PHYSICS AND CHEMISTRY OF CLOUDS

Clouds affect our daily weather and play key roles in the global climate. Through their ability to precipitate, clouds provide virtually all of the fresh water on Earth and are a crucial link in the hydrologic cycle. With ever-increasing importance being placed on quantifiable predictions – from forecasting the local weather to anticipating climate change – we must understand how clouds operate in the real atmosphere, where interactions with natural and anthropogenic pollutants are common.

This textbook provides students – whether seasoned or new to the atmospheric sciences – with a quantitative yet approachable path to learning the inner workings of clouds. Comprehensive treatments are given of the mechanisms by which cloud droplets form and grow on soluble aerosol particles, ice crystals evolve into diverse shapes, precipitation develops in warm and cold clouds, trace gases and aerosol particles are scavenged from the atmosphere, and electrical charge is separated in thunderstorms.

Developed over many years of the authors' teaching at Penn State University, *Physics and Chemistry of Clouds* is an invaluable textbook for advanced students in atmospheric science, meteorology, environmental sciences/engineering, and atmospheric chemistry. It is also a useful reference text for researchers and professionals.

DENNIS LAMB is Professor Emeritus of Meteorology at the Pennsylvania State University. He has been fascinated with clouds ever since growing up in the Midwestern United States, and the true nature of clouds and the processes that form them have been gradually revealed to him through years of formal training and self study. Professor Lamb worked as a researcher for nearly 14 years at the Desert Research Institute (Reno) before embarking on a teaching career at Penn State University. With more than 40 years of observational and laboratory research experience, and more than 20 years teaching cloud physics and atmospheric chemistry at both the undergraduate and graduate levels, he now realizes that the best path toward understanding clouds is to understand water itself, at the molecular level. The deeper the understanding, the greater becomes the appreciation of clouds as gate keepers in the water cycle and energy budget of Earth. This book is the culmination of his career studying the physics and chemistry of water and clouds.

HANS VERLINDE is an Associate Professor of Meteorology at the Pennsylvania State University. He is an observational meteorologist who has studied clouds in the Antarctic, at the equator, and in the Arctic. He is currently the site scientist for the US Department of Energy Atmospheric Radiation Measurement Program Climate Research Facility at Barrow on the North Slope of Alaska, and he teaches classes in atmospheric thermodynamics, cloud physics, mesoscale meteorology, and radar meteorology at Penn State University. After a start in meteorology as a weather forecaster, he developed a passion for clouds, and likes to look at clouds from all sides now, from up and down, from inside out. He enjoys seeing the surprised looks on student faces when, on the first day of every semester, he asks "Who can tell me what clouds are in the sky today?" which inevitably leads to the next question, "Who can tell me what the sky looks like today?"

PHYSICS AND CHEMISTRY OF CLOUDS

DENNIS LAMB
Pennsylvania State University

JOHANNES VERLINDE
Pennsylvania State University

CAMBRIDGE
UNIVERSITY PRESS

CAMBRIDGE
UNIVERSITY PRESS

University Printing House, Cambridge CB2 8BS, United Kingdom

Cambridge University Press is part of the University of Cambridge.

It furthers the University's mission by disseminating knowledge in the pursuit of education, learning and research at the highest international levels of excellence.

www.cambridge.org
Information on this title: www.cambridge.org/9780521899109

First published 2011

A catalogue record for this publication is available from the British Library

ISBN 978-0-521-89910-9 Hardback

To our families

Pat and Julie

Terri, Anneke, Kris, Katrina, and Maarten

Contents

Preface

Clouds contribute to our lives in both direct and indirect ways. Clouds are at once the most visible elements of the sky and the dominant contributors to the weather we experience every day. Less apparent, but perhaps even more important, are the roles clouds play in the global energy and water budgets that determine the climate of Earth. Through their ability to precipitate, clouds provide virtually all of the fresh water on Earth and a crucial link in the hydrologic cycle. Clouds are also the most effective agents cleansing the atmosphere, although some terrestrial and aquatic ecosystems pay the price for anthropogenic emissions of chemicals into the air. With ever-increasing importance being placed on quantifiable predictions, whether to forecast the local weather or to anticipate changes in global climate, we must learn how clouds operate in the real atmosphere, where two-way interactions with natural and anthropogenic pollutants are common.

Clouds have been the subject of observation for centuries, but serious systematic investigations began only a few decades ago. For all practical purposes, the study of clouds can be traced back to Luke Howard, the English pharmacist who began, around 1803, the system of naming cloud types that we still use today (see Appendix A). Speculation about the composition and nature of clouds persisted for many years. Direct observations from balloons and aircraft helped greatly to develop a base of empirical knowledge upon which the research community could later build testable hypotheses. With ongoing improvements in instrumentation and measurement techniques, the invention of cloud chambers, and an ability to test hypotheses quantitatively, the research community of atmospheric scientists has gradually developed a broad and quantitative understanding of clouds.

The atmosphere, with all its dynamical and chemical complexity, is the environment in which clouds form. We cannot understand clouds divorced from that parental setting. The atmosphere is a mixture of a huge number of chemical compounds, some gaseous, some particulate in nature. Indeed, water is just one of those myriad components, but the only one of note that changes phase under ordinary conditions. The atmosphere is far more than "dry air" and water vapor, so any modern treatment of clouds must deal with this mixture head-on. Indeed, cloud droplets form on the more soluble subset of the particulate matter, and they subsequently absorb some of the trace gases. The microphysical properties, even the macrophysical forms, of clouds are significantly affected by the chemicals in the air. In turn, many of those chemicals are altered and removed from the atmosphere by clouds and the precipitation they produce. The physics and chemistry of the atmosphere go hand in hand when developing a complete picture of clouds and how they behave in the atmosphere.

The need for a new textbook arose partly out of our frustrations with identifying textbooks suitable for teaching cloud physics, mainly at the graduate level, but also as part of the undergraduate curriculum that meteorology majors take. Many of our first-year graduate students come from disciplines other than meteorology or the atmospheric sciences. They typically have strong backgrounds in science, engineering, or math, but few have had any formal course dealing with clouds. The graduate course offered at Penn State University therefore starts out with few assumptions other than that the students are bright and skillful in mathematics and scientific reasoning. An ideal text parallels the course structure by presenting the atmospheric context and showing how fundamental scientific principles lead to the phase changes of water we know of as clouds. We hope that we have met the challenge with this textbook by emphasizing the basic disciplines of physics and chemistry.

This textbook offers students, whether seasoned or new to the atmospheric sciences, a quantitative, yet approachable path to learning the inner workings of clouds. Comprehensive treatments are given of the mechanisms by which cloud droplets form and grow on soluble aerosol particles, ice crystals evolve into diverse shapes, precipitation develops in warm and cold clouds, trace gases and aerosol particles get scavenged from the atmosphere, and electrical charge becomes separated in thunderstorms. Overall, the book emphasizes how clouds ultimately depend on the molecular properties of matter.

The book is broken down into five parts to allow the reader to focus separately on the several main areas. Part I provides background material of use to those either unfamiliar with the atmospheric sciences or wishing a brief refresher of concepts and terminology. This first part could serve as the basis for a survey course in physical meteorology. Alternatively, it may be skimmed or skipped by readers with strong backgrounds in atmospheric physics and chemistry. Part II shows how transformations, both physical and chemical, come about in nature when a system deviates from equilibrium. Special attention is paid to the concepts of equilibria pertinent to phase changes because of their central role in theories of cloud formation. Part III discusses clouds from a macroscopic point of view. At this level, one need not know much about the composition of clouds other than that they are composed of water in condensed form. Part IV elucidates the processes responsible for the microstructure of clouds, the phases, sizes, and shapes of the individual particles making up a cloud. Part V brings the reader back to the cloud scale and the effects of large populations of cloud particles.

The units used for dimensioned quantities in this book are based on the International System of Units (Système International d'Unités, SI). The SI uses decimal units of measure, with the base units being the meter [m], kilogram [kg], second [s], ampere [A], kelvin [K], and mole [mol]. A standard set of prefixes are used to allow quantities to be specified with convenient values. Some accommodation has been made in this book for nonstandard units that are still in common use. For instance, hPa is the SI equivalent of mbar [or mb], the unit of pressure so commonly used by meteorologists. It is expected that readers are familiar with or have access to the complete guidelines of SI symbols and usage.

This textbook is intended to augment lectures in upper-division and graduate courses in physical meteorology or cloud physics. Introductory material in each major section is intentionally descriptive in nature in order to present students with a qualitative feel for the subject. Where the subject matter is presented from a theoretical viewpoint, mathematics through vector calculus and differential equations is employed. It is intended that students in a graduate course would read further into each subject than would undergraduates. The instructor is best able to decide the dividing points. Each chapter ends with a bibliography for further reading and a set of problems to emphasize certain points and give opportunities for further learning. The instructor should feel free to modify or expand on problems to meet the particular goals of his/her course.

Authors of textbooks invariably depend on the hard work of and discussions with many others. The research literature has been heavily exploited, but citations have been limited to sources of graphical material in order to ease the reading of technical material by beginning students. We are indebted to the many authors of prior texts and reference books, upon which we have relied to develop material for teaching the subjects of cloud physics and atmospheric chemistry over many years. Among the books we have relied on most are those by Pruppacher and Klett (*Microphysics of Clouds and Precipitation*, 2nd edn.), Rogers and Yau (*A Short Course in Cloud Physics*, 3rd edn.), Seinfeld and Pandis (*Atmospheric Chemistry and Physics*), Cotton and Anthes (*Storm and Cloud Dynamics*), and Houze (*Cloud Dynamics*). We think this book should find a niche somewhere between the short course by Rogers and Yau and the extensive reference book of Pruppacher and Klett. We would also like to think of it as complementing the comprehensive work on atmospheric chemistry by Seinfeld and Pandis.

Special acknowledgment is graciously extended to the many individuals who have, in one way or another, enabled us to carry out this ambitious project. At the onset, Raymond Shaw graciously provided a quality environment in the Physics Department at Michigan Technological University, where the first author spent a sabbatical leave in 2006 and began the actual process of writing. Many subsequent discussions with him and other colleagues there and elsewhere have contributed enormously over the years to the development of the material presented in this book. Among those individuals, the authors wish to thank (in alphabetical order) Alex Avramov, Chad Bahrman, William Brune, Will Cantrell, J.P. Chen, Eugene Clothiaux, William Cotton, Graham Feingold, Jose Fuentes, Barry Gardiner, Jerry Harrington, Alex Kostinski, Zev Levin, Nathan Magee, Paul Markowski, Alfred Moyle, Yvette Richardson, Lindsay Sheridan, Nels Shirer, Ariel Stein, and Huiwen Xue. The first author also benefited greatly from the opportunity, offered by Huiwen Xue and Chunsheng Zhao, to teach part of a graduate course in cloud physics in the Department of Atmospheric Sciences at Peking University in 2009. The feedback on various chapters of the book that was received from the students there and at Penn State University has helped us greatly during revisions of the text. We are especially grateful to Eugene Clothiaux,

Barry Gardiner, and Nels Shirer for their careful and critical reading of parts of the manuscript at various stages. The publication process was made relatively painless by the able assistance provided by Laura Clark and Matt Lloyd at Cambridge University Press. Finally, we are ever thankful for the patience of and encouragement given by our respective families, to whom this book is dedicated.

Part I

Background

1
Introduction

1.1 Importance of clouds

What would our world be like without clouds? Unimaginable – quite literally – for clouds are essential for our lives on earth. Humans, and for that matter most other land-dwelling species, would simply not exist, let alone thrive in the absence of the fresh water that clouds supply. The favorable climate we have enjoyed for thousands of years might also not exist in the absence of atmospheric water and clouds. A world without clouds would be different indeed.

Clouds contribute to the environment in many ways. Clouds, through a variety of physical processes acting over many spatial scales, provide both liquid and solid forms of precipitation and nature's only significant source of fresh water. Under extreme circumstances, however, clouds and precipitation may not form at all, leading to prolonged droughts in some regions. At other times and places, too much rain or snow falls, giving rise to devastating floods or blizzards. Liquid rain drops bring usable water directly to the surface, while simultaneously carrying many trace chemicals out of the atmosphere and into the ecosystems of the Earth. Chemical wet deposition thereby supplies nutrients (and sometimes toxic compounds) to both terrestrial and aquatic lifeforms, as well as the weak acids responsible for the weathering of the Earth's crust. The solid forms of precipitation contribute in additional ways to the world as we know it. Snow, for instance, forms the winter snowpacks that dramatically affect the radiation balance and climate of high latitudes on a seasonal basis. In mountainous regions, snow simultaneously yields a long-lasting supply of water and lucrative opportunities for human recreation. Snow that accumulates from one year to the next gives rise to glaciers that carve out valleys as they slowly flow downhill under their enormous weight. Atmospheric clouds and the precipitation they yield are responsible for much of the world that we take for granted.

Many aspects of weather revolve around the presence or absence of clouds. The weather systems that routinely pass through the mid-latitudes transform invisible water vapor into sometimes beautiful, sometimes dreary clouds of many sizes and types. These clouds affect the radiation balance of the region and hence the temperature of the air and exposed surfaces. The precipitation they generate removes the water and trace chemicals from the sky, serving simultaneously to dry and cleanse the air. Forecasting the meteorological events of the next day or of the coming season is becoming ever more crucial to our individual lives and to the economy of our society. Being able to anticipate the amount and nature of the clouds with assurance is an important skill of every forecast meteorologist, for which one needs a thorough understanding of the atmosphere and the processes responsible for cloud formation.

Less apparent than the weather we see and feel, but which is nevertheless important to the workings of the atmosphere, are the roles clouds play in the atmospheric energy budget. Clouds reflect incoming solar radiation back to space, thus helping regulate the overall input of solar energy and its distribution around the world. Clouds also intercept infrared radiation emitted from the surface that would otherwise be lost to space; reradiation of infrared radiation by the same clouds helps warm the surface. Energy deposited in the oceans helps evaporate water and provides the primary ingredient for cloud formation, water vapor. The transformation of water vapor into the many liquid and solid particles that compose clouds necessarily results in a warming of the air. This energy consequence of a physical change of phase determines in part the macroscopic shape and behavior of clouds, whether they are convective or stratiform in nature. On a much larger scale, the thermal energy "released" by the phase transformations in the large convective clouds of the tropics becomes an important component of the Earth's energy balance. Major circulation patterns in the atmosphere are thus spawned, helping redistribute the surplus energy from the tropical regions to other parts of the world, where less energy is received from the Sun than is lost to space by thermal infrared radiation.

The composition of the atmosphere, especially regarding its trace gases and particulate matter, is greatly influenced by clouds. The precipitation resulting from clouds serves as a carrier of the material taken up by the cloud and precipitation particles. Cloud and precipitation "scavenging" thus serves as a remarkably efficient mechanism by which the atmosphere is cleansed of the diverse gases and particles continually emitted into the air, thereby preventing the build-up of natural and anthropogenic pollutants. At the same time that air quality is improved by precipitating clouds, the precipitation itself becomes correspondingly fouled, leading to such ecosystem problems as acidic rain, for instance. Even in the absence of precipitation, clouds offer several important opportunities for transforming trace components of the atmosphere into other compounds. In the lower atmosphere, such in-cloud reactions oxidize sulfur and nitrogen compounds, leading to enhanced summertime hazes across industrial regions. In the middle atmosphere, chemical balances may be altered rather profoundly by reactions occurring in or on the surfaces of aerosol and cloud particles, causing ozone to be lost. The chemistry of clouds thus becomes as important as the physics of clouds toward the workings of the atmosphere.

The study of clouds offers rich opportunities for applying our understanding of physics and chemistry to real-world phenomena. Clouds give direct evidence of changes taking place in the atmosphere. By our conventional ways of categorizing the disciplines of science, we would say that some of these changes are physical in nature, some are chemical in nature. Nature, of course, knows no such distinction, so we need to realize that dividing the atmospheric sciences into "physical" and "chemical" domains is largely a matter of convenience. Physics, the science of matter, energy, and their interactions, is the discipline used in traditional cloud physics to understand the fundamentals of cloud and precipitation formation, the microscale structure of clouds, cloud electrification, and the impacts of clouds on climate. The conventional restriction to the physics of clouds, however, ignores several important chemical attributes of clouds. Chemistry, the science of the composition, structure, and properties of substances and their transformations, is needed to understand the very nature of water itself, as well as how water interacts with aerosol particles to permit clouds to form under atmospheric conditions. Chemistry is also needed to understand how those aerosol particles that serve as the sites of condensation came to exist in the first place. The atmospheric phenomena of acidic haze formation, acid rain, stratospheric ozone depletion, and some aspects of the natural biogeochemical cycles and climate can be understood only via the traditional discipline of chemistry. Throughout this book, we will find frequent occasion to jump between physical and chemical concepts, often without mention. It is only important to recognize that the fundamental principles of science guide us aptly as we try to understand atmospheric clouds in their natural, complex setting.

1.2 Observed characteristics of clouds

1.2.1 Overview

Careful observations of the atmosphere reveal much about clouds. Some, "macroscopic" characteristics of clouds are readily seen with the unaided eye, whereas other "microscopic" properties require elegant instrumentation. At all levels, we find it natural and helpful to give names to phenomena, properties, and concepts. The nomenclature and jargon of the science become the means by which we communicate effectively with one another and so must be learned along with the scientific concepts. Some attention is therefore given to the proper use of terminology throughout the text.

Cloud formation requires moisture, aerosol particles, and a process for cooling the air. The abundances of moisture and aerosol particles determine the total mass of condensate and the number concentration, respectively, and they affect the ability of clouds to produce precipitation. These two components also regulate the radiative properties of clouds and how we perceive them visually. The necessary cooling to form a cloud may arise from any one or combination of processes: radiative cooling, turbulent mixing of air across moisture/temperature gradients, or expansion of air during forced ascent or free convection.

The observed characteristics of a cloud depend on how the atmosphere organizes itself to provide these key ingredients. Atmospheric moisture, derived from evaporation of surface water or transpiration of plants, originates at or near the Earth's surface. The fact that most clouds are observed well above the surface suggests that surface moisture must be transported upward by atmospheric motions. The surface is likewise the dominant source for aerosol particles, although these particles may also be formed in the atmosphere through gas-to-particle conversion. The processes responsible for upward moisture transport also results in a cooling of the air, the other requirement for cloud formation. The type of vertical motion is the major determinant of the cloud forms we commonly see. Slow, large-scale ascent results in broad, featureless clouds, whereas rapid ascent of smaller parcels of air result in cloud turrets. The mixing of warm, moist air with cooler air leads to transient clouds of limited spatial extent.

Once formed, a cloud changes in response to the relative rates of the processes responsible for condensate formation and loss. Continued cooling from the net loss of radiation, adiabatic expansion, and/or moisture advection adds condensate; conversely, radiative heating, adiabatic compression, mixing with drier atmospheric air, and precipitation all remove condensate. The radiative heating rates are determined by a combination of the macroscopic (large-scale) and microscopic (small-scale) characteristics of the cloud. The vertical distribution of diabatic heating and cooling plays a further role in changing the atmospheric stability profile, the impact of which is to enhance (suppress) vertical motions.

The vertical velocities in a cloud determine the time an air parcel spends inside the cloud. This in-cloud duration limits the time available for condensate to grow to sizes large enough to fall against the updraft and remove condensate. Spatial variations in vertical velocities induced by turbulence result in spatial variations of microscopic properties, and are responsible for lumpiness in the visually observed cloud outline. Turbulence further serves to mix drier atmospheric air into the cloud, leading to the loss of condensate through evaporation. The interactions between the macroscopic air motions and the microscale processes ultimately determine the characteristics of the clouds we observe.

1.2.2 Macroscopic forms

The sky is rich in information about the state of the atmosphere and the diverse processes that bring about changes. One needs to learn how to read the sky much as one does to read a book. In both cases, we depend on our sense of vision to recognize patterns (the shape of a cloud, or the words on a page) and on our mind to interpret what we see. "Sky reading", the art of interpreting observed properties of the sky in terms of categories and processes, is practiced by many, amateurs and professionals alike, as a way of understanding atmospheric phenomena and foretelling weather events. By combining the ever-changing visual clues presented by clouds with an understanding of physical processes, much can be gleaned about the current or anticipated weather. We start with basic terminology and categorization before explaining the processes that bring clouds into being and eventually to their demise.

Clouds can be seen at various times from virtually every point on the Earth's surface, but it is often challenging to know what to call them or how they evolve. Despite difficulties in determining the sizes and altitudes of clouds with precision, level in the atmosphere is good for telling one cloud type from another. We can usually distinguish low-lying clouds from those higher in the atmosphere, for instance. Thus, clouds may be categorized as "low" (up to about 2 km above the surface), "mid-level" (2 to 7 km), or "high" (above 7 km). Clouds confined to distinct levels often take on a "stratiform" appearance, one exhibiting dimensions in the horizontal that are substantially greater than those in the vertical. Stratiform clouds form in air that is thermodynamically stable, meaning that small vertical displacements have little effect on the overall air motions. "Cumuliform" clouds, by contrast, tend to extend farther in the vertical than in horizontal directions. Cumuliform clouds are sometimes said to be convective because they form in air that is locally unstable, meaning that small vertical displacements lead to further displacements and convective overturning of the air. The stability of the air is determined by how rapidly the temperature and humidity change with altitude. Large cumuliform clouds can span all altitude categories, from "low" to "high". The conventional names given to the various cloud forms are derived from visual observations taken at the ground (see Appendix A for a summary).

The observed forms of clouds differ substantially within a given category. Often, the shape of a cloud itself best reveals its type. A few photographic examples illustrate the diverse forms clouds can take in various settings. As Fig. 1.1 shows, the view from the ground often reveals multiple cloud types at one time. The cloud elements near the bottom of the figure are individually cumuliform in nature and indicative of a turbulent, nearly well-mixed boundary layer. Such disconnected clouds are classified as cumulus (Cu). If the edges of the cloud elements were touching, the deck would be classified as stratocumulus (Sc). The particles constituting these clouds are liquid water droplets, the result of vapor condensation at relatively high temperatures. The clouds toward the top of Fig. 1.1 have a fibrous appearance indicative of high cirrus clouds (Ci), perhaps the result of a jet aircraft. The cloud particles are ice crystals, evidence for which is the coloration (bright spots left and right of center) arising from the refraction of sunlight through adjacent facets of the crystal (causing a "circumhorizontal arc" in this case). Another example of cirrus, in this case Ci uncinus, is shown in Fig. 1.2. Such high clouds are made of relatively large

Figure 1.1 Diverse cloud forms over eastern Oregon. The two bright spots in the upper third are from a circumzenithal halo, which forms when sunlight refracts through hexagonal ice crystals. Photo by D. Lamb.

Figure 1.2 Cirrus uncinus clouds over Victoria, British Columbia. Photo by D. Lamb

ice particles that sediment rather rapidly. Ci uncinus in effect, represent snow that never reaches the ground.

Mid-level clouds are often relatively thin and take on a variety of sub-forms. Perhaps the most diverse types occur with altocumulus (Ac). Figure 1.3a shows an example of Ac undulatus, cumuliform elements that formed in a relatively thin layer where moisture accumulated preferentially. A distinctly different form of Ac is shown in Fig. 1.3b. In this case, stable air was forced over an upstream mountain, forming a gravity wave with clouds in the crest. Such a cloud is commonly called a wave cloud, although the scientific designation is Ac lenticularis because of the lens-like shape. Figure 1.3c shows the edge of an altostratus deck (As), a continuous deck of clouds at mid levels. Where precipitation is falling out, the cloud is called nimbostratus (Ns). It is hard to tell if the precipitation is rain or snow here, but any precipitation that evaporates before reaching the ground is termed virga.

Clouds sometimes look tall and vertically extended relative to their horizontal dimension. Such cumuliform clouds arise when moist air rises rapidly in an unstable atmosphere. Examples of cumulus clouds in the trade winds of the Pacific Ocean are shown in Fig. 1.4. The flat bases indicate that each of the clouds was derived from boundary-layer air having common thermodynamic properties. The towering nature of these clouds shows the importance of buoyancy generated by the release of the latent heat of condensation. Particularly impressive cumulus and cumulonimbus clouds (Cb) can develop when air moves upwards rapidly in moist air that it unstable over large depths. The cloud shown in Fig. 1.5, for instance, towered over the Grand Canyon in Arizona and was just about to develop an anvil and begin raining at the time the photograph was taken. Such a cloud with hints of an anvil is called Cb calvus. A fully developed Cb is shown Fig. 1.6. Note the active convection on the upshear (left-hand) side and the extended anvil on the downshear side of

Figure 1.3 Examples of mid-level clouds. a. Altocumulus (Ac) undulatus over Pennsylvania. b. Ac lenticularis over Alberta. c. Nimbostratus (Ns). Photos by D. Lamb.

Figure 1.4 Trade-wind cumuli off the shore of Kauii. Photo by D. Lamb.

Figure 1.5 Towering cumulus cloud over the Grand Canyon. Photo by D. Lamb.

Figure 1.6 Cumulomnimbus cloud with long anvil along east coast of the United States. Photo by D. Lamb.

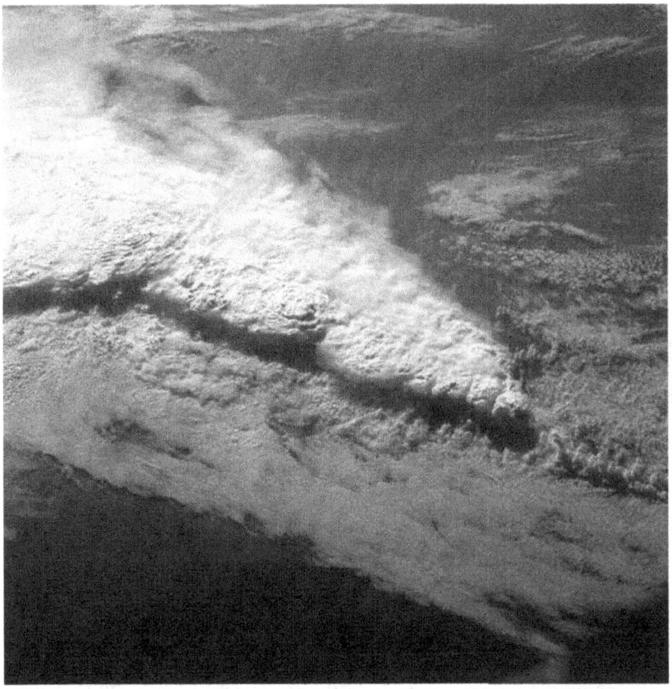

Figure 1.7 Squall line over the middle United States. Image from NOAA satellite.

the central cloud. Large cumulonimbi can develop airflow patterns that actively suppress the development of neighboring clouds and so appear isolated. Cumuliform clouds sometimes organize themselves into well-defined circulations, which we recognize at different times as squall lines (Fig. 1.7) or hurricanes (Fig. 1.8). Still photographs show impressive macro-features of clouds, but they cannot do justice to the evolution and microphysical properties of clouds.

1.2.3 Microscopic properties

The individual particles that make up a cloud are not generally visible to the human eye. We see the macroscopic, overall form that the cloud takes in the sky, the net effect of sunlight being scattered by the many water droplets and ice crystals within the boundaries of the cloud, but the cloud particles themselves are simply too small to be resolved from outside the cloud. Raindrops and snowflakes falling near us may be seen individually, but not the vast majority of cloud particles, which are truly microscopic. We seek here to expose the properties of those many minute particles that make up the cloud "microstructure".

A large cloud can look imposing, a potential hazard were one to venture into its interior. However, one would not encounter a wall of water as one would upon jumping into a swimming pool. Rather, entering a cloud would be akin to leaving one's house and

Figure 1.8 Satellite view of hurricane Katrina in the Gulf of Mexico in 2005. Image from NOAA satellite.

walking into a morning fog. You might find the scene eerie perhaps, but you certainly would find little to impede your progress, save the limited visibility. Clouds are not much different. Indeed, fogs are actual clouds, just ones that contact the ground. We must appreciate the fact that clouds are mostly air, the many particles being dispersed widely and more or less randomly throughout the cloud interiors.

The microstructure of a cloud can be appreciated at a qualitative level by a consideration of scales, the relative sizes of the various constituents that make up a cloud. Figure 1.9 shows a "telescoping" view of a cloud. The cloud itself (top panel) forms on the cloud macroscale, which overlaps the meteorological mesoscale (kilometers to hundreds of kilometers) from moisture carried aloft by large-scale air motions. Supersaturations are developed as the air cools by adiabatic expansion and allow condensation to take place on aerosol particles. Any small part of this cloudy air contains many liquid water droplets of diverse sizes (second panel). It is on this microscale that the particles compete for the available water vapor, causing the interstitial supersaturation to be less than that determined thermodynamically. Further extension of our telescoping view (third panel) would show an individual droplet surrounded by the vapor from which it grows by condensation. Gradients of both vapor concentration and temperature would be found on this particle scale because of the resistances to mass and heat transport imposed by the air. The droplet mass increases during condensational growth, when the interstitial vapor concentration exceeds the equilibrium value, and it decreases during evaporation, when the vapor concentration

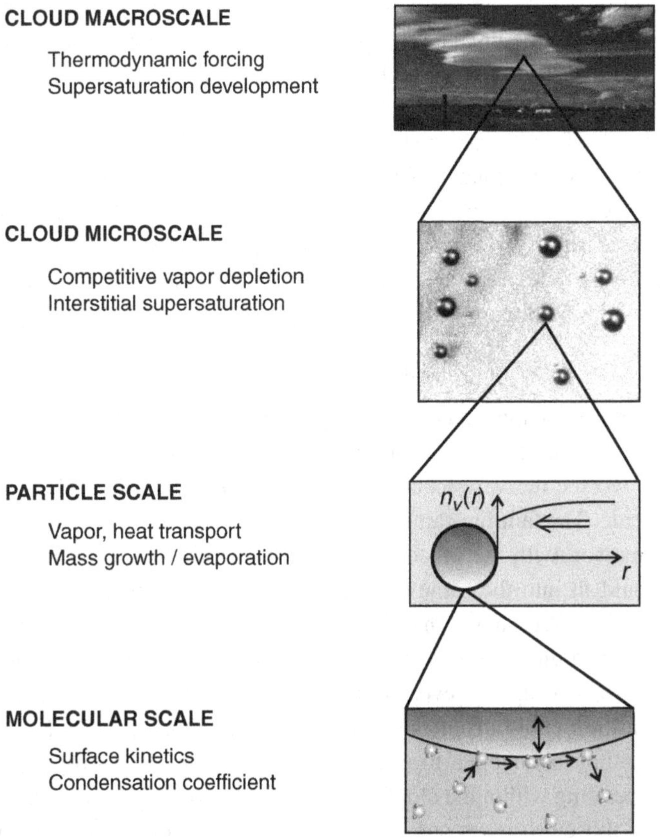

Figure 1.9 Relative scales of clouds and particles.

falls below this threshold value. If we were to ever see the true molecular-scale processes near the droplet surface (bottom panel), we would be amazed at the rapidity and complexity of action. The water molecules continually scurry about just above the surface, being impacted randomly by the molecules that make up the air (mostly nitrogen and oxygen). The vapor molecules that collide with the surface of the droplets may stick for a short while where they landed, then move around on the surface a bit, before they either escape back into the vapor phase or enter the bulk liquid and add their mass to the droplet. All of the many possibilities for molecular "incorporation" into the condensed phase are summed up in the "condensation coefficient", a parameter used in theoretical models to account for the many molecular-scale processes we can only speculate about. The science of cloud physics must deal with a wide range of scales (from molecules to whole cloud systems), a range encompassing many factors of ten (called "orders of magnitude").

The particles making up a cloud may be solid, liquid, or a mixture of both states of matter. Clouds made solely of liquid droplets are termed "warm" clouds, whereas those containing ice particles are said to be "cold" clouds. These terms probably arose from our general realization that liquid water exists at relatively high temperatures (above 0 °C) and that ice can form only at lower temperatures. However, the distinction between "warm" and "cold" clouds hinges on the phase of the particles, not on the temperature. Some parts of "cold" clouds may also contain liquid drops, which are termed "mixed-phase" regions. Later, we will come to appreciate why the lower parts of deep clouds tend to be "warm", the middle parts are often "mixed phase", while the upper-most parts may be all ice. We will also talk about "warm-cloud" microphysical processes, those involving liquid drops; ice particles may be present, or not, but they do not influence how the liquid drops interact. "Cold-cloud" processes, by contrast, can involve liquid drops, as during hail formation, although only ice particles may be involved in other situations, such as during crystal aggregation and snow formation.

Warm-cloud microstructure

"Warm" clouds often contain liquid droplets of many different sizes. An appreciation of the diversity of sizes can be gained from diagrams that compare the sizes directly, such as Fig. 1.10. Here, we see the various categories of cloud particles and their representative sizes drawn to scale. The familiar raindrop (only half of which is shown to conserve space) is huge in comparison with the other categories of particles. Something like a thousand drizzle drops could fit into the same volume occupied by a single raindrop. (Remember that the particles are spherical and that the volume of a sphere increases as the cube of the diameter.) Likewise, a thousand cloud droplets of 10 μm diameter could fit into the volume of a single drizzle drop. So, every raindrop that falls to earth represents the effective gathering of a million cloud droplets or more. It is important to note, however, that the cloud droplets do not just happen to gather together to form a raindrop. Specific mechanisms must be operating within the cloud to allow raindrops to grow, via processes covered later. One hint of those processes can be seen, however, by contrasting the concentrations of the respective categories. Note that many fewer raindrops exist than do drizzle drops or cloud droplets. In fact, the ratio of concentrations is just the inverse of the ratio of sizes, suggesting that the large drops indeed form from the smaller drops, conserving the overall mass of liquid water in a given volume of cloudy air. The large drops gain mass at the expense of the smaller drops, but the process by which this "coalescence" takes place is anything but simple.

At the small end of the range of droplet sizes we see a different pattern emerging. Haze droplets are yet another order of magnitude smaller in size, but their concentrations are comparable to those of the cloud droplets. If the cloud droplets formed from the haze droplets, then clearly the mass of liquid water was not conserved in the process. The mass of water added to a given haze droplet to form a cloud droplet must have come from somewhere other than the already-formed liquid phase. Indeed, it is the process of condensation, whereby individual water molecules leave the gas phase and enter the liquid phase, that causes this mass increase. It is, therefore, not a coincidence that the concentrations of haze and cloud droplets are similar, just as it is no coincidence that the even smaller "cloud condensation nuclei" (CCN) are found in comparable concentrations. We will come to realize that condensation preferentially takes place on existing surfaces, and it is the CCN that

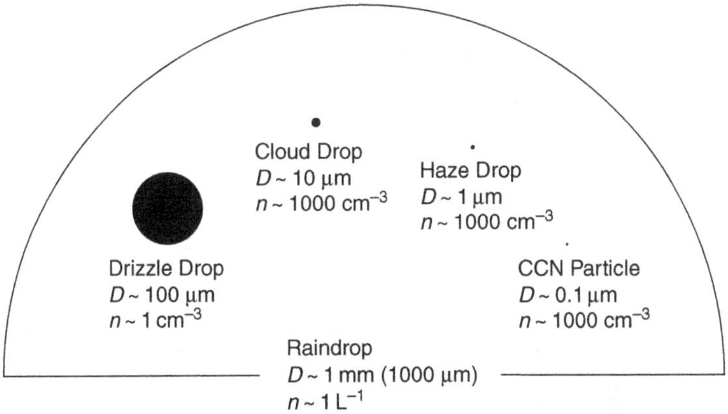

Figure 1.10 Relative sizes of various liquid drops found in clouds. Only half of the raindrop is shown to conserve space.

serve as the sites upon which condensation takes place. In fact, the microstructure of warm clouds, as well as the efficiency with which clouds develop rain, depends strongly on the concentrations of CCN, a most important subset of the atmospheric aerosol.

At a quantitative level, the microstructure of a warm cloud is described by the way in which the cloud droplets are distributed in size. Such "drop spectra" tell us how many of what size drops are present. Drop spectra are often determined empirically, for instance, by flying through a cloud and sampling the droplets with special instrumentation. An example of what one would find by letting the droplets impinge for a known length of time onto a sticky surface that leaves a permanent impression (via a Formvar replicator) is shown in the top part of Fig. 1.11. Each of the circles here represents the region where the momentum of impact pushed aside some of the sticky coating, leaving a visible crater. What we see therefore are the impact craters, not the droplets themselves. Nevertheless, a bit of work behind the scenes allows one to relate the crater diameters to the sizes of the droplets causing the craters (compare scales along the bottom axis). One must also compute the volume of cloudy air sampled in order to convert the numbers of droplets sampled to concentrations or number densities (vertical axis).

What results from performing such a sampling procedure? First, note that all of the droplets here (Fig. 1.11) may be correctly classified as "cloud droplets", but that they nevertheless range considerably in size. Droplets vary more or less continuously in size, but we must always define small subcategories or "bins" of well-defined size intervals before arriving at a true size distribution. We are concerned only with the number of droplets within a given size interval because we can never keep track of every droplet in a cloud, only some statistical measure of the droplet population. When we count the number of droplets in each category from the sample shown in Fig. 1.11, for instance, we obtain the "histogram" at the bottom of the figure, in which the ordinate specifies the droplet concentration per bin (even when the labeling is not so specific). Here, we see

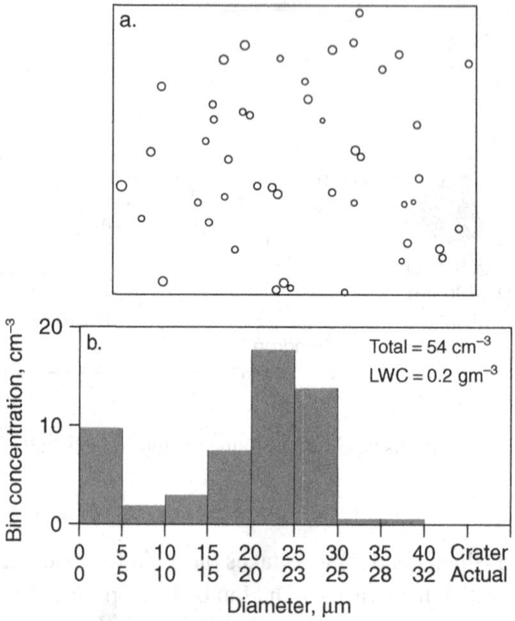

Figure 1.11 Example of in-cloud data from a Formvar replicator. a. Craters made in the Formvar. b. Histogram derived from the drops shown in (a). (LWC = Liquid water concentration.) From John Hallett, used with permission.

that most of the cloud droplets fall into the 20- to 23-μm category. We typically take the mid-point (21.5 μm) of the bin containing the most droplets to be the "modal" diameter of the sample. The variation in the number densities of sampled droplets across the various size bins gives us some sense of how the droplet population in the cloud varies in size.

Other statistical measures can be obtained from drop spectra. For instance, if we simply add up the number densities in all bins, we would get the total concentration of droplets in the sampled air. We generalize counting within a drop spectrum by labeling the bins sequentially and expressing the sum mathematically. Thus, if we let the bins be numbered from $j = 1$ at the small end to $j = N_{bins}$ at the large end of the size axis, then the total concentration is given by the relationship,

$$n_{tot} = \sum_{j=1}^{N_{bins}} n_j. \tag{1.1}$$

Similarly, we obtain the mean diameter of the sample by adding up all the diameters (D) and dividing by the total number of drops (n_{tot}). Mathematically, we express this mean diameter as

$$\bar{D} = \frac{1}{n_{tot}} \sum_{j=1}^{N_{bins}} D_j n_j, \tag{1.2}$$

where D_j is the mid-point value of bin j. The collective surface area in a unit volume of cloudy air can also be calculated readily by noting that each small drop closely approximates the geometry of a sphere. So, if we use A_{tot} [possible units $\mu m^2\, m^{-3}$] to designate the combined surface area concentration of the drops in the sample and recall that the area of a single sphere of diameter D is $A = \pi D^2$, we find

$$A_{tot} = \sum_{j=1}^{N_{bins}} n_j A_j = \pi \sum_{j=1}^{N_{bins}} D_j^2 n_j. \tag{1.3}$$

The surface area of a drop or aerosol population is sometimes needed to calculate the interaction of clouds with atmospheric pollutants. A particularly useful statistical measure for cloud physics is the liquid water concentration [LWC, symbol ω_L, typical units $g\, m^{-3}$], the collective mass of liquid water in the sample. Here, again, we simply add up the masses of water in each bin, recognizing that the mass of a single drop of diameter D and mass density ρ_L is $m_D = \rho_L v_D$, where $v_D = \pi/6 \cdot D^3$ is the drop volume. Thus, the liquid water concentration of the sample is given by

$$\omega_L = \frac{\pi \rho_L}{6} \sum_{j=1}^{N_{bins}} D_j^3 n_j. \tag{1.4}$$

Note the similarity of the expressions being summed in each of the previous equations: each contains n_j and D_j raised to some power. We use the exponent of D_j to designate the type of statistic derived from the given drop spectrum. Thus, the statistical measures of drop concentration (Eq. (1.1)), mean diameter (Eq. (1.2)), surface area concentration (Eq. (1.3)), and liquid water concentration (Eq. (1.4)) are said to come, respectively, from the zeroth, first, second, and third moments of the distribution. The sixth moment of the drop size distribution is important for radar studies.

The microstructure of a cloud varies with location within the cloud, as well as with time. Typically, the properties vary most in the vertical direction, in part because pressure decreases rapidly with height, as do the thermodynamic variables that cause cloud to form in the first place. An example of how the microstructure of a small cumuliform cloud varies with distance above cloud base is presented in Fig. 1.12. The set of curves along the right-hand side show the drop spectra as measured by instrumentation aboard an aircraft that flew at five different heights. The spectra are seen to broaden (span a wider range of sizes) with increasing altitude and shift toward larger sizes. Care must be taken when viewing and interpreting these curves, for the concentrations are plotted on a logarithmic scale and refer to the number densities within each bin. The total concentration of drops at each level would be computed via Eq. (1.1), whereas the mean diameter of the drops sampled at each level would be calculated from Eq. (1.2). The variation with height of the total concentration is shown in the left-hand side of Fig. 1.12. We clearly see that the number density of drops decreases with height (at least above 0.4 km) at the same time that the average size of the drops increases.

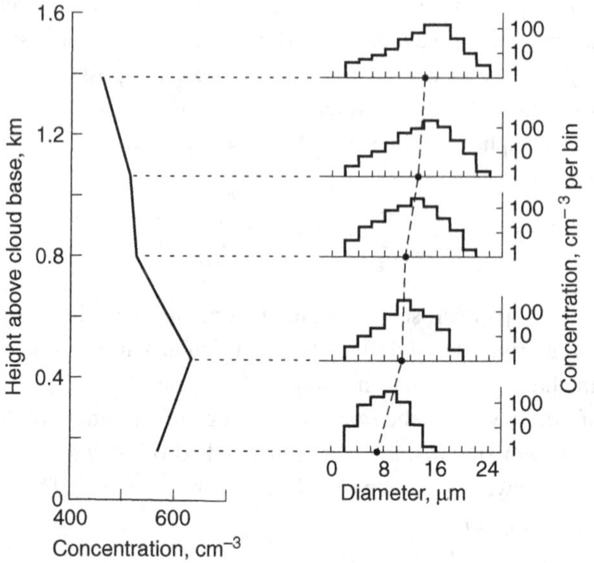

Figure 1.12 The microstructure of a small cumulus cloud. Left: Total concentration of droplets at various altitudes above cloud base. Right: Droplet spectra at each level. The dots connected by a dotted line show the average diameters of the droplets at each level. This figure, published in *A Short Course in Cloud Physics* by Rogers and Yau (1989), page 69 (copyright Elsevier 1989) was derived from a report by Schemenauer *et al.* (1980). Printed with permission of the authors. Copyright Environment Canada (1980).

An important function of cloud physics research is to understand the causes of various observed microstructures. Why do the spectra of this cumulus cloud broaden with height? What causes the drops to be larger near the top of the cloud, but fewer in number? Some additional information, shown in Fig. 1.13, may help us answer such questions. Computations of the maximum and average liquid water concentration show the total mass of water to increase with height above cloud base, as might be expected from continuous growth by condensation as the droplets rise in the atmosphere. The maximum observed LWC approached the "adiabatic" value, the maximum possible for the given thermodynamic conditions, but the average LWC was only about half of the adiabatic values. One explanation consistent with these data is based on "entrainment", the mixing in of drier environmental air surrounding the cloud. Some parts of the cloud at each level below 1 km may have represented cloudy air that ascended undiluted from cloud base, but much of the cloud experienced the effects of dilution. Some of the droplets in the mixed air would probably have disappeared completely from the population, causing the number concentration to be lowered. At the same time, it appears that those droplets that did survive were able to continue growing larger, perhaps even faster than if entrainment had not occurred. We will see later how fewer droplets in a given parcel of air can give rise to faster growth rates. Real clouds are always complicated, even when the only condensed phase present is liquid.

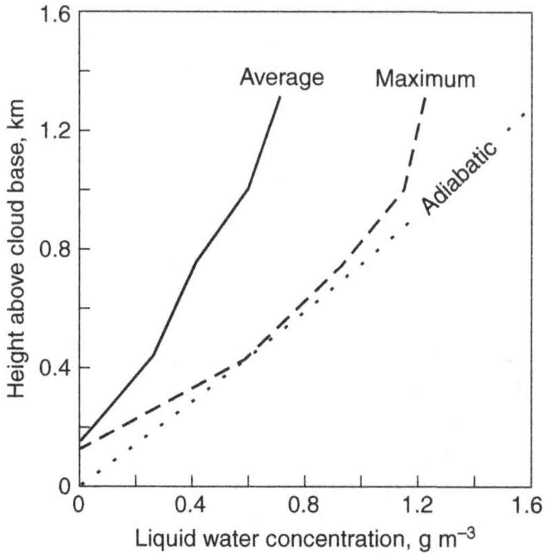

Figure 1.13 Liquid water concentrations at various heights in a small cumulus cloud. Solid curve: average measured LWC at each level; Dashed curve: maximum measured LWC; Dotted curve: computed adiabatic LWC. This figure, published in *A Short Course in Cloud Physics* by Rogers and Yau (1989), page 68 (copyright Elsevier 1989), was derived from a report by Schemenauer *et al.* (1980). Printed with permission of the authors. Copyright Environment Canada (1980).

Cold-cloud microstructure

By definition, "cold" clouds contain ice, the solid phase of water. Ice particles are seldom spherical, as the rigidity of the crystalline state prevents them from responding to surface tension, the force that causes small liquid drops to conform to the spherical shape. The microstructures of cold clouds are therefore more complicated than are those of warm-clouds. In addition to specifying the sizes of the particles, we must describe their shapes. Particle shape or form becomes a new variable, a new descriptor of cloud properties.

A few examples illustrate the wide diversity of ice forms found in the atmosphere. A collection of snow crystals gathered over several hours in Antarctica is shown in Fig. 1.14. We see, from a single event, both "columns" (long, pencil-like forms) and "plates" (thin, flat hexagons) of various sizes. The crystals that overlap in this image most likely formed individually in the atmosphere by vapor deposition and then landed on top of one another before the photograph was taken. Snow falling in more temperate climates typically develops more complicated shapes, such at the crystal shown in Fig. 1.15. The "habit" of this crystal is also a "plate", but we see a tendency for the corners of the hexagon to develop more than the sides. We also see multiple hexagons, as if one formed on top of another. Indeed, microscopic inspection of many crystals originating at temperatures in the -10 to $-12\,°C$ range show plates separated by small distance. As the diagram in Fig. 1.16 suggests, plates form on opposite sides of a cloud droplet after it freezes into a single ice crystal. The two plates align perfectly with one another because each arises (via vapor

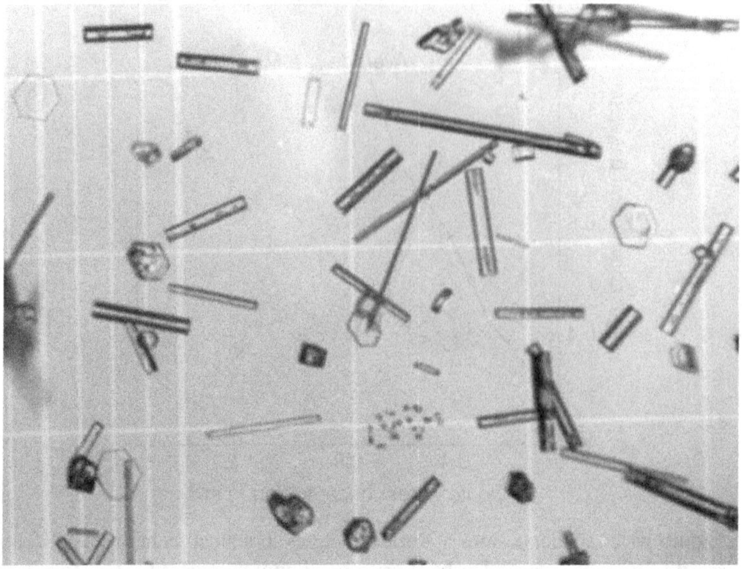

Figure 1.14 Diverse snow crystals at the South Pole. The air temperature near the surface was −60 °C, but probably increased to −40 °C near the top of the inversion. Photo by S. Warren, used with permission.

Figure 1.15 Photo of a sector plate that may be a double plate. Photo by D. Lamb.

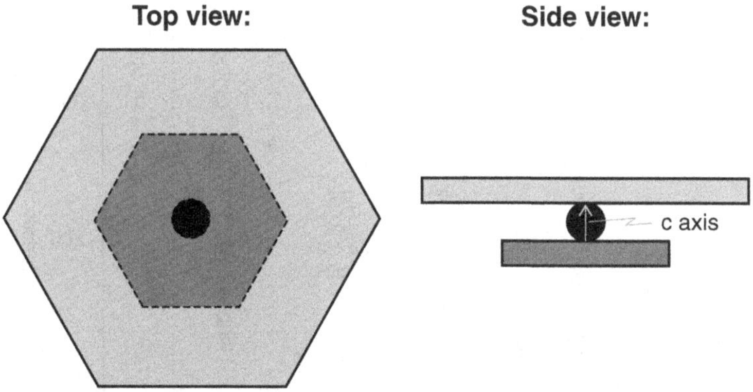

Figure 1.16 Schematic showing the geometry of a double plate. Left: view from the top. Right: view from the side showing the c-axis perpendicular to the basal plane.

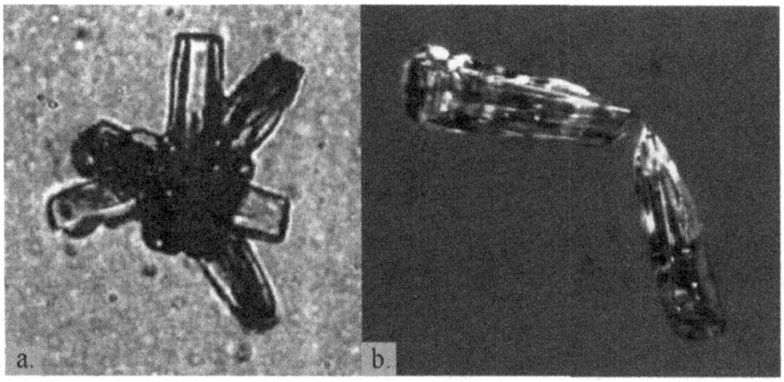

Figure 1.17 Examples of polycrystals. a. Bullet rosette. Photo by A. Heymsfield, used with permission. b. Two bullets still attached. Photo by D. Lamb.

deposition) from the same initiating crystal (the frozen droplet). At lower temperature, such as that found in cirrus clouds, each droplet freezes into a "polycrystal", a single ice particle containing several crystalline grains formed during a common freezing event. Each of the individual crystals in the particle subsequently grows by vapor deposition into either a plate or a column depending on the temperature. Figure 1.17 shows examples. Figure 1.17a shows a bullet rosette formed as the crystals grew as columns. The term "bullet" comes from the shape of the individual crystals emanating from the central frozen droplet, as shown in Fig. 1.17b. One can readily appreciate how complex ice particles can arise.

Many attempts have been made over the years to systematize the complicated patterns of the ice crystals found in the atmosphere. The most important variables controlling the shape or "habit" of a crystal formed by vapor deposition are temperature and the humidity (often expressed as an amount of water vapor above saturation). One summary for single snow crystals that grew at relatively high temperatures is provided schematically in

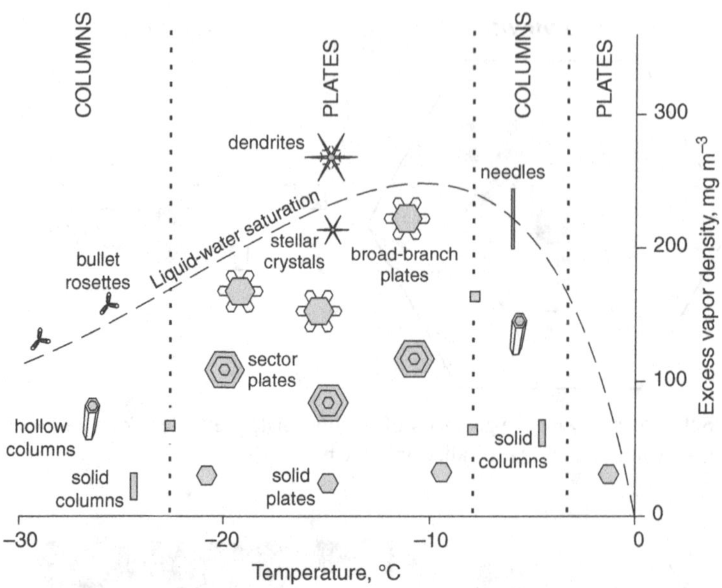

Figure 1.18 Categorization of snow crystals by temperature and excess vapor density.

Fig. 1.18. Because of the complexity in shapes (habits), we need to simplify our descriptions as much as possible. First, note that all single crystals can be classed as either "plates" or "columns", giving us the first-order category for crystal shape called the "primary habit". Observations in the atmosphere and in laboratory experiments show that the primary habit of vapor-grown ice crystals is determined mainly by the growth temperature, in the manner shown in Fig. 1.18. Plate-like crystals are found to grow under the warmest conditions, within about 3 °C below the melting temperature, but they are most commonly found in the lower temperature range of about −8 to −22 °C. Columnar crystals, by contrast, are found in the temperature ranges of −3 to −8 °C and below −22 °C. The primary habit is thus seen to alternate with temperature, with transition temperatures at approximately −3, −8, and −22 °C.

The complexity of crystal shape goes well beyond the designation of primary habit. Characteristic features superimposed on the primary habits are called "secondary habits". For instance, the indentations along the edges of plates, seen earlier in Fig. 1.15, subdivide each face into sectors, so such crystals are called "sector plates". Plate-like crystals often grow preferentially at the corners of the hexagons, because of variations in the concentration of vapor in the vicinity of the crystal. When this tendency for growth on the corners is carried to an extreme, "dendrites" result, an example of which is shown in Fig. 1.19. The branches of dendritic crystals seem to sprout randomly along each spine, but research suggests that they develop from fluctuations in the ambient humidity (degree of supersaturation). Columns, too, can generate secondary habits, but the range of variations

Figure 1.19 A dendritic snow crystal. Photo by D. Lamb.

is less extensive than that of plates. Mostly, columns vary in the length-to-width ratio, leading from compact (short) columns to needles (very long columns). Columns may also grow with hollowed features that are sometimes called "sheath" crystals. Particularly interesting crystals are combinations of columns and plates, such as the "capped column" of Fig. 1.20.

The range of possibilities for crystal shape is vast, especially at high humidities and low temperatures. As seen in Fig. 1.18, the crystal morphology becomes more intricate as the excess vapor density increases. The secondary habits are much more dependent on the humidity than they are on the temperature at the time of growth. Even the distinction between plates and columns is less clear cut at low temperatures, as shown in Fig. 1.21. Some of the crystal habits found by scientists may arise from the manor in which the studies have been carried out, but it seems clear that nature has found many ways to confound our attempts to understand how so many different forms of ice crystals can arise by vapor deposition.

Once crystals grow large enough to sediment, that is, fall relative to the local air in which they are embedded, they can collide with other crystals or with cloud droplets. Individual ice crystals often collide and stick to one another, in which case an "aggregate" or "flake" forms (Fig. 1.22). (The common usage of the term "flake" for single (usually dendritic) crystals of snow is a misnomer.) Typical aggregates can contain anywhere from several to hundreds of individual crystals. On the other hand, collisions with supercooled (liquid) water droplets that freeze on contact is termed riming. At the earliest stage of riming, one finds lightly rimed crystals, such as that shown in Fig. 1.23a. The underlying crystal is

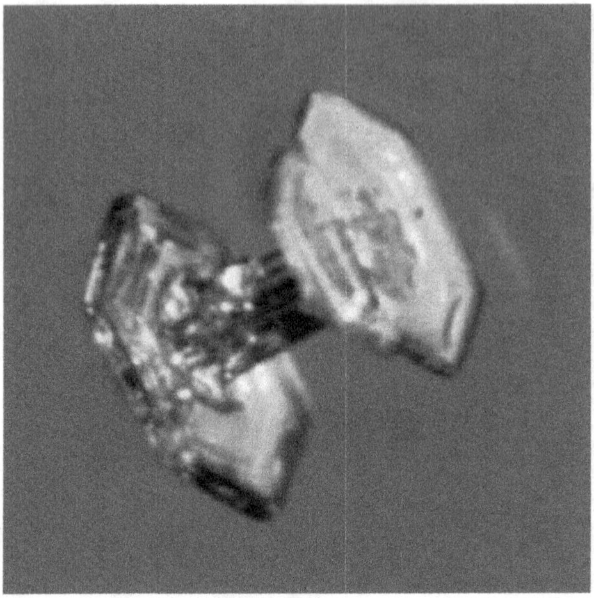

Figure 1.20 Capped column, also known as a tsusumi crystal. Photo by D. Lamb.

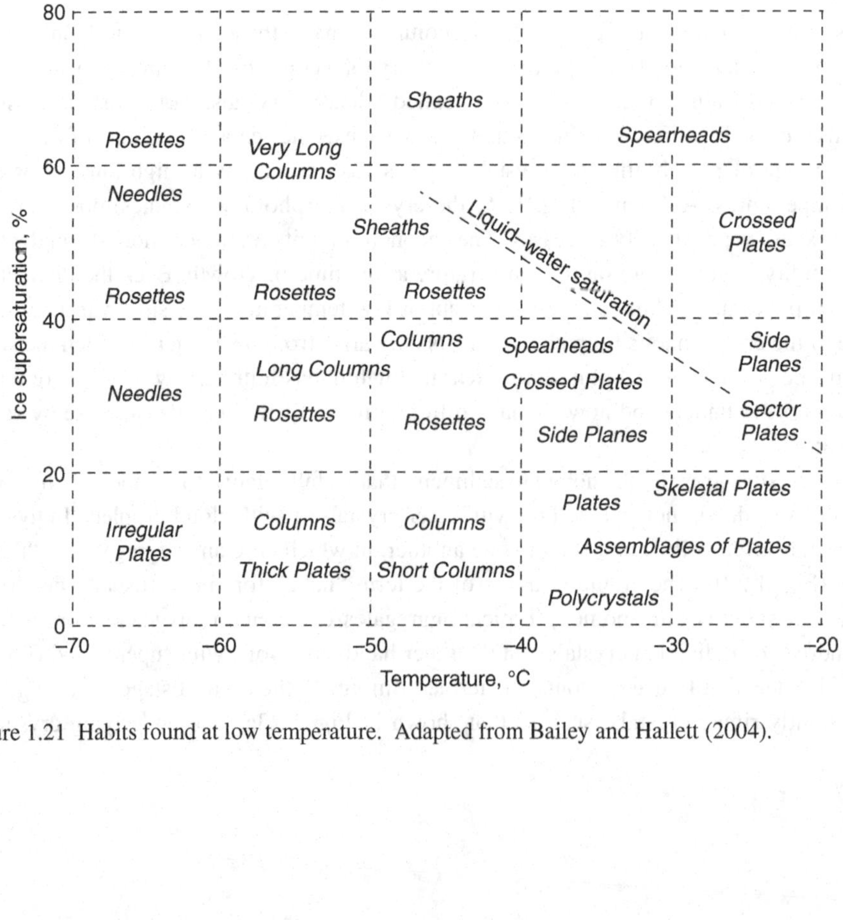

Figure 1.21 Habits found at low temperature. Adapted from Bailey and Hallett (2004).

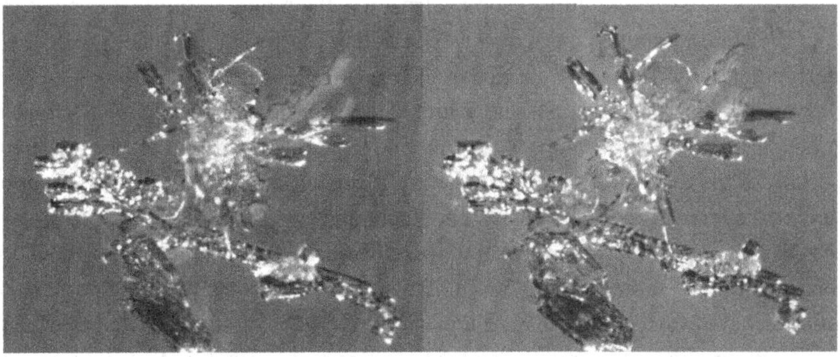

Figure 1.22 A complex snow crystal, viewable in stereo by the cross-eyed method. Photo by D. Lamb.

Figure 1.23 Stages in the riming of snow crystals. a. Light riming. Photo by N. Magee, used with permission. b. Heavy riming, sufficient to obscure the underlying crystal. Photo by D. Lamb c. Wet accretion of supercooled water on a hailstone.

clearly visible, as are the individual droplets frozen at the points of original contact. As the process of collection continues, progressively more of the crystal becomes obscured, until the underlying structure is no longer apparent (Fig. 1.23b). Such a heavily rimed particle, called “graupel”, appears roundish and white. The accretion of supercooled cloud water onto an ice particle can, in principle, continue indefinitely, leading to ever larger particles of ice. “Hail” is the term used to describe an ice particle that exceeds a diameter of 5 mm and exhibits a layered structure of alternating clear and opaque ice (Fig. 1.23c).

The sizes of ice particles found in the atmosphere vary enormously, from just a few micrometers across to more than 10 cm in diameter. Not all clouds contain particles of all sizes, of course, but taken collectively they lead to the kind of size distribution illustrated in Fig. 1.24. When the concentrations and sizes range widely, one finds it most convenient to plot data logarithmically, as done here. On average, one finds many more small particles (frozen fog or cloud droplets) than large particles (e.g., snow aggregates or hail). As a rule, with only so much condensed matter available in a given volume of cloud, it takes many small particles to make the few large particles. As suggested by the shaded area in Fig. 1.24, the concentrations of ice particles vary naturally. Much of the variability in the sizes and shapes of the ice particles found in cold clouds depends on the altitude at which the ice formed. Low altitudes are typically associated with high temperatures, whereas high altitudes imply low temperatures. Considering clouds in a collective sense, we find smaller ice particles at lower temperatures (higher altitudes) and larger particles at higher temperatures (lower altitudes). This pattern (small particles high up and large particles lower down) is the reverse of that found in warm clouds (discussed in connection with Fig. 1.12). Many ice particles defy ready classification and so are termed “irregular”. Cold clouds are very complicated and challenge our research capabilities.

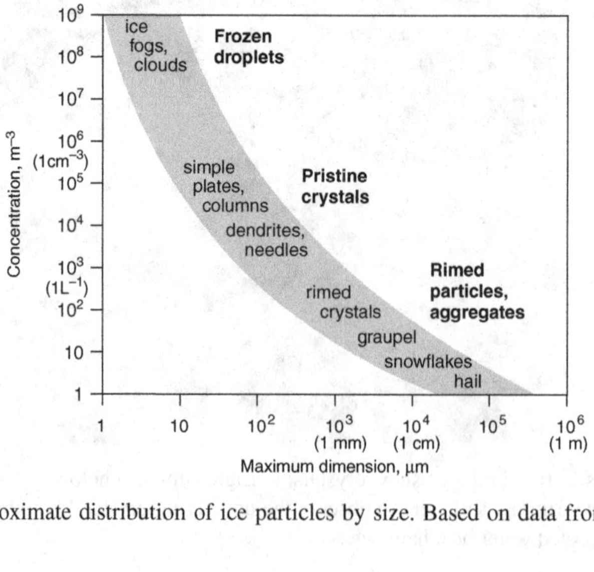

Figure 1.24 Approximate distribution of ice particles by size. Based on data from J. Hallett, used with permission.

The many past studies of the macroscopic forms of clouds and their microscopic properties have yielded a vast amount of information. With such detailed descriptions, we can begin our exploration of the physical and chemical processes that lead to the observed features. First, however, it is important that we understand the environment in which clouds form and dissipate.

1.3 Further reading

Fletcher, N.H. (1962). *The Physics of Rainclouds*. Cambridge: Cambridge University Press, 386 pp. Chapters 1 (cloud forms) and 5 (cloud microstructure).

Greenler, R. (1980). *Rainbows, Halos, and Glories*. Cambridge: Cambridge University Press, 195 pp. Chapter 2 (halos).

Houze, R.A. (1993). *Cloud Dynamics*. San Diego, CA: Academic Press, 570 pp. Chapter 1 (cloud identification).

Lehr, P.E., R.W. Burnett, and H.S. Zim (1987). *Weather*. New York: Golden Press, 160 pp.

Ludlam, D.M. (1991). *National Audubon Society Field Guide to North American Weather*. New York: Alfred A. Knopf, 655 pp.

Parviainen, P. and H. Fountain (2004). *Clouds*. New York: Barnes and Noble, 319 pp.

Pruppacher, H.R. and J.D. Klett (1997). *Microphysics of Clouds and Precipitation*. Dordrecht: Kluwer Academic Publishers, 954 pp. Chapter 1 (cloud microstructure).

Rogers, R.R. and M.K. Yau (1989). *A Short Course in Cloud Physics*. New York: Pergamon Press, 293 pp. Chapter 5 (cloud properties).

Schaefer, V.J. and J.A. Day (1981). *A Field Guide to the Atmosphere*. Boston, MA: Houghton Mifflin, 359 pp.

1.4 Problems

1. Draw a "slash diagram" of "Issues/Concerns in Atmospheric Physics". Categorize them appropriately, recognizing that the greater the variety, the better.

2. Clouds consist of moist air and particles of condensed water, either in the liquid or solid state. Here, consider an idealized, but nevertheless "typical" cloud that contains only liquid water droplets of uniform radius $r_d = 15\ \mu\text{m}$. The droplet number concentration is $n_d = 100\ \text{cm}^{-3}$, the temperature $T = 0\,°\text{C}$, and the pressure $p = 600\,\text{hPa}$. Calculate the following cloud properties.
   (a) The approximate distance between the centers of the droplets. How many droplet radii does this spacing represent? (It is sufficient to assume that the drops are uniformly spaced for this problem.)
   (b) What is the "volume fraction" of liquid water in the cloudy air? (i.e., What fraction of any given volume of the cloudy air, say 1 m^3, is occupied by the liquid drops?). What is the dominant phase of the cloud (by volume)?
   (c) The "liquid water concentration" ω_L of the cloud, in units of g m^{-3}.

3. For an entire semester, look at the sky at least once each day and keep a daily log of your sky observations. You may choose the time of day when you find it most convenient to

make the observations, but try to stick to that time within about an hour. An observation is to be made and recorded every day, weekends and holidays included. The log need not be complicated or verbose; it is more important that you look up routinely and assess what you see in a systematic way. (The log, a notebook or a collection of individual sheets of paper stapled together, will be collected toward the end of the semester and graded as a term project. Your log will be returned unmarked, so make it neat and something to retain for your personal records.)

The minimal pieces of information to include in the Sky Log are the date, time, and location of the observation, the general state of the sky (e.g., partly cloudy, overcast and raining) and the type(s) of clouds present. For cloud type, use the international abbreviations given in Appendix A. Remarks about any unusual or interesting conditions may be included as appropriate.

4. You may have noticed that the drops in some rain events are larger than in others. This activity gives you the opportunity to quantify your observations of rain. Over the course of the semester, develop and use a technique for measuring the sizes of individual raindrops. Then, analyze your data in terms of a proper size distribution and write up your methods and findings in a report.

   Questions to consider and discuss in your report:

   - What is the range of drop diameters that your method can potentially accommodate? That is, what are the diameters of the smallest drop and the largest drop that your method could reliably measure?
   - How large is the uncertainty of your individual size measurement? What is the basis for determining this uncertainty value?
   - How could you improve on the method, were you to have a future opportunity?
   - Which kinds of clouds or weather systems tend to form the largest drops? Explain.

2

The atmospheric setting

Clouds form in an atmosphere that is rich in water vapor and diverse chemical compounds. These components reside in the Earth's gaseous envelope and so become organized is ways that respond to gravitational attraction and to the steady input of energy from the Sun. If we are to understand the properties and behavior of atmospheric clouds, we must first understand the environment in which clouds form.

2.1 Composition

What actually makes up or constitutes the Earth's atmosphere? The answer to such a question depends on time, for the composition of the atmosphere has been and still is changing. For some compounds the changes are slow, for others they are rapid. Here, we explore how the atmosphere evolved to its present form, the fundamental nature of the substances making up the atmosphere, and how those substances are currently distributed throughout the atmosphere.

2.1.1 History of the atmosphere

Our atmosphere has evolved dramatically since the Earth formed some 4.6 billion (4.6×10^9) years ago. Initially, while the Earth was still hot from the gravitational accumulation of matter in the solar nebula and from the heavy bombardment of rocky planetesimals, the atmospheric composition was more akin to that of the Sun. It is during this early heavy-bombardment period that the Moon most likely formed and helped stabilize the Earth's rotational axis. The primary or proto-atmosphere contained much hydrogen and helium, reflecting the dominant composition of the stellar matter. However, other elements were present, too, for the Sun is thought to be a second- or third-generation star, meaning it arose from the debris of earlier star explosions (supernovae) that generated higher atomic-number elements during nuclear fusion. The Earth's proto-atmosphere could not have lasted long without an effective way to shield the atmosphere from the solar wind, the flux of electrically charged particles (electrons, protons and ions) released from the Sun.

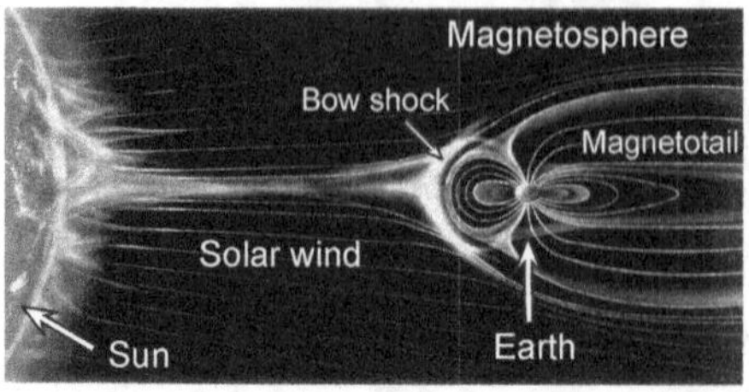

Figure 2.1 The dominant connections between the Sun and the Earth. Not to scale. Rendition from NASA.

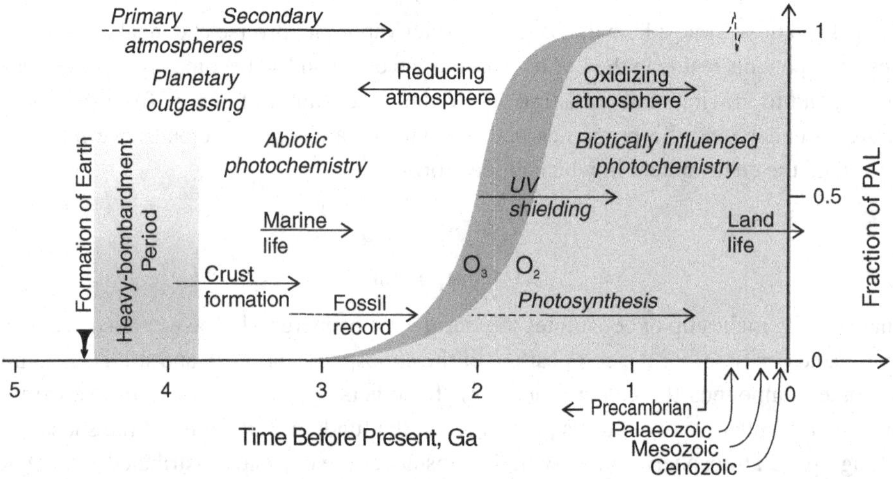

Figure 2.2 Time line of atmospheric evolution. Time units are Ga, 10^9 years. Shaded regions suggest the build-up of oxygen (O_2) and ozone (O_3). PAL is the abbreviation for present atmospheric level.

A secondary atmosphere subsequently evolved under the protection of a magnetosphere (Fig. 2.1) that formed while the Earth was still hot. Iron and nickel settled into the core of the planet, where rotational motion generated a magnetic field that extends well out into space. The charged particles in the solar wind respond to this magnetic field by following the magnetic lines of force. Some solar-wind particles can sneak into the atmosphere near the poles (and cause auroras), but most skirt around the Earth, missing the atmosphere entirely. The magnetosphere is an important protector of the atmosphere we have today.

The early evolution of the secondary atmosphere was characterized by the build-up of gases released from the cooling planet. Figure 2.2 illustrates how the atmosphere may have evolved over the course of Earth's history. Water vapor injected into the atmosphere by volcanic activity and general outgassing condensed in and precipitated from clouds, giving rise to great oceans. Over time, carbon dioxide (CO_2) dissolved into the oceans and

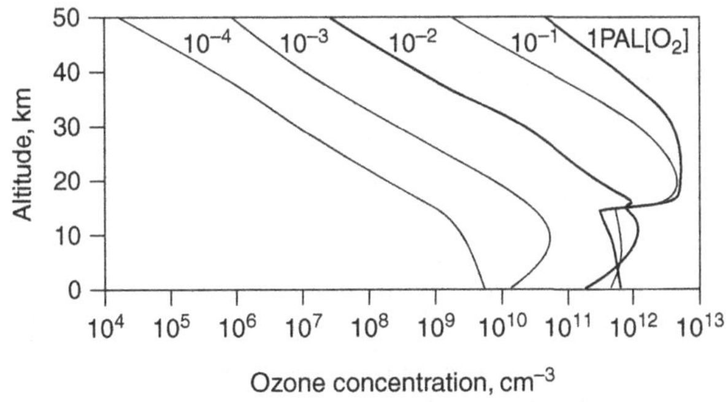

Figure 2.3 Evolution of ozone in the Earth's atmosphere. PAL = present atmospheric level. From Levine (1982) used with kind permission from Springer Science + Business Media: *J. Mol. Evol.*, The photochemistry of the paleoatmosphere, vol. **18**, 1982, pp. 161–172, J.S. Levine, Fig. 2.8.

combined with calcium ions (Ca^{2+}) to form carbonate rocks. Diatomic nitrogen (N_2), with its strong triple bond, is relatively unreactive and so was left to become the major gaseous constituent of the atmosphere, then as now.

The initial volatiles introduced into the atmosphere were exposed to harsh sunlight and high temperatures. Some water vapor, along with carbon dioxide, was photolyzed by ultraviolet (UV) radiation to produce some of the first oxygen (O_2). Such abiotic photochemical reactions also led to the formation of formaldehyde (H_2CO), an important precursor of life. Some of the atmospheric nitrogen was likely "fixed" (made reactive) by lightning, leading to the formation of cyanic acid (HCN), another precursor of life. The chemistry of the early atmosphere was complicated, even before the appearance of life.

Oxygen levels were held down in the early atmosphere because O_2 readily combined with iron-based minerals to form oxides that sequestered the oxygen in the Earth's crust. The early atmosphere was mildly reducing until such sinks of O_2 became saturated and new sources of oxygen arose. In the presence of ultraviolet light, O_2 itself photolyzes, producing in the process small amounts of ozone (O_3), which acts to shield the surface from harsh ultraviolet radiation. Until the concentration of ozone built up sufficiently, life was confined to deeper parts of the oceans. Results from modeling the chemistry of the early atmosphere, shown in Fig. 2.3, indicate that ozone can become an effective shield of UV radiation well before molecular oxygen reaches a significant fraction of its present atmospheric level (PAL).

Details about how and when a reducing atmosphere changed to an oxidizing one are unclear, but the transition is likely tied to the evolution of oxygen-producing bacteria and more complex life forms in the oceans and/or lakes. The earliest organisms capable of photosynthesis had to face the dual problem of needing sunlight, while at the same time avoiding the deadly UV radiation that penetrated through the atmosphere and into the upper levels of the ocean before the ozone shield was in place. Nevertheless, once photosynthesis became the dominant means for converting sunlight into organic matter, oxygen levels rose dramatically to the levels ($\sim 20\%$) we observe today, save for a time of rapid fluctuation

Table 2.1 *Abundances of gases in the present atmosphere.*

Gas	Relative Abundance	Mean Residence Time
	Major Constituents	
Nitrogen (N_2)	78%	10^6 years
Oxygen (O_2)	21%	10 years
Argon (Ar)	~1%	(accumulates)
	Variable/Minor Constituents	
Water (H_2O)	~0.5% varies 10 ppm – 5%	5–12 days
Carbon dioxide (CO_2)	380 ppm (increasing)	10 ++ years
Methane (CH_4)	2 ppm (increasing)	7 years
Ozone (O_3)	1–10 ppm (stratosphere) 10–100 ppb (troposphere)	~ months
Nitrous oxide (N_2O)	300 ppb	10 years
Ammonia (NH_3)	0.1–1 ppb	20 days
Sulfur dioxide (SO_2)	10–100 ppt	40 days
Nitric oxide (NO·)	1 ppt–10 ppb	1 day
Hydroxyl (OH·)	ppt	<1 second
Chlorofluorocarbon (CFC-12)	500 ppt	100 years

some 350 million years ago (during the late Paleozoic era; see dashed curve at top of Fig. 2.2). As the atmosphere shifted from a reducing to an oxidizing environment, around two billion years ago, the existence of the protective ozone layer was ensured and life could gradually evolve into ever-more complex forms in the oceans and eventually on land. Oxygen, produced by life, gives an evolutionary advantage to the forms of life we recognize today.

The delicate balance that now exists between the biosphere and the atmosphere has prompted the formulation of the "Gaia Hypothesis", a suggestion that life not only adapts to, but actively adjusts its environment for its own good. Whether the interactions between the biosphere and the atmosphere are purposeful or not, we do well to recognize that the composition of the atmosphere is strongly coupled to biological processes and photochemistry. Clouds, by providing the fresh water needed to sustain terrestrial biota, by reusing the same evapotranspired water, and by modulating the climate, play key roles in the biospheric–atmospheric interchanges.

The composition of the present atmosphere reflects its evolutionary history. A concise summary of the so-called permanent and variable gases is presented in Table 2.1. We recognize the permanence of nitrogen (N_2), certainly, but we must also realize that this relatively stable gas is nevertheless gradually being formed and destroyed. It is just that the turn-over time, the time required to replace the current set of N_2 molecules with a new set, is millions of years, so N_2 appears permanent from our human perspective. Oxygen (O_2), also seems permanent, and indeed its concentration in the atmosphere remains stable over very long periods. However, unlike N_2, O_2 is both produced and destroyed rapidly. The average time an O_2 molecule resides in the atmosphere is only about 10 years. The steady abundance of O_2, it has been argued, arises from the balance between its production by photosynthetic plants and its consumption by animals, to some extent, and wild fires. Other permanent gases include the noble gases, compounds unable to react at all. They are either hold-overs from the proto-atmosphere or arise from radioactivity and simply accumulate (e.g., argon). Most other gases are variable, in the sense that their concentrations depend on location and circumstances. The atmospheric residence times of these gases is usually short compared with the times needed for air to move around the world or between hemispheres. Carbon dioxide (CO_2) and methane (CH_4) are two of the most important gases associated with global climate. Both are produced naturally, although humans have contributed to their abundances. These greenhouse gases are currently being introduced into the atmosphere at rates exceeding their removal, meaning that their concentrations are increasing with time. The residence time for CO_2 is difficult to estimate and so has been specified with a low number (10 years, reflective of removal by the biosphere), followed by plus signs, to suggest that other removal processes act over much longer time periods (hundreds to thousands of years for absorption into the oceans; tens to hundreds of thousands of years for weathering of rocks). Water is also an effective greenhouse gas, but its concentration is limited by its ability to change phase and precipitate out of the atmosphere. Water vapor is perhaps the most variable of the gaseous compounds because of its ability to condense into liquid and solid particles. On average, water vapor enters the atmosphere in relatively warm regions and condenses in cooler regions, inside clouds.

In general, the characteristics and behaviors of the various gases depend in large measure on their molecular structures and abilities to react with other compounds. Large abundance of a compound does not always signify importance in an atmospheric-chemistry sense. In fact, many of the most reactive compounds (e.g., OH) are found in the least abundance, simply because they are consumed by chemical reactions almost as rapidly as they are produced. Water will be the focus of subsequent discussion because of its ability to change phase and form clouds.

2.1.2 Molecular nature of matter

Matter is composed of atoms and molecules. Molecules, groupings of atoms bound together by electrical forces, constitute the common substances of our everyday world. Atoms are made of various subatomic particles, mainly electrons, protons, and neutrons. Most of the mass of an atom is contained in the protons and neutrons that constitute the nucleus, the tiny center of the atom. The electron, the carrier of negative charge, revolves around the nucleus, positively charged because of the protons. Protons are bound to the uncharged neutrons and to each other inside the nucleus by the so-called "strong nuclear force". A short-range "weak nuclear force" also exists and accounts for radioactivity, but neither of these nuclear forces needs to be considered further for our understanding of atmospheric physics. Of the fundamental forces of nature, only the "electromagnetic force" and the "gravitational force" are important. Of these two, the electromagnetic force is by far the strongest and accounts for almost all of the bonding between atoms in molecules, as well as between molecules in the condensed phases of matter (liquids and solids). Gravity affects the macroscopic physical behavior of matter, but molecular-scale electrical forces determine the chemical properties of matter.

The nature of matter is best understood by considering the electronic structure of atoms and molecules in some detail. A schematic representation of the atoms that make up the water molecule is shown in Fig. 2.4. As with all atoms, the hydrogen and oxygen atoms each have a positively charged nucleus, around which the negatively charged electrons revolve. It is tempting to think of the electrons in an atom as analogous to the planets orbiting the Sun in our solar system. However, important differences exist. Whereas planets are attracted to the Sun by the gravitational force, electrons are attracted to the nucleus by the electromagnetic force. Planets follow well-defined orbits, all nearly in a common

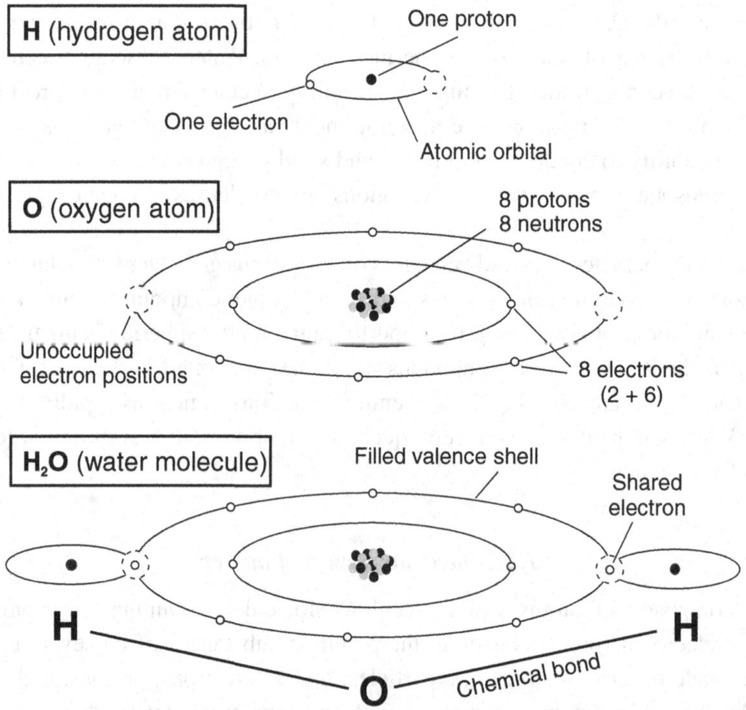

Figure 2.4 Schematic of the atoms that combine to make water. Note that atomic and molecular orbitals are not confined to a single plane. From page 73, Figure 3.19 in Earth under Siege by R.P. Turco (1997), used by permission of Oxford University Press, Inc.

plane, whereas electrons surround the nucleus in ill-defined shells called orbitals. The laws of classical physics describe the motions of planets around the Sun; the laws of quantum mechanics are needed to understand the behavior of electrons in atoms and molecules. The properties of water arise in large measure from the quantum mechanical interactions between the oxygen and hydrogen atoms in the water molecule.

How atoms combine with each other chemically to form molecules can be understood in part through elementary concepts of quantum mechanics. Atomic orbitals have discrete energy levels, which means that the electrons, even though their positions are uncertain, exist in well-defined energy levels. Electrons in low-level energy states are physically close to the nucleus and so are bound tightly to it. Outer electrons, those in higher energy levels, are further removed from the nucleus and are moreover partially shielded by the inner electrons. These outermost, or "valence" electrons feel the electrostatic pull of the nucleus less and so interact with other atoms relatively easily.

Beyond requiring discrete energies, quantum mechanics places restrictions on the interactions of elementary particles with each other. The Pauli exclusion principle, for instance, asserts that no two identical particles can occupy the same quantum state at the same time. The quantum state of an atom is described by a set of four discrete quantum numbers: the principal quantum number, the angular momentum quantum number, the magnetic quantum number, and the spin quantum number. Of these, we need concern ourselves mainly with the first two, but we must recognize that the spin quantum number allows two electrons of opposite spin to coexist in otherwise the same quantum state. In fact, electrons of opposing spins readily pair up.

The quantum-mechanical rules set limits on the number of electrons the shells can accommodate, and they help us understand how water arises from hydrogen and oxygen. As shown in Fig. 2.4, the first, inner-most shell (principal quantum number 1) may contain at most two electrons (one with spin up, the other with spin down); the neutral hydrogen atom contains only one electron, so another electron could be accommodated in principle. The next shell out from the nucleus (principal quantum number 2) can accommodate eight electrons; oxygen by itself contains only six valence electrons, so it has room for two additional electrons, as suggested by the dashed circles in Fig. 2.4. The water molecule is formed when two hydrogen atoms combine with a single oxygen atom, in the process satisfying the full-occupancy rule for each shell. The two missing positions in the valence shell of the oxygen atom are filled by the electrons from the two hydrogen atoms, while at the same time the lone electron of each hydrogen atom gains an opportunity to pair up with an electron of opposite spin from the oxygen. The net result is a stable water molecule, comprised of two hydrogen atoms and one oxygen atom, with all of the electrons appropriately paired up. Water is therefore H_2O. Quantum mechanical principles let us understand the composition of the water molecule, but we must delve deeper. If we stopped here, we would still have an insufficient basis upon which to understand the properties of water and the formation of clouds.

The geometrical and electrical structure of water, not just its composition, proves to be crucially important. Once the hydrogen and oxygen combine in the proper proportions,

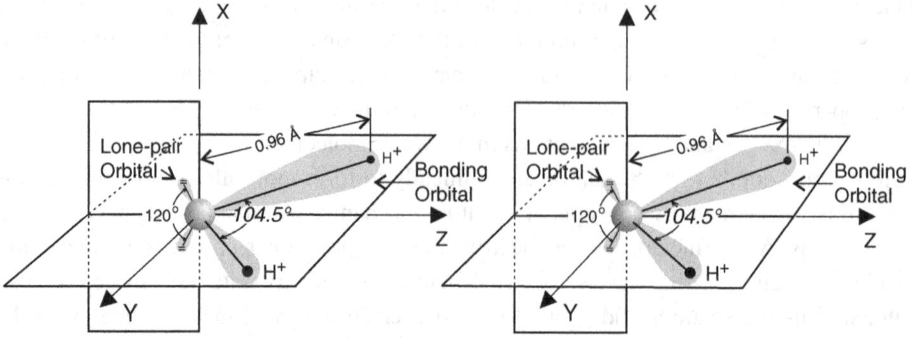

Figure 2.5 Stereographic view of the water molecule (using the cross-eye method).

the electrons arrange themselves into molecular (versus atomic) orbitals. The electrons in effect rearrange themselves to minimize the total energy of the molecule. The resulting structure for water is shown schematically in Fig. 2.5. We see that the water molecule has a distinct three-dimensional character. Branching out from the oxygen atom in the center are two orbitals that bond the two hydrogen nuclei (protons, H^+) with a bond angle of 104.5 °. Each bond is just shy of 1 Å (10^{-10} m) in length and contains two of the eight valence electrons (shaded region). In a plane perpendicular to that formed by the bonding orbitals are two non-binding or "lone-pair" orbitals. These short orbitals are oriented at an angle of 120 ° from each other and also contain two electrons each. The eight valence electrons are thus arranged in four directed orbitals, the two bonding orbitals in one plane and the two lone-pair orbitals in an orthogonal plane.

This arrangement of orbitals gives water its basic physical and chemical properties. The bonds between the O atoms and H^+ are strong ($\sim$ 430 kJ mol^{-1}), so water is chemically stable. (Liquid water is, after all, an effective fire-fighting substance.) As we shall see, the bonds are also polar, so liquid water is a good, almost universal solvent. The polarity of its bonds gives water a strong permanent dipole moment (6.1×10^{-30} C m), so the water molecule is readily agitated when exposed to microwave radiation; liquid water can be heated in a microwave oven, and weather radar can "see" clouds. The polarity of the bond also lets water molecules attach strongly to other water molecules, so clouds form under conditions that would not permit other substances to condense.

The bond holding each hydrogen atom to the oxygen atom inside the water molecule is said to be a "polar covalent" bond. It is covalent in the same way that many strong chemical bonds are: an electron pair is shared among two neighboring atoms, in the process holding the adjoining, positively charged nuclei together and acting as electrostatic "glue". However, the sharing between unlike atoms is not equal. Rather, it is unbalanced, one atom hoarding more of the electronic charge than the other. Every atom hangs on to its valence electrons with different strengths. The lone outer electrons in the alkali metals (lithium, sodium, potassium, etc.), for instance, are only weakly held to their respective nuclei; the outermost electron in each case is not only relatively far

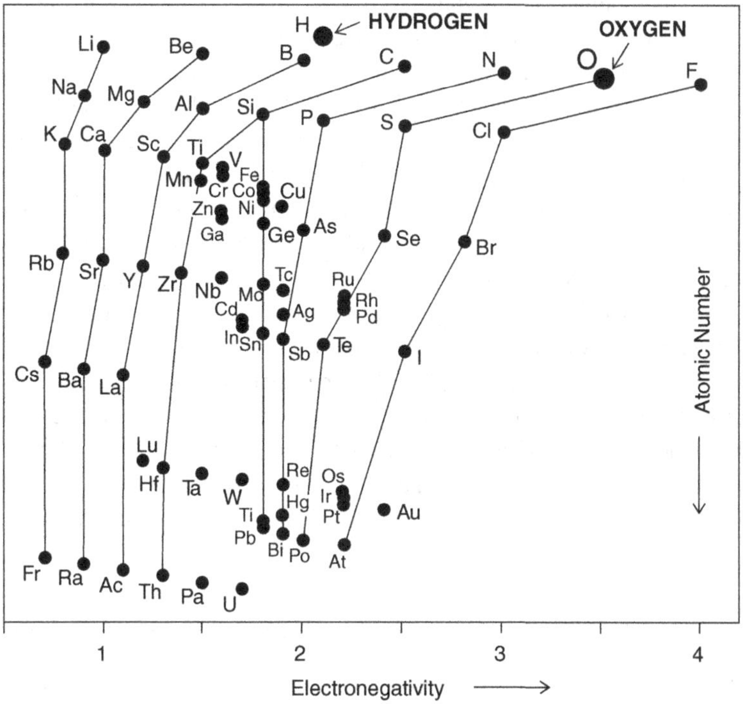

Figure 2.6 Electronegativities of the atoms. After Pauling (1960).

removed from the nucleus, but it is also shielded from the attracting positive charge of the nucleus by the intervening electrons. By contrast, the nuclei of atoms at the other side of the Periodic Table, such as fluorine and oxygen, exert a strong hold on their outer electrons. One measure of the strength with which a nucleus attracts valence electrons is the electronegativity of the atom. As Fig. 2.6 shows, the most electronegative elements are fluorine, oxygen, nitrogen, and chlorine; the least electronegative are the alkali metals. When atoms of strongly opposing electronegativities combine, electrons may be transferred completely from the atoms of small electronegativities to those of large electronegativities. Substances such as HF and NaCl are bound electrostatically by such ionic bonds. Hydrogen, by contrast, has an intermediate electronegativity, so when it is combined with oxygen, as in the water molecule, the electrons forming the O-H bond are still shared, but they are biased toward the oxygen molecule. That is, the center of negative charge is shifted toward the oxygen atom, leaving the bond polar (i.e., with positive and negative ends). The modest difference in electronegativities of hydrogen and oxygen give the covalent O-H bonds in the water molecule partial ionic character. The polar-covalent bonds of the bonding orbitals give each water molecule its permanent dipole moment and account for water's chemical stability and ability to condense readily under atmospheric conditions.

The strength of any bond is measured in terms of the energy required to break the bond and set the neighboring atoms or molecules free from one another. We may like to think of bonds as strong or weak in terms of force, but energy provides a more unique and useful measure of bond strength. The relationship between force and energy can be understood through the physical concept of work, the amount of energy required to move an object against an attracting force. Think of lifting a heavy object, a bowling ball, for instance. The Earth's gravitational force tries to pull the ball downward at the same time that you are lifting it. The work you put into the job depends on how high you lift the ball vertically; horizontal displacements do not matter because gravity has no component of force in those directions. In the molecular world, the dominant force is electrical, not gravitational, and the important directions are radially outward from an atom or center of charge.

Calculations of work need account only for components of displacement along the direction of force. Mathematically, we take such factors into account by recognizing that forces and displacements are vectors and by taking the dot product in order to isolate the component of displacement along the lines of force. Thus, for a small displacement **dr** within a force field $\mathbf{F}(r)$, the work V is given by integrating over the distance traveled $(r_2 - r_1)$:

$$V = -\int_{r_1}^{r_2} \mathbf{F}(r) \cdot \mathbf{dr}. \tag{2.1}$$

The negative sign tells us that **F** and **dr** act in opposing senses, letting V be positive when work is done on the object. The work so done is in effect stored in the field because it can be utilized later. This stored energy is called the potential energy of the field and depends only on the position of the object in the field. As shown by Eq. (2.1), only the integral of force, not the detailed dependence of the force on distance, $\mathbf{F}(r)$, determines the potential energy. Calculating the integral, however, does require detailed knowledge of how force varies with distance.

A key concept to understand with any type of bonding is the competing effects of electrostatic attraction (between charges of opposite sign) and electrostatic repulsion (between charges of like sign). Attractions can arise in a variety of ways between the electrons and nuclei of atoms, but they always result in a lowering of the potential energy as the relative distance between atoms decreases (lower, dashed curve in Fig. 2.7). Repulsions, on the other hand, arise from the interactions between the nuclei and cause the potential energy to increase as the atoms approach closely to one another (upper, dotted curve in Fig. 2.7). The net result, the sum of the separate potential-energy curves (shown as the solid curve), invariably has a minimum, which is where one atom resides when bound stably to another. The one atom (B) sits at the base of the potential well formed by the other (A). We can now appreciate why energy is a more meaningful measure of bond strength than is force. Force varies in complicated ways with separation distance, and the net force at the balance position is zero. However, the energy to break the bond is the integral of the net force from the balance point to infinity, which is just the depth of the potential well.

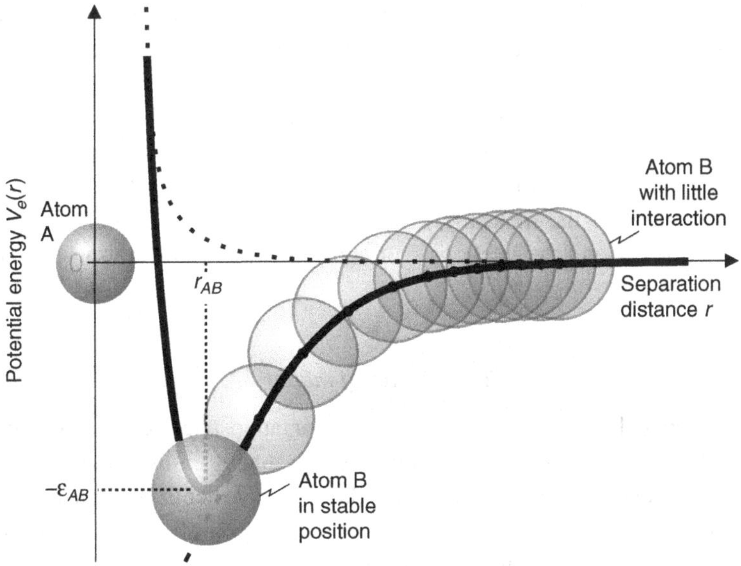

Figure 2.7 The interaction potential between two atoms viewed as a "molecular slide".

The energy of a bond manifests itself in the heat liberated or released as the bond is made. The position of atom B in Fig. 2.7 is near the zero of potential energy when B is relatively far from the stable, bound position. Initially, as atom B first responds to the attractive force of atom A, it accelerates slowly. The force causing the acceleration of B toward A is electrostatic and equal to the gradient of the electrostatic potential energy V_e:

$$\mathbf{F} = -\nabla V_e. \tag{2.2}$$

The gradient is small at large separation distances, so the attractive force is initially weak. As atom B approaches A, the potential energy decreases ever more sharply, causing an ever stronger acceleration of atom B toward atom A (schematically illustrated by the progressively greater distances between the centers of atom B along the solid curve in Fig. 2.7). Atom B literally picks up speed as it gets near the bottom of the potential well, much as a bicycle would traveling down a steep hill. Once beyond the potential minimum, however, atom B increasingly responds to the repulsive force of the nuclei. The potential energy curve increases sharply and causes atom B to slow down. The considerable momentum gained by atom B in the attractive part of the curve is conveyed forcefully to atom A during the repulsive part of the curve. The potential energy has been converted into kinetic energy, which becomes shared between the two atoms. Unless the kinetic energy can be dissipated to other atoms in the vicinity, the kinetic energy will stay in the bond, causing atom B to reverse its motion and return along its initial path to the unbound state. Generally, however, the kinetic energy imparted to atom A during collision is dissipated one way or another, allowing atom B to settle into its stable position at the bottom of the potential well. Atom B is then trapped by A and we say that the bond between A and B has been formed. The decrease in potential energy has manifested itself in an increase in molecular kinetic energy of the neighboring atoms (or molecules). The so-called heat of reaction in chemistry, or the latent heat of phase change in physics, is the thermal energy generated during bond formation. When atoms or molecules combine, molecular potential energy is converted into molecular kinetic energy (the random motions of the molecules), which we observe as a heating of the system. Both chemical energy and the latent heat of condensation follow the same principles of physics.

2.1.3 States of matter

The diverse chemical compounds that make up our atmosphere exist in distinct states, either as gases, liquids, or solids. The atmosphere is mainly in the gaseous state, that is as individual molecules (mostly N_2 and O_2). Gas molecules collide frequently with each other, but they do not stick together. Gases are thus unconstrained and tend to fill the space available. Suspended in this gaseous mixture are many particles, bits of condensed matter that may be either in the liquid or solid state. The molecules in condensed states of matter are bound to each other by electrical forces of the type just discussed. Particulate matter is relatively dense ($\sim$ 1 to 2 $\mathrm{g\,cm^{-3}}$), so only small particles (up to about 100 μm in diameter) remain suspended for significant times.

Any suspension of liquid or solid particles in a gaseous medium is called an aerosol. Because clouds contain liquid droplets and solid ice particles suspended in air, clouds may be thought of as an aerosol. However, we often distinguish the atmospheric aerosol from a cloud based on the degree of water condensation that has taken place. The nonaqueous component of the atmospheric aerosol is pervasive, but clouds exist only where active condensation is thermodynamically allowed, that is, where the relative humidity exceeds 100%. Some of the aerosol particles become cloud particles as air becomes supersaturated and the particles serve as preferred sites of condensation.

The physical properties of all ordinary matter depend on the state of matter and the conditions under which the matter is exposed. For example, most materials compress somewhat under pressure and expand as the temperature rises. An increase in pressure causes the molecules to move closer together, making the material denser; higher temperature causes molecular motions to increase, making the molecules move farther apart and the material less dense. Instead of density, a measure of how concentrated the matter is, one may use specific volume, the physical volume occupied by a specified amount of matter. Specific volume and density are inversely related. Thus, if we let v represent the volume occupied by a mole of material [units $\mathrm{m^3\,mol^{-1}}$] and n the molar density [$\mathrm{mol\,m^{-3}}$], then $v = 1/n$. Meteorologists tend to use mass-based units, in which case the mass-specific volume α [units $\mathrm{m^3\,kg^{-1}}$] $= 1/\rho$, where ρ is the mass density [$\mathrm{kg\,m^{-3}}$]. Regardless of how we choose to express the amount of material, the volume of matter in a particular state depends on the environmental temperature T and pressure p.

Common measures of how strongly the volume changes with given changes in pressure and temperature are expressed by coefficients of compression and expansion, respectively. The isothermal compressibility is the fractional change in volume per change in pressure while the temperature is held constant:

$$\kappa_T \equiv -\frac{1}{v}\frac{\partial v}{\partial p}\bigg|_T . \tag{2.3}$$

(The inverse of the compressibility is called the bulk modulus, a measure of resistance to compression.) The isobaric coefficient of thermal expansion, sometimes called the expansivity, is

$$\alpha_p \equiv \frac{1}{v}\frac{\partial v}{\partial T}\bigg|_p . \tag{2.4}$$

These response functions are important ways of characterizing matter in different states.

A quantitative relationship between pressure, temperature, and the density (or specific volume) of any given substance is known as an equation of state. Pressure and temperature are independent state variables that affect the material density or specific volume. Each state of matter behaves differently and so is often specified by a unique equation of state. However, efforts are under way to develop equations of state that describe substances generally, without restriction to a particular state. We must distinguish between an empirical relationship between the state variables, meant to describe real substances, and an idealized relationship derived from a conceptual or molecular model. Analytical equations of state can be good representations of matter, but we must recognize that they are only approximations of reality.

Each of the three states of matter have different equations of states and exhibit different properties under a given set of environmental conditions. The unique features of each state are discussed below. Clouds represent changes in state (e.g., from vapor to liquid water, or from liquid to ice), so one cannot understand clouds without a clear understanding of each state separately.

Gases

A gas is a volume-filling fluid that is both highly compressible and elastic. An external pressure applied to a gas causes the volume occupied by a given amount (mass or mole) of the gas to shrink; its specific volume decreases with increasing pressure at any given temperature. When the external pressure is removed, the specific volume returns to its former value; the gas springs back to its original state with no permanent change in its properties.

The equation of state of a gas has been thoroughly explored through experimentation and theory over the past couple of centuries. In diverse experiments, the magnitudes of the state variables were varied systematically, demonstrating that ordinary air obeys the following proportionality to good approximation:

$$\frac{pV}{T} \propto N, \tag{2.5}$$

where p is the pressure, T is the thermodynamic (absolute) temperature, and N is the amount of gas in volume V. Refined experiments show that real gases deviate slightly from the form given by Eq. (2.5), but many purposes are satisfied by this simple proportionality, which forms the basis of the ideal gas law. An ideal gas is one made of atoms or molecules that have neither significant volumes nor any attractions toward one another; the molecules of an ideal gas may be thought of as point masses. The proportionality coefficient in Eq. (2.5) is obtained from experimental data. If N is specified in moles, then the data yield the universal gas constant $R = 8.314\ \mathrm{J\,mol^{-1}\,K^{-1}}$. One version of the ideal gas law, the extensive form used by physical chemists working with well-defined volumes of gas, is $pV = NRT$. Atmospheric scientists, however, because they work with ill-defined volumes of air, use the simpler, intensive form of the ideal gas law:

$$p = nRT, \tag{2.6}$$

where $n \equiv N/V$ is the molar concentration of the gas, the number of moles per unit volume (usually taken to be 1 $\mathrm{m^3}$). Frequently, one needs to express the gas concentration as the number of individual molecules per unit volume, in which case the ideal gas law may be written

$$p = n_1 k_B T, \tag{2.7}$$

where n_1 is the molecular number concentration, $k_B = R/N_{Av} = 1.38 \times 10^{-23}$ J molec^{-1} K^{-1} is the Boltzmann constant and $N_{Av} = 6.02 \times 10^{23}$ molec mol $^{-1}$ is Avogadro's number.

The simplicity of the ideal gas law reflects the nature of gases at relatively low pressures and high temperatures. We can conceptualize an ideal gas as a region of space having many individual atoms or simple molecules darting around and interacting with each other only through elastic collisions (those conserving total kinetic energy). Viewed from the center of a gas, an arbitrary molecule would see the same average molecular concentration in all directions and at all distances, for the motions are everywhere equally random. The random motions also mean, however, that the local concentration of molecules will vary from the mean value on short time scales; for fleeting moments, a greater number of molecules may just happen to occupy one region of space more than some other region. Only after many molecular collisions do the random fluctuations average out and the concentrations settle down to the mean value. The gas molecules themselves occupy only a tiny fraction of the total volume, so a gas is mostly empty space, and the number concentration (or density) of the gas is small compared with that of the condensed states of matter. The large compressibility of a gas indeed arises from the big void spaces between the molecules, so an externally applied pressure simply causes the molecules to move closer together. The larger the applied pressure, the smaller the volume occupied by a given number of molecules. This connection between pressure and volume is readily seen by rewriting the ideal gas law in the form

$$pv = RT, \tag{2.8}$$

where now $v \equiv V/N = 1/n$ is the molar volume, the volume occupied by Avogadro's number of gas molecules. At any given temperature, pressure and volume are seen to be inversely related to each other, yielding one version of Boyle's law: $pv = const$. At constant volume, the pressure is seen to vary in direct proportion to the temperature, giving the law of Gay-Lussac (or Charles): $p \propto T$. The compressibility of an ideal gas is readily calculated to be $\kappa_T \equiv -(1/v)\,(\partial v/\partial p)_T = 1/p$, showing that a gas at low pressure is more compressible than one at high pressure. The beauty of the ideal gas law is its insensitivity to any property of the atoms or molecules constituting the gas. We base intensive variables on the mole, rather than on mass, because only numbers of molecules are relevant.

Another useful property of ideal gases is expressed by Dalton's law of partial pressures for gaseous mixtures. As long as the molecules maintain large average separations and do not attract each other, the total pressure of the gas is equal to the sum of the partial pressures of the individual components of the gas. In effect, each component (generically called j) acts as if it were the only gas occupying the volume, so total pressure, that measured by a barometer, is given by

$$p = \sum_j p_j, \tag{2.9}$$

where p_j is the partial pressure of component j. The ideal gas law applies to each component separately: $p_j = n_j RT$, as well as to the mixture. The composition of a gaseous mixture can be characterized by specifying the partial pressures, alternatively the partial molar concentrations n_j. In a unit volume, the total number of moles of gas is simply the sum

$$n = \sum_j n_j. \tag{2.10}$$

If we next divide this equation by n, we get a convenient relationship between the individual mole fractions $y_j \equiv n_j/n$:

$$\sum_j y_j = 1. \tag{2.11}$$

This relationship must be true if all components have been accounted for. Application of the ideal gas law to the definition of mole fraction gives us a convenient way of specifying this measure of composition in terms of pressures:

$$y_j \equiv \frac{n_j}{n} = \frac{p_j/RT}{p/RT} = \frac{p_j}{p}. \tag{2.12}$$

Expressing mole fraction in terms of pressures simplifies expressions for Henry's law, which will be discussed later in the context of gaseous interactions with cloud water. The mole fraction y_j, also called a molar (or volume) mixing ratio, is often used to express the relative concentration of a trace gas as so many parts per million (1 ppm $= 10^{-6}$), parts per billion (1 ppb $= 10^{-9}$), parts per trillion (1 ppt $= 10^{-12}$), etc. Dalton's law is useful in many atmospheric applications, but it is strictly valid only for gases that exhibit ideal behavior.

Real gases deviate from ideal-gas behavior at high pressures or low temperatures. The simplifying assumptions that underpin the ideal gas law (e.g., Eq. (2.8)) break down when the volumes of the molecules cannot be ignored, or when the temperature falls sufficiently to allow condensation. The so-called compressibility factor, $Z \equiv pv/RT$, which is by definition unity for ideal gases, then varies with conditions and the composition of the gas. Deviations from ideal-gas behavior are sometimes accounted for by including higher order terms involving density or molar volume. The link between p, v, and T for real gases is often given by the virial equation of state,

$$Z \equiv \frac{pv}{RT} = 1 + \frac{B_2(T)}{v} + \frac{B_3(T)}{v^2} + \cdots \simeq 1 + \frac{B_2(T)}{v}. \tag{2.13}$$

With rare exceptions, one need account only for the second virial coefficient (B_2), as the higher terms contribute little to Z. The dependence of B_2 on temperature can be obtained from experiments or statistical–mechanical theories. Even real gases can exhibit ideal-gas behavior under narrow sets of conditions, as when $B_2 \to 0$. The temperature at which a real gas behaves ideally is called the Boyle temperature T_B, for there $B_2(T_B) = 0$ and thus $Z = 1$.

An early attempt was made by the Dutch chemist van der Waals to develop an equation of state that explicitly incorporates molecular properties. Molecules are, after all, real entities of matter that have finite (albeit small) volumes and that interact with each other with real forces of attraction and repulsion. The collective volume of all the molecules in a gas is accounted for by defining an excluded volume b that is subtracted from the macroscopic volume. The volume to use is thus $(v - b)$, not v. The molecular-volume parameter (b) depends on the sizes of the molecules, which is ultimately defined by the repulsive nature of the molecules during collisions (as suggested by the upper dotted curve in Fig. 2.7). The attractions of molecules for one another that occur during collisions (Fig. 2.7) act to reduce the pressure. All molecular collisions occur in pairs, so the opportunities for interactions varies in proportion to the square of molecular concentration (i.e., as $an^2 = a/v^2$, where a is a parameter related to attraction). Combination of the volume and pressure effects gives the van der Waals equation of state:

$$\left(p + \frac{a}{v^2}\right)(v - b) = RT. \tag{2.14}$$

The general behavior of this equation of state is shown as a family of isotherms in Fig. 2.8. The "gorge", a mathematical artifact of the equation, is in reality crossed by condensation (arrow). Nearly ideal, Boyle's-law behavior is apparent above the critical temperature T_c, the point distinguishing gases (non-condensible) from vapors (condensible). The atmosphere acts as an ideal gas because the temperature never falls close to the critical point of the main components, nitrogen and oxygen. Trace amounts of vapor, such as water, also act nearly ideally because the dominant interactions are with gases (N_2 and O_2), not other vapors (other H_2O molecules).

Adding energy to a system typically causes its temperature to rise. The precise relationship between the energy of a system and its temperature depends on the conditions under

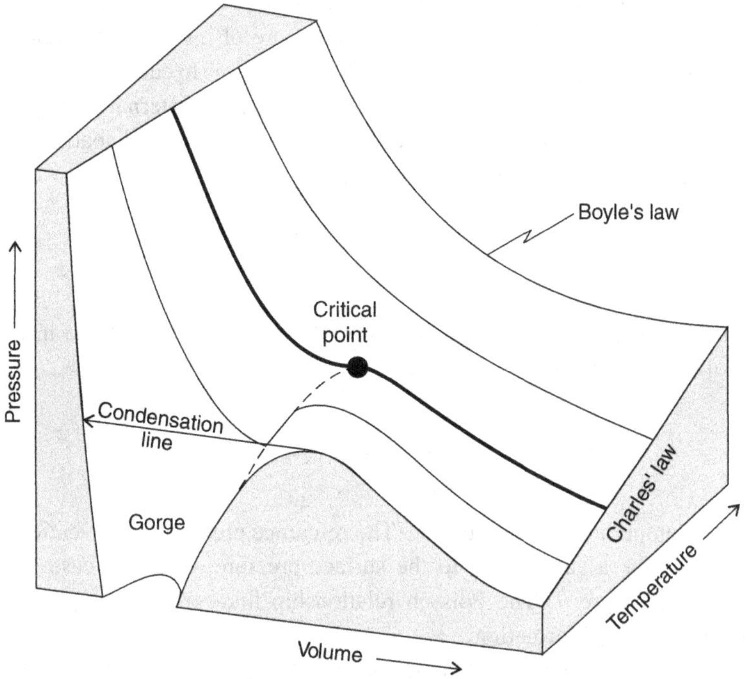

Figure 2.8 Perspective view of the van der Waals equation of state. From page 30, Figure 1.8(a) in *Physical Chemistry* by P.W. Atkins (1986), used by permission of Oxford University Press, Inc.

which the interaction takes place between the system and its environment. Two limiting situations arise in practice, one in which the volume is fixed, the other in which the pressure is constant. As shown by basic treatments of thermodynamics, the relationship (for an ideal gas) between the specific heat at constant pressure (c_p) and that at constant volume (c_v) is

$$c_p = \left(\frac{\partial u}{\partial T}\right)_p = c_v + R, \tag{2.15}$$

where u is the internal energy. The specific heat at constant pressure is always larger than that at constant volume simply because some of the added energy goes toward the work of expanding the system when the volume can change. It then takes more energy to accomplish a given change in temperature.

The heating of a gas in general leads to changes in both temperature and pressure. The enthalpy-form of the first law of thermodynamics is the most convenient way for atmospheric scientists to relate the heating rate (q) and the simultaneous rates of change of temperature and pressure:

$$q = c_p\frac{dT}{dt} - v\frac{dp}{dt}. \tag{2.16}$$

Were the pressure in any given situation to be fixed, as in processes confined to the Earth's surface, the heating rate would increase the temperature of the gas at the rate $dT/dt = q/c_p$. The most common application of Eq. (2.16), however, is to calculate the temperature change as an air parcel rises toward lower pressure without any external heating, i.e., when $q = 0$, the condition for an adiabatic change. The temperature of an adiabatic parcel is thus found to change at the following logarithmic rate:

$$\frac{d \ln T}{dt} = \frac{R}{c_p}\frac{d \ln p}{dt}, \tag{2.17}$$

where we have used the ideal gas equation to replace the specific volume in Eq. (2.16). Integration of this equation results in the so-called Poisson relationship:

$$\frac{T}{T_0} = \left(\frac{p}{p_0}\right)^{R/c_p}, \tag{2.18}$$

where T_0 is the temperature at pressure p_0. The reference pressure is conventionally taken to be $p_0 = 1000\,\text{hPa}$, a value close to the surface pressure, in which case T_0 is called the potential temperature θ. The Poisson relationship finds frequent application in the thermodynamics of cloud formation.

The connection of energies to the molecular world is easily seen through the specific heat at constant volume, which is given by

$$c_v = \left(\frac{\partial u}{\partial T}\right)_v. \tag{2.19}$$

Specific heat c_v is essentially a measure of the change in internal energy u for a given change in temperature while the volume is fixed. The internal energy of an ideal gas composed of simple atoms (e.g., argon) exists entirely in the translational kinetic energies of the atoms. Thus, if the average kinetic energy of each atom is $\bar{\varepsilon}_k$, then the total internal energy of a mole of gas is just $u = N_{Av}\bar{\varepsilon}_k$. The kinetic-molecular theory of gases shows that the average kinetic energy of an atom in an ideal gas is related to temperature by $\bar{\varepsilon}_k = (3/2)\,k_B T$, so the specific heat $c_v = (3/2)\,k_B N_{Av} = (3/2)\,R$ ($12.5\ \text{J}\,\text{mol}^{-1}\,\text{K}^{-1}$). An increase in temperature thus increases the kinetic energies of the atoms. Each atom moves faster on average at a higher temperature.

By recognizing that the atoms in an ideal gas move independently in any of the three spatial directions, we see that the molar energy per direction is on average $(1/2)\,RT$. In fact, the equipartition theorem asserts generally that energy is shared equally by all possible modes, not just those due to translation. If a gas is composed of diatomic molecules, such as N_2 and O_2, then the rotational degrees of freedom must likewise be taken into account. Diatomic molecules are similar to dumbbells, which can rotate around two orthogonal axes about the line joining the centers of the atoms. With these two degrees of freedom added to the three translational degrees of freedom, we find that the internal energy of the gas is $u = (5/2)\,RT$, which gives a specific heat $c_v = 5/2R$ (20.8 J mol^{-1} K^{-1}). At this level of analysis, theory and measurements agree very well, so we gain important confidence that our picture of the gaseous state is largely correct.

In any population of molecules, deviations from averages are common. In fact, these deviations, also called statistical fluctuations, become important for our deeper understanding of matter and changes of state. Temperature provides a robust measure of the mean kinetic energy of the molecules constituting a system. However, many of the molecules at a given temperature will have energies larger than the mean, and many will have energies less than the mean. The distribution of molecular speeds was shown by Maxwell and Boltzmann to be

$$n_v \equiv \frac{dn}{dv} = 4\pi n_v v^2 \left(\frac{m}{2\pi k_B T}\right)^{3/2} \exp\left(-\frac{mv^2}{2k_B T}\right), \qquad (2.20)$$

where n_v is the number of molecules (out of the population of n molecules each of mass m) having a speed between v and $v + dv$. An even more general relationship emerges upon dividing Eq. (2.20) through by the total number n. We then obtain the probability density function, the likelihood of finding molecules having speeds between v and $v + dv$ (see Fig. 2.9):

$$f_v = 4\pi v^2 \left(\frac{m}{2\pi k_B T}\right)^{3/2} \exp\left(-\frac{\epsilon_k}{k_B T}\right). \qquad (2.21)$$

Note that the molecular kinetic energy appears (as $\epsilon_k = \frac{1}{2}mv^2$) in the argument of the exponential function, known as the Boltzmann factor. Only a small fraction of the molecules

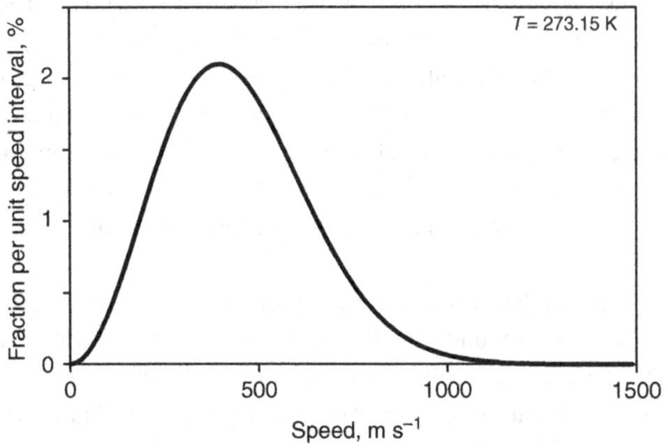

Figure 2.9 Maxwell–Boltzmann distribution of molecular speeds at 0 °C.

have large kinetic energies, those appearing in the tail of the distribution (toward the right in Fig. 2.9). Nevertheless, these energetic molecules, those with $\epsilon_k >> k_B T$, are responsible for phase changes and chemical reactions.

Liquids

A liquid is a fluid that maintains a free surface when exposed to a gas. The distinction between liquid and gas is one of density, alternatively the volume occupied by a given number of molecules. The relatively large density of a liquid is due to the fact that the molecules of a liquid are cohesively bound to each other by electrical forces. However, the intermolecular attractions in a liquid are not as strong as those in a solid of the same substance. As with gases, the molar volume v of a liquid is determined by the temperature, pressure, and composition. Unlike gases, however, liquids are difficult to compress, because the molecules are already in close contact and have little free space in which to move before they experience the repulsive forces of neighbors. The isothermal compressibility, $\kappa_T \equiv -(1/v)(\partial v/\partial p)_T$, is not zero, however, as all substances in a condensed phase do compress slightly with increased pressure. Most liquids also expand upon heating, which is to say that the molar volume increases with increasing temperature; the isobaric coefficient of thermal expansion $\alpha_p = (1/v)(\partial v/\partial T)_p$ is normally positive.

Liquid water is anomalous in several important respects. Whereas normal liquids monotonically expand with increasing temperature, water at low temperatures and low pressures actually contracts with increasing temperature. Careful measurements show that the density reaches a maximum at 4 °C under atmospheric pressures (see Fig. 2.10). The temperature of maximum density (TMD, dotted curve along which $\alpha_p = 0$) decreases as the pressure increases, but it disappears at high pressures. The behavior of water becomes more like that of normal liquids when the pressure exceeds about 1000 atm at ordinary temperatures. Liquid water at temperatures below the TMD, however, compresses more readily as the temperature decreases. That is, $\partial\kappa_T/\partial T = (1/v)\left(\partial^2 v/\partial T^2/(dp/dT)_{TMD}\right) < 0$ because $(dp/dT)_{TMD} < 0$. The pattern of compressibility in this region is shown in Fig. 2.11. Compressibility and the coefficient of expansion reflect changes in molecular-level behavior of water in response to changes in temperature and pressure. Before we explain how the molecular-level behavior of water impacts its compressibility, we will first look at a few other macroscopic properties dependent on the same environmental variables.

Measurements of the enthalpies associated with changes of state exhibit interesting patterns among various compounds, including water. If we consider the change from the liquid to vapor states at the boiling point of various hydrides, for instance, we find that the enthalpy of vaporization depends both on the chemical family (Group) and the Series (Period) of the Periodic Table (see Fig. 2.12). Within any given family (e.g., carbon), the enthalpy of vaporization (energy needed to evaporate a given amount of liquid) decreases systematically as the atomic number decreases (from right to left in Fig. 2.12).

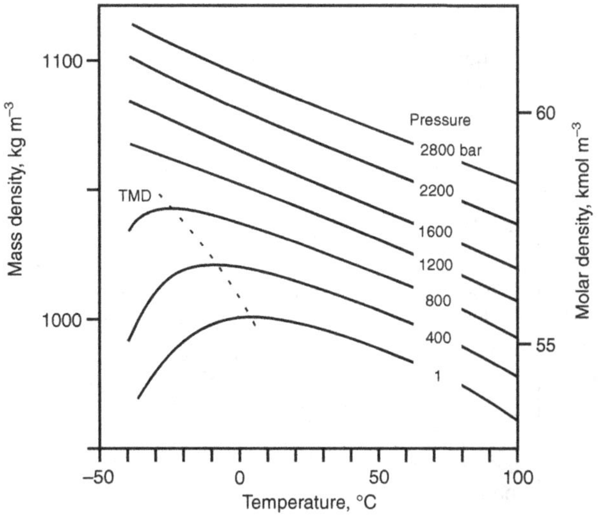

Figure 2.10 Density of liquid water as functions of temperature at selected pressures, as determined by the bond-mixture model (see text). The temperature of density maximum (TMD, dotted curve) decreases with increasing pressure. Adapted from Jeffery and Austin (1997).

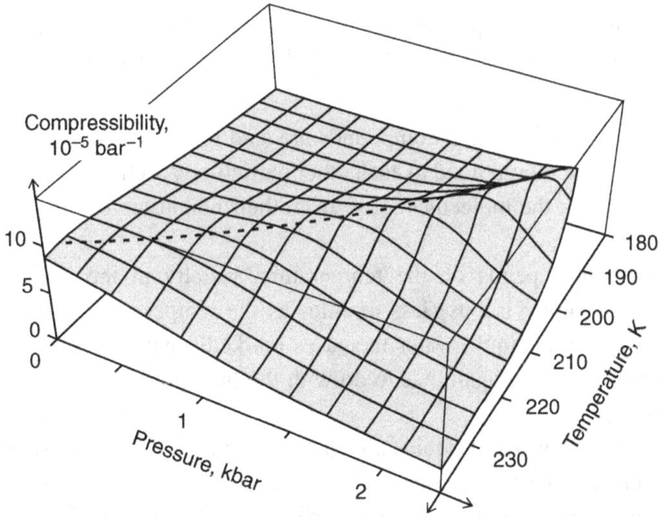

Figure 2.11 Compressibility of supercooled liquid water. Dashed curve identifies the maxima. Adapted from Baker and Baker (2004).

This pattern is interrupted, however, in the nitrogen, fluorine, and oxygen families at the lightest-weight elements; the respective liquids (ammonia, NH_3; hydrogen fluoride, HF; and water, H_2O) all have anomalously large enthalpies of vaporization. In the case of water, for instance, the enthalpy of vaporization is more than twice the expected magnitude based

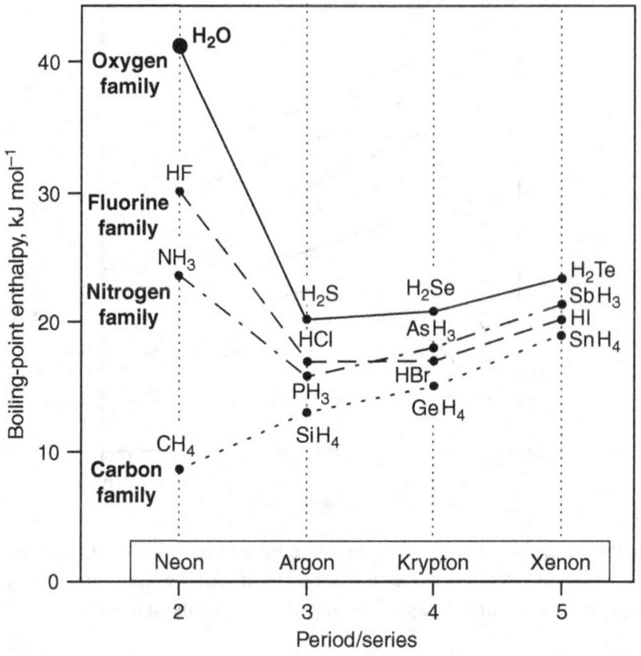

Figure 2.12 Boiling-point enthalpies of vaporization arranged by position (Period/Series and Group/Family) of the main element in the Periodic Table. Adapted from Pauling (1960).

on extrapolation to Period 2. Such compounds are said to be associated liquids, meaning that the molecules in the liquid bond to each other particularly well; it is difficult (i.e., takes lots of energy) to break the molecules loose from their neighbors and send them into the vapor phase.

The viscosity, too, depends on the temperature, usually in the sense that the liquid becomes increasingly viscous (less mobile) as the temperature drops. As shown in Fig. 2.13, the viscosity of liquid water increases markedly with decreasing temperature. The reason for this tendency again has to do with the bonding between the molecules, as we shall see shortly.

As a rule, the specific heats of liquids are larger than those of the corresponding vapors. The liquid is a state in which molecules are bound to each other, so it takes extra energy to raise the temperature of the liquid a unit amount compared with that of the vapor; some of the energy goes into increasing molecular potential energies, such as those associated with vibrations of the molecules relative to their neighbors and with the breaking or rearrangement of intermolecular bonds (configurational changes). Liquid water is anomalous in that a relatively large amount of energy is needed to raise the temperature of the liquid by 1 K. The solid curve in Fig. 2.14 shows how the measured specific heat varies with temperature in each state of water. We clearly see the large values for the liquid state compared with

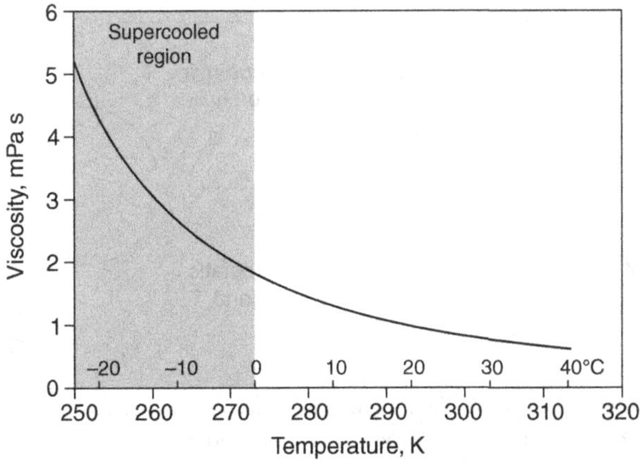

Figure 2.13 The dependence of the viscosity of liquid water on temperature. Based on data from Hallett (1963) for supercooled water.

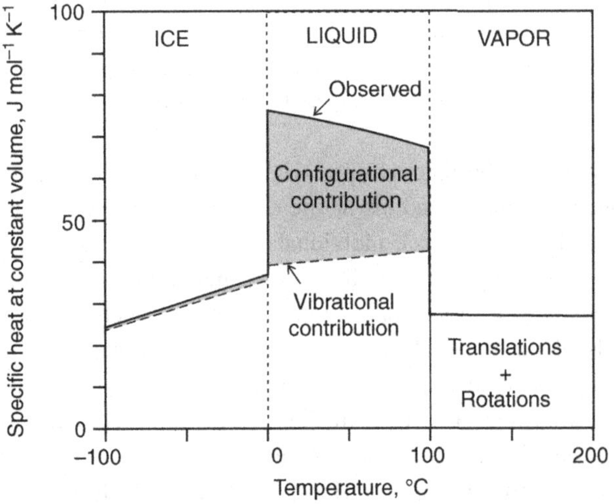

Figure 2.14 Specific heat of water at constant volume. The shaded region is the contribution due to structural rearrangements with respect to neighboring molecules. Vibrational contributions in liquid and ice refers to intermolecular vibrations constrained by H bonds. The heat capacity of vapor is due to molecular translations and rotations. Adapted from Eisenberg and Kauzmann (1969).

those for the solid (ice) and vapor states. The contribution to the specific heat from the intermolecular vibrations of the molecules in liquid water, shown by the dashed curve, is found to be only about half that observed in experiments. The remaining portion arises from the changing configurations of the molecules (that is, the geometrical and energetic arrangements of molecules relative to one another) with changing temperature. The molecules in

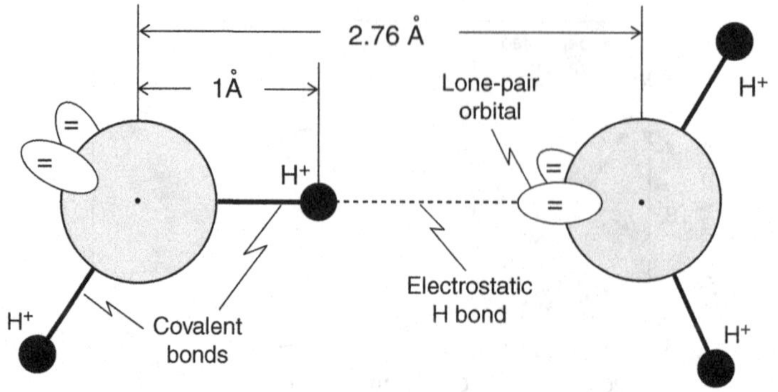

Figure 2.15 Cartoon identifying the hydrogen bond (dashed line) as the electrostatic attraction between the positive hydrogen atom of one molecule and the negative lone-pair orbital of a neighboring molecule.

ice vibrate relative to one another, but they lack the freedom to move around and form new molecular arrangements with neighbors; the specific heat of ice is determined mainly by intermolecular vibrations. The specific heat of ice is therefore only about half that of liquid water, but it is nevertheless larger (at least at temperature $T > -70\,°C$) than that of the vapor, which offers mainly molecular translations and rotations to its specific heat.

In order to account for the macroscopic properties of water, those that are measured in experiments, we must understand the nature of the bonding between molecules. Recall that the covalent bond holding each individual water molecule together is polar, so the molecule as a whole is also polar; H_2O has a permanent dipole moment. The electronegative nature of the oxygen atom concentrates the negative charge near itself, leaving the hydrogen ends of the molecule positive. Thus, the positive (H) end of each bonding orbital is easily attracted to one of the negatively charged lone-pair orbitals of a neighboring water molecule, as shown in Fig. 2.15. The attraction of two electronegative atoms (here oxygen) mediated by an intervening hydrogen atom (with partial positive charge) is termed a hydrogen bond (or H bond). When fully formed (as in ice), the H bond holds the centers of two adjacent water molecules an optimal distance of 2.76 Å (0.276 nm). About 1 Å of that separation is due to the polar covalent bond within one of the molecules, the remaining distance is accounted for by the electrostatic attraction between the H^+ and the lone-pair orbital of the other molecule. In the case of water, the H bond is only about 5% the strength of the covalent bond, so the H bonds between molecules break more readily than do the covalent bonds within molecules. Water thus evaporates from the liquid as intact molecules. We note at this point that the maximum number of H bonds that any one water molecule can form is two. Each water molecule has four attachment points (at the two bonding orbitals and at the two lone-pair orbitals), but each bond is shared among two neighboring molecules. Two bonds must therefore be broken on average for every water molecule that evaporates from the condensed phase.

The fundamental force causing the H bond is electromagnetic. A positively charged H atom (a proton, H^+) of one molecule is electrostatically attracted to the negative charge in one of the lone-pair orbitals of a neighboring molecule. As a first approximation, one may think in terms of simple electrostatics (as describe by Coulomb's law), or "static cling" (causing clothes to stick together) when trying to understand the cohesion of water in the liquid state. However, the bonding between water molecules is complicated, and it is in the complications that the properties of water are best understood. The H bond is actually the net effect of two separate contributions, a relatively strong, directed component and a weaker, non-directional component. Sometimes, these components are respectively called the strong and the weak H bond. The overall strength of the bond between two water molecules depends on the distance between the molecules and the orientation they make with one another. The H bond is strongest when the covalent bond of the one molecule lines up closely (within about 20 °) with the lone-pair orbital of the adjoining molecule (as in Fig. 2.15). Bent (misaligned) bonds are weaker than straight bonds, but the weaker, non-directional bonding helps hold the molecules together.

Both the strength of the intermolecular bonds and the structure (molecular configuration) of the liquid depend on temperature. Thermal energy induces molecular vibrations (linear motions along the bonds) and librations (twisting motions relative to the optimal alignment) that affect the separation distance and the orientations of the molecules. If a molecule becomes excessively separated (by more than about 15%) or significantly misaligned relative to its nearest neighbors, the strengths of the bonds are compromised. Any given molecule is also sensitive, although to a lesser extent, to the presence of the next-nearest neighbors, indeed to all other molecules in its vicinity, even those to which it is not directly bonded. Such cooperative effects give rise to long-range order, which is especially prominent in a solid, but which also prevails somewhat within a liquid at low temperatures.

The molecular structure of a bulk phase can be revealed in an average sense by using X-ray diffraction. The mean positions of the atoms relative to each other are expressed as radial distribution functions, defined as the ratio of density $\rho(r)$ at a specified distance r from an hypothetical central atom to the bulk (average) density ρ_{bulk}:

$$g\,(r) = \frac{\rho(r)}{\rho_{bulk}}. \tag{2.22}$$

Examples of $g\,(r)$ for various phases are shown in Fig. 2.16. The dotted curve at $g\,(r) = 1$ shows the limiting case for a perfectly isotropic, featureless medium, such as an ideal gas composed of point masses; the local density is the same as the bulk density at all distances. The thin solid line represents the distribution of atoms in a non-interactive van der Waals gas or a gas of hard-shell spheres, one in which the finite size of the atoms limits

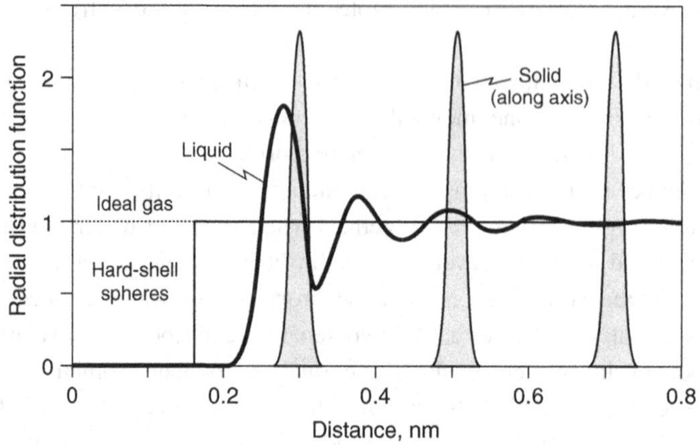

Figure 2.16 Radial distribution functions of different states of matter. Adapted from Goodstein (1985, p. 236).

how close the atomic centers can get. Real substances invariably reveal more complicated structures, however. The bold solid curve shows that the oxygen atoms in liquid water tend to maintain a separation distance of about 0.28 nm (2.8 Å), which is the distance to the nearest neighbors and consistent with the tetrahedral coordination of water. The local maximum in $g\,(r)$ at $r \simeq 0.38$ nm is thought by some to be evidence of interstitial water molecules, those lying between other H-bonded molecules among the next-nearest neighbors. Nevertheless, any evidence of structure in liquid water disappears with distance; liquids exhibit short-range order, but relatively little long-range order, such as that in solids (shown by the repetitive shaded peaks).

The configuration of molecules in liquid water at any given instant is determined in complicated ways by the interplay of the directional and non-directional components of the H bonding. To go beyond the average geometrical arrangements of molecules revealed by X-ray diffraction, we must resort to conceptual models, representations of molecular patterns that help us visualize a liquid at the molecular level. An accurate model would be one that is consistent with known physics and yields the same macroscopic properties that are observed. Many models of liquid structure have been devised over the years, but our purpose is served if we focus on a classical model, modified as needed to fit recent findings. As with any model, we must recognize that the individual water molecules, with their oxygen and hydrogen atoms bound tightly together by polar covalent bonds, retain their shape and size, so it is the arrangement of the intact water molecules relative to neighbors that establishes the structure of the liquid. The intermolecular (hydrogen) bonds, unlike the intramolecular (covalent) bonds, are comparable in strength to the thermal energy that causes neighboring molecules to vibrate and librate relative to each other. Thus, it should not be surprising to learn that some fraction of the intermolecular bonds break and become temporarily ineffective. Each water molecule, however, has four nearest neighbors, so the breaking of one bond does not mean the molecule in question is really free. It may rotate a bit more freely, perhaps, and it may find new neighbors with which to bond, but a molecule in the bulk material cannot escape. Also, because the intermolecular bonding is complicated by having directional and non-directional contributions, the breaking of a directional bond leaves the non-directional component more or less intact. The directional bonds are favored at lower temperatures and when the density is reduced in the local region around the molecule. Later, we will see that molecular-scale fluctuations in density help account for the initiation of ice in supercooled water.

The "flickering-cluster" model, introduced by Frank and Wen in the 1950s, helps us visualize the molecular structure of liquid water. The cooperative nature of the strong hydrogen bonding encourages the formation of clusters, groupings of several water molecules that act as a unit for a short period of time. These clusters tend to have a more open structure than do molecules bound by the weaker hydrogen bonds. The breaking of one or more bonds weakens the coherent arrangement, causing the cluster to disintegrate as such, only to be reformed later but in a different arrangement and likely with other molecules. The average lifetime of a cluster is thought to be on the order of 10 ps (1 ps $= 10^{-12}$ s), which may seem to be a short time until one realizes that the period for molecular vibrations is only about 0.1 ps. Thus, for ~ 100 vibrational periods, each of the many clusters scattered throughout the liquid is held together in a loose, wobbly association. Then, relatively suddenly (but asynchronously) those structures break down and form anew. The liquid, were its structure possible to see in high temporal and spatial resolution, would appear to consist of many molecular clusters flickering into and out of existence.

The molecular structure of liquid water envisioned by the flickering-cluster model allows us to understand some observed properties. The minimum in specific volume (or maximum in density) at 4°C, for instance, arises from the interplay of the directional and non-directional components of the H bonding between molecules. At relatively high temperatures, or high pressures, the molecules cannot easily maintain alignment with one another, so the non-directional bonding tends to dominate, leading to distorted bonds and specific volumes that increase with increasing temperature, as with normal liquids. As the temperature drops, the thermal energy of the molecules decreases, allowing them to settle into more stable arrangements with each other. Below the temperature of maximum density (minimum specific volume), the directional bonding becomes increasingly more important; the stronger and straighter directional component of the H bonds nudges the molecules farther apart and thereby reduces the density. At still lower temperatures, the compressibility reaches a maximum (Fig. 2.11) where the microstructure of water shifts from one still influenced by the weak, non-directional bonds to one dominated by the strong, directional H bonds. This shift in the nature of bonding leads to open, ice-like clusters. Neither a structure dominated by compact molecules bound by weak H bonds (as at high temperatures), nor one composed of rigid open clusters connected by strong H bonds (as at low temperatures), is as compressible as the mixture of the two types, which occurs near the compressibility maximum.

Other properties of liquid water are similarly amenable to molecular interpretation. The anomalously large latent heats of vaporization of associated liquids, as well as the large viscosity of liquid water, are due to the relatively strong attractions afforded by hydrogen bonding. Evaporation requires bonds to be broken completely, whereas the mobility of molecules relative to each other demands that some bonds break before new ones can form. The enthalpy of vaporization decreases with increasing temperature because the H bonds progressively weaken as thermal agitation becomes more pronounced and the bonding becomes stretched and distorted. Similarly, the large specific heat of liquid water is a consequence of the rearrangements of molecular structure that take place whenever the temperature rises. Each addition of thermal energy causes some of the bonds to weaken, allowing the molecules to find new neighbors. Overall, many of the properties of water we take for granted are determined by the average behavior of the molecules in relation to their neighbors.

Numerous attempts have been made in recent years to develop an equation of state for liquid water that is consistent with its known properties over a wide range of temperatures and pressures. The task is challenging for water because of the complicated bonds connecting the molecules. The most success thus far has been achieved by extending the concept of a van der Waal's equation of state to take the effects of hydrogen bonding into account. The intermolecular bonding in water is envisioned to consist of a mixture of van der Waal's bonds (arising from dispersion forces, which we may liken to the weak H bond) and of (strong) hydrogen bonds. The van der Waal's bonds give rise to a background pressure p_0, while the H bonds gives rise to a separate pressure component p_{HB}. The total pressure is then $p = p_0 + 2p_{HB}$, where the factor of two arises because two H bonds can form around each molecule. The background pressure p_0 is obtained from the van der Waal's equation of state, whereas p_{HB} is derived from an appropriate Helmholtz free energy A_{HB}, as follows. The Helmholtz function is defined in general as $A \equiv U - T\Phi$, where U is the internal energy and Φ is the entropy of the system. Application of the first and second laws of thermodynamics gives the differential relationship $dA = -pdV - \Phi dT$. Thus, p_{HB} can be calculated at constant T from the Maxwellian relationship,

$$p_{HB} = -\left.\frac{\partial A_{HB}}{\partial V}\right|_T . \tag{2.23}$$

The total H-bond free energy is a mixture of the strong and weak types, so the bond-mixture model gives

$$- A_{HB} = f A_{HB,strong} + (1 - f)\, A_{HB,weak}, \tag{2.24}$$

where f is the fraction of strong bonds. The strong-bond fraction (f) is a complicated function of both the density and temperature of the liquid, but it is thought to have a form involving a Gaussian centered around a molar volume corresponding closely to that of ice:

$$f \sim \exp\left[-\left(\frac{v - v_{ice}}{\sigma}\right)^2\right]. \tag{2.25}$$

Here, σ is a measure of the width of the distribution, basically the range of v over which the strong H bonds contribute significantly to the liquid structure. Equation (2.25) is a mathematical way of accounting for the random fluctuations in local density that occur naturally in any liquid.

The free energies of the strong and weak H bonds, used in Eq. (2.24), have been derived with the aid of statistical mechanics. The contribution of the strong H bonds to the free energy is $A_{HB,strong} = -RT \ln Z_{HB,strong}$, where $Z_{HB,strong} = \Omega_0 + \exp(-\epsilon_{HB}/k_B T)$ is the partition function for strong H bonds, Ω_0 is the weak-bond configuration number (ways the weak H bonds can be obtained), which is related to the thermodynamic entropy Φ_0 through Boltzmann's relationship: $\Phi_0 = k_B \ln \Omega_0$. The energy of a strong H bond is estimated to be $\epsilon_{HB} \simeq -14$ kJ mol^{-1}. It is assumed that the weak H bonds contribute no energy, so the partition function for weak bonds is $Z_{HB,weak} = \Omega_0 + 1$, and $A_{HB,weak} = -RT \ln Z_{HB,weak} = -RT \ln(\Omega_0 + 1)$. With appropriate estimates for the weak and strong components to the free energy (for Eq. (2.24)), as well an empirical relationship for the strong-bond fraction f, the effect of the H bonds on the pressure is determined from Eq. (2.23). The total pressure ($p = p_0 + 2p_{HB}$) is then calculated and related to temperature and molar volume to give an effective equation of state. One set of results for water was shown earlier as isobars in Fig. 2.10. The liquid phase of water is extremely complicated and still the subject of active research. Having an accurate equation of state for liquid water is an important first step toward being able to calculate the onset of ice in supercooled clouds.

Solids

A solid is the condensed state of matter that maintains its shape and size, even in the presence of modestly strong external forces. Solids resist being deformed or compressed. As with liquids, the molecules of a solid are bound together cohesively by electrical forces. However, in the solid state, unlike in the liquid state, the molecules are arranged regularly in a crystalline lattice. It is this repetitive pattern that gives solids long-range order, a characteristic missing in liquids and gases (see Fig. 2.16). The geometrical arrangement of molecules in a solid persists throughout individual crystals, whereas in liquids the molecules deviate from set positions relative to one another in unpredictable ways. Even though the molecular arrangement is regular in solids, the pattern often differs along each crystallographic axis, so the macroscopic properties of many solids depend on direction; solids are typically anisotropic (not uniform in all directions). A solid will always have a characteristic density, for instance, but the packing density of molecules, the number per unit distance, may vary in different directions through the solid. The conduction of heat through a solid may also depend on crystallographic direction for any given temperature gradient. The solid phase has many properties not shared by the liquid or gas phases.

The solid phase of water is usually thought of as ice, but the familiar ice of everyday life is just one of many possible solid phases, or polymorphs, of water. The various solid phases of water differ in the way the H_2O molecules are arranged relative to one another. As shown in Fig. 2.17, many of these polymorphs are stable only under pressures or temperatures not found naturally on Earth. Note that the melting points of some of the high-pressure ices are well above 0 °C. Experimental scientists, with specialized apparatus to form the crystals and analyze them with X-ray or electron diffraction, have so far found more than 10 distinct polymorphs. The high-pressure polymorphs (ices II through VIII) form only once the pressure exceeds several thousand atmospheres. They all have relatively low specific volumes (high densities), but they nevertheless maintain an open molecular structure compared with close-packed solids. It is the open spaces between the molecules that allows the structure to compress when the applied pressure is increased. As is the case with liquid water, the molecules in all ices are held to one another by combinations of strong (highly directional) and weak (non-directional) hydrogen bonds. The tendency of strong H bonds to be linear is what gives the ice polymorphs an open structure. The bonds in ices I, VII, and VIII tend to be more linear than those in ices II, III, V, and VI, which have bonds that are distorted (partially bent) and so are more influenced by weak H bonding.

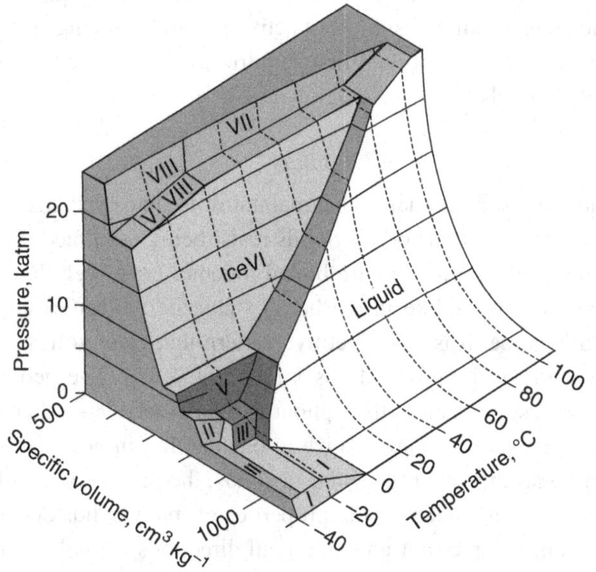

Figure 2.17 Perspective view of the high-pressure phases of water ice. Adapted from Eisenberg and Kauzmann (1969, p. 80).

The low-pressure polymorphs of solid water exist at temperatures below 0 °C and include two types of ice I (called Ih and Ic) and vitreous ice. Vitreous or amorphous ice, which forms only at very low temperatures (below about −135 °C), is not technically a crystalline solid, as the molecules are not arranged in regular patterns. Vitreous ice is more like a glass or a very viscous liquid. Ices Ih (called hexagonal ice) and Ic (cubic ice) are regular solids with relatively straight H bonds. As the names imply, the H_2O molecules in ice Ih are arranged in a hexagonal pattern, whereas those of ice Ic are arranged in a cubic pattern (as are the carbon atoms in diamond). Ice Ic is thought to form only at temperatures below about −70 °C, and it transforms irreversibly to ice Ih upon warming. Ice Ic is said to be metastable with respect to ice Ih. All of the ice we encounter in everyday life, regardless of the external form (i.e., the morphology), is ice Ih. This "ordinary ice" is henceforth the focus of discussion.

The basic building block for constructing any macroscopic lattice, including that of ice Ih, is called the unit cell. This group of molecules is analogous to the bricks that make up a building. Each unit cell contributes to the construction of the lattice through repetitive translations, much as a wall is made by laying bricks side by side, then repeating the operation at the next level. The actual formation of a crystalline solid is more complicated, however, in that individual molecules, not unit cells, incorporate into the lattice via processes to be considered later. Nevertheless, the unit cell, by representing the smallest grouping of molecules that can repeat ad infinitum, serves the useful purpose of describing the fundamental properties of the lattice.

The unit cell of ice Ih has a shape like that of a prism having four sides perpendicular to a base in the shape of a rhombus. Figure 2.18 provides both a perspective view of the unit cell, as well as a projection onto the bottom, or basal plane. The strong covalent bonds holding each H atom (small dot) to its parent oxygen atom (open circle) are shown as solid lines, while the weaker hydrogen bonds are shown as dashed lines. The number of water molecules contained in each unit cell of ice Ih is four: Two of these are contained completely inside the unit cell, while the other two molecules appear along the edges or at the corners that are shared equally with adjacent unit cells. (We calculate the number of molecules in the unit cell of ice Ih as 2 whole molecules + 4 edge molecules/4 edge-sharing cells + 8 corner molecules/8 corner-sharing cells $= 2 + 1 + 1 = 4$.) The size of the unit cell is given by the lattice parameters, the dimensions along the edges: $a = 0.452$ nm, $c = 0.737$ nm at 0 °C, consistent with an intermolecular distance $d = 0.276$ nm. Each H_2O molecule is seen to have four nearest neighbors symmetrically arranged as in a tetrahedron, so all angles (β) between the bonds are equal. In space, the tetrahedral angle is $\beta = 109.5\,^\circ$, but the projection of this angle onto a plane gives an angle $\theta = 360\,^\circ/3 = 120\,^\circ$. It is the simple tetrahedral geometry that gives pristine snow crystals their hexagonal shapes.

The relationship between the unit cell and a macroscopic crystal of ice Ih is shown in Fig. 2.19, a projection onto the basal plane. Note that the repetitious translations of the unit cell yield an unending array of hexagons, much as one can find in the arrangement of

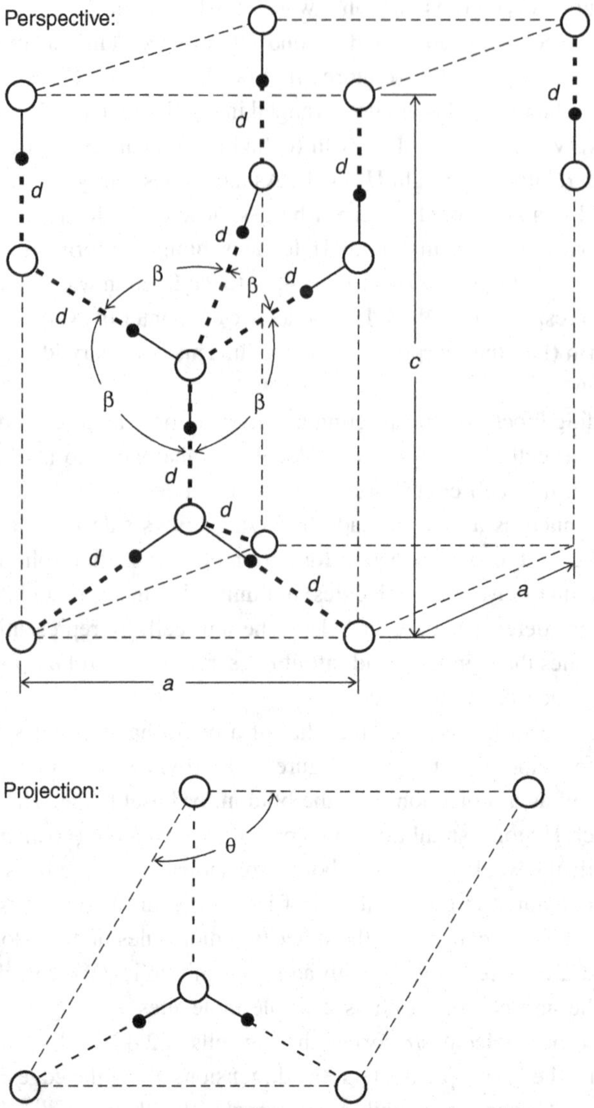

Figure 2.18 The unit cell of ice Ih. Top: Perspective view showing the intermolecular spacing *d* and the tetrahedral angle β. Bottom: Projection of the unit cell onto the basal plane, showing the hexagonal angle θ.

hexagonal tiles on some floors. No crystal extends for ever, of course, so the crystal (shown shaded) terminates along directions that severe the fewest number of bonds.

Directions in ice Ih are specified most naturally in the hexagonal coordinate system. Figure 2.19 shows the three a-axes lying in the basal plane; the c-axis is normal to the

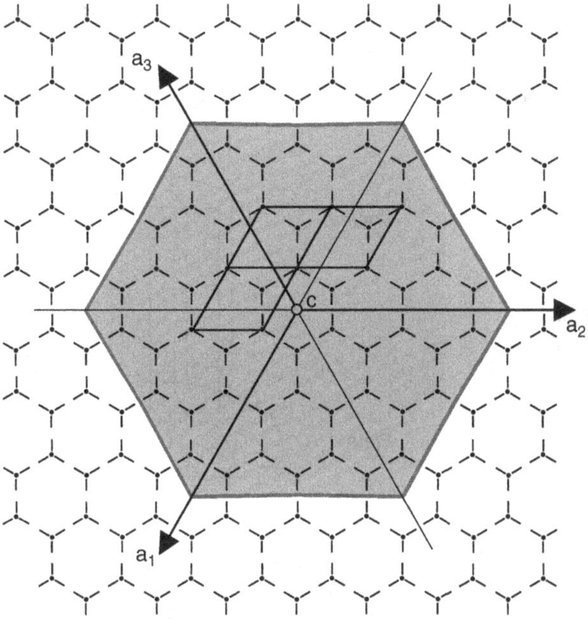

Figure 2.19 The lattice of ice Ih projected onto the basal plane. The c-axis points out of the image, whereas the lateral a-axes lie in the image plane. Projections of three unit cell are shown near the center by closed solid lines. The shaded region identifies one possible crystal formed by breaking bonds along its edge.

basal plane. The position of any point in a hexagonal crystal is identified by choosing the distances along these four axes and writing them within brackets: $[a_1\ a_2\ a_3\ c]$. A plane within the crystal is defined as the positions at which the plane intersects the axial coordinates: One first takes the reciprocal of each axial coordinate, where it intersects the plane, then reduces the set of four values to the lowest integers that maintain the ratios. The resulting values for designating planes are set in parentheses. The so-called low-index planes have small values: The basal plane, for instance, is designated (0 0 0 1), whereas the prism planes are variants of $(1\ 0\ \bar{1}\ 0)$. (The overbar designates a negative value; the first three values must sum to zero; the fourth position is always zero for planes parallel to the c-axis.) Figure 2.20 shows the relationship of the crystallographic axes and the simplest forms of ice found in the atmosphere. The primary habit (plate or column) of ice Ih reflects the hexagonal symmetry of the molecular arrangement.

The molecules in an ideal crystal of ice Ih are arranged in rather specific ways. Recall that the covalent bond in water is polar, so the partial positive charge of each hydrogen atom (acting as a proton, H^+) is attracted electrostatically to the negative lone-pair orbital of a neighboring molecule. Extensive research has established how the H atoms are arranged in the Ih lattice. The description is provided by the so-called ice rules, attributed to Bernal, Fowler, and Pauling:

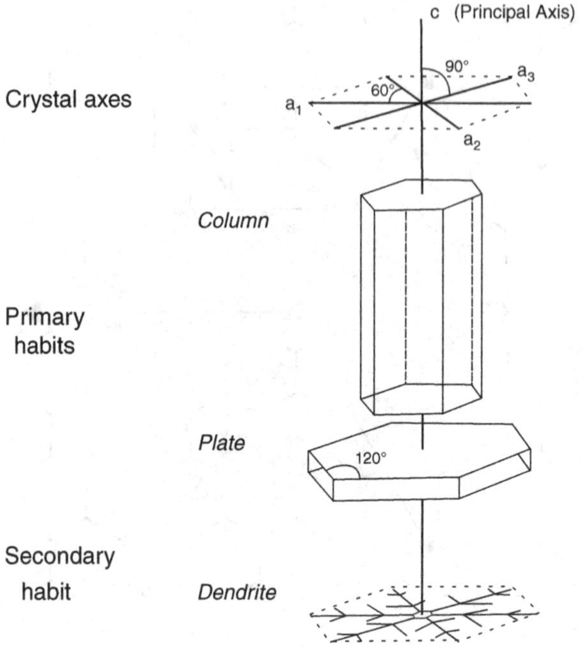

Figure 2.20 Schematic of the basic geometry of snow. Top: Hexagonal axes. Middle: The two primary habits of ice Ih. Bottom: An example of a secondary habit. Adapted from LaChapelle (1969, p. 6).

(a) Two H atoms are bound (strongly) to each O atom, thus forming discrete molecular units of H_2O;
(b) The two H atoms in an H_2O molecule are bound (weakly) to two of the four neighboring molecules, thus making a regular tetrahedron about the central molecule;
(c) Only one H atom is present on each intermolecular bond.

These ice rules, sometimes called the Bernal–Fowler rules for the arrangement of protons in an ideal lattice, state that ice is a molecular solid consisting of permanent dipoles (the polar water molecules) oriented tetrahedrally with respect to nearest neighbors. In other words, the structural units of ice are intact water molecules (H_2O), and each is linked to the four neighbors by hydrogen bonds (provided by the single H atom on each linkage). As suggested by the ice rules and shown on the left side of Fig. 2.21, ice is crystalline (having an orderly, repeating pattern) in a molecular sense, but the orientations of the molecules are irregular (glass-like, non-repeating) over larger distances. As long as the ice rules are followed locally, the H atoms may point in various, quasi-random directions throughout the solid (in the absence of an external electric field). Strong evidence for proton disorder comes from measurements of a finite (non-zero) entropy at very low temperatures (~ 0 K).

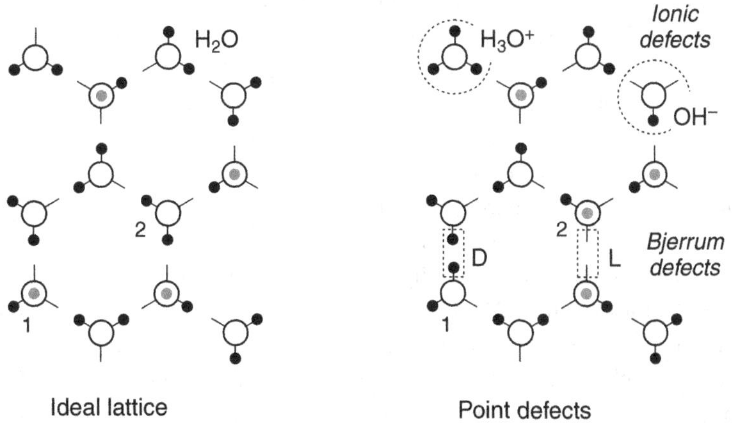

Figure 2.21 Schematic of the ice Ih lattice projected onto the basal plane. Open circles are oxygen atoms, solid circles are H atoms roughly in the plane, and shaded circles are H atoms on bonds pointing out of the plane. Left: One of several possible proton arrangements in an ideal lattice. Right: Common types of protonic point defects.

Crystals in nature are never perfect. The idealized hexagonal arrangement of molecules in ice Ih, for instance, can be upset by a number of point and line defects. Such defects account for important processes (discussed later), such as the growth of ice from the vapor, chemical interactions with the environment, and cloud electrification. Ice may also contain various planar defects, such as the two-dimensional interfaces between individual grains of polycrystalline ice, as well as faults in the way one molecular layer stacks on top of another. Grain boundaries are disordered regions that tend to accumulate foreign molecules that cannot substitute for water in the lattice. Stacking faults give rise to several non-prismatic morphologies of snow.

Point defects in a lattice are diverse in their form and effects. We categorize them as molecular or protonic. Among molecular point defects, individual lattice sites may occasionally lack a molecule altogether. Such a vacancy, or hole in the lattice, may get filled as a neighboring molecule jumps into that empty spot, but then the vacancy just reappears at the original location of the molecule. Vacancies may thus move through the lattice, in the process serving as a mechanism for self-diffusion of water molecules in the solid. The open nature of the ice Ih lattice also allows the possibility that extra molecules may fit in between the regular lattice sites. Such interstitial molecules would still experience attractive forces to neighboring molecules, but they nevertheless move about and so also contribute to solid-state diffusion. Interstitials and vacancies tend to originate at the crystal surface, but an interior molecule at a regular lattice site could jump into an interstitial position, thus forming a vacancy and an interstitial simultaneously. By contrast, an interstitial encountering a vacancy eliminates both. The number concentrations of these two types of molecular defects increase with the temperature because energy is needed for their formation.

Other types of point defects involve the protons rather than whole molecules. The right-hand side of Fig. 2.21 shows common protonic point defects relative to an ideal lattice (left-hand side). Protons (H^+) have net positive charge, so the addition of an H^+ to a molecule converts it into the positive hydronium ion (H_3O^+, top of figure), whereas the removal of H^+ from a neutral molecule results in a negative hydroxyl ion (OH^-, top right). Hydronium and hydroxyl ions (dashed circles in Fig. 2.21) are examples of ionic defects, molecules having net electric charge at regular lattice sites. Protons occasionally move to other bonds on a given water molecule, in effect causing the molecule to rotate or turn on one of its axes and resulting in electrically neutral point defects in the proton distribution (two dashed rectangles in Fig. 2.21). Note, for the example shown, how the rotation of the molecule labeled 1 results in two H atoms on one bond, a clear violation of the ice rules. Such a defect, in which two H atoms occupy one bond, is called a D-defect (the D referring to double, or doppelt in German). In the other example, the rotation of the molecule labeled 2 leaves a bond with no H atoms. Such a defect, an empty bond, is termed an L-defect (L for leer, the German word for empty). Orientational defects of the D and L types are called Bjerrum defects.

Local violations of the protonic ice rules are common and result from the thermal agitation that all molecules experience at temperatures well above absolute zero (0 K). As we just learned, the turning of a molecule results in the formation of Bjerrum defects. A proton can also hop from one molecule to another, in effect sliding along the common bond and simultaneously forming a positive and a negative ionic defect. The protons in ice may thus diffuse, when mediated by a combination of Bjerrum and ionic defects, about the lattice at rates that depend on the temperature. In fact, protons diffuse through solid water (ice) much faster than through pure liquid water, likely a result of the long-range order allowed by hydrogen bonding.

Protonic point defects are key to understanding the electrical properties of ice. Ice lacks the free electrons of a metal that respond readily to the presence of an electric field. Nevertheless, experimental data show that ice can conduct electricity. Ice, the experiments show, is a protonic conductor, meaning that protons (H^+) rather than electrons carry charge through the solid (in the general direction of the applied electric field).

Many scientists have studied the electrical properties of ice and have come up with a number of explanations. Figure 2.22 illustrates the basis behind the so-called Grotthuss mechanism. In panel I, the introduction of a proton (small arrow at top) into an otherwise ideal lattice causes an ionic defect (H_3O^+) to form. The extra proton on this molecule encourages another proton on the same molecule to slide along a bond and thus hop to the next molecule on the side toward lower electric potential. The original (top) molecule is now back to normal (except for having a different orientation), but the next lower molecule has now become ionic. That molecule in turn lets an extra proton hop to the next molecule lower on the chain. The sequence of hops continues in the general direction of the electric field and releases a proton at the end (bottom-most arrow). A positive current has flowed briefly, but no new protons can enter the top until the top-most molecule rotates and frees up the up-field bond. Facilitated by the force from the field, this molecule

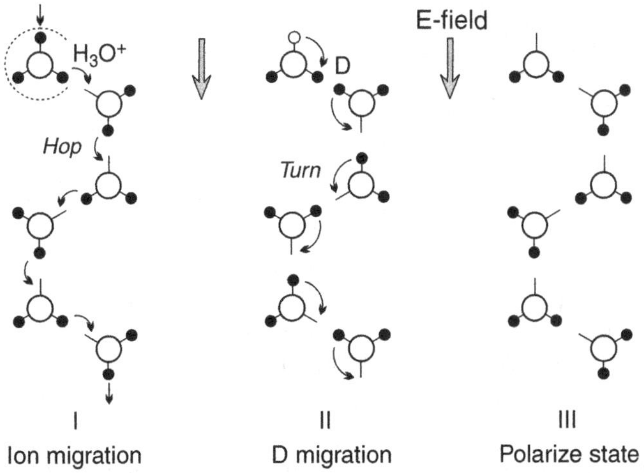

Figure 2.22 Schematic illustration showing the effects of an electric field (broad arrows) on the protons attached to water molecules in the lattice of ice Ih. I: Hopping of a proton from one molecule to the next, resulting in the transfer of charge from high to low potential (small arrows at top and bottom). II. Turning of a molecule, which moves a D-defect down the potential gradient (starting from initial orientation shown by the small open circle at top). III: A polarized state in which protons are at the lowest potential and unable to conduct current until a new proton enters (top left arrow). Adapted from Fletcher (1970, p. 213).

reorients (panel II) and forms a D-defect on the down-field side. The next molecule in the chain then turns and puts a proton on the next lower bond. The ensuing sequence of molecular turnings causes the D-defect to migrate, but no actual charge to flow. The lattice has become polarized (panel III), because of the electric field, and now sits at the lowest possible electric potential energy. However, the down-gradient migration of the D-defect has freed up the upper-most bond and made it receptive to the introduction of a new proton (upper arrow in panel I) and repetition of the process. Both ionic and orientational defects are needed for electricity to pass continuously through ice via this hop-turn mechanism.

Line defects, called dislocations, also appear in real ice. As shown in Fig. 2.23, dislocations are of two types, screw and edge. We discuss each separately, although combined forms are common. An edge dislocation (right-hand side) represents the linear boundary of an extra plane of molecules in the lattice. The extra plane is suggested by the dots over the dashed line terminating at the dislocation (aligned perpendicular to the plane and shown by the T-like symbol). Screw dislocations (left-hand side) represent a shearing of the lattice up to the line defect itself (shown as the vertical dashed line). Molecules in the dislocation core are stressed relative to neighbors, but away from it, the molecules all line up as in an ideal lattice. The adjective "screw" comes from the fact that the lattice planes perpendicular to the dislocation are helical about the dislocation; following a path that keeps the screw dislocation on the right leads to progressively lower lattice planes (suggested by the gray

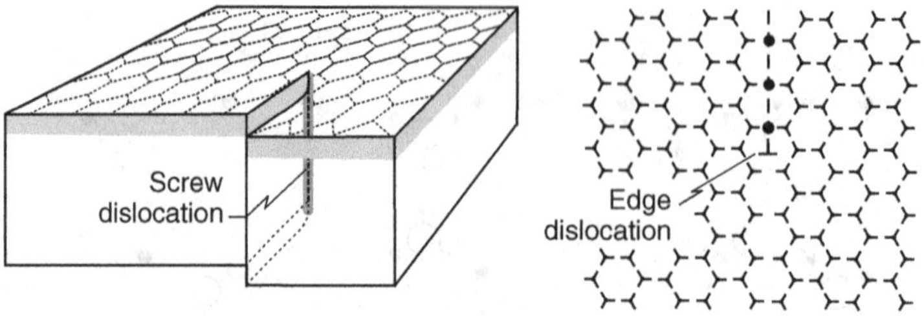

Figure 2.23 Two tyes of dislocations in a hexagonal lattice. Left: Screw dislocation, resulting from the partial displacement of the lattice. The gray layer shows how one plane spirals around the dislocation (heavy vertical dashed line). Right: Edge dislocation, caused by the presence of an extra plane of molecules (suggested by dots along the dashed line). The dislocation itself is oriented perpendicular to the image plane.

band in the figure). Screw dislocations always introduce a step into the surface and so play an important role in the growth of ice crystals in clouds, as we learn in Chapter 8.

2.1.4 The atmospheric aerosol

Solids and liquids exist in the atmosphere as particles, bits of condensed matter suspended in air. A collection of such suspended particles is termed an aerosol. One also speaks generically of an aerosol as particulate matter (PM), a term suggesting that the liquids and solids are dispersed, separated from each other within the air (in contrast to substances aggregated together in a macroscopic, bulk form. A glass of water, for instance, is water in bulk, but that same water sprayed into the air becomes dispersed.) Individual particles of an aerosol typically contain non-aqueous substances and are tiny by human standards. The molecules in each particle are bound together by electrical forces in a single cohesive mass of microscopic dimensions. Atmospheric particles are not to be confused with the elementary particles (electrons, protons, etc.) of atomic and high-energy physics, nor with any of the gas molecules making up the atmosphere. Except for the blue sky itself, it is the atmospheric aerosol, condensed matter in particulate form, that we see and recognize as smoke, haze, fog, or cloud. Atmospheric particles, as we shall learn later, frequently undergo changes in physical and chemical properties during their residence in the atmosphere. Aerosol particles play key roles in clouds because some particles serve as the precursors of cloud particles, while others help ice to form. The atmospheric aerosol is intimately intertwined with clouds and so is discussed here at length.

Classifications

Aerosol particles enter and leave the atmosphere by several mechanisms, which may be put into broad categories, as shown in Fig. 2.24. Particles may enter the air directly, or they may form *in situ*, in the air itself. Direct injections of particles into the atmosphere are primary sources, in contrast to particles resulting from chemical reactions in the atmosphere,

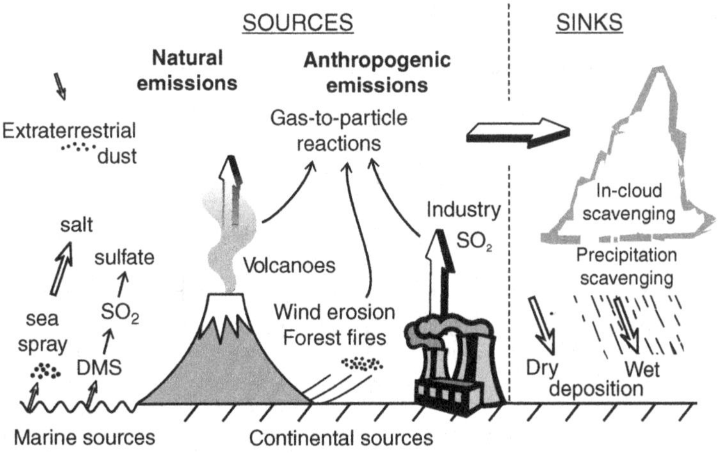

Figure 2.24 Cartoon of aerosol sources (left side) and sinks (right side).

which constitute secondary sources. Examples of primary sources are sea spray, which introduces sea salts into maritime air, wind erosion, which injects dust into the air over continents, and meteoric dust, which brings in particles from outside the Earth. Examples of secondary sources include particulate sulfate formed by gaseous reactions involving SO_2 emitted by industry (coal burning) or formed from oxidation of dimethyl sulfide (DMS) over some oceanic regions. Volcanoes and forest fires simultaneously release particles and reactive gases, so they are sources of both primary and secondary particulate matter. We also categorize sources by location, labeling them marine (or maritime), continental, or extraterrestrial. Sources are either natural or anthropogenic (human-caused), but the distinction is vague in some cases (e.g., human-set forest fires). The terminology used need only be appropriate for the discussion at hand. The sinks of aerosol particles are categorized on the basis of whether clouds are involved or not. Aerosol particles may leave the surface by dry deposition (e.g., sedimentation), but the most important sink for many particles is scavenging by clouds and precipitation (discussed in Chapter 13).

Communication about aerosols is made through the proper use of terminology for classification (see Fig. 2.25). One aerosol is typically distinguished from another by the properties of the individual particles. A cloud of chalk dust, for instance, is clearly different than cigarette smoke; the former is made of large, irregularly shaped particles, simple fragments of the chalk itself, whereas the latter is made of very small particles that result from the condensation of volatile organic substances released during the burning of tobacco. Thus, whereas chemical composition is one way to classify aerosols, particle shape and size are others. Composition, while important for a number of physical and chemical reasons, is complicated by the variety of the mixtures that are possible in individual particles and by the difficulties of analyzing particles that are often only a tiny fraction of the size of a sand grain. The shapes of the particles constituting different aerosols often differ, but

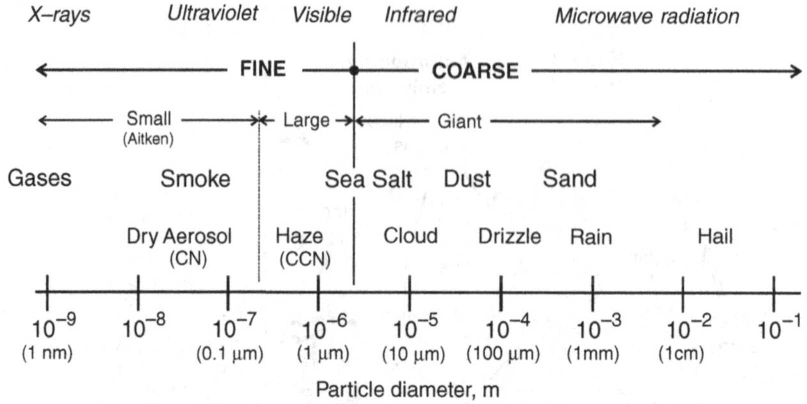

Figure 2.25 Categories of particulate matter found in the atmospheric aerosol. Radiation bands are shown at the top for comparison.

this characteristic is difficult to determine without tedious techniques. Size, on the other hand, is good for characterizing aerosols, as it can be measured with modest effort, and size has predictable consequences: Chalk dust settles out quickly because the particles are relatively large, not because of their composition. Cigarette smoke hangs around and gets into our lungs because the particles are tiny.

Particle size

The size of a particle is meant to convey a measure of how large it is. Some people, especially those concerned with the theory of aerosol formation, specify size as the radius of a sphere having the same volume as that of the particle (volume-equivalent radius). Others, such as those who measure particle sizes, use diameter or maximum dimension. Regardless, one picks a convention and makes computations consistent with it. Here, we use the volume-equivalent diameter D_p.

The sizes of particles in the atmospheric aerosol vary enormously. Figure 2.25 shows that individual particles can have diameters anywhere from nanometers (clusters of molecules) to several centimeters (hailstones). We use a logarithmic scale in such presentations in order to display the broad range of possibilities on one graph. A natural subdivision occurs in the atmospheric aerosol at a diameter of roughly 2 μm (air-quality standards use 2.5 μm) because of a persistent minimum in particle concentration there. The fine particles, those smaller than this threshold, typically come from different sources than do the coarse particles. Even though we occasionally use the term "large" toward the right-hand end of the "fine" range, we note that such particles are still very small. Haze is typically associated with this range of particles because the diameters are comparable to the wavelengths of visible light. The "small" sub range of fine particles is often called the Aitken range after the person who developed an instrument for measuring the concentration of aerosol particles. An Aitken counter develops a large supersaturation and so measures particles of all sizes by their ability to act as condensation nuclei (CN). The association of Aitken particles (or nuclei) with "small" arises because his method found these particles first. In the "coarse" range, we see familiar names, such as dust, cloud, rain or snow, and hail. Most cloud particles are coarse particles in the sense of size, but we often maintain a distinction, for convenience, between cloud particles and aerosol particles on the basis of composition (cloud particles are mostly water, unlike aerosol particles that are dominated by non-aqueous substances). Simple categorization of particles by size is helpful, but we must find more quantitative ways to characterize cloud particles and the atmospheric aerosol.

The number of particles within a given size range must be established if we are to complete the specification of an aerosol. The aerosol arising from a dust storm, for instance, may contain many coarse particles, but few fine particles. On the other hand, hazes and smokes typically contain mostly fine particles. A specification of the concentration of particles according to size is called a size distribution (or size spectrum, plural spectra). Particle size spectra tell us how the particles in an aerosol or cloud are distributed across the range of possible sizes. In effect, we ask ourselves how many particles have what sizes? When devising and interpreting size distributions, it proves important to distinguish empirical displays (as derived from data) from mathematical functions meant to approximate data or represent particle sizes derived from theory. Nevertheless, all size distributions serve the purpose of specifying how many particles in the population are contained in a given size range. The size distribution is an important characteristic of a population of particles because it enables us to distinguish one aerosol from another and to shed light on likely mechanisms of formation.

Statistical methods are employed when determining size spectra because the number of particles contained in even a small volume (e.g., 1 cm^3) of air can be large (thousands). A long list of particle diameters, while providing all possible size information about the population, serves little useful purpose; it is simply too much information, and the information is not organized in any useful way. Statistics provides the means of summarizing the main characteristics of the population in concise terms.

A size distribution is typically specified in terms of concentration (number of particles in a unit volume of air) per given size interval (termed a size category or bin). Let N be the total number of particles of all sizes in the unit volume of air and N_j be the concentration in size category j. The distribution of the N particles by size is obtained by following several steps, as summarized in Table 2.2. The histogram resulting from Step 1 yields a bar graph, one showing the particle number concentration versus diameter interval or bin number. The heights of the bars in the histogram, representing the category values N_j, may look uneven when the bin widths are not uniform; wide bins tend to contain more particles than do narrow bins. Division of the concentration in each bin by the bin width (Step 2) yields a discrete size distribution, $n_j = n(D_{p,j})$, where $D_{p,j}$ is the diameter at the mid-point of size interval j. Discrete size distributions from Step 2 look smoother than the histogram from Step 1. Size distributions have dimensions of concentration per unit size interval; typical units might be $cm^{-3}\ \mu m^{-1}$ (assuming the interval for standardization was 1 μm).

Table 2.2 *Operations for developing a size distribution.*

(1) Group the data:	Conceptually put each particle into its appropriate size category (or bin) by adding a count to the number already in that category. The choice of size intervals (bin widths) is arbitrary and need not be uniform, but the intervals must be contiguous to ensure that all particles are counted (i.e., that no particle falls between the "cracks"). Particles having a size exactly the same as one of the bin limits are, by convention, grouped into the larger of the two possible size bins. This operation yields a table of counts (or frequency) versus bin number (or size class). A histogram results when the frequency is plotted against the size class (represented by the mid-point of each bin).
(2) Standardize the grouped data:	Divide the number in each size class by the size interval (bin width). This operation allows the information in one bin to be comparable to those in the other bins. A discrete size distribution results when this count per unit size interval is plotted against the class size.
(3) Normalize the distribution:	Divide the standardized values in each bin by the number of particles in the entire sample. This operation yields a fraction per unit size interval, the frequency or likelihood that particles in the sample fall into a unit size interval about a specified size. This relative (or fractional) size distribution results when this fraction per unit size interval is plotted against the class size.

A discrete size distribution may be approximated by one of several mathematical functions, so yielding a continuous size distribution $n(D_p)$. Frequency or probability distribution curves (pdf from Step 3) are often plotted as continuous curves and labeled as a fraction per unit size interval. Probability distributions express the relative importance of various size ranges, but for the atmospheric sciences it suffices to use standardized distributions (those resulting from Step 2).

Mathematical representations of size distributions help us refine our view of an aerosol. When we let $n\left(D_p\right)$ represent the continuous size distribution function, we are specifying the number of particles in a unit volume having diameters between D_p and $D_p + dD_p$. With units of $\text{cm}^{-3}\,\mu\text{m}^{-1}$, the distribution $n\left(D_p\right)$ gives the concentration of particles within a 1-μm size interval about diameter D_p. With a countable number of size ranges, the number of particles within a finite, but small size range ΔD_p is approximately $n\left(D_p\right)\Delta D_p$; we simply multiply the distribution function by the size range to gain the concentration of particles in that interval. The total number of aerosol particles in the unit volume of air, N, is obtained by adding up the particles in all of the size intervals:

$$N = \sum_{j=1}^{N_{bins}} n(D_{p,j})\Delta D_{p,j}, \tag{2.26}$$

where the subscript j identifies one of the N_{bins} size categories, the one having the mean size $D_{p,j}$. The total aerosol concentration is not directly dependent on particle size, so the units of N in this case are simply cm^{-3} (note the cancellation of size units). When the size interval is made infinitesimally small, the size distribution becomes a continuous function and the total concentration is calculated by integration (infinite summation):

$$N = \int_0^{\infty} n(D_p)dD_p. \tag{2.27}$$

Reversal of this procedure shows that a size distribution is a differential function: $n(D_p) = dN/dD_p$. If we divide the differential number concentration by the total number concentration, we get the fraction of particles contained within this differential size interval: $f(D_p) \equiv n(D_p)/N$. Integration of the differential frequency function over all possible sizes gives us the expected normalization property,

$$\int_0^{\infty} f(D_p)dD_p = 1. \tag{2.28}$$

In other words, all particles are accounted for once the full size range is covered.

The atmospheric aerosol is often characterized in terms of size distributions. An example of a size distribution determined from data taken in clean air over the middle of the North American continent is shown by the dashed curve in Fig. 2.26. We see immediately that the concentration of particles varies enormously over the diameter range, so logarithmic axes are needed to show the details. The wiggles in the distribution function prevent $n(D_p)$ from being characterized by simple analytical functions.

Dealing with the wide range of particle concentrations and the wiggles in measured size distributions can be facilitated by transforming $n(D_p)$ into the logarithmically based function $n(\log D_p) \equiv dN/d\log D_p$. (Note that these two functions differ both numerically and dimensionally.) In any given infinitesimal size interval the number dN of particles is determined physically by the aerosol sample and not by our choice of coordinates. Thus, we make use of the continuity relationship $dN = n(D_p) \cdot dD_p = n(\log D_p) \cdot d\log D_p$. Furthermore, from the properties of logarithms, $\log x = \ln x / \ln 10$ and $d\ln x = dx/x$, we get the convenient conversion formula $n(\log D_p) = D_p \cdot n(D_p) \cdot \ln 10$. The result of making this transformation is shown by the solid curve in Fig. 2.26. The ordinate, $n(\log D_p) = dN/d\log D_p$, has the units [$cm^{-3}$] and represents the concentration of particles in the range from $\log D_p$ to $\log D_p + d\log D_p$. This curve is flatter than $n(D_p)$ (because of the weighting by D_p), and the wiggles are now more pronounced.

Other distributions, beyond those based on number concentration, further help us characterize an aerosol. The collective surface area an aerosol exposed to the air (in a unit volume)

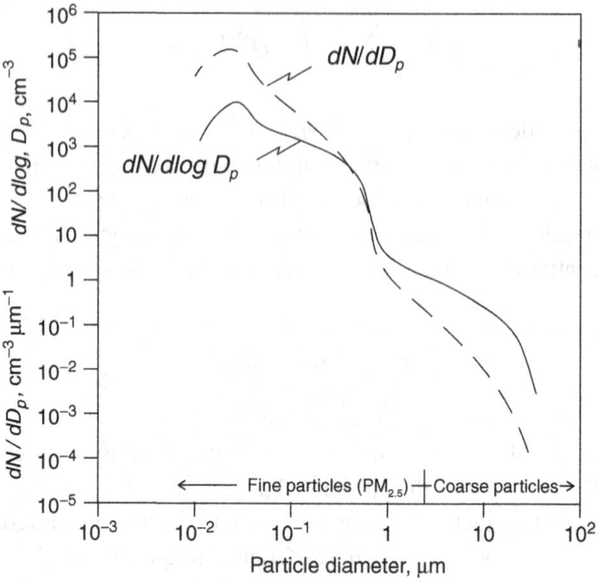

Figure 2.26 Distributions of a continental background aerosol by size. The ordinate is interpreted differently for each curve. Particles with diameters smaller than 2.5 μm are called fine based on the air quality standard for particulate matter, $PM_{2.5}$; larger particles are called coarse. Based on Whitby (1972).

is important for understanding physical and chemical transformations, as well as for computing radiative interactions. The collective volume occupied by the aerosol is closely related to mass concentration and the lifetimes of particles in the atmosphere. We can compute these other distributions by recalling that the area of a spherical particle is $A_p = \pi D_p^2$ and its volume is $V_p = \pi D_p^3/6$. We thus gain the surface and volume distributions (area or volume per unit size interval) via multiplication of the number distribution by the respective particle properties:

$$\frac{dS}{d\log D_p} \equiv n_S(\log D_p) = A_p \cdot n(\log D_p) = \pi D_p^2 \cdot n(\log D_p) \tag{2.29}$$

$$\frac{dV}{d\log D_p} \equiv n_V(\log D_p) = V_p \cdot n(\log D_p) = \frac{\pi}{6} D_p^3 \cdot n(\log D_p). \tag{2.30}$$

Graphs of distributions based on $\log D_p$ are convenient because the areas under the curves are proportional to the respective total concentrations. The total surface-area concentration $S = \int_{-\infty}^{\infty} n_S(\log D_p) d\log D_p = \pi \int_{-\infty}^{\infty} D_p^2 \cdot n(\log D_p) d\log D_p$, and the total volume concentration $V = \int_{-\infty}^{\infty} n_V(\log D_p) d\log D_p = \frac{\pi}{6} \int_{-\infty}^{\infty} D_p^3 \cdot n(\log D_p) d\log D_p$. The surface area and volume represent, respectively, the second and third moments of the size distribution. The first moment of the distribution is the mean

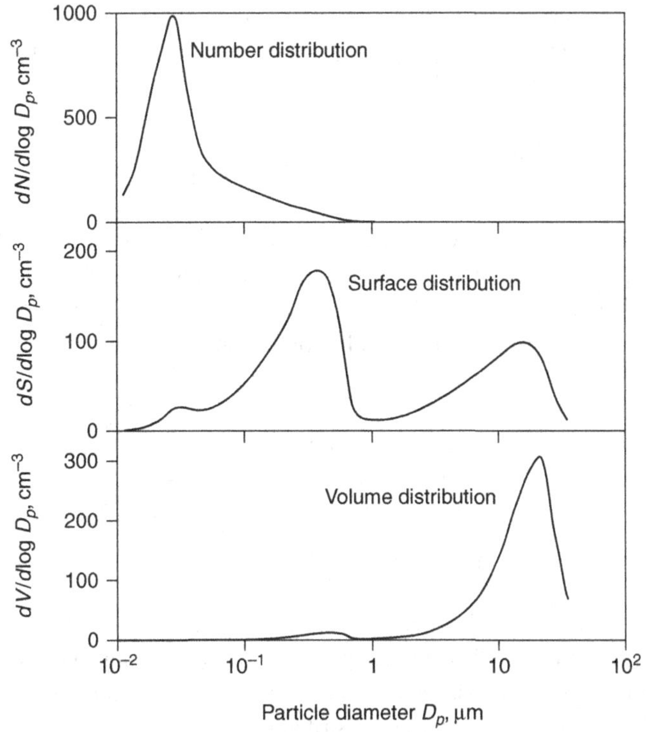

Figure 2.27 Distributions of aerosol properties by size. Adapted from Whitby (1972).

diameter $\bar{D}_p = \frac{1}{N}\int_{-\infty}^{\infty} D_p \cdot n(\log D_p)d\log D_p$, while the total number concentration $N = \int_{-\infty}^{\infty} n(\log D_p)d\log D_p$ is the zeroth moment. Note the difference between the limits of integration for linear and logarithmic functions.

The various size distribution functions are compared in Fig. 2.27, where we note the multi-modal nature of the atmospheric aerosol. Whereas the number distribution has a single mode (maximum number), the surface-area and volume distributions often have two or more modes each. Each function describes the same aerosol (that represented in Fig. 2.26), but the higher moments weight the concentrations in the larger sizes more. Each mode is given a name: the maximum in the Aitken range is the "nucleation mode", the main maximum in the surface-area distribution is termed the "accumulation mode", and the mode above 2 μm is the "coarse mode". Even though all three modes seldom show up in any one of the distribution functions, it is common to refer to the atmospheric aerosol as being trimodal.

The individual modes in the atmospheric aerosol can be described, at least approximately, by analytical functions. Each mode generally acts independently of the others, so we describe the composite size distribution as the sum of three separate functions. As a first approximation, one might expect that a given mode could be represented by a normal distribution, that is, one in which the independent variable (D_p) is distributed randomly about some mean value ($\bar{D}_p$) with a standard deviation σ_D. A normal distribution about $\bar{D}_p$ would then be represented by the function

$$n(D_p) = \frac{N}{\sigma_D \sqrt{2\pi}} \exp\left[-\frac{(D_p - \bar{D}_p)^2}{2\sigma_D^2}\right], \tag{2.31}$$

which describes a bell-shaped or Gaussian curve. However, this function would admit negative values of particle diameter, which is clearly unrealistic. A slight variation of the normal distribution solves this problem and accounts nicely for real aerosol size distributions. In place of D_p, we use the logarithm of D_p and obtain the so-called log-normal distribution:

$$\begin{aligned} n(\log D_p) &= \frac{N}{\log \sigma_g \sqrt{2\pi}} \exp\left[-\frac{(\log D_p - \log \bar{D}_{pg})^2}{2\log^2 \sigma_g}\right] \\ &= \frac{N}{\log \sigma_g \sqrt{2\pi}} \exp\left[-\frac{\log^2 (D_p/\bar{D}_{pg})}{2\log^2 \sigma_g}\right]. \end{aligned} \tag{2.32}$$

Here, $\bar{D}_{pg}$ is the geometric (rather than the arithmetic) mean diameter, and σ_g is the geometric standard deviation of this mode. In effect, we are saying that the logarithm of particle diameter is normally distributed. The composite, three-mode distribution is then described by the sum of three log-normal distributions:

$$n(\log D_p) = \sum_{i=1}^{3} n_i(\log D_p) = \sum_{i=1}^{3} \frac{N_i}{\log \sigma_{g,i} \sqrt{2\pi}} \exp\left[-\frac{\log^2 (D_p/\bar{D}_{pg,i})}{2\log^2 \sigma_{g,i}}\right]. \tag{2.33}$$

Each mode (index i) is characterized by its own set of parameters, N_i, $\bar{D}_{pg,i}$, and $\sigma_{g,i}$. Observed distributions are often fitted to the three-mode log-normal distribution. Mathematical descriptions of the atmospheric aerosol help us relate measurements with theory.

The differential distributions we have been considering can be integrated to form cumulative distributions, especially those that find applications in air-quality regulations and cloud formation. Often, we simply need to know the number or mass of particles smaller than some specified size. The particulate matter taken up by our lungs, for instance, has been found to be in the "fine" category of sizes, specifically below 2.5 μm in diameter (see Fig. 2.26). Such fine particles have been deemed unhealthy to breathe, so a so-called $PM_{2.5}$ standard may exist in some regions. (The US EPA set the standard for fine aerosol mass concentration in 2006 to be 35 $\mu g\,m^{-3}$ in any 24-hour period or 15 $\mu g\,m^{-3}$ when averaged over one year.) The number concentration of particles in the $PM_{2.5}$ category is calculated from the size distribution by integration:

$$N_{2.5} = \int_0^{2.5} n(D_p) dD_p. \tag{2.34}$$

The mass concentration [$\mu g\,cm^{-3}$] is calculated from the third moment of the distribution, given knowledge of the mass density ρ_p of the individual particles:

$$M_{2.5} = \frac{\pi \rho_p}{6} \int_0^{2.5} D_p^3 \cdot n(D_p) dD_p. \tag{2.35}$$

Here, we have assumed that the particle density (often taken to be $2\,g\,cm^{-3}$) varies little with particle size. Fine-particle mass concentrations are frequently measured and compared with the relevant air-quality standard. Also of importance is the total suspended particulate matter:

$$TSP = \frac{\pi \rho_p}{6} \int_0^{\infty} D_p^3 \cdot n(D_p) dD_p. \tag{2.36}$$

Note that information about the size distribution is not obtainable from size-integrated measurements.

Composition of particulate matter

The chemical composition of the atmospheric aerosol depends on the origins of the individual particles. As indicated earlier in Fig. 2.24, some particles may enter the atmosphere naturally by wind erosion, while others arise from chemical reactions within the air. The particles from each such aerosol have distinctive compositions reflective of their sources. Wind erosion might yield soil particles, themselves complex mixtures of organic and inorganic compounds. By contrast, *in situ* chemical reactions involving SO_2 and NH_3 would result in partially neutralized ammonium sulfate and possibly semi-volatile organic compounds. Individual particles may each contain a variety of compounds, but a clear distinction between the two sources would nevertheless be evident through detailed sampling and analysis. Particles from a given source that have related composition, but which are themselves composed of a variety of compounds are said to be internally mixed. As the aerosols from different sources blend through normal turbulent transport with the wind, the aerosol becomes externally mixed, a collection of the individual particles from the different upwind sources. The atmospheric aerosol tends to be an external mixture because of the multitude of sources within any one region. Specialized sampling methods are needed to determine the internal mixture of individual particles; analyses based on bulk sampling (e.g., filtration) cannot distinguish an internal from an external mixture.

The composition of the atmospheric aerosol is complicated by the large number (thousands) of possible compounds that the particles might contain. It is therefore useful to define broad classes of compounds and consider individual substances only as needed. Compounds are usually classified as either inorganic (without carbon) or organic (containing carbon), but exceptions exist in the case of carbonates (oxides of carbon that are considered to be inorganic despite containing carbon). The dry aerosol over continents tends to be roughly half (by mass) organic (hydrocarbons, oxidation products of volatile organic compounds, or VOC) and half inorganic (sulfates, nitrates, and various metallic compounds). Over the remote oceans, the dominant compounds in the aerosol are sulfates and sea salts.

The chemical properties of an aerosol affect how the individual particles behave in and out of clouds. Many inorganic compounds, especially mineral acids and salts, are hygroscopic, which means they interact readily with water, even to the point of dissolving completely in liquid water. Hygroscopic particles that absorb water from the atmosphere and help initiate cloud formation are termed cloud condensation nuclei (CCN). The uptake of water alters the composition, as well as the size distribution of the aerosol. Many organic compounds, on the other hand, tend to be hydrophobic and insoluble. Soot (or elemental carbon), which results from some types of combustion (e.g., diesel engines), interacts only weakly with water, but such substances may absorb radiation and indirectly affect cloud properties. Some organic substances may be surface active, meaning that they reside preferentially in the surface of liquid water. Such surfactants change the surface tension of the liquid and so also influence cloud formation. Crustal matter (e.g., clay) does not readily dissolve, but it can sometimes facilitate the freezing of supercooled liquid water. Insoluble particles that aid the formation of ice are termed ice nuclei (IN). It is worth emphasizing that the various processes that occur during atmospheric transport and aging act on the individual particles, rather than on the aerosol as a whole. Thus, an increase in humidity causes the individual particles to take up water and swell to the degree characteristic of each internal mixture. However, although each particle responds separately, the collective action may nevertheless manifest itself in some complicated, but observable way (e.g., reduced visibility with increasing humidity). Resolving the responses of the atmospheric aerosol to changing conditions remains an outstanding challenge.

Geographical distributions of aerosols

Aerosol particles exist because of processes that put them into the air in one place, but they are removed from the atmosphere by other processes in other regions. Each source enriches the atmosphere with its characteristic compounds, which subsequently get transported and transformed downwind. Important sources of the atmospheric aerosol are condensation of semi-volatile vapors (a secondary source) and mechanical lifting of surface particles by the wind (a primary source). Such sources are widespread, in contrast to some anthropogenic sources, such as combustion in electric generation plants, which may be considered virtual point sources. The plume from a smoke stack is initially well defined, but it does eventually disperse through turbulent transport and mix with other plumes and the background aerosol of the region. Industrial plumes contain both primary particulate matter and reactive gases that serve as precursors for other gases and new particles. Gas-to-particle conversion produces fine aerosol particles that stay suspended for long times and so can travel great distances. By contrast, the mechanical lifting of sea spray, dust, and vegetation debris produces mostly coarse aerosol particles that may fall out relatively close to the source region.

Clouds greatly augment the removal of particles by providing effective transport and transformation mechanisms. Aerosol particles participate in the nucleation of cloud droplets by acting as the sites of condensation, and they may be scavenged by existing cloud particles inside the cloud. Cloud-active particles are frequently removed from the atmosphere through precipitation processes. As a result, regions with high cloud and precipitation frequencies often have low particle concentrations.

The concentration of aerosol particles thus depends not only on the relative strength of the sources and losses, but also on the locations of those sources and sinks. Higher concentrations are found closer to strong source regions; conversely, lower concentrations are found close to strong sinks. Continents, with all their human and other biological activities, have an abundance of sources, but the removal mechanisms, mainly by clouds enriched in particulate matter, are modest. By contrast, the oceans have relatively weak sources, but relatively active removal opportunities in clean clouds that precipitate efficiently. As a result, aerosol concentrations are much higher over continents than over the oceans. Figure 2.28 compares the size distribution of a typical marine aerosol to that of a continental aerosol. We see that marine aerosols have lower concentrations at all size ranges, and that the mode of the distribution, the size with highest concentration, occurs at a larger size than that in a continental aerosol.

Removal mechanisms operate throughout the depth of the atmosphere, so we should expect concentration variations in vertical, as well as horizontal directions. Figure 2.29 shows profiles of Aitken particle concentrations measured over several Russian cities. The high concentrations at low altitudes suggest that sources predominate near the surface, while sinks dominate aloft. The surface (planetary boundary) layer is typically well mixed by turbulence, so these fine particles exhibit relatively constant profiles throughout the lower 2 km. Above the boundary layer, concentrations fall off systematically with height, an average effect of removal by precipitation over long time and distance scales. Clouds are one of natures best cleansing mechanisms.

The concentration profiles of "large" aerosol particles show an important difference. As shown in Fig. 2.30, a distinct maximum in concentration is observed in the lower stratosphere. This stratospheric aerosol layer, called the Junge layer after the scientist

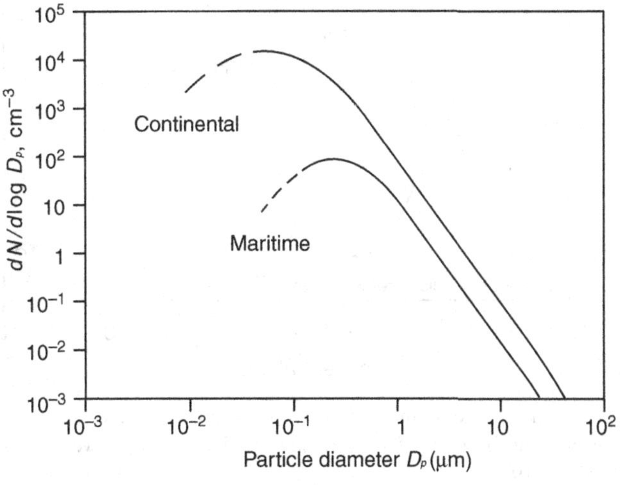

Figure 2.28 Comparison of aerosol size distributions from continental (upper curve) and maritime (lower curve) air masses.

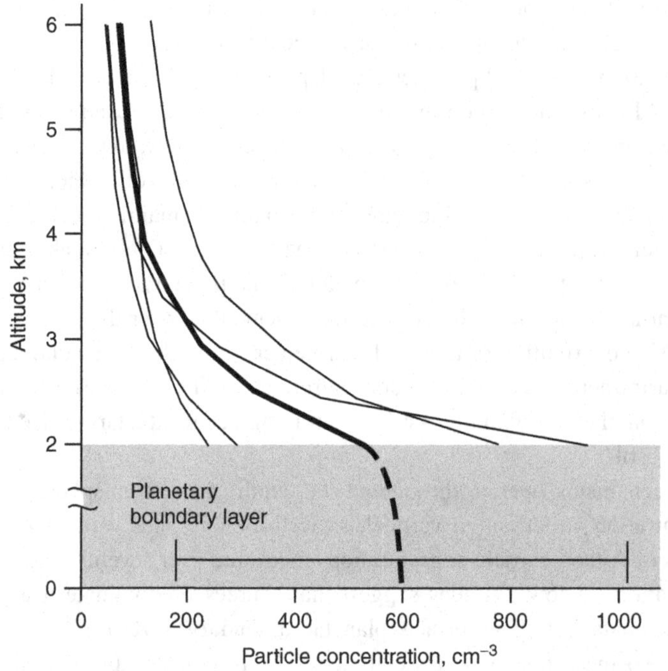

Figure 2.29 Profiles of Aitken particles (CN, all sizes) over several Russian cities. Adapted from Selezneva (1966).

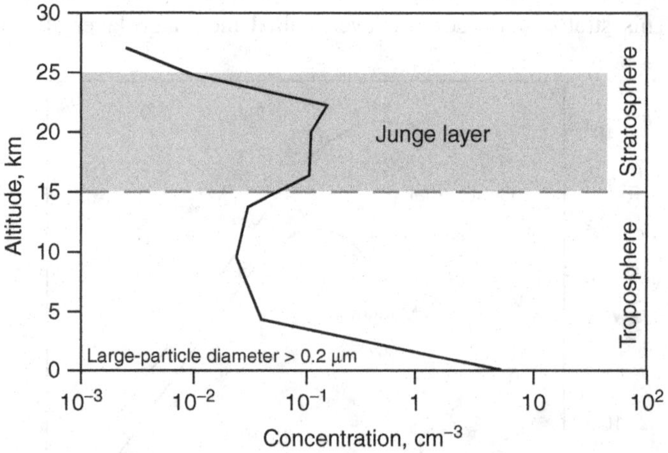

Figure 2.30 Concentration of large (diameters > 0.2 μm) particles versus altitude. The Junge layer is a region of enhanced concentration of sulfuric-acid particles in the lower stratosphere. Adapted from Chagnon and Junge (1961).

who discovered it, is persistent and found around the globe. The Junge layer results from periodic injections of sulfur-containing gases into the stratosphere by volcanoes. The reduced sulfur compounds (e.g., SO_2) become oxidized to sulfuric acid, which readily condenses into small, highly acidic droplets. Fresh injections of sulfurous gases by sporadic volcanic activity temporarily enhance the concentrations of the sulfate particles to levels high enough to disturb the Earth's energy balance for a year or two.

2.2 Energy in the atmosphere

Energy is as much a part of the atmosphere as is the matter it contains. Without the steady supply of energy from the Sun, nothing in our world would change; the winds would not blow, clouds would not form, rain would not fall. We need to recognize the various forms of energy that bring about the changes we see everyday. On average, as much energy must return to space as was gained from the Sun, or the Earth would warm indefinitely. So, we must also understand the transformations and redistributions of energy that allow the energy flows to be balanced and our climate to remain reasonably stable.

Energy appears in various forms and is the subject of important thermodynamic principles. Work done on a system may heat the system and cause its internal energy to increase, but overall the total energy must be conserved (first law of thermodynamics). The reverse process, conversion of internal energy into useful work, can also occur (subject to the second law of thermodynamics), but again the total energy must be conserved. Energy may change forms, but it can neither be formed anew nor destroyed.

The forms that energy takes are seemingly endless. We use various terms for the different forms (e.g., gravitational, electromagnetic, chemical, latent heat), but all forms fit into just a couple of categories. Energy is fundamentally either kinetic (KE) or potential (PE) in nature, and it operates in both our everyday world (macroscale) and inside objects (microscale). A ball lifted above one's head puts it at a higher macroscopic potential energy (gravitational PE). Dropping the ball causes it to accelerate and generate macroscopic kinetic energy at the expense of its potential energy. We can readily visualize macroscopic energy transformations, but interchanges between potential and kinetic energies also occur among the molecules that make up matter. Once the dropped ball hits the ground and the macroscopic motions stop, the motions of the molecules inside the ball and in the ground are increased; the macroscopic kinetic energy has been transformed into microscopic kinetic energy. The molecules in the ball and ground have been agitated just as much as if they had been heated. Increases in the internal energy of an object may also cause the molecules to move farther apart or vibrate more rapidly relative to the neighbors to which they are bound, in which case the potential energies of the molecules increase as well. Whether at the macroscopic or microscopic level, energy transformations may occur subject to the principle that an increase in one form of energy brings about a decrease in other forms. Overall, energy must be conserved.

The internal energy of a system (an object or a part of the atmosphere) may be sensed, or it may be hidden (not sensed). At the macroscopic level, a system may feel hot or cold to the touch, but we are sensing a microscopic property, namely the motions of the molecules making up the system. The system when hot has much sensible energy and would cause a thermometer to show a high reading. Temperature is a measure of the concentration of sensible energy, the mean KE of the molecules. Molecular KE is here referred to as thermal energy, although some authors use this term (thermal energy) as an alternative to internal energy, which is the sum total of sensible and hidden energy. The hidden component of internal energy is not readily sensed, but it is just as real; it is hidden, or latent, possibly appearing later when it has been "released", converted into molecular KE. The latent forms of energy represent molecular PE, energy possessed by the molecules because of the forces they exert on one another. As molecules move away from one another against these binding forces, they acquire molecular PE, much as a ball acquires gravitational PE when lifted against gravity. At the molecular level, the dominant forces are derived from electromagnetism, not gravitation.

Two systems may exchange internal energy by various mechanisms. Some of the sensible energy of a system may be transferred to another system if that system has a lower temperature. If two objects of differing temperature make physical contact, some of the molecular KE of the hotter object may be transferred by conduction, the process by which molecules with high KE give up some energy to those of lower KE during molecular collisions. Thermal energy may also be transferred by radiation, the process by which two non-touching systems exchange electromagnetic energy. Energy always flows spontaneously, by both conduction and radiation, from the hotter to the colder system. Energy transfer by radiation is especially important in the atmosphere.

2.2.1 Radiation

Radiation in the context of the atmospheric sciences refers to the electromagnetic energy that flows from one system to another; we are not concerned here with any type of nuclear radiation. Sunlight is perhaps the most familiar form of electromagnetic radiation, but invisible forms of radiation exist and are just as important. Infrared radiation helps transfer energy from warm bodies to cold bodies, ultraviolet radiation brings about chemical reactions in the atmosphere, and microwave radiation helps us detect precipitation with weather radar. Each category of electromagnetic radiation brings about a specific action that alters the atmosphere in some way or provides information about the state of the atmosphere via remote sensing.

Nature of electromagnetic energy

Electrical and magnetic forms of energy were found long ago to have the common capability of causing action at a distance by the propagation of waves. The electric and magnetic fields in an electromagnetic wave interact with each other and cause the energy to propagate through empty space at a constant speed, the speed of light $c \simeq 3 \times 10^8\ \mathrm{m\,s^{-1}}$. As shown in Fig. 2.31, electromagnetic waves are transverse in free space because the electric and magnetic fields vary normal to the direction that the wave travels (arrow toward right).

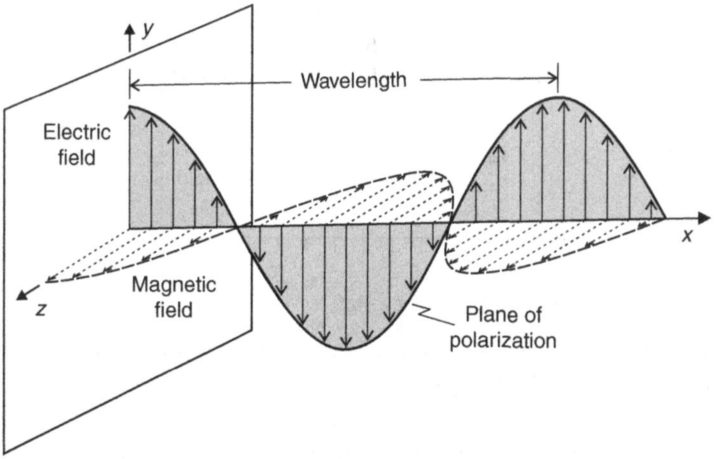

Figure 2.31 Schematic depiction of an electromagentic wave penetrating a plane normal to the direction of propagation. Adapted from Weidner and Sells (1960, p. 21).

The strength of the electric field E varies sinusoidally in sync with that of the magnetic field H, with time dependences described by the following equations:

$$\mathbf{E}(t) = \widehat{y}E_0 \sin\left[2\pi\left(\frac{x}{\lambda} - \upsilon t\right)\right],$$
$$\mathbf{H}(t) = \widehat{z}H_0 \sin\left[2\pi\left(\frac{x}{\lambda} - \upsilon t\right)\right], \tag{2.37}$$

where x is distance in the direction of propagation, while $\widehat{y}$ and $\widehat{z}$ are unit vectors in the orthogonal directions. Note that these fields are vectors, with the amplitudes of the electric and magnetic waves being E_0 and H_0, respectively. The transport of radiant energy is described by the Poynting vector, $\mathbf{S} = \mathbf{E} \times \mathbf{H}$, and the orientation of the electric field (the rotational angle of $\widehat{y}$ about $\widehat{x}$) defines the polarization of an electromagnetic wave. In the absence of a specific process causing polarization, radiation is often unpolarized, meaning it is a mixture of waves having all possible rotational angles. The distance between crests of the waves is the wavelength λ, and the frequency of the radiation is ν. The relationship between ν and λ is given by $c = \lambda\nu$, so we see that wavelength and frequency are inversely related.

The spectral (wavelength- or frequency-dependent) properties of electromagnetic waves are important for understanding the various roles that radiation plays in the atmosphere. As shown in Fig. 2.32, the possible wavelengths and corresponding frequencies vary by many orders of magnitudes. It is therefore convenient to give names to various ranges (or bands) of the electromagnetic spectrum. Visible light, for instance, is that portion to which the human eye is sensitive, namely radiation with wavelengths between about 0.4 μm and 0.7 μm. Note by the shading in Fig. 2.32 how narrow the visual range is. Specialized equipment is needed to detect radiation in other spectral bands. Solar radiation overlaps

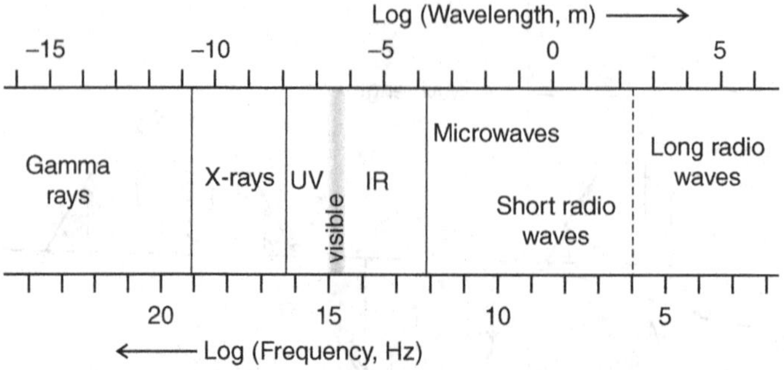

Figure 2.32 Spectrum of electromagnetic radiation expressed both in terms of wavelength (top axis) and frequency (bottom axis). Adapted from Weidner and Sells (1960, p. 23).

the visible band and extends into the infrared (IR) and ultraviolet (UV) bands, which bound the visible band on the high- and low-wavelength sides, respectively. IR radiation is particularly important for transporting energy from Earth back into space and maintaining a stable climate. UV radiation comes mainly from the Sun and causes photochemical reactions in the atmosphere.

The modern view of radiant energy extends the classical wave picture by accounting for observations that show light to exhibit some characteristics of particles. Diffraction phenomena show light to be wavelike, but the photoelectric effect strongly suggests that light acts as if it were particles of specific energy. The emergence of quantum mechanics in the early 1900s helped resolve the paradox, which came to be known as the wave–particle duality. Through the pioneering research of Planck, Einstein, and Millikan, we have come to accept that electromagnetic radiation is quantized, made up of photons, packets of radiant energy having discrete energies. The energy of a photon of frequency ν is $E_{photon} = h\nu$, where $h = 6.6\times10^{-34}$ J s is the Planck constant. The concept of photons can be appreciated by noting that a truly monochromatic wave would have to be infinitely long and contain an incredible amount of energy. Heisenberg showed that limits exist on the certainty of knowing two related variables simultaneously. If the wavelength of a wave is known precisely, for instance, then its position (or length) is uncertain. As shown in Fig. 2.33, a photon may be viewed as a wave packet derived from the superposition of waves having wavelengths differing from the central wave by small amounts. Where the waves are nearly in phase, constructive interference gives rise to enhanced amplitudes, whereas when the waves are out of phase, they interfere destructively. The net result is an envelope of electromagnetic energy having finite extent (on the order of the wavelength) and slightly uncertain position. Electromagnetic radiation acts as if the waves were encapsulated in little particles that move at the speed of light until they encounter matter. Short-wavelength radiation is sometimes conceptualized as quanta, whereas longer-wave radiation (including visible light) can often be treated as if it were only waves. The best way to visualize electromagnetic radiation depends on the application, but remember that all radiation has both wave-like and particle-like characteristics.

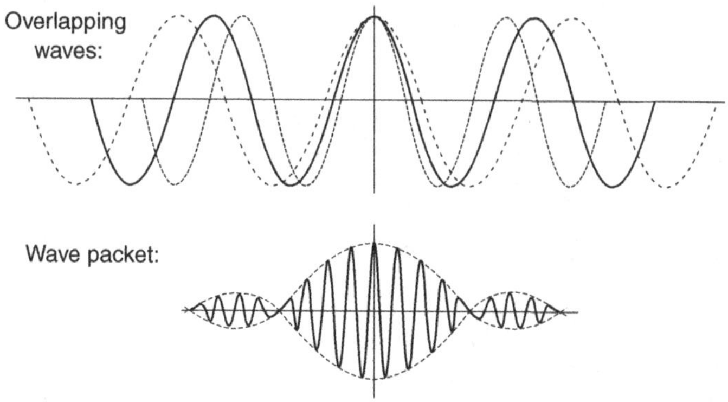

Figure 2.33 Cartoon depiction of electromagnetic waves leading to the concept of photons. Top: Overlapping waves that interfere constructively and destructively. Bottom: A wave packet resulting from interference of waves about the principal wave. Adapted from Weidner and Sells (1960, pp. 489, 490).

Sources of radiation

The source of most usable energy on Earth is the Sun. The energy from the Sun drives atmospheric circulations, stimulates chemical reactions, and causes water to evaporate from the oceans. Water in vapor form eventually condenses in clouds and returns to the surface as precipitation. The never-ending hydrologic cycle itself is driven by energy from the Sun.

The radiation emitted by the Sun, derived indirectly from nuclear fusion in its core, is distributed over a range of wavelengths centered in the visible band. Radiation, generated by the hot plasma (ionized gases) near the Sun's surface, travels outward into space, where some of it is intercepted by the Earth. The spectrum of solar radiation irradiating the top of the atmosphere is shown by the solid curve in Fig. 2.34, where it is compared with radiation from an ideal blackbody (dashed curve) having a temperature of 6000 K. Some solar radiation entering the atmosphere is absorbed by atmospheric gases and particles before reaching the Earth's surface. The shaded part of Fig. 2.34 shows the radiant energy reaching sea level on a cloud-free day with the Sun overhead. Note that atmospheric gases cause radiation to be absorbed strongly in relatively narrow ranges of wavelengths. The energies of electrons in matter are quantized, so specific photon energies are required for transitions from one energy state to another. Photons having just the right energy for a molecular transition give up their electromagnetic energy, in the process adding to the internal energy of the gas molecules.

The Sun acts, to reasonable first approximation, as a blackbody with an effective temperature of 5777 K. The effective blackbody temperature is found by integrating the solar spectrum to obtain the total power emitted and then solving for the temperature a blackbody would need in order to emit that same power. The total (wavelength-integrated) flux of energy irradiating the top of the Earth's atmosphere per unit time and area is

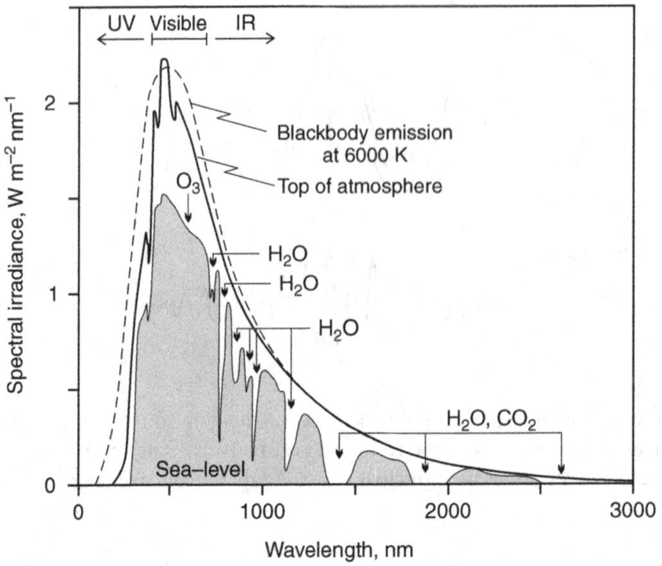

Figure 2.34 Spectrum of solar radiation at the top of the atmosphere (solid curve) and at sea level on a cloudless day with the Sun overhead (shading). Some of the compounds responsible for absorption are labeled with arrows. The radiation from a blackbody at 6000 K is shown by the dashed curve. Adapted from Salby (1996, p. 201).

the solar irradiance (previously called the solar constant), $S_0 = 1367\ \mathrm{W\,m^{-2}}$. The solar irradiance is an average value that accounts for the annual variation of the Sun–Earth distance, and it varies slightly with time because of natural variations in solar output. However, many calculations are made as if S_0 were constant. Thinking of the Sun as a hot body that radiates electromagnetic energy by virtue of its temperature is a good starting place for understanding the role of radiation in the atmosphere and the climate of Earth.

The theory of blackbody radiation was shown by Planck in the early 1900s to require taking the quantum nature of electromagnetic radiation into account. A blackbody is by definition one that absorbs all incident radiation, but such a body must also emit radiation with the same efficiency. However, not all photons are generated with the same probability during thermal emission. Short-wavelength photons each require a lot of energy and so are less likely to be emitted than are photons with longer wavelengths; there is simply not enough thermal energy in a body to generate many high-energy photons. A proper calculation of how blackbody radiation depends on wavelength takes the limited emissions of both short- and long-wavelength photons into account and yields the so-called Planck function:

$$B_\lambda(T) = \frac{2\pi hc^2}{\lambda^5\left(e^{\frac{hc}{\lambda k_B T}} - 1\right)}. \tag{2.38}$$

The Planck function specifies the rate at which electromagnetic energy within a given wavelength interval is emitted by a unit surface area of a blackbody at temperature T into a hemisphere; commonly used SI units of $B_\lambda(T)$ are W m^{-2} nm^{-1}. The wavelength at which the radiative power from a blackbody is a maximum is given by Wien's displacement law:

$$\lambda_m = \frac{a}{T}, \tag{2.39}$$

where $a = 2898$ μm K (for wavelengths expressed in μm). The higher the temperature, the smaller the wavelength λ_m. Integration of the Planck function over all wavelengths yields the Stefan–Boltzmann equation, the total radiative power emitted by a unit surface area of the blackbody:

$$F_B = \sigma T^4, \tag{2.40}$$

where $\sigma = \left(2\pi^5 k_B^4\right) / \left(15c^2h^3\right) = 5.67 \times 10^{-8}$ W m^{-2} K^{-4} is the Stefan–Boltzmann constant. Note the strong dependence of emitted power on temperature.

Radiation absorbed by the Earth's surface or in the atmosphere is converted into thermal energy, which may be converted again into radiation at some later time. By virtue of the molecular agitation in any substance, be it a gas, liquid, or solid, radiation will be given off continuously. The main difference between an object like the Sun and ordinary objects on the Earth is the temperature (nearly 6000 K on the Sun, closer to 250 K on Earth). A direct comparison of the relative power radiated from a blackbody at these two temperatures is shown in Fig. 2.35. The spectrum is shifted so much by the temperature difference that radiation from the Earth overlaps that from the Sun very little. Thus, radiation with wavelengths greater than about 4 μm is often called longwave (LW) radiation, whereas that of smaller wavelengths is called shortwave (SW) radiation. More appropriate terminology refers to the source of radiant energy; thus, LW radiation is really terrestrial

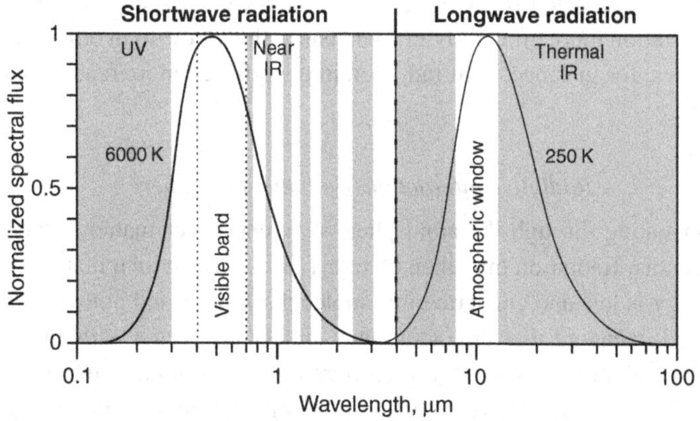

Figure 2.35 Separation of radiation into shortwave (solar) and longwave (terrestrial) bands. The shortwave and longwave bands have been separately normalized to their respective maxima. Adapted from Fleagle and Businger (1963, p. 148).

radiation, and SW radiation is solar radiation. The infrared band is subdivided into a near-IR band that is really part of the solar spectrum and a thermal-IR band that overlaps the LW band (shading on right half of Fig. 2.35). Clear regions of the diagram (including that labeled atmospheric window) indicate the ranges of wavelengths in which the atmosphere is reasonably transparent. As we saw in the theory of blackbody radiation, the wavelength of maximum radiative power varies inversely with temperature (Eq. (2.39)). Thus, whereas solar radiation peaks in the visible range ($\lambda_m \sim 0.5\ \mu\text{m}$), terrestrial radiation peaks in the thermal infrared ($\lambda_m \sim 10\ \mu\text{m}$). The clear separation in the spectral properties of solar and terrestrial radiation allows us to simplify discussions of the roles played by radiation in various atmospheric processes.

All objects emit radiation when thermal energy is converted into electromagnetic energy, but real objects seldom emit as effectively as does an idealized blackbody at the same temperature. We take the non-ideal behavior of real objects into account by introducing the emissivity, a simple ratio of the actual energy emitted by a real object to the idealized maximum emitted by a blackbody. The emissivity of an object is a dimensionless quantity characteristic of the material. The spectral emissivity depends on wavelength and so can be written

$$\epsilon_\lambda \equiv \frac{F_\lambda(T)}{B_\lambda(T)}, \tag{2.41}$$

where $F_\lambda(T)$ is the actual emitted power at wavelength λ per unit surface area of the object at temperature T. The radiant power emitted in a unit interval about wavelength λ by an object having emissivity ϵ_λ and temperature T is therefore $F_\lambda(T) = \epsilon_\lambda B_\lambda(T)$. Emissivity can be viewed as the efficiency with which an object radiates energy. Conservation principles force us to realize that an object that emits radiation with a certain efficiency must also absorb radiation with that same efficiency, so the terms absorptivity (α_λ) and emissivity may be interchanged (i.e., $\alpha_\lambda = \epsilon_\lambda$). It is often convenient to use an overall emissivity, one representing an average over a broad range of wavelengths, even though objects emit (and absorb) radiation more efficiently in some bands than they do in others. Certain radiatively active gases, for instance, emit radiation in the infrared more readily than they do in the visible band.

Radiative interactions with the atmosphere

Radiation propagating through the atmosphere may encounter matter, either in the form of gases or aerosols. Radiation may then be either absorbed by such matter, in which case the radiant energy is lost and converted into molecular kinetic and potential energies, or it may be scattered, meaning that the radiation is simply redirected without loss of radiant energy. Whether radiation is absorbed or scattered depends in part on the energies of the radiation compared with the energies of the bonds in the matter they encounter. Matter is composed of molecules that contain negatively charged electrons that are electrically bound to positively charged protons in the nuclei of the constituent atoms, so all matter interacts with electromagnetic energy to some extent. At the most basic level, we may think of atoms and molecules as composed of oppositely charged masses attached to each other by springs. The masses naturally oscillate relative to one another with a resonant frequency that depends on the stiffness of the spring (strength of the bond). Electromagnetic radiation incident on a molecule thus exerts forces on the charges that induce new motions of the electrons relative to the much more massive nuclei. The radiation-induced oscillations of electrical charges in turn generate electromagnetic waves that propagate away. As long as the frequency of incident radiation is well removed from natural (resonant) frequencies of the molecule, the radiation will be scattered. But, when the frequencies of radiation and resonance are comparable, the radiation is likely to be absorbed, causing the energy of the molecule to increase, perhaps even to the point of changing the molecular structure.

Most real molecules are more complicated than simple electric dipoles. A molecule to consider in some detail regarding interactions with radiation is water (H_2O). Water exists in all three common states of matter, and it is the main ingredient of clouds. We have already seen that the H_2O molecule has a three-dimensional structure that is permanently polar, meaning that the center of negative charge is spatially separated from the center of positive charge. Electromagnetic radiation exerts forces on these charges that cause the isolated molecules to vibrate and rotate in complicated ways.

The water molecule has several natural modes of vibration, each of which can be excited by radiation at specific wavelengths in the IR band. The polar covalent bonds between the oxygen atom and the two protons (H^+) can stretch asymmetrically (when $\lambda = 2.66\ \mu m$) or symmetrically ($\lambda = 2.73\ \mu m$); alternatively, the bonds may bend away from the normal bond angle of 104.5 ° (when $\lambda = 6.27\ \mu m$). The specificity of the wavelengths of radiation being absorbed arises from the fact that the energies of molecular vibrations are discrete, a consequence of the quantum nature of matter. Possible energies in a molecule exist as steps called energy levels, so radiant energy can be absorbed only if the incident energy induces a complete change in vibrational energy from one level to the next; fractional changes in energy levels cannot occur.

Once radiation is absorbed by one molecule, however, the energy can be dissipated to other molecules through collisions. In effect, the absorption of radiation by one atmospheric component (e.g., water vapor) serves to heat the entire gas. Molecular collisions rapidly distribute energy among all available internal-energy modes (vibrations, rotations, translation) and thereby maintain local thermodynamic equilibrium. The H_2O molecule, because of it three-dimensional structure, also rotates about each of its three axes of symmetry (see right-hand side of Fig. 2.36). The molecule is free to rotate in many possible ways, but any given rotation can be resolved into the three principal components shown in the figure. As with vibrations, molecular rotations are quantized. However, the separation between rotational energy levels is relatively small compared with that between vibrational energy levels. Water vapor can therefore, by rotating faster, absorb photons of relatively low energy, even those in the microwave band. Interactions between the rotational and vibrational modes cause the absorption spectra to be complicated in water and other radiatively active molecules. Additional complexity arises in condensed phases due

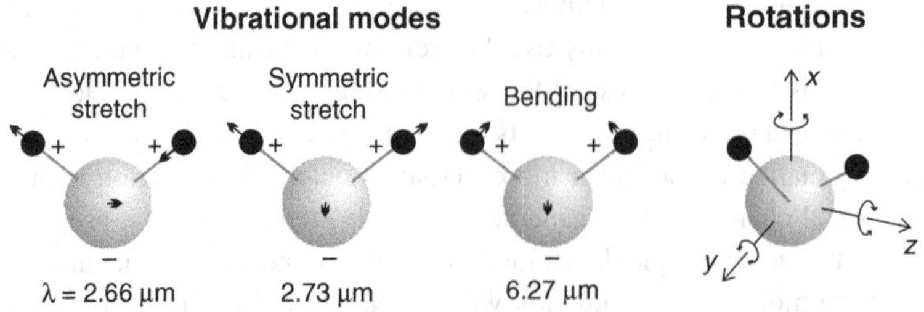

Figure 2.36 Schematic of the normal modes of vibrations and rotations of the water molecule.

to the intermolecular bonding, causing clouds, for instance, to absorb and radiate across broad, almost continuous bands in the IR, but little at visible wavelengths. At the other extreme, when photons are energetic, as with UV radiation, absorption can result in electronic transitions and ionization (loss of electrons). Away from the absorption bands, water acts as any other molecule in simply scattering radiation into new directions.

Gas molecules, aerosol particles, and clouds may all scatter radiation, as well as emit and absorb it. Scattering occurs fundamentally at the atomic scale because of the electrons and protons atoms contain. This individual unit of scattering may be viewed as an electric dipole, one forced into oscillation by the incident field. The incident radiation forces the charges in the dipole to accelerate, which in turn generates a new field that interferes constructively and destructively with the incident field. It is the interference of the incident radiation with the induced radiation that leads to the sometimes complicated pattern of scattered radiation. A bound grouping of atoms, as in molecules and small aerosol particles, can interact coherently with the incident radiation, as if it were a single scattering center and all of the dipoles respond to the incident radiation more or less in unison. On the other hand, widely and randomly spaced particles, such as in a cloud, act independently and scatter radiation incoherently, circumventing the need to consider the phase of the radiation. When the scattering by many particles significantly alters the original beam, the radiation incident on each particle is itself derived from radiation scattered by other particles. Such multiple scattering becomes geometrically complicated and dominant once the optical thickness for scattering, $\tau_s \simeq \beta_s s$, exceeds unity. We next discuss single-particle scattering, leaving the effects of multiple scattering for the next subsection.

Scattering of radiation by individual particles requires precise characterization of the interaction environment. Scattering is a transient interaction with matter that results in a change of direction, but not in energy (or frequency) of the radiation. We therefore define the geometry of scattering relative to the direction that the incident radiation travels, as shown in Fig. 2.37. The scattering angle θ is the plane angle that the scattered radiation makes with the incident radiation; it measures how much the particle caused the radiation to deviate from its original direction. Scattering by angles less than $90°$ ($\theta < \pi/2$) is termed forward scattering, whereas larger scattering angles indicate back scattering. The relative amount of radiation scattered within unit solid angle (Ω) centered

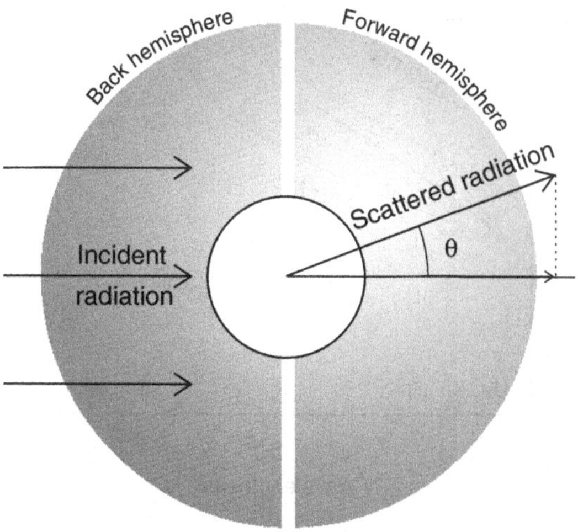

Figure 2.37 Geometry of electromagnetic scattering from a particle (center circle).

about direction θ is called the scattering, or phase, function $P(\theta)$. The phase function is a probability distribution, so it can be normalized by integration over all 4π solid angles: $\oint P(\theta)\, d\Omega = 1$ simply states that all scattering has been accounted for. Multiplying $P(\theta)$ by the cosine of the scattering angle, $P(\theta)\cos\theta$, gives the projection of the phase function onto the incident direction, so the integral $g \equiv \oint P(\theta)\cos\theta d\Omega$ measures how asymmetrically distributed the scattered radiation is. The asymmetry parameter g is zero when the radiation is scattered equally in all directions, positive when scattering is primarily into the forward hemisphere, and negative when most scattering is into the back hemisphere. The phase function and the asymmetry parameter vary greatly from particle to particle in a given radiation field.

The phase function $P(\theta)$ depends mainly on the size of the particle in relation to the wavelength λ of radiation. Often, one assumes that a particle is spherical and defines a dimensionless size parameter x to be the ratio of the circumference of the particle to the wavelength:

$$x \equiv \frac{2\pi r_p}{\lambda}, \tag{2.42}$$

where r_p is the particle radius. The general theory of scattering is complicated and beyond the scope of this work. It suffices here to recognize that scattering falls into various regimes, as shown in Fig. 2.38. Note that the categories depend only on x (slanted lines), and not on the absolute magnitude of the particle radius nor on the wavelength. Polar plots of the phase functions characteristic of each regime have been added to illustrate the dramatic influence the size parameter has on the scattered radiation. In the Rayleigh regime, when a particle is small compared with the wavelength ($x << 1$), the scattering is weak and symmetrically distributed, with equal amounts of radiation being scattered into the forward

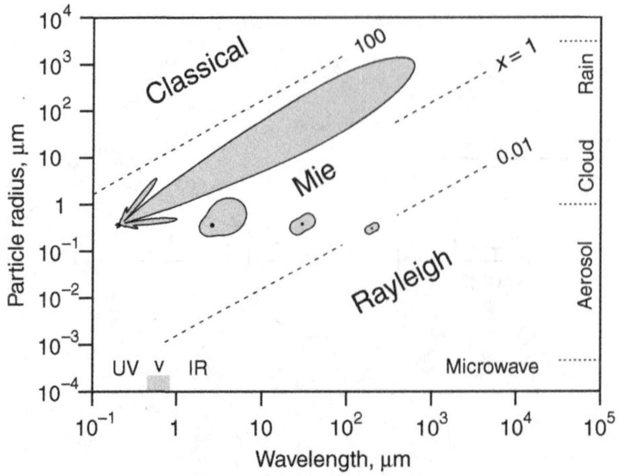

Figure 2.38 Regimes of scattering from spherical particles based on the non-dimensional size parameter x. The polar plots are oriented with the forward-scattering direction upward and parallel to the x isopleths. Adapted from Petty (2004 pp. 339, 360).

and back hemispheres. In the Mie regime, when the particle size is comparable to the wavelength ($x \sim 1$; aerosol particles), forward scattering is favored, giving positive values to the asymmetry parameter. A characteristic feature of scattering from larger particles is the strong forward-scattering lobe, caused by radiation inside and outside the particle remaining in phase. (The effects of forward scattering are readily apparent when viewing thin, bright-looking clouds in the general direction of the Sun. The same clouds viewed in the opposite direction look distinctly duller.) As x increases within the Mie regime, lobes at specific angles show up, indicating that the radiation resonates inside the particle and propagates preferentially in specific directions (giving the lobes on a polar diagram). In the classical regime, when the particle size is much larger than the wavelength ($x >> 1$; large cloud and raindrops), scattering is predominantly in the forward direction, and the principles of geometric optics apply. (Reflection and refraction phenomena are seen as rainbows and halos.)

The total amount of radiation scattered by a particle in all directions depends (for a given medium) on its composition and size. Composition dictates the index of refraction m, which for many applications regarding clouds varies little around that for water at visible wavelengths, $m = 1.33$. The size of a particle enters calculations through the dimensionless size parameter x, so only size relative to the wavelength is relevant, and we should expect the scattered power to depend on the regime (Fig. 2.38). The total power W_{scat} (SI units, W) scattered by a particle must be proportional to the incident flux of radiation F ($\mathrm{W\,m^{-2}}$), so we use the following relationship to define the proportionality constant as the scattering cross section $\sigma_{scat} = \sigma_{scat}(x)$:

$$W_{scat} = \sigma_{scat} F. \tag{2.43}$$

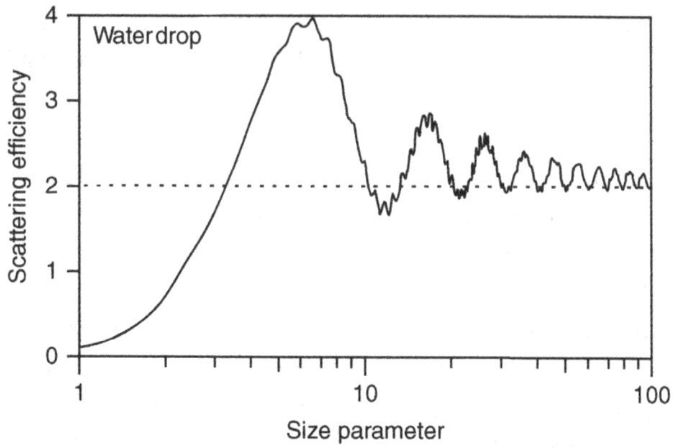

Figure 2.39 Efficiency for scattering by water drops as a function of the size parameter (x). Adapted from Hansen and Travis (1974).

The scattering cross section is an effective area [m^2] intercepting the incident radiation, but it is only loosely related to the physical cross-sectional area of the particle, $A_p = \pi r_p^2$. Optical cross sections are sometimes normalized by the physical cross section to yield dimensionless parameters called efficiency factors. The efficiency factor for scattering is thus

$$Q_{scat} \equiv \frac{\sigma_{scat}}{A_p}. \tag{2.44}$$

The variation of Q_{scat} with x for a water drop is shown in Fig. 2.39. Scattering by molecules and tiny particles is very inefficient when $x << 1$ (Rayleigh regime, not shown), but it increases rapidly with x (because of increasing particle size or decreasing wavelength). In this regime, Q_{scat} increases as x^4. Air molecules exposed to sunlight fall into this regime, so we see that the natural atmospheric scattering varies as λ^{-4}; blue light is scattered much more efficiently than red light. As x exceeds ~ 1 (Mie regime), the scattering becomes efficient, so much so that the scattering cross section can be as much as a factor of four greater than the cross-sectional area of the particle itself when $r_p \simeq \lambda$. The broad maxima and minima at $x > 1$ arise from structural resonances of the radiation inside the particle, and the fine-scale ripple structure is due to higher-order interference patterns. Toward $x >> 1$ (classical regime), the scattering efficiency approaches an asymptote with a value $Q_{scat} = 2$, a counter-intuitive result arising from the edges of the particle causing diffraction and a slight bending of the peripheral rays away from the forward direction.

The atmosphere is a medium that offers many opportunities for interactions with radiation. The collective effects of radiation–particle interactions may be appreciated by considering a beam of photons propagating in a certain direction (as with sunlight). If a photon is absorbed by a molecule, an aerosol particle, or a cloud drop, it is lost from the beam. If a photon is scattered, it is also lost from the beam. Both absorption and scattering represent losses that partially extinguish the radiant energy of a beam. Extinction is just the net result of absorption and scattering. The beam becomes progressively depleted as more photons are lost by either process, and fewer photons are available for subsequent interactions with the atmosphere. Many interactions cause the beam to weaken rapidly.

The rate with which the beam becomes diminished with distance depends on the amount of material in its path. If a unit volume of air contains N monodisperse particles, each of which offers a radiation cross section σ, then the power of a beam irradiating that volume will be attenuated by the fractional amount $N\sigma$ upon passing a small distance through the volume. Think of each encounter as an opaque disk of area σ_e blocking the path. The larger N is, the greater the number of disks that block the radiation. To be mathematically precise, we let F_λ be the spectral power (per unit wavelength interval) passing through unit area at the start of its passage along a small path ds of air, in which case the power will be diminished by the amount $F_\lambda \beta_e ds$, where $\beta_e = \beta_a + \beta_s$ is the extinction coefficient, a measure of the amount of absorption ($\beta_a = N\sigma_a$) and scattering ($\beta_s = N\sigma_s$) taking place in that small length of path. The law of attenuation can thus be written as

$$-\frac{dF_\lambda}{F_\lambda} = \beta_e ds, \tag{2.45}$$

from which we see that β_e must have units of inverse length. The extinction coefficient is related to a distribution of particles $n(r_p)$ having extinction efficiencies $Q_{ext}(r_p)$ by the relationship

$$\beta_e = \pi \int_0^\infty r_p^2 Q_{ext}(r_p) n(r_p) dr_p. \tag{2.46}$$

Similar relationships can be written separately for absorption and scattering. The scattering coefficient, for instance, is

$$\beta_s = \pi \int_0^\infty r_p^2 Q_{scat}(r_p) n(r_p) dr_p. \tag{2.47}$$

Recall the strong size dependence implied by the scattering efficiency, $Q_{scat}(r_p)$, for radiation of given wavelength (see Fig. 2.39 leading up to the maximum). This strong size dependence in optical efficiency correlates with that of the accumulation-mode aerosol concentration, $n(r_p)$, which causes a large peak in scattering to occur in the visible range. Indeed, hazy skies result when high concentrations of pollutant particles reside in the accumulation mode because such particles have sizes comparable to the wavelength of solar radiation ($\lambda \simeq 0.5$ μm) and so scatter sunlight very efficiently into our eyes. The optical effect is enhanced during periods of high relative humidity as the same haze particles swell by the uptake of atmospheric water vapor.

The net effect of the many particles interacting with radiation along a finite path length s can be determined by integration of Eq. (2.45):

$$\frac{F_\lambda}{F_{\lambda 0}} = \exp\left(-\int_0^s \beta_e ds'\right) \tag{2.48}$$

in the general case that the extinction coefficient varies along the path. In a homogeneous medium, in which β_e is uniform, we obtain the simpler equation,

$$\frac{F_\lambda}{F_{\lambda 0}} = \exp(-\beta_e s) . \tag{2.49}$$

The fraction of radiation surviving passage through the medium, $F_\lambda/F_{\lambda 0}$, is called the transmissivity. These exponential relationships between attenuation and distance have been attributed to various individuals, but often to Beer. Beer's law is important for linking the physics of absorption and scattering to beam attenuation by a collection of particles.

The concept of optical path stems directly from Beer's law. Thus, Eq. (2.49) can be written as

$$\frac{F_\lambda}{F_{\lambda 0}} = \exp(-\tau) , \tag{2.50}$$

where $\tau \equiv \int_0^s \beta_e ds \simeq \beta_e s$ is the optical path, a dimensionless measure of distance through an optically active medium. Optical path is related to distance, but it is not a distance in the ordinary sense. Rather, the optical path of the atmosphere, or a portion of it, is a measure of how effective the medium is in attenuating radiation. For instance, air that absorbs and scatters little radiation at a given wavelength requires a large physical distance to achieve the same attenuating effect that a shorter, but more strongly absorbing air mass would.

Penetration of radiation through a medium is often defined as the physical distance s_1 needed to attenuate the beam by the factor $1/e \simeq 0.37$, corresponding to an optical path of unity. With $\tau = 1$ in the formula $s = \tau/\beta_e$, the critical distance becomes $s_1 \equiv 1/\beta_e$ (for a homogeneous medium). Approximately two-thirds of the photons in a beam can penetrate air with an optical path of unity. A cloud of unity optical path would let the Sun and Moon be visible as disks, but these objects would look dim. Optical path is also related to horizontal visibility, but human perception of contrast must then be considered. The daytime visual range in meteorology, L_V, is the distance that a black target can just be distinguished from a white background. The human threshold for apparent contrast in brightness is typically taken to be $C_{R,th} = 0.02$, so we calculate visual range from the formula

$$C_{R,th} = 0.02 = \exp(-\beta_e L_V) , \tag{2.51}$$

which gives the so-called Koshmieder relationship

$$L_V = -\frac{1}{\beta_e} \ln(0.02) = \frac{3.91}{\beta_e} . \tag{2.52}$$

Both visual range and penetration distance are inversely related to the extinction coefficient.

The vertical penetration of radiation through a horizontal layer of the atmosphere is expressed in terms of optical depth or thickness, rather than optical path. Optical depth accounts for the specific angle that the radiation makes with the vertical. The optical depth is the vertical component of the optical path through the layer. Consider sunlight shining onto a layer of clouds at a solar zenith angle θ. If the layer has a physical (vertical) thickness Δz, then the sunlight travels along the longer slant path of length $\Delta s = \Delta z/\cos\theta$. Vertical distances are usually taken to be positive upward, whereas sunlight always points downward. This directional distinction is accounted for by noting that $\cos(\pi - \theta) = -\cos\theta$. Equation (2.45) is thus modified to be applicable to a plane-stratified atmosphere:

$$\frac{dF_\lambda}{F_\lambda} = \beta_e \sec\theta dz. \tag{2.53}$$

The transmissivity of the layer is calculated by integrating Eq. (2.53) over the thickness of the layer, from z_{bot} to z_{top}:

$$\frac{F_\lambda(z_{bot})}{F_\lambda(z_{top})} = \exp\left(-\sec\theta \int_{z_{bot}}^{z_{top}} \beta_e dz\right). \tag{2.54}$$

Note the reversal of the order of integration.

When we want to calculate how much sunlight penetrates the atmosphere to any specified altitude z, we integrate Eq. (2.53) from that altitude to the top of the atmosphere $z_{TOA} \to \infty$. The transmissivity at altitude z is thus

$$\frac{F_\lambda(z)}{S_{0\lambda}} = \exp\left(-\sec\theta \int_{z}^{\infty} \beta_e dz\right), \tag{2.55}$$

where we use the monochromatic solar irradiance as the incident radiant flux at the top of the atmosphere. In a cloudless atmosphere, the dominant loss of solar radiation is due to absorption by the various atmospheric gases, so we expect the optical depth to depend on the wavelength, and we replace the extinction coefficient by the absorption coefficient: $\beta_e \simeq \beta_a(z) = \sum_j \sigma_{abs,j}(z) n_j(z)$, where the absorption coefficient varies with altitude and is the sum over all absorbing species (j). This approach requires information about the absorption cross sections of the gases, $\sigma_{abs,j}$, as well as the dependence of concentration with altitude, $n_j(z)$. The particular altitude z_1 to which solar radiation penetrates the atmosphere is calculated from Eq. (2.55) with these modifications and setting the optical path to unity:

$$\tau \equiv \sec\theta \int_{z_1}^{\infty} \beta_a(z) dz = 1. \tag{2.56}$$

Note that the desired quantity appears as a limit in the integration. The consequence of the many interactions of solar radiation with the atmosphere is shown schematically in Fig. 2.40. X-rays and photons with wavelengths less than about 100 nm have enough energy to ionize molecules and atoms in the upper-most parts of the atmosphere. This part of the thermosphere is therefore called the ionosphere (or electrosphere); it is electrically conducting because of the relatively high concentrations of ions (e.g., O_2^+, O^+, and NO^+). Photons of longer wavelengths have less energy, interact with the atmospheric gases less, and so can penetrate farther into the atmosphere. Radiation with wavelengths $200 \le \lambda \le 300$ nm may not be able to ionize atoms, but it nevertheless has enough energy to break some molecular bonds. Such photodissociation initiates numerous secondary

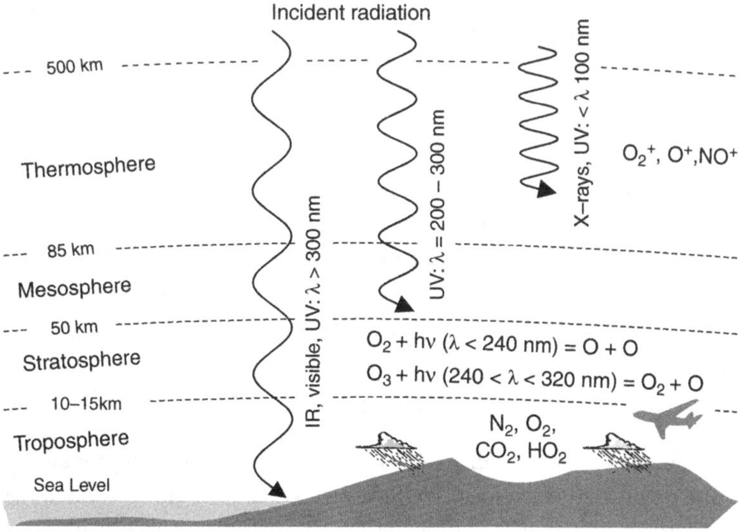

Figure 2.40 Cross section of the atmosphere showing the wavelength-dependent penetration of solar radiation into the atmosphere. Adapted from Manahan (1993).

reactions of importance in the middle atmosphere (mesosphere and stratosphere). Ozone (O_3) is produced in the stratosphere as oxygen (O_2) is photolyzed by radiation having wavelengths less than about 240 nm. Mostly, visible light, UV radiation with $\lambda \geq 300$ nm, and some IR radiation can penetrate the entire depth of the atmosphere without significant loss.

The penetration of terrestrial radiation, by contrast, is severely hampered by interactions of IR radiation with atmospheric constituents. Fortunately, the atmosphere is relatively transparent to radiation in the IR range 8 μm $< \lambda <$ 12 μm, so radiation emitted from the surface can propagate into space and compensate for the absorbed solar radiation. Concerns over climate change stem largely from the increase in radiatively active gases (e.g., CO_2, CH_4, O_3, and H_2O) that cause this so-called atmospheric window to become less transparent with time.

The combined effect of solar radiation penetrating relatively easily to the surface, but terrestrial radiation being absorbed by the atmosphere gives rise to the so-called greenhouse effect. We may never get away from using this term, but we should at least recognize that the effect of the atmosphere on the radiation balance has little to do with the energy balance of human-made greenhouses. The atmospheric effect is caused by IR emissions from radiatively active gases and clouds in the atmosphere, as we can see by considering a simple two-layer model of the Earth–atmosphere system. Such a model ignores horizontal and vertical variations in temperature, as well as the abundances of atmospheric constituents, but it nevertheless conveys the concept adequately. Let the temperature of the atmosphere be T_{atm} and that of the surface be T_{sfc}. Looking at the system from above, one notes that a certain net flux of solar (SW) radiation S_0 enters the system and an equal flux of terrestrial (LW) radiation leaves it. The energy balance could thus be written as (net SW energy in = LW energy out)

$$\pi R_E^2 S_0 = \left(\epsilon\sigma T_{atm}^4 + (1-\epsilon)\sigma T_{sfc}^4\right) 4\pi R_E^2, \tag{2.57}$$

where ϵ is the average LW emissivity of the atmosphere and we assume complete penetration of solar radiation to the surface (where all radiation is assumed to be absorbed). The first term on the right-hand side represents the outgoing LW radiation from the atmosphere itself, the second term the LW radiation from the surface that can penetrate through the atmosphere. The energy balance of the atmosphere can itself be written as (energy gained = energy lost)

$$\epsilon\sigma T_{sfc}^4 = 2\epsilon\sigma T_{atm}^4. \tag{2.58}$$

The factor 2 accounts for the fact that LW energy is lost by radiation propagating in both the upward and downward directions. We have two equations with two unknowns, so we determine the surface temperature in terms of the physical parameters by eliminating T_{atm} and solving for T_{sfc}:

$$T_{sfc} = \left[\frac{S_0}{4\,(1-\epsilon/2)\,\sigma}\right]^{1/4}. \tag{2.59}$$

We thus see, as we should expect, that the effective surface temperature depends on the net solar irradiance (S_0), as well as on the emissivity of the atmosphere (ϵ). In the limiting case that $\epsilon = 0$, $T_{sfc,0} = (S_0/\sigma)^{1/4}$, meaning that in the absence of any emissions from the atmosphere, the surface temperature would be just the radiative equilibrium temperature of an atmosphere-free planet. More generally, we find the ratio of the temperatures with and without an atmosphere to be

$$\frac{T_{sfc}}{T_{sfc,0}} = \frac{2}{2-\epsilon}. \tag{2.60}$$

Note that the surface temperature with an emitting atmosphere (T_{sfc}) is always greater than that in the atmosphere-free case ($T_{sfc,0}$). The greater the concentration of substances the atmosphere contains that can radiate energy downward toward the surface, the larger the value of ϵ, and the warmer the surface becomes. We are fortunate that the atmosphere does provide some greenhouse effect (largely through CO_2), but the climate invariably becomes warmer with every addition of gases that can absorb and radiate in the IR band.

2.2.2 *Effects of clouds*

Clouds introduce a level of complexity into radiative interactions beyond those of gases because clouds both absorb and scatter radiation over broader ranges of wavelengths. Clouds absorb a small amount of short-wave (solar) radiation, but most absorption and hence production of thermal radiation occurs in the infrared parts of the electromagnetic spectrum. Clouds effectively scatter solar radiation, but it is multiple scattering of clouds that causes them to appear white in the backscattering direction. Multiple scattering in clouds also enhances the amount of actinic (chemically important) radiation available for photochemical reactions.

Solar radiation encountering clouds may be scattered back to space and contribute little to the warming of the Earth. Sometimes we think of clouds as reflecting sunlight, but reflection in the case of objects composed of many independent particles (as in an aerosol or cloud) differs from that of a mirror. In a cloud, the radiation is elastically scattered (altered in direction without loss of energy) multiple times by the cloud particles and leaves it in arbitrary directions. Thick clouds act as if they send a significant fraction of the incident shortwave radiation back into space, so it is common to speak as though clouds reflect sunlight. Reflectivity then is the ratio of radiant power reflected (or scattered back) to the incident power.

Reflected solar radiation is not absorbed and so has the effect of reducing the input of energy into the Earth–atmosphere system. A simple energy balance illustrates the effect of clouds on the effective temperature of the Earth. Let R_E be the Earth's radius and α_E its solar-band albedo (the overall reflectivity averaged across all wavelengths in the solar spectrum). The solar energy intercepted and absorbed by the Earth (in unit time) is thus $E_{in} = \pi R_E^2 S_0(1 - \alpha_E)$. If the effective temperature of the Earth is T_E, then it sends IR radiation back to space from all over the surface at the rate $E_{out} = 4\pi R_E^2 \overline{\epsilon} \sigma T_E^4$, where $\overline{\epsilon}$ is the average IR emissivity of the Earth. On average, we expect steady-state conditions, so we equate the energies into and out of the system. The flux of solar radiation at the top of the atmosphere, averaged over the length of a day, is $\overline{S_0} = S_0/4 \simeq 342$ W m^{-2}, and the effective temperature is calculated as

$$T_E = \left(\frac{S_0(1 - \alpha_E)}{4\overline{\epsilon}\sigma} \right)^{1/4}. \tag{2.61}$$

The planetary albedo has been estimated to be about $\alpha_E \simeq 0.3$, so for a blackbody Earth ($\overline{\epsilon} = 1$) $T_E \sim 255$ K. In the absence of clouds, the albedo would be about half what it is, in which case the effective temperature would be higher by about 13 K. Clouds, because of their strong ability to scatter shortwave radiation, thus offset some of the input of solar energy.

A more detailed radiation budget helps us refine our understanding of the roles clouds play in the Earth's climate. Figure 2.41 identifies the various energy paths as a percentage of the average solar irradiance ($\overline{S_0} = 342$ W m^{-2}). Energy enters the Earth system as shortwave radiation from the Sun, and energy leaves it as longwave radiation. Of the incoming shortwave radiation, approximately 30% is reflected back to space by a combination of clouds, aerosol, and the surface. Of the 70% of solar radiation that is absorbed, some heats the atmosphere itself (mostly in the near IR by trace gases and clouds), but most makes it through the atmosphere, thereby heating the surface (oceans and land). The heated surface causes some direct heating of the atmosphere by sensible and latent heat transfer, but it also radiates much energy upwards. An important fraction of this thermal IR radiation escapes

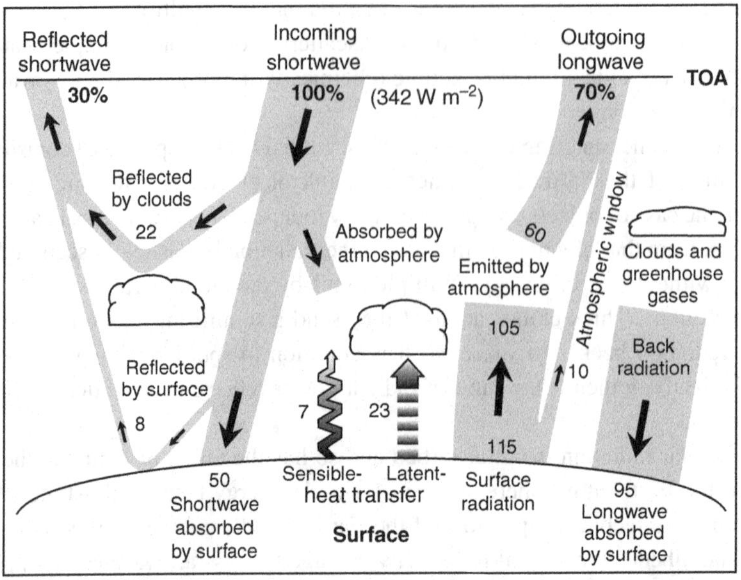

Figure 2.41 Cartoon showing the Earth's radiation budget as approximate percentages of the mean solar input (342 W m^{-2}). The widths of the arrows suggest the importance placed on individual pathways. Adapted from Trenberth *et al.* (2009).

directly to space through the atmospheric window (the spectral band between about 8 μm and 12 μm) when skies are clear. The presence of clouds, however, reduces the amount of shortwave radiation reaching the surface and contributes, along with the greenhouse gases, to additional IR radiation being send toward the surface. The atmosphere itself (including clouds and gases) radiates most of the energy to space to balance the net solar radiation absorbed by the Earth–atmosphere system.

The effects of clouds on the radiation field have long been studied as a way of learning how the atmosphere responds to various forcings. Climate scientists are keenly aware of the uncertain roles clouds play in determining the radiation budget of the Earth, so they devise special methods for dealing with the issue. The planetary albedo depends to a large extent on the extent of cloud cover (cloud fraction), but it also depends on the optical thickness of those clouds, as shown in Fig. 2.42. Clouds are important to consider in all discussions of climate because their albedos can vary widely, from less than 10% to more than 90%. Cloud drops exposed to solar radiation scatter in the classical regime, so the optical depth of a cloud is roughly equal to two times the sum of the cross-sectional areas of the drops in a column of cloudy air. Clouds with many small drops scatter more radiation than those with fewer drops, other factors being the same. Thus, low clouds in polluted regions tend to be optically thick and brighter (when viewed from space) because of the high drop concentration. Such clouds partially offset the warming effects of radiatively active gases. On the other hand, high clouds tend to be optically thin in visible wavelengths, so they

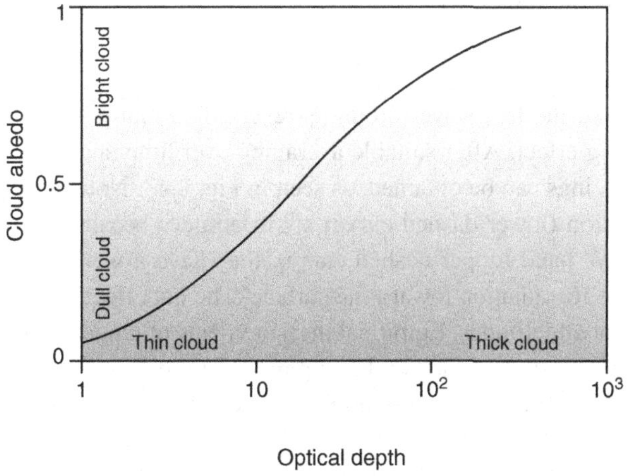

Figure 2.42 Albedo of clouds of various optical depths. Adapted from Twomey (1974).

let solar radiation penetrate to the surface. Despite the low temperatures of high clouds, they act much as do the radiatively active gases and so contribute significant amounts of thermal IR radiation to the surface. High clouds augment atmospheric LW emissions to the surface and so warm the surface. Quantification of the competing effects of clouds on climate can be accomplished by looking at how clouds force, or bring about, changes in the radiation entering and leaving the Earth–atmosphere system.

Empirical measures of cloud forcing make use of satellite data to unravel the different types of interactions clouds have on radiation in the shortwave and longwave bands. An image of the Earth taken from a satellite shows the patchy distributions of clouds around the globe. Some parts of any broad image show clouds, while other parts show the surface unhindered by clouds. The radiation received by the cameras aboard the satellite is filtered into various channels to distinguish the shortwave and the longwave effects. A climatology is developed by averaging all the cloudy regions and all the clear-sky, separately and together. The effect of clouds is then determined by comparing the total-scene (cloudy plus clear-sky) regions with the clear-sky regions in the SW and LW bands. Mathematically, one calculates the overall (net) cloud forcing by considering both SW and LW bands together:

$$\Delta F_{net} = F_{total} - F_{clear}, \tag{2.62}$$

where $F_{total} = F_{SW} - F_{LW}$ is the net radiation crossing the top of the atmosphere (here, $F_{SW} = (1 - \alpha_E)\, S_0$ is the solar radiation entering the system; F_{LW} is the measured outgoing LW radiation). Positive values represent more radiation entering the system than leaving it (hence causing a warming effect). The net cloud forcing is resolved into the separate effects clouds have in the SW and LW bands as follows. The shortwave cloud forcing is

$$\Delta F_{SW} = F_{SW} - F_{SW,clear}, \tag{2.63}$$

and the longwave cloud forcing is

$$\Delta F_{LW} = F_{LW,clear} - F_{LW}. \tag{2.64}$$

Note that the sign of the longwave forcing is reversed so that positive values of ΔF_{LW} represent a warming effect. After suitable averaging over time and space, a global climatology of cloud forcings can be obtained, as seen in Fig. 2.43. Note that clouds reduce the input of SW radiation (lower dashed curve) at all latitudes because of the albedo effect. However, in the LW band (upper dashed curve), they have a warming effect because of their ability to emit IR radiation toward the surface. The net effect of clouds (solid curve) is a slight cooling at all latitudes. Empirical measurements of cloud forcings are consistent with our general expectations based on basic physics of radiative interactions, but modeling the effects of clouds depends on our ability to understand the processes responsible for the observed properties of clouds.

2.2.3 Global distribution of energy

The steady flow of radiant energy from the Sun to the Earth is not distributed evenly around the globe. The Earth is round, and its rotation axis is inclined about 23.4 ° to the normal of the ecliptic (the plane in which the Earth orbits). Even after averaging the incident sunlight over a complete rotation of the Earth (i.e., one day), the Sun's radiation varies strongly with latitude and season. The solar irradiance (S_0) itself, defined as the radiant power incident on a unit area of a plane oriented perpendicular to the Sun's rays, is relatively constant and uniform at the distance of the Earth's orbit. However, that same irradiance gets spread over a larger area on a plane oriented in any other direction. The radiant power per unit area incident on a horizontal plane (one locally parallel to the Earth's surface) is called

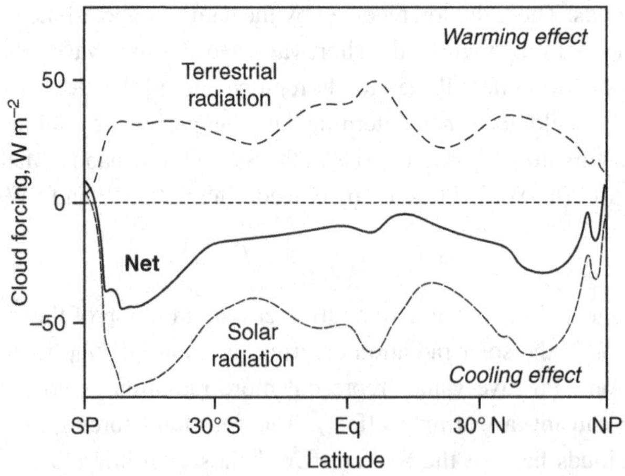

Figure 2.43 Zonal averages of measured forcings of solar and terrestrial radiation by clouds. Adapted from Hartmann (1993).

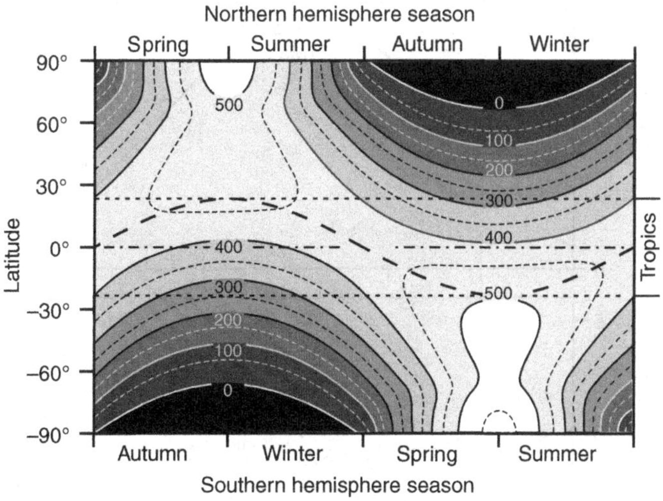

Figure 2.44 Insolation (W m^{-2}) at the top of the atmosphere (isopleths). Dashed curve: solar declination, bounded by the Tropic of Cancer (upper dashed line) and the Tropic of Capricorn (lower dashed line). Adapted from Incredio (2010).

the insolation (a blended term derived from INcident SOLar radiATION). The result of computing the insolation at the top of the atmosphere (TOA), taking the main geometrical factors into account and averaging over individual days, is shown graphically in Fig. 2.44. The roundness of Earth causes the strong dependence of averaged insolation on latitude (ordinate), while the axial tilt (also known as obliquity) is the main cause of variations with time of year, hence the seasons (abscissa). The summer season starts in each hemisphere when the solar declination (the apparent path of the Sun in the sky at noon) has its maximum absolute value in the respective hemisphere. The TOA insolation varies strongly by season at high latitudes, but only weakly in the tropics (defined as the latitude range between the Tropics of Cancer and Capricorn). The large insolation near the poles in summer results from the large fractions of each day that the Sun is above the horizon. Note the strong north–south gradient in insolation in the winter season of each hemisphere. It is this uneven input of energy that drives the complicated atmospheric and oceanic circulations.

Annual averages of measured radiation at the top of the atmosphere confirm the importance of latitude in the radiation budget. Earth-orbiting satellites that separately measure the incident and outgoing radiation in the shortwave and longwave bands monitor the two-way fluxes of radiant energy passing through the virtual sphere defining the top of the atmosphere. As shown in Fig. 2.45 (top), the net input of solar radiation (the amount actually absorbed by the Earth–atmosphere system) has a different dependence on latitude than does the outgoing terrestrial radiation. Both the solar and terrestrial components are balanced when taken across all latitudes (and amount to about 240 W m^{-2}), but the solar radiation varies by a factor of three or more from poles to equator, whereas the terrestrial

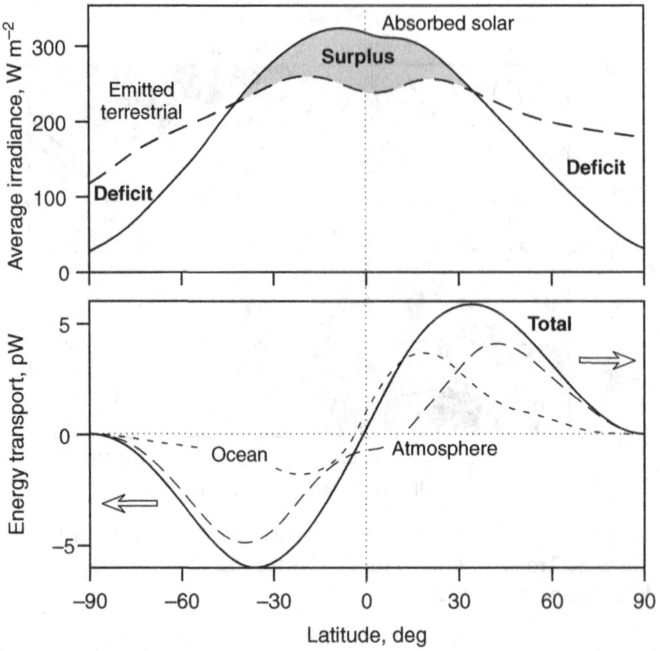

Figure 2.45 Latitude dependence of (top) solar and terrestrial radiation and (bottom) the energy transported poleward (arrows) by the oceans and atmosphere. Top figure adapted from Vonder Haar and Suomi (1971); bottom figure adapted from Zhang and Rossow (1997).

radiation varies by less than a factor of two over the range of latitudes. The steady input of solar radiation into the tropics is significantly greater than that into the polar regions (despite the large input during high-latitude summers; see Fig. 2.44). Terrestrial radiation, by contrast, responds to the mean temperatures of the Earth and atmosphere and so varies relatively little with latitude. As shown by the shading in Fig. 2.45, a surplus of radiant energy pours into the latitude zone between $\pm 40\,°$, whereas a deficit in the radiation balance exists poleward of the equatorial zone.

Any imbalance in the zonal radiation balance of a steady-state system must be compensated for by non-radiational flows of energy. The continual surplus of radiation entering the low-latitude zone of Earth does not lead to a continual warming in that region because the excess energy is transported to higher latitudes by oceanic currents and atmospheric winds. The dependences of these flows on latitude are shown by Fig. 2.45 (bottom). The total northward flowing energy is shown by the solid curve, the components due to the oceans and atmosphere are shown by short- and long-dashed curves, respectively. Clouds contribute significantly to the atmospheric component because of latent heating (transport of water vapor that condenses at high latitudes), thereby demanding less of the oceanic component. The atmosphere is an important part of the Earth system in its continual efforts to restore the imbalances imposed by the uneven distributions of radiant energy.

2.3 Structure and organization

The atmosphere is a complicated environment, one driven by many competing processes. Nevertheless, the average behavior of the atmosphere yields an apparent structure and a sense of organization that can be understood in a broad sense upon considering the physics. Throughout the discussions that follow it proves helpful to maintain the perspective that the atmosphere is really a very thin and tenuous component of the Earth. The entire atmosphere, out to about 500 km, represents less than 10% of the Earth's radius. More striking still, the ratio of the vertical extent of the atmosphere to the available horizontal distance (the circumference of the Earth) is about 1/100. Cross sections of the atmosphere often exaggerate the vertical direction for clarity, but the atmosphere is in reality thin, analogous to the skin of an onion.

2.3.1 Vertical structure

The strongest variations in the atmosphere typically occur in the vertical direction. Climb a mountain just a kilometer in height and you find the average temperature to decrease noticeably, by some 5 to 10 °C. To find an equivalent change in a horizontal, north–south direction, you would have to travel many hundreds of kilometers. Some of the most important atmospheric properties, such as temperature, pressure, and density, decrease rapidly with increasing altitude (in the direction opposite that of the Earth's gravity vector). It is the gravitational force acting on the atmospheric constituents that causes strong vertical gradients of many properties

Temperature

Temperature, the common measure of "hotness", varies in complicated, but systematic ways with altitude. The dynamical behavior of the atmosphere is strongly affected by how strongly the temperature decreases with altitude, so the sign of the vertical temperature gradient is traditionally used to define the various layers of the Earth's atmosphere. As illustrated on the left-hand side of Fig. 2.46, the mean temperature first lapses (decreases with altitude) in the layer called the troposphere. Temperature then increases with altitude in the stratosphere, before again lapsing in the mesosphere. The temperature again increases in the thermosphere (sometimes referred to as the "upper atmosphere"), all the way into the exosphere (not shown), the outermost layer of the atmosphere about 500 km above the Earth's surface. The stratosphere and mesosphere are together referred to as the "middle atmosphere", whereas the troposphere is the "lower atmosphere". Each layer terminates in a thin zone called its respective "pause".

The causes of the complicated thermal structure of the atmosphere can be understood in a general way. Not surprisingly, the three warmest regions of the atmosphere tend to get heated by the Sun more so than do the other parts of the atmosphere. It is in the upper part of the thermosphere that solar radiation first encounters a significant amount of matter

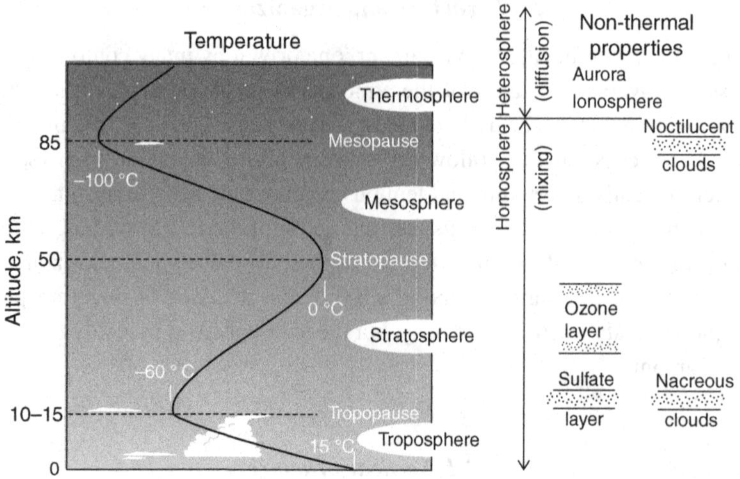

Figure 2.46 Vertical structure of the atmosphere in terms of temperature, dominant transport mechanisms, and composition.

after leaving the Sun. The most energetic radiation, that in the far ultraviolet part of the electromagnetic spectrum, is absorbed by the gas molecules there, leading to daytime temperatures exceeding 1000 K. Less energetic ultraviolet radiation is able to penetrate through the thermosphere and most of the mesosphere before being absorbed by oxygen and ozone in the upper stratosphere. The temperature maximum in the middle atmosphere arises from this extra input of solar energy directly into that part of the atmosphere. The relatively high temperatures found near the surface of the Earth are due to the insolation with which we are all familiar. It is the heating of the ground at the lowest levels of the atmosphere that indirectly heats the atmosphere and leads to the convective overturning of air in the troposphere.

The temperature profiles of other planetary atmospheres differ from that of the Earth in rather specific and instructive ways. Figure 2.47 shows how temperature varies with altitude on the three terrestrial planets, Venus, Earth, and Mars. All of these planets have relatively warm surfaces and high daytime thermospheres. What is conspicuously absent on Venus and Mars, however, is the temperature maximum in the middle atmosphere that is so prominent on Earth. Recall that it is oxygen and its photochemical reaction product, ozone, that readily absorb solar radiation in the near-ultraviolet range, so planetary atmospheres without oxygen lack a mechanism for warming the middle atmosphere. The temperatures on such planets lapse from the surface right through to the respective thermospheres. Because of the strong heating of our middle atmosphere, the troposphere is relatively thin, only some 10 to 15 km, depending on latitude. We should appreciate the fact that a chemical phenomenon (ozone formation) profoundly influences the physical structure of the lower atmosphere and limits the sizes of convective clouds.

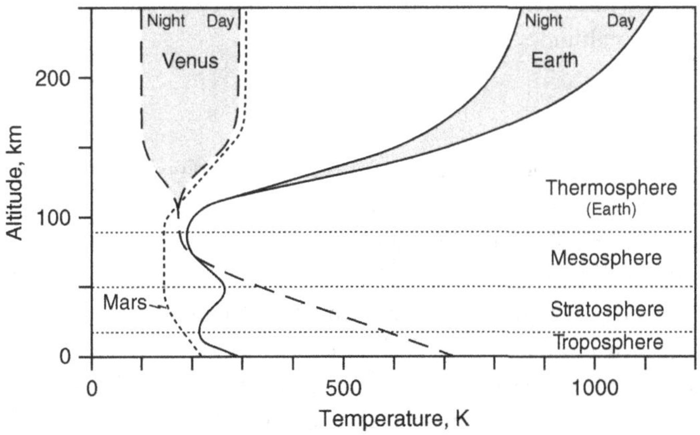

Figure 2.47 Temperature profiles of the terrestrial planets: Venus (dashed curves), Earth (solid curves), and Mars (dotted curve). Shaded regions indicate the diurnal ranges of temperatures in the upper atmospheres of Venus and Earth. The layers identified along the right side apply to Earth. From Pollack (1982) in *The New Solar System*, 2nd edition, J. Kelly Beatty (ed.); reprinted with permission.

The vertical structure of the atmosphere can be defined in other terms, such as those schematically represented on the right-hand side of Fig. 2.46, but temperature is nevertheless linked to several non-thermal properties. Most clouds, for instance, form in cooler regions. Where water vapor is abundant, as in the lower troposphere, clouds can form at relatively low altitudes despite the higher temperatures. But, as clouds precipitate and remove water from the air, clouds form at progressively lower temperatures. Nacreous clouds can even form in the dry lower stratosphere of wintertime polar regions when the temperature gets sufficiently low. Noctilucent clouds, visible at high latitudes after sunset, form at the other prominent minimum in temperature, that near the mesopause. Unlike the temperature minimum at the tropopause, which is coldest in wintertime, the minimum at the mesopause appears in the summertime, when the heating of the ozone layer in high latitudes drives strong mesospheric circulations. The chemical make-up of the atmosphere lets us recognize other layers as well. For instance, the lower stratosphere is characterized by a layer of small (submicron) aerosol particles composed mostly of sulfuric acid. The ionosphere is really part of the thermosphere and upper mesosphere, but the separate terminology is useful for defining the region containing relatively high concentrations of electrons and ions, which are generated when high-energy photons from the Sun interact with gas molecules. The aurora is a visible phenomenon arising from collisions of charged particles in the solar wind with atmospheric gases in the thermosphere of polar regions. We also find it convenient to classify the lower and middle atmospheres (from the surface to nearly 100 km) as the homosphere, where turbulent mixing makes the composition relatively uniform. The thermosphere may also be termed the heterosphere, because molecular diffusion dominates turbulent mixing there, causing the composition to favor the lighter compounds at higher altitudes.

Pressure

Pressure is the force acting on a unit area of surface, either physical or virtual. As with any force, pressure can cause objects to move or at least to strain under its influence. A greater pressure acting on one side of an object than on the other will tend to push it toward the lower-pressure side. The pressure exerted by air is a fundamental variable of atmospheric physics, one that varies with location, time, density, and temperature. The accepted global mean pressure at sea level p_0 is 101.325 kPa (equivalent to 1013.25 mb), but pressure decreases steadily with increasing altitude, by about six orders of magnitude in the first 100 km (near the base of the thermosphere). Pressure varies most rapidly in the vertical direction because of gravity, although the much smaller variations along horizontal planes suffice to move air and generate wind.

Unlike temperature, pressure in a hydrostatically balanced atmosphere decreases monotonically with altitude. Pressure at any one altitude is always greater than it is higher up. The justification for making such a bold assertion can be found in the mathematical relationship between pressure and altitude. Consider Fig. 2.48 and the forces acting on a small volume element, a slab of air somewhere in the atmosphere. Let the bottom of the slab have an area A and be located at altitude z [m]; the top has the same area, but is located slightly higher, at an altitude $z + \Delta z$. (Think of Δz being about 1 m, although the magnitude is not critical.) The volume of this rectangular slab is $V = A\Delta z$. The air contained within the slab has mass density ρ [kg m^{-3}], so the slab has a mass $m = V\rho$ and a weight $F_g = mg = V\rho g = A\Delta z\rho g$. If the air is not accelerating, the forces must balance in all

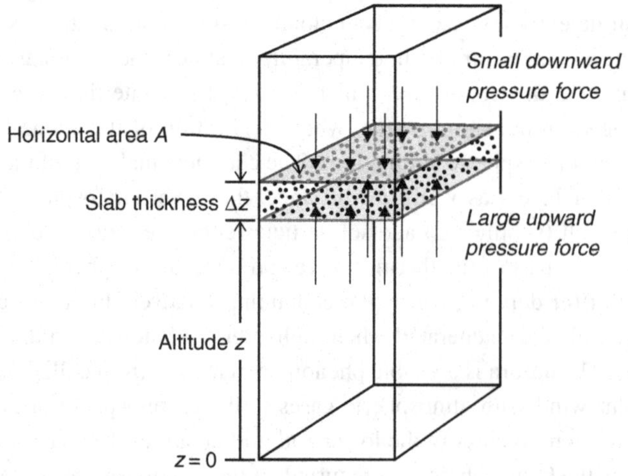

Figure 2.48 Schematic showing the geometric basis for deriving the variation of pressure with altitude. Adapted from Wallace and Hobbs (2006, p. 68).

directions. In the vertical direction, the weight of the parcel is borne by the difference in atmospheric pressure, $p_2 - p_1 = \Delta p < 0$, acting on the top area relative to that acting on the bottom area. The pressure is greater on the bottom than on the top in order to support the weight of the air in the slab, so Δp is a negative quantity. The pressure force, $F_p = -\Delta p A$, acts upward, toward lower pressure and in the positive z direction. At the same time, the weight acts downward. For hydrostatic balance, the net force $F_{net} = F_p - F_g = 0$, implying

$$\begin{aligned} F_p &= F_g \\ -\Delta p A &= A \Delta z \rho g. \end{aligned} \tag{2.65}$$

Once we cancel the area appearing on both sides of the equation and let $\Delta z \to 0$, we gain the hydrostatic equation, a differential equation for the rate of variation of p with z:

$$\frac{dp}{dz} = -\rho g. \tag{2.66}$$

Note that ρ and g are physical variables with positive values, so the vertical pressure gradient is invariably negative, meaning that the hydrostatic pressure must always decrease with increasing altitude. Pressure decreases monotonically with altitude, and its magnitude at any one level represents the weight of the air above that level in a column of unit cross-sectional area.

An explicit dependence of pressure on altitude can be obtained by combining the hydrostatic equation with the ideal gas law. Recall a valuable attribute of the gas laws: the behavior of an ideal gas is independent of its composition and depends only on the number of molecules present in a given volume, not on any other property, not on the shapes of the molecules nor on their masses. However, when the Earth's gravitational force acts significantly on a gas, as it does in the atmosphere, then we do need to consider the masses of the molecules. Gravity acts on mass, not on number alone. Because the atmosphere is relatively homogeneous below about 100 km, we can use the mean molecular mass of the molecules, $\bar{m}$, to compute the local gas density as $\rho = n\bar{m}$, where n is the local number concentration of molecules in the air as given by the ideal gas law, $n = p/k_B T$. Thus, $\rho = p\bar{m}/k_B T$, and the hydrostatic equation becomes

$$\frac{dp}{dz} = -\frac{\bar{m}g}{k_B T} p. \tag{2.67}$$

This equation, now representing the influence of the gravitational field on an ideal gas, can be solved for pressure as a function of altitude. The usual variation of temperature with altitude makes the integration difficult, but it sometimes suffices to use an average temperature $\bar{T}$ and treat the atmosphere as if temperature were constant. In the case of such a nearly isothermal atmosphere, pressure varies log-linearly with altitude:

$$\ln \frac{p}{p_0} = -\frac{\bar{m}g}{k_B \bar{T}} z, \tag{2.68}$$

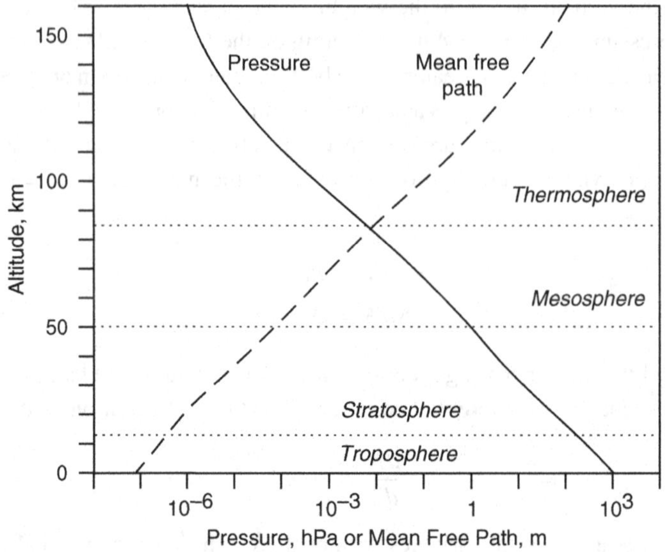

Figure 2.49 Variations of pressure and mean free path with altitude. Adapted from Wallace and Hobbs (2006, p. 9).

where p_0 is the pressure at altitude $z = 0$, which is usually taken as the Earth's surface (but the reference altitude can be chosen arbitrarily). Figure 2.49 shows the dependence of pressure on altitude and the nearly linear relationship between the logarithm of pressure and altitude throughout the lower and middle atmospheres. (Above about 100 km, the assumption of a fixed mean molecular weight breaks down.) Thermodynamic charts, such as the skew T-log p diagram, also make use of this relationship between pressure and altitude by scaling pressure logarithmically to make altitude more or less linear along that same axis.

Scale height

The decrease of pressure with increasing altitude reflects the decreasing weight of air with increasing altitude. Progressively less and less air exists in the atmosphere as one travels to greater and greater heights, and the density of that air becomes likewise smaller and smaller. Less mass and fewer molecules exist in each higher slab of air than in those below. Less mass means that a smaller difference in pressure is needed between the top and bottom of the slab to counteract the smaller weight. Pressure thus decreases with altitude at an ever-decreasing rate. We can see this pattern in the mathematics by exponentiating Eq. (2.68):

$$p(z) = p_0 \exp\left(-\frac{\bar{m}g}{k_B \bar{T}} z\right). \tag{2.69}$$

Pressure decreases exponentially with increasing altitude. As z gets larger, the argument of the exponent becomes progressively more negative, depending on the magnitude of the factor in front of z. Dimensional analysis shows that this grouping of variables ($\bar{m}g/k_B\bar{T}$ = force/energy) has dimensions of reciprocal distance, so we make the connection between altitude and this prefactor more explicit by defining the atmospheric scale height

$$H \equiv \frac{k_B\bar{T}}{\bar{m}g}. \tag{2.70}$$

In terms of scale height, the pressure relative to its reference value thus varies with z as

$$\frac{p(z)}{p_0} = \exp\left(-\frac{z}{H}\right). \tag{2.71}$$

At altitude $z = H$, the pressure is just $1/e$ of its reference value (p_0); at $z = 2H$, the pressure has fallen another factor of e, that is to $1/e^2$ of p_0, and so on. Scale height provides a concise measure of the rate with which pressure decreases with altitude; pressure decreases by a factor of e for each increment of H in altitude. As illustrated in Fig. 2.50, scale height also represents the full depth an hypothetical atmosphere would have were it to have the same density uniformly distributed from the surface to altitude H. The areas under the two curves are equal.

We also see in the scale height (Eq. (2.70)) the physical factors responsible for this systematic variation of pressure with altitude. Larger scale heights (and slower pressure decreases) arise with warmer conditions (larger $\bar{T}$), lighter gases (smaller $\bar{m}$), weaker gravitational attraction (smaller g), or any suitable combination of these variables. The scale height of our atmosphere, averaging about 8 km, varies slightly with the season because

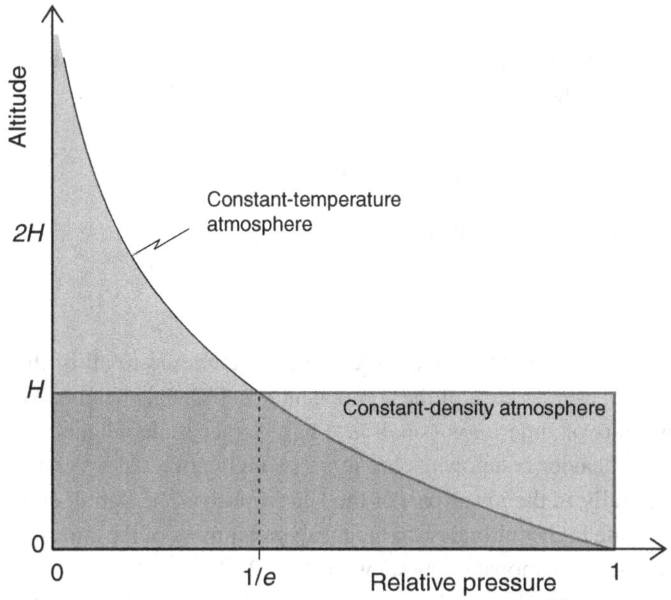

Figure 2.50 Idealizations of atmospheric pressure in terms of scale height.

of variations in the mean temperature; the other two factors remain relatively constant. However, the scale heights of the atmospheres on other planets (e.g., Venus and Mars) differ considerably from that on Earth because of different compositions and/or different gravitational attractions.

The atmosphere is not strictly isothermal, nor is the mean molecular mass of the air exactly constant with altitude. The scale height therefore varies somewhat with the altitude-dependence of temperature and with atmospheric composition, but the percentage deviations from the average state are generally small in the lowest 100 km (i.e., homosphere). However, significant deviations from average conditions do occur in the heterosphere, where the temperature rises rapidly and the blend of molecules changes significantly with altitude. The variation of pressure with altitude is more complicated in the heterosphere, because each component has its own scale height. According to Dalton's law, the total pressure p_{tot}, that which could be measured with a barometer, is the sum of the partial pressures p_j arising from the individual components (generically designated as j):

$$p_{tot} = \sum_j p_j, \tag{2.72}$$

where the summation accounts for all components. In the heterosphere, each component varies with altitude independently, as if it were the only substance present in the atmosphere. In hydrostatic balance, we thus find the partial pressure of component j to vary with altitude as

$$p_j(z) = p_{j,HP} \exp\left(-\frac{m_j g}{k_B T}(z - z_{HP})\right), \tag{2.73}$$

where we take the reference pressure $p_{j,HP}$ to be that at the homopause $z_{HP} \doteq 100\,\text{km}$. Expressed in terms of the scale height, the relative pressure changes as

$$\frac{p_j(z)}{p_{j,HP}} = \exp\left(-\frac{z - z_{HP}}{H_j}\right), \tag{2.74}$$

where the component-specific scale height is

$$H_j(z) = \frac{k_B T(z)}{m_j g}. \tag{2.75}$$

We now have to recognize that temperature, which is common to all of the constituents, depends on altitude, as does each of the component-specific scale heights. We also see that gaseous components of little mass (small m_j) have larger scale heights than do those of greater mass. The heavier components are more strongly attracted by gravity and so tend to settle preferentially in the lower part of the heterosphere. The lighter components preferentially occupy the higher altitudes, so the mean molar mass of the atmosphere decreases with altitude above the homopause, as shown in Fig. 2.51.

Despite the greater complexity of atmospheric structure in the thermosphere, it is worth reminding ourselves that the pressure at any given level is still simply the weight of the

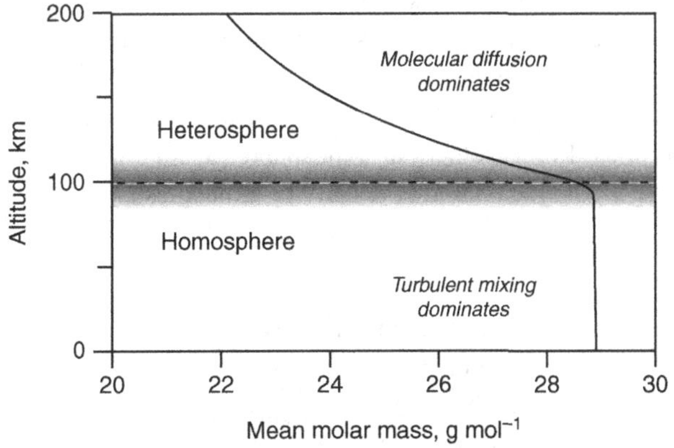

Figure 2.51 Variation of mean molar mass with altitude. The altitude separating the heterosphere from the homosphere (the homopause) is approximate, as suggested by the shading at about 100 km. Adapted from Torr (1985).

air in a column of unit cross-sectional area. The total mass M_j of compound j in this unit column above level z is just the local mass density times the component-specific scale height:

$$M_j(z) = \rho_j(z) H_j(z) = \left(\frac{m_j p_j(z)}{k_B T(z)}\right)\left(\frac{k_B T(z)}{m_j g}\right) = \frac{p_j(z)}{g}. \tag{2.76}$$

The altitude-dependence of temperature affects both the density (through the ideal gas law) and the scale height, but in opposing senses. Thus, we again see that the partial pressure $p_j = M_j g$, the force per unit area exerted by component j, is just the weight $(M_j g)$ of that component above level z. The total mass of air is the sum of the component masses,

$$M_{tot} = \sum_j M_j = \frac{1}{g}\sum_j p_j = \frac{p_{tot}}{g}, \tag{2.77}$$

so we again see that the total, or barometric pressure, represents the composite weight of the air in the unit column.

Density

The mass of air contained in a unit volume has been seen above to play a key role in the hydrostatic equation and the variation of pressure with altitude. It is this mass density [SI units: kg m^{-3}] that gives air its weight and the need for the pressure to decrease with altitude. But, through the ideal gas law, density also varies with the pressure; the larger the pressure, the greater the density: $\rho = p\bar{m}/k_B T$. To first approximation, the composition of the lower and middle atmospheres changes little (keeping $\bar{m}$ constant), so at a given temperature, density $\rho \propto p$. Air is readily compressible, because of the void spaces between its molecules, so the greater pressures at low altitudes in effect squeeze the gas molecules closer together. Density has also been shown to be the mean mass of the molecules times the number density n: $\rho = n\bar{m}$, so we see that it is the greater packing of the molecules at higher pressure, not their individual masses, that causes the mass density to be greater at higher pressures and lower altitudes.

The molecular number density (or concentration) varies with altitude in much the same way that pressure does. In an isothermal atmosphere, again characterized by a column-average temperature $\bar{T}$, the local number density of molecules is given by

$$n = \frac{p}{k_B\bar{T}}. \tag{2.78}$$

Variations in pressure are reflected directly in variations in number density: $n \propto p$. We may develop the dependence of n on z, as we did for pressure, thus obtaining the function

$$n(z) = n_0 \exp\left(-\frac{\bar{m}gz}{k_B\bar{T}}\right). \tag{2.79}$$

Here, n_0 is the number density at the surface, where we take $z = 0$. As z increases, $n(z)$ decreases exponentially, with approximately the same scale height as pressure did. However, in the form presented here, we gain new insight into the vertical organization of the atmosphere. Note that the numerator in the argument of Eq. (2.79) is just the ordinary expression for the gravitational potential energy of an object of mass $\bar{m}$: $\Delta V = \bar{m}gz$. Here, the object of interest is a representative molecule of the air and z is the height of this molecule above the Earth. We may now rearrange Eq. (2.79) to reflect how the relative number density varies with the potential energy:

$$\frac{n}{n_0} = \exp\left(-\frac{\Delta V}{k_B\bar{T}}\right). \tag{2.80}$$

In this form, we see that the number density n of molecules decreases exponentially with the magnitude of the potential energy difference ΔV.

Potential energy, an energy of position, represents the work done by an object to move it against a force. In the atmosphere, the applicable force is the gravitational attraction of a molecule toward the Earth, but any force that can be represented by a potential energy would similarly affect the molecular number density n relative to n_0 (that density when the potential energy is zero). Equation (2.80) is thus general. Molecules in liquids and solids, for instance, interact predominantly with each other via electromagnetic forces, so V there represents an electrical type of potential energy. Whether resulting from gravitational or electrical forces, the potential energy, V, relative to molecular kinetic energy, k_BT, is what determines the concentration ratio of molecules, n/n_0. Equation (2.80), known as Boltzmann's law, gives the probability of finding molecules in energy state V relative to the "ground" state $V = 0$. This fundamental principle of statistical mechanics is applicable to numerous macroscopic and microscopic situations, including cloud formation.

Column abundance

The amount of material in a vertical column of the atmosphere is a quantity that arises in various physical and chemical applications. We have already seen that the pressure at the surface is a measure of the total mass of air in a unit column. As another example, atmospheric ozone measured from space represents the total ozone in a column without regard for the altitude dependence of the ozone concentration. Satellites often sense electromagnetic radiation coming from the Earth's surface, the atmosphere as a whole (along the optical path) acting as a filter that lets some rays through, but blocking others. Radiation from the Sun interacts in complicated ways with the various gases in the atmosphere; some wavelengths can penetrate only so far before that radiation is absorbed and lost. Such examples depend on the integrated amount of a substance in the atmosphere more so than the actual concentration of that substance at any given altitude.

Column abundance represents a generalization of column weight discussed earlier. We can calculate the column abundance by adding up the amounts in the many thin slabs of air constituting the column, much as one would count the number of people in a tall building by simply adding up the numbers on each of the floors: Let the number of people on the k-th floor be n_k, then the population N of a building having K floors is

$$N = \sum_{k=1}^{K} n_k. \tag{2.81}$$

The analogy breaks down for atmospheric applications in the sense that buildings have a finite number of discrete floors, whereas an atmospheric column has an infinite number of infinitesimally thin slabs, so we need to use calculus. Let the concentration [e.g., molec m^{-3}] of substance j be $n_j(z)$ at level z. Then, the total number of j molecules in an atmospheric column of unit cross-sectional area is

$$N_j = \int_0^\infty n_j\left(z'\right) dz'. \tag{2.82}$$

If we care only about the number per square meter above a given altitude z, then

$$N_j\left(z\right) = \int_z^\infty n_j\left(z'\right) dz'. \tag{2.83}$$

Note that the independent variable (z) appears as a limit of the integration, and a dummy variable (z') is used in the integrand. One sometimes needs to calculate the amount of material in a path aligned off the vertical, in which case straightforward trigonometric methods aid the calculations.

2.3.2 Horizontal structure

The structure of the atmosphere along planes perpendicular to the gravity vector is more subtle than it is in the vertical direction. Nevertheless, the relatively small variations in pressure, density, and temperature on horizontal planes are sufficient to drive the winds and the overall atmospheric circulation.

Zonal averages

The average variations of atmospheric properties in the north–south direction give us a good sense for how the atmosphere organizes itself in horizontal directions and how the seasons affect that organization. When the temperature at selected latitudes and altitudes is averaged around individual latitude belts (zones), we see how temperature varies around the globe. Figure 2.52 shows the mean distribution of air temperature when it is summer in the southern hemisphere and winter in the northern hemisphere. The isotherms bulge upward near the Earth's surface at low latitudes because of the strong surface insolation there throughout the year. The surface heating is shifted toward the summer hemisphere, leaving the winter hemisphere relatively cold. The lower stratosphere is cold, particularly near the winter pole and at low latitudes because of strong tropical convection. By contrast, the temperature maximum in the middle atmosphere, arising from the absorption of solar radiation by ozone, is shifted far toward the summertime pole. In response to this strong middle-atmospheric heating, convection in the summer polar mesosphere forces air to rise, expanding and cooling as it moves upward toward lower pressure. The result is the extreme temperature minimum some 30 to 40 km higher in the atmosphere. The

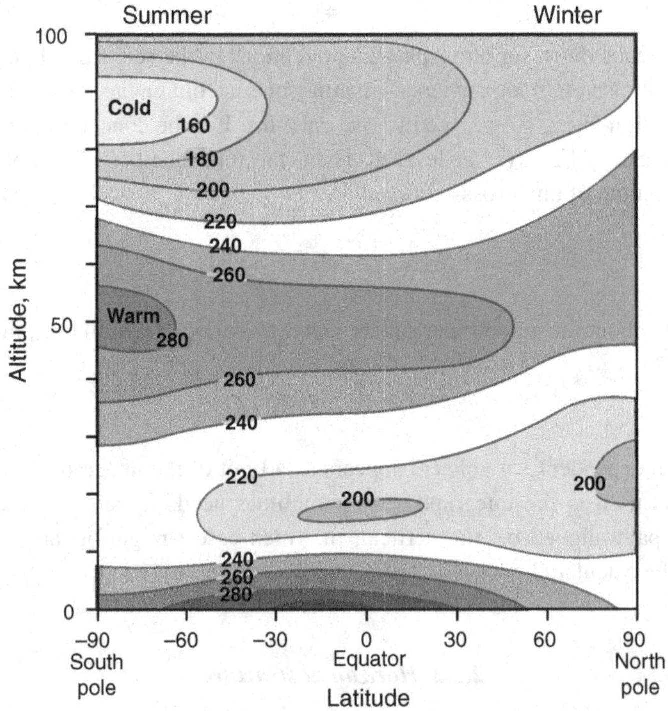

Figure 2.52 Vertical structure of zonally averaged temperatures during Northern Hemisphere winter. Curves are isotherms, lines of constant temperature (labeled in K). From Brasseur and Solomon (1986, p. 34).

temperatures at the summertime polar mesopause are the lowest found anywhere in the atmosphere. Note that the juxtaposition of low-temperature above regions of high temperature is not a coincidence; rather, the low temperatures exist precisely because of the warm region lower down and the consequent convective overturning of air. It should not be surprising to learn that noctilucent clouds form in the cold pocket near the summer mesopause, while polar stratospheric and nacreous clouds form in the winter part of the polar stratopause. Clouds form when moist air is lifted and cools.

Air masses

An air mass is a large-scale body of tropospheric air with more or less uniform properties. With time, the air acquires some characteristics of the surface in which it is in contact. So, for example, air having spent significant time in polar regions during the winter months is likely to be cold and dry. Air in tropical regions, by contrast, can be expected to be warm and moist.

Temperature and humidity are two important meteorological variables used by climatologists to characterize air masses. A summary of the major categories of air masses around the world is shown in Fig. 2.53. Generally, cold polar air (designated by P) arrives from higher latitudes, whereas warm tropical air (T) comes from the lower latitudes. Wind from the north is generally cooler than that from the south in the northern hemisphere. A further subdivision arises upon considering the moisture content of the air. Air having spent considerable time over the oceans tends to be more humid than air over dry land. Thus, the prefix "m" designates air with a maritime character, whereas the prefix "c" identifies a drier continental air mass. The four main air mass types, maritime polar (mP), continental polar (cP), maritime tropical (mT), and continental tropical (cT), suffice for many climatological purposes, but in cloud physics we also need to recognize that maritime air is typically cleaner than continental air. Relatively few sources of particulate matter exist over

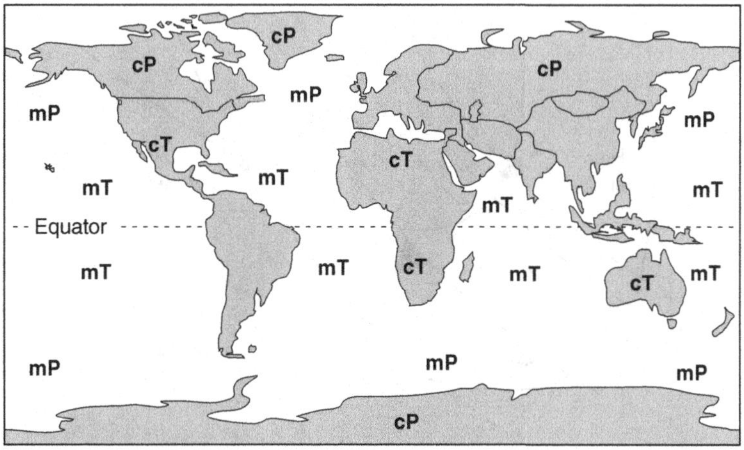

Figure 2.53 World map showing the principal air masses.

the oceans, and precipitating clouds themselves act to cleanse the air. The microphysical properties of clouds forming in Maritime air masses thus tend to differ substantially from those that form in continental air masses.

The dynamical behavior of an air mass is affected mainly by the mass density of the air near the surface, which in turn depends on the temperature and pressure of the air. Recalling the approximate equation of state for moist air,

$$p \doteq \rho R_d T, \tag{2.84}$$

we see that density ρ and temperature T are inversely related when the pressure is constant (as along an isobar). The adage "cold air is denser than warm air" is seen to be true, but only for a given pressure. Cold continental air is typically denser than warm maritime air. Density differences between air masses become important for cloud formation, especially when neighboring air masses make contact.

Fronts

A weather front is a quasi-two-dimensional structure that separates one air mass from another. The properties of the air change significantly over relatively short distances across a front, which can extend upward from the surface all the way to the tropopause. A curve designating a front on a surface weather chart (see Fig. 2.54) represents the intersection of

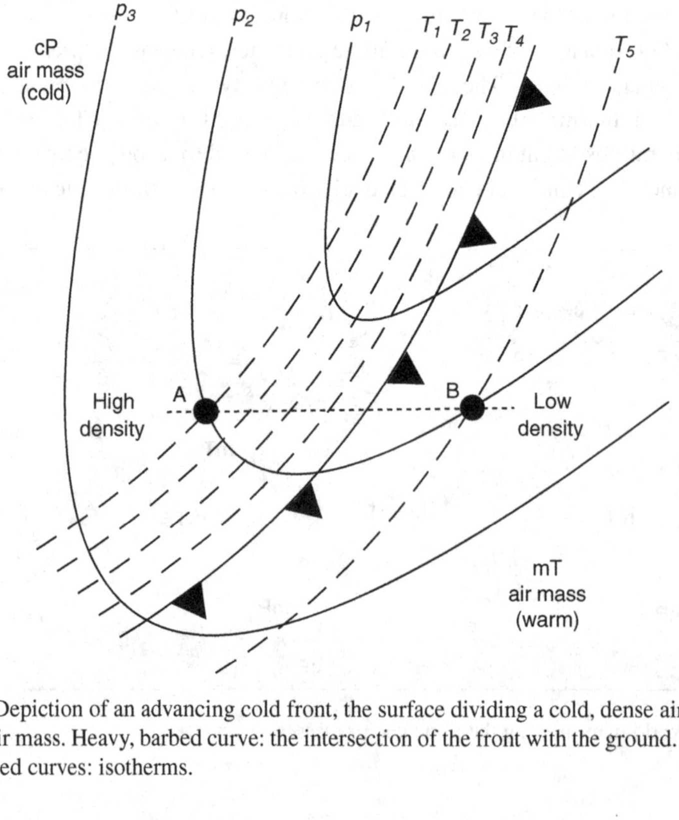

Figure 2.54 Depiction of an advancing cold front, the surface dividing a cold, dense air mass from a warm, light air mass. Heavy, barbed curve: the intersection of the front with the ground. Solid curves: isobars. Dashed curves: isotherms.

the front itself with the surface of the Earth. As the colder cP air mass advances upon the warmer mT air mass, the isotherms (dashed curves) pack together on the cold side of the front, giving rise to a strong density gradient in the frontal zone. One can appreciate the difference in density that must be present across the front by considering the two dark spots joined by the dotted line in the figure. Both spots (A and B) are on the same isobar (p_2), so the pressures must be equal at the two locations. Spot A is, however, colder than spot B ($T_5 > T_1$). Thus, by the equation of state (Eq. (2.84)), the density at A must be greater than that at B. This density difference across the front causes the pressures to decrease with altitude at different rates (according to the hydrostatic equation, Eq. (2.66)) on either side of the front. The pressure pattern set up in turn determines the motions of the air in the vicinity of the front. Of particular importance for cloud formation, the density difference causes the warmer and usually moister air to ride over the the denser air below the advancing front.

2.3.3 Planetary-scale patterns

On the scale of the Earth, the atmosphere organizes itself in a way that promotes the effective transfer of energy from low to high latitudes. Low-latitude regions, those within about 30 ° of the equator, receive more energy from the Sun on average than do polar latitudes (review Section 2.2.3). The oceans and atmosphere help redistribute this excess energy by moving the warmer fluids poleward and the colder fluids equatorward. Whereas oceanic circulations are restricted by the land masses of the Earth, the atmosphere moves freely around the globe. The warmed air of the tropics rises in response to its lowered density, forms clouds, and travels poleward near the tropopause. Were the Earth to be a non-rotating sphere, this poleward motion would continue all the way to the poles.

The Earth's rotation introduces angular momentum into the air and causes the atmospheric circulation to be complicated. The poleward movement of air therefore develops a westerly component (air from the west) that inhibits the direct transport of the air into higher latitudes. As shown in Fig. 2.55, the circulation on a rotating spheroid breaks up into a set of interconnected, toroidal circulations, called cells. Three distinct cells form in each hemisphere. Two of these cells are thermally direct in the sense that air rises where it has been heated and sinks where it has been cooled. The thermally direct cell in the tropics is called the Hadley cell, that near the poles, the Polar cell. In mid-latitudes, a less-well defined circulation, the Ferrel cell, is driven indirectly by the motions of the thermally direct cells. Clouds typically form along the rising branches of the cells.

Large-scale air motions develop in response to the cellular circulations. The flow of air in the upper troposphere within each cell brings air with distinct properties to the regions where the cells meet. The large spatial gradients in temperature and pressure that result cause locally strong westerly winds, called jet streams (shown in Fig. 2.55 as broad curvy arrows). The subtropical jet stream tends to form where the Hadley cell meets the Ferrel cell, whereas the polar jet forms near the junction of the polar and Ferrel cells. These jets meander around the globe in quasi-horizontal waves called planetary or Rossby waves. The wavelengths of Rossby waves can exceed thousands of kilometers, so only a few

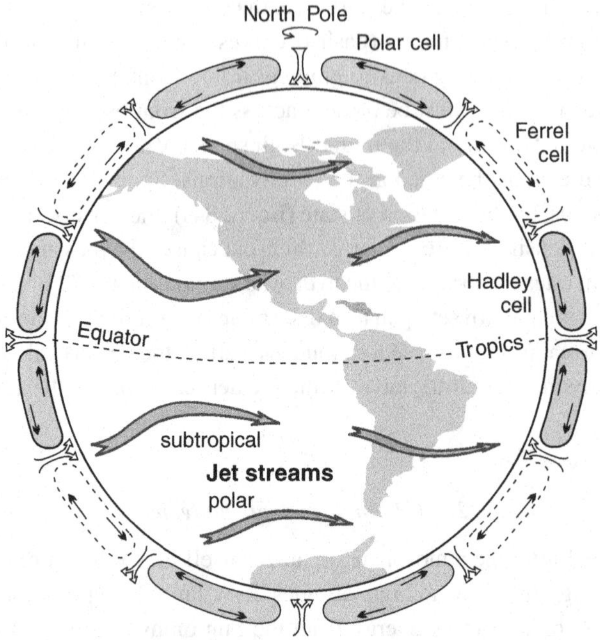

Figure 2.55 Cartoon of atmospheric organization on a global scale. Of the three circulation cells in each hemisphere, two (shaded and named Hadley and Polar) are direct and one (unshaded, Ferrel) is indirect, as suggested by the small arrows. Jet streams (curvey arrows) form in the upper troposphere near the convergence of neighboring cells.

ridges and troughs form around the globe. Rossby waves are important for cloud formation because cyclones with fronts form and generate weather near the poleward branch of each wave.

2.4 Further reading

Andreae, M.O., D.A. Hegg, and U. Baltensperger (2009). Sources and nature of atmospheric aerosols. In *Aerosol Pollution Impact on Precipitation*, ed. Z. Levin and W.R. Cotton. Berlin: Springer, pp. 45–89.

Bohren, C.F. and E.E. Clothiaux (2006). *Fundamentals of Atmospheric Radiation*. Weinheim: Wiley VCH, 472 pp. Chapters 1 (emission), 2 (absorption), and 3 (scattering).

Curry, J.A. and P.J. Webster (1999). *Thermodynamics of Atmospheres and Oceans*. London Academic Press, 467 pp. Chapters 12 (energy, radiation), 13 (climate), and 14 (planetary atmospheres).

Eisenberg, D. and W. Kauzmann (1969). *The Structure and Properties of Water*. Oxford: Oxford University Press, 296 pp. Chapters 1 (water molecule), 3 (ice structure), and 4 (liquid structure).

Fleagle, R.G. and J.S. Businger (1963). *An Introduction to Atmospheric Physics*. New York: Academic Press, 346 pp. Chapters I (basics), II (gases), and IV (radiation).

Fletcher, N.H. (1970). *The Chemical Physics of Ice*. Cambridge: Cambridge University Press, 271 pp. Chapters 1 (water molecule), 2 (ice structure), 4 (liquid), 7 (point defects), and 9 (electrical properties).

Hobbs, P.V. (1974). *Ice Physics*. Oxford: Oxford University Press, 837 pp. 1 (ice structure), 2 (electrical properties), 4 (mechanical properties), and 10 (atmospheric ice).

Hobbs, P.V. (2000). *Introduction to Atmospheric Chemistry*. Cambridge: Cambridge University Press, 262 pp. Chapters 1 (atmospheric history, Gaia), 3 (atmospheric composition), 4 (radiation), and 6 (aerosols).

Petrenko, V.F. and R.W. Whitworth (1999). *Physics of Ice*. New York: Oxford University Press, 373 pp. 1 (water molecule, H bond), 2 (ice Ih), 4 (electrical properties), 6 (point defects), and 7 (dislocations).

Petty, G.W. (2004). *A First Course in Atmospheric Radiation*. Madison, WI: Sundog Publishing, 443 pp. Chapters 2 (radiation basics), 3 (EM spectrum), 6 (emission), and 12 (scattering).

Pruppacher, H.R. and J.D. Klett (1997). *Microphysics of Clouds and Precipitation*. Dordrecht: Kluwer Academic Publishers, 954 pp. Chapters 3 (water), and 8 (gases, aerosol).

Salby, M.L. (1996). *Fundamentals of Atmospheric Physics*. San Diego, CA: Academic Press, 624 pp. Chapters 1 (composition), 2 (gases), and 8 (radiation).

Seinfeld, J.H. and S.N. Pandis (1998). *Atmospheric Chemistry and Physics*. New York: John Wiley and Sons, 1326 pp. Chapter 7 (aerosols).

Turco, R.P. (1997). *Earth under Siege*. New York: Oxford University Press, 527 pp. Chapters 3 (basics) and 4 (Earth history); Appendix A (units of measure).

Wallace, J.M. and P.V. Hobbs (2006). *Atmospheric Science: An Introductory Survey*. Amsterdam: Academic Press, 483 pp. Chapters 1 (atmospheric structure), 2 (Earth history), 3 (gas laws, thermodynamics), and 8 (weather systems).

2.5 Problems

1. The number density of molecules in air enters into many calculations in atmospheric physics. Recognizing that air as a whole, and each of its constituents, behaves as an ideal gas, estimate the number of water molecules in an SI unit volume (1 m^3) of dry air to one significant figure at standard temperature and pressure (STP) when the water vapor occupies 1% of the air by volume. [Do not use a calculator for the first part of the exercise – use your head!] What is the percentage error in your estimation of vapor density compared with the true, high-precision value obtained by using a calculator? What importance do you place on estimating (versus calculating) variables?

2. The following exercise is intended to help you "visualize" molecular scales.
   (a) Compute the total number of molecules in an atmospheric column that has a surface pressure of 1013 hPa. [Units: cm^{-2}]
   (b) Assume the scale height is 8 km. Calculate the average molecular concentration. [Units: cm^{-3}]
   (c) Assume the effective diameter of an air molecule is 0.5 nm. Compute the volume of this molecule. [Units: cm^3]

(d) From parts (b) and (c), calculate the ratio of the volume of space occupied by an average molecule to the actual volume of that particle.

(e) From kinetic theory, the molecular mean free path can be defined as $\lambda_m = \left(\sqrt{2}\pi d^2 N\right)^{-1}$ where d is the molecular diameter and N is the number of molecules per unit volume. Compute this mean free path using the information from (b) and (c).

3. By assuming constant temperature, we are able to integrate the hydrostatic equation combined with the ideal gas law to obtain an expression for the exponential decrease of pressure with height. But suppose that temperature decreases uniformly with height. That is, $-dT/dz = \Gamma$, where Γ is called the temperature lapse rate.
   (a) On physical grounds, you should be able to guess if pressure decreases with height more or less rapidly when temperature decreases with height than it does in an isothermal atmosphere.
   (b) Derive the mathematical expression of how pressure decreases with height in this atmosphere where temperature decreases with height.
   (c) The scale height is defined as the height where the pressure has decreased to $1/e$ times its surface value. Calculate the scale height for this atmosphere.
   (d) How does this scale height compare to that calculated using the average temperature of the layer? Discuss.
   (e) Sketch a plot of the pressure drop-off with z for an isothermal atmosphere. Now, on top of this plot, draw the pressure profile for our constant lapse rate atmosphere. You may assume that the isothermal atmosphere has a temperature exactly the mean value of the constant lapse rate atmosphere, and that the surface pressures are the same in each case.

4. The tetrahedral nature of the water molecule is extremely important, as it determines many of the physical properties of water that we take for granted. You have probably known from early on that the mass density of ordinary ice $\rho_{ice} = 0.917\ \mathrm{g\,cm^{-3}}$. With this knowledge and an understanding of the geometrical arrangement of the water molecules in the unit cell of ice, calculate the spacing (d) between the centers of neighboring molecules in ice. You will probably find it easiest to approach the problem by pretending that you know the magnitude of d, then calculating the following in the order given:
   (a) the obtuse angle θ of the unit cell for ice Ih,
   (b) the tetrahedral angle β (the bond angle plus a bit), (try working with four vectors radiating from the center of a sphere),
   (c) the dimensions a and c of the unit cell,
   (d) the volume V of the unit cell.

5. The ozone layer lies between the altitudes of 14 km and 42 km and contains 4 molecules of ozone for every million molecules of "air", i.e., its volume mixing ratio

is 4 ppmv (parts per million by volume). Assume that the atmosphere is isothermal at 240 K and that the scale heights for pressure and density are constant at 7 km.

(a) Calculate the total number density of molecules at the surface under typical conditions.
(b) What is the number density of ozone at 21 km?
(c) The (vertical) column density is defined as the total number of molecules per square centimeter overhead. How is the column density of a molecular species related to the vertical structure of its number density?
(d) What is the column density of ozone for the information given above?
(e) If the vertical column of ozone were brought down to the surface, assuming the same typical conditions you used in (a), how thick would this layer be?

6. Suppose the Earth has an albedo of 0.4. What will be Earth's effective temperature if the solar irradiance is 1370 $\mathrm{W\,m^{-2}}$? If the albedo could be increased to 0.6, what would be the new equilibrium temperature?

7. A container contains two gases separated by a partition. Assume that the volumes containing each gas are equal, and that volume A contains helium and volume B oxygen. Initially both gases are at the same temperature and pressure.
   (a) What is the ratio of the number densities in the two volumes?
   (b) What is the ratio of the average molecular speeds in the two volumes?
   (c) If the partition is removed, initially will the pressure in B increase, decrease, or remain the same relative to A? Explain!

8. A black horizontal disk is exposed to direct solar radiation (zenith angle 30°), diffusely reflected radiation from an infinite horizontal surface with an albedo of 0.3, and long-wave radiation emitted by the horizontal surface as a blackbody of 300 K. Compute the equilibrium temperature of the disk when the solar irradiance is 1370 $\mathrm{W\,m^{-2}}$.

9. A column of 1 km height contains moist air of temperature T with an initially uniform water mixing ratio r_{v0}, where $r_v = \rho_v/\rho_d$. Derive an expression for the vertical distribution of the mixing ratio after diffusive equilibrium has been established under the assumption that the temperature remains unchanged.

10. Refer to the table below and perform the indicated tasks. Explain each step of your work carefully, using equations as appropriate. Always specify appropriate SI units of all values.
    (a) Complete the table by calculating the following discrete distributions for each bin: number distribution n_j, surface-area distribution $n_{s,j}$, and volume distribution $n_{v,j}$. Specify the correct SI units in the blank cells at the top of the respective columns.
    (b) Plot each of these three distributions on separate, correctly labeled graphs, using logarithmic abscissas and linear ordinates.
    (c) Calculate the total concentrations of number (N), surface area (S), and volume (V).

Bin j	Diameter D_j [μm]	Interval ΔD_j [μm]	Concentration N_j [cm^{-3}]	n_j	$n_{s,j}$	$n_{v,j}$
1	1.12E-02	2.59E-03	1.33E+02			
2	1.41E-02	3.26E-03	2.77E+02			
3	1.78E-02	4.10E-03	5.80E+02			
4	2.24E-02	5.17E-03	8.19E+02			
5	2.82E-02	6.50E-03	1.01E+03			
6	3.55E-02	8.19E-03	6.36E+02			
7	4.47E-02	1.03E-02	3.57E+02			
8	5.62E-02	1.30E-02	2.59E+02			
9	7.08E-02	1.63E-02	2.10E+02			
10	8.91E-02	2.06E-02	1.79E+02			
11	1.12E-01	2.59E-02	1.52E+02			
12	1.41E-01	3.26E-02	1.27E+02			
13	1.78E-01	4.10E-02	1.05E+02			
14	2.24E-01	5.17E-02	8.57E+01			
15	2.82E-01	6.50E-02	6.50E+01			
16	3.55E-01	8.19E-02	4.50E+01			
17	4.47E-01	1.03E-01	2.71E+01			
18	5.62E-01	1.30E-01	1.18E+01			
19	7.08E-01	1.63E-01	1.63E+00			
20	8.91E-01	2.06E-01	4.93E-01			
21	1.12E+00	2.59E-01	3.04E-01			
22	1.41E+00	3.26E-01	2.15E-01			
23	1.78E+00	4.10E-01	1.71E-01			
24	2.24E+00	5.17E-01	1.36E-01			
25	2.82E+00	6.50E-01	1.10E-01			
26	3.55E+00	8.19E-01	9.19E-02			
27	4.47E+00	1.03E+00	7.13E-02			
28	5.62E+00	1.30E+00	5.54E-02			
29	7.08E+00	1.63E+00	4.10E-02			
30	8.91E+00	2.06E+00	3.04E-02			
31	1.12E+01	2.59E+00	2.20E-02			
32	1.41E+01	3.26E+00	1.56E-02			
33	1.78E+01	4.10E+00	9.62E-03			
34	2.24E+01	5.17E+00	5.17E-03			
35	2.82E+01	6.50E+00	1.52E-03			
36	3.55E+01	8.19E+00	2.91E-04			

Part II

Transformations

3
Equilibria

Transformations, changes in the structure or composition of something, are common in our everyday world. Lakes freeze in the winter; ponds dry up on hot summer days. Salt melts ice and dissolves in water. Such examples may be classified as either physical or chemical, if you wish, but many natural phenomena represent blends of both disciplines. Of primary interest is the notion that the entity in question (here, water or salt) has undergone a change of one sort or another. Learning how transformations proceed in nature is a fundamental goal of science, so we are often concerned with "processes" and "mechanisms", the sequence of discrete events that leads to a particular outcome. Clouds owe their existence to particular transformations: Water vapor changes into liquid droplets, and those droplets later freeze when the temperature becomes low enough. Clouds would never form (or dissipate) were water not to change, be transformed from one state to another. We simply cannot understand much about clouds without dealing with transformations in detail. In fact, it is often useful to view clouds as processes, rather than as objects or entities. This part of the book lays out the guiding principles upon which to build our deeper understanding of atmospheric processes leading to clouds.

Natural transformations take place when a system deviates in some way from its most balanced or equilibrium state. Clouds develop, and ordinary substances transform from one state to another, when the conditions in the local environments are "out of whack", not appropriate for keeping them stable. An ice cube taken from your refrigerator melts if you keep it in your hand or put it into a warm beverage. Ice is simply not stable at temperatures above its melting point; it gradually loses all rigidity and becomes liquid. Similarly, clouds form by condensation only when the concentration of water vapor exceeds an equilibrium value. Should the vapor concentration fall below this threshold, the cloud dissipates. The equilibrium condition, even though not often present in the atmosphere, is an important reference state, in effect a "cusp" in the range of conditions that separates processes going in one direction (e.g., condensation) from those going in the opposite direction (evaporation). We cannot predict whether net condensation or evaporation of a cloud droplet will take place unless we can determine the state of equilibrium with some quantitative precision.

The equilibrium state may be thought of as the set of environmental conditions that allows the system to remain as it is without observable change. These equilibrium conditions, often dictated by specific values of environmental temperature, pressure, or composition, let the system (whether an ice cube, water in a container, or a cloud droplet) persist indefinitely. Ice at temperatures below the melting point, for instance, will remain ice; liquid water will remain liquid at modest temperatures above the melting point (in a suitably moist environment). Such observations come from everyday experience, and they exemplify the common concept of equilibrium that underpins much of our science.

Equilibrium may be categorized in various ways. Some equilibria are "stable", whereas others are "unstable", the distinction revolving around the propensity of the system to return to its original state, or not, following an externally imposed perturbation. An example of a stable equilibrium is a marble in a bowl: Push the marble away from its resting position in the deepest part of the bowl, then release it; the marble falls back toward the bottom of the bowl. It may roll back and forth before all apparent motion ceases, but it will invariably come to rest close to its initial point of rest. As a counter example, try balancing a new wooden pencil upright on its flat end; it might stay put for a short while, but even a gentle shove causes the pencil to topple and move irreversibly away from its initial vertical orientation. The pencil on end represents a system in an unstable (possibly metastable) state. These examples of macroscopic objects in a gravitational field illustrate mechanical equilibria. Later, we will explore ways of specifying thermal and chemical equilibria, as well. First, we develop the concept of equilibrium from the viewpoint of molecules, the fundamental units of matter.

3.1 Molecular interpretation of equilibrium

We may justifiably view the equilibrium state as one in which objects are stationary and non-varying with time (as a marble sitting peacefully in a bowl), but such steadfastness exists only at the macroscopic, directly observable level. Most of the systems with which we need to deal operate about a state of dynamic equilibrium. The system overall may appear static by every normal measure, but the individual molecules constituting the system are anything but still. We may never directly see them, but the molecules are real, and it is their incessant motions within the seemingly static system that give rise to the paradox associated with the term "dynamic equilibrium". Whereas the word equilibrium implies no change, the word dynamic connotes rapid, even forceful change. You can reconcile the paradox by realizing that the separate terms refer to different views of the system. The molecules are indeed constantly and randomly changing positions and orientations, but their average behavior, which is what ultimately determines the macroscopic properties, remains constant. The equilibrium state is fixed, or stationary, only in a statistical sense. We also say a system in equilibrium is in "steady state" at the molecular level.

Consider an example closely related to clouds. The molecules in liquid water are perpetually jostled around by one another, and some of those in the surface break away altogether and enter the gas or vapor phase; they evaporate (see Fig. 3.1). At the same time, vapor molecules frequently strike the surface and enter the liquid; they condense. When the liquid is in equilibrium with its environment (in this case, the surrounding gas phase),

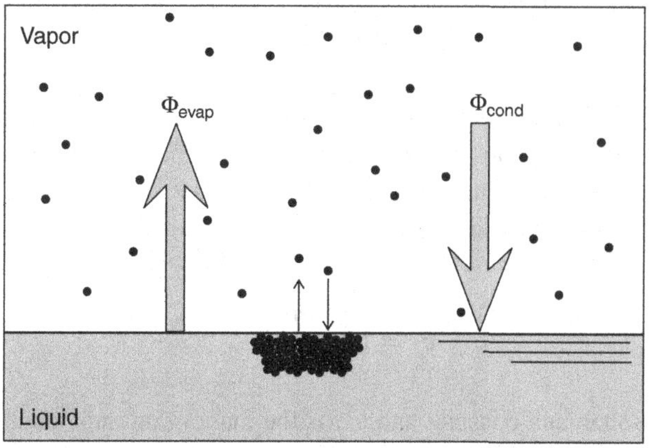

Figure 3.1 Schematic of the molecular fluxes leaving and entering the surface of a liquid.

the relative amounts of liquid and vapor remain fixed, invariant with time, even though the molecules are exchanged rapidly across the interface. As suggested by the equal lengths of the arrows in Fig. 3.1, equilibrium between two phases is characterized by a balance of the molecular fluxes. Such a physical statement of equilibrium may be expressed mathematically by equating the fluxes of molecular condensation Φ_{cond} and of molecular evaporation Φ_{evap}:

$$\Phi_{cond} = \Phi_{evap}. \tag{3.1}$$

The number of water molecules entering the liquid through unit area (1 m^2) in a unit interval of time (1 s) is the same as the number leaving through that same area in the same time interval.

Under such steady-state, flux-balanced conditions, no net condensation or evaporation occurs; the amount of liquid (the observable depth of water in a container, or the size of a water drop) remains perfectly constant, despite a high rate of molecular exchange across the liquid–vapor interface. Note the dual use of the words here for explaining equilibrium in molecular terms. On the one hand, we speak of "condensation" and "evaporation" as molecular-scale processes; whereas the same words are also used to describe the macroscopic effect of those microscale processes. Usually, the context allows us to distinguish a process from its effect, but adding the term "net" for macroscopic descriptions will usually eliminate any ambiguity. Liquid cloud droplets experience no net condensation under equilibrium conditions, and so they neither grow larger nor shrink with time. Only if the rate of molecular condensation were to exceed the rate at which molecules evaporate and leave the liquid, would the droplets grow.

How frequently do vapor molecules actually contact the liquid surface? Fortunately, the magnitude of a molecular flux can be calculated with the help of the kinetic theory of gases. In cloud physics, we are naturally interested in the exchange of water vapor molecules across the liquid–air interface, but the mathematical treatment works equally well for any type of gas molecules impinging onto a surface. To keep our treatment general, we will think of a generic gas, which we call "j" (an algebraic symbol that can be replaced by a specific compound, such as H_2O, SO_2, or O_3, as needed). For now, in the case of water, we identify this generic "impingement flux" with the condensation flux, but later we will find that not all impinging molecules automatically end up in the condensed phase. Consider a unit area of surface exposed to the gas at some temperature T. The mean thermal speed of the gas molecules is shown by kinetic theory to be

$$\bar{c}_j = \left(\frac{8k_BT}{\pi m_j}\right)^{\frac{1}{2}}, \tag{3.2}$$

where k_B is the Boltzmann constant and m_j is the mass of an individual j molecule. The mean speeds of gas molecules are typically several hundreds of meters per second. If the number density of j molecules (the molecular concentration) immediately above the surface is n_j, then the impingement flux of gas j is simply

$$\Phi_{imp,j} = \frac{1}{4}n_j\bar{c}_j. \tag{3.3}$$

Note that the product of a molecular concentration (molecules per unit volume) and a speed (distance per unit time) has the dimensions of a molecular flux (molecules per unit area per unit time). The leading fraction (1/4) arises from the fact that only half of all the randomly translating molecules are moving toward the surface (and so have a chance to collide at all with the surface); the other factor of two arises when one accounts for the speeds and directions of motion toward the surface. (Molecules moving nearly parallel to the surface, for instance, approach it only very slowly.) Using values at 0 °C for water vapor concentration ($n_{H_2O} \sim 2 \times 10^{23}$ m^{-3}) and molecular speed ($\bar{c}_{H_2O} \sim 500$ m s^{-1}), we calculate $\Phi_{imp,H_2O} \sim 10^{25}$ molec m^{-2} s^{-1}. Such a flux amounts to something like 10^{19} water molecules impinging each second onto every square millimeter of liquid surface. In general, molecular impingement fluxes are always enormous, even without any net condensation occurring.

Our understanding of the molecular impingement flux is refined if we make use of other properties of the gas. Fortunately, virtually all atmospheric gases behave nearly ideally, so an adequate approximation to the equation of state for our gas j is given by the ideal gas law:

$$p_j = n_jk_BT, \tag{3.4}$$

where p_j is the partial pressure of gas j. If we now solve Eq. (3.4) for n_j and insert the resulting expression, along with that for c_j from Eq. (3.2), into Eq. (3.3), we obtain the practical equation

$$\Phi_{imp,j} = \frac{p_j}{(2\pi m_jk_BT)^{1/2}}. \tag{3.5}$$

The mathematics very clearly shows us that, at a given temperature, the impingement of any gas onto any surface occurs at a rate that is directly proportional to the partial pressure of the gas in question (i.e., $\Phi_{imp,j} \propto p_j$). The impingement flux depends on temperature, but only weakly.

We are now in a position to develop criteria for phase equilibrium based on vapor concentration and partial pressure. In so doing, we will see key connections between basic physical principles and later applications in meteorology and the atmospheric sciences. Equation (3.1) states that equilibrium occurs when the molecular fluxes into and out of the liquid surface are equal, whereas Eq. (3.3) relates the molecular flux colliding with the surface to the vapor concentration. The ideal gas law (Eq. (3.4)) further shows how the concentration of vapor is related to its partial pressure, which then gives us Eq. (3.5) and a direct link between the molecular fluxes and the pressure exerted by the gas.

Let us now restrict attention to water as the substance of interest. Partly out of tradition and partly to avoid cumbersome subscripts on mathematical variables, we use the following identities:

Impingement flux of water vapor:	$\Phi_{imp,H_2O} \equiv \Phi_{cond}$
Concentration of water vapor:	$n_{H_2O} \equiv n_v$
Partial pressure of water vapor:	$p_{H_2O} \equiv e.$

With this terminology in place, we see that the rate with which water molecules condense onto a unit area of liquid water is

$$\Phi_{cond} = \frac{1}{4} n_v \bar{c}_v = \frac{e}{(2\pi m_v k_B T)^{1/2}} \tag{3.6}$$

where now $\bar{c}_v$ is the mean thermal speed of water molecules in the gas phase at temperature T and m_v is the mass of an individual water molecule. To reiterate an important point, the rate with which water molecules contact the surface (and presumed for now to enter the liquid phase) at a given temperature is proportional to the concentration (n_v), alternatively to the partial pressure (e) of water vapor just above the surface: $\Phi_{cond} \propto n_v \propto e$. The greater the concentration of water vapor molecules above a liquid water surface, the faster they enter the condensed phase.

Under the restriction that equilibrium prevails, we can go a step further and determine the outgoing, or evaporation flux. When, and only when, the liquid is in equilibrium with its vapor is it true that

$$\Phi_{evap} = \Phi_{cond,eq} = \frac{1}{4} n_{v,eq} \bar{c}_v = \frac{e_s}{(2\pi m_v k_B T)^{1/2}}. \tag{3.7}$$

Note the subtle distinction in terminology used in Eqs. (3.6) and (3.7): The special value of vapor concentration under equilibrium conditions is explicitly so designated (by subscript "eq"), and the equilibrium or "saturation" value of the water partial pressure is given the subscript "s" (by convention). The criteria for phase equilibrium in terms of macroscopic vapor concentration and partial pressure are $n_v = n_{v,eq}$ and $e = e_s$. These equilibrium values ($n_{v,eq}$ and e_s) are the reference states (or "cusps") alluded to earlier. They vary with temperature (primarily) and total pressure (slightly), but not with the magnitudes of n_v or e. They give us the conditions under which the molecular fluxes into and out of the condensed phase are equal: $\Phi_{cond} = \Phi_{evap}$.

Distinguishing between the macroscopic and molecular viewpoints, as we have been doing, is useful, often necessary, in science. Whether we consider the atmosphere as a whole, a cloud, or a single water droplet, each system can be studied either as a macroscopic entity (with our senses or suitable instrumentation), alternatively as a collection of individual molecules that interact with each other in diverse and complex ways.

3.2 Thermodynamic perspective of phase equilibrium

Thermodynamics, that branch of science concerned with energy in its various forms, regards nature from the macroscopic point of view. The temperature of an object, for instance, tells us how hot or cold it is, either by touch or as measured by a thermometer. Many molecules, through their incessant and random motions, contribute to the single measurement of temperature at a given instant; we do not speak of the temperature of individual molecules. The discipline of statistical mechanics, however, allows us to make connections between the actions of the molecules in a system and its macroscopic, thermodynamic properties. We emphasize that the thermodynamic temperature derived from measurements and used in calculations is closely related to the average kinetic energies of the molecules in the system. Thermodynamic principles are relevant whenever temperatures or energies are involved. Atmospheric motions, especially those resulting in changes of altitude, lead to changes in temperature and relative humidity, and so directly affect clouds. Changes in the state of matter also have energy consequences, which affect the rates of droplet growth and the heating of cloudy air parcels. Our intent here is to establish criteria for equilibrium in terms of various thermodynamic variables.

Let us again consider the two-phase system depicted in Fig. 3.1, but recognize that the conclusions we draw from this idealized example will apply equally well to water droplets in a cloud. In our scenario, liquid water is assumed to be in contact with the air, which contains water vapor that exchanges rapidly with the water in the liquid. We assume, for the time being, that the system is in equilibrium. The molecular fluxes are therefore balanced, so the water partial pressure is constant in time with the value $e = e_s$. The total pressure (p), too, is everywhere constant in time, so the system is also in mechanical equilibrium. The temperatures of the liquid and the air are both the same (namely T), so the system is in thermal equilibrium, as well. Liquid water in equilibrium with its vapor is a system at constant temperature and pressure.

The principles of thermodynamics are particularly useful for dealing with systems at constant temperature and pressure. The first law of thermodynamics states that energy overall is conserved, even as it changes from one form to another, so let us conceptually track a group of water molecules as they move from one part of a system to another, say from the liquid phase to the vapor phase. (At the same time, of course, an equal number of molecules moves in the opposite direction.) The number of molecules tracked is not important as long as it is large enough to yield statistically meaningful results and remains the same throughout the analysis. We choose to follow an Avogadro's number of molecules (one mole), which amounts to about 6×10^{23} molecules. Let this mole of water molecules occupy a physical volume v [units, $\mathrm{m^3\,mol^{-1}}$] and have an internal energy u [$\mathrm{J\,mol^{-1}}$]. The first law can therefore be written in energy form as

$$q = \frac{du}{dt} + e_s \frac{dv}{dt}, \tag{3.8}$$

where q [units, $\mathrm{J\,s^{-1}\,mol^{-1}}$] is the rate with which thermal energy (often called "heat") is added to the system (here, the group of molecules we are tracking). This heat transfer is related to the change in entropy (φ) of the system through the defining relationship

$$\frac{d\varphi}{dt} \equiv \frac{q}{T}, \tag{3.9}$$

where it is understood that the transfer is performed reversibly. Thus, we may write

$$T\frac{d\varphi}{dt} = \frac{du}{dt} + e_s \frac{dv}{dt}. \tag{3.10}$$

This relationship tells us that any thermal energy transferred into the molecules, thus causing the entropy to increase (LHS of Eq. (3.10)), will manifest itself partly as an increase in the internal energy (first term on RHS) and partly as work to expand the volume occupied by the molecules (second term on RHS).

Complete movement of our group of molecules from the liquid state to the vapor state can be represented mathematically by integration of Eq. (3.10):

$$T\int_L^V d\varphi = \int_L^V du + e_s \int_L^V dv. \tag{3.11}$$

Here, we have represented the liquid and vapor states by the letters L and V, respectively. Performing the integration, using the fact that T and e_s are constant during the integration, gives

$$T(\varphi_V - \varphi_L) = u_V - u_L + e_s(v_V - v_L). \tag{3.12}$$

If we algebraically regroup the variables by phase, we come up with an important relationship:

$$(u + e_s v - T\varphi)_V = (u + e_s v - T\varphi)_L\,. \tag{3.13}$$

Remarkably, the particular combination of thermodynamic variables inside the parentheses remains the same as our group of molecules moves from one phase to another. The individual values of internal energy (u), occupied volume (v), and entropy (φ) all change during the evaporation (or condensation) process, but the algebraic combination remains exactly the same (as long as the process occurs under equilibrium conditions).

The combination of variables shown on each side of Eq. (3.13) retains its overall value during any process for which temperature and pressure are fixed, although condensation is clearly the most important application for our study of clouds. The thermodynamics of constant-pressure processes shows that the first two terms of the combination represent the system enthalpy, which for our mole of water molecules is

$$h \equiv u + e_s v. \tag{3.14}$$

Thus, Eq. (3.13) could equally well have been written as

$$(h - T\varphi)_V = (h - T\varphi)_L\,, \tag{3.15}$$

which says that any increase in the enthalpy of the molecules upon going from one phase to another is offset by a commensurate increase in their entropy (times temperature). Recall, too, that the parenthetical quantity of Eq. (3.15) is defined as the Gibbs function

$$g \equiv h - T\varphi, \tag{3.16}$$

so we may equally well write

$$g_V = g_L. \tag{3.17}$$

As our mole of water molecules moves from the liquid state to the vapor state, the Gibbs function, in effect the energy available or "free" to do work, retains its value under equilibrium conditions.

It is this conservation principle, the constancy of this grouping of variables during constant-temperature, constant-pressure transformations, that gives importance to such abstract concepts as the Gibbs function. The equality of the Gibbs function in each phase, as specified mathematically in Eq. (3.17), is our simplest and most robust way of stipulating the criterion for phase equilibrium in a one-component system. For systems containing multiple components, the same concepts apply to the chemical potential (μ) of each component j: $\mu_j \equiv \partial G/\partial n_j$, where G is the total free energy of the system and n_j is the number of moles of j. Chemical potential, as the partial molar free energy, is more general than the Gibbs function, but they are equivalent when only a single compound (e.g., water) is present. Whenever the chemical potential of a substance is the same in any two phases, the system is in equilibrium with respect to that component. How systems respond away from equilibrium is discussed in the next chapter.

3.3 Phase relationships

Any one phase is always related to the other phases in which it is in contact in some physical sense. In effect, the one phase has to "know" that another phase exists. It is the physical contact or exchange of molecules at a phase boundary that constitutes the main mechanism for this transfer of "information". Molecules evaporating at the liquid–vapor interface, say, transfer not only their masses to the neighboring gas phase, but also some of their kinetic energies and momenta. We have already seen that the rates of molecular exchange are enormous under normal atmospheric conditions, so it is not surprising that a system like that portrayed in Fig. 3.1 is able to settle into a stable, equilibrium state within a relatively short period of time.

Let us continue our conceptual tracking of a single mole of water molecules as it moves from one phase to another. The discussion in the last section emphasized the power of thermodynamics to give us a concise criterion for equilibrium: when the Gibbs function (g) of a group of molecules remains constant during the phase change, then the system must be in equilibrium and the molecular fluxes across the interface are balanced. It is important to note, however, that the particular value of g that yields equilibrium is not fixed. The Gibbs function has the value it has only under that particular set of conditions. Under another set of conditions, g will have a different value.

The variation of the Gibbs function with conditions can be determined readily from the thermodynamic principles, and it then gives us new insight into the relationships between phases. The definitions of the Gibbs function ($g \equiv h - T\varphi$) and enthalpy ($h \equiv u + e_s v$) may be combined to give an expression for g in terms of basic thermodynamic variables:

$$g = u + e_s v - T\varphi, \tag{3.18}$$

where we use $e_s = p_{H_2O}$ because we are still restricting attention to water in phase equilibrium. In general, any or all of the variables on the RHS of Eq. (3.18) can vary with time, so we take the total derivative of g with respect to time t to determine the overall effect:

$$\frac{dg}{dt} = \frac{du}{dt} + e_s\frac{dv}{dt} + v\frac{de_s}{dt} - T\frac{d\varphi}{dt} - \varphi\frac{dT}{dt}. \tag{3.19}$$

However, the variations in one variable are not really independent of the variations in the other variables, so we must impose constraints on the relative behavior of the variables. The system cannot violate the principle of energy conservation, for instance. The first two terms on the RHS of Eq. (3.19) may be seen to be q, the rate of heat transfer (Eq. 3.8), which in turn is related to the rate of change of entropy through Eq. (3.10). The first two terms in Eq. (3.19) thus cancel with the fourth term, leaving us with a relatively simple expression for the change in g:

$$\frac{dg}{dt} = v\frac{de_s}{dt} - \varphi\frac{dT}{dt}. \tag{3.20}$$

This expression in effect summarizes the first and second laws of thermodynamics applied to a closed, one-component system experiencing changing partial pressure and temperature.

3.3.1 Effect of temperature on vapor pressure

The purpose of this analysis is to establish how strongly the partial pressure in equilibrium with liquid water (i.e., the "vapor pressure") increases with temperature. The only criterion that must be met to maintain equilibrium during a temperature increase is the equality of the free energies of the two phases. That is, as the molar free energy of the liquid phase increases, that of the vapor phase must increase at the same rate. We express this stipulation mathematically by the equality,

$$\frac{dg_V}{dt} = \frac{dg_L}{dt}. \tag{3.21}$$

It follows then, when Eq. (3.20) is applied respectively to the vapor (V) and liquid (L) phases, that

$$v_V \frac{de_s}{dt} - \varphi_V \frac{dT}{dt} = v_L \frac{de_s}{dt} - \varphi_L \frac{dT}{dt}. \tag{3.22}$$

Combining terms and solving for the differential ratio, de_s/dT, gives the rate at which the vapor pressure increases with temperature. We thus gain the so-called Clausius–Clapeyron equation:

$$\frac{de_s}{dT} = \frac{\varphi_V - \varphi_L}{v_V - v_L} \equiv \frac{\Delta\varphi}{\Delta v}. \tag{3.23}$$

Note that the vapor pressure increases with temperature at a rate that is directly proportional to the ratio of the changes in molar entropy ($\Delta\varphi$) and molar volume (Δv).

Phase changes of atmospheric interest invariably imply shifts in system properties. A mole of water molecules in the vapor phase is scattered widely and occupies a large volume, whereas those same molecules in the liquid phase are bound to each other and so occupy a relatively small volume of space. Thus, we can treat the vapor as ideal and use (in Eq. (3.23)) the approximation

$$\Delta v \equiv v_V - v_L \doteq v_V = \frac{RT}{e_s}. \tag{3.24}$$

The molar volume of vapor is large compared with that of liquid, in fact, about a factor of 10^5 larger. This approximation can be used in Eq. (3.20) to show that under constant-temperature conditions the Gibbs function varies simply as

$$\frac{dg}{dt} \doteq RT \frac{d \ln e_s}{dt}. \tag{3.25}$$

This relationship between free energy and partial pressure is useful for quantifying the formation and growth of cloud particles.

Just as the physical state of our mole of water molecules has changed upon evaporating from liquid to vapor, so have their energies relative to one another. Physical changes of state have energy consequences, quite simply because molecular-scale forces of attraction are involved. The molecules in the liquid phase, unlike those in the vapor phase, are trapped within a small volume because they are bound to each other, constrained in their movement by the electrostatic forces existing between neighboring molecules (see Section 2.1). Molecules in the liquid phase must overcome the forces of attraction to their neighbors before they can evaporate. In other words, thermal energy must be added to the system, not to increase its temperature, rather to boost the molecules from their bound, liquid state to their unbound, vapor state. The amount of energy required to evaporate a unit quantity (e.g., one mole) of molecules from the liquid is the enthalpy of vaporization, or the latent heat of vaporization, l_v. Evaporation is an endothermic process, because energy must be added, the heating rate (q) and entropy change ($d\varphi/dt$) being positive. The heat transfer is reversed for condensation (an exothermic process), so we get the general result,

$$q = T\frac{d\varphi}{dt} \begin{cases} < 0, \text{ for condensation} \\ > 0, \text{ for evaporation} \end{cases} . \tag{3.26}$$

Integration of the heating rate for a mole of evaporating molecules is the latent heat: $l_v \equiv \int_L^V q dt = h_V - h_L$, where h_V and h_L are the molar enthalpies of the vapor and liquid, respectively. The molar entropy change for vaporization is therefore

$$\Delta\varphi \equiv \varphi_V - \varphi_L = \frac{l_v}{T}. \tag{3.27}$$

Molecules in the vapor phase have larger entropy than those in the liquid phase, so $\Delta\varphi > 0$.

The equilibrium vapor pressure of liquid water increases with temperature because of molecular bonding. A higher temperature means that the molecules have greater kinetic energies, vibrate more rapidly, and so find it easier to escape the bondage to their neighbors. Substances, such as water, have relatively large bonding energies and therefore have vapor pressures that increase strongly with increasing temperature; the water molecules in the liquid need fairly high temperatures in order to escape from the surface. The actual rate of increase is obtained quantitatively by inserting the expressions for the changes in molar volume (Eq. (3.24)) and molar entropy (Eq. (3.27)) into Eq. (3.23). We thus arrive at the approximate form of the Clausius–Clapeyron equation for liquid water,

$$\frac{de_s}{dT} \doteq \frac{l_v e_s}{RT^2}, \tag{3.28}$$

or (because $d \ln x = dx/x$ for any positive value of x),

$$\frac{d \ln e_s}{dT} \doteq \frac{l_v}{RT^2}. \tag{3.29}$$

The actual vapor pressure at any specified temperature is found by integration of Eq. (3.29). The process of integration is straightforward when l_v is kept constant, an approximation that is often reasonable for the range of temperatures of atmospheric interest. The constant of integration must come from measurement, the most convenient value being the vapor pressure $e_0 = 611$ Pa in equilibrium with a mixture of pure liquid water and ice at a total pressure of 1 atm, which occurs at the so-called "ice point" $T_0 = 273.15$ K (0 °C). Integration of the Clausius–Clapeyron equation from T_0 to any other T, where the vapor pressure is $e_s(T)$, gives

$$e_s(T) \doteq e_0 \exp\left(\frac{l_v}{R}\left(\frac{1}{T_0} - \frac{1}{T}\right)\right). \tag{3.30}$$

Sometimes, it is convenient to specify the difference in temperature $\Delta T \equiv T - T_0$, in which case

$$e_s(\Delta T) \doteq e_0 \exp\left(\frac{l_v}{RT_0} \cdot \frac{\Delta T}{T}\right). \tag{3.31}$$

Either of these forms gives the relationship between the saturation vapor pressure and the temperature. More precise expressions for $e_s(T)$ can be obtained by taking into account the dependence on temperature of latent heat itself and other parameters. A convenient and reasonably accurate formula for calculating the vapor pressure of liquid water within about 20 °C of T_0 is the Magnus equation,

$$e_s(\Delta T) \doteq e_0 \exp\left(A_L \cdot \frac{\Delta T}{T - B_L}\right), \tag{3.32}$$

where $A_L \simeq 17.2$ and $B_L \simeq 36$ K. For theory, we typically designate the ideal, perfectly accurate function as $e_s(T)$.

All first-order phase changes, those involving changes in system energy, can be described by the same fundamental principles. Simply by replacing the enthalpy of vaporization (l_v) with the enthalpy of sublimation (l_s) in Eq. (3.30), we obtain the equilibrium vapor pressure with respect to ice, $e_i(T)$:

$$e_i(T) = e_0 \exp\left(\frac{l_s}{R}\left(\frac{1}{T_0} - \frac{1}{T}\right)\right). \tag{3.33}$$

Or, in terms of the temperature deviation from the ice point (T_0),

$$e_i(\Delta T) = e_0 \exp\left(\frac{l_s}{RT_0} \cdot \frac{\Delta T}{T}\right). \tag{3.34}$$

Note that water molecules are on average bound more tightly in the ice phase than in the liquid phase, so the enthalpy of sublimation is larger than that of vaporization. In particular, because energy must be conserved,

$$l_s = l_v + l_f, \tag{3.35}$$

where l_f is the enthalpy of fusion. Thus, the equilibrium vapor pressure with respect to ice increases with temperature more rapidly than does that with respect to the liquid. The Magnus equation for ice (for $\Delta T < 0$) is

$$e_i(\Delta T) \doteq e_0 \exp\left(A_I \cdot \frac{\Delta T}{T - B_I}\right), \tag{3.36}$$

where $A_I \simeq 21.9$ and $B_I \simeq 7.6$ K.

We can use the expression from Eq. (3.35) to obtain a relationship between the respective vapor pressures in equilibrium with the liquid and solid phases (when the temperature is below the melting point). Start by taking the ratio of the vapor pressures and note that e_0 cancels out. We then obtain

$$\frac{e_s(T)}{e_i(T)} = \exp\left(\frac{l_f}{R}\left(\frac{1}{T} - \frac{1}{T_0}\right)\right). \tag{3.37}$$

When the temperature is below the melting point, it is common to define the "supercooling", the negative departure of the temperature from the ice point: $\Delta T_s \equiv T_0 - T$ $(= -\Delta T)$. Expressed in terms of ΔT_s, the ratio of vapor pressures becomes

$$\frac{e_s}{e_i} = \exp\left(\frac{l_f}{RT_0} \cdot \frac{\Delta T_s}{T}\right). \tag{3.38}$$

For small supercoolings, the exponential function may be expanded in a Taylor series, leading to

$$\frac{e_s}{e_i} \doteq 1 + \frac{l_f}{RT_0} \cdot \frac{\Delta T_s}{T}. \tag{3.39}$$

Note that the vapor pressure with respect to liquid water is always greater than that with respect to ice, a reflection of the fact that the molecules in liquid are less tightly bound than those in ice. This distinction between the vapor pressures over liquid water and ice is key to understanding precipitation formation in "cold" clouds.

Improved approximations to the vapor-pressure functions take the temperature dependence of latent heats into account. Our realization that latent heat of vaporization, for instance, is a measure of the mean energy bonding molecules together in liquid water helps us understand this additional temperature dependence. As temperature increases, the molecules in the liquid vibrate more relative to one another, causing them to be trapped higher up on the interaction potential (Fig. 2.7 in Chapter 2). The effective depth of the potential well, the difference in energy between the bound (liquid) state and the unbound (gaseous) state, thus decreases with increasing temperature. This decreasing difference in the potential energies of the molecules in the different phases is just the latent heat, so we calculate the rate of change of latent heat by recalling the definitions of latent heat as a difference in enthalpies ($l_v \equiv h_V - h_L$) and of specific heats as the rates of change of enthalpies ($c_{pv} \equiv dh_V/dT;\ c_l \equiv dh_L/dT$):

$$\frac{dl_v}{dT} = \frac{dh_V}{dT} - \frac{dh_L}{dT} = c_{pv} - c_l. \tag{3.40}$$

The specific heats are only weak functions of temperature (they decrease by less than 1% over the range of atmospheric temperatures), so Eq. (3.40) yields a good approximation to the rate of change of l_v with T. Thus, by assuming constant specific heats, we can integrate Eq. (3.40) and obtain an expression for the temperature dependence of the latent heat of vaporization:

$$l_v = l_{v0} + \left(c_{pv} - c_l\right)(T - T_0), \tag{3.41}$$

where l_{v0} is the value at T_0 (typically taken to be the ice point, 273.15 K). Similar relationships hold for all phase transitions, so we obtain

$$e_s(T) \doteq e_0 \exp\left(\frac{l_{v0} + \left(c_l - c_{pv}\right) T_0}{R}\left(\frac{1}{T_0} - \frac{1}{T}\right) - \frac{c_l - c_{pv}}{R} \ln\frac{T}{T_0}\right) \tag{3.42}$$

and

$$e_i(T) \doteq e_0 \exp\left(\frac{l_{s0} + (c_i - c_{pv})\, T_0}{R}\left(\frac{1}{T_0} - \frac{1}{T}\right) - \frac{c_i - c_{pv}}{R} \ln \frac{T}{T_0}\right). \tag{3.43}$$

These expressions improve calculations involving vapor pressures, but we must realize that they are still only approximations to the perfect functions, $e_s(T)$ and $e_i(T)$.

The theory leading to the Clausius–Clapeyron equation and the resulting vapor-pressure functions, $e_s(T)$ and $e_i(T)$, underpins much of atmospheric physics. As we have seen, thermodynamics lets us quantify the average behavior of molecules bound to their neighbors in the liquid (and ice) state. Given sufficient thermal energy, a molecule in the surface of a condensed phase can occasionally break its bonds and escape into the vapor phase. The higher the temperature, the stronger the vibrations, the weaker the bonding, and the greater are the chances that the molecule can escape the condensed phase and enter the vapor phase. Equilibrium vapor pressures must always be larger at higher temperatures in order to compensate for this greater escaping tendency. Similar reasoning lets us realize that the vapor pressure of supercooled liquid water must always be larger than that of ice at any given temperature (below T_0).

3.3.2 Pressure melting and the phase diagram

An equilibrium also exists between the liquid and solid phases and so must also be considered. Even when the vapor phase is not explicitly present, as at the interface of ice and liquid water, we can still use vapor pressures because of the zeroth law of thermodynamics. In this case, we use the symbol e_m for the hypothetical vapor pressure of the melt water in equilibrium with ice. However, we cannot make the approximation used earlier to derive the Clausius–Clapeyron equation for liquid water (Eq. (3.28)). The volumes occupied by a given number of water molecules in the liquid and solid phases differ by only about 8%. Thus, we must start with the full differential form of the Clausius–Clapeyron equation:

$$\frac{de_m}{dT} = \frac{l_f}{T\,(v_L - v_S)}, \tag{3.44}$$

where v_S is the molar volume of the solid. Water is unusual in the sense that it expands upon freezing (ice floats), so $v_L < v_S$. This volume difference alone causes the slope of e_m versus T to be negative and e_m to decrease with increasing temperature. The ice point (at 1 atm) is thus about 0.01 K lower than the triple point of water ($T_t = 273.16\,\text{K}$), which is a small, but important magnitude. Ice melts in response to an increase in pressure, although the applied pressure must be large to make the effect noticeable. The fact that ice floats on its melt has important climatic consequences.

The various phases of water are all related to each other in systematic ways, as summarized in a phase diagram. Thermodynamic theory allows us to apply fundamental principles

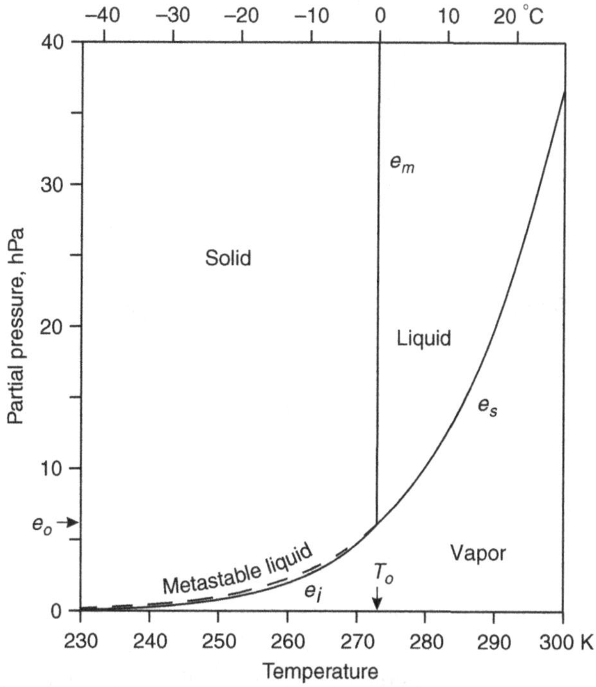

Figure 3.2 Phase diagram of water.

of nature and determine how the equilibria depend on temperature and pressure. The result is the various forms of the Clausius–Clapeyron equation. It also proves useful to plot each of the vapor-pressure functions against temperature, as shown by the phase diagram in Fig. 3.2. Both the e_s and the e_i curves increase exponentially with temperature, an attribute caused by thermally induced vibrations that break the bonds holding water in the condensed phases. When drawn to scale, we see little difference between the e_s and e_i curves for $T < T_0$, but the difference has profound effects on cloud microphysics. Also, the e_m curve looks to be vertical, but it slopes gently "backward" and identifies the pressure-melting effect. The phase diagram of water graphically summarizes the important relationships between the phases of this unique substance. We will make extensive use of the phase diagram throughout this book.

3.4 Interfaces

Interfaces, the physical boundaries between phases, require special attention. Not only do the molecular properties of interfaces differ from those in the bulk, but all phase transformations take place at interfaces. The various interfaces of interest are schematically portrayed in Fig. 3.3. Note that the solid phase itself can have internal interfaces, such as

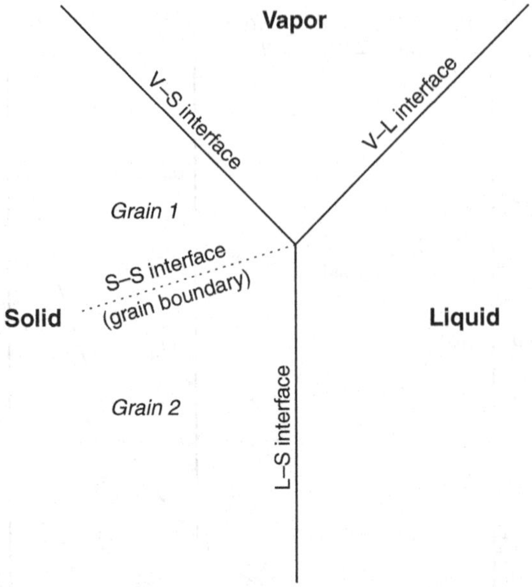

Figure 3.3 Schematic depiction of the various phases and interfaces of water.

the boundaries between individual grains in polycrystalline ice. Our aim here is develop a relationship (the Kelvin equation) that is used frequently in cloud physics.

Cloud droplets are spherical because of surface tension, the asymmetric force acting at the liquid–gas interface. Surface tension is real, a result of the differing attractions between molecules in two adjacent phases. The molecules in the interface lack the symmetry of forces experienced by interior molecules. The collective action of the molecules along an interface manifests itself as a tension that pulls the molecules in toward the condensed phase. If a water droplet were to become distorted, for instance, by becoming elongated in one direction relative to another, the surface forces would exert a greater inward force along the elongated direction and so restore the droplet to the spherical shape. We see the effect of surface tension when water spills onto a hydrophobic (e.g., waxed) surface and beads up. However, other forces, such as those due to gravity or molecular attractions with another substance, are often present, so large drops and puddles are seldom spherical. Surface tension nevertheless serves to minimize the surface area and gives rise to a surface free energy. Surface tension may be expressed either as a force per unit distance [$\mathrm{N\,m^{-1}}$] or equivalently as an energy per unit surface area [$\mathrm{J\,m^{-2}}$].

3.4.1 Effect of curvature on vapor pressure

Curvature, measured by the reciprocal of radius, is the deviation from a flat surface. Curvature effects are important in cloud physics because all liquid droplets are round and so require a larger partial pressure to keep them from evaporating away.

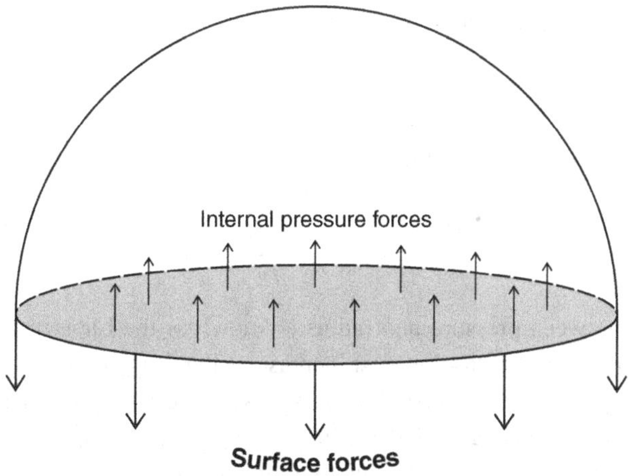

Figure 3.4 Cross section of a liquid drop showing the forces involved in maintaining mechanical equilibrium. Bold arrows identify tangential forces arising from surface tension; thin arrows indicate the internal pressure forces acting on the cross-sectional area (shaded).

An explanation of the curvature effect can be appreciated by considering molecular interactions further. Anything that weakens the intermolecular bonds holding the liquid together lets the molecules escape more readily. Curvature affects the way surface tension acts on the bulk liquid. As we shall see shortly, the greater the curvature, the greater the pressure on the bulk liquid and the higher the vapor pressure.

Thermodynamic principles give us the quantitative tools needed to determine the effect of curvature on vapor pressure. The most important application for cloud physics is the increased vapor pressure over small liquid droplets, so we restrict attention now to liquid water in contact with air containing water vapor. The surface tension arising from the liquid–vapor interface acts to increase the total pressure inside the drop by an amount Δp_{drop} above that needed to balance the normal atmospheric pressure. This extra internal pressure is not the same as the increase in the vapor pressure that we seek, but these two pressures are nevertheless related.

The excess internal pressure Δp_{drop} arising from a curved surface can be calculated by considering the forces needed to maintain a drop in mechanical equilibrium. A drop is stable when the internal pressure force (pushing outward) balances the surface-tension force (pulling the surface inward). We analyze the situation by viewing a liquid droplet as two identical hemispheres, one of which is shown in Fig. 3.4. The internal pressure acts over the entire cross-sectional area A_c of the drop, whereas the surface tension σ_{LV} acts in an opposing sense on the circumference L_c of the drop. These forces must balance, so

$$\Delta p_{drop} \cdot A_c = \sigma_{LV} \cdot L_c. \tag{3.45}$$

For a spherical drop, we therefore have

$$\Delta p_{drop} \cdot \left(\pi r_d^2\right) = \sigma_{LV} \cdot (2\pi r_d). \tag{3.46}$$

Upon solving for Δp_{drop}, we find the pressure increment due to the liquid–vapor surface tension to be

$$\Delta p_{drop} = \frac{2\sigma_{LV}}{r_d}. \tag{3.47}$$

This relationship between pressure and radius is known as the Laplace equation. We see that the internal pressure of a droplet must increase as the droplet size decreases.

An increase in pressure on a system, regardless of the cause, impacts its thermodynamic properties by increasing the chemical potential of all components in the system. The chemical potential μ for a pure, one-component system is just the free energy per mole (g), so the change in chemical potential, when temperature remains constant, is related to pressure change by

$$\frac{\partial \mu}{\partial t} = v\frac{\partial p}{\partial t}, \tag{3.48}$$

where v is the molar volume of the substance (cf., Eq. (3.20)). A fundamental property of chemical potential is its equality in any two phases that are in equilibrium with each other. Thus, any change in chemical potential in the one phase must be the same as that in the other. Applying the thermodynamic relationship (Eq. (3.48)) to each phase, we obtain

$$v_L\frac{\partial p}{\partial t} = v_V\frac{\partial e_{eq}}{\partial t}. \tag{3.49}$$

Note that p is the total pressure acting on the liquid (from whatever cause), while e_{eq} is the equilibrium vapor pressure. We thus gain the general result, called the Poynting equation:

$$\frac{\partial e_s}{\partial p} = \frac{v_L}{v_V}. \tag{3.50}$$

The Poynting equation describes the effect of total pressure on the equilibrium vapor pressure. The volume v_L occupied by a mole of water molecules in the liquid phase is much smaller than the volume v_V occupied by the same number of molecules in the vapor phase, so the dependence of vapor pressure on total pressure is relatively weak (about one part in 10^5). Nevertheless, we shall see that this small effect has profound consequences on the formation of cloud droplets.

In order to calculate the increase in chemical potential in a liquid drop of radius r_d, we identify v in Eq. (3.48) with the molar volume of liquid water, the inverse of the molar density (n_L): $v_L = 1/n_L$. Liquid water is relatively incompressible, so the integration of Eq. (3.48) amounts to simple multiplication:

$$\Delta\mu = v_L \Delta p_{drop}. \tag{3.51}$$

Thus, the chemical potential of the liquid water in a drop of radius r_d is, from Eq. (3.47),

$$\Delta\mu = \frac{2\sigma_{LV}}{n_L r_d}. \tag{3.52}$$

It is this increment of chemical potential, brought about by the surface tension σ_{LV}, that determines how much the vapor pressure exceeds "saturation", that value in equilibrium with a flat water surface (e_s).

On the vapor side of the liquid–vapor interface, this extra chemical potential manifests itself as an increase in the partial pressure of water vapor needed to prevent the droplet from either evaporating or growing. The equilibrium vapor pressure e_r over a pure droplet of radius r_d is found by applying the same thermodynamic principle used to calculate the extra chemical potential due to the surface tension, only now we employ the volume and pressure variables appropriate to the vapor (rather than to the liquid). The volume (v in Eq. (3.48)) is now the molar volume of vapor, v_V, and the pressure is the equilibrium vapor pressure, initially in equilibrium with a flat surface (e_s), then in equilibrium with the droplet (e_r). Vapor, unlike a liquid, is compressible, so the increase in vapor pressure (arising from the finite curvature of the droplet) causes the molar volume to decrease, according to the ideal gas law: $v_V = RT/e$. A full integration of Eq. (3.48) is therefore necessary to obtain the overall change in chemical potential:

$$\Delta\mu = \int_{e_s}^{e_r} v_V dp = RT \int_{e_s}^{e_r} \frac{dp}{p}. \tag{3.53}$$

Thus,

$$\Delta\mu = RT \ln\left(\frac{e_r}{e_s}\right). \tag{3.54}$$

We now need only to equate the chemical potential increments from Eqs. (3.52) and (3.54):

$$\frac{2\sigma_{LV}}{n_L r_d} = RT \ln\left(\frac{e_r}{e_s}\right), \tag{3.55}$$

then solve for the vapor-pressure ratio to obtain the Kelvin–Thomson equation:

$$S_r \equiv \frac{e_r}{e_s} = \exp\left(\frac{2\sigma_{LV}}{n_L RT r_d}\right). \tag{3.56}$$

We thus see how surface tension (σ_{LV}) quantitatively affects the equilibrium vapor pressure of a droplet of specified radius (r_d). The equilibrium saturation ratio must increase, because of the pressure put on the liquid by the surface forces, as the size of the droplet decreases. The exact relationship between S_r and r_d given by Eq. (3.56) is plotted in Fig. 3.5. The saturation ratio of submicron droplets can be many times that of a flat surface. As long as the droplet radius is not too small, the exponential factor can be linearized, yielding the convenient approximation,

$$S_r = \frac{e_r}{e_s} \doteq 1 + \frac{A_K}{r_d}, \tag{3.57}$$

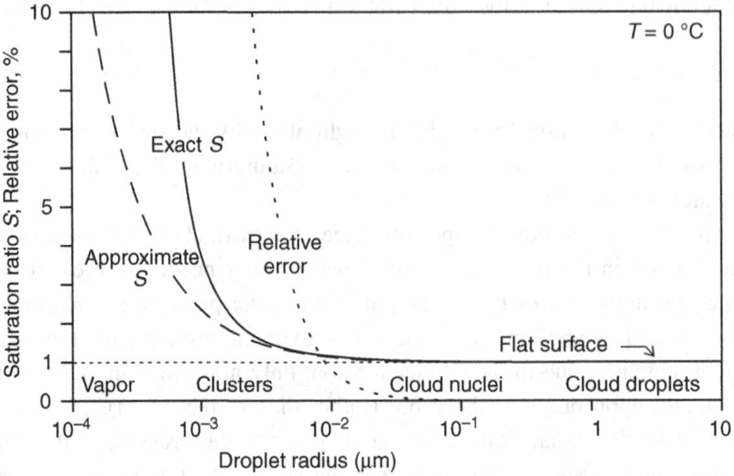

Figure 3.5 Exact and approximate dependences of saturation ratio S on droplet radius at temperature $T = 0\,°\mathrm{C}$.

where $A_K \equiv 2\sigma_{LV}/n_L RT$. The approximate calculation gives reliable estimates (fractions of a percent, dotted curve) for cloud droplets and cloud nuclei, but the exact equation (3.56) is needed for clusters.

3.4.2 Surface melting

The solid–liquid system, too, is affected by the same principles that cause the equilibrium vapor pressure of small water droplets to be larger than that of a flat surface of water. Consider Fig. 3.6, a representation of the junction of three grain boundaries in a mass of polycrystalline ice (such as in a hailstone or a snow pack). Each of the grains of ice is in contact with the liquid water in the vein, a tortuous channel extending some tens of meters in length for every cubic centimeter of compacted snow. Here, equilibrium means the tendency for the ice phase to melt as rapidly as the liquid phase freezes along each interface. At a temperature of $T_0 = 0\,°\mathrm{C}$, the thermodynamic melting point of bulk ice, the solid–liquid interface would have to be flat (meaning, of course, that three such grains could not meet as depicted in the figure). At all temperatures $T < T_0$, the solid–liquid interface must be curved (as depicted), with the interface being convex toward the liquid phase (just the opposite of the liquid–vapor interface). A curved surface of ice melts more readily (i.e., at a lower temperature) than does a flat ice surface. Note that the liquid water in contact with the ice grains is not supercooled in the usual thermodynamic context despite temperatures below $0\,°\mathrm{C}$; water in veins does not freeze, for if it did, the interface of the ice grains would become more sharply curved and simply melt back to their equilibrium forms. The liquid and solid phases at grain boundaries and along veins maintain a stable equilibrium with each other. The precise nature of the equilibria is, however, sensitive to the

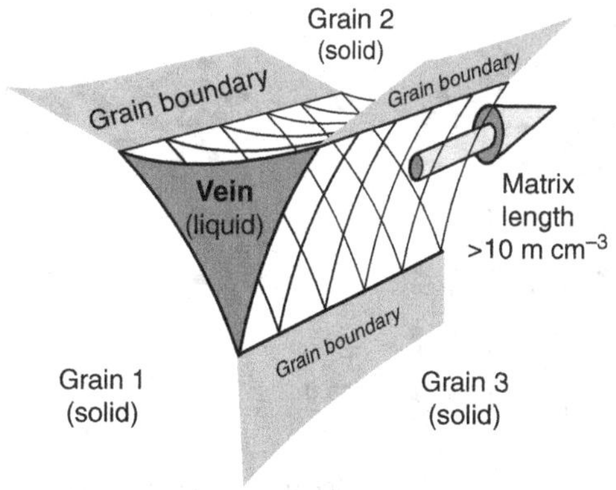

Figure 3.6 Perspective of grain boundaries at a triple junction in polycrystalline ice. Adapted from Mader (1992).

presence of trace chemicals. Veins, in particular, have been thought to harbor substances that may be soluble in liquid water, but which cannot enter the ice lattice directly.

The solid–vapor interface, that between ice and air, has been actively studied, largely because of its importance in the formation of snow in clouds and the electrification of clouds. Early observations by Faraday and by Tyndall suggested that the surface of ice and snow must be more liquid-like than solid. Recall how snow at temperatures just below the melting temperature can be compacted easily into snowballs, but not when the snow is really cold. Cold snow retains its powdery nature despite attempts to compact it. Warm snow, by contrast, seems to solidify as soon as it touches another snow mass, as if the surfaces simply refreeze on contact. This tendency for one solid substance to stick to another piece of the same material is called sintering. Sintering is increasingly favored as the temperature approaches the melting point (from the cold side). The surface of snow acts as if it has a liquid-like coating as long as the surface is exposed to air, but returns to the solid state on contact with another surface. The surface of ice is said to be covered by a quasi-liquid layer (QLL), the thickness of which increases markedly as the melting point is approached.

The bare surface of ice is disordered relative to the bulk. The long-range order that so characterizes a crystal breaks down near the surface. Various hypotheses have been put forth to explain the phenomenon, but the most compelling ideas are based on the relative strengths of the intermolecular bonds. Molecules in the bulk find neighbors in every direction, whereas those at the surface are missing bonding opportunities on the vapor side. Interior molecules are more tightly bound than surface molecules. At any given temperature within a few degrees of the melting point, the surface molecules readily break bonds and rotate, possibly even migrate away from their normal lattice positions. The well-defined order of the lattice breaks down near the surface, as shown by Fig. 3.7. Deep in the interior, well away from the surface, the molecules are locked into their normal lattice positions, as molecular bilayers that parallel the basal plane (left) and as the hexagonal

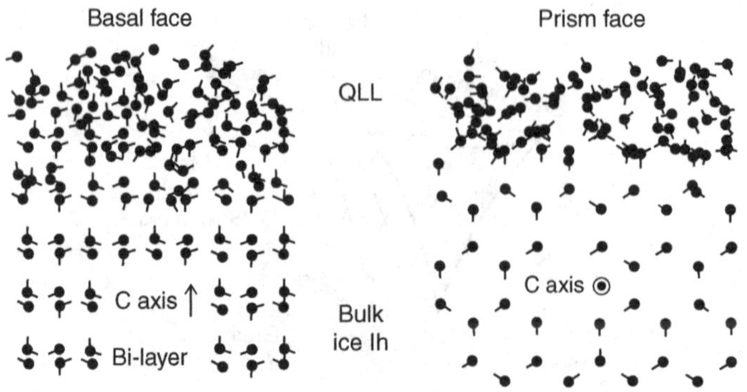

Figure 3.7 Results of molecular simulations of ice Ih showing molecular disorder at the solid–vapor interface. Left: Cross section parallel to the c axis (arrow up). Right: Cross section perpendicular to the c axis (arrow head on). Adapted with permission from Furukawa and Nada (1997). Copyright 1997 American Chemical Society.

pattern characteristic of ice Ih (right). Closer to the surface (upward in the figure), the instantaneous positions become progressively less aligned and subject to vibrations of greater and greater amplitude. The uppermost molecules have lost most of their long-range order and so act almost liquid-like. The depth of the quasi-liquid layer, under the conditions imposed for these calculations, is thicker on the basal face than on the prism face. However, the relative thicknesses of the QLL varies differently with temperature on these two main faces, so generalizations are to be made cautiously. Nevertheless, the molecular restructuring of the ice surfaces at least partially accounts for the observed alternation with temperature of the primary habits of snow crystals (see Section 1.2). Macroscale phenomena commonly have roots in the behavior of molecules.

3.5 Multicomponent systems

Water in our natural world is never pure. The oceans are noted for their saltiness, and ground water is often "hard" because of the dissolved minerals it contains. However, even the cleanest rainwater contains "solute", trace substances picked up from the air and dissolved into the cloud water at some point during its existence in the atmosphere. Perhaps the cleanest form of atmospheric water is "pristine" snow, the light, fluffy kind one finds on cold winter days in remote regions. Even then, careful measurements reveal a plethora of trace substances contained within the snow. All forms of atmospheric water contain many components in addition to water itself.

Multicomponent systems containing water are termed "aqueous" solutions. Even the air itself can be considered an aqueous solution because water vapor is always associated with many other compounds, those that constitute the atmosphere. However, we will here restrict our attention to liquid solutions, those in which solutes affect the properties of liquid water and hence the formation and properties of clouds.

The need to consider the non-aqueous components of the atmosphere can be appreciated through consideration of common phenomena. For instance, the white hazy sky that often prevails in summertime over industrialized regions is caused by the scattering of sunlight by the many small particles constituting the aerosol in the atmospheric boundary layer; pure water droplets could not exist under such subsaturated conditions. Those aerosol particles contain complex combinations of mineral acids (e.g., sulfuric and nitric acids) and organic compounds dissolved in the suspended haze droplets. Those same aerosol particles may later be lifted into a cloud, where they may serve as the sites on which active condensation takes place. It is the profound effects of the solute on the properties of liquid water that give multicomponent systems such importance in cloud physics.

3.5.1 Effect of solute on vapor pressure

The effect a solute has on the equilibrium properties of liquid water depends primarily on the amount of solute present, only slightly on the nature of the solute itself. Salts, sugar, or acids dissolved in water all lower the melting point of ice more or less to the same degree. This effect of a solute on the melting point is, for instance, relied upon when making homemade ice cream the old-fashioned way (i.e., without refrigeration); the salt lowers the melting point and lets the ice–water–salt mixture cool the ingredients enough to let the cream freeze. Of particular relevance to clouds, however, the presence of solute in water lowers the equilibrium vapor pressure of the water at any given temperature. These seemingly independent effects have a common origin at the molecular level.

The magnitude of impact a solute has on the properties of liquid water depends on how much is present. A small amount of solute in a given amount of water (i.e., a dilute solution) has a weak effect, whereas a large amount of solute in the same amount water has a stronger impact. Such a tendency was first put into quantitative terms by Raoult towards the end of the nineteenth century. He noted that the equilibrium vapor pressure of a solvent is reduced in proportion to the mole fraction of solute contained in the solution. (The solute mole fraction is the ratio of solute molecules to total molecules in solution.)

We can express Raoult's proportionality in the following mathematical way. Let the solution be composed of water (the solvent; subscript "w") and a generic non-volatile solute (subscript "s") that is present in fractional amount (solute mole fraction) x_s. Thus, if we continue to use e_s to represent the equilibrium vapor pressure of pure bulk water at a particular temperature and let the symbol Δe express the amount that the vapor pressure is reduced by the specified amount of solute, then Raoult's law becomes

$$\frac{\Delta e}{e_s} = x_s. \tag{3.58}$$

We see immediately from this equation that the magnitude of vapor pressure reduction varies directly with the amount of solute in the liquid. If no solute is present ($x_s = 0$), then no effect is observed (i.e., $\Delta e = 0$). At the other (though unrealistic) extreme, were the solute mole fraction to be unity (meaning the solution was all solute and contained no water), the depression of the vapor pressure would equal the vapor pressure itself (i.e., the water vapor pressure would vanish). Between these extremes, the equilibrium vapor pressure of the water in the solution decreases linearly, according to this law, as the fraction of non-water molecules (or ions) increases.

Raoult's law can be put into practice with a little mathematical manipulation. Let us start by being precise about the meaning of the vapor pressure depression Δe. The word "depression" implies a difference, as does the symbol delta (Δ). Thus, we define

$$\Delta e \equiv e_s - e_{sol}, \tag{3.59}$$

where the symbol e_{sol} is used to represent the equilibrium vapor pressure of water over the solution. It is important to recognize that $e_s = e_s(T)$ is the vapor pressure the liquid would have at temperature T, were it to be pure and the surface flat. The liquid is not pure, however, so e_s is to be interpreted as a reference value, one that can always be calculated even when the pure state does not exist. The only real, measurable vapor pressure is that over the solution, e_{sol}. Substitution of Eq. (3.59) into Eq. (3.58), followed by a bit of algebra, yields the equilibrium saturation ratio (or relative humidity),

$$S_{sol} \equiv \frac{e_{sol}}{e_s} = 1 - x_s. \tag{3.60}$$

This form of Raoult's law lets us calculate the saturation ratio required to keep the solution stable, neither increasing nor decreasing in mass. The equilibrium vapor pressure over an ideal solution can now be expressed in terms of temperature and solute mole fraction. From Eq. (3.60) we get the functional form,

$$e_{sol}(x_s, T) = (1 - x_s)\, e_s(T). \tag{3.61}$$

Furthermore, because the solution is comprised of only two kinds of substance, water and not-water (i.e., solute), the water mole fraction $x_w = 1 - x_s$ and Eq. (3.61) can also be written as

$$e_{sol}(x_w, T) = x_w e_s(T). \tag{3.62}$$

This form of Raoult's law, in which we see that the equilibrium vapor pressure of water over an aqueous solution is directly proportional to the mole fraction of water, proves useful for generalizing our treatment of solute effects in non-ideal solutions. For now, note that, whereas the equilibrium vapor pressure of pure water depends only on the temperature, the equilibrium vapor pressure of a solution depends on both the temperature and the relative amount of solute in solution. Solutions are more complicated than pure substances.

Equation (3.61) may be plotted to illustrate the combined effects of temperature and solute mole fraction graphically, as shown in Fig. 3.8. The vertical axis is the vapor pressure calculated from Eq. (3.61) for the temperature and solute mole fraction specified on

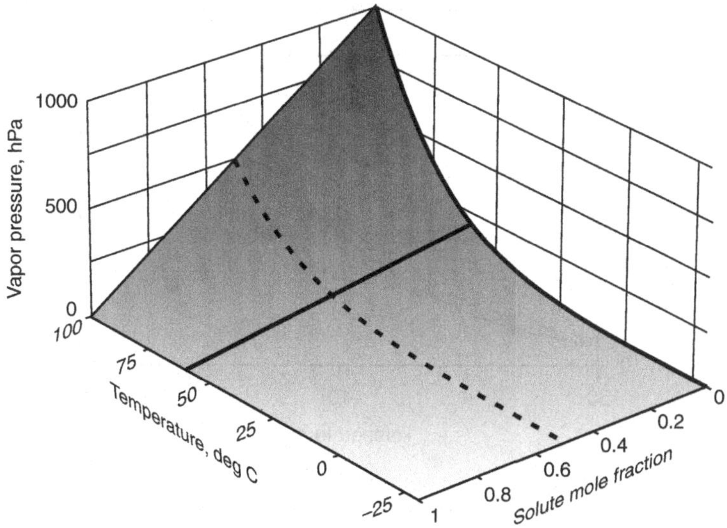

Figure 3.8 Effects of solute and temperature on the equilibrium vapor pressure of liquid water.

the abscissas. The vapor pressure over a solution is now seen to be a surface because it depends on two independent variables (T and x_s). Along the "back" side of the graph, where $x_s = 0$, we see the exponential rise of vapor pressure with temperature for pure water. In this limiting case, the system contains only water, so the function reverts back to a curve (our familiar Clausius–Clapeyron relationship). So, too, whenever the solute amount is fixed, but the temperature is allowed to vary, the dependence of vapor pressure on temperature over a solution maintains the same one-dimensional form, but all values are reduced by the factor $(1 - x_s = x_w)$. Thus, as shown by the dashed curve in Fig. 3.8, if half the molecules in the solution were solute, the equilibrium vapor pressure would be just half of its pure-water value. At any given temperature, on the other hand, the vapor pressure follows the linear trend shown by the straight line (at about 60 °C). The greater the solute concentration (larger x_s), the more the vapor pressure over the solution is suppressed. The mathematical relationship (Eq. (3.61)) and the graphical depiction both provide the same information, but each offers its own viewpoint.

3.5.2 *Effect of solute on melting*

The same solute that lowers the equilibrium vapor pressure of a solution also lowers the melting point of ice. We can see how the two phenomena are related by considering the phase diagram in the vicinity of the nominal melting point of ice, that is, near $T = T_0 = 0\,°\mathrm{C}$. Figure 3.9 shows curves of vapor pressure versus temperature for various constant-solute mole fractions. The upper curve (solid above 0 °C, dashed below) shows the standard relationship for pure water, $e_s(T)$, whereas the lower solid curve gives the

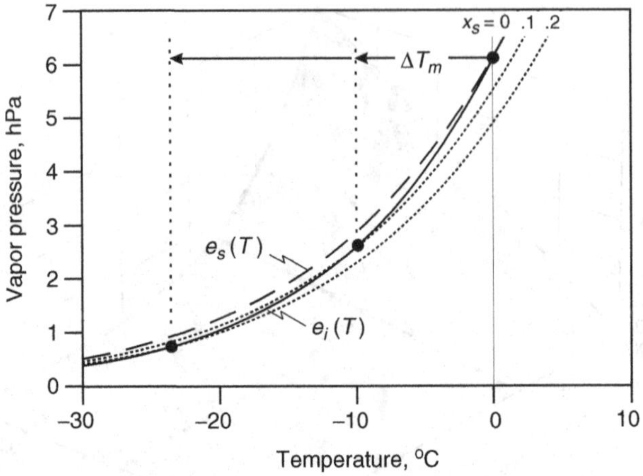

Figure 3.9 Concept behind the depression of the melting point of ice. The solute in solution lowers the equilibrium vapor pressure in proportion to the solute mole fraction x_s.

vapor pressure of ice, $e_i(T)$. The dotted curves below $e_s(T)$ give the vapor pressure for two hypothetical solutions ($x_s = 0.1$ and 0.2). Note, that each of the solution curves maintains its proportionality with respect to the pure-water curve, in accordance with Eq. (3.61); remember that solute resides in the liquid, not in the ice. Where the solution curves intersect the ice curve (at the large dots), new solid–liquid equilibria are established; there, both the ice and the solution have the same vapor pressures and the same free energies; the intersections represent equilibria between the solid and liquid phases. The deviation of these melting temperatures from the pure-water melting point is the so-called melting-point depression ΔT_m. Calculations of ΔT_m start by equating the vapor pressures over solution and ice.

The melting point varies continuously with the concentration of solute in the liquid. The concept of the melting-point depression illustrated in Fig. 3.9 is readily generalized by connecting the loci of solid–solution equilibrium points (such as the large dots in the example). Note that the melting points follow the vapor-pressure curve for ice, which lies below that for the pure liquid. The ice and pure-liquid vapor pressures are thermodynamically related to each, so we can easily construct a general melting curve by taking the ratios of the vapor pressures, $e_i(T)/e_s(T)$, as shown in Fig. 3.10. The ordinate is explicitly the vapor-pressure ratio, but it also expresses the effective mole fraction of water in the liquid solution. As more solute is added to a solution, the vapor pressure of water in the solution is suppressed, forcing the freezing point to get depressed further. Vapor-pressure ratios provide a convenient measure of the "activity" of water in solution, an effective mole fraction (as implied by Eq. (3.62)). Water activity is an important concept for cloud physics, as it offers a general way of accounting for the complexities of real aqueous solutions.

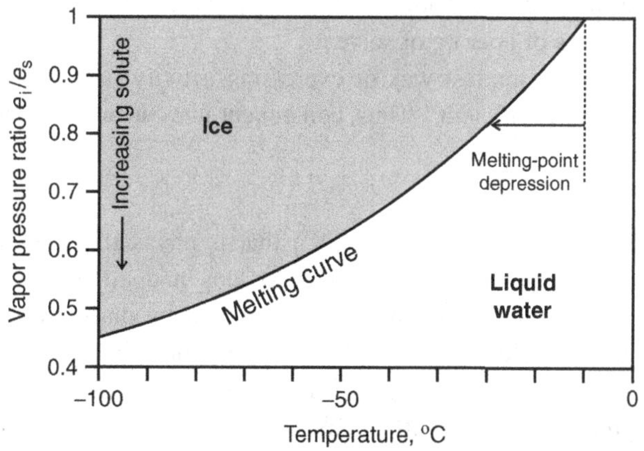

Figure 3.10 The depression of the melting point as a function of the activity of water. Solute concentration increases toward the right. Adapted from Koop *et al.* (2000).

3.5.3 *Non-ideal solutions*

Most real solutions found in nature are not ideal in the sense considered so far. The solute molecules in real liquids often interact with the water molecules in complex ways. Fortunately, by considering aqueous systems semi-empirically, we can avoid dealing with the complexity in detail. In fact, the principles introduced by Raoult's law can be preserved by simply redefining how to specify the amount of solute in solution. Instead of using the actual mole fraction of a component (e.g., solute or water) by itself, we replace the mole fraction (x) with the activity (a), which acts as the mole fraction that gives the physical or chemical effect. Thus, in place of x_s, the mole fraction of solute, use a_s, the "activity" of solute; instead of x_w, use the "activity" of water a_w. Raoult's law (Eq. (3.62)), for instance, is more generally written

$$e_{sol}(x_w, T) = a_w(x_w) \cdot e_s(T), \tag{3.63}$$

where we understand that the activity of water depends on (is a function of) the actual mole fraction of water (x_w), which in turn depends on the mole fraction of solute (x_s) in the solution.

Activity can be related to the amount of solute in solution in several ways. At the outset, recognize that the purpose of all descriptions is to let us calculate macroscopic properties (e.g., equilibrium vapor pressures) from knowledge of the environmental conditions (especially temperature) and the amounts of various solutes present in the liquid phase, even when the solution properties vary with composition and temperature in highly non-linear ways. We also need to recognize that several conventions exist for specifying the physical amount of any given substance (which we generically call "j") in a known amount of solution. In addition to mole fractions (moles of j per total moles of solution), some scientists use molar concentration (moles of j per liter of solution; units designated "M"), or molality (moles of j per kg of solvent).

The first, conceptually simplest way of expressing activity is as the product of mole fraction and an "activity coefficient". Thus, component j has an activity defined as

$$a_j \equiv \gamma_j x_j. \tag{3.64}$$

Note that x_j specifies the relative amount of j that is physically present in the liquid, whereas the activity coefficient (γ_j) accounts for any and all non-linear behavior. If the solution were to be ideal (few are), then γ_j would be unity and the activity of j would simply equal the mole fraction of j. In real solutions having several components, $\gamma_j = \gamma_j(x_j, x_{k\neq j}, T, p)$, which is a potentially complicated function of composition, temperature, and pressure. The activity coefficient is dimensionless, so Eq. (3.64) lets us appreciate the viewpoint that the activity may be thought of as an "effective" mole fraction. Unlike mole fractions, activities can sometimes be greater than unity. The activity acts as a mole fraction, but one modified (made larger or smaller) to account for the presence of other solutes and possibly strong interactions with the water. The activity of water is written as $a_w = \gamma_w x_w$, where values of γ_w are determined experimentally over a wide range of conditions and published in the open literature.

A second way of accounting for complicated solution-phase interactions is the use of the so-called van't Hoff factor (i) within a modified form of mole fraction. If a unit volume of solution contains n_s moles of solute and n_w moles of water, then the activity of water is given as

$$a_w \equiv \frac{n_w}{n_w + i n_s}. \tag{3.65}$$

As with activity coefficient, values of i are determined empirically using this defining relationship. Some solutes are strong electrolytes in the sense that they dissociate (i.e., break up) completely into the ions of which they are composed. Ordinary salt, for instance, dissolves in liquid water as pairs of ions (Na^+ and Cl^-), not as neutral NaCl molecules; the van't Hoff factor would then simply reflect this "degree of dissociation", $\nu = 2$. Strong mineral acids, too, dissociate completely in water, so for nitric acid (HNO_3), $\nu = 2$, but for sulfuric acid (H_2SO_4), $\nu = 3$. In general, simple strong electrolytes yield van't Hoff factors equal to the integer degree of dissociation: $i = \nu$. Complex mixtures and weak electrolytes, those that do not completely dissociate, generally yield non-integer values of i. Dilute solutions, those for which $i n_s << n_w$, typically act linearly once the degree of dissociation is taken into account. Then, Eq. (3.65) may be approximated as $a_w \doteq 1 - i x_s$.

A third way of expressing activity is through the effect of solutes on osmotic pressure. The mere presence of solutes in an aqueous solution reduces the concentration of water, and those solutes may further suppress the activity of water through strong molecular interactions. Water molecules pass through a semi-permeable membrane separating pure water from the solution until the hydrostatic pressure builds up sufficiently to counteract the flow.

The so-called osmotic coefficient of j (Θ_j) accounts for the non-linear behavior of j over and above that already taken into account by the idealized degree of dissociation (ν_j). The activity of water in a binary mixture with solute j is expressed as

$$a_w \equiv \exp\left(-\frac{\nu_j n_j}{n_w}\Theta_j.\right) \tag{3.66}$$

Here, the factor n_j/n_w is the molar ratio of solute j and water in the solution; it must not be confused with the mole fraction of j, $x_j = n_j/\left(n_w + n_j\right)$. If several different solute species are present at the same time, then a more complicated approach is needed. Regardless of how it is expressed, activity is best viewed as a dimensionless measure of the effect that solutes have on the macroscopic properties of water.

3.5.4 Köhler theory for non-volatile solutes

New cloud droplets form on individual aerosol particles, each of which can be viewed as a multicomponent system. The presence of soluble substances in a particle lowers the equilibrium vapor pressure of liquid water sufficiently to allow water vapor to condense readily, often even when the relative humidity is less than 100% (as in hazes). Soluble aerosol particles (especially those containing sulfates and other non-volatile solutes) thus serve as the sites of condensation and the precursors of cloud droplets. It is precisely because of the reductions in vapor pressure caused by solutes that the atmospheric aerosol serves as one of the three main requirements for cloud formation (moisture, aerosol, cooling; acronym MAC). Without aerosol particles, water vapor in the atmosphere would have difficulty condensing, even if the moist air were cooled greatly. Köhler theory (developed in 1921) provides the foundation upon which we can quantify the effects of solute on the equilibria of small solution droplets. We start by assuming the solute to be non-volatile and confined to the particle. In this section, we assume that water is the only substance that exchanges between the droplet and the surrounding air.

Two fundamental attributes characterize every liquid droplet in the atmosphere: its solute content and its curvature. Liquid water resides within aerosol particles that usually contain hygroscopic substances (e.g., mineral acids and salts), so all liquid particles (including cloud droplets) are aqueous solutions. In addition, every particle has limited size, a finite radius of curvature, meaning that an interface exists between the liquid solution and the gas phase.

The dual attributes of solute content and curvature compete with each other in their influence on the water in the particle. As illustrated in Fig. 3.11, droplet curvature (middle) increases the vapor pressure of the water in a particle (relative to a flat surface, left) by increasing the internal pressure and so making the molecules less tightly bound to each other. On the other hand, the solute reduces the equilibrium vapor pressure of the water (right). Think of solute ions (represented by black dots) as molecular "anchors" that hold on to neighboring water molecules; the greater the number of such anchors, the greater

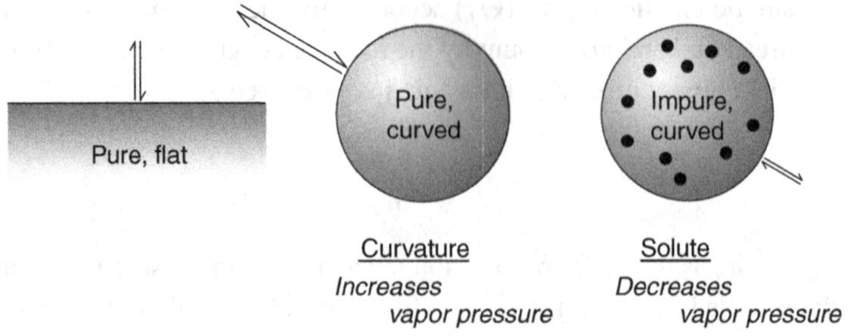

Figure 3.11 Illustration of the independent effects of curvature and solute on equilibrium vapor pressure.

the number of water molecules held tightly and retained in the liquid. Curvature raises the vapor pressure, solute lowers it.

The net effect of solute and curvature yields a quantitative relationship between vapor pressure and droplet radius that is nonlinear and rich in interpretation. Imagine a hypothetical intermediate state, a droplet of pure water having the same radius r_d as that of the solution droplet (middle of Fig. 3.11). The relationship between the radius of curvature and the vapor pressure was shown earlier through the Kelvin–Thomson equation to be

$$\frac{e_r}{e_s} = \exp\left(\frac{2\sigma}{n_L RT r_d}\right) > 1. \tag{3.67}$$

The vapor pressure e_r is inversely related to r_d. At the same time, the effect of solute was given by Raoult's equation in its most general form as

$$\frac{e_{eq}}{e_r} = a_w < 1. \tag{3.68}$$

Both of these relationships must be satisfied simultaneously for a given droplet, so we multiply the two expressions together: $(e_{eq}/e_r) \cdot (e_r/e_s) = e_{eq}/e_s \equiv S_K$. We thus arrive at the general Köhler-theory equation:

$$S_K = a_w \cdot \exp\left(\frac{2\sigma}{n_L RT r_d}\right). \tag{3.69}$$

It must be stressed that the Köhler-theory saturation ratio, $S_K = S_K(r_d)$, represents an equilibrium between the vapor and the solution droplet. The droplet will grow or evaporate only if the actual vapor saturation ratio, $S = e/e_s$, differs from this equilibrium value.

Several assumptions are commonly made for applications of the Köhler function to cloud physics. When the solution is dilute, one can make the approximation $a_w \doteq 1 - ix_s$, and if the droplet radius is not too small, the exponential may be expanded in a Taylor series and approximated as $1 + A_K/r_d$, where $A_K \equiv 2\sigma/n_L RT$. Upon making these substitutions, multiplying the factors together and ignoring the cross terms, we find a simple expression for the equilibrium saturation ratio over a solution droplet compared with that over a flat surface of pure water:

$$S_K \doteq 1 + \frac{A}{r_d} - ix_s. \tag{3.70}$$

The second term on the right-hand side gives the effect of curvature explicitly in terms of the droplet radius. However, the last term changes significantly as the droplet grows or shrinks: the effective solute mole fraction (ix_s) is an implicit function of the droplet radius. Even though the mole fraction of solute is not constant, the total number of moles of solute in the droplet is. (Recall that we are here dealing with a non-volatile solute.) Note that this "solute content" N_s is not the same as the solute concentration n_s (amount per unit volume). These two variables are related for a given droplet of volume V_d by $N_s = n_s V_d$. The magnitude of the solute effect depends on its concentration, but it is the total solute in solution that is conserved as the droplet grows. The solute mole fraction can be related explicitly to the size of the droplet through the approximate relationships

$$ix_s \doteq \frac{in_s}{n_L} = \frac{iN_s}{n_L V_d} \equiv \frac{BiN_s}{r_d^3}, \tag{3.71}$$

where $B_K \equiv 3/(4\pi n_L)$ varies only weakly with temperature (through the liquid density n_L). The solute mole fraction and the equilibrium saturation ratio thus vary explicitly with the droplet radius as

$$S_K \doteq 1 + \frac{A_K}{r_d} - \frac{B_K iN_s}{r_d^3}. \tag{3.72}$$

Now, if we subtract unity from each side, the left-hand side becomes a "supersaturation":

$$S_K - 1 \equiv s_K \doteq \frac{A_K}{r_d} - \frac{B_K iN_s}{r_d^3}. \tag{3.73}$$

The symbol s_K, called the Köhler function, represents the supersaturation (fractional deviation from equilibrium with respect to a flat surface of pure water) needed to keep a solution droplet from either growing or shrinking in size. Note that in the limit of extremely large droplet size ($r_d \to \infty$), both the curvature and solute terms vanish, leaving $s_K \to 0$ and $e_{eq} \to e_s$, the vapor pressure of flat, pure water.

The dependence of the equilibrium supersaturation on droplet size is most easily viewed graphically. Figure 3.12 shows a set of "Köhler curves", plots of Eq. (3.73) for selected values of the effective solute content (iN_s). Note that each curve represents the equilibrium behavior of an individual droplet having the specified amount of solute. Each curve exhibits a maximum, called the critical supersaturation. The larger the solute content, the lower is the equilibrium saturation ratio (the curves shift downward on the graph) and the critical supersaturation. The limiting case for a droplet of pure water ($iN_s = 0$), shown as the dashed curve, represents the effect of curvature only (as calculated from the Kelvin equation). Köhler curves are much used in cloud physics as the reference states for droplets containing non-volatile solutes.

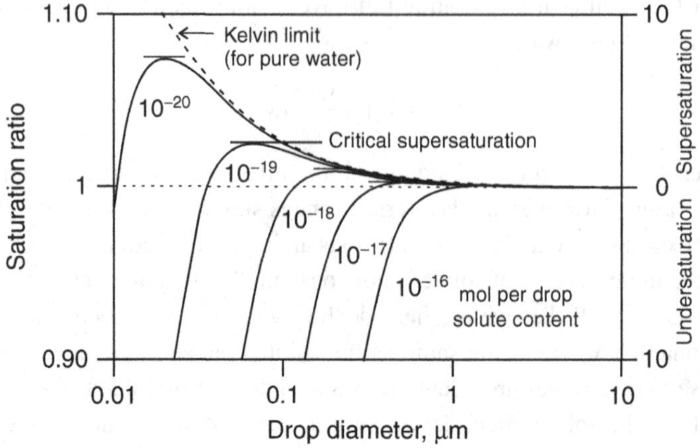

Figure 3.12 A set of Köhler curves, giving the equilibrium saturation ratios of solution drops containing the indicated solute contents.

3.5.5 Volatile solutes

When both the solvent and the solute are volatile, we must account for the vapor pressures of each. The same formalism that we applied to water could also be applied to any other component in the solution, but it is found more convenient to use the convention of Henry's law for volatile solutes, those that exchange between phases. Our goal remains, however, to obtain a quantitative relationship between the concentration of any non-aqueous, but volatile substance ("A") in the liquid and gas phases. Just as $e_{sol} \equiv p_{H_2O,sol}$ is a measure of the concentration of water vapor in equilibrium with water in a liquid solution, so too does $p_{A,sol}$ represent the gas-phase concentration of substance A in equilibrium with A in the aqueous solution. Both e_{sol} and $p_{A,sol}$ depend on the physical and chemical properties of the solution.

It is useful to reiterate our approach and anticipate the main outcome of this analysis: Think of A as a specific soluble gas that is commonly found in the atmosphere, carbon dioxide (CO_2), for instance. We first calculate how much of this gas dissolves into cloud water. Then, we estimate the impact this gas has on the equilibrium vapor pressure of water and on cloud microstructure.

We calculate the uptake of trace gas A by considering its equilibrium state. Chemical equilibria are conveniently expressed in terms of reactions, from which we can derive a useful relationship between the concentrations of solute in two phases. A multiphase equilibrium reaction is expressed as

$$A(g) \rightleftarrows A(aq). \tag{3.74}$$

As indicated by the right-pointing arrow, molecules of substance A in the gas phase (g) pass at some rate into the aqueous phase (aq), while at the same time other A molecules escape from the aqueous phase and enter the gas phase (left-pointing arrow). The flux of gas molecules entering the liquid depends directly on the concentration of A molecules immediately above the surface, just as the condensation flux of water molecules depends on the water vapor concentration. Similarly, the escaping flux of A is the same as the impingement flux of A when equilibrium prevails. During equilibrium, the rates of absorption and desorption are equal, and the concentration of A molecules in the aqueous phase is steady and directly proportional to the partial pressure ($p_{A,sol}$) of A in the gas phase. One form of Henry's law is thus:

$$[\mathrm{A(aq)}] \equiv H_A p_{A,sol}. \tag{3.75}$$

We follow the convention of chemistry and use brackets to indicate "concentration of" whatever compound is inside the brackets (in order to minimize the levels of subscripting). The proportionality constant H_A [common units, mol L^{-1} atm^{-1}] is called the Henry's law coefficient of A and is interpreted as the solubility of gas A in the aqueous solution, the number of moles of A dissolved in unit volume (typically 1 L) of solution per unit partial pressure (typically 1 atm). One seldom finds the second subscript ("sol") on the pressure symbol, but we will continue using it for now to emphasize the fact that Henry's law is a statement of equilibrium between the gas and solution. Note that Eq. (3.75) is a defining relationship, so it is also used in laboratory experiments to determine the values of Henry's law coefficients for many trace gases from measured values of partial pressure and aqueous-phase concentration. These empirically determined Henry's law coefficients are available in the literature as needed, so Eq. (3.75) allows us to calculate the concentration of any trace gas dissolved in cloud water (our aqueous solution) given its partial pressure in the gas phase.

Dissociating gases

The behavior of some gases is more complicated than that depicted thus far. Chemically stable substances (e.g., N_2, Ar) retain their molecular forms when dissolved in water, whereas more reactive substances (CO_2, SO_2, HNO_3) interact with the water and change their forms. Such gases dissociate (are pulled apart) upon coming in contact with liquid water, and some of these substances (called electrolytes) generate ion pairs in solution, making the solution electrically conducting in the process. The conventional Henry's law coefficient (H_A) that we have considered up to now applies strictly to the physically dissolved component of a dissolved gas, that is, to the molecular form preserved in both the aqueous and gas phases. We will soon see the need for deriving an overall or total Henry's law coefficient $\left(H_A^*\right)$, one that takes dissociation into account. We may correctly anticipate that if a trace gas is physically taken up by cloud water and additionally forms new species in the aqueous solution, the overall Henry's law coefficient must be larger than the Henry's law coefficient based on the physically dissolved portion alone.

The process of dissociation and how it affects gas–liquid equilibria can be understood best by first thinking about how water molecules interact with themselves in the liquid phase. Liquid water, because of the polar nature of the water molecule, is a weak electrolyte, a substance that occasionally dissociates partially into ions. The self-dissociation of water can be expressed in the language of chemistry by the following reaction:

$$H_2O + H_2O \rightarrow OH^- + H_3O^+. \tag{3.76}$$

Normally, two neighboring water molecules pull on each other, the positive (proton) end of one being attracted to the negative (lone-pair) end of the other, thus forming the hydrogen bond. Occasionally, however, the thermal energy in the bond is strong enough to let the proton break away from the one molecule and jump across to the other molecule, leaving the first molecule with a net negative charge, the second with a net positive charge. The two initially neutral molecules form an ion pair simply through the act of transferring a proton from one molecule to another. The negative ion (OH^-) is called the "hydroxyl" ion, and the positive ion (H_3O^+) is called the "hydronium" ion. (Note: The hydroxyl ion should never be confused with the neutral OH radical, which is formed via photochemical reactions; the properties and behavior of these two odd-hydrogen species differ greatly.) The hydronium ion, by convention, is often designated by H^+, but protons never reside freely in solution; they are always associated with water. Thus, we may write hydronium variously as $H_3O^+ \equiv H^+ \cdot H_2O \equiv H^+(aq) \equiv H^+$.

Ionic equilibrium is established in liquid water when the rate of dissociation is exactly countered by the rate of the reverse process, ionic recombination or association. We may write the equilibrium reaction for water in the convenient form,

$$H_2O \overset{K_{water}}{\rightleftarrows} OH^- + H^+, \tag{3.77}$$

where the variable over the double arrows designates the equilibrium constant, defined as the concentration ratio of products (species on the right-hand side) to reactants (species on the left):

$$K_{water} \equiv \frac{\left[OH^-\right]\left[H^+\right]}{[H_2O]}. \tag{3.78}$$

The concentrations of ions in a weak electrolyte such as water are extremely small compared with that of the water itself ($[H_2O] = n_L = 55.5\,\text{M}$; $\text{M} \equiv \text{mol}\,\text{L}^{-1}$), so it is conventional to fold the relatively constant denominator of Eq. (3.78) into the equilibrium constant itself. For water, we therefore use the so-called "ion product",

$$K_w \equiv K_{water} \cdot [H_2O] \equiv \left[OH^-\right]\left[H^+\right], \tag{3.79}$$

to express the extent of dissociation. The magnitude of K_w depends on temperature, as given by the following equation:

$$\log K_w = -\frac{4470.99}{T} + 6.0875 - 0.01706T, \tag{3.80}$$

where the units of K_w are M^2 when T is expressed in Kelvin (K). At a temperature of 298.15 K (25 °C), $K_w = 1.0 \times 10^{-14}\,\text{M}^2$, which is small, but nevertheless significant.

The relative concentrations of the positive and negative ions in an electrolytic solution can be calculated using the fundamental principle of charge conservation. Just as neither mass nor energy can be created or destroyed, so too is it impossible to create or destroy electric charges via ordinary chemical reactions or physical processes. Reactions can generate ions, to be sure, but they always appear in pairs (equal numbers of positive and negative ions), and those charges were already present inside the atoms (as electrons and protons). Reactions simply release or separate pre-existing charges. The principle of charge conservation lets us calculate the relationship between "cations" (positive ions, q_+) and "anions" (negative ions, q_-) because solutions that are electrically neutral before a reaction takes place remain electrically neutral after ions are released into the solution. In its most generic form, the electroneutrality equation can be written as

$$\sum [q_+] = \sum [q_-]. \tag{3.81}$$

Operationally, we simply sum up the concentrations of all of the cations and equate that sum to the sum of all anionic concentrations. We see this principle of charge conservation acting most simply in the dissociation of ordinary water through the formation of equal numbers of hydronium and hydroxyl ions. Pure water at a temperature of 25 °C contains, based on Eqs. (3.79) and (3.81), $[H^+] = [OH^-] = K_w{}^{1/2} = 1 \times 10^{-7}\,M^{-1}$. Such a concentration is not likely to be preserved in solutions containing other dissociating compounds, but the principle of calculation remains valid.

Many substances dissociate into ionic components when dissolved in water, and in so doing they may alter the balance of hydronium and hydroxyl ions in solution. Adding a "neutral" electrolyte, such as ordinary salt (NaCl), does nothing to this balance, as can be seen by considering the following generic and specific reactions,

$$\begin{aligned} &BA \rightarrow B^+ + A^- \\ &NaCl \rightarrow Na^+ + Cl^-. \end{aligned} \tag{3.82}$$

Equal numbers of positive and negative ions are produced by the dissociation of a strong electrolyte; neither the H^+ nor the OH^- ions are involved, so their concentrations remain unchanged. However, if we add an acid (a substance that donates H^+), we find a different situation. Let the generic acid be designated HA and the specific example be hydrochloric acid (HCl). Now, the reactions are

$$\begin{aligned} &HA \rightarrow H^+ + A^- \\ &HCl \rightarrow H^+ + Cl^-. \end{aligned} \tag{3.83}$$

The number of hydronium ions in the solution has changed by the addition of acid, but by how much?

The concentration of hydronium ions in a solution can be calculated from the ionic equilibria and the electroneutrality equation. In the case of a generic acid, the equilibrium constant for the reaction (Eq. (3.83)) is defined as

$$K_A \equiv \frac{[\mathrm{H}^+][\mathrm{A}^-]}{[\mathrm{HA}]}. \tag{3.84}$$

The magnitude of K_A for a "strong" acid (one that dissociates completely) is large, whereas that for a "weak" acid (one that only partially dissociates) is small. The terms large and small are relative and take on meaning only after the analysis is complete. The hydrogen ions from the acid are added to those already present from the partial dissociation of the water (as well as that of any other substances), but the addition is not linear. Once new ions are added, the equilibria of all other ions shift. The easiest way to account for such complex interactions is to set up all relevant equilibria as equations involving the equilibrium constants. The set of equations is then solved simultaneously for the unknown quantities, the concentrations of species in solution. One finds that the number of equations so established is one short of the number needed to solve the set completely. The final equation needed is the electroneutrality equation, which imposes the constraint of charge conservation and allows us to calculate $[\mathrm{H}^+]$.

The procedure for calculating $[\mathrm{H}^+]$ in the case of a generic acid follows:

(a) Set up equilibrium reactions and defining equations:

$$\mathrm{HA} \overset{K_A}{\rightleftarrows} \mathrm{H}^+ + \mathrm{A}^-,$$

$$K_A \equiv \frac{[\mathrm{H}^+][\mathrm{A}^-]}{[\mathrm{HA}]} \Rightarrow [\mathrm{A}^-] = \frac{K_A[\mathrm{HA}]}{[\mathrm{H}^+]} \tag{3.85}$$

$$\mathrm{H_2O} \overset{K_w}{\rightleftarrows} \mathrm{H}^+ + \mathrm{OH}^-$$

$$K_w \equiv [\mathrm{OH}^-][\mathrm{H}^+] \Rightarrow [\mathrm{OH}^-] = \frac{K_w}{[\mathrm{H}^+]}. \tag{3.86}$$

(b) Write the electroneutrality equation, inserting appropriate expressions from (a):

$$[\mathrm{H}^+] = [\mathrm{OH}^-] + [\mathrm{A}^-] = \frac{K_w}{[\mathrm{H}^+]} + \frac{K_A[\mathrm{HA}]}{[\mathrm{H}^+]}. \tag{3.87}$$

(c) Combine terms and solve for $[\mathrm{H}^+]$:

$$[\mathrm{H}^+]^2 = K_w + K_A[\mathrm{HA}] \Rightarrow [\mathrm{H}^+] = \{K_w + K_A[\mathrm{HA}]\}^{1/2}. \tag{3.88}$$

Thus, knowledge of the concentration of acid, [HA], and use of the correct values for the equilibrium constants (found in the literature for specific temperatures), one gains a value for $[\mathrm{H}^+]$ (from Eq. (3.88)). Note that the concentration of hydronium in an acidic solution is always greater than it is in pure water (when $[\mathrm{HA}] = 0$). The solution becomes ever more acidic, and the concentration of hydronium increases as more acid is added.

The pH of an aqueous solution is a convenient measure of its acidity, the concentration of hydronium. The defining relationship between pH and $[H^+]$ is

$$pH \equiv -\log [H^+]. \tag{3.89}$$

The pH is expressed on a logarithmic scale, so a change of a single pH unit amounts to a factor of ten change in $[H^+]$. (The "p" in pH stands for the (negative) power of ten representing the concentration of "H".) By convention, $[H^+]$ is specified in molar units [mol L^{-1}], which is one reason for using the liter (rather than the cubic meter) as the unit of liquid volume in cloud chemistry. Thus, we find the pH of pure water (at 25 °C) to be

$$pH|_{\substack{pure \\ water}} \equiv -\log [H^+] = -\log K_w^{1/2} = -\frac{1}{2}\log\left(10^{-14}\right) = 7. \tag{3.90}$$

Solutions with a pH of 7 are said to be "neutral", with the concentrations of H^+ and of OH^- then being equal. An acid added to pure water increases $[H^+]$ (and decreases $[OH^-]$), so the pH of an acidic solution is less than 7; a base (e.g., NaOH) added to pure water decreases $[H^+]$, so the pH of a basic (or alkaline) solution is greater than 7. This pattern can be summed up as follows:

$$\begin{aligned} \text{Acid: } & [H^+] > [OH^-] \quad \Rightarrow \quad pH < 7 \\ \text{Base: } & [H^+] < [OH^-] \quad \Rightarrow \quad pH > 7. \end{aligned} \tag{3.91}$$

An acronym for remembering which side of 7 the solution pH lies is LAMB = Less, Acid; More, Base.

Consider the specific example of gaseous CO_2 taken up by otherwise pure water. We may want to know, for instance, the pH of the resulting solution and the total amount of CO_2 dissolved in the water. We start with the premise that the partial pressure of CO_2, p_{CO_2}, is known, as well as the values of the relevant equilibrium constants. We next apply the general procedure outlined above for finding the hydronium-ion concentration, including the phase equilibrium and the second ionization:

(a) Equilibria (example):

Phase equilibrium:

$$CO_2(g) \overset{H_{CO_2}}{\rightleftarrows} CO_2(aq), \quad H_{CO_2} \equiv \frac{[CO_2(aq)]}{p_{CO_2}} \tag{3.92}$$

$$\Rightarrow \quad [CO_2(aq)] = H_{CO_2} p_{CO_2}. \tag{3.93}$$

First dissociation:

$$CO_2(aq) + H_2O \overset{K_{1C}}{\rightleftarrows} H^+ + HCO_3^-, \quad K_{1C} \equiv \frac{[H^+][HCO_3^-]}{[CO_2(aq)]} \tag{3.94}$$

$$\Rightarrow \quad [HCO_3^-] = \frac{K_{1C}[CO_2(aq)]}{[H^+]}. \tag{3.95}$$

Second dissociation:

$$\mathrm{HCO_3^-} \overset{K_{2C}}{\rightleftarrows} \mathrm{H^+} + \mathrm{CO_3^{2-}},\ K_{2C} \equiv \frac{[\mathrm{H^+}]\left[\mathrm{CO_3^{2-}}\right]}{[\mathrm{HCO_3^-}]} \tag{3.96}$$

$$\Rightarrow \left[\mathrm{CO_3^{2-}}\right] = \frac{K_{2C}\,[\mathrm{HCO_3^-}]}{[\mathrm{H^+}]}. \tag{3.97}$$

Water dissociation:

$$\mathrm{H_2O} \overset{K_w}{\rightleftarrows} \mathrm{H^+} + \mathrm{OH^-},\ K_w = [\mathrm{OH^-}][\mathrm{H^+}] \tag{3.98}$$

$$\Rightarrow [\mathrm{OH^-}] = \frac{K_w}{[\mathrm{H^+}]}. \tag{3.99}$$

(b) Electroneutrality relationship (example):

$$[\mathrm{H^+}] = [\mathrm{OH^-}] + [\mathrm{HCO_3^-}] + 2\left[\mathrm{CO_3^{2-}}\right]. \tag{3.100}$$

(Note that the factor 2 is needed to account for the charge on each carbonate ion.)

(c) Solving for $[\mathrm{H^+}]$ (example):

$$[\mathrm{H^+}]^3 - \left(K_w + K_{1C}H_{CO_2}p_{CO_2}\right)[\mathrm{H^+}] - 2K_{2C}K_{1C}H_{CO_2}p_{CO_2} = 0. \tag{3.101}$$

The last term can be shown to be small for realistic values of p_{CO_2}, so we gain the reasonable approximation

$$[\mathrm{H^+}] \doteq \left(K_w + K_{1C}H_{CO_2}p_{CO_2}\right)^{1/2}. \tag{3.102}$$

After inserting the appropriate values for the constants, we find $[\mathrm{H^+}] = 3 \times 10^{-6}$ M when $p_{CO_2} = 400$ ppm, yielding a pH of about 5.5. Thus, we find that even the purest of water is acidic when in equilibrium with atmospheric CO_2.

The total concentration of CO_2 dissolved in cloud water is calculated by counting all of the carbon atoms in a unit volume of water. If we designate total carbon by C_{tot}, then we simply sum the concentrations of the individual components containing carbon:

$$[\mathrm{C_{tot}}] = [\mathrm{CO_2(aq)}] + [\mathrm{HCO_3^-}] + \left[\mathrm{CO_3^{2-}}\right]. \tag{3.103}$$

(Note that the carbonate ion is here counted only once, as we are tracking the carbon atoms, not the electrical charges.) We must also count the physically dissolved component, even though it has no charge. Inserting the relevant expressions into this equation gives us

$$[\mathrm{C_{tot}}] = \left(1 + \frac{K_{1C}}{[\mathrm{H^+}]}\left(1 + \frac{K_{2C}}{[\mathrm{H^+}]}\right)\right) H_{CO_2}p_{CO_2}. \tag{3.104}$$

We can factor out the CO_2 partial pressure, and so form an expression that closely resembles the conventional Henry's law coefficient:

$$[C_{tot}] = H^*_{CO_2} p_{CO_2}. \tag{3.105}$$

The pH-dependent variable $H^*_{CO_2}$ is called the "overall Henry's law coefficient" because it acts as the conventional Henry's law coefficient does, but it accounts for the dissociation of the trace gas as well as its physical uptake. The overall Henry's law coefficient represents the complete solubility of the gas in a liquid of specified pH.

The range of gas solubilities varies enormously with the type of gas involved. The solubilities of several important trace gases are represented in Fig. 3.13. Note that the vertical axis is the specific and overall Henry's law coefficient plotted as a function of the solution pH for a number of important gases. The pH of a real solution is, of course, determined by the composite influence of all dissolved substances, but here we assume that the pH is given for the sake of simplicity. The typical range of pH values expected in cloud water over

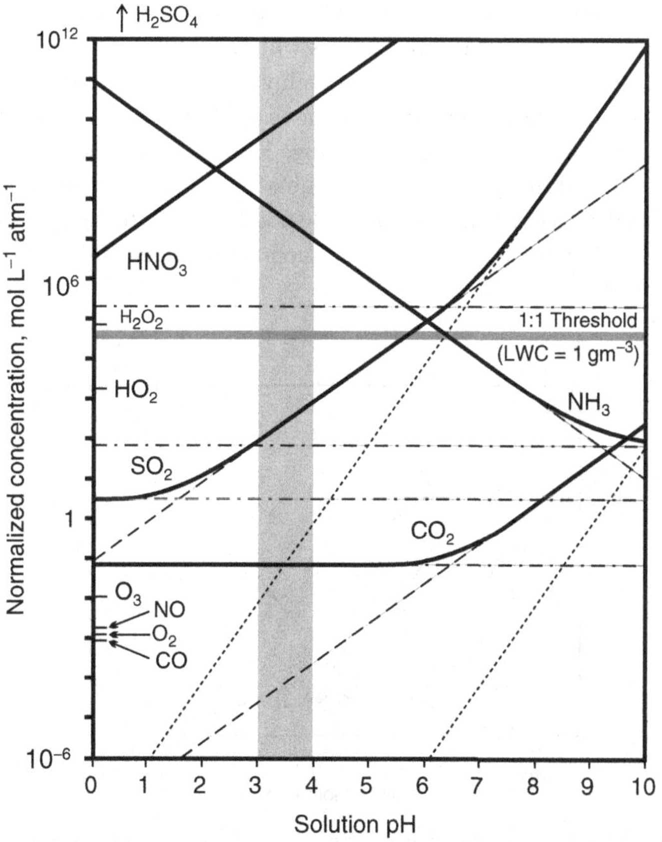

Figure 3.13 Plots of the Henry's law coefficients for selected gases as a function of the solution pH.

industrial regions is shown by the vertical band. In cleaner environments, the cloud-water pH is generally higher, perhaps above 5. Recall from our earlier discussion of reactive gases that the pH of otherwise pure water in equilibrium with atmospheric CO_2 is 5.5, which is still acidic and well below the neutral pH of 7 for ideally pure water.

The solid curves in Fig. 3.13 represent the overall or total solubilities of the specified gas. Acidic gases (e.g., HNO_3, SO_2) become more soluble as the solution pH increases, whereas alkaline gases (mainly NH_3) become less soluble with increasing pH. The contributions to the total solubilities are shown as the various types of dashed curves in Fig. 3.13. The horizontal (dash-dot) lines give the Henry's law coefficients for the physical, undissociated form of the dissolved gas. Note that the Henry's law coefficients for unreactive components are independent of solution pH because they are electrically neutral and so do not change the balance of ions in solution. The longer-dashed lines represent the pH dependence of the first dissociation product of the reactive gases, whereas the shorter-dashed lines give that for the second dissociation product (if it exists). Note that each first dissociation of an acidic gas generates a single hydronium ion, which causes the concentration of this product to vary inversely with $[H_3O^+]$ (a dependence that shows up on a log–log plot as a upward straight line with a slope of unity). The second dissociation yields two hydronium ions and so the concentration of its product gives curves with slopes of 2. The sums of the various components at any given pH give the total solubility of the parent gas (solid curves).

The reason for using Henry's law for trace gases and Raoult's law for solutes dissolved in water can be understood with the help of Fig. 3.14. Plotted there are the activities of water and of nitric acid (representative of a soluble trace gas) as functions of the solution composition specified in terms of the solute mole fraction (x_A). The heavy curves show that the activities of both water and nitric acid are suppressed relative to Raoult's law (dashed

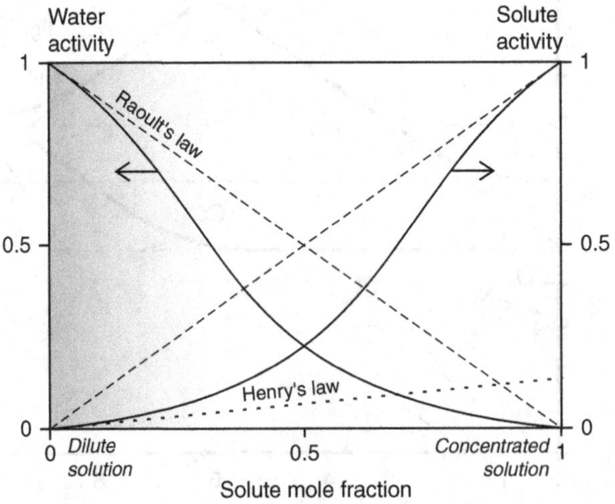

Figure 3.14 Activities of a binary solution as a function of solute mole fraction. Based on the water–nitric acid system.

lines), a reflection of the strong intermolecular attractions between H_2O and the nitrate ion (NO_3^-). Concentrated solutions are represented toward the right-hand side, where x_A and the solute activity are relatively large. Dilute solutions are represented toward the left-hand side, which is shaded to remind us of cloud-water characteristics. In the dilute range, the water activity is large (because cloud water is mainly water) and initially follows the linear Raoult's law behavior as solute is added. By contrast, the activity of the solute stays low in dilute solutions (because of the strong molecular-scale attractions). Note that the solute activity initially increases linearly with x_A, but with a slope considerably smaller than that for Raoult's law. It is this linear dependence of solute activity on x_A (lower dashed line) that is represented by Henry's law. Henry's law describes the behavior of a volatile solute, whereas Raoult's law describes the behavior of the solvent (H_2O) in dilute solutions.

Let us now make the connection between Henry's law when specified by solute concentration and partial pressure (Eq. (3.75)) to Henry's law as shown graphically in Fig. 3.14. We seek an expression that relates the solute activity to the mole fraction (the variables shown on the graph). Start with Eq. (3.75) and note that the partial pressure of any substance in equilibrium with a solution is always related to the corresponding activity (analogous to Eq. (3.63) for water) as a dimensionless ratio:

$$\frac{p_{A,sol}}{p_{A,pure}} = a_A. \tag{3.106}$$

Thus, $p_{A,sol} = a_A p_{A,pure}$, and Eq. (3.75) becomes

$$\left[\text{A(aq)}\right] = \left(H_A p_{A,pure}\right) \cdot a_A. \tag{3.107}$$

Now, working with the left-hand side, express the solute concentration as a mole fraction:

$$\left[\text{A(aq)}\right] \doteq n_w x_A, \tag{3.108}$$

where the approximation is justified when the solution is dilute ($n_A << n_w$). Finally, combine Eqs. (3.107) and (3.108) to gain the desired expression:

$$a_A = \left(\frac{n_w}{H_A p_{A,pure}}\right) x_A. \tag{3.109}$$

Note that the quantity in parentheses is the slope of the lower dashed line in Fig. 3.14. We see that this slope is inversely related to the Henry's law coefficient (H_A), hence to the trace-gas solubility, and to the equilibrium vapor pressure of the solute in the hypothetical pure state. We see from this expression that a highly soluble gas (large H_A) has a very low activity for any given concentration (represented by x_A). Low activity means that only a small partial pressure is needed to maintain equilibrium with the solution, which makes sense in view of our expectation that soluble substances are those that bond well to water. The solute molecules tend to stay in the solution because of their strong attachments to the water molecules in the liquid.

A dimensionless form of the Henry's law coefficient can also be derived, one that provides additional insight into the nature of equilibria between two phases. Most trace gases obey the ideal gas law, so the concentration of A in the gas phase ($n_{A(g)} \equiv [A(g)]$) is related to the partial pressure (p_A) by

$$p_A = [A(g)]\,RT. \tag{3.110}$$

Substitution of this expression for the partial pressure into the defining form of Henry's law (Eq. 3.75) yields

$$[A(aq)] = H_A RT\,[A(g)] \equiv \hat{H}_A\,[A(g)], \tag{3.111}$$

from which we see that the dimensionless Henry's law coefficient ($\hat{H}_A$) is the ratio of concentrations in the aqueous and gaseous phases:

$$\hat{H}_A = \frac{[A(aq)]}{[A(g)]}. \tag{3.112}$$

The greater the solubility of a gas, the higher is its concentration in the aqueous phase relative to that in the gas phase. The molecules of a highly soluble gas are thus tightly packed together in the liquid solution, by virtue of their strong attachments to the water molecules, compared with those in the gas phase, where intermolecular bonding is virtually absent.

Gas uptake in confined systems

From the point of view of gaseous absorption, each parcel of cloudy air acts as an independent system, one in which a finite amount of soluble gas interacts with a finite amount of liquid cloud water. Mass continuity limits the amount of gas that can be dissolved into the cloud water, for as gas is taken up by the cloud water, the concentration in the gas phase decreases and shifts the equilibrium state. For a given concentration of liquid (say $\omega_L \sim 1\ \mathrm{g\,m^{-3}}$), the uptake of trace gas depletes the gas and lowers its partial pressure. The lowered partial pressure in turn means weaker subsequent uptake of dissolved gas by the cloud water. The process is inherently non-linear and depends on the amount of liquid present. We should anticipate that when the liquid water concentration is small, most of the trace substance is in the gas phase, interstitially between the cloud drops. As the amount of liquid water becomes large, however, progressively more of the compound resides in the aqueous phase, leaving less in the gas phase.

The most appropriate measure of the liquid water concentration is the volume v_L of liquid contained in a unit volume of cloudy air, in effect the volume fraction. A volume fraction is normally dimensionless, of course, but it is often convenient to carry specific units along as a way of distinguishing the liquid volume from the air volume. By convention, liquid volumes are given in units of liter [L] or milliliter [mL], whereas gas volumes are specified as cubic meter [m^3]. The most convenient units of liquid-water volume concentration (v_L) are [mL m^{-3}]. The relationship between this volumetric concentration and the conventional mass concentration (or "content" ω_L) of cloud water is $v_L = \omega_L/\rho_L$, where ρ_L is the mass density of bulk liquid water. Note that v_L and ω_L are approximately equivalent numerically when ω_L is specified in units of [g m^{-3}] and density is given in units of [g mL^{-1}].

The partitioning of a soluble trace gas between the liquid and gas phases in a cloudy air parcel can be calculated from the principle of mass conservation. Within a unit volume of air, molecules of trace gas A can either be in the aqueous phase or in the gas phase, not both at the same time. The sum is thus conserved: $n_A|_{total} = n_A|_{air} + n_A|_{aqueous}$. We assume that gaseous A obeys the ideal gas law, so the molar concentration of A in the gas phase is related to its partial pressure p_A at any given temperature T by $n_A|_{air} = p_A/RT$. The collective number of moles of A dissolved in all of the cloud drops in the unit air volume is simply the product of the liquid-volume concentration v_L and the aqueous-phase concentration $[\mathrm{A}_{tot}]$. Thus,

$$n_A|_{total} = \frac{p_A}{RT} + [\mathrm{A}_{tot}]v_L. \tag{3.113}$$

Expressing the aqueous-phase concentration in terms the overall Henry's law coefficient, we get

$$n_A|_{total} = \frac{p_A}{RT} + H_A^* p_A v_L. \tag{3.114}$$

If we multiply this equation through by RT, we see that the left-hand side is just the partial pressure in the absence of liquid cloud water. Let us represent this clear-air partial pressure as $p_A(0)$ and the partial pressure of A in liquid volume v_L [mL m^{-3}] as $p_A(v_L)$. Then,

$$p_A(0) = p_A(v_L) + \hat{p}_A, \tag{3.115}$$

where $\hat{p}_A = RTH_A^* v_L \cdot p_A(v_L)$ is termed the "potential pressure", the additional pressure that A would exert if all of the dissolved forms of A were to be released from the aqueous solution. Equation (3.115) provides one way of expressing how the A molecules are distributed among the gaseous and liquid phases.

The ratio of partial pressures is another way of showing the partitioning of A between phases. The interstitial partial pressure at any given value of liquid volume v_L is thus

$$\frac{p_A(v_L)}{p_A(0)} = \frac{1}{1 + RTH_A^* v_L}. \tag{3.116}$$

At "low" liquid water concentrations, that is when $RTH_A^* v_L << 1$, we see that $p_A(v_L) \doteq p_A(0)$, meaning that too little cloud water is available to deplete A from the gas phase appreciably. At the other extreme, when the liquid water concentration is "high", such that $RTH_A^* v_L >> 1$, we find

$$\frac{p_A(v_L)}{p_A(0)} = \frac{1}{RTH_A^* v_L} \propto \frac{1}{v_L}. \tag{3.117}$$

When relatively much liquid water is present, the A molecules find many opportunities to reside in the aqueous phase (dispersed as cloud drops). The trace gas is then depleted inversely proportional to the concentration of cloud water. The full range of possibilities is

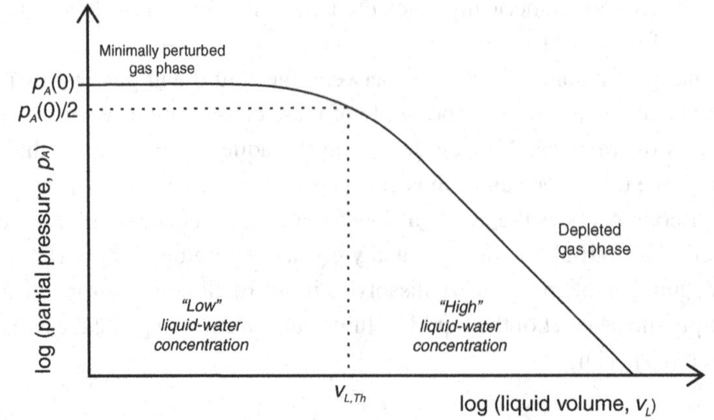

Figure 3.15 Dependence of the partial pressure of a trace gas in equilibrium with its aqueous solution on the liquid-water concentration.

depicted graphically in Fig. 3.15. The break-even point between "low" and "high" liquid-water concentrations, when one half of the substance is in the gas phase and the other half is dissolved in the cloud water, defines a threshold

$$v_{L,Th} = \frac{1}{RTH_A^*}. \tag{3.118}$$

Note that the threshold value depends inversely on H_A^*. For a cloud of known liquid water concentration, one can alternatively define the threshold for equal distribution between the two phases in terms of the solubility:

$$H_{A,Th}^* = \frac{1}{RTv_L}. \tag{3.119}$$

For a typical liquid water concentration of $1\ \text{g}\,\text{m}^{-3}$ at a temperature of 25 °C, $H_{A,Th}^* = 4.6 \times 10^4$ mol L^{-1} atm^{-1}. This value is shown on Fig. 3.13 as the gray horizontal bar identified as the 1:1 threshold. Any trace gases with solubilities larger than this value will reside preferentially in the liquid phase; otherwise, they pass through the cloud relatively uninfluenced by the presence of cloud water. Keep in mind that the magnitude of $H_{A,Th}^*$ depends mainly on the concentration of liquid water, relatively little on the temperature.

Effect of volatile solutes on Köhler theory

When a soluble trace gas is taken up by very small droplets, such as the submicron particles in a summertime haze, one must consider the effect of curvature, as well as the effect of the solute on vapor pressures. Köhler theory is applicable, but the Köhler curves discussed earlier no longer apply to fixed solute contents, rather to solute contents that vary with droplet size and with the collective influence of all other particles in the aerosol.

The uptake of the trace gas by one particle depends on the concentration of the interstitial gas, which in turn depends on how much gas was taken up by the other particles. Each particle in effect competes with all of the others in the population for the same trace gas. Also, particles would not exist in the first place were it not for the presence of non-volatile solutes, so we must deal with ternary, not just binary solutions. We are left with a rather complicated situation when trying to calculate equilibria in hazes and clouds.

The way to solve such complex problems is, as we have been doing, to lay out the various components of the problem systematically and find the solution that simultaneously satisfies all of the constraints. Here, because we want to determine how the equilibria change in a parcel of air that is hypothetically rising adiabatically in the lower atmosphere, we must simultaneously consider all of the following:

(a) Phase equilibria (Henry's law, Clausius–Clapeyron equation)
(b) Aqueous-phase interactions (dissociation into ions, charge conservation)
(c) Solute and curvature effects (Köhler theory)
(d) Distribution of constituents among particles and gas (conservation of mass).

Each aspect of the problem is represented by one or more equations that can be combined simultaneously with the others to yield a mathematical solution. The result of this procedure will be a set of computed relationships between the ambient saturation ratio (for water vapor) and the sizes of solution droplets for a given set of conditions. Often, the relationships are complicated, so numerical computations may be needed, and the results are best presented graphically.

Consider the following simplified example to illustrate the effect a trace gas could have on equilibria in a haze. Our parcel of air is composed of a monodispersed-phase aerosol, the particles of which each contain the same amount of sulfuric acid (representing the non-volatile component, N_{SO_4}), a given number concentration of droplets (n_d), and a specified mixing ratio of nitric acid vapor before condensation ($y_{HNO_3} = p_{HNO_3,0}/p_{tot}$). The following equations represent this situation:
Phase equilibrium:

$$\mathrm{HNO_3(g)} \overset{H_{HNO_3}}{\rightleftarrows} \mathrm{HNO_3(aq)},$$

$$H_{HNO_3} \equiv \frac{[\mathrm{HNO_3(aq)}]}{p_{HNO_3}}. \tag{3.120}$$

Acid dissociation:

$$\mathrm{HNO_3(aq)} + \mathrm{H_2O} \overset{K_{1N}}{\rightleftarrows} \mathrm{H^+} + \mathrm{NO_3^-},$$

$$K_{1N} \equiv \frac{[\mathrm{H^+}][\mathrm{NO_3^-}]}{[\mathrm{HNO_3(aq)}]}. \tag{3.121}$$

Water dissociation:

$$H_2O \overset{K_w}{\rightleftarrows} H^+ + OH^-,$$

$$K_w \equiv [OH^-][H^+]. \tag{3.122}$$

Charge balance:

$$[H^+] = [OH^-] + [NO_3^-] + 2\left[SO_4^{2-}\right]. \tag{3.123}$$

Köhler-theory relationship:

$$\frac{e_{sol}}{e_s} = a_w \cdot \exp\left(\frac{2\sigma}{n_L R T r_d}\right) \tag{3.124}$$

$$a_w \doteq 1 - x_s$$

$$x_s \doteq [\text{solute}]/[H_2O].$$

Mass conservation:
Total solute concentration:

$$[\text{solute}] = \left[SO_4^{2-}\right] + [N_{tot}]. \tag{3.125}$$

Sulfate concentration:

$$\left[SO_4^{2-}\right] = N_{SO_4}/V_d, \text{ individual drop volume } V_d = \frac{4}{3}\pi r_d^3. \tag{3.126}$$

Nitrogen concentration:

$$[N_{tot}] = [HNO_3(aq)] + [NO_3^-] = \frac{H^*_{HNO_3} y_{HNO_3} p_{tot}}{1 + RT H^*_{HNO_3} v_L}. \tag{3.127}$$

Trace-gas partial pressure:

$$p_{HNO_3} = \frac{y_{HNO_3} p_{tot}}{1 + RT H^*_{HNO_3} v_L}, \text{ collective liquid volume } v_L = n_d V_d. \tag{3.128}$$

Using representative values for the various coefficients, we calculate the equilibrium saturation ratio for a full range of droplet radii (r_d) at a temperature of 0 °C, as shown in Fig. 3.16. The solute content is here assumed to be $N_{SO_4} = 10^{-17}$ mol per droplet. The four solid curves represent the results for different initial mixing ratios of nitric acid, $y_{HNO_3} = 0,\ 1,\ 2,$ and 3 ppb; each solid curve has a corresponding lower dashed curve, which isolates the effect of the solute (the sum of sulfate and nitrate) on the equilibrium saturation ratio. The uppermost dashed curve gives the effect of droplet curvature (assuming

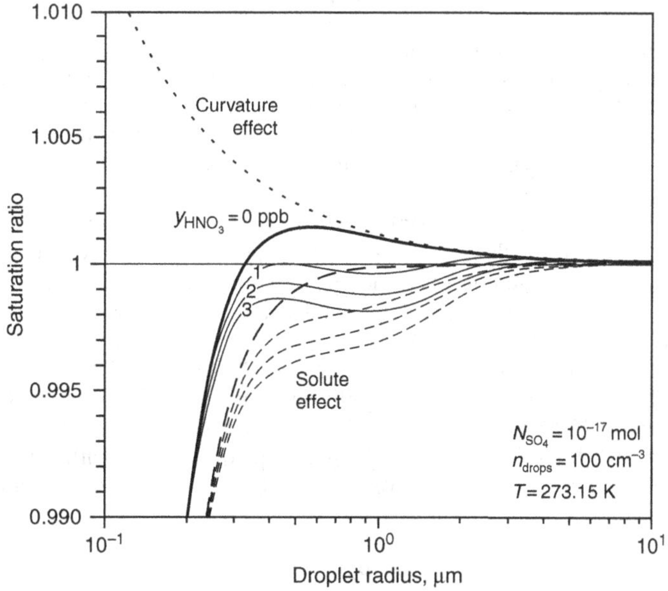

Figure 3.16 Saturation ratio of water vapor in equilibrium with a population of ternary solution droplets in the presence of various mixing ratios of nitric acid vapor.

fixed surface tension). To good approximation, the solute and curvature effects are additive and yield the net effect shown by the respective solid curves. When no nitric acid is present in the air parcel ($y_{HNO_3} = 0$), the upper solid curve gives the same curve that conventional Köhler theory would yield for the same non-volatile solute content. By the merging of the several solid curves toward the left-hand end, we see that nitric acid has little impact on the properties of the smallest droplets; nitric acid cannot compete with concentrated sulfuric acid. On the other hand, as the droplets grow larger, by adding water, the sulfate becomes diluted, allowing some nitric acid to be absorbed. The solute content added by the nitric acid progressively suppresses the evaporation of water as the droplets grow. The first maxima in saturation ratios are clearly reduced in magnitude compared with that when only the non-volatile solute is present. However, as the droplets in the population collectively absorb nitric acid, the supply of nitric acid vapor diminishes and the effectiveness of subsequent uptake is reduced. The nitric acid in effect becomes diluted by the increased amount of liquid water in the aerosol, so the equilibrium saturation ratio again rises beyond the local minima. Toward infinite dilution, the droplets respond no differently from pure water droplets, whether nitric acid is present or not. The presence of a soluble trace gas in the atmosphere adds an interesting level of complexity to the equilibria of cloud droplets, but we have all the mathematical tools available for quantifying its influence. Observational evidence suggests that some fogs in the Po River Valley of Italy are especially persistent because of this effect of nitric acid in the polluted air.

3.6 Further reading

Curry, J.A. and P.J. Webster (1999). *Thermodynamics of Atmospheres and Oceans*. London: Academic Press, 467 pp. Chapter 4 (molecular structure, phase equilibria).

Hobbs, P.V. (2000). *Basic Physical Chemistry for the Atmospheric Sciences*. Cambridge: Cambridge University Press, 209 pp. Chapters 1 (chemical equilibria), 2 (thermodynamics, free energy), 4 (aqueous-phase solutions and equilibria), and 5 (acids, bases, pH).

Pruppacher, H.R. and J.D. Klett (1997). *Microphysics of Clouds and Precipitation*. Dordrecht: Kluwer Academic Publishers, 954 pp. Chapters 3 (water and ice structures), 4 (aqueous-phase equilibria, activity, chemical potential, Clausius–Clapeyron equation), 5 (surface properties), 6 (Köhler theory), 8 (gases, aerosols), and 17 (soluble trace gases, dissociation).

Salby, M.L. (1996). *Fundamentals of Atmospheric Physics*. San Diego, CA: Academic Press, 624 pp. Chapter 4 (chemical equilibria, Clausius–Clapeyron equation).

Seinfeld, J.H. and S.N. Pandis (1998). *Atmospheric Chemistry and Physics*. New York: John Wiley and Sons, 1326 pp. Chapters 6 (aqueous-phase equilibria, dissociation, Henry's law) and 9 (chemical potential, Kelvin equation, trace-gas equilibria).

3.7 Problems

1. From a meteorological perspective, water is one of the most important components of the atmosphere, so the sources of atmospheric moisture should be well understood. Geographically, the tropics tend to be the source region since the tropics are noted for having particularly high absolute humidities. What is the main reason for the strong latitudinal variation in absolute humidity that we find on Earth? Is it because
   (a) the clouds tend to be thicker in the tropics than in extra-tropical regions,
   (b) the average temperatures are higher in the tropics, or
   (c) the fact that the winds converge at low levels in the tropics to generate the intertropical convergence zone?

   Explain your answer carefully in terms of the cause and effect relationships. In addition, explain why the other two choices are not correct explanations.

2. Answer the following questions regarding equilibrium vapor pressure over liquid water:
   (a) How is the latent heat of vaporization l_v related to the entropy of phase change $\Delta\varphi$? (Think of latent heat as an energy per mole of water transferred from the liquid to the vapor phase.)
   (b) Assume that water vapor is described by the ideal gas law and derive an approximate form of the differential relationship

   $$\frac{de_s}{dT} = \frac{\Delta\varphi}{\Delta v},$$

   where Δv is the change in molar volume during phase change.
   (c) Integrate this approximate form (assuming that l_v is constant) from the triple point to some general temperature T to get the function $e_s(T)$.

(d) Put your result in a form that includes the Boltzmann factor, $\exp\left(-\frac{V}{k_B T}\right)$, where V is a potential energy. Which of the variables in your expression for $e_s(T)$ is best associated with a potential energy? Explain the potential energy of what with respect to what?

3. We all know that the equilibrium vapor pressure over liquid water is greater than that over ice at any given temperature less than 0 °C. Without resorting to numbers, prove that this has to be the case. Start with only the knowledge that the equilibrium vapor pressures are equal at 0 °C.
   (a) Calculate the ratio of the equilibrium vapor pressure over liquid water to that over ice at a temperature of −10 °C.
   (b) Provide a molecular-scale interpretation for the difference in vapor pressure over water and ice.

4. Water is anomalous in that the melting point of ice at atmospheric pressure is 0.01 °C *lower* than its triple point temperature at 273.16 K. How would our everyday world differ if the melting point of ice were instead 0.01 °C *higher* than the triple point?

5. This exercise is designed to help you work through the concepts of condensation and evaporation from a molecular perspective. At the same time, you should gain familiarity with the magnitudes of some important variables.
   (a) Start with the kinetic-theory formula for the molecular condensation flux, $\Phi_{cond} = 1/4 n_v \overline{c_v}$. Derive an expression for calculating the evaporation flux in terms of the saturation vapor pressure of water at any given temperature T. (The vapor may be treated as an ideal gas.)
   (b) What is the mass of the water molecule [units, kg]?
   (c) How fast do water vapor molecules travel on average at 0 °C?
   (d) Calculate the magnitude of the physical evaporation flux for water at 0 °C [both in units of molec $m^{-2}\,s^{-1}$ and in units of mol $m^{-2}\,s^{-1}$].
   (e) What is the net evaporation flux if the relative humidity is 50% immediately over the water surface?
   (f) How long would it take for a pan of liquid water 1 cm deep to evaporate at the net rate found in (e)? How realistic is this evaporation rate? What physical process has not been considered here that would account for a slower evaporation rate?

6. Solutes are non-aqueous substances dissolved in liquid water. Water bodies in nature, whether an ocean, a lake, or rain, are never pure; they all contain solutes of one kind or another and in varying concentrations. Usually, the amount of solute is more important than the type, so we often use "salt" as a representative, generic solute. The mere presence of solute in liquid water causes the equilibrium vapor pressure to be reduced, the freezing point to be depressed, and the boiling point to be raised. For the questions below, express the dependence on temperature T of pure-water vapor pressure $e_s(T)$ using the approximate relationship

$$e_s(T) = A \exp\left(-\frac{B}{T}\right),$$

where $A = 2.53 \times 10^{11}$ Pa and $B = 5.42 \times 10^3$ K.

(a) Plot the equilibrium partial pressures of water over the range $-20\,°\mathrm{C}$ to $40\,°\mathrm{C}$ for the following:
   (i) pure liquid water;
   (ii) water with solute mole fraction $x_s = 0.03$ (as in the ocean);
   (iii) water with solute mole fraction $x_s = 0.3$ (found in the Dead Sea).

(b) Hurricanes tend to form best over warm ocean water when the surface temperature exceeds a threshold value $T_s = 27\,°\mathrm{C}$. How much higher would the threshold temperature have to be if the ocean were as salty as the Dead Sea? Briefly explain the difference you find.

7. Free energy is a concept that helps us predict the behavior of systems in nature. In cloud physics, we are most interested in phase changes, a class of applications that occur at constant pressure and temperature. We therefore restrict attention to the Gibbs free energy g. The laws of thermodynamics show that g changes in a general way at a rate given by

$$\frac{dg}{dt} = v\frac{dp}{dt} - \varphi\frac{dT}{dt}$$

where v is the molar volume and φ is the molar entropy of the substance in question.

Apply this general concept to the specific case of liquid water in equilibrium with water vapor. Assume that the temperature is uniform and constant at T, but that the partial pressure e of vapor exceeds the equilibrium value e_s by an amount Δe. Show that for small supersaturations $s \equiv \Delta e/e_s$ the free energy difference between the vapor and liquid, defined as $\Delta g \equiv g_L - g_V$, can be approximated by $\Delta g \simeq RTs$. This relationship justifies the common use of the term "supersaturation" in cloud physics as a measure of the thermodynamic driving force behind phase changes in the atmosphere.

8. Ammonia (NH_3) is the principal basic gas in the atmosphere. Absorption of NH_3 in liquid H_2O (Henry's law coefficient $H_{NH_3} = 62\,\mathrm{M\,atm^{-1}}$ at 298 K) leads to

$$\mathrm{NH_3(g)} \overset{H_{NH_3}}{\rightleftarrows} \mathrm{NH_3}(aq),$$

$$\mathrm{NH_3(aq) + H_2O} \overset{K_{1N}}{\rightleftarrows} \mathrm{NH_4^+ + OH^-},$$

with $K_{1N} = 1.7 \times 10^{-5}$ M.

(a) Calculate $[\mathrm{NH_4^+}]$ in the solution if the partial pressure of ammonia in the gas phase is 1 ppb.

(b) Calculate the total dissolved ammonia $[\mathrm{NH_3^T}]$ in solution at equilibrium.

(c) Derive the expression for the overall Henry's law coefficient.

4
Change

Change occurs in nature when a system is not in equilibrium. Natural systems are frequently thrust out of balance, into a disequilibrium state that cannot persist in the long run. Sunlight shining on a puddle, for instance, heats the ground and forces the water to evaporate. Radiative transfer constitutes one way change and transformations are brought about. Atmospheric systems (e.g., clouds) routinely interact with their environments through exchanges of matter and energy, and they respond to external forces (e.g., pressure gradients). Environmental interactions shift the mechanical, thermal, and chemical balances that allowed a system to be at equilibrium in the first place, forcing the system to change, be transformed to another state. The atmosphere is never at equilibrium, so it is continually changing in one way or another. Indeed, clouds ultimately owe their existence to the disequilibrium forced by solar heating of the surface.

4.1 Deviations from equilibrium

Deviations from equilibrium drive changes in the observable properties of a system. A ladder that is kicked may suddenly topple, an example of mechanical forces that become imbalanced. A turkey placed in an oven gradually gets hot, an example of energy transfer because of a temperature difference. A mixture of air and natural gas (methane, CH_4) exposed to a spark explodes, an example of one set of compounds (oxygen and methane) transforming to another (carbon dioxide and water) because of differences in the chemical potentials of the compounds involved. Indeed, liquid droplets form and grow in clouds because of the chemical-potential differences that exist between the phases of water.

Thermodynamics helps us anticipate what to expect when a system becomes unbalanced. In general, an unbalanced system, one pushed away from equilibrium, responds by moving back toward equilibrium. Equilibrium is the stable state, the one that can exist over the long haul, so the ultimate goal of change is the restoration of equilibrium. At an even more fundamental level, a transformation occurs spontaneously if it results in an increase in the entropy of the Universe. Such an abstract statement, though true, is not helpful unless it can be related to measurable quantities associated with the system of interest (say, a water droplet in a cloud). We already showed in Section 3.2 that the free energy g of any system is a useful quantity because it is preserved during transformations at constant temperature and pressure.

How can we show that changes in the free energy of a system are related to changes in the entropy of the Universe? Recall the definition of the Gibbs function (free energy) and recognize that the specific quantities involved (internal energy u; volume v; pressure p; enthalpy $h \equiv u + pv$; entropy φ) pertain to the system (not to the surroundings):

$$g_{sys} = h_{sys} - T\varphi_{sys}. \tag{4.1}$$

When a process is isothermal (temperature T remains constant), the system free energy changes in response to changes in these other variables according to

$$\frac{dg_{sys}}{dt} = \frac{dh_{sys}}{dt} - T\frac{d\varphi_{sys}}{dt}. \tag{4.2}$$

Integration of this equation from the beginning (t_0) to the end (t_{end}) of the process yields

$$\int_{g_{sys}(t_0)}^{g_{sys}(t_{end})} dg_{sys} \equiv \Delta g_{sys} = \Delta h_{sys} - T\Delta\varphi_{sys}. \tag{4.3}$$

This equation is general and relates the overall change in system free energy to the net changes in enthalpy and entropy from the process. Next, divide this equation through by T:

$$\frac{\Delta g_{sys}}{T} = \frac{\Delta h_{sys}}{T} - \Delta\varphi_{sys}. \tag{4.4}$$

Note that Δh_{sys} is a state variable that represents the net heat transferred into the system. By the conservation principle, energy transferred into a system must equal that transferred out of the surroundings, so $\Delta h_{sys} = -\Delta h_{sur}$. Thus, the entropy transferred is $\frac{\Delta h_{sys}}{T} = -\frac{\Delta h_{sur}}{T} \equiv -\Delta\varphi_{sur}$, and Eq. (4.4) becomes

$$\frac{\Delta g_{sys}}{T} = -\Delta\varphi_{sur} - \Delta\varphi_{sys} = -\left(\Delta\varphi_{sur} + \Delta\varphi_{sys}\right) = -\Delta\varphi_{universe}. \tag{4.5}$$

So, the system variable $-\Delta g_{sys}/T$ is exactly the entropy change of the universe, the universal driver of change. A process that decreases the system free energy ($\Delta g_{sys} < 0$) increases the entropy of the universe and can occur spontaneously. By calculating changes in system free energy, we avoid all complications associated with determining entropy changes of the universe.

The abstract concepts that arise from thermodynamics are beautifully general, but we need to apply them to familiar and measurable quantities. In Section 3.2, we showed that changes in the free energy of a liquid-vapor system vary with changes in vapor pressure and temperature as

$$\frac{dg}{dt} = v\frac{de}{dt} - \varphi\frac{dT}{dt}. \tag{4.6}$$

At constant temperature, this relationship reduces to $dg/dt = v de/dt$, where the molar volume v is related to temperature and partial pressure of vapor via the ideal gas law ($ev = RT$). We thus see how a change in the vapor partial pressure causes the free energy to change:

$$\frac{dg}{dt} = \frac{RT}{e}\frac{de}{dt} = RT\frac{d\ln e}{dt}. \tag{4.7}$$

Integration of this equation, from vapor in equilibrium with liquid water to vapor that deviates from equilibrium, yields

$$g_V - g_L \equiv \Delta g_{VL} = RT\ln S, \tag{4.8}$$

where $S \equiv e/e_s$ is the saturation ratio (the same as fractional relative humidity in the US convention). The partial pressure of water vapor can be determined from measurements of the dew-point temperature T_d: $e = e_s(T_d)$, so the entropy change of the Universe, should it be needed, could be calculated as $\Delta\varphi_{universe} = -\Delta g_{VL}/T = R\ln\left[e_s(T)/e_s(T_d)\right]$. Measurable variables, used within an appropriate theory, let us calculate an abstract quantity.

Applications in cloud physics typically use a relative deviation from equilibrium, as given by the supersaturation $s \equiv S - 1 = \frac{e}{e_s} - 1 = \frac{e - e_s}{e_s}$. When the partial pressure of water exceeds the equilibrium value, perhaps because a moist parcel of air is forced to rise in the atmosphere, supersaturation and excess vapor pressure $\Delta e = e - e_s$ both become positive, and the entropy of the universe temporarily decreases. The system, initially forced out of equilibrium, responds by condensing water vapor into a liquid state. Net condensation occurs spontaneously as the initially positive deviation in free energy Δg_{VL} relaxes back toward equilibrium. When s is small, Eq. (4.8) yields the useful approximation $\Delta g_{VL} = RTs$. If the condensed phase is ice, rather than liquid water, one simply replaces e_s with e_i.

Deviations from equilibrium occur commonly in liquid–solid systems, too. Clouds at temperatures below the melting point of ice (T_0), for instance, contain liquid drops that are often supercooled relative to the ice phase. The thermodynamic driver for change is still a difference in free energy, but now it is more convenient to relate the driver to the supercooling, $\Delta T_s \equiv T_0 - T$, the deviation of the temperature from the ice point. Integration of the free-energy differential equation (4.6) from a state in which the vapor is in equilibrium with the solid to one in equilibrium with the supercooled liquid yields the desired free-energy change:

$$g_L - g_S \equiv \Delta g_{LS} = RT\ln\frac{e_s}{e_i}. \tag{4.9}$$

The saturation ratio of liquid water with respect to ice was shown in Section 3.3 to be

$$S_{i,L} \equiv \frac{e_s}{e_i} = \exp\left(\frac{l_f}{RT_0}\cdot\frac{\Delta T_s}{T}\right), \tag{4.10}$$

so we find

$$\Delta g_{LS} = l_f \frac{\Delta T_s}{T_0}. \tag{4.11}$$

The abstract thermodynamic driver (Δg_{LS}) is directly proportional to the easily measurable supercooling (ΔT_s). Throughout our study of clouds we will use the measurable deviations from equilibrium, but we know that they are just convenient surrogates for the true drivers of change, namely the entropy increase of the universe.

Graphical methods may help solidify the concept of disequilibrium. Phase relationships were shown in Section 3.3 to be conveniently summarized on a phase diagram. Here, we want to add deviations from equilibrium to such a diagram and develop the concept of a property plot. Every point on a phase diagram represents a state of the system, which we here take to be a parcel of moist air. As shown in Fig. 4.1, the abscissa is the temperature, a representative measure of the energy content of the air, whereas the ordinate is the partial pressure of vapor, a measure of moisture, one component of the air in the parcel. Energy and composition are two essential properties that distinguish one parcel from another. These two types of properties are independent of each other, for we can find parcels of any given composition having any temperature. A property plot is a convenient way of representing a moist parcel as a point located at the intersection of the orthogonal variables (temperature and partial pressure). Parcel properties are two dimensional because they have two degrees of freedom. Stipulation that two phases are in equilibrium, by contrast, adds a constraint and loss of one of the two degrees of freedom; equilibria have only one degree of freedom, because once one variable (e.g., temperature) is specified, the other (partial pressure) is determined by the Clausius–Clapeyron relationship (Section 3.3). The triple point offers no degrees of freedom because nature uniquely defines the temperature and partial pressure at which all three phases coexist in equilibrium. The phase diagram,

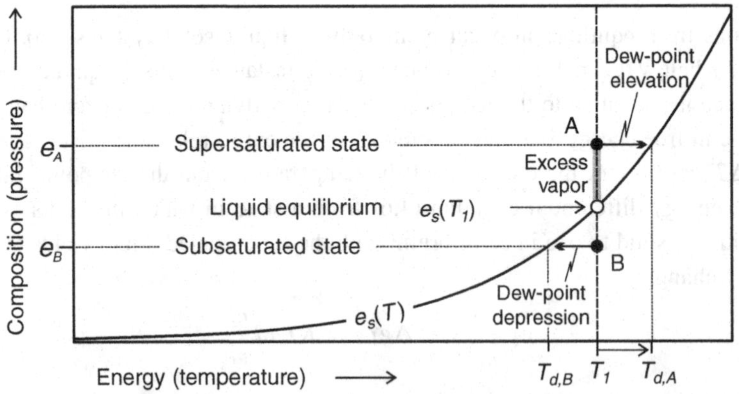

Figure 4.1 Property plot showing liquid–vapor equilibrium (solid curve) and deviations from equilibrium at temperature T_1. Vertical gray bar identifies the excess vapor associated with state A, which gives rise to a dew-point elevation indicated by the upper arrow.

especially when viewed as a property plot, is a powerful tool for interpreting states and processes applicable to clouds.

Consider next a specific example of the phase diagram of water used as a property plot. The two black dots shown in Fig. 4.1 represent two disequilibrium states at arbitrary temperature $T_1 > T_0$. Property point A identifies a parcel having a partial pressure greater than that required for equilibrium with respect to liquid water at that temperature: $e_A > e_s(T_1)$. Parcel A is supersaturated with respect to liquid water ($S_A = e_A/e_s(T_1) > 1$). Parcel B, on the other hand, identifies a different parcel, one that is subsaturated ($S_B = e_B/e_s(T_1) < 1$) because $e_B < e_s(T_1)$. Both parcels are out of equilibrium (with respect to the liquid) and will attempt to readjust back toward equilibrium by either condensing water vapor ($V \rightarrow L$; parcel A) or evaporating any liquid present ($L \rightarrow V$; parcel B). Note that in each case the direction of anticipated mass transfer is from high to low partial pressure, from high to low free energy (via Eq. 4.8). Relative humidity, which we take to be the saturation ratio, can be seen graphically as the fractional distance from the point in question to the equilibrium point. For instance, parcel B has a relative humidity about $0.75 = 75\%$. Parcel A is supersaturated by about 50%. Dew point alternatively specifies the moisture content of a parcel as the temperature T_d needed to maintain the liquid in equilibrium at the given partial pressure. Thus, if parcel B has vapor partial pressure e_B, then its dewpoint temperature is given by $e_B = e_s(T_{d,B})$, whereas parcel A has dew point specified by $e_A = e_s(T_{d,A})$. Subsaturated parcels have dew points less than the temperature and require isobaric cooling to reach saturation, whereas supersaturated parcels have dew points larger than the temperature and require isobaric warming to reach saturation. Deviations of the dew point from the temperature serve as alternative measures of deviations from equilibrium. How much any system deviates from equilibrium, regardless of the measure used, yields information about how the system is likely to respond.

4.2 Rates of change

Change occurs because a system is out of equilibrium, but how the change occurs, and at what rate, requires disciplines beyond thermodynamics. As a general rule, the greater the deviation from equilibrium, the greater the impetus for change. And, we may logically anticipate that the rate of change will be proportional to the deviation. Beyond such generalizations, one must invoke kinetic arguments.

4.2.1 Molecular kinetics

Any imbalance in the fluxes of molecules exchanged between a liquid and the surrounding vapor causes the liquid to either increase or decrease in volume. Refer again to Fig. 3.1 and note the relative magnitudes of the condensation and evaporation fluxes, respectively Φ_{cond} and Φ_{evap}. We saw that equilibrium prevails when $\Phi_{cond} = \Phi_{evap}$, meaning that water vapor molecules condense into the liquid phase as fast as they leave it; no net change in liquid volume occurs at equilibrium, despite the rapid exchange of molecules between the two phases. What would happen if the condensation flux were to exceed the evaporation flux by an amount $\Delta\Phi$? A greater number of molecules would enter the liquid than leave it. The liquid volume would have to increase because of this flux imbalance, while vapor molecules would be removed from the gas phase. The disequilibrium quantity $\Delta\Phi$ could equally well have a negative value, in which case the condensation flux would be insufficient to compensate for the evaporation flux at the given temperature. The liquid volume would then decrease, but the air would become richer in vapor. The difference in the molecular fluxes into and out of the liquid determines the net transport between the two phases.

We can quantify the rate of liquid gain or loss using the same mathematical analysis we applied to the equilibrium situation (review Section 3.1 as needed). In terms of partial pressure e, the net flux is

$$\Delta\Phi \equiv \Phi_{cond} - \Phi_{evap} = \frac{e - e_s}{(2\pi m_v k_B T)^{1/2}} \equiv k_{cond}\Delta e, \tag{4.12}$$

where $\Delta e \equiv e - e_s$ is the excess vapor pressure at the phase boundary and $k_{cond} \equiv (2\pi m_v k_B T)^{-1/2}$ is the proportionality constant, a kinetic coefficient for condensation. The net flux is proportional to the deviation from equilibrium (Δe), but quantification of the rate requires knowledge of the coefficient, which comes from the kinetic theory of gases. In cloud physics, in which it is common to express deviations from equilibrium in terms supersaturation $s \equiv \Delta e/e_s$ (a relative deviation from equilibrium), one could use

$$\Delta\Phi = k_{cond} e_s(T) \cdot s \tag{4.13}$$

to calculate the net flux. At any given temperature, net condensation is now proportional to supersaturation as the driver of change, but one gains a false sense of security because the constant of proportionality $k'_{cond} \equiv k_{cond} e_s(T)$ takes on a much stronger temperature dependence. We will later express rate equations in terms of supersaturation, but it pays to be aware of the distinction between absolute and relative measures of disequilibrium.

Rules for calculating rates of change in all systems follow that discussed for the specific case of water evaporation/condensation. We will later calculate the transport of water, other trace substances, and energy to and from cloud particles. In every case, we need to identify the drivers of change, the deviations from equilibrium, and find ways to calculate the relevant kinetic coefficients. We also need to distinguish between discrete differences across a well-defined interface (e.g., between liquid and vapor) and gradients of the driver within a continuous medium (such as air).

4.2.2 Budgets

The diverse substances and phases contributing to the atmosphere are continually entering and leaving it at various rates. Sometimes a substance enters the atmosphere directly from a source near the ground, while at other times the substance is formed in the atmosphere itself by chemical reactions or changes in phase. Directly introduced substances are primary components, whereas substances formed *in situ* are secondary components. The appearance of a new, secondary substance or phase results in the disappearance of some other substance or phase, because of mass conservation. The loss of a substance is termed a sink, and the balance or imbalance that exists at any moment between the sources and sinks is the budget. Chemical budgets tell us whether substances are building up or diminishing in the atmosphere. Budgets also help us understand how substances and phases are distributed throughout the atmosphere.

The simplest, and also the most useful, framework for developing a chemical budget is the box model. A box model tracks the abundance of a substance, say the number of moles of a given compound X in a specified volume (Fig. 4.2). This volume may be the whole atmosphere or just a portion of it (i.e., the planetary boundary layer or a column above a given area of the surface). In any case, the concentration is assumed to be uniform throughout the volume. The abundance of X in the volume, represented as $N(t)$, increases by chemical production (P), primary emissions (E), and advection of X into the box (F_{in}), whereas deposition onto surfaces (D), losses by chemical processes (C), and advection out of the box (F_{out}) reduce the abundance. Using the concept of mass conservation, we find the time rate of change of the abundance to be equal to the difference between the collective sets of sources ($S = F_{in} + P + E$) and sinks ($L = F_{out} + D + C$):

$$\frac{dN(t)}{dt} = S - L. \tag{4.14}$$

Possible scenarios include increasing N ($dN/dt > 0$), when the strengths of the sources exceed those of the sinks, and decreasing N ($dN/dt < 0$), when the sinks exceed the sources. When the sources and sinks just balance ($L = S$), $dN/dt = 0$ and the abundance remains constant at the steady-state value N_{ss}.

Consider a simple application of the box model when the sources are independent of the abundance in the box (e.g., emissions from a smoke stack, which are clearly not dependent upon what is already in the atmosphere). Loss rates typically depend on the abundance, because if the substance were absent, there would be nothing to lose. The loss rate can

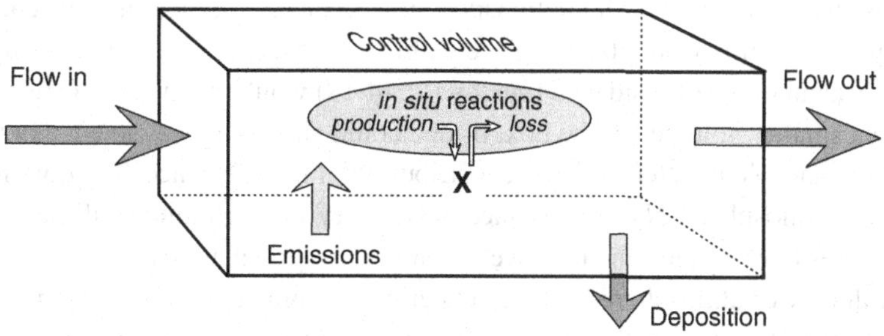

Figure 4.2 Schematic of a hypothetical control volume for defining atmospheric budgets.

be a complicated function of the abundance, but for many applications it is sufficient to consider a first-order loss rate, one proportional to the abundance: $L = kN(t)$. We then get the material-balance equation in the form

$$\frac{dN}{dt} = S - kN. \tag{4.15}$$

If we choose the initial condition $N(t = 0) = 0$, then the solution of Eq. (4.15) is

$$N(t) = \frac{S}{k}\left(1 - \exp\left(-kt\right)\right). \tag{4.16}$$

Note that the abundance initially increases at the rate S and asymptotes exponentially to the steady-state value $N_{ss} = S/k$ as $t \to \infty$.

The residence time, or lifetime, of a constituent is defined as the average time that a compound remains in the box. The lifetime τ can be seen in Eq. (4.15) as the time for the abundance to come within $1/e$ of its final value (N_{ss}): $\tau = 1/k$. We gain more physical insight into the nature of residence time by considering the depletion of X by first-order losses from any initial abundance N_0 after all sources have been shut off. Solving Equation (4.15) for $S = 0$ shows N to decrease exponentially with time:

$$N(t) = N_0 \exp(-kt) = N_0 \exp\left(-\frac{t}{\tau}\right). \tag{4.17}$$

So, we again see that the residence time is inversely related to the reciprocal of the exponential decay constant (i.e., $\tau = 1/k$). A practical way of computing the residence time is to assume that the substance in question is in steady-state ($dN/dt = 0$), in which case Eq. (4.15) yields

$$\frac{1}{k} = \tau = \frac{N_{ss}}{S}. \tag{4.18}$$

The steady-state abundance of X is obtained by measuring its concentration in the atmosphere and multiplying by the volume of the box. The collective sources of X entering the atmosphere (S) are determined from published emissions inventories.

Our substance X has been assumed to be uniformly distributed within the box. Such an assumption is based on the rapid transport of X within the box. If, on the other hand, the lifetime τ is short compared with the transport time scale τ_{trans}, then the concentration of X may vary significantly from place to place within the box, and X would not be uniformly distributed. Calculations of residence time by Eq. (4.18) would be in error. One could perhaps define a smaller volume for the box, or one could simply acknowledge that transport plays a significant role in determining the distribution of the substance. The dominant constituent of the atmosphere, N_2, for instance, has a very long lifetime (millions of years), so N_2 molecules have ample time to move around in the global atmosphere (τ_{trans} months to years in the lower atmosphere) without experiencing much loss. At the other extreme, the OH radical is highly reactive and so lasts just a second or less; it may travel only some tens of meters through the air before a reaction partner destroys it. Substances with lifetimes comparable to or significantly less than the transport time can be expected to vary appreciably throughout the atmosphere.

4.3 Microscale transport

The properties of the atmosphere change and are transported from one place to another because of interactions between matter and energy. On the largest scales, that of the Earth, radiant energy from the Sun heats the surface and atmosphere, forcing air to expand and circulate by differences in pressure. On smaller scales, clouds form as moist air moves upward, cools, and becomes supersaturated. Within clouds, vapor is transported from the interstitial spaces between the particles toward their surfaces, where actual condensation takes place. At still smaller scales, new substances form in the gas phase as molecules of one type collide and react with those of another type. At all scales, matter, energy, and momentum are transported, by one mechanism or another, in response to differences in atmospheric properties. Here, we briefly outline the types of transport processes needed for later use in our study of cloud formation.

The mechanisms by which mass, energy, and momentum are transported within the atmosphere are diverse and dependent on the property being transported. Whole parcels of air move with the wind in response to pressure gradients and carry, or advect, the properties intact in the direction of the wind (vector velocity $\mathbf{v}$). The flux $\mathbf{\Phi}_\chi$ of a specific property χ having local concentration n_χ is just the product: $\mathbf{\Phi}_\chi = n_\chi \mathbf{v}$. If the concentration were everywhere the same, no net change would occur; the property would move to new locations, only to be replaced by air with exactly the same property. Only spatial gradients in the magnitude of the property can lead to change. Thus, the contribution to the Eulerian time rate of change in property χ, that at a fixed location, is given by

$$\frac{\partial n_\chi}{\partial t} = -\mathbf{v} \cdot \nabla n_\chi, \tag{4.19}$$

where the dot product is used because we need the component of the gradient in the direction the wind is blowing; the minus sign is needed so that a wind blowing down the gradient (meaning higher concentrations are coming) correctly leads to a positive derivative and increasing n_χ. Advection is sometimes resolved into horizontal and vertical components. Clouds, because they form in upward-moving air, are important for advecting atmospheric properties vertically.

Properties are also transported by gradients inside a given parcel. As with advection, without gradients, no net change can occur. Some of the most important small-scale transport mechanisms occur in relationship to the growth of cloud particles. We will treat the growth processes in detail later, but here we introduce some terminology and describe the microscale phenomena underlying the different mechanisms of transport. We are interested primarily in the transport of mass, energy, and momentum to a given cloud particle through the surrounding air. Transport on the microscale is brought about by the random motions of the gas molecules, and relatively little by coherent motions of the air.

Molecules in any gas are continually in motion, being agitated because of thermal energy. Collisions with neighbors are frequent, and those collisions send the molecules in arbitrary directions. From the kinetic theory of gases, we know that the average distance between collisions for any molecule, is given by the mean free path,

$$\lambda_{air} = \frac{1}{\sqrt{2}\pi d_g^2 n_{air}}, \tag{4.20}$$

where d_g is the mean diameter of the gas molecules and n_{air} is the molecular density of the air. At standard temperature and pressure $\lambda_{air} \sim 0.1\,\mu\text{m}$. When this uninterrupted travel distance is combined with the mean molecular speed ($\bar{c} = \left[(8k_B T)/(\pi m)\right]^{1/2}$), we can estimate the collision frequency as $c/\lambda = 500\ \text{m}\,\text{s}^{-1}/10^{-7}\ \text{m} \sim 10^9\ \text{s}^{-1}$. These frequent collisions cause the molecules to move relative to each other along random paths, a spreading-out process known as diffusion.

The transport of vapor near a surface occurs via molecular diffusion. Individual water molecules, being constantly jostled about by neighboring air molecules, are forced to move in completely undirected ways. The molecules may each behave randomly, yet the collective behavior yields a coherent, predictable effect (net migration of the vapor). Net transport requires a large population of molecules and a gradient in the average property (a constituent concentration, for instance). We do not need to keep track of individual molecules, as the overall effect is described phenomenologically by Fick's first law of diffusion:

$$\mathbf{\Phi}_j = -D_j \nabla n_j, \tag{4.21}$$

where $\mathbf{\Phi}_j$ is the net flux of constituent j, D_j is the coefficient of diffusion, or simply the diffusivity of j-type molecules in air. Fick's law reflects the difference in molecular impingement onto an imaginary surface oriented perpendicular to the gradient from the high and low sides. The diffusivity of water vapor in air is approximately $D_v \doteq 2 \times 10^{-5}\ \text{m}^2\,\text{s}^{-1}$ near sea level, but it varies inversely with atmospheric pressure. Transport is proportional to the concentration gradient, the driver of transport, while the coefficient D_v provides the kinetic information (note time in the units).

Thermal energy is also transported through the air by molecular motions, only now mass per se need not be transferred, rather just the energy that each molecule carries. In the presence of a temperature gradient, the average energy of the molecules differs with distance, so an energy flux is transported down-gradient, as described by the Fourier law of conduction:

$$\mathbf{\Phi}_T = -k_T \nabla T, \tag{4.22}$$

where k_T is the thermal conductivity. The conductivity of air, about $2.4 \times 10^{-2}\ \text{J}\,\text{m}^{-1}\ \text{s}^{-1}\,\text{K}^{-1}$, varies little with pressure because of offsetting influences: the smaller molecular number concentration at lower pressure is compensated by a greater mean free path.

The transport of momentum is more complicated than that of mass or energy, for both the bulk motions and the molecular nature of the fluid are involved. When molecular interactions are strong, the fluid acts more sticky, or viscous, than one with weaker interactions, and differences in fluid motion are readily damped out. Viscosity is a property of the fluid, a type of resistance to fluid motion; it is the internal friction due to the motion of one part of the fluid relative to another. In the presence of a gradient in velocity (a velocity shear), momentum is transferred from faster-moving fluid toward slower-moving fluid because of molecular collisions. A force (the time rate of change of momentum) is imparted to the slow-moving fluid from the fast-moving fluid, the result being a tendency for the shear to be reduced. This force acts on a unit area parallel to the shear and is called the shear stress τ; a force per unit area (as with pressure, but shear stress acts parallel to the surface, not perpendicular to it). The driver for transport is a gradient of velocity. Velocity is a vector quantity, so a complete specification of shear stress requires a tensor, rather than a vector. However, we gain sufficient insight into the effects of differential fluid motion by considering the simple case of shear normal to flow in one direction. If air moves at speed u along the x direction, for instance, and varies in an orthogonal direction z with gradient du/dz, then the viscous stress is given by

$$\tau = \mu \frac{du}{dz}, \tag{4.23}$$

where μ is the coefficient of dynamic viscosity, or just viscosity. The viscosity of air $\mu_{air} \simeq 1.7 \times 10^{-5}$ Pa s varies with temperature, but little with pressure because the changes in molecular concentration are offset by the changes in mean free path. For many purposes, the dynamic viscosity is replaced by the kinematic viscosity $\nu = \mu/\rho$, where ρ is the mass density of the fluid. Kinematic viscosity has the same SI units as diffusivity, $m^2\,s^{-1}$.

4.4 Formation of new substances

Molecular interactions are responsible for chemical transformations, as they are for microscale transport processes. Each collision between molecules in air, for instance, has one of two possible outcomes: either the collision is elastic or it is inelastic. During elastic collisions the molecules simply exchange kinetic energy, such that the total kinetic energy is conserved. The individual identities of the molecules are retained during elastic collisions, so the net result is a redistribution of kinetic energies among the gas molecules and maintenance of local thermodynamic equilibrium. During inelastic collisions, by contrast, the total kinetic energies of the molecules is not conserved; rather, some of the initial kinetic energy is converted into molecular potential energy. Two molecules colliding inelastically may undergo internal rearrangements and exit the collision with altered identities. The composition of the atmosphere changes because of chemical reactions brought about by inelastic collisions.

In this section we develop the methodology for understanding the various processes that control the concentrations of atmospheric constituents and their influences on clouds. We note that reactions may occur in the gas phase, within liquid droplets, or on the surfaces of particles. We are especially concerned about how trace gases form compounds of low volatility that lead to new particulate matter and cloud condensation nuclei. We first consider gas-phase reactions, then heterogeneous media, such as aerosols and clouds. Stratospheric clouds are especially sensitive to the chemical state of the atmosphere, and they contribute to the loss of the Earth's protective ozone shield. Tropospheric clouds are strongly influenced by the secondary aerosol that results from gas-phase reactions, and the precipitation they release is often acidified by reactions occurring in the cloud water.

How do the collisions between molecules account for the types and rates of reactions observed in nature or the laboratory? Chemical kinetics is the discipline used to explain observable changes in composition in terms of molecular interactions. The subject of chemical kinetics is vast, so our treatment is restricted to gas-phase reactions of importance to clouds in one way or another. Reactions are usually classified as either thermal (or dark, not needing light) or photochemical (initiated by solar radiation), but the atmosphere is a complex environment in which both types can occur simultaneously.

4.4.1 Thermal reactions

Chemical reactions are frequently presented as empirical, net results involving several reactive compounds. The general reaction of a molecules of compound A with b molecules of B and c molecules of C to produce d molecules of D and e molecules of E is written as

$$aA + bB + cC \rightarrow dD + eE, \tag{4.24}$$

with the rates [molecules cm^{-3} s^{-1}]

$$= -\frac{1}{a}\frac{d[A]}{dt} = -\frac{1}{b}\frac{d[B]}{dt} = -\frac{1}{c}\frac{d[C]}{dt} \tag{4.25a}$$

$$= \frac{1}{d}\frac{d[D]}{dt} = \frac{1}{e}\frac{d[E]}{dt} \tag{4.25b}$$

$$= k[A]^l[B]^m[C]^n. \tag{4.25c}$$

Rates are specified relative to the species of interest, but, in the end, the units of the coefficient k must be consistent with the final rate equation, which is called the rate law. Although an overall reaction may look simple, the exponents in its rate equation (l, m, and n) can not be deduced from the reaction itself (Equation (4.24)); rather, they must be determined empirically through laboratory experimentation. We can write the time tendencies for the various constituents in a simple reaction involving two reactants as follows:

$$-\frac{d\,[A]}{dt} = -\frac{d\,[B]}{dt} = k[A][B] = k \times [\text{reactants}]\,. \tag{4.26}$$

Overall, or apparent, reactions and their corresponding rate laws represent useful summaries of complex processes, but our purpose here is best served by focusing on the individual reactions that combine to provide apparent reactions.

All reactions, no matter how complex they may appear, arise from elementary steps involving either one, two, or at most three molecules. Each reaction starts with reactants,

Table 4.1 *Three orders of elementary-step reactions and their rate laws*

Reaction order	Generic reaction	Rate law r_{Rx}	Coefficient units
Unimolecular	$A + h\upsilon \rightarrow C + D$	$\frac{d[C]}{dt} = j[A]$	$\left[s^{-1}\right]$
Bimolecular	$A + B \rightarrow C + D$	$\frac{d[C]}{dt} = k[A][B]$	$\left[cm^3\ molec^{-1}\ s^{-1}\right]$
Termolecular	$A + B + C \rightarrow D + E$	$\frac{d[D]}{dt} = k[A][B][C]$	$\left[cm^6\ molec^{-2}\ s^{-1}\right]$

the molecules before reaction, and ends with products, the molecules that result from reaction. In a most general sense, we can write reactions in the generic form

$$\text{Reactants} \overset{rate\ coeff}{\rightarrow} \text{Products.} \tag{4.27}$$

The arrow shows the direction of the reaction, always from reactants to products, and the symbol over the arrow identifies the coefficient that specifies the speed of the reaction. Sometimes the arrow points toward the left, in which case the molecules on the right are the reactants. A double arrow identifies an equilibrium reaction, one in which the products, once formed, become the reactants for the reverse reaction. Elementary-step reactions involving a single reactant are termed unimolecular; elementary-step reactions with two reactants are bimolecular; and reactions involving three reactants are termolecular. The molecularity, or order of reaction, determines the specific rate law and the units of the rate coefficient. Unimolecular reactions are mostly associated with photochemistry and so will be discussed later. Bimolecular and termolecular reactions are common ways that new compounds form in the atmosphere without the need for light. A summary of the reaction types, their respective rate laws, and the units of the rate coefficients is given in Table 4.1.

Some reactions are reversible:

$$\begin{aligned} A + B &\overset{k_f}{\rightarrow} C + D \\ C + D &\overset{k_b}{\rightarrow} A + B \end{aligned} \tag{4.28a}$$

or

$$A + B \underset{k_b}{\overset{k_f}{\rightleftarrows}} C + D, \tag{4.28b}$$

where the rate coefficient in the forward direction, k_f, in general differs in magnitude and dimensionality from the rate coefficient for the reverse reaction, k_b. Such a system reaches "chemical equilibrium" when the forward and reverse reactions occur at the same rate and the concentrations of all compounds remain constant. At chemical equilibrium of the generic bimolecular reaction, we define the equilibrium constant to be

$$K \equiv \frac{k_f}{k_b} = \frac{[C]\,[D]}{[A]\,[B]}. \tag{4.29}$$

A bimolecular reaction may result when two molecules collide inelastically. Such a reaction can be represented generically as

$$A + B \overset{k_{AB}}{\rightarrow} C + D. \tag{4.30}$$

Reactant molecules A and B collide with a loss of kinetic energy and form product molecules C and D. The symbol k_{AB} refers to the second-order rate coefficient, which will be evaluated later. A specific atmospheric example of a bimolecular reaction is

$$NO + O_3 \overset{k_{NO-O_3}}{\rightarrow} NO_2 + O_2. \tag{4.31}$$

Nitric oxide (NO, emitted by automobiles and other high-temperature combustion operations) reacts with ozone (O_3, a secondary oxidant) to form nitrogen dioxide (NO_2, a compound sensitive to UV light) and ordinary oxygen (O_2). This reaction is fast and so tends to reduce ozone concentrations in urban areas, at least temporarily.

The speed of a reaction is determined by the details of the collision process. Molecules must first come into contact with each other, then the atoms must rearrange to form different atomic combinations (i.e., the new compounds). Breaking the process into two steps (collision and internal rearrangement) allows us to quantify the speed of the generic reaction (4.30) as

$$r_{Rx} = Z_{AB} P_{CD}, \tag{4.32}$$

where Z_{AB} is the collision frequency [$m^{-3}\,s^{-1}$] and P_{CD} is the probability that A and B will rearrange properly to yield the products C and D during the short period of contact. Note that probability is dimensionless, a number between zero and one, so the reaction speed has the same dimensions and units as Z_{AB}.

The collision frequency is calculated from the kinetic theory of gases by visualizing a mathematical cylinder passing through the gas. This so-called collision tube represents the path that molecule A takes through the air in unit time. The tube is not straight, as the non-B molecular collisions send A in many directions; it only matters that we can estimate the total length of the tube in unit time as the relative speed ($\overline{c}_{rel} = \left[(8k_BT)\,/\,(\pi\mu_{AB})\right]^{1/2}$, where μ_{AB} is the reduced mass). Knowing the cross-sectional area ($A_{AB} = (\pi/4)\,(d_A + d_B)^2$) and the tube length allows us to calculate the volume swept out in unit time ($V_A = A_{AB}\overline{c}_{rel}$). Then, knowing the concentration of B molecules (n_B), we calculate the individual collision frequency as $Z_1 = V_A n_B$. Many such collision tubes, one for each A molecule, exist in a unit volume of air, so the total collision frequency is

$$Z_{AB} = \frac{1}{2} Z_1 n_A = \frac{1}{2} A_{AB} \overline{c}_{rel} n_A n_B. \tag{4.33}$$

The factor $1/2$ prevents overcounting of the collisions, because A colliding with B is the same event as B colliding with A.

The speed with which A and B molecules react is next calculated as the product of the collision rate (Eq. (4.33)) and the reaction probability P_{CD} (according to Eq. (4.32)). We thus gain the rate law

$$r_{Rx} = \frac{1}{2} A_{AB}\bar{c}_{rel} P_{CD} n_A n_B \equiv k_{AB} n_A n_B. \tag{4.34}$$

In conventional chemical nomenclature, the bimolecular rate law is written as $r_{Rx} = k_{AB}[A][B]$. Note that reaction rates are in general proportional to the concentrations of reactants, which serve as the drivers of chemical change. The rate coefficient, $k_{AB} \equiv \frac{1}{2} A_{AB}\bar{c}_{rel} P_{CD}$, contains all the details about how fast the given products can be transformed into the designated products and is readily interpreted in terms of the collision rate and the probability of reaction. The rate coefficient for any given reaction contains all the information about the reaction except for the concentrations of the reactants. Reactions described by Eq. (4.34) are said to follow second-order kinetics.

Determining the probability of reaction is difficult, as accessing the inner workings of molecules during the few femtoseconds (10^{-15} s) that a reaction occurs is still an active area of research. Nevertheless, it is expected that an energy barrier must be overcome before A and B can react; old bonds must be broken before new bonds can form. Weak collisions may not offer enough energy to exceed this so-called activation energy. Slow collisions between A and B are not likely to result in new products. Strong collisions, on the other hand, could send the reactants over the barrier and give a higher chance of success. The kinetic energies of molecules, it will be recalled from the Maxwell–Boltzmann distribution, fall off exponentially with energy, so relatively few molecules in a population have the requisite energy for reaction. The fraction of molecules in a gas having energies exceeding the activation energy ϵ_{act} is proportional to $\exp(-\epsilon_{act}/k_B T)$, so temperature becomes a controlling variable. Thus the bimolecular rate coefficient also varies exponentially with temperature, as $k_{AB} \propto \exp(-\epsilon_{act}/k_B T)$. Reactions with large activation energies have strong temperature dependencies and so proceed slowly unless the temperature is elevated. The activation energy can be determined experimentally by measuring the reaction rate over a range of temperatures and interpreting the data in terms of the empirical Arrhenius relationship

$$k_{AB} = A \exp(-\epsilon_{act}/k_B T), \tag{4.35}$$

where A is the pre-exponential factor, which is related to the free energy and entropy of product formation.

We can visualize the reaction process better by hypothesizing the existence of an intermediate stage, which can be shown by rewriting the bimolecular reaction as

$$\mathrm{A} + \mathrm{B} \rightleftarrows \mathrm{AB}^* \overset{k_{AB}}{\rightarrow} \mathrm{C} + \mathrm{D}. \tag{4.36}$$

The species AB*, called the activated complex, is formed by collision of A and B (when sufficient energy is available). It is at this stage that some of the kinetic energy of collision is converted into molecular potential energy. This intermediate compound, the excited addition product (adduct) of A and B, contains extra energy (which we denote by the asterisk) and so is unstable; it may simply fall apart and return back into the reactants (left-pointing arrow), or it may indeed complete its mission and form the products (single arrow to the right). The fate of the complex depends in part on the collision energy and in part on the geometrical alignment of the two colliding molecules. We envision many attempts and few successes at product formation, so the formation of AB^* is shown as a quasi-equilibrium reaction (double arrows); the single arrow suggests that the products form irreversibly.

A ternary reaction involves three reactant molecules, but only two molecules actually collide at any one instant. The third molecule typically collides slightly later, while the first two are still in the process of reacting. Termolecular reactions are found to depend only weakly on temperature, but they depend strongly on pressure and have variable reaction orders. Under some conditions (low pressure), termolecular reactions exhibit third-order kinetics, whereas under other conditions (high pressure), the same reaction set exhibits second-order kinetics. Termolecular reactions are relatively complicated, but they are common in the atmosphere and so deserve additional attention.

The most common type of termolecular reaction is the gas-phase association reaction, the joining of two simple molecules in the presence of a third. We may write an association reaction in the following generic way

$$A + B + M \rightarrow AB + M. \tag{4.37}$$

The molecules A and B are the molecules being combined into molecule AB; AB is the adduct, or addition product. Molecule M is any third molecule, which in the atmosphere is most frequently N_2 and O_2, simply because they are most common. Molecule M serves as a catalyst to the reaction; it is required for AB to become stable, but it reappears after the reaction. One may also write the reaction as

$$A + B \underset{M}{\rightarrow} AB. \tag{4.38}$$

Showing M below the arrow signifies that the catalyst M is required for A and B to combine into AB. An example of a common association reaction in the atmosphere is the formation of ozone (O_3):

$$O + O_2 + M \rightarrow O_3 + M. \tag{4.39}$$

Here, an oxygen atom combines with ordinary oxygen to form O_3. Ozone is a strong oxidant that stimulates the formation of many other reactive compounds. Oxidation in the atmosphere tends to make molecules more soluble and less volatile, properties important for the formation of secondary aerosols, in general, and cloud condensation nuclei, in particular.

The need for a catalyst in association reactions arises from the energy released during bonding. As we learned in Chapter 2, the energy of bond formation must be dissipated, or the colliding molecules simply return to the dissociated, unattached state. The catalyst M collides with the excited adduct (while it still contains the original collision energy) and carries away the excess energy. The catalyst enters the reaction with ordinary thermal energy, but it leaves hot, having gained additional kinetic energy from bond formation. The energy extracted from the excited adduct allows it to settle into a stable configuration (with AB sitting near the minimum of the interaction potential). The catalyst has served its purpose of stabilizing the new compound (AB), while at the same time heating the surrounding gas through energetic elastic collisions.

The kinetics of association reactions can be understood by recognizing the various stages of the process. Let us break the termolecular reaction into three separate reactions as follows:

$$\mathrm{A} + \mathrm{B} \overset{k_c}{\rightarrow} \mathrm{AB}^* \tag{4.40a}$$

$$\mathrm{AB}^* + \mathrm{M} \overset{k_s}{\rightarrow} \mathrm{AB} + \mathrm{M} \tag{4.40b}$$

$$\mathrm{AB}^* \overset{k_d}{\rightarrow} \mathrm{A} + \mathrm{B}. \tag{4.40c}$$

The first subreaction (Step 1) brings the two primary molecules together, but they still contain the original collision energy, which is too much energy for stable bond formation (hence the asterisk). This first reaction is the combination step, which proceeds with bimolecular rate coefficient k_c. Step 2 is the primary stabilization reaction, the one that lets the stable adduct form as a new species (with bimolecular coefficient k_s). It is this step that contributes additional kinetic energy to M. Step 3 represents the failed attempt to form a new compound; the energetic adduct simply falls apart if M does not appear in time (unimolecular coefficient k_d). The last two subreactions are competitive with each other because they offer alternative mechanisms for dealing with the excess energy.

The overall effect of the three steps required for associative combination is determined by considering them as a consecutive set and writing the appropriate rate laws. Product is formed in Step 2 (stabilization) according to the bimolecular rate law,

$$\frac{d\,[\mathrm{AB}]}{dt} = k_s\,[\mathrm{M}]\left[\mathrm{AB}^*\right]. \tag{4.41}$$

We can calculate the rate of product formation if we know $\left[\mathrm{AB}^*\right]$. Changes in the concentration of this excited adduct are found by considering the mass balance:

$$\frac{d\left[\mathrm{AB}^*\right]}{dt} = k_c\,[\mathrm{A}]\,[\mathrm{B}] - k_s\,[\mathrm{M}]\left[\mathrm{AB}^*\right] - k_d\left[\mathrm{AB}^*\right]. \tag{4.42}$$

The first term on the right-hand side represents the rate of formation (Step 1); the second term represents loss of AB* by stabilization (Step 2); and the third term represents loss by decomposition (Step 3). Note that contributions to the formation of the species in question (AB*) appear as positive terms, whereas losses appear as negative terms. All loss terms must also contain the concentration of the substance being tracked. We next factor out the common concentration and assume that the source terms match the loss terms. The steady-state concentration of AB* is thus

$$\left[AB^*\right]_{ss} = \frac{k_c [A][B]}{k_s [M] + k_d}. \tag{4.43}$$

We next substitute this concentration into Eq. (4.41) to gain the overall rate law:

$$\frac{d[AB]}{dt} = \frac{k_c k_s}{k_s [M] + k_d} [M][A][B]. \tag{4.44}$$

Comparison of this equation with that for the general third-order rate law, $d[AB]/dt = k^{III}[M][A][B]$, shows that the third-order rate coefficient is

$$k^{III} = \frac{k_c k_s}{k_s [M] + k_d}. \tag{4.45}$$

The overall rate coefficient depends on the coefficients for the individual steps in the sequence, as well as on the concentration of the catalyst.

The pressure dependence to the kinetics of the association reaction is assessed by analyzing Eq. (4.45). First, recognize that pressure p manifests itself mainly by changing the catalyst concentration via the ideal gas law: $[M] = p/(k_B T)$. Atmospheric pressure changes strongly in the vertical dimension, so [M] varies significantly with altitude. Next, look at the limiting relationships implied by Eq. (4.45) while recognizing that we are still dealing with the generic reaction, so terms like high and low are relative. We gain the low-pressure limit by letting $[M] \to 0$, which makes the first term in the denominator disappear. The rate coefficient in the low-pressure limit is thus

$$k_0^{III} = \frac{k_c k_s}{k_d}, \tag{4.46}$$

and the low-pressure rate law,

$$\left.\frac{d[AB]}{dt}\right|_0 \doteq \frac{k_c k_s}{k_d}[M][A][B] = k_0^{III}[M][A][B] \propto p, \tag{4.47}$$

exhibits third-order kinetics because it depends on all three reactants. As long as $k_s[M] << k_d$, the product AB forms at a rate that depends on pressure, because M is a limiting reagent. At the other extreme, when pressure is relatively high, $k_s[M] >> k_d$ and the high-pressure rate coefficient is

$$k_\infty^{III} = \frac{k_c k_s}{k_s [M]} = \frac{k_c}{[M]} \propto \frac{1}{[M]}. \tag{4.48}$$

The high-pressure rate law is therefore

$$\left.\frac{d[AB]}{dt}\right|_\infty \doteq \frac{k_c}{[M]}[M][A][B] = k_c [A][B]. \tag{4.49}$$

At high pressures, the reaction is independent of pressure and exhibits second-order kinetics with rate coefficient $k_\infty^{II} \equiv k_\infty^{III}[M] = k_c$. At high pressures, plenty of catalyst is always available, so the limiting factor in the formation of AB is the bimolecular combination step. For any specific reaction, one determines the pressure threshold between high and low by comparing the terms in the denominator of Eq. (4.44): $[M] = k_d/k_s \Longrightarrow p = k_B T\ (k_d/k_s)$. Over the entire pressure range, both the dependence on pressure (hence altitude) and the reaction order of association reactions change. Thermal reactions of various molecularities are numerous in the atmosphere, but they ultimately depend on chemical compounds whose reactivities were spawned by photochemical reactions.

4.4.2 Photochemistry

Photochemical reactions play important roles in atmospheric chemistry because they release reactive compounds that subsequently initiate many thermal reactions. A photolytic reaction results when a molecule absorbs electromagnetic radiation and dissociates into its atomic and molecular components. The energy of a photon of frequency ν, $E_\nu = h\nu$, does the work needed to move the atoms against the electrical forces holding the molecule together. Photolysis, chemical decomposition induced by radiation, requires that the photon energy (E_ν) exceed the interatomic bond energy E_b. Chemical bonds are strong, so ultraviolet (UV) radiation is typically needed to break trace gases into fragments that then drive the rest of atmospheric chemistry.

The Sun provides a broad spectrum of radiation that serves as an external source of energy to drive atmospheric chemistry. The Sun, at a temperature of approximately 6000 K, provides the chemically active (actinic) radiation, photons having wavelengths in the UV part of the spectrum. As we learned in Chapter 2, the upper atmosphere removes much of the far-UV radiation. However, near-UV radiation ($290 < \lambda < 400$ nm) penetrates into the stratosphere and troposphere, where it drives much of the photochemistry relevant to clouds.

Photolytic reactions are typically first-order in the reactant, the species experiencing photolysis. A common way to write a photolytic reaction involving generic molecule A is

$$\mathrm{A} + h\nu\ \ (\lambda < \lambda_{th}) \xrightarrow{j} \mathrm{C} + \mathrm{D}, \tag{4.50}$$

where λ_{th} is the threshold wavelength needed for the reaction to produce fragments C and D. The threshold wavelength is seldom sharp, so detailed quantification requires the wavelength-dependent quantum yield $\phi(\lambda)$, the fraction of absorbed photons that actually produce the designated product. The rate coefficient for photolysis, j, sometimes called the photodissociation coefficient, is used in the photolytic rate law, which for production of C is

$$\frac{d\,[\mathrm{C}]}{dt} = j_C\,[\mathrm{A}]\,. \tag{4.51}$$

By way of example, nitrogen dioxide (NO_2) photolyzes in the presence of sunlight to form nitric oxide and an O atom:

$$\mathrm{NO_2} + h\nu\ \ (\lambda < 400\,\mathrm{nm}) \xrightarrow{j_O} \mathrm{NO} + \mathrm{O}. \tag{4.52}$$

A photon having a wavelength less than about 400 nm has sufficient energy to break one of the N-O bonds and release an O atom from the parent NO_2 molecule. The rate law for O formation by NO_2 photodissociation is therefore

$$\frac{d\,[\mathrm{O}]}{dt} = j_O\,[\mathrm{NO_2}]\,. \tag{4.53}$$

Note that photolytic rate coefficients are like first-order rate coefficients for thermal reactions; they represent a frequency and so have dimensions of inverse time [e.g., s^{-1}]. The photolytic rate coefficient is independent of reactant concentration, but it does depend on the availability of photons having sufficient energy for the specified reaction to occur. Photolytic coefficients therefore depend indirectly on altitude, season, and the time of day.

Calculation of the rate coefficient to produce product C by photolysis of parent compound A must take several factors into account simultaneously. The chemically active radiation flux, the number of photons available for photolysis, depends on wavelength (λ), so it is written $I(\lambda)$ and called the spectral actinic flux [SI units, $m^{-2}\,s^{-1}\,nm^{-1}$]. The parent molecule A intercepts the radiation with an effective cross-sectional area $\sigma_A(\lambda)$, which combines with the actinic flux to give the spectral absorption rate coefficient $k_\lambda = \sigma_A(\lambda)\,I(\lambda)$. Factoring in the quantum yield for forming product C at this wavelength, $\phi_C(\lambda)$, then provides the wavelength-specific rate coefficient, $j_{C\lambda} = \phi_C(\lambda)k_\lambda$. We care only about the overall effect of photolysis on the production of C, so we integrate the spectral coefficient over all pertinent wavelengths (λ_{min} to λ_{max}):

$$j_C = \int_{\lambda_{min}}^{\lambda_{max}} \phi_C(\lambda)\sigma_A(\lambda)\,I(\lambda)\,d\lambda. \tag{4.54}$$

The photolytic rate coefficient contains all relevant information about the available radiation field ($I(\lambda)$), as well as the molecular properties of the reactant ($\sigma_A(\lambda)$) and the efficiency for forming the product ($\phi_C(\lambda)$). It is important to distinguish between the absorption of radiation and the ability of that absorbed energy to form the stated product.

Application of these principles to the photodissociation of NO_2 is shown in Fig. 4.3. NO_2 absorbs solar absorption at a rate ($k_\lambda = \sigma_A(\lambda)\,I(\lambda)$) that peaks in the blue end of the visible spectrum (thus causing high concentrations of NO_2 to look reddish brown), but only that portion of the spectrum that overlaps with non-zero quantum yields is able to liberate O atoms. The product $\phi_O(\lambda)k_\lambda$ is the overlap function (bold curve) for this reaction, and its integral (shaded area under the curve) is the NO_2 photodissociation coefficient,

$$j_O = \int_{\lambda_{min}\simeq 300\,\mathrm{nm}}^{\lambda_{max}\simeq 440\,\mathrm{nm}} \phi_O(\lambda)\sigma_{NO_2}(\lambda)\,I(\lambda)\,d\lambda. \tag{4.55}$$

Only a portion of the absorbed radiation serves to produce O atoms, in which case the photon energy is expended mainly in raising the O atoms to higher molecular potential energy. Radiation that is absorbed, but not chemically productive is dissipated by heating surrounding molecules via elastic collisions. The rate of O-atom production is thus calculated by the specific rate law, Eq. (4.53) with j_O from Eq. (4.55). Hidden in the photodissociation

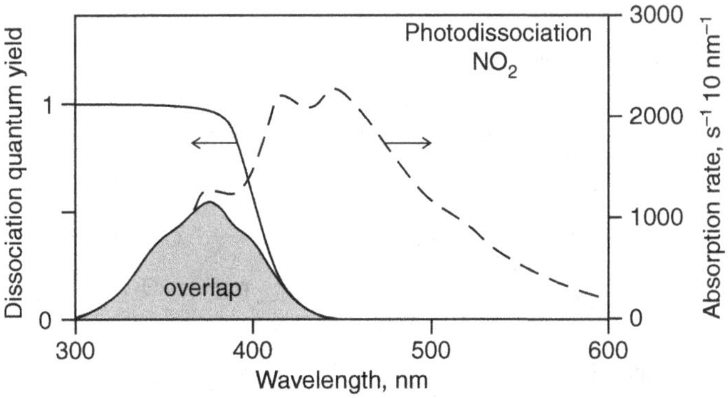

Figure 4.3 Dependence of NO_2 photodissociation on wavelength of incident radiation. Based on data from DeMore *et al.* (1997) and Roehl *et al.* (1994).

rate coefficient is the spectral actinic flux, $I(\lambda)$, so the production of O atoms by NO_2 photolysis is driven by the available sunlight, which is maximum near noon on sunny days during summer. NO_2 photolysis is a key step in the production of odd-oxygen (O, O_3) and hence in the formation of ozone in the troposphere (via reaction (4.39)).

4.4.3 Oxidant chemistry

Ozone forms indirectly by photochemistry and is itself photochemically sensitive. It forms in both the troposphere and stratosphere, and it stimulates the formation of other oxidants and many of the aerosol particles (CCN) that influence cloud properties. Ozone in the troposphere is a secondary pollutant that is harmful to both plants and animals. Stratospheric ozone, by contrast, is beneficial to life because it intercepts and filters out harmful ultraviolet radiation well above the surface. Here, we highlight key elements of oxidant chemistry relevant to CCN formation.

Only one elementary step exists for the production of ozone, in both the troposphere and the stratosphere, namely the association of atomic oxygen with molecular oxygen (via the termolecular reaction (4.39)). Molecular oxygen is readily available in both the troposphere and stratosphere, but the O atoms have distinctly different sources. A summary of the pertinent reactions in each regime is provided in Fig. 4.4. The oxidant chemistries of the troposphere and stratosphere are distinct and need to be studied separately. The origin of the distinction is the difference in the wavelengths of available solar radiation. The short-wavelength radiation needed to photolyze O_2 is available in the stratosphere, but it does not penetrate significantly into the troposphere.

Tropospheric oxidants

Ozone in the troposphere is a common oxidant resulting from photochemistry involving radiation with wavelengths 290 nm $< \lambda <$ 400 nm, the range we call the near-UV. Ozone is a secondary compound because it is formed *in situ* (not emitted). The pathways from

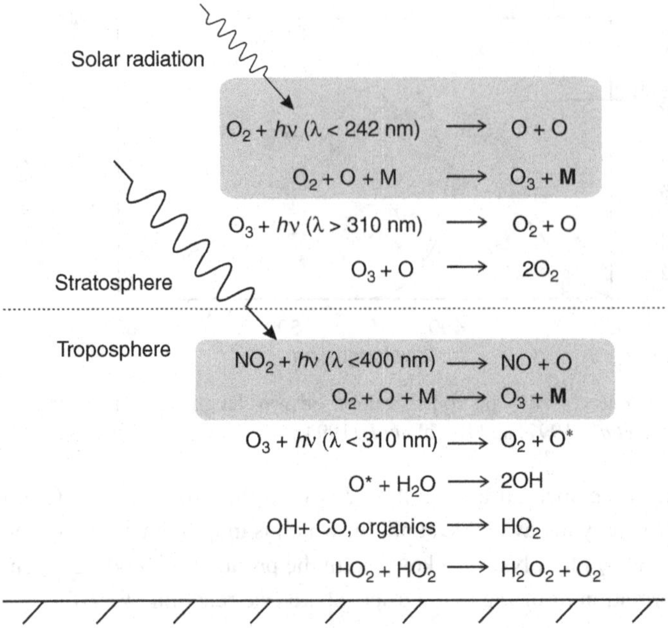

Figure 4.4 Summary of oxidant chemistry in the stratosphere and troposphere. The primary mechanisms for ozone formation are highlighted in gray. Shown for the stratosphere is the classic Chapman cycle, whereas for the troposphere additional reactions are included to illustrate production of other oxidants.

primary precursor gases (those emitted) to ozone are relatively complicated, so we restrict attention to continental air masses, where high concentrations of many pollutants exist. Sunlight initiates photolysis of a few compounds, which then leads to numerous secondary compounds that in turn react thermally to produce still other compounds. Some of these secondary compounds are themselves photosensitive, and some have low volatilities and a propensity to form secondary aerosol particles.

Photochemical smog is the common term for the chemical mixture that arises in urban areas under sunny skies (such as Los Angeles), but the word smog, a contraction for smoke and fog, is best reserved for the particulate matter, the visible part of urban pollution. With rare exceptions, the gases are invisible. Photochemical smogs are highly oxidizing and form during warm, sunny weather. Smogs are also, and more appropriately, associated with sulfurous pollution, the type that forms under cool, dank, and foggy conditions in conjunction with the burning of sulfur-bearing coal. Sulfurous smogs, most notably the killer-smog event of December 1952 in London, can be highly toxic despite the lack of strong oxidants. We focus here on oxidizing situations, because oxidation favors the formation of compounds of high solubility and low volatility, properties important for the production of secondary aerosol particles, many of which become CCN.

Consider the Los Angeles basin as a representative urban environment for our photochemical smog example. Under stagnant meteorological conditions, the boundary-layer air flowing in from the sea is trapped vertically by a strong temperature inversion and horizontally by a ring of mountains inland from the coast. Large concentrations of primary pollutants, such as NO, CO, SO_2, and many volatile organic compounds (VOC), accumulate from industrial and vehicular emissions. Nature also contributes to the VOC abundance through the emissions of terpenes (by coniferous trees) and isoprene (deciduous trees). Such biogenic VOC can dominate the anthropogenic VOC. Ironically, the same meteorological conditions responsible for the popular California sunlight also stimulate the photochemical processes that lead to pollution and visibility degradation. Because of the huge number of pollutant species, the resulting chemistry is complex. We therefore limit the presentation here to a few essential chemical reactions.

Oxidant production starts in the troposphere with the photolysis of NO_2, the only significant source of the O atoms needed for ozone formation. The starting sequence of reactions includes

$$\begin{aligned} NO_2 + h\nu\ (\lambda < 400\,\mathrm{nm}) &\xrightarrow{j_O} NO + O \\ O_2 + O + M &\xrightarrow{k_2} O_3 + \mathbf{M} \\ O_3 + NO &\xrightarrow{k_3} NO_2 + O_2. \end{aligned} \tag{4.56}$$

Note that the sum of all reactants is equal to the sum of all products, so this sequence is a null cycle, meaning that each species is destroyed soon after it is formed. No build-up of ozone is possible by this reaction sequence. Nevertheless, the sequence is essential, so let us analyze it further. The solar UV radiation absorbed by the NO_2 ends up as enhanced thermal energy of the catalyst M (shown bold in the sequence for emphasis) at the end of the ozone-formation step. The O atoms are removed at a rate that roughly balances the rate of production, so one may assume a quasi-steady-state concentration of O, from which one calculates the O_3 concentration:

$$[O_3]_{ss} = \frac{j_O}{k_3}\frac{[NO_2]}{[NO]}. \tag{4.57}$$

When realistic numbers are substituted into Eq. (4.57), the calculated ozone concentration is well below observed values. Therefore, other chemical reactions must be involved, especially those that can convert NO to NO_2 and so increase the $[NO_2]$-to-$[NO]$ ratio.

These other reactions often involve free radicals (atoms or molecules with unpaired electrons) and VOC. NO and NO_2 are themselves radicals, but even more reactive radicals are formed by the photolysis of O_3 from UV radiation with $\lambda < 310$ nm. Once some ozone has formed, new photochemistry opens up to generate still more ozone. Careful investigations have revealed that ozone can photodissociate along either of two pathways:

$$\begin{aligned} O_3 + h\nu &\rightarrow O_2 + O, \quad \lambda > 310\,\mathrm{nm} \\ &\rightarrow O_2 + O^*, \quad \lambda < 310\,\mathrm{nm}. \end{aligned} \tag{4.58}$$

Most of the photolysis events produce ordinary, ground-state atomic oxygen ($O(^3P)$, designated O), but some produce electronically excited $O(^1D)$, which we designate as O^*. An

O^* atom is so energetic that it can break the ordinarily stable polar covalent bonds in the water molecule. Water vapor now becomes a chemical reactant that unleashes the radical chemistry. The radical-initiating reaction is

$$O^* + H_2O \rightarrow OH + OH. \tag{4.59}$$

The hydroxyl radical (OH) is highly reactive and plays a central role in the oxidation of many tropospheric trace gases. Reactive species typically have short lifetimes, because they are so rapidly destroyed in the act of forming still other compounds. OH, in particular, serves as an important cleanser of the atmosphere by increasing the solubility of many compounds and making them vulnerable to removal by clouds.

Reactions involving the OH radical are numerous and best portrayed through bubble diagrams, graphical summaries of alternative chemical pathways. As we see in Fig. 4.5, the OH radical is one member of a chemical family, a group of related compounds that interact more rapidly among themselves than with chemicals outside the family. The rapid interconversion between the two dominant members of the HO_x family, OH and HO_2 (hydroperoxyl radical), is called the HO_x cycle. During one half of the cycle, carbon monoxide (CO) is oxidized indirectly by OH, in the process forming HO_2 via the two sequential reactions,

$$OH + CO \rightarrow H + CO_2$$
$$H + O_2 \rightarrow HO_2,$$

which combine to form the net reaction,

$$OH + CO + O_2 \rightarrow HO_2 + CO_2. \tag{4.60}$$

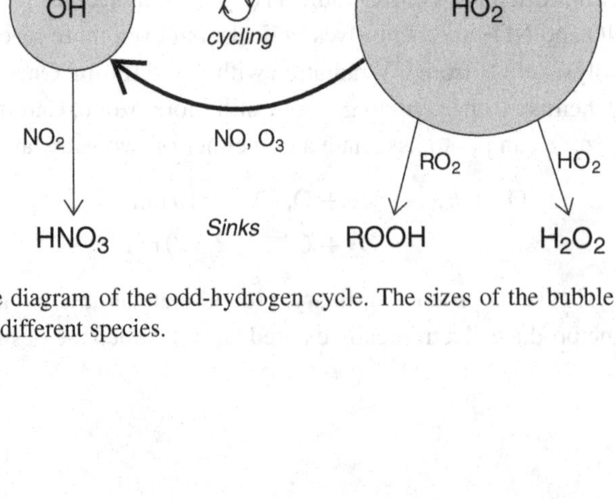

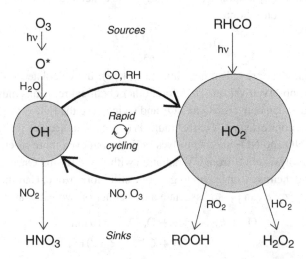

Figure 4.5 Bubble diagram of the odd-hydrogen cycle. The sizes of the bubble reflect the relative abundances of the different species.

Carbon monoxide (CO) offers an important control on the atmospheric concentration of OH and opens up reactions involving HO_2. The other half of the intra-family conversion involves reaction of HO_2 with NO:

$$HO_2 + NO \rightarrow OH + NO_2. \tag{4.61}$$

Note that this reaction simultaneously completes the HO_x cycle and oxidizes NO to NO_2, the main source of the O atoms needed for ozone formation (recall reaction (4.52)).

The odd-hydrogen family interacts more slowly with chemicals outside the family. As we have seen, HO_x species are introduced mainly through the reaction of O* with H_2O, but aldehydes (RHCO) also provide HO_2 through photolysis. An important sink of HO_x includes the reaction of OH with NO_2 to form nitric acid vapor (HNO_3), via the termolecular reaction,

$$NO_2 + OH \underset{M}{\rightarrow} HNO_3. \tag{4.62}$$

Nitric acid is an important trace gas in both the troposphere and stratosphere, as it is a highly soluble acid and provides a reservoir for NO_x (retrieved following photolysis of HNO_3). Gaseous organic acids and peroxides also represent losses of HO_x, by reactions with HO_2, and play important roles on their own. Hydrogen peroxide (H_2O_2), for instance, is not only an important odd-hydrogen termination product via the self reaction

$$HO_2 + HO_2 \rightarrow O_2 + H_2O_2, \tag{4.63}$$

it is also responsible for much acidification of clouds and precipitation. H_2O_2 is both highly soluble and a strong oxidizing agent in liquid water, especially toward dissolved SO_2. Reaction (4.63) thus provides an important link between tropospheric gas-phase photochemistry and heterogeneous chemistry in clouds.

Hydrocarbons and other VOC also contribute to the HO_x cycle, in the process serving to oxidize NO to NO_2 and generate ozone. Reactive hydrocarbon compounds (often designated for simplicity by RH) are readily oxidized by OH because of the abundance of H atoms they contain; OH acts as a water molecule seeking to regain a lost member (an H atom). RH oxidation follows a sequence of reactions having the general form

$$\begin{aligned} RH + OH &\rightarrow R + H_2O \\ R + O_2 + M &\rightarrow RO_2 + M \\ RO_2 + NO &\rightarrow RO + NO_2. \end{aligned} \tag{4.64}$$

Note that this hydrocarbon sequence can be summarized by the single net reaction,

$$RH + OH + O_2 + NO \rightarrow RO + H_2O + NO_2. \tag{4.65}$$

From the point of view of oxidant production, we thus see that VOC, through interactions with the HO_x family, raises the $[NO_2]$-to-[NO] ratio and enhances ozone formation (per Eq. (4.57)). In the simplest of terms, the effect of organic compounds on NO oxidation can be written as

$$NO \underset{VOC}{\rightarrow} NO_2, \quad (4.66)$$

and their stimulating effects on ozone production as

$$VOC + NO_x + h\nu \rightarrow O_3. \quad (4.67)$$

NO and VOC ultimately stimulate the formation of ozone, as long as the VOC are present to ensure the conversion of NO to NO_2. Biogenic VOC, because they often contain double bonds between neighboring carbon atoms, are susceptible to attack by ozone and become even more reactive. Oxidant formation in polluted regions is non-linear, increasing dramatically with the concentrations and types of pollutants. At the other extreme, in clean (e.g., maritime) air masses, the concentrations of all VOC and NO are low, in which case ozone is consumed instead via the competitive reaction,

$$HO_2 + O_3 \rightarrow OH + 2O_2. \quad (4.68)$$

Ultimately though, polluted continental regimes, with their rich sources of organics and nitrogen oxides, are sure to experience frequent episodes of oxidant pollution on warm, sunny days.

Stratospheric ozone

Ozone (O_3) in the stratosphere is considered good, protective ozone, so considerable concern arises whenever its abundance decreases. Depleted abundances of stratospheric ozone have been linked to skin cancers in humans and even to changes in the depths of the troposphere and convective clouds through the impact on the thermal structure of the stratosphere. The chemistry of ozone formation and destruction in the stratosphere involves both gas-phase reactions and heterogeneous interactions with aerosols and clouds.

A mechanism for the formation and maintenance of stratospheric O_3 was proposed by Chapman in 1930. The essential steps in his mechanism include the following sequence:

$$O_2 + h\nu\,(\lambda < 242\,\text{nm}) \xrightarrow{j_1} 2O \quad (4.69)$$

$$O + O_2 + M \xrightarrow{k_2} O_3 + M \quad (4.70)$$

$$O_3 + h\nu \xrightarrow{j_3} O_2 + O \quad (4.71)$$

$$O_3 + O \xrightarrow{k_4} 2O_2. \quad (4.72)$$

Molecular oxygen (O_2) is initially photodissociated by radiation having wavelengths $\lambda < 242$ nm, thereby producing atomic oxygen, an important member of the odd-oxygen family (O, O_3). The O atoms then combine with O_2 to form O_3 in the termolecular association reaction. Ozone itself photodissociates, thus temporarily destroying itself as a molecule, but not the abundance of odd oxygen (sum of [O] and [O_3]). The only real loss of odd oxygen occurs in the fourth step of the Chapman mechanism, when O_3 and O combine and reform stable O_2.

Calculations based on the original Chapman mechanism yield ozone concentrations significantly in excess of those observed in the stratosphere. Moreover, the calculated distribution of O_3 with altitude also disagrees with observations. Such comparisons imply that either the source term for odd oxygen in the Chapman mechanism was too strong, or that important loss mechanisms were not being accounted for. The discrepancies between model results and measurements remained unresolved for decades, until 1970, when Crutzen proposed the role of nitrogen oxides in the destruction of O_3. Odd-nitrogen compounds (mainly NO and NO_2, called NO_x), as do odd-hydrogen (HO_x) compounds, serve as catalysts in the termination of odd oxygen via a chain reaction of the following general form:

$$O_3 + X \rightarrow O_2 + XO \tag{4.73}$$

$$O + XO \rightarrow O_2 + X \tag{4.74}$$

$$\text{Net} \quad O_3 + O \underset{X}{\rightarrow} 2O_2 \tag{4.75}$$

Molecule X (NO when the catalytic cycle is driven by odd nitrogen, or OH in the case of odd hydrogen) abstracts one of the O atoms from O_3 and donates it in turn to the free O atom, thus forming molecular oxygen. Note how catalyst X acts as a transfer agent shuttling an O atom from the oxygen-rich compound (O_3) to the oxygen-poor compound (O), but then returns unscathed at the end of each cycle to initiate another round of odd-oxygen destruction. The catalytic cycle shown above operates effectively in the middle and upper stratosphere, where O is abundant. In the lower stratosphere, however, this mechanism is too slow. The following catalytic cycle circumvents the need for free O atoms:

$$O_3 + X \rightarrow O_2 + XO \tag{4.76}$$

$$O_3 + XO \rightarrow 2O_2 + X \tag{4.77}$$

$$\text{Net} \quad 2O_3 \underset{X}{\rightarrow} 3O_2. \tag{4.78}$$

Some chlorine compounds, too, are effective catalysts of ozone destruction, but cycles based on Cl_x offer other possibilities. When O is plentiful, Cl atoms serve as X in the first catalytic cycle above, but when O is lacking, the following cycle can be effective:

$$O_3 + Cl \rightarrow O_2 + ClO \tag{4.79}$$

$$ClO + ClO + M \rightarrow (ClO)_2 + M \tag{4.80}$$

$$(ClO)_2 + h\nu \rightarrow ClO_2 + Cl \tag{4.81}$$

$$ClO_2 + M \rightarrow O_2 + Cl + M. \tag{4.82}$$

The first step, abstraction of an O atom from O_3, is similar in each of the cycles. The second step, a termolecular self reaction, forms the ClO dimer, $(ClO)_2$, which serves as a Cl reservoir because it releases Cl atoms in the presence of sunlight. The last step represents the thermal decomposition of the intermediate product of dimer photolysis,

ClO_2. When catalytic mechanisms were added to the Chapman mechanism, they yielded estimates of ozone concentration close to observed values. Catalytic cycles, as diverse as they are, all effectively speed up the fourth step in the Chapman mechanism, the termination of odd oxygen. Because of the large effect these catalysts have on ozone concentrations, any changes in catalyst availability will potentially produce large changes in ozone concentrations.

Sources of reactive radicals in the stratosphere arise indirectly from the third step in the Chapman mechanism (ozone photolysis). With the realization that two ozone-dissociation branches exist, depending on the wavelength of radiation (review reaction (4.58)), we now know that photons with $\lambda < 310$ nm generate energetic O atoms (O*). The dominant precursor of reactive nitrogen in the stratosphere, nitrous oxide (N_2O), is itself a source of O* through its photolysis in the UV-rich stratosphere:

$$N_2O + h\nu\ (\lambda < 220\,\text{nm}) \rightarrow N_2 + O^*. \tag{4.83}$$

Regardless how it is formed, O* is highly reactive toward water (reaction (4.59)), producing OH, as in the troposphere. O* also reacts with as-yet unphotolyzed N_2O molecules to form NO via

$$O^* + N_2O \rightarrow 2NO. \tag{4.84}$$

Again, the excess energy in the excited O atom (put there by solar energy) suffices to break the relatively strong bonds in N_2O, an otherwise stable trace gas. Reaction (4.84) is the primary mechanism by which NO_x is introduced into the stratosphere. Reactive chlorine is introduced naturally into the stratosphere when methyl chloride (CH_3Cl, from the biosphere) diffuses through the tropopause and photodissociates:

$$CH_3Cl + h\nu\ (\lambda < 215\,\text{nm}) \rightarrow CH_3 + Cl. \tag{4.85}$$

Both Cl and NO arise naturally and contribute to the loss of stratospheric ozone.

An even more dominant source of stratospheric Cl is the upward migration of man-made chlorofluorocarbon (CFC) compounds, gaseous substances with little reactivity until they photodissociate in the stratosphere. The potential damage resulting from this artificial source of Cl raised concerns among scientists in the 1970s, when it was found that CFCs were widespread. Chlorofluorocarbons were manufactured to be inert and therefore useful in many industrial and household applications. But, that same characteristic gives CFCs long residence times in the atmosphere. Such molecules have adequate time to mix globally and enter the stratosphere, where the high-energy UV photons release Cl atoms from the parent molecules. By way of example, one of these CFC compounds, CFC-12, used extensively as a refrigerant until banned because of its role in ozone destruction, photodissociates once in the stratosphere as follows:

$$CF_2Cl_2 + h\nu\ (\lambda < 220\,\text{nm}) \rightarrow CF_2Cl + Cl. \tag{4.86}$$

These newly freed Cl atoms add to the natural abundance of Cl and disturb the delicate balance of ozone production and destruction in the stratosphere. The first evidence for such an impact was found in the 1980s over Antarctica, where ozone concentrations dropped drastically during each austral spring (especially in October). All of the reactive radicals (especially NO_X and Cl_X) individually contribute to the catalytic loss of odd oxygen in the stratosphere, but they never act in isolation.

The effects of multiple free-radical families coexisting at a given location in the stratosphere are complicated by chemical interactions. The NO_X family, for instance, can alone destroy odd oxygen, as can the Cl_X family. However, certain members of these two families react with each other and limit the catalytic destruction of odd oxygen. The effects of the families are not additive, rather they are competitive and hold each other in check, at least until the stalemate is upset by cloud and aerosol particles. This larger view of family interactions is diagrammed in Fig. 4.6. The left half shows the reactive-nitrogen family, whereas the right half shows the reactive-chlorine family. The key to understanding the limits imposed by interacting families is the formation of relatively inactive compounds, called reservoir species (shown by bold rectangles). NO radicals entering the nitrogen cycle from N_2O photolysis destroy some odd oxygen, but other reactions (left side) produce NO_3 and N_2O_5, which in turn leads to the reservoir compound, HNO_3. The conversion of N_2O_5 to HNO_3 occurs via heterogeneous reactions on sulfuric acid aerosol particles (lower left):

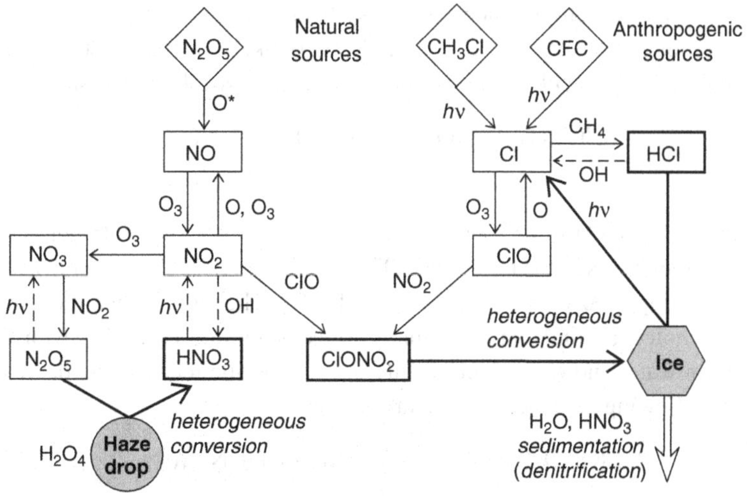

Figure 4.6 Bubble diagram of reactive nitrogen and chlorine in the stratosphere. Gray symbols represent aerosol and cloud particles on which heterogeneous conversion takes place (bold lines). Bold boxes identify reservoir species. Chlorine is activated and nitric acid is sequestered by polar stratospheric ice particles.

$$N_2O_5(g) \rightleftharpoons N_2O_5(aq) \quad (4.87)$$

$$N_2O_5(aq) \rightleftharpoons NO_2^+(aq) + NO_3^-(aq) \quad (4.88)$$

$$NO_2^+(aq) + H_2O(aq) \rightarrow 2H^+(aq) + NO_3^-(aq) \rightleftharpoons 2HNO_3(g) \quad (4.89)$$

$$\text{Net} \quad N_2O_5(g) + H_2O(aq) \rightarrow 2H^+(aq) + 2NO_3^-(aq) \rightleftharpoons 2HNO_3(g). \quad (4.90)$$

The first step is the Henry's law equilibrium of the physically dissolved N_2O_5 in the aqueous solution (aq). The second step is the splitting of $N_2O_5(aq)$ into the ions, NO_2^+ and NO_3^-. Then, NO_2^+ reacts with water to form nitric acid in solution, which tries to maintain a Henry's law equilibrium with gaseous HNO_3. Overall, this reaction sequence that gaseous N_2O_5 is converted into gaseous HNO_3, not directly, rather mediated by liquid aerosol particles: $N_2O_5(g) + H_2O(aq) \rightarrow 2HNO_3(g)$. Nitric acid formation removes NO_2 and thereby limits the effect of odd nitrogen on the destruction of odd oxygen. Chlorine atoms freed from their sources (CH_3Cl and CFC) also tend to get tied up in gaseous reservoirs, especially hydrochloric acid (HCl, upper right of diagram) and chlorine nitrate ($ClONO_2$, center). This latter species locks up both odd nitrogen and odd chlorine and so is particularly effective in hindering the actions of each catalytic cycle.

The chemical impasse, especially in the southern hemisphere, is broken by aerosol and cloud particles. Air within the Antarctic polar vortex is cold and so offers an environment suitable for cloud formation in the lower stratosphere. The sulfate aerosol particles and polar stratospheric clouds (PSC) provide surfaces for heterogeneous reactions. The gaseous reservoir species, HCl and $ClONO_2$, adsorb onto the surfaces of ice crystals and interact with one another in ways not possible in the gas phase. The nature of the interaction is heterogeneous because both a condensed phase (solid ice) and the gas phase coexist at the surface. Following adsorption, HCl and $ClONO_2$ likely interact first with the water molecules of the ice, then with each other on the surface, possibly in two-steps:

$$ClONO_2 + H_2O(ice) \rightarrow HOCl + HNO_3 \quad (4.91a)$$

$$HOCl + HCl \rightarrow Cl_2 + H_2O(ice). \quad (4.91b)$$

Chlorine nitrate is thought to be a planar molecule, which initially hydrogen-bonds with the polar water molecules in the ice surface. The chlorine atom is abstracted from $ClONO_2$ by the water and replaced with a proton, giving HNO_3. The HOCl, once it encounters an adsorbed HCl molecule, finds an opportunity to exchange the H in HCl with the Cl in HOCl to reform water and Cl_2, which readily desorbs. While mechanistic details may be uncertain, the following net reaction does takes place:

$$ClONO_2(s) + HCl(s) \rightarrow Cl_2(g) + HNO_3(s). \quad (4.92)$$

Heterogeneous reactions of this type cause the reservoir species to be dismantled, in the process releasing Cl_2 gas, but sequestering the HNO_3 on the ice surface. In addition to catalyzing the heterogeneous reaction (4.92), the ice particle also removes odd nitrogen from the atmosphere by retaining the HNO_3 in the ice, thereby limiting the return of ClO to the reservoir species $ClONO_2$. Under suitably cold conditions, the ice particles become large enough to sediment to lower altitudes, bringing its burden of HNO_3 along with it. This denitrification of the air permits Cl_2 gas to build up during the dark wintertime. However, with the first return of sunlight, atomic Cl atoms emerge that can reinitiate the O_3-destroying catalytic cycle unencumbered by reactive nitrogen and its tendency to tie up chlorine in reservoirs. Some studies also suggest that chlorine activation can occur on liquid aerosol particles, thus relaxing the temperature constraint for effective ozone depletion. Regardless of the details, it is clear that the so-called ozone hole (region of depleted O_3) that forms over Antarctica each Austral spring is a chemical phenomenon, enhanced by the photolytic breakdown of man-made chlorine-containing compounds and heterogeneous chemistry on aerosol and cloud particles.

4.5 Aerosol formation

The atmospheric aerosol at any instant is the result of many complex, simultaneously acting processes. Some processes are physical, others are chemical in nature. As suggested by Fig. 4.7, processes may be broadly classified as molecular and mechanical. New coarse particles are continually entering the atmosphere mechanically by, for instance, resuspension of dust, release of pollen by trees, volcanic eruptions, and bubble bursting from wave action on the oceans. At the other end of the size spectrum, large numbers of fine particles

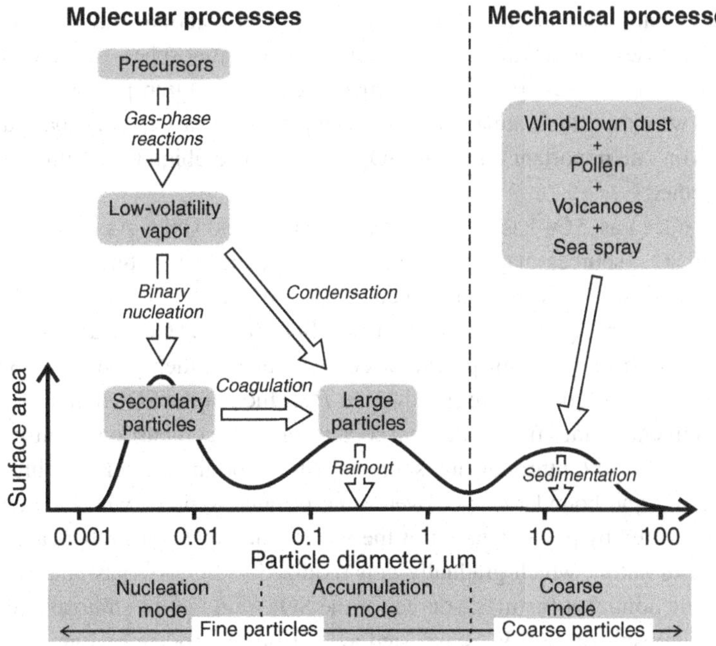

Figure 4.7 Processes that produce the various modes of the atmospheric aerosol, represented schematically in terms of relative surface-area concentration. Adapted from Whitby (1978).

are formed by the molecular process of gas-to-particle conversion. Gas-phase chemical reactions generate new compounds with low vapor pressures that then condense and form new secondary particles. Some of the reaction products may alternatively condense onto larger particles in the accumulation mode, contributing to the aerosol mass concentration, but not to the number concentration of particles. During their residence in the atmosphere, aerosol particles grow either by condensation of vapors or by colliding and coagulating with other particles. Particles are eventually removed by either dry deposition (e.g., sedimentation of the largest particles) or wet deposition (involvement in clouds leading to rainout). The focus here is on the molecular processes leading to new secondary aerosols. The formation of secondary aerosols requires gaseous precursors, the reactants that combine chemically to form new compounds with reduced equilibrium vapor pressures.

4.5.1 Precursors

The precursors of a secondary aerosol are typically trace gases that may have either primary or secondary sources. These precursors may be emitted from point sources (e.g., industrial operations) or from many widely distributed sources (e.g., wetlands). Alternatively, the gases may be the products of prior chemical reactions in the atmosphere. The precursor gases may be relatively simple inorganic compounds, or they may be complex organic molecules. Precursors in the context used here have the common attribute of being the reactants, the starting material, of the gas-phase chemical reactions that lead to particle formation. Trace gases have a wide range of lifetimes in the atmosphere. For the most part, the lifetimes of reactive gases are short or comparable to typical transport times within the atmosphere, so we expect their atmospheric concentrations to vary in both space and time. Here, we focus on one important trace gas, SO_2, because it so ubiquitously impacts clouds one way or another.

As with any trace gas, SO_2 has its own distinct sources and sinks. As shown in Fig. 4.8 for the case of SO_2, sources may be natural (i.e., volcanoes) or anthropogenic (derived from human activities; i.e., electric power generation). We can further distinguish primary sources (emissions E in Eq. (4.14)), those that emit SO_2 directly into the atmosphere, from secondary sources, those that emit precursor compounds that then produce SO_2 through chemical reactions in the atmosphere (production P). Much of the SO_2 found in the atmosphere over continents comes from industrial sources through combustion of sulfur-bearing fuels, such as oil or coal (anthropogenic sources). An important secondary source of SO_2 is found in the marine boundary layer over some oceanic regions, where dimethyl sulfide (DMS) is emitted by phytoplankton in the water. The precursor in this case is DMS, a reduced form of sulfur, which gradually gets oxidized to SO_2 via gas-phase reactions. Volcanoes are the dominant (natural) source for the SO_2 found in the stratosphere.

Once in the atmosphere, SO_2 is transported away from the source region and becomes subject to a variety of loss processes. Gas-phase chemical reactions may transform SO_2 to particulate sulfate (SO_4^{2-}), a gas-to-particle conversion process that reduces the

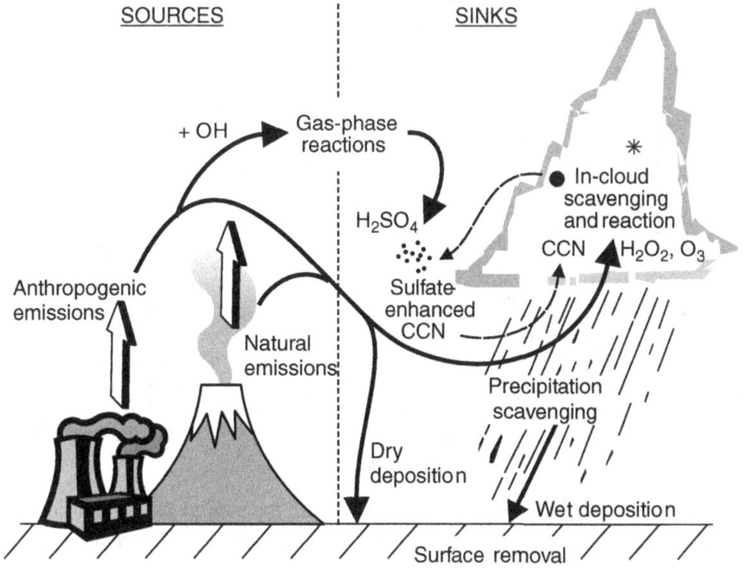

Figure 4.8 Sources and sinks of SO_2.

concentration of SO_2, but which simultaneously increases the sulfate concentration. Chemical processes change the concentrations of compounds, but they do not by themselves eliminate the atoms constituting those compounds. Real losses result from deposition (D in Eq. (4.14)) of compounds onto the surface. Deposition can take either of two pathways: dry deposition, which includes the decomposition or sorption of a compound on a surface, and wet deposition, the process by which the compound is taken up by cloud particles and carried out of the atmosphere in precipitation. Clouds are instrumental in taking up SO_2 and providing a medium in which the sorbed gas gets oxidized to sulfate (SO_4^{2-}) within the cloud drops. Sulfur atoms, and indirectly SO_2, are subsequently lost from the atmosphere through precipitation (wet deposition). Alternatively, when sulfur-containing cloud drops are detrained from clouds and evaporate, sulfur-enriched particles may be produced that serve as effective cloud condensation nuclei (CCN).

A chemical budget for SO_2 requires that all significant sources and sinks be quantified. When the rate of input exceeds the rate of removal, noticeable increases in concentration occur. As shown in Fig. 4.9, the anthropogenic emissions of SO_2 have increased dramatically over the past ~ 150 years, so much so that they now surpass the estimated natural fluxes of all reduced forms of sulfur. This increase in SO_2 emissions is largely the result of increased combustion of coal for electric power generation and home heating. Regional-scale controls over the release of SO_2 from power generation, when they exist, do result in decreased concentrations of this pollutant in the region. A global-scale budget for a trace gas like SO_2 is challenging to determine because of the variable nature of the SO_2 lifetime relative to atmospheric transport times.

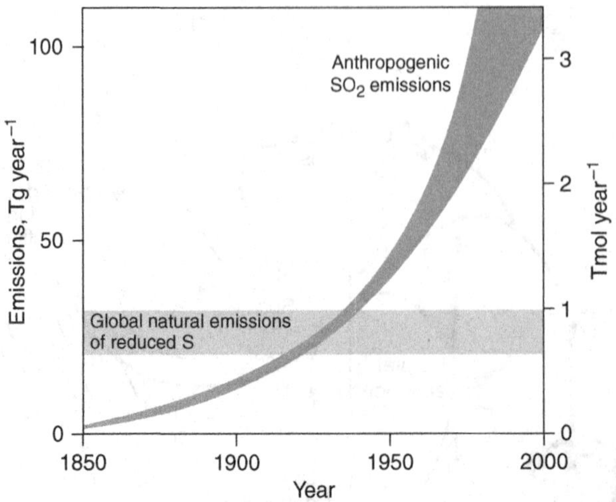

Figure 4.9 Emissions of antrhopogenic SO_2 compared with the natural global emissions of reduced S. Widths of curves reflect the ranges of estimated values. Adapted from Charlson *et al.* (1992).

The concentrations of other gases, especially carbon dioxide (CO_2) and methane (CH_4), have also increased rapidly over the last century. Such trends show that the composition of our atmosphere is not constant, rather that it is continually changing. However, these gases are minor trace gases with low concentrations. Do we mean to imply that these variations in concentration are not important for the atmosphere? Unfortunately not. Even small increases in the concentrations of greenhouse gases (e.g., water vapor, CO_2, CH_4) contribute to a warming of the climate. Some trace gases (e.g., CO, NO_2, O_3), too, are chemically reactive and lead to potentially harmful compounds. Our concern then is more with what trace gases do, rather than just how much exists in the atmosphere. The impacts and importance of a substance depend both on its abundance and on its reactivity. Reactive trace gases often involve oxidation, which makes the reaction products more attractive to water and therefore susceptible to forming secondary aerosols.

Organic compounds, too, can act as precursors of secondary aerosol particles and therefore of cloud condensation nuclei (CCN). Literally hundreds, if not thousands of organic compounds are released into the atmosphere, either directly through primary emissions or indirectly through *in situ* chemical reactions. Many of these compounds are volatile and reside in the gas phase until oxidized. Some trees are prolific sources of reactive hydrocarbons, such as isoprene (from deciduous trees) and terpenes (from coniferous trees). Isoprene and other alkenes contain double carbon bonds that are readily attacked by ozone, making them highly reactive and good sources of HO_x and ultimately secondary organic aerosols (SOA). The source strengths of biogenic VOC emissions typically vary with location, the season, and local meteorological conditions. Formaldehyde (CH_2O) has both primary and secondary sources, as it is released by industry and is an intermediate by-product of methane oxidation. Organic solvents are commonly emitted into the atmosphere from evaporation of industrial and household products. Both natural and anthropogenic VOC serve as precursors of secondary aerosols and CCN.

4.5.2 Secondary aerosols

The conversion of gases into particles starts with a gas-phase chemical reaction and ends with new particulate matter suspended in the air. The reactants of the chemical reactions are some of the trace-gas precursors discussed in the previous subsection. The reactions themselves take place among the reactants through inelastic collisions of the precursor molecules. The products of those reactions are substances that have new physical properties. The most appropriate products for the formation of secondary-aerosol formation exhibit a tendency to bond well to water and other compounds that may aid in the formation of a condensed phase (e.g., liquid). Sometimes the condensed phase must be nucleated, that is formed only after a critical concentration has been reached (because an energy barrier must be overcome), but sometimes it appears that the particles form almost as rapidly as the condensible vapors come together. The former case of aerosol formation is nucleation-limited and thermodynamically driven, whereas the latter case is kinetically limited; both are sensitive to the concentrations of the involved compounds.

The formation of secondary sulfate aerosols starts with the conversion of SO_2 into sulfate (SO_4^{2-}), the product that readily condenses into particles. Recall that sulfur dioxide is a modestly soluble trace gas that will not, by itself, form particles under atmospheric conditions; it must first be oxidized. The chemistry associated with SO_2 oxidation under summertime conditions is schematically illustrated by the bubble diagram in Fig. 4.10. Two basic pathways exist for chemical conversion (SO_2 to SO_4^{2-}), gas-phase reactions involving OH and aqueous-phase (in-cloud) reactions involving H_2O_2. Note that we cannot ignore the HO_x cycle, even when our primary concern here is SO_2 oxidation. The abundance of HO_x is stimulated by ozone photolysis in the presence of water vapor, so SO_2 oxidation proceeds most effectively during the summer season. The aqueous pathway takes place inside liquid cloud drops via the net reaction

$$H_2O_2(aq) + SO_2(aq) \rightarrow H_2SO_4(aq). \tag{4.93}$$

The sulfuric acid formed by the aqueous pathway is already inside cloud droplets, so this pathway adds new solute to the droplets, but it does not generate new particles per se. The added solute content in the droplets indirectly contributes to the aerosol if and when the droplets evaporate, releasing the solute back into the particulate phase. On the other hand, the gas-phase oxidation of sulfur dioxide results in a vapor and takes place as follows:

$$\begin{aligned} SO_2 + OH + M &\rightarrow SO_2 \cdot OH + M \\ SO_2 \cdot OH + O_2 &\rightarrow SO_3 + HO_2 \\ SO_3 + H_2O + M &\rightarrow H_2SO_4 + M. \end{aligned} \tag{4.94}$$

Note that this reaction sequence is initiated by the termolecular reaction of SO_2 with OH, but that HO_2 is formed in the second, bimolecular step. Gas-phase oxidation of SO_2 by

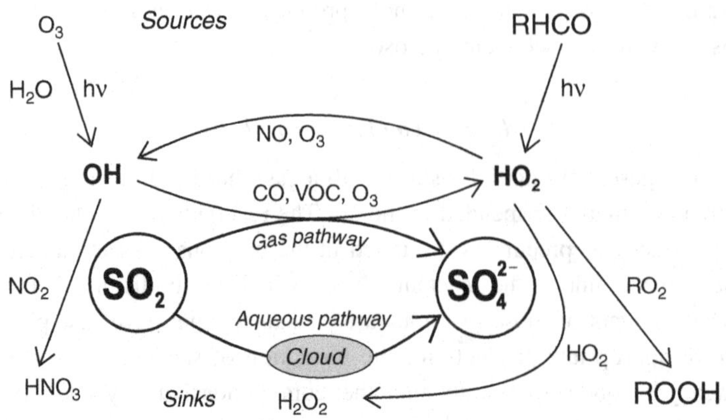

Figure 4.10 Bubble diagram of sulfur oxidation. From Stein and Lamb (2002), Chemical indicators of sulfate sensitivity to nitrogen oxides and volatile organic compounds. *J. Geophys. Res*, **107** (D20), 4449. Copyright 2002 American Geophysical Union. Reproduced by permission of American Geophysical Union.

OH preserves the total abundance of odd hydrogen, so the HO_x cycle can continue its effectiveness in atmospheric chemistry. The last, termolecular step combines SO_3 with water vapor to form sulfuric acid vapor.

The conversion of sulfuric acid (H_2SO_4) vapor into liquid particles of aqueous sulfuric acid occurs by heteromolecular homogeneous nucleation. The term heteromolecular indicates that the process involves two or more chemical species, here H_2SO_4 and H_2O, sometimes also NH_3; the word homogeneous suggests that the nucleation of the new phase (liquid) occurs in one (gas) phase. Here, we restrict attention to binary nucleation involving the vapors H_2SO_4 and H_2O in otherwise clean air. Binary nucleation is favored over single-component (homomolecular) nucleation because each component helps the other component condense. For the same reason that sulfuric acid is highly soluble in liquid water, namely the strong bonding between sulfate ions $\left(SO_4^{2-}\right)$ and H_2O, is the same reason that the combination of H_2SO_4 and H_2O gives lower vapor pressures and a stronger propensity to condense than either component alone (refer to the discussion of binary nitric acid–water solutions in Chapter 3).

The rate of new-particle formation depends strongly on temperature and the concentrations of the condensing vapors (generically A and B). Nucleation of a condensed phase from multiple vapors occurs when enough of the A and B vapor molecules collide and stick together to overcome the free-energy barrier for particle formation. This barrier arises because a liquid particle, unlike a gas, has a surface in which the molecules of the condensed phase are missing neighbors on the vapor side. The surface free energy (same as surface tension) acts as a cost, a penalty to form the new phase. Even pure substances have such a cost, but the cost to form binary solutions is less than those of the pure systems individually. The strong mutual attraction between the A and B molecules reduces the

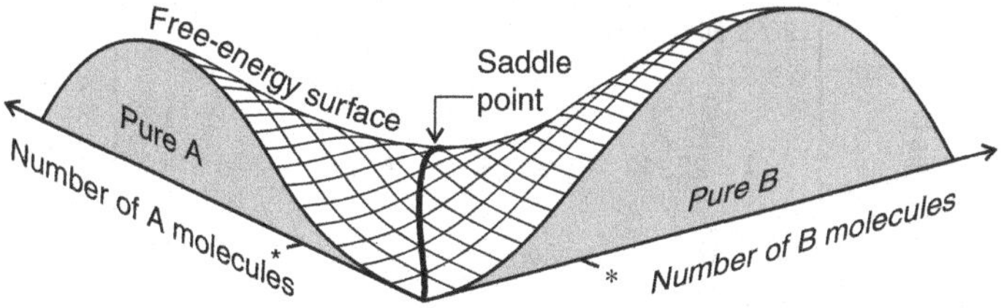

Figure 4.11 Free-energy surface for a generic binary system of A and B. The shaded regions identify the free energy to form particles of the pure compounds. The heavy curve passing through the saddle point represents a possible pathway up and over the free-energy barrier. The tick marks (*) along the axes identify the numbers of molecules of each type in the critical nucleus.

equilibrium vapor pressure. Once vapor pressure comes into play, then the process must be dependent on temperature (review Chapter 3). The free energy to form liquid droplets from a binary vapor also depends on the number of molecules in the incipient particle, hence also on the size of the particle. With two independent variables (numbers of A and B in the particle), the free energy can be viewed as a two-dimensional surface in the shape of a saddle, as seen in Fig. 4.11. Note that the curve is everywhere lower than the maxima of the pure components. The rate of binary nucleation is exponentially related to the height of the free-energy surface at the saddle point, ΔG^*_{AB}, as follows:

$$J_{AB} = C \exp\left(-\frac{\Delta G^*_{AB}}{k_B T}\right), \tag{4.95}$$

where the pre-exponential factor C is a coefficient that reflects the kinetics of molecules entering the particle and climbing the energy barrier (through the saddle point). The binary-nucleation rate depends strongly on the vapor concentrations because of their ability to suppress the free energy of the particle. The situation applied to sulfuric acid–water system is shown in Fig. 4.12. The nucleation rate is strongly dependent on the acid concentration at any given temperature. Water vapor is always in great abundance compared with sulfuric acid vapor, so the nucleation rate is limited by the availability of acid molecules. High rates of acid production from the gas-phase reactions (4.94) favor rapid nucleation and the formation of new secondary aerosol particles interstitially between the existing particles. Lower chemical rates of acid production limit the nucleation rate and let the acid vapors deposit preferentially onto the surfaces of existing particles. Note that new particles can form directly from binary vapors even when the individual vapors are subsaturated with respect to the pure liquids.

Secondary aerosols form frequently in the atmosphere, and the resulting particles have important consequences, such as contributing to hazes and helping initiate clouds. The example of sulfuric acid generation from gas-phase oxidation of SO_2 in the presence of water vapor serves as an example that applies equally well in the stratosphere as in the troposphere. SO_2 may be injected directly into the stratosphere by powerful volcanoes and there oxidized to sulfuric acid vapor by the same chemical mechanisms discussed above

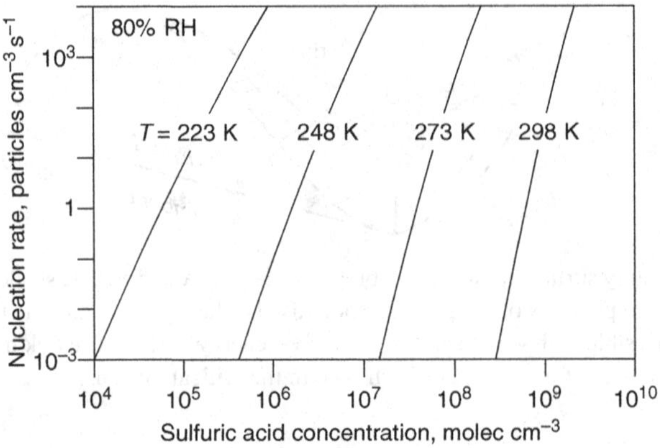

Figure 4.12 Rate of binary nucleation in the sulfuric acid–water system at a relative humidity of 80% with respect to pure water. Reprinted from *Atmos. Environ.*, vol. **23**, Jaecker-Voirol and Mirabel. Heterogeneous nucleation in the Sulfuric acid–water system, 2053–2057. Copyright (1989), with permission from Elsevier.

(4.94). The acid vapor then condenses into the sulfate particles that form the so-called Junge layer. This sulfate layer surrounds the globe in the lower stratosphere, persists for a couple of years following significant eruptions, and affects the global energy budget by scattering solar energy back to space. In the troposphere, sulfuric acid vapors are produced when the SO_2 emitted from coal burning, for instance, gets oxidized. The resulting H_2SO_4 vapor condenses into new particles that contribute to summertime hazes (at least in the eastern United States) and sulfate episodes in the boundary layer. Secondary particles, being submicron in size and likely combined with other substances, readily enter the lungs and circulatory systems of humans. Beyond possible health effects, secondary aerosols result in visibility degradation and reductions in solar insolation at the surface.

Organic compounds contribute to the secondary aerosol, too, by adding the oxidized fragments of VOC released from vegetation. Biogenic VOC commonly contain double bonds between some carbon atoms (making them olefins), which are easily attacked by ozone. The molecular fragments resulting from the severing of the double bonds are often reactive and lead to several additional reactions. Even in the absence of sulfuric acid and other mineral acids (e.g., nitric acid), the oxidation of biogenic VOC may contribute to visible effects, known as the blue haze. Some researchers have suggested that molecular clusters (~ 1 nm) of inorganic sulfate serve as embryos upon which organic vapors condense, allowing the particles to grow into true, stable aerosol particles. This hypothesis suggests that nucleation and growth are decoupled. Many questions remain about the precise chemical and physical mechanisms responsible for the appearance of new particles in the atmosphere.

The growth of secondary particles from molecular clusters to particles large enough to impact clouds requires both condensation and coagulation. Condensation results when individual vapor molecules collide with and stick to a particle, in the process adding its mass to the particle. Coagulation is the process of particles moving about and randomly colliding with other particles in the vicinity. Fewer particles, but larger ones, result from coagulation. These two processes are interactive, as condensation causes the particles to grow and become more massive and less mobile, making them less likely to coagulate with neighboring particles. Coagulation, for the same reason, is self-limiting, because the resulting, enlarged particles move more slowly and find fewer new partners within a given time interval. Nevertheless, both processes invariably move a population of particles up the size spectrum, from small to large, from the nucleation mode into the accumulation mode (refer again to Fig. 4.7). Particles in the accumulation mode are massive by comparison with those in the nucleation mode, so coagulation effectively stops, and new condensation of vapors contributes relatively little additional mass. These large particles literally accumulate in the size range around 0.1 nm; they are neither tiny, nor huge, but they nevertheless represent an important subset of the atmospheric aerosol. Accumulation-mode particles contain relatively much soluble matter, which helps them gather water vapor from the surrounding air; they swell up to sizes comparable to the wavelength of light and so are able to scatter radiation more efficiently than when they were smaller. We will come to appreciate the microscale processes in greater detail later, but for now just recognize that the particles on which cloud drops form ultimately stem from the chemical reactions and nucleation processes that lead to secondary aerosols.

4.6 Further reading

Bohren, C.F. and B.A. Albrecht (1998). *Atmospheric Thermodynamics*. New York: Oxford University Press, 402 pp. Chapter 7 (transport phenomena).

Brasseur, G. and S. Solomon (1986). *Aeronomy of the Middle Atmosphere*. Dordrecht: D. Reidel, 452 pp. Chapter 2 (chemical thermodynamics and kinetics, photochemistry).

Curry, J.A. and P.J. Webster (1999). *Thermodynamics of Atmospheres and Oceans*. London: Academic Press, 467 pp. Chapters 3 (transport processes) and 4 (phase-change thermodynamics).

Hobbs, P.V. (2000). *Introduction to Atmospheric Chemistry*. Cambridge: Cambridge University Press, 262 pp. Chapters 2 (chemical lifetimes), 5 (tropospheric chemistry), and 6 (aerosol sources).

Jacob, D.J. (1999). *Introduction to Atmospheric Chemistry*. Princeton, NJ: Princeton University Press, 266 pp. Chapters 3 (box model), 8 (aerosols), 9 (chemical kinetics), 10 (stratospheric ozone), 11 (tropospheric oxidants), and 12 (ozone pollution).

Pruppacher, H.R. and J.D. Klett (1997). *Microphysics of Clouds and Precipitation*. Dordrecht: Kluwer Academic Publishers, 954 pp. Chapters 8 (aerosol formation), 11 (aerosol dynamics), and 13 (diffusion, conduction).

Rogers, R.R. and M.K. Yau (1989). *A Short Course in Cloud Physics*. New York: Pergamon Press, 293 pp. Chapter 2 (free energy change).

Seinfeld, J.H. and S.N. Pandis (1998). *Atmospheric Chemistry and Physics*. New York: John Wiley and Sons, 1326 pp. Chapters 2 (chemical cycles, lifetimes), 3 (photochemistry), 4 (stratospheric chemistry), 5 (tropospheric chemistry), 6 (aqueous-phase chemistry), 8 (aerosol dynamics), 9 (aerosol thermodynamics), 10 (aerosol nucleation), and 23 (chemical-transport models).

Turco, R.P. (1997). *Earth under Siege*. New York: Oxford University Press, 527 pp. Chapters 5 (air pollution), 6 (urban pollution), and 13 (stratospheric ozone).

Wayne, R.P. (1985). *Chemistry of Atmospheres*. New York: Oxford University Press, 361 pp. Chapters 3 (chemical kinetics, photochemical principles), 4 (stratospheric ozone), and 5 (tropospheric chemistry).

Wayne, R.P. (1988). *Principles and Applications of Photochemistry*. New York: Oxford University Press, 268 pp. Chapter 1 (photochemical principles).

4.7 Problems

1. Show how the impingement flux ($\Phi = (1/4)\,\overline{c}n$) can be used to obtain an expression for the diffusion coefficient D in terms of mean free path λ and mean thermal speed $\overline{c}$ of molecules in a gas. Let the concentration $n = n(x)$ vary only in the x-axis and have the gradient dn/dx.

2. Water enters the atmosphere a vapor via evaporation near the Earth's surface, and it is removed by precipitation. For this exercise, assume that water enters the atmosphere at a constant rate $E = 10^{15}$ kg d^{-1} and that the atmosphere can be treated as a single well-mixed reservoir. Let m be the collective mass of water (in all forms) in the atmosphere at any given instant of time t. Water is removed from the atmosphere (via precipitation) at a rate D that is proportional to the atmospheric abundance (m), let k be the proportionality constant.
   (a) Set up the differential equation that describes the instantaneous rate of change of m.
   (b) Integrate this material balance equation, under the conditions that E and k remain constant, to obtain an explicit equation for describing the build-up of water as a function of time starting with $m = 0$ at $t = 0$ (i.e., the atmosphere starts out dry).
   (c) What is the abundance of water in the atmosphere once steady state has been reached?
   (d) What is the steady-state precipitation rate?
   (e) What is the average residence time of water in the atmosphere?

3. Calculate the change in thickness of thin ice covering a fresh water lake during
   (a) a 12-h night
   (b) a 12-h day

   if the latitude is 60 °, the atmospheric transmissivity is 0.7 for solar radiation, long wave radiation from the atmosphere is 200 W m^2 and the ice temperature remains constant at 0 °C.

   (Hint: Because the ice is very thin, assume that the energy received at the interface will immediately be felt throughout the ice layer.)

4. Professor Dummkopf, a brilliant student of atmospheric physics, carelessly left a beaker of concentrated sulfuric acid open and exposed to the room air on his lab bench while he went on a long summer vacation. Upon his return, he was surprised to find that the level of liquid in the beaker had not changed appreciably. He had fully expected,

because sulfuric acid is hygroscopic and a good desiccant, that the solution would have absorbed large quantities of water vapor from the humid room air. Should Professor Dummkopf have been so surprised, or did he overlook some important process?

First, qualitatively explain the phenomenon of the slowly changing level of liquid in the beaker in terms of fundamental physical processes. Then, set up equations that would allow one to calculate the rate of rise of the liquid level, being careful to relate the mathematical relationships to the physics discussed initially. Treat the liquid–air system as a one-dimensional, constant temperature problem and look for quasi-stationary conditions. Consider the initial depth of liquid and the concentration of the acid to be known, and let the ambient relative humidity at the mouth of the beaker be known and invariant.

5. In cloud physics, common use is made of the concept of supersaturation, defined here as $s = \Delta e/e_s$, where $\Delta e = e - e_s$ is the excess partial pressure of water vapor in the air, i.e., the difference between the actual partial pressure e and the reference value e_s needed to maintain equilibrium with a flat surface of pure water. Show here, through the use of basic thermodynamic principles, that the supersaturation is a direct measure of the thermodynamic driving potential for the vapor-to-liquid phase change. In particular, show that $s \simeq \Delta\mu/RT$ when $|s| \ll 1$, where $\Delta\mu$ is the difference in the chemical potential between the vapor and liquid phases, R is the universal gas constant, and T is the temperature of the isothermal system.

6. We will compare some features of the Chapman mechanism (Eqns. (4.69) through (4.72)) at two different altitudes: 20 km ($T = 200$ K, $n_{air} = 1.8 \times 10^{18}$ molec cm^{-3}, $n_{O_3} = 3 \times 10^{12}$ molec cm^{-3}) and 45 km ($T = 270$ K, $n_{air} = 4.1 \times 10^{16}$ molec cm^{-3}, $n_{O_3} = 0.1 \times 10^{12}$ molec cm^{-3}).
   (a) Identify the sink reactions for atomic oxygen and ozone, respectively.
   (b) Which molecule is M in Eq. (4.70) most likely to be?
   (c) Which reaction is the dominant sink for each species?
   (d) What is the residence time of O at 20 km and 45 km? How do these times compare to atmospheric motion time scales?
   (e) Assuming steady-state concentrations, calculate the $[O]/[O_3]$ concentration ratio at 20 km and at 45 km altitude. (Hint: What does steady-state concentration imply?)
   (f) Show that the mass balance equation for odd oxygen $[O_x] = [O_3] + [O]$, ignoring transport terms, can be written as

$$\frac{d\,[O_x]}{dt} = P - k\,[O_x]^2$$

   where $P = j_1$ and $k = 2k_3k_4/k_2 y_{O_2} n_{air}^2$.
   (g) Express the lifetime of O_x as a function of k and $[O_x]$, then calculate the lifetime of O_x at 20 km and 45 km altitude.
   (h) Based on your answer for (f), in what part of the stratosphere would you expect $[O_x]$ to be closer to being in steady state?

7. A pan with a small amount of dry salt is placed inside a sealed enclosure, the bottom of which is covered with a layer of pure water. The pan with salt is not in direct contact with the water (it is suspended above the water), and the entire system is at constant temperature.

   Explain what you expect to happen over time. Justify your expectation in terms of physical processes, then interpret the expected result in thermodynamic and mathematical terms. How can you use this simple demonstration to show that chemical potential is a free energy?

Part III

Cloud macrophysics

5

Cloud thermodynamics

5.1 Overview

Seeing an isolated thunderstorm from a distance can be awe-inspiring, partly because of its size, partly because of the apparent sense of organization. At this macroscopic level, we may see distinct turrets and sharp, bumpy edges along one side of the storm, evidence of turbulent motions and the rapid penetration of moist, cloudy air into the dry surroundings. The other side of the storm, by contrast, may look diffuse and wispy, evidence of gentler air motions and less abrupt distinctions between cloudy and clear air. These visible macroscale features of mature storms evolved from smaller convective elements in response to the effects of energy conversions on the air motions. We realize that cloudy air is composed of many small, subvisible particles that eventually become precipitation, but it is the macroscale structure of the storm as a whole that compels us to understand the relevant energy conversions and the connections between the microphysics of condensate formation and the macrophysics of cloud development.

The conversion of energy from potential to kinetic is of fundamental importance to cloud formation and evolution. Sometimes this conversion occurs on the atmospheric mesoscale (as when air motions respond to pressure readjustments), sometimes at the microscale (as in the release of latent heat during condensation). We come to understand energy transformations, regardless of scale, through the discipline of thermodynamics. The fundamental principles of thermodynamics are here applied to clouds. We focus on the macroscale, but we must realize that some of the energy utilized by whole clouds arises at the molecular level during condensation. The latent heat of phase transformation is intimately linked to storm energetics and organization, the macrostructures we see in the sky.

5.2 Characterization of a moist atmosphere

A cloud, being composed of water in both condensed and vaporous states, naturally responds to the concentration of water in the air. Energy conversions, in particular, depend on the relative amounts of water in the various phases. Cloudy air is a multicomponent, heterogeneous system, consisting of moist air and condensate (the liquid and solid particles). Moist air itself is the binary mixture of dry air and water vapor, where dry air is the

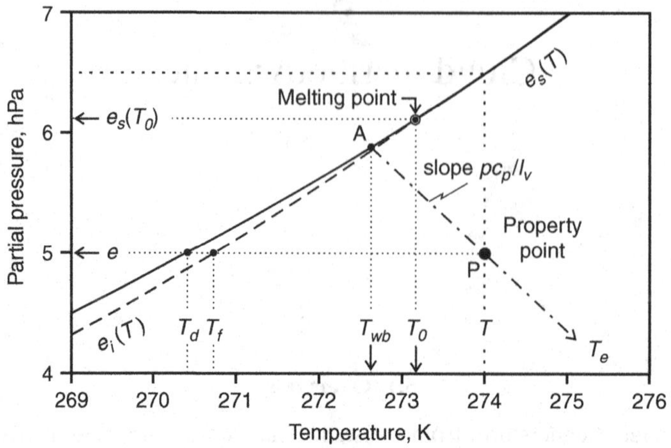

Figure 5.1 A partial phase diagram of water to illustrate relationships between relevant moisture variables.

mixture of all atmospheric gases except water vapor; the partial pressure of dry air is here represented by p_d. Important variables that help us quantify the concentration of water vapor include the partial pressure of vapor e, the molar density of vapor n_v, and the dewpoint temperature T_d. The dewpoint temperature, defined by $e = e_s\,(T_d)$ is the temperature to which a moist air mass must be cooled (or warmed in the case of supersaturated air) at constant pressure in order to bring the air to saturation. This process is graphically illustrated in Fig. 5.1 for moist air, the property point of which is identified by the large black dot (labeled P). When the dewpoint temperature is below the ice point (T_0), the dewpoint temperature is found where the horizontal e line intersects the curve (light solid) defining equilibrium with supercooled liquid water. A frost-point temperature T_f may then also be defined, where e intersects the ice-saturation curve (dashed), $e_i(T)$.

The relevant thermodynamic variables are related to one another through the equation of state. Moist air acts much as an ideal gas does, so we use the ideal gas equation, which is most simply written in molar form:

$$p = p_d + e = nRT, \tag{5.1}$$

where $n = n_d + n_v$ is the molar density of the moist air.

The microphysical processes that lead to cloud particles are often coupled to the dynamics that drive air motions in a cloud, where gravity (acting on mass) is the dominant force. Thus, it is expedient to develop the thermodynamical relationships with mass-specific variables. We can rewrite the ideal gas equation in terms of mass density ρ_{air} by using the mean molar mass (molecular weight), $M_{air} = \sum_j y_j M_j = y_d M_d + y_v M_v$, where y_j is the mole fraction of component j, and M_j is the molar mass, respectively, of dry air (j = d) and water vapor (j = v). To aid interpretation, we often represent the molar mass of the gas explicitly in each appropriate equation to emphasize when mass is needed. Because the mole fraction of water vapor is highly variable and often separately tracked when evaluating or modeling atmospheric processes, it is convenient to express the equation of state in forms showing the vapor content explicitly. With the humidity expressed as a mole fraction, $y_v = 1 - y_d$, we obtain

$$p = \rho_{air} \frac{R}{M_{air}} T = \rho_{air} \frac{R}{M_d} \frac{M_d}{M_{air}} T = \rho_{air} R_d \left(\frac{1}{1 - y_v(1-\varepsilon)} \right) T, \tag{5.2}$$

where $\varepsilon = M_v/M_d = 0.622$ is the ratio of molar masses of water and "dry" air, respectively, and in which the ratio R/M_d has been replaced by the dry-air gas constant R_d. A mass-based expression for the amount of water vapor in the air is the specific humidity q_v, defined as

$$q_v \equiv \frac{\rho_v}{\rho_{air}} = \frac{n_v M_v}{n_d M_d + n_v M_v} = \frac{\varepsilon y_v}{1 - y_v(1-\varepsilon)} \simeq \varepsilon y_v, \tag{5.3}$$

where we used $n_d = n - n_v$ to show its relationship to vapor mole fraction (y_v). Another expression for humidity, commonly used in the meteorological community, is the (mass) mixing ratio of water vapor r_v, defined as

$$r_v \equiv \frac{\rho_v}{\rho_d} = \frac{n_v M_v}{n_d M_d} = \frac{\varepsilon n_v}{n_d} = \frac{\varepsilon e}{p_d} = \frac{\varepsilon y_v}{1 - y_v} = \frac{q_v}{1 - q_v}. \tag{5.4}$$

Here, we have used the ideal gas law and the previous relationships to obtain equivalent expressions for r_v in terms of the other moisture variables. Now, we can replace y_v in Eq. (5.2) with q_v (using Eq. (5.3)) to obtain the alternative mass-based equation of state,

$$p = \rho_{air} R_d \left(1 + q_v \left(\frac{1-\varepsilon}{\varepsilon} \right) \right) T = \rho_{air} R_d T_v, \tag{5.5}$$

where we have introduced the virtual temperature $T_v = (1 + q_v (1/\varepsilon - 1)) T \simeq (1 + 0.61 q_v) T$. The virtual temperature is the temperature a perfectly dry parcel would have were it to have a density equivalent to that of the moist parcel at the same physical temperature. The virtual temperature is always greater than the physical temperature, with the biggest deviations in a humid boundary layer (see Fig. 5.2). The virtual temperature may also be defined in terms of the other moisture variables, giving

$$T_v = \frac{T}{1 - y_v(1-\varepsilon)} = \left(\frac{1 + r_v/\varepsilon}{1 + r_v} \right) T. \tag{5.6}$$

Thus, we can rewrite Eq. (5.2) in terms of r_v (using Eq. (5.4)), obtaining

$$p = \rho_{air} R_d \left(\frac{1 + r_v/\varepsilon}{1 + r_v} \right) T = \rho_{air} R_d T_v. \tag{5.7}$$

Describing an ideal gas (which the atmosphere well approximates) in terms of the masses of its constituents adds mathematical complexity and undermines the inherent simplicity of the universal (molar-based) gas laws, but such is the sacrifice some atmospheric scientists make when developing mathematical models of the atmosphere.

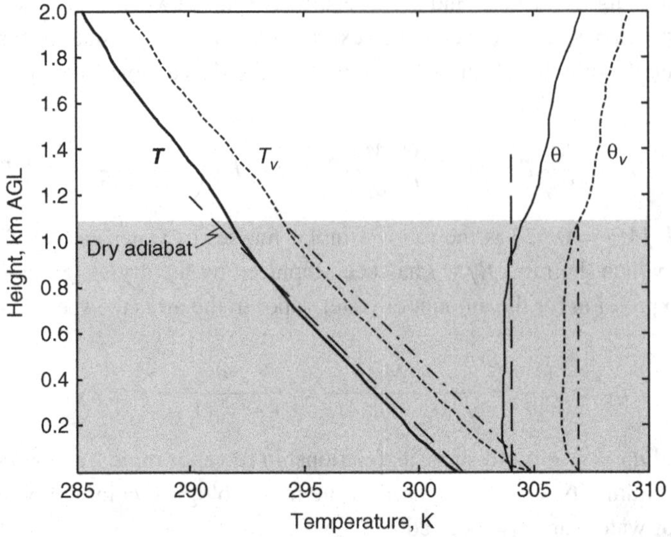

Figure 5.2 Vertical profiles of temperature T, virtual temperature T_v, potential temperature θ, and virtual potential temperature θ_v. The shaded region represents the atmospheric boundary layer, defined by the intersection of the dashed-dot lines with the virtual-temperature profiles. The dashed curves represent profiles for the dry-adiabatic reference process.

5.3 Reference processes

Clouds result from changes in the phases of water. Vapor gets transformed into liquid, for instance, by condensation onto water droplets. Any such change in the physical state of a system implies an energy transformation, as well, so at least some thermodynamic analysis is needed. We here examine two possible reference processes, idealized scenarios by which the phase changes in a representative parcel of air. No real process meets the criteria of a reference process, but understanding the idealization helps us interpret natural phenomena. In the first scenario, the parcel remains at a given height, whereas in the other scenario, the parcel moves vertically. We assume that both processes are adiabatic, although the first is also isobaric (pressure remains constant). In the isobaric case, we assume a pre-existing deviation from equilibrium to prevail, whereas when the parcel moves vertically, the equilibrium state is continually disturbed by the changing state of the system.

Mathematical tools are developed below to determine both the equilibrium amount of water in each phase and the impact of the phase change on the system properties. Heating rates (q) are most conveniently linked to the simultaneous rates of change of temperature and pressure through the enthalpy form of the first law of thermodynamics.

5.3.1 Adiabatic isobaric processes

Many atmospheric processes are well approximated by the adiabatic ($q = 0$) and isobaric ($dp/dt = 0$) reference process. An example is the mixing of two air parcels or the evaporation of liquid at a fixed height. Enthalpy is conserved for such a process, so for our multi-phase system

$$h = y_d h_d + y_v h_v + y_l h_l, \tag{5.8}$$

where the subscript l represents the liquid phase. Defining y_w as the mole fraction for total water (all phases) and applying the conservation principles, the enthalpy change for our process is calculated from the total derivative of Eq. (5.8) as

$$\begin{aligned} \frac{dh}{dt} &= y_d \frac{dh_d}{dt} + y_v \frac{dh_v}{dt} + y_l \frac{dh_l}{dt} + h_v \frac{dy_v}{dt} + h_l \frac{dy_l}{dt} \\ &= y_d c_{pd} \frac{dT}{dt} + \frac{d}{dt} \left(y_v \left(h_v - h_l \right) \right) + y_w c_l \frac{dT}{dt}. \end{aligned} \tag{5.9}$$

With the definition of latent heat of vaporization ($l_v \equiv h_v - h_l$), the conservation of enthalpy requires that

$$\left(y_d c_{pd} + y_w c_l \right) \frac{dT}{dt} = -\frac{d}{dt} (l_v y_v). \tag{5.10}$$

This expression suggests a definition of the specific heat capacity for the cloudy system as

$$c_p = y_w c_l + y_d c_{pd} \, . \tag{5.11}$$

We use Eq. (5.10) to determine the so-called wet-bulb temperature T_{wb} to which a volume of air will be cooled by evaporating water into it at constant pressure until saturation is reached. We find the expression for the wet-bulb temperature by integration over the process, which yields (assuming that the heat capacity and latent heat do not depend on temperature)

$$c_p \left(T - T_{wb} \right) \simeq l_v \left(y_{vs} \left(T_{wb} \right) - y_v \right). \tag{5.12}$$

The final vapor mole fraction is the saturation value $y_{vs} = e_s/p$. This isobaric adiabatic process is illustrated in Fig. 5.1, where the property point of initial (gaseous) state of the volume is indicated by the large dot at (T, e). As liquid evaporates, the temperature decreases (because of conversion of thermal energy into latent heat) and the partial pressure of vapor increases along the process line AP, the slope of which is given by pc_p/l_v, as required by Eq. (5.12). The temperature at which the process line intersects the saturation vapor pressure line is the (isobaric) wet-bulb temperature (T_{wb}). We can also calculate the final, equivalent temperature (T_e) of the parcel under the assumption that all vapor in the parcel condenses out and is removed. Integration of Eq. (5.10), from the initial state to the final state (no water), yields

$$T_e = T + y_v \frac{l_v}{c_p}. \tag{5.13}$$

Graphically, the equivalent temperature is determined by following the same process line (AP) used to determine the wet-bulb temperature, but now toward the right in Fig. 5.1. The temperature at which the partial pressure is zero is the (isobaric) equivalent temperature T_e. The difference between the equivalent and physical temperatures ($T_e - T$) is another useful measure of the concentration of moisture in a parcel. This temperature increase represents the potential energy that exists between water molecules in the vapor phase (see discussion regarding Fig. 2.7). Likewise, the decrease in temperature toward the wet-bulb temperature results from the conversion of molecular kinetic energy into molecular potential energy of the now-broken intermolecular bonds.

5.3.2 Adiabatic expansion or compression

Although our primary interest is in multi-phase heterogeneous systems, we also need to understand the effects of a vertical displacement of moist, but subsaturated air. An example of subsaturated vertical motion is the flow of air up the side of a mountain, or the rise of a thermal in the boundary layer. Here, we consider unsaturated vertical displacement before considering a complete multi-phase system. Our departure point is the enthalpy form of the first law of thermodynamics:

$$q = c_p \frac{dT}{dt} - v \frac{dp}{dt}. \tag{5.14}$$

Here, $v = M_{air}/\rho_{air}$ is the molar volume of the moist air. When we exchange the time derivative for a height derivative, we obtain the temperature lapse rate under the assumption that the air parcel rises toward lower pressure without heating (i.e., $q = 0$):

$$-\left.\frac{dT}{dz}\right|_{dry} = -\frac{M_{air}}{c_p \rho_{air}} \frac{dp}{dz} = \frac{M_{air} g}{c_p} \equiv \Gamma_d, \tag{5.15}$$

where we have used the hydrostatic-balance equation ($dp/dz = -\rho_{air} g$). The mean molar mass of the parcel air (M_{air}) appears in the equation because we use molar-specific units and like to show mass explicitly when gravitational attraction is important. In some treatments, M_{air} is combined with c_p to give the mass-based specific heat $c'_p = c_p/M_{air}$. In either case, the mean specific heat should be adjusted for the different humidities of the air.

The dry-adiabatic lapse rate Γ_d, called the dry adiabat because no condensation occurs within the parcel, is closely associated with an important conservative property. Let us rearrange Eq. (5.14), again using the ideal gas equation, to obtain

$$\frac{d \ln T}{dt} - \frac{R}{c_p} \frac{d \ln p}{dt} = \frac{d}{dt}\left(\ln T + \ln p^{-R/c_p}\right) = 0, \tag{5.16}$$

from which we get an expression for the relationship between temperature and pressure,

$$T p^{-R/c_p} = const, \tag{5.17}$$

for a dry-adiabatic displacement. This relationship is used to define potential temperature, a variable that lets us compare temperatures at different pressure levels in the atmosphere. The comparison is made by adiabatically moving all parcels, from any level, to the common pressure level of $p_0 = 1000$ hPa. The potential temperature θ is thus defined by the relationship

$$\theta = T\left(\frac{p_0}{p}\right)^{R/c_p}. \tag{5.18}$$

Potential temperature is conserved for dry-adiabatic displacements of unsaturated, but moist air, so each dry adiabat is associated with a unique value of θ. For applications in the boundary layer (as in Fig. 5.2), where the effect of water vapor on air density is greatest, the preferred (nearly) conserved temperature variable for a dry-adiabatic process is the virtual potential temperature, defined as

$$\theta_v \equiv T_v\left(\frac{p_0}{p}\right)^{R/c_{pd}}. \tag{5.19}$$

Here, we have negated the need to include the contribution of water vapor to the heat capacity of the air in Eq. (5.18) by replacing the temperature with the virtual temperature. A comparison of virtual potential temperatures from two different air masses at the same pressure level yields a comparison of their virtual temperatures, and hence densities.

Variations of measured and calculated variables with altitude are often depicted graphically. Each altitude or pressure level in the atmosphere has a certain temperature and a measurable vapor concentration. The upper part of Fig. 5.3 shows the results of a sounding, the air temperature (jagged solid curve) and the dewpoint temperature (dash curve) measured during the ascent of a balloon released from the ground. Note how the temperature decreases steadily in the boundary layer (up to the 860-hPa level), then increases slightly before decreasing irregularly all through the free troposphere. Potential temperature, too, may be calculated for each pressure level, so it is frequently convenient to represent the atmospheric temperature profile on a θ–z, rather than T–p graph (lower part of Fig. 5.3). As the expanded view of this sounding shows in the boundary layer (Fig. 5.2), θ is relatively constant in the lowest kilometer or so, a reflection of the mixing that allows the air to be approximated by a dry adiabat.

With these relationships for unsaturated ascent/descent defined, we are now ready to look at a multi-phase system. Consider a parcel, consisting of moist air and liquid water (as droplets), that is displaced upward. Both pressure and temperature decrease, which in turn causes the saturation vapor pressure to decrease and the concentration of liquid to increase. If we assume no external heating for the parcel $q = 0$, we say that such a process is wet adiabatic (in contrast to dry adiabatic). In a wet-adiabatic process, the air temperature changes not only because of expansion, but also because of vapor condensation. If we assume that the parcel is a closed system, the process is reversible and both entropy ($\varphi = y_d\varphi_d + y_{vs}\varphi_v + y_l\varphi_l$) and total water ($y_w = y_{vs} + y_l$) must be conserved. Inside the cloud, we assume that the vapor mole fraction has the saturation value, $y_{vs} = e_s(T)/p$. The specific entropy for our multi-phase system is calculated as

$$\varphi = y_d\varphi_d + y_{vs}\varphi_v + y_l\varphi_l, \tag{5.20}$$

the total change of which is given by

$$\frac{d\varphi}{dz} = y_d\frac{d\varphi_d}{dz} + y_v\frac{d\varphi_v}{dz} + y_l\frac{d\varphi_l}{dz} + \varphi_v\frac{dy_{vs}}{dz} + \varphi_l\frac{dy_l}{dz}. \tag{5.21}$$

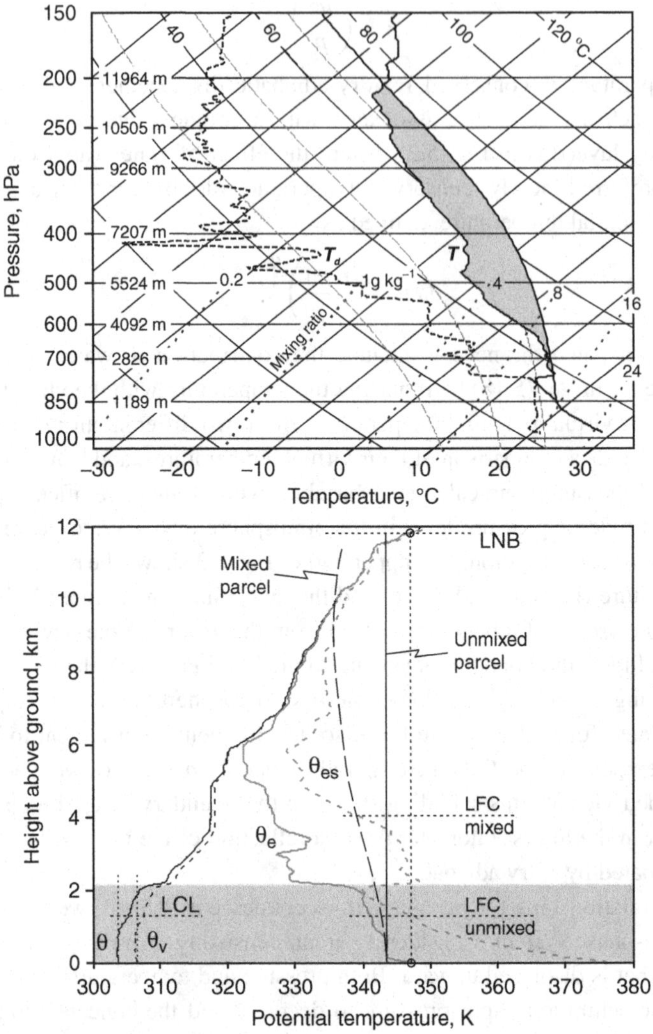

Figure 5.3 Radiosonde profiles from the DOE-ARM site at Lamont, OK, 2:30 pm June 5, 2001, when tornadic storms were observed in the area. Top: Skew-T profiles with convective available potential energy (CAPE) indicated by shading. Bottom: Profiles of conserved variables, showing the surface-parcel lifting condensation level (LCL) and the levels of free convection (LFC), and the level of neutral buoyancy (LNB). The solid vertical line may apply to a more realistic parcel. Also indicated are characteristic values for a parcel experiencing mixing during ascent (long-dashed curve). Some of the features introduced here are discussed further in Chapter 6.

From the definition of entropy Eq. (3.9) and Eq. (5.14), we can write the entropy changes with height for the dry air as

$$\frac{d\varphi_d}{dz} = \frac{c_{pd}}{T}\frac{dT}{dz} - \frac{R}{p_d}\frac{dp_d}{dz} \tag{5.22}$$

and for the water vapor as

$$\frac{d\varphi_v}{dz} = \frac{c_{pv}}{T}\frac{dT}{dz} - \frac{R}{e_s}\frac{de_s}{dz} = \frac{c_{pv}}{T}\frac{dT}{dz} - \frac{R}{e_s}\frac{de_s}{dT}\frac{dT}{dz} = \frac{1}{T}\frac{dT}{dz}\left(c_{pv} - \frac{l_v}{T}\right), \tag{5.23}$$

where we have used the Clausius–Clapeyron equation (Eq. (3.28)) to express the saturation vapor pressure dependence on temperature. The entropy change of the liquid is approximated well by

$$\frac{d\varphi_v}{dz} = \frac{c_l}{T}\frac{dT}{dz}. \tag{5.24}$$

Substituting these expressions into Eq. (5.21) and applying conservation of water gives

$$\frac{d\varphi}{dz} = y_d\left(\frac{c_{pd}}{T}\frac{dT}{dz} - \frac{R}{p_d}\frac{dp_d}{dz}\right) + \frac{y_{vs}}{T}\frac{dT}{dz}\left(c_{pv} - \frac{l_v}{T}\right) + \frac{y_l c_l}{T}\frac{dT}{dz} + (\varphi_v - \varphi_l)\frac{dy_{vs}}{dz}, \tag{5.25}$$

which may be simplified further by recalling how entropy is related to latent heating: $\varphi_v - \varphi_l = l_v/T$. Thus,

$$\frac{d\varphi}{dz} = \frac{1}{T}\frac{dT}{dz}\left(y_d c_{pd} + y_{vs} c_{pv} + y_l c_l - \frac{y_{vs} l_v}{T}\right) - \frac{y_d R}{p_d}\frac{dp_d}{dz} + \frac{l_v}{T}\frac{dy_{vs}}{dz}. \tag{5.26}$$

Using the product rule of calculus, we can write the last term as

$$\frac{l_v}{T}\frac{dy_{vs}}{dz} = \frac{d}{dz}\left(\frac{y_v l_{vs}}{T}\right) + \frac{y_v l_{vs}}{T^2}\frac{dT}{dz} - \frac{y_{vs}}{T}\frac{dl_v}{dT}\frac{dT}{dz}, \tag{5.27}$$

which, after replacing the latent heat dependence on temperature (dl_v/dT) with its appropriate relationship (Eq. (3.40)), yields

$$\frac{d\varphi}{dz} = \frac{1}{T}\frac{dT}{dz}\left(y_d c_{pd} + (y_l + y_{vs})\, c_l\right) - \frac{y_d R}{p_d}\frac{dp_d}{dz} + \frac{d}{dz}\left(\frac{y_{vs} l_v}{T}\right). \tag{5.28}$$

For an isentropic (reversible, adiabatic) process we get

$$\frac{d\varphi}{dz} = 0 = \frac{d}{dz}\left(\frac{l_v y_{vs}}{T}\right) + c_p \frac{1}{T}\frac{dT}{dz} - \frac{y_d R}{p_d}\frac{dp_d}{dz}, \tag{5.29}$$

where the specific heat for the multi-phase system is defined as before (Eq. (5.11)). Moreover, we have assumed that the heat capacities are not functions of temperature (it varies by less than 1% over the typical atmospheric range) and that $v_v \gg v_l$ in the Clausius–Clapeyron equation (Eq. (3.28)). We can write Eq. (5.29) in a physically more revealing form by substituting for dry air pressure $p_d = p - e_s$ in the last term, which allows us to rewrite it as

$$\frac{y_d R}{p_d}\frac{dp_d}{dz} = \frac{n_d}{n}\frac{R}{p_d}\left(\frac{dp}{dz} - \frac{de_s}{dz}\right)$$
$$= \frac{n_d R}{np_d}\frac{pM_{air}g}{RT} + \frac{n_d R}{np_d}\left(\frac{l_v e_s}{RT^2}\right)\frac{dT}{dz}$$
$$= M_{air}\frac{g}{T} + \frac{y_{vs} l_v}{T^2}\frac{dT}{dz}, \tag{5.30}$$

after substitution of the hydrostatic equation (Eq. (2.66)), the ideal gas law (Eq. (5.2)) and the Clausius–Clapeyron equation (Eq. (3.28)). Substituting Eq. (5.30) back into Eq. (5.29) and solving for the temperature lapse rate yields

$$-\left.\frac{dT}{dz}\right|_{saturated} = \frac{M_{air}g}{c_p} + \frac{1}{c_p}\frac{d}{dz}\left(y_v l_v\right). \tag{5.31}$$

The form of Eq. (5.31) lends itself readily to physical interpretation. The first term on the right is essentially the dry adiabatic lapse rate (Γ_d, Eq. (5.15)), with only a small difference in the definition of the specific heat capacity. If a cloudy parcel, initially at equilibrium, is suddenly displaced vertically through a distance Δz, it will expand and cool, resulting in net condensation on the existing condensate, or $d\left(y_v l_{vs}\right)/dt < 0$ because the vapor mixing ratio decreases. The resulting temperature lapse rate of the parcel is less than the dry adiabatic lapse rate. The same underlying physics uncovered in our discussion of the equivalent temperature is at play here, only in this case a real physical process exists for converting the molecular potential energy to mean kinetic energy of the molecules in the parcel (heating that offsets some of the temperature decrease). One can now see how the process is related to the equivalent temperature: The lifting of a parcel to low temperatures, where the saturation mole fraction of vapor is close to zero, causes virtually all of the vapor to be removed before the parcel is brought back (dry adiabatically) to its original level.

A more common use of the *reversible adiabatic process*, described in Eq. (5.29), is to define another conserved quantity, the equivalent potential temperature θ_e. We can define a potential temperature analogous to Eq. (5.18) as

$$\theta_d = T\left(\frac{p_0}{p_d}\right)^{y_d R/c_p} \tag{5.32}$$

such that the spatial derivative of its natural log equals the last two terms in Eq. (5.29). Note that θ_d differs from θ, the ordinary potential temperature (Eq. (5.18)), in that the total pressure in the denominator is replaced by the dry-air partial pressure, and the specific heat capacity is that of the multi-phase system $c_p = y_w c_l + y_d c_{pd}$. Combining Eq. (5.32) and Eq. (5.29) gives

$$c_p\frac{d\ln\theta_d}{dz} = -\frac{d}{dz}\left(\frac{l_v y_{vs}}{T}\right), \tag{5.33}$$

which can be integrated from its initial state θ_d at T and p_d to its final state θ_{df} with no vapor (i.e., $y_{vs} = 0$). With $\theta_{df} = \theta_e$, the equivalent potential temperature is

$$\theta_e = \theta_d \exp\left(\frac{l_v y_{vs}}{c_p T}\right). \tag{5.34}$$

Here, we have neglected the temperature dependence of l_v and c_p.

In the derivation of Eq. (5.34), we assumed a closed system in which all the liquid remains in the parcel. In a cloud, however, one may expect some of the condensate to be removed from the parcel via precipitation. Therefore, we also consider the *pseudo-adiabatic process*, the limiting case in which all condensate is immediately removed from the parcel. This assumption is implemented by replacing the total water mole fraction y_w with the saturation vapor mole fraction y_{vs} in the multi-phase heat capacity ($c_p = y_{vs}c_l + y_d c_{pd}$). The resulting temperature lapse rate for this pseudo-adiabatic process is higher than that for the reversible process because the heat capacity is slightly smaller in the pseudo-adiabatic process. In a real cloud, one would expect a process somewhere between a reversible and pseudo-adiabatic process. The difference between a reversible and pseudo-adiabatic equivalent potential temperatures is small for many applications. In practice, one can calculate θ_e by first lifting a parcel to saturation and then calculating θ_e from Eq. (5.34), but using the temperature of the parcel at the point of saturation (i.e., the cloud-base temperature).

We can use Fig. 5.3 to explore the behavior of these variables graphically. We defined θ_e in the subcloud layer by lifting the parcel to saturation. This dry-adiabatic process is best visualized on the skew-T log-p graph (top) by realizing that, for this dry ascent, potential temperature θ and the water vapor mixing ratio r_v are conserved. The parcel becomes saturated when the decreasing saturation mixing ratio $r_s(T, p) = r_{sfc}$, the mixing ratio of the parcel ascending from the surface. The height where this equality is reached is called the lifting condensation level (LCL), and it is found by following the dry adiabat through the surface temperature to the intersection with the mixing-ratio line through the surface dew-point temperature. The smooth solid curve in the sounding of Fig. 5.3 (top) follows the moist adiabat from cloud base (near pressure level 860 hPa). The value of θ_e is determined by reading the maximum potential temperature reached along this moist adiabat (at the top of the skew-T diagram). The level of free convection (LFC) is the height at which the parcel first becomes warmer than its environment, or where $\theta_e(z_{sfc}) = \theta_{es}(z_{LFC})$. The parcel will accelerate upward through the atmosphere as long as its θ_e exceeds θ_{es} at each level. We note in Fig. 5.3 (bottom) that a thin layer exists near 2.2 km within which $\theta_{es}(z) > \theta_e$. The skew-$T$ profile reveals that this layer corresponds to the top of the boundary-layer inversion, in which the parcel temperature becomes colder than its environment. When the parcel enters this layer, its acceleration becomes downward, although its upward momentum may carry it through the layer if thin enough. Once the parcel has penetrated through this layer, it ascends unimpeded to the level of neutral buoyancy ($\theta_{es}(z_{LNB}) = \theta_e$), above which $\theta_{es}(z) > \theta_e$.

We next consider saturated descent to the reference pressure level p_0 and the concept of wet-bulb potential temperature θ_w. In concept, water may have to be added to the parcel as it descends in order to maintain saturation all the way to p_0. The temperature attained at p_0 by this reference process is θ_w. The wet-bulb potential temperature is determined from the same saturated-adiabat process as used for the equivalent potential temperature, so an expression for θ_w may be obtained by integrating Eq. (5.33) from its initial saturated state θ_d at T and p_d to the pressure level p_0, where the temperature is θ_w (and the parcel is still saturated):

$$\theta_w = \theta_d \exp\left(\frac{l_v}{c_p}\left(\frac{y_{vs}(T)}{T} - \frac{y_{vs}(\theta_w)}{\theta_w}\right)\right). \tag{5.35}$$

Because water may have been added to maintain saturation, the parcel represents an open system, and the process is pseudo-adiabatic. We observe, therefore, from its definition in Eq. (5.11), that the specific heat changes slightly as water is added. Because θ_e of a parcel is determined from the same process line as θ_w, θ_e uniquely determines θ_w, and vice versa. By their respective definitions, both quantities are conserved in both dry-adiabatic and moist-adiabatic processes, and they are quasi-conservative with respect to evaporation of rain.

Finally, the definition of potential temperature (Eq. (5.18)) may be used to calculate adiabatic versions of the equivalent temperature T_{ae} and of the wet-bulb temperature T_{aw}. These psuedo-adiabatic temperatures differ from their isobaric counterparts by the introduction/removal of liquid water, which changes the total energy of the parcel. The relationship of the various temperatures to one another may be summarized by the following inequalities (for unsaturated air):

$$T_d < T_{aw} < T_w < T < T_v < T_e < T_{ae}. \tag{5.36}$$

Note that the only directly measurable temperatures are T (thermometer), T_d (dewpoint hygrometer), and T_w (sling psychrometer); all the others are computed quantities.

The amount of condensate produced in a cloud is often of great interest to cloud physicists. In a reversible adiabatic process, the adiabatic condensate mixing ratio r_{ad} at each altitude z is easily calculated from the difference

$$r_{ad}(z) = r_{sfc} - r_s(z), \tag{5.37}$$

where $r_{sfc} = r(z_{cldbase})$ is the vapor mass mixing ratio at the surface, which is conserved below cloud base $z_{cldbase}$, and where the saturation mixing ratio at altitude z is

$$r_s(z) = r_s(T(z), p(z)) = \frac{\varepsilon e_s}{p_d} = \frac{\varepsilon e_s(T(z))}{p(z) - e_s(T(z))}. \tag{5.38}$$

The temperature to use for calculating the saturation (i.e., equilibrium) vapor pressure of liquid water (alternatively ice) is that realized when the parcel is lifted from the surface to altitude z in reversible adiabatic ascent. The adiabatic condensate mixing ratio is often used as a reference value for diagnostic studies of cloud processes. Departures from adiabatic values may result from mixing or precipitation.

5.4 Stability

The concept of stability may be introduced generally by asking the following question: for a system initially at equilibrium, does a small perturbation cause it to return to or depart from that equilibrium state? Within the context of atmospheric stability, the system is a closed parcel of air, and we consider the perturbation to be a vertical, adiabatic displacement away from its original altitude. A parcel is in stable air if the displacement up (or down) makes the parcel colder (or warmer) than its surroundings, for then the parcel returns to its original level. The parcel is in unstable air if the same displacement reverses the temperature difference and causes the parcel to accelerate even farther away from the original level. The static stability of the atmosphere at a given place is determined by the local lapse rate of temperature ($-dT/dz$), which in turn becomes an indicator for convective (unstable case) or stratiform (stable case) air motions.

5.4.1 Buoyancy

Evaluation of stability criteria requires an understanding of force balances and buoyancy. Consider a parcel to be a fixed volume V of air having density ρ, enclosed by a virtual surface of area A, and located at some height z in the atmosphere. This parcel is in mechanical equilibrium when the sum of all forces acting on it is zero. The force exerted by the pressure p over all A is then balanced by the weight of air within V. This force balance is expressed mathematically as

$$-\int_A p\mathbf{n}dA + \int_V \rho\mathbf{g}dV = -\int_V \nabla p dV + \int_V \rho\mathbf{g}dV = 0, \tag{5.39}$$

where $\mathbf{n}$ is the outward-directed, unit normal vector to A, $\mathbf{g} = -g\mathbf{e}_z$ is the acceleration due to gravity ($\mathbf{e}_z$ is the unit vector pointing vertically upward), and where we have used the divergence (or Gauss') theorem of calculus to convert from an area to a volume integral. This equation must be satisfied for arbitrary V, so the condition for mechanical equilibrium in the vertical direction is

$$\nabla_z p = -\rho g, \tag{5.40}$$

where $\nabla_z p$ is the vertical component of the total pressure gradient. The atmosphere is locally in hydrostatic balance when Eq. (5.40) applies.

Buoyancy is the net force acting on a body as a consequence of a density difference between the body and its fluid environment. If we let the density of the air in the parcel be ρ_p and that of the surrounding air be ρ, then the net vertical force is

$$\begin{aligned} \mathbf{F}_B &= \int_V \rho_p \mathbf{g} dV - \int_A p\mathbf{n}dA \\ &= \int_V (\rho_p - \rho)\, \mathbf{g} dV \\ &\approx -(\rho_p - \rho)\, gV\mathbf{e}_z, \end{aligned} \tag{5.41}$$

where we have again employed the divergence theorem. The approximation applies for suitably small volumes. The parcel will be accelerated either upward or downward depending on the sign of the density difference. Buoyancy, in contrast to net force, which depends on parcel volume, is best viewed as a net force per unit mass. Then, using the ideal gas law for moist air (Eq. (5.2)), buoyancy B may be approximated as

$$\begin{aligned} B &\equiv \frac{F_B}{\rho_p V} \\ &= -g\frac{(\rho_p - \rho)}{\rho_p} \\ &= g\left(\frac{T_{v_p} - T_v}{T_v}\right), \end{aligned} \tag{5.42}$$

where the vertical direction is implied (positive upward). If the density of the parcel exceeds that of its environment, that is if $\rho_p > \rho$, or $T_{v_p} < T_v$, the buoyancy force is directed downward ($B < 0$), and the parcel will experience a downward acceleration. Conversely, if $\rho_p < \rho$, or $T_{v_p} > T_v$, the volume will experience an upward acceleration. Finally, we can, applying Eq. (5.18) to Eq. (5.42), express the buoyancy also in terms of potential temperatures as

$$B = g\frac{(\theta_p - \theta)}{\theta}. \tag{5.43}$$

This form is convenient for evaluating buoyancy for large displacements. The potential temperature in Eq. (5.43) may be replaced by the appropriate form (virtual, equivalent) for the process considered.

The condensate in cloud also impacts the parcel density. The density of the parcel is

$$\rho_p = \frac{m_d + m_v + m_c}{V} = \rho_{air}\left(1 + \frac{\rho_c}{\rho_{air}}\right) = \rho_{air}\left(1 + r_c\right), \tag{5.44}$$

where m_c is the mass of condensate in the parcel, and r_c is the condensate mass mixing ratio. For a fixed amount of condensate, the parcel density decreases with increasing amounts of water vapor, while for a fixed air density, the parcel density increases with the amount of condensate. If we use Eq. (5.44) for the parcel density in our expression for the buoyancy (Eq. (5.42)), then

$$B = -g\frac{(\rho_{p,air}(1 + r_c) - \rho)}{\rho_{p,air}(1 + r_c)} \approx -g\frac{(\rho_{p,air} - \rho)}{\rho_{p,air}} - gr_c = g\frac{\Delta T_v}{T_v} - gr_c. \tag{5.45}$$

We see the physical effects of condensed water on the buoyancy: The formation of condensate causes warming, which increases ΔT_v and thus buoyancy (due to latent-heat release). At the same time, the condensate adds mass r_c to the parcel, which decreases the buoyancy (realized through the drag forces on the particles).

5.4.2 Static stability and lapse rates

No condensation

With our understanding now of buoyancy, the issue of stability may be addressed. The local temperature lapse rate of the atmosphere ($\gamma \equiv -\partial T/dz$) is determined by the local weather and is not equal in general to the dry-adiabatic lapse rate (Γ_d) even in a cloudless sky. For example, mid-level warm air advected ahead of an approaching trough causes γ to become smaller (closer to isothermal). Alternatively, subsidence below a mid-level high-pressure system may cause a temperature inversion, in which case the lapse rate is negative (T increases with z).

The atmosphere is considered to be unstable if a parcel of air, displaced from its original equilibrium state at height z to $z + \Delta z$, experiences an acceleration in the same direction as the displacement. We first consider a dry displacement, that is, a displacement without any net condensation. The temperature change of the displaced parcel may reasonably be assumed to be adiabatic, because the thermal energy transferred between the parcel and its environment is through diffusion or radiation, both of which are slow processes. Therefore, we can rewrite Eq. (5.42), using first-order Taylor expansions for the temperature at $z + \Delta z$, as

$$B = g\left(\frac{T_v(z) + \Gamma_d \Delta z - T_v(z) + \gamma \Delta z}{T_v}\right) = g\Delta z\left(\frac{\Gamma_d - \gamma}{T_v}\right). \tag{5.46}$$

We see that the difference between the dry-adiabatic and the local atmospheric lapse rate determines the sign and the magnitude of the buoyancy. If the temperature decreases more slowly than the dry-adiabatic lapse rate, the sign of the acceleration is in the same sense as the displacement, and the atmosphere is unstable.

We may view this same displacement in terms of potential temperature. Taking the natural logarithm of Eq. (5.18), then differentiating with respect to height z yields

$$-\frac{T_v}{\theta_v}\frac{d\theta_v}{dz} = \Gamma_d - \gamma \tag{5.47}$$

and thus

$$B = -\frac{g\Delta z}{\theta_v}\frac{d\theta_v}{dz} \approx -g\frac{\theta_v - \theta_{vp}}{\theta_v}. \tag{5.48}$$

A decrease in the potential temperature with height corresponds to an acceleration in the same direction as the displacement and hence indicates an instability. This expression for buoyancy leads to a very intuitive interpretation of potential temperature. Adiabatic displacement of a parcel conserves potential temperature. The potential buoyancy of a displaced parcel, and thus its stability, can be determined by comparing its potential temperature to that of the environment at any height. We conclude the discussion of static stability for displacements without condensation by summarizing the three main cases:

$$\frac{dT_v}{dz} > -\Gamma_d, \text{ or } \frac{d\theta_v}{dz} > 0 \quad \text{(stable)} \tag{5.49}$$

$$\frac{dT_v}{dz} = -\Gamma_d, \text{ or } \frac{d\theta_v}{dz} = 0 \quad \text{(neutral)} \tag{5.50}$$

$$\frac{dT_v}{dz} < -\Gamma_d, \text{ or } \frac{d\theta_v}{dz} < 0 \quad \text{(unstable).} \tag{5.51}$$

The static stability is fundamentally determined by how the density (or virtual temperature) of the environment changes with height compared with that of the parcel following an adiabatic reference process.

With condensation

The situation changes when a parcel contains condensate (is saturated) and condensation (or evaporation) takes place as a result of the displacement. Energy released from the phase change impacts the temperature lapse rate (i.e., Eq. (5.31)), the consequence of which is that the temperature lapse rate now depends on how much water changes phase. The atmosphere is considered conditionally unstable if the environmental lapse rate is greater than the moist adiabatic lapse rate, the condition being unstable for saturated ascent. Unfortunately, the amount of water changing phase is itself a function of temperature (height), and therefore the moist adiabatic lapse rate is variable, making a direct comparison to the environmental lapse rate difficult. It is therefore convenient to introduce a new variable to make the stability analysis for wet displacements easier. We define the saturated equivalent potential temperature of an unsaturated parcel as the equivalent potential temperature it would have if it were saturated, that is

$$\theta_{es} = \theta_d \exp\left(\frac{l_v y_{vs}}{c_p T}\right). \tag{5.52}$$

If the environmental temperature in a layer decreases at the moist adiabatic lapse rate, θ_{es} will be constant with height, and the density of a saturated parcel ascending through the layer will remain equal to that of its environment. On the other hand, if θ_{es} decreases with height, the parcel density will be less; and the layer is conditionally unstable. Conversely, if θ_{es} increases with height, the layer is absolutely stable. The definition of θ_{es} allows for easy evaluation of the static stability for any displacement, dry or saturated, as

$$\frac{d\theta_v}{dz} < 0 \quad \text{(absolutely unstable)} \tag{5.53}$$

$$\frac{d\theta_v}{dz} < \frac{d\theta_{es}}{dz} < 0 \quad \text{(conditionally unstable)} \tag{5.54}$$

$$\frac{d\theta_{es}}{dz} > 0 \quad \text{(absolutely stable).} \tag{5.55}$$

For a graphical representation, refer again to the bottom of Fig. 5.3, where the environmental potential temperature profiles are plotted. The θ_e of the displaced parcel is represented by the solid vertical line. As long as the environmental θ_{es} is less than the parcel θ_e, the displaced parcel will experience an upward acceleration above cloud base.

Stability, when applied to clouds, is often expressed in terms of the energy available for driving updrafts. Parcels in the atmosphere frequently experience forced displacements through finite distances, even though initially located in a stable layer. We can think of low-level convergence, for instance, as forcing a local uplift of the air. Alternatively, air may be undercut by a denser outflow boundary from a storm, forcing moist air ahead of the gust front upward. The stability of an atmospheric layer is evaluated by determining whether the displaced parcel acquires positive buoyancy during the displacement. For example, if a saturated parcel is lifted in a conditionally unstable environment beyond the level of free convection, it will attain positive buoyancy and continue accelerating upward. The parcel will accelerate freely upward as long as $B > 0$, regardless of the static stability of the layers through which it is moving. The instability is that of the layer (possibly much of the troposphere) and expressed in terms of the available energy over the layer in which the buoyant acceleration is in the same direction as that of the displacement. A common form of available energy used by scientists studying deep convective storms is the convective available potential energy (CAPE), the total energy available from the environment between the LFC and the level of neutral buoyancy (LNB):

$$\mathrm{CAPE} = \int_{LFC}^{LNB} B dz = g \int_{LFC}^{LNB} \frac{T_{vp} - T_v}{T_v} dz. \tag{5.56}$$

The shaded region in Fig. 5.3 is a graphical presentation of the CAPE. (How CAPE is related to atmospheric motions is discussed in Chapter 6.) The work required to lift the parcel to its LFC is called the convective inhibition, or CIN, which is defined as

$$\mathrm{CIN} = -\int_{0}^{LFC} B dz = -g \int_{LFC}^{LNB} \frac{T_{vp} - T_v}{T_v} dz. \tag{5.57}$$

Storms often develop only after some large-scale lifting mechanism forces air through the CIN layer into a region of positive CAPE.

5.5 Mixing

In our discussion thus far we have assumed that the parcel does not mix with its environment. In reality, when a parcel rises or sinks through an environment with different properties, it typically entrains some of and mixes with the surrounding air. The parcel gains mass and its properties change. If the parcel is cloudy (contains condensate) and entrains unsaturated (dry) air, it may cool substantially as condensate evaporates, moistening the entrained air and bringing it toward saturation. The buoyancy of cloudy parcels may thereby be reduced when entraining dry air. Entrainment mixing is not reversible, but we may nevertheless bring the conservation properties of equivalent potential temperature and total water content to bear on the problem.

Mixing of two masses of air, each characterized by different properties, results in a new mass with blended properties. Consider a generic conserved property P, such that the one mass m_1 has property P_1 and the other mass m_2 has property P_2. The masses are simply additive ($m_{mix} = m_1 + m_2$), and the property of the final mixture, P_{mix}, is determined by forming a linear combination of the initial properties:

$$P_{mix} = \chi P_1 + (1 - \chi) P_2, \tag{5.58}$$

where $\chi = m_1/(m_1 + m_2) = m_1/m_{mix}$ is the mass mixing fraction, the mass of parcel 1 ending up in the mixture. The final property, P_{mix}, is said to be the mass-weighted average of the original properties. Applying this principle to an entraining air parcel lets us calculate the thermodynamic state of the mixture as

$$\theta_e = \chi \theta_{e,p} + (1 - \chi) \theta_{e,s} \tag{5.59}$$

$$r_t = \chi r_{t,p} - (1 - \chi) r_{t,s}, \tag{5.60}$$

where the second subscript identifies the initial parcel (p) and the surrounding, entrained air (s), and where r_t is the total water mass mixing ratio. The effect of mixing is always to bring the parcel closer to the environmental state through which the parcel is moving.

We next explore the consequences of mixing for the case of deep convection. The bottom of Fig. 5.3 shows θ_e profiles for an idealized unmixed parcel (vertical solid line), as well as for mixed parcels (long dash). As a parcel ascends through the unsaturated boundary layer (in which θ and $\theta_{e,s}$ are approximately constant), $r_{t,s} = r_{v,s}$ decreases with height, resulting in an LCL higher than that of an unmixed parcel. Much more pronounced for this particular sounding is the impact on the LFC, the level at which convection becomes free. As the cloudy air now ascends into the much drier surrounding air, with lower $\theta_{e,s}$ and $r_{t,s}$ values, entrainment further increases the negative buoyancy of the ascending parcel, in turn increasing the required work (CIN) to lift that air through the stable layer. Only those plumes with the strongest forcing will manage to break through the stable layer. Once the LFC is reached the parcel becomes positively buoyant, but the CAPE of the mixed parcel is substantially reduced compared to that of the unmixed parcel. Not only is the temperature difference between the cloud and its environment smaller, but the vertical range of positive buoyancy has been reduced. Mixing reduces the impact of the CAPE calculated from a sounding, thus emphasizing the fact that CAPE represents the theoretical maximum potential energy that can be converted into kinetic energy.

Clouds do not consist of updrafts only; downdrafts play an important role in the structure and evolution of clouds and storm systems. The buoyancy forcing of these downdrafts may result from two sources, the redistribution of condensate into precipitation (Eq. (5.44)) and through mixing and the consequent cooling from rain evaporation. Mixing at cloud top may produce air that is negatively buoyant relative to the cloud, causing it to sink into the cloud as a penetrative downdraft. However, the same mixing that produced the negative buoyancy also reduced the condensate mixing ratio. As the downdraft descends into the cloud and warms along the moist adiabat, condensate will decrease unless replenished by further mixing with other cloudy air. Once the condensate is depleted, the downdraft warms dry adiabatically, quickly losing its negative buoyancy. Above the mid-tropospheric θ_{es} minimum, penetrating downdrafts will remain warmer than their cloud-free environments, and thus remain cloudy.

Below the mid-tropospheric θ_{es} minimum, moist downdrafts may sink into environments where they may maintain their negative buoyancy. Furthermore, if the downdraft coincides with a precipitation shaft, the continuous reinforcement of the condensate loading along with evaporative cooling may allow the downdraft to descend all the way to the surface. From this simple discussion we may conclude that outflow from convective clouds must have originated in the lower parts of the troposphere.

Mixing between clouds and their environments can be analyzed conveniently using conserved-variable diagrams (e.g., a $r_t - \theta_e$ diagram, in which r_t is the total water mixing ratio). From Eqs. (5.59) and (5.60), we see that mixing between the environment and the cloud should fall on the mixing line, a straight line between the thermodynamic coordinates of the two air masses contributing to the mixed parcel (Fig. 5.4). Other physical processes active in the cloud may cause deviations from the mixing line. For example, precipitation into or out of the volume will change r_t, while leaving θ_e unchanged. Non-adiabatic heating or cooling from radiation will change θ_e while leaving r_t unchanged. Such conserved-variable diagrams are particularly useful for studying mixing processes in stratocumulus and small, isolated cumulus clouds, where the non-adiabatic processes are weak and may be ignored. Measurements of conserved variables for volumes of air at any altitude inside these clouds are closely packed around the mixing line between the

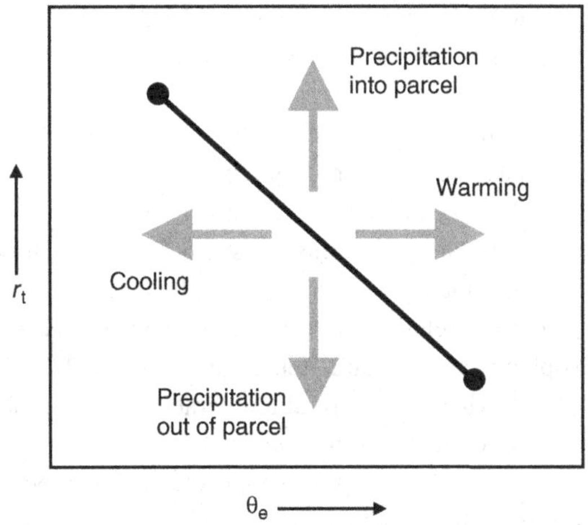

Figure 5.4 Schematic diagram of conserved variables, showing a mixing line and the directions of non-adiabatic processes.

cloud-base and cloud-top thermodynamic coordinates, suggesting that the cloudy air at any level is a result of mixing between the cloudy air that has risen adiabatically from cloud base with air entrained near the cloud top.

5.6 Further reading

Bohren, C.F. and B.A. Albrecht (1998), *Atmospheric Thermodynamics*. Oxford: Oxford University Press, 402 pp. Chapters 3 (dry adiabatic processes, stability and buoyancy), 5, and 6 (equivalent potential temperature, moist adiabatic lapse rate, thermodynamic diagrams, mixing).

Markowski, P.M. and Y.P. Richardson (2010), *Mesoscale Meteorology in Mid-latitudes*. Chichester: John Wiley & Sons, Ltd, 406 pp. Chapters 2.1 (moist thermodynamics) and 2.7 (thermodynamic diagrams).

Rogers, R.R. and M.K. Yau (1989). *A Short Course in Cloud Physics*. New York Pergamon Press, 293 pp. Chapters 1 (thermodynamics of dry air), 2 (moist air, adiabatic liquid water content), 3 (parcel buoyancy and stability), and 4 (mixing of air masses).

Salby, M.L. (1996). *Fundamentals of Atmospheric Physics*. San Diego, CA: Academic Press, 624 pp. Chapters 2 (general thermodynamics), 3 (second law), and 5 (moist-air processes).

Wallace, J.M. and P.V. Hobbs (2006). *Atmospheric Science: An Introductory Survey*. Amsterdam: Academic Press, 483 pp. Chapter 3 (atmospheric thermodynamics).

Young, K.C. (1993). *Microphysical Processes in Clouds*. New York: Oxford University Press, 427 pp. Appendix 1 (thermodynamics).

5.7 Problems

1. Use Table 2.1 plus outside resources for the following:
   (a) Calculate the mean molar mass (molecular weight) of "dry" air composed of the gases listed. How well does your computed value compare with the accepted value of 28.97 g mol^{-1}?
   (b) Assume now that the atmosphere contains 4% water vapor (by molecules, not mass). Calculate the mole fractions of the three major components of the permanent gases (N_2, O_2, and Ar) in this moist air.
   (c) Calculate the mean molar mass of this moist air (from (b)). How much change does the addition of this amount of water vapor to the air make? (Specify your answer as a percentage change relative to the accepted value of dry air of 28.97 g mol^{-1}.)
   (d) For an atmosphere at standard temperature ($T = 273.15$ K) and pressure ($p = 1013.25$ hPa), what is the virtual temperature of this moist air?
   (e) What is the specific humidity of this moist air?
   (f) Suppose that a 1 m^3-parcel of this moist air were to be embedded in a completely dry and static (i.e., calm) atmosphere under the same temperature and pressure conditions. How would you expect the parcel to move? What will be its initial acceleration?

2. One of the authors likes to drink tea, but unlike most people in the US, he never makes tea by putting a tea bag in his cup. Rather, he makes tea in a teapot. To get the best results, he puts tea leaves in the pot, then fills it almost to the top with boiling water. After that, he places a lid on the pot to let the tea steep. However, a few seconds after the lid goes on, he notices tea spilling out of the spout. This continues for a few seconds, and then stops. Please explain qualitatively why this happens. Then, provide a quantitative answer. Please estimate the magnitude of any contributing factors. Assume that the initial air temperature was 20 °C and that the water temperature was 100 °C.

3. The hydrological cycle on Earth may be viewed as a thermodynamic heat engine that does work on the atmosphere. Follow a typical water molecule through this cycle, from evaporation of liquid water at the Earth's surface, through transport in the atmosphere and condensation in clouds, to its return to the original liquid state as precipitation on the ground, and
   (a) sketch the path on a pressure–volume diagram,
   (b) sketch the path on a temperature–entropy diagram,
   (c) calculate an approximate magnitude of the work done by 1 g of water upon going through the cycle.

   Label the diagrams clearly and provide adequate explanations of the physical processes involved and any assumptions made.

4. Helium balloons are commonly used in meteorology to carry instruments (the payload) that can probe the vertical structure of the atmosphere. For this problem, consider two independent balloons, each with the same amount and initial density of helium and weighing exactly the same in total (i.e., payload, balloon itself, and gas). The only difference between the two is the flexibility of the balloon material. Balloon "R" is made of a material (e.g., mylar) that does not stretch at all; that is, it is Rigid. The material of balloon "F", by contrast, is perfectly Flexible (no such material exists, but this is a thought experiment). Answer the following questions under the assumption that each balloon is launched under identical conditions:
   (a) Which balloon will ascend higher into the atmosphere? Explain.
   (b) If each helium balloon were to ascend without exchanging energy with the surrounding atmosphere and a thermometer were placed in the gas, what would be the change in helium temperature in each case after an ascent of 1 km.
   (c) Under the conditions of (b), how does the buoyancy of balloon F compare with that of balloon R
      (i) at launch and
      (ii) at 1 km above the launch site?

      Justify your answers through mathematical reasoning.

5. In this problem we will consider a hot air balloon. In contrast to the balloons discussed in Problem 4, such a balloon has an opening in the bottom through which the air inside

the balloon may be warmed. We will address the physics behind what keeps the balloon floating. The easy answer is that it is less dense than the environment, which, though it is partially correct, misses some important understanding.

(a) What is the impact of the opening in the balloon on the thermodynamic state inside the balloon? How does the state of the air in this balloon differ from that in the previous problem?

(b) Use the concept of scale height to discuss the pressure distribution inside the balloon and environment.

(c) Based on your discussion, what do you conclude about what is keeping the balloon floating?

(d) Now, consider a balloon with a 25 m diameter. Estimate what the minimum temperature differential between the balloon and environment in order to lift a payload of 2000 kg.

6. Figure 5.5 below shows a vertical profile of potential temperature for the atmosphere. Consider two parcels with potential temperatures θ_1 and θ_2, respectively, at a height z_1 (indicated) in this environment:

(a) In which direction will each parcel accelerate? Provide full motivation.

(b) Show graphically what the final height of each parcel would be.

(c) Now sketch a temperature profile for this potential temperature profile. You should first draw a line for the dry adiabatic lapse rate on your T–z diagram, and then show the atmospheric lapse rate represented by the potential-temperature profile relative to the dry adiabatic lapse rate.

7. An atmospheric layer below cloud base is moistened by rain falling from a cloud. Assume that the initial temperature of the layer (before the rain started falling into the layer) was 20 °C and the wet-bulb temperature was 10 °C. If the mixing ratio at 900 hPa increases by 3 g kg^{-1} during this rain event, what will be the temperature when the rain ceases? What is the initial and final relative humidity?

8. A parcel of air at 900 hPa has a temperature of 19 °C and a mixing ratio of 11.5 g kg^{-1}. The parcel is lifted to 600 hPa by passing over a mountain, during which 50% of

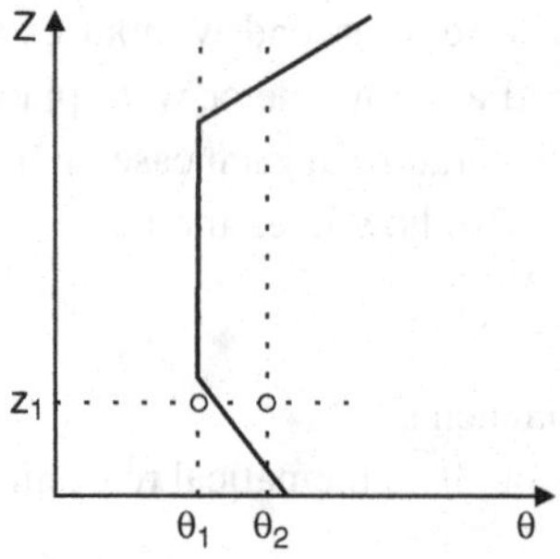

Figure 5.5 Figure for problem 6.

the water vapor condensed in the ascent is removed by precipitation. Determine the temperature, potential temperature, vapor mixing ratio, and wet-bulb potential temperature of the parcel after it has descended to the 950 hPa level on the other side of the mountain.

9. A few years ago, a famous golfer died in an aircraft crash. The first people to reach the accident site found the inside of the aircraft to be coated with ice, which must have formed as the temperature in the aircraft dropped to very low temperatures. Later, investigators determined that the aircraft frame had been punctured by a piece of the engine. How were the low temperatures realized inside the aircraft? Explain why the ice formed. How would the ice formation itself affect any calculation you were asked to perform? You may assume that the aircraft was flying at an altitude of ~ 150 hPa.

10. Consider an atmosphere with no aerosol particles. Describe (with approximate equations) how you would estimate cloud base pressure under these conditions with the surface temperature T_0, pressure p_0, and mixing ratio ω_0 respectively. Assume air is being lifted to cloud base. Be sure to explain why you are doing what you are doing. Now, let $T_0 = 19\,°\mathrm{C}$, $p_0 = 1000$ hPa, and $r_0 = 8\ \mathrm{g\,kg^{-1}}$. What would be the cloud base pressure in this case?

6
Cloud formation and evolution

Clouds form when atmospheric conditions become appropriate. The necessary ingredients for cloud formation include an adequate supply of water vapor, aerosol particles, and a mechanism for cooling the air. Water, in condensed form, is, of course, the primary component of any cloud. The aerosol particles provide the sites for condensation by offering places where water vapor can adhere. Cooling lets condensation take place by causing the physical temperature to fall below the dew point and the concentration of water vapor to exceed the equilibrium value.

The properties of the resulting cloud depend on both macro- and micro scale processes. The macroscale air motions, driven by atmospheric pressure gradients, move the requisite water vapor and aerosol particles upward toward lower pressures, thereby cooling the air and generating excess water vapor. The microphysical processes, driven initially by aerosol abundance, determine how the excess vapor is utilized within the cloud. This chapter offers an overview of the macroscale processes that produce environments conducive to cloud formation. Subsequent chapters focus on the evolution of the microphysical properties.

6.1 Cooling mechanisms

A cloud, the visible aggregation of liquid or ice particles suspended in the atmosphere, requires cooling so that the partial pressure of vapor, $p_{H_2O} \equiv e$, can exceed the equilibrium vapor pressure of the condensate, $e_{eq}(T)$, a function that increases monotonically with temperature T. Note, however, that partial pressure varies relatively little in typical situations; rather, it is the temperature that changes most. As T decreases, $e_{eq}(T)$ also decreases and eventually falls below e. With $e > e_{eq}(T)$, the condensation flux exceeds the evaporation flux and net condensation occurs. It is the cooling, the lowering of temperature and vapor pressure, that brings the air into a state of disequilibrium and permits cloud to form.

Cooling mechanisms may be classified as uplift, radiation, conduction, or mixing. Most commonly, the requisite cooling for supersaturation results from adiabatic expansion of the air during rising motions (uplift), but cloud can also form at constant pressure (isobarically) by radiation, conduction, or mixing. Diabatic processes involve exchange of energy with the environment by virtue of a temperature difference, adiabatic processes do not.

6.1.1 Isobaric diabatic cooling

Air cools isobarically and diabatically whenever energy leaves faster than it arrives. Radiation was covered in Chapter 2, but in the context of cloud formation we must realize that some entity, either a gas molecule or the surface of a liquid or solid, is responsible for radiating energy away. When more radiant energy leaves a parcel than it receives from its surroundings, the air cools. Often, the cooling of air occurs indirectly, such as when the Earth's surface (not the air) radiates energy to space under clear skies at night. The air in contact with the cooled surface then gives up its thermal energy by conduction and in turn cools neighboring air parcels by mixing. The moist, boundary-layer air may eventually cool below the dew-point temperature, giving rise to radiation fog. The condensed water in the fog droplets can also radiate energy away, further cooling the air and enhancing the fog. Alternatively, if warm, moist air travels over a colder surface, such as a cold lake, an advection fog may result. Most isobaric cooling scenarios combine radiation, conduction, and mixing. Details of the mixing process help elucidate isobaric processes of cloud formation.

6.1.2 Isobaric adiabatic mixing

Conditions conducive to cloud formation may arise when two distinct, subsaturated air masses mix together. It is the mixing process itself that causes a parcel to cool and generate a supersaturated state. The process is termed isobaric because it does not depend on any change in altitude, and it is adiabatic because no other exchanges of energy are needed. The most obvious mixing clouds arise when the contrasts in temperature and vapor concentration are most pronounced. For instance, when an aircraft flies in the upper troposphere, a linear cloud often forms behind the aircraft. Such condensation trails, or contrails, result from the mixing of warm, vapor-rich air from the aircraft engine with the cold ambient air. Another example is one's own foggy breath that forms on a cold day.

Mixing always blends the properties of the joining parcels. We can understand the generation of supersaturation by considering the mathematics of mixing. Initially, the parcels have distinct properties, as shown by the dots in the property plot of Fig. 6.1. Each parcel has its own temperature and its own partial pressure of water vapor. As the two parcels begin to mix together, a new mixed parcel forms with properties that represent a linear combination of the original properties. The relative proportion of the two parcels in the mixture, termed the mixing fraction, varies continuously throughout the mixing process. For instance, if the mass fraction of parcel 1 in the mixture at any instant is expressed as $\chi \equiv m_1/(m_1+m_2)$, χ will range from near 1 (parcel 1 intact initially) to near 0 (parcel 1 highly diluted by parcel 2). When expressed in terms of mole fractions, the mixing mass fraction may be approximated as

$$\chi = \frac{M_{air,1}N_{air,1}}{M_{air}N_{air}}\left(\frac{M_{d,1}}{M_{air,1}}y_{d,1} + \frac{M_{v,1}}{M_{air,1}}y_{v,1}\right) \approx \frac{N_{d,1}}{N_{d,1}+N_{d,2}}, \tag{6.1}$$

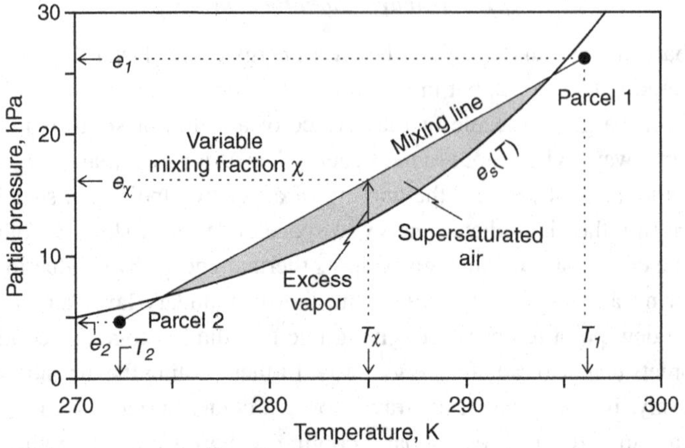

Figure 6.1 Property plot showing the mixing of two air parcels (straight line) in relationship to the equilibrium vapor pressure. Shaded region identifies the range of mixing fractions that give rise to supersaturated air.

where N is the number of moles and we have assumed that $y_v \ll y_d \approx 1$, from which we see that we may ignore the second term in the bracket and that $M_{air} \approx M_d$. The vapor mole fraction in any mixture is thus well approximated by

$$y_{v,\chi} = \chi y_{v,1} + (1 - \chi)\, y_{v,2}, \tag{6.2}$$

or, using $y_v = e/p$, the partial pressure of the mixture is

$$e_\chi = \chi e_1 + (1 - \chi)\, e_2. \tag{6.3}$$

Conservation of enthalpy over the mixing process yields the temperature of the mixture:

$$\begin{aligned} T_\chi &= \frac{M_{air,1}N_{air,1}c_p}{M_{air,1}N_{air,1}c_p + M_{air,2}N_{air,2}c_p} T_1 + \frac{M_{air,2}N_{air,2}c_p}{M_{air,1}N_{air,1}c_p + M_{air,2}N_{air,2}c_p} T_2 \\ &= \chi T_1 + (1 - \chi)\, T_2. \end{aligned} \tag{6.4}$$

Both equations (6.3) and (6.4) depend on χ in the same way, so mixing is represented as a straight line on a phase diagram having linear axes of e and T.

Supersaturation is shown in Fig. 6.1 by the shaded region, where the partial pressure of the mixture exceeds that in equilibrium with liquid water, $e_s(T)$. Mixing yields a linear combination of properties, whereas the equilibrium vapor pressure is an exponential function of temperature (as derived in Chapter 3). Thus, ranges of mixing fractions can exist within which the partial pressure exceeds the equilibrium vapor pressure. Of course, not every mixing scenario yields supersaturation in the mixture, but when appropriate conditions exist, condensate forms and a visible mixing cloud appears.

6.1.3 Adiabatic cooling

Up to now we considered processes constrained to a single level. However, most clouds in the atmosphere arise from the lifting of moist air toward lower pressure. As the air is lifted, the total pressure and (by Dalton's law) the partial pressure of vapor both decrease. As a consequence, the temperature decreases by adiabatic expansion, and the partial and equilibrium vapor pressures decrease (but at different rates). Cloud forms because the temperature-dependent equilibrium vapor pressure decreases faster than does the partial pressure of vapor, resulting in an increase of the saturation ratio with height. The base of a cloud is located at the lifting condensation level, where the saturation ratio becomes unity and net condensation begins.

Lifting condensation level

The lifting condensation level (LCL; refer back to Fig. 5.3) is the altitude of the thermodynamic cloud base, the height a parcel of moist air must rise adiabatically in the atmosphere to reach moisture saturation. In typical situations, a parcel is subsaturated near the surface of the Earth. As it rises, the parcel experiences lower pressure; the air cools, lowering the equilibrium vapor pressure of the condensate, which in turn raises the relative humidity of the air. As the parcel passes through the LCL, the relative humidity reaches 100%; then, with continued lifting, the air becomes supersaturated, and cloud formation becomes possible.

The height of the LCL can be computed from known values of the air temperature T and dewpoint temperature T_d at the surface. The defining feature of the LCL is phase equilibrium, where the partial and equilibrium vapor pressures are equal. Thus, $e = e_s(T_{LCL})$. By definition of the dew point $e = e_s(T_d)$, so it follows that $T_{d,LCL} = T_{LCL}$. Our task is to calculate the vertical distance z_{LCL} needed to achieve this equality. It is important to realize that the temperature and dew point in an adiabatic parcel each decrease with increasing altitude at distinct rates.

The temperature of an adiabatic parcel changes with altitude z at the dry adiabatic lapse rate Γ_d (Eq. (5.15)), so we can write

$$T(z) = T_{sfc} - \Gamma_d z. \tag{6.5}$$

The dew point, on the other hand, decreases at a smaller rate, one dependent on the variation in pressure during ascent. In the absence of vapor loss due to condensation, the vapor mole fraction, $y_V = e/p$, remains constant. The partial pressure varies directly with p as $e = y_V p$, so

$$\frac{de}{dz} = y_V \frac{dp}{dz}. \tag{6.6}$$

But, because $e = e_s(T_d)$, we use the chain rule of calculus to obtain

$$\frac{de}{dz} = \frac{de_s(T_d)}{dz} = \frac{de_s(T_d)}{dT_d}\frac{dT_d}{dz} = y_V \frac{dp}{dz}. \tag{6.7}$$

By the Clausius–Clapeyron equation, the rate the equilibrium vapor pressure varies with temperature is

$$\frac{de_s(T_d)}{dT_d} = \frac{l_v e_s(T_d)}{RT_d^2}, \tag{6.8}$$

and the equation for hydrostatic balance describes the rate that pressure decreases with altitude:

$$\frac{dp}{dz} = -\rho g = -n_{air} M_{air} g, \tag{6.9}$$

where n_{air} is the molar density and M_{air} the molar mass of air. The air density ρ is related to the temperature and pressure via the ideal gas law ($p = n_{air} RT$), so

$$\frac{dp}{dz} = -\frac{p}{RT} M_{air} g. \tag{6.10}$$

From Eqs. (6.7), (6.8), and (6.10), we gain the dew-point lapse rate

$$\Gamma_{dew} \equiv -\frac{dT_d}{dz} = -\left(\frac{RT_d^2}{l_v e_s(T_d)}\right)\left(\frac{e}{p}\right)\left(-\frac{pM_{air}g}{RT}\right) = \frac{M_{air} g T_d^2}{l_v T}. \tag{6.11}$$

For small dewpoint depressions, T and T_d cancel approximately, giving

$$\frac{d \ln T_d}{dz} \doteq -\frac{M_{air} g}{l_v}. \tag{6.12}$$

Integration of Eq. (6.12) yields the dependence of the dew-point temperature on altitude:

$$T_d(z) \doteq T_{d,sfc} \exp\left(-\frac{M_{air} g}{l_v} z\right). \tag{6.13}$$

Note that $l_v / M_{air} g$ has the dimensions of distance and so represents an effective dew-point scale height $H_{dew} \sim 160$ km. Because the dew point decreases slowly, a further approximation may be made for small z:

$$T_d(z) \doteq T_{d,sfc}\left(1 - \frac{M_{air} g}{l_v} z\right). \tag{6.14}$$

The LCL can be determined by setting the air temperature (Eq. (6.5)) and dew point (Eq. (6.14)) equal to each other and solving for z_{LCL}:

$$z_{LCL} \doteq \frac{T_{sfc} - T_{d,sfc}}{\Gamma_d - \dfrac{T_{d,sfc}}{H_{dew}}}. \tag{6.15}$$

The second term in the denominator offsets the dry adiabatic lapse rate (9.8°C km^{-1}) by less than 2 °C km^{-1}, so a convenient application of the theory is obtained by letting the denominator equal 8 °C km^{-1}, so allowing us to estimate the LCL in units of km from the surface dew-point depression: $\delta T_{d,sfc} \equiv T_{sfc} - T_{d,sfc}$:

$$z_{LCL}[\,\text{km}\,] \doteq \frac{\delta T_{d,sfc}\,[°\text{C}]}{8\,°\text{C}\,/\,\text{km}}. \tag{6.16}$$

The closer the dew point is to the temperature, the higher the relative humidity and the lower are the bases of boundary-layer clouds.

6.2 Adiabatic supersaturation development

Supersaturation develops adiabatically when moist air rises rapidly enough for exchanges of thermal energy with surroundings to be negligible over the time period of interest. All energy conversions occur inside the rising parcel. As we have seen, rising air may be subsaturated initially, but the equilibrium vapor pressure, $e_s(T)$, decreases faster than does the vapor pressure, e, so the saturation ratio, $S \equiv e/e_s(T)$, increases with time as the parcel approaches cloud base (LCL, where $S = 1$). As the parcel continues its rise above cloud base, the air becomes supersaturated (supersaturation $s \equiv S - 1 > 0$) and net condensation takes place. Supersaturation continues to develop above cloud base, but condensation causes vapor to be lost and slows the rate of supersaturation increase. Eventually, loss of vapor by condensation exceeds the rate at which excess vapor is generated by adiabatic uplift, and the supersaturation begins to decrease. These competing effects on the supersaturation in a parcel of air experiencing adiabatic uplift can be treated mathematically by considering the vapor budget.

The budget of water vapor in a cloudy parcel of air rising adiabatically forms the conceptual basis for deriving an expression for the rate of change of supersaturation with time. A graphical display of how the partial pressure of water vapor varies during uplift near cloud base is shown in Fig. 6.2, using temperature as a surrogate measure of time (or altitude). The partial pressure naturally decreases during expansion (moving toward lower temperature), but it decreases even faster above cloud base because of condensation. We seek a differential equation that describes the time rate of change of partial pressure,

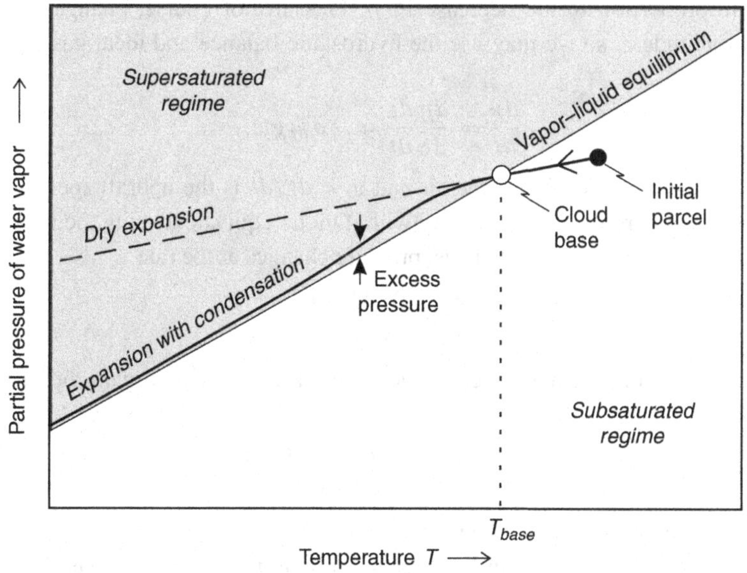

Figure 6.2 A portion of the phase diagram of water, showing the trajectory of a rising parcel in e–T space.

alternatively supersaturation, in an isolated parcel being lifted at a constant updraft speed w through cloud base. We start by recalling that supersaturation, $s = \Delta e/e_s$, represents the relative excess vapor pressure, $\Delta e = e - e_s$, and noting the relationship between supersaturation s and saturation ratio S: $s = S - 1$. Thus,

$$\frac{ds}{dt} = \frac{dS}{dt} = \frac{d}{dt}\left(\frac{e}{e_s}\right) = \frac{1}{e_s}\frac{de}{dt} - \frac{e}{e_s^2}\frac{de_s}{dt}. \tag{6.17}$$

This equation relates the change in supersaturation with the changes in the partial and equilibrium pressures.

The mathematical derivation of ds/dt involves numerous algebraic operations, but it systematically incorporates the essential physical principles: conservation of water, first law of thermodynamics, ideal gas law, Clausius–Clapeyron equation, and the hydrostatic-balance equation. The treatment is facilitated by using molar quantities, in particular specifying the amount of condensed (liquid) water as a mole fraction, $y_L \equiv$ mol-liquid/mol-air. Changes in water vapor mole fraction, $y_V = e/p$, are then simply governed by mass conservation: $dy_V/dt = -dy_L/dt$. Water gained by the condensate is lost from the vapor. Each term on the right-hand side of Eq. (6.17) is then considered separately before combining terms for the final result.

The partial pressure (e) in a rising parcel decreases both because of the expansion itself (a geometrical effect) and because water vapor is lost to any condensate that may form. We see these effects together when e is expressed in terms of mole fraction: $e = y_V p$, which implies the rate $de/dt = p dy_V/dt + y_V dp/dt$. The first term here reflects the loss of vapor due to condensation, whereas the second term shows that e decreases in proportion to the decrease in p. Pressure, of course, changes as the parcel changes altitude z, so we may use the hydrostatic-balance and ideal-gas equations to obtain

$$\frac{dp}{dt} = \frac{dp}{dz}\frac{dz}{dt} = -\rho_{air} g w, \tag{6.18}$$

where ρ_{air} is the mass density of the air and $w = dz/dt$ is the updraft speed (following the parcel). We may replace mass density with it molar equivalent using the ideal gas law: $\rho_{air} = n_{air} M_{air} = p M_{air}/(RT)$. Thus, pressure changes at the rate

$$\frac{dp}{dt} = -\frac{p M_{air} g}{RT} w. \tag{6.19}$$

By mass conservation, we may interchange water variables and so gain the first term of Eq. (6.17):

$$\frac{1}{e_s}\frac{de}{dt} = -\frac{p}{e_s}\frac{dy_L}{dt} - \frac{p M_{air} g}{RT} w. \tag{6.20}$$

We will use this expression again later.

Considering the second term in Eq. (6.17), we note that $e_s = e_s(T)$ is the Clausius–Clapeyron function, which describes the dependence of equilibrium vapor pressure on temperature. This function depends only indirectly on time, so we use the chain rule of calculus to get the rate, $de_s/dt = (de_s/dT)(dT/dt)$. The slope of the Clausius–Clapeyron function was derived in Chapter 3 as $de_s/dT = l_v e_s/\left(RT^2\right)$, where l_v is the enthalpy of vaporization (molar units). The temperature change in the closed parcel has two components, one due to the adiabatic expansion itself and another due to the release of latent heat because of condensation. The appropriate expression to use comes from the enthalpy form of the first law of thermodynamics, $q = c_p dT/dt - v dp/dt$, where $q = l_v dy_L/dt$ is the heating rate due to condensation, c_p is the specific heat at constant pressure, and $v = RT/p$ is the molar volume of the air. After a bit of algebra, we find the temperature in the parcel to change at the rate

$$\frac{dT}{dt} = \frac{l_v}{c_p}\frac{dy_L}{dt} - \frac{M_{air}g}{c_p}w, \tag{6.21}$$

where we have again used the hydrostatic-balance equation to gain the rate of pressure change. The first term in Eq. (6.21) represents the heating due to condensation, the second term the cooling due to expansion. The second term in Eq. (6.17) is therefore found to be

$$\begin{aligned} -\frac{e}{e_s^2}\frac{de_s}{dt} &= -\frac{e}{e_s^2}\left(\frac{l_v e_s}{RT^2}\right)\left(\frac{l_v}{c_p}\frac{dy_L}{dt} - \frac{M_{air}g}{c_p}w\right) \\ &= -\frac{e}{e_s}\left(\frac{l_v^2}{c_p RT^2}\frac{dy_L}{dt}\right) + \frac{e}{e_s}\left(\frac{l_v M_{air}g}{c_p RT^2}w\right). \end{aligned} \tag{6.22}$$

The first term here reflects the loss of supersaturation because of condensate formation, whereas the second term the formation of excess vapor as the equilibrium vapor pressure decreases with rise time.

With all the relevant physics included, we now combine terms algebraically to gain the desired rate of supersaturation development in a wet parcel. Note that two of the terms in Eqs. (6.20) and (6.22), taken together, contain the updraft speed (w) as a factor and the other two terms contain the rate of condensate formation (dy_L/dt). Grouping the respective terms gives us the result we seek, the so-called adiabatic supersaturation development equation:

$$\frac{ds}{dt} = Q_1 w - Q_2\frac{dy_L}{dt}, \tag{6.23}$$

where

$$Q_1 = \frac{e}{e_s}\left(\frac{l_v}{c_p T} - 1\right)\frac{M_{air}g}{RT} \tag{6.24}$$

$$Q_2 = \frac{e}{e_s}\frac{l_v^2}{c_p RT^2} + \frac{p}{e_s}. \tag{6.25}$$

As with any budget, the variable of interest (here, s) changes at a rate equal to the difference between production ($P = Q_1 w$) and loss ($L = Q_2 dy_L/dt$): $ds/dt = P - L$. Uplift produces excess vapor (hence supersaturation), whereas condensation consumes it. Below cloud base, where vapor is not consumed, the supersaturation increases in direct proportion to the updraft speed. Once the parcel reaches cloud base, nucleation and growth of cloud particles starts consuming excess vapor, slowing the rate of increase in supersaturation. A maximum supersaturation is reached some distance above cloud base (on the order of tens of meters), after which the supersaturation decreases and eventually achieves a quasi-steady-state value deeper in cloud, where the production term (updraft) is roughly balanced by constant consumption of vapor. The ramifications of this evolution of supersaturation are discussed in Chapter 10.

The principle upon which we derived the differential equation (6.23) describing the evolution of supersaturation in a warm (all-liquid) cloud may be adapted for use in mixed-phase clouds. Recognize that the diabatic heating rate (q) resulting from vapor depositing on both liquid and ice is just

$$q = l_v \frac{dy_L}{dt} + l_s \frac{dy_I}{dt}, \tag{6.26}$$

which, after appropriate substitutions, yields

$$\left. \frac{ds}{dt} \right|_{mixed} = Q_1 w - Q_2 \frac{dy_L}{dt} - Q_3 \frac{dy_I}{dt}, \tag{6.27}$$

where

$$Q_3 = \frac{p}{e} + \frac{l_v l_s}{c_p R T^2}. \tag{6.28}$$

The physical principles are robust and can be applied in a variety of situations to yield useful relationships between supersaturation, updraft speed, and cloud microphysical properties.

6.3 Cloud dynamics

Clouds form and evolve in ways that reflect the motions of the air in which the phase changes of water take place. Cloud dynamics are concerned with the fundamental forces that drive atmospheric motions on the scale of the clouds themselves. Large-scale motions, those associated with the common cyclonic storms of mid-latitudes, often lead to warm fronts, which spawn stratiform clouds having uniform properties over broad areas. Cold fronts similarly generate clouds over large distances, but cold-frontal clouds tend to be more convective, often separated from each other. The intrusion of cold air into warmer regions typically leads to unstable, overturning air, making cloud properties highly variable in space and time. Radiatively driven, smaller scale circulations often exist within large-scale systems, as when nighttime radiative cooling and/or cloud-top entrainment produce negative buoyancy at cloud top, producing upside-down convection. Smaller-scale motions also arise from turbulence in the boundary layer, or from air being forced up over mountains or other terrain features. Waves of both large and small scales form frequently in the atmosphere and cause clouds to form over a large range of scales. Regardless of the specific situation, most clouds form during rising air motions, which are generated by gradients in the pressure field and which generally follow adiabats, both outside (dry adiabats) and inside clouds (wet adiabats). We focus here on cloud-scale convective motions and seek to gain physical insight into the major factors that determine the environment in which the cloud microphysics play out.

6.3.1 Principles

The principles underlying dynamic meteorology are based on the conservation of momentum, mass, and energy. All accelerations (changes of motion) of the air are described by Newton's laws of motion. However, Newtonian physics strictly applies to an inertial coordinate system, so adjustments must be made for applications in a rotating coordinate system, such as the Earth. Newton's second law of motion preserves its form if we add to the real (physical) forces an apparent force due to the rotating body. The real forces of interest arise from gradients in pressure p, from gravity, and from friction. The apparent force is the Coriolis force, which causes a moving body to move toward the right in the northern hemisphere and toward the left in the southern hemisphere. Here, we consider the motion of an air parcel having unit mass.

The momentum equations appropriate for meso- and cloud-scale motions may be written in scalar form as

$$\frac{du}{dt} = \frac{\partial u}{\partial t} + \mathbf{v} \cdot \nabla u = -\frac{1}{\rho}\frac{\partial p}{\partial x} + fv + F_u \tag{6.29}$$

$$\frac{dv}{dt} = \frac{\partial v}{\partial t} + \mathbf{v} \cdot \nabla v = -\frac{1}{\rho}\frac{\partial p}{\partial y} - fu + F_v \tag{6.30}$$

$$\frac{dw}{dt} = \frac{\partial w}{\partial t} + \mathbf{v} \cdot \nabla w = -\frac{1}{\rho}\frac{\partial p}{\partial z} - g + F_w, \tag{6.31}$$

where $\mathbf{v} = u\widehat{\mathbf{i}} + v\widehat{\mathbf{j}} + w\widehat{\mathbf{k}} \equiv (u,v,w)$ is the velocity of the parcel, f is the Coriolis parameter, and $\mathbf{F} = (F_u, F_v, F_w)$ represents the frictional forces. The scalar speeds in the x, y, and z directions are respectively u, v, and w, the acceleration due to gravity is g, and the mass density of the air is ρ.

Convection in the atmosphere results when gravitation acts upon spatial variations in air density. The resulting atmospheric motions will in turn induce pressure variations that oppose the motions. In order to gain better physical intuition of convective motions, it is expedient to rewrite the vertical momentum equation in terms of a buoyancy force and a vertical perturbation pressure-gradient force. Let our real atmosphere be represented by a hypothetical base-state atmosphere, one in hydrostatic balance, upon which small anomalies in pressure and density are superimposed. Neglecting for now the frictional forces in the vertical momentum equation (Eq. (6.31)) and subtracting the base-state hydrostatic balance, we can write the perturbation form of the vertical momentum equation as

$$\rho \frac{dw}{dt} = -\frac{\partial p}{\partial z} - \rho g - \frac{\partial \overline{p}(z)}{\partial z} - \overline{\rho}(z) g$$
$$= -\frac{\partial p'}{\partial z} - \rho' g, \tag{6.32}$$

or

$$\frac{dw}{dt} = -\frac{1}{\rho}\frac{\partial p'}{\partial z} - \frac{\rho'}{\rho} g, \tag{6.33}$$

where the first term on the right is the vertical perturbation pressure-gradient force and the second term is the buoyancy force B. We see that accelerations in the vertical direction result from imbalances between the vertical perturbation pressure gradient force and the buoyancy force. The vertical perturbation-pressure gradient may result from hydrostatic processes (horizontal advection of density into or out of the column, horizontal divergence into or out of the column, diabatic or adiabatic temperature changes in the column) or it may have nonhydrostatic origins associated with velocity gradients or density anomalies resulting from vertical motions. In its simplest application (Eq. (6.33)), the vertical perturbation pressure may be understood as resulting from the nonhydrostatic high (low) pressure that must form above (below) the rising parcel to force the environment to accommodate its upward movement. Overlaying air must be pushed out of the parcel's path, while environmental air fills the wake, the space just vacated by the rising parcel.

For most applications, the buoyancy force (B) may be expressed in terms of temperature and pressure perturbations (denoted by primes) by making suitable approximations. First, we replace the density in the denominator of Eq. (6.33) with that of the mean-state atmosphere. Next, recognize that perturbations in one variable are related mathematically to the perturbations in the other variables upon which it depends. Density (ρ) is related to pressure (p) and virtual temperature (T_v) through the equation of state (which, for an ideal gas, is $\rho = p/(R_d T_v)$), so we find the relative density perturbation with the help of Taylor's expansion:

$$\frac{\rho'}{\overline{\rho}} = \left(\frac{\partial \ln \rho}{\partial p}\right) p' + \left(\frac{\partial \ln \rho}{\partial T_v}\right) T_v', \tag{6.34}$$

which yields the following approximate relationship for buoyancy:

$$B \equiv -\frac{\rho'}{\rho} g \approx \left(\frac{T_v'}{\overline{T_v}} - \frac{p'}{\overline{p}}\right) g. \tag{6.35}$$

This expression may be simplified further by realizing that for most atmospheric conditions $p'/\overline{p} << T_v'/\overline{T_v}$. This approximation is possible because pressure adjusts to induced changes much faster (at the speed of sound) than does temperature. (This assumption would fail only if the speed of the process could approach that of the speed of sound.) It is also customary to take the environmental virtual temperature profile as the reference state, such that the virtual-temperature perturbation is the difference between that of the parcel and that of the environment at a given level. We can therefore write

$$B \approx \left(\frac{T_{v,parcel} - T_{v,env}}{T_{v,env}} \right) g. \tag{6.36}$$

A parcel warmer than its environment will therefore experience a positive buoyancy force and an upward acceleration. Inside cloud, we also need to take account of the condensate. Using Eq. (5.44), we can then write the buoyancy as

$$B \approx \left(\frac{T_{v,parcel} - T_{v,env}}{T_{v,env}} - r_c \right) g. \tag{6.37}$$

Examination of this equation reveals that 3 g kg^{-1} of condensate (r_c) corresponds to approximately a 1°C temperature perturbation. For most applications, the contribution of the condensate to buoyancy is small and may be ignored. However, redistribution of condensate within a thunderstorm by precipitation can significantly shift the centers of positive and negative buoyancy in deep convective clouds, from regions of updraft to regions of downdraft.

6.3.2 Parcel theory

The simplest dynamical representation of a cloud is the parcel model. The cloud is viewed as a parcel of buoyant air that freely moves under the effects of gravity (free convection) without disturbing its environment. The parcel has uniform properties throughout, and its pressure adjusts instantaneously (at the speed of sound) to that of its environment. The only relevant governing equation is the vertical momentum equation, but under the assumptions of the parcel model vertical accelerations arise solely from the buoyancy force. Multiplying Eq. (6.33) by $w = dz/dt$ and ignoring the pressure perturbation term, we obtain

$$w\frac{dw}{dt} = B\frac{dz}{dt}, \tag{6.38}$$

which yields, upon integration from height z_0 to z,

$$w^2(z) = w^2(z_0) + 2\int_{z_0}^{z} B dz. \tag{6.39}$$

Using the hydrostatic equation and the ideal gas law (Eq. (5.2)), we obtain the integral

$$\int_{z_0}^{z} B dz = \int_{p_0}^{p} R_d \left(T_{v_p} - T_{v_{env}} \right) (-d \ln p), \tag{6.40}$$

which is the area on a thermodynamic diagram bounded by the adiabatic process curve of the parcel temperature and the environmental temperature profile between the pressure levels corresponding to heights z_0 and z. This area is proportional to the increase in kinetic energy as the parcel moves from height z_0 to z. If we now consider z_0 as the level of free convection (LFC), where we assume $w^2 = 0$, and z, the level of neutral buoyancy (LNB), then the vertical velocity at the LNB is $w(z_{LNB}) = \sqrt{2 \cdot CAPE}$, where $CAPE$ is the convective available potential energy, the area on a thermodynamic diagram bounded by the process curve and the environmental temperature profile. Theoretically, this vertical velocity represents an upper limit for the vertical velocity in buoyant convection.

Despite it simplicity, the parcel model yields useful results, and it is widely used in convective-storm forecasting (e.g., to calculate convective-cloud forecasting indices, such as CAPE, the Showalter index (SI), and Lifted Index (LI), among others.

Limitations of parcel theory

In reality, a buoyant parcel does interact with its environment in two ways that are evident from the vertical momentum equation, through the vertical pressure-gradient force and through frictional forces. When the parcel moves, it has to push the environmental air out of its path and refill the space it vacated with environmental air. As a result, a relatively high (low) pressure tends to develop above (below) a warm parcel, and a relatively low (high) pressure above a cold parcel, as shown in Fig. 6.3. These pressure perturbations cause the environmental air to move around the parcel. The high pressure above a rising warm parcel is the force behind the horizontal divergence; likewise, the low pressure below the rising parcel forces convergence. But, other effects of these pressure perturbations also exist. The pressure perturbations above and below the rising parcel indicate that a vertical perturbation-pressure gradient opposes the upward directed buoyancy force; hence, by Eq. (6.33), the vertical acceleration experienced by the parcel is less than predicted by parcel theory. Yet another effect of the upward (downward) motion of the parcel arises. By continuity, compensating subsidence (convection) must exist in the environment, the effect of which is to warm (cool) the environment and diminish the parcel buoyancy.

These simple conceptual arguments reveal an important way of understanding convective cloud motions. If a density anomaly is of small horizontal extent, the work required to push the environmental air laterally out of the parcel path is small, and a small force

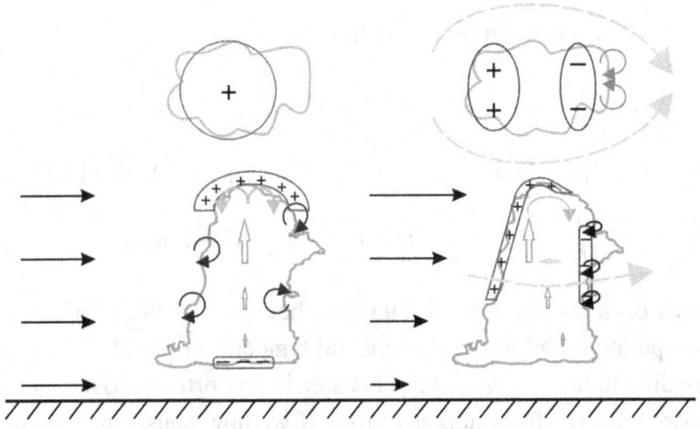

Figure 6.3 Plan views (top) and cross sections (bottom) of perturbation pressure fields in a growing cumulus cloud. Shown are clouds in a no-shear (left) and sheared (right) environment.

(from the pressure perturbation) suffices; conversely, if the density anomaly is large in extent, much air needs to be pushed over larger distances, requiring a much larger pressure perturbation. As a result, the vertical perturbation-pressure gradient scales with the width of the parcel and increasingly balances the buoyancy force as the parcel size increases. For sufficiently large anomalies, the opposing vertical pressure-gradient force balances the buoyancy force, resulting in no acceleration: the atmosphere is then in hydrostatic balance. It becomes clear that in one-dimensional (upright) convection, large, intense deep convection is not optimal and will not occur often. We will expand upon this idea later.

The frictional effects in the momentum equation result from the turbulent exchange of momentum, water vapor, condensate, and temperature between the parcel and its environment. This mixing process is called entrainment, which is easily discernible in convective clouds by the textured outline of the cloud. Alternatively, look closely at a smoke-stack plume on a cold winter morning. Entrainment dilutes the plume, thereby reducing parcel buoyancy and vertical momentum, with the effect, on both accounts, of slowing the vertical motion of the parcel. Mixing takes place only at the interface between the parcel and its environment, so a large parcel surface-to-volume ratio makes the parcel more susceptible to the effects of entrainment. The cores of bigger plumes/clouds are better protected against the effects of entrainment.

We next explore effects of mixing on the buoyancy with the aid of another thermodynamic diagram. Only in this case, warmer, condensate-laden (hence saturated) cloudy air mixes with colder, subsaturated environmental air. As with the mixing of two unsaturated parcels, the final state of the air depends on the mass fraction χ of cloudy air in the mixture, but with the added complexity of evaporational cooling as condensate evaporates to raise the saturation ratio of the environmental air. The state of the resultant mixture then depends on χ, the environmental saturation ratio, and r_c, the condensate mass fraction of the cloudy air. Fig. 6.4 shows the variation of the final virtual temperature of the mixture as a function

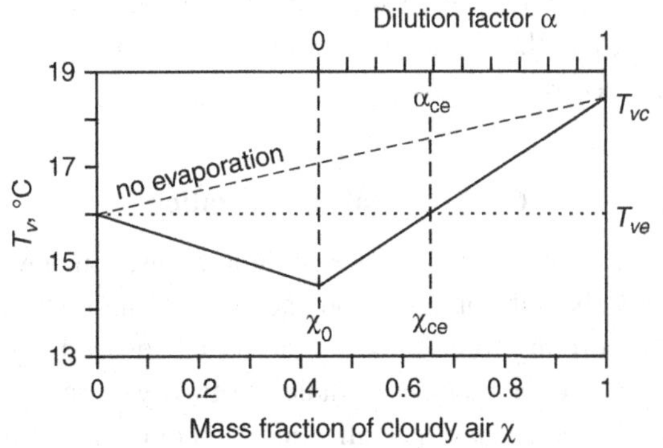

Figure 6.4 Thermodynamic mixing diagram of the resulting virtual temperature T_v as a function of the mass fraction of cloudy air χ and liquid-water dilution factor α. Adapted from Grabowski (1993) and Burnet and Brenguier (2007).

of χ for a specific environmental saturation ratio and r_c. For reference, we included the mixing line (dashed) for the case of no evaporation of condensate. A peculiar outcome is that the resulting mixture may be colder than either of the contributing air masses because evaporation takes energy from the air. The extent of the deviation of the mixing line below the reference case and the minimum mixing fraction χ_0 that will produce a cloudy parcel depend on the environmental saturation ratio and the amount of condensate. Drier environments mixing with higher condensate-content cloudy air cause greater deviations because of the larger potential for evaporative cooling, and they are more likely to produce negatively buoyant air, even with respect to the environment. In such conditions, we observe downward motion in both clear and cloudy air mixtures, provided that the mixing fraction of cloudy air is below a critical value χ_{ce}, corresponding to a critical dilution ratio α_{ce}. Only mixtures with cloud mass fractions $\chi > \chi_{ce}$ remain positively buoyant with respect to the environment, even though all mixtures are negatively buoyant with respect to the original cloudy air.

One sees competing demands on the characteristics of a parcel from the two additional terms in the vertical momentum equation. The vertical perturbation pressure gradient term favors infinitesimally narrow updrafts in deep convection, but such updrafts are highly susceptible to the effects of entrainment, which homogenizes the parcel and its environment. On the other hand, very large density anomalies are well protected against the effects of entrainment, but they tend more toward hydrostatic balance. Neither of these limits are thus conducive to sustained convection. These arguments suggest that there is an optimal size range that protects the parcel against the effects of entrainment, yet allows the buoyancy force to exceed the vertical perturbation pressure gradient force.

Finally, and closer to the subject matter of this book, parcel theory neglects contributions of hydrometeors to the parcel buoyancy (Eq. (6.37)). If all the condensate remains suspended in the updraft, the buoyancy may be substantially reduced, weakening the updraft, whereas if precipitation removes condensate from the updraft, it will be strengthened. When we realize that hydrometeors always fall relative to the air in which they are imbedded, we see a delicate interaction between the microphysical processes that determine the microstate of the clouds and the dynamical processes that provide the larger-scale forcing of the microphysical processes.

6.4 Mesoscale organization

Clouds often organize themselves into various configurations, which we call cloud or storm systems. When thinking about the organization of cloud systems on the mesoscale, we must think of buoyancy as a driving force (convection, up or down), buoyancy as a restoring force (gravity waves), horizontal buoyancy gradients (gravity currents), and vorticity (wind shear). All mesoscale phenomena are specializations and/or combinations of these physical phenomena. Of these, the vertical shear of the horizontal wind is commonly the dominant determinant of convective organization.

6.4.1 Convection systems

Deep convective clouds are driven by strong buoyancy forces. Such systems therefore occur only in environments where sufficient CAPE exists. However, although CAPE is a necessary condition for deep convection, it is not always a sufficient condition. Convection is closely associated with processes that force low-level horizontal convergence, which is necessary to lift boundary-layer air up to the LFC. Low-level convergence can arise in a number of ways, most notably from orographic forcing, but more often from air mass boundaries in the boundary layer. Once a storm is initiated, the storm itself can provide the density boundary along which subsequent lifting occurs, as when evaporatively cooled outflow air from the precipitation spreads out along the surface underneath the storm.

In order to understand the organization of a storm, it is necessary to consider the interaction of the storm with the flow field in which it is embedded. The cloud itself consists mostly of air lifted from the boundary layer, so the updraft carries with it the momentum of the boundary-layer air. If the wind speed and direction throughout the depth of the troposphere are constant, the updraft will be vertical, even when the cloud is moving relative to the surface. From a storm-relative framework, low-level air will converge into the base of the updraft and diverge from the updraft at the level of neutral buoyancy (left-hand side of Fig. 6.3). On the other hand, if the updraft ascends into an atmosphere where the wind speed or/and direction changes with altitude (vertical shear of the horizontal wind), there will be net convergence (divergence) on opposite flanks of the updraft (right-hand side Fig. 6.3). The shear vector, the vector difference between the wind at altitude and that in the boundary layer, plays an important role in the storm dynamics. The relative in-cloud and environmental flow fields produce convergence (divergence) with associated high (low) non-hydrostatic pressure perturbations on the upshear (downshear) side of the storm. The perturbation high pressure on the upshear side of the storm is necessary to divert the environment flow around the storm, whereas the low perturbation pressure is necessary to force the convergence of the environmental flow downshear of the storm. As a consequence, the updraft of the storm is protected from the environmental air by the relative high, whereas the downshear side of the storm experiences significant entrainment.

The impact of shear on storm development and organization may be viewed from two perspectives. Shear across the depth of the boundary layer is important for the role it plays in determining the efficiency of lifting boundary-layer air to its level of free convection. The deep-layer shear, the shear across the depth of the precipitation-producing layer, determines where precipitation is deposited relative to the location of the updraft. The relative role that each type of shear plays depends on the environment in which the storm develops. The organization of convection is typically described in three distinct classes: single-cellular, multicellular, and supercellular convection, with environmental deep-layer shear the dominant factor in determining the more likely storm-type. Updrafts in single-cell storms are mostly driven by buoyancy, while vertical pressure-gradient forces play increasingly more important roles in supercellular convection. However, it is important to realize that such distinctions exist exclusively in textbooks. Clouds in nature exhibit a continuum of storm types.

Single-cellular convection

Single-cellular or ordinary convection occurs in weak-shear environments, those in which the deep-layer shear (typically taken as the 0 to 6-km layer) is less than $\sim 10\ \mathrm{m\,s^{-1}}$. Weak-shear environments are typically regions of weak synoptic-scale forcing, such that storm initiation depends strongly on the diurnal cycle of the boundary-layer value of θ_e, which is a strong function of the temperature and water vapor mole fraction. Increasing the boundary-layer θ_e reduces CIN while increasing CAPE.

The lifecycle of an ordinary cell may be viewed as consisting of three stages. During the first or towering-cumulus stage (cumulus congestus) most of the cell exists as an updraft (Fig. 6.5). The mature stage is reached when precipitation has grown large enough to fall through the updraft. The redistribution of the precipitation loading increases buoyancy in regions from which the precipitation falls, strengthening the updraft there, while reducing buoyancy lower in the cloud. As the precipitation settles into the subcloud layer evaporation of the hydrometeors further reduces buoyancy and induces downdrafts. The downdraft is just as important a part of the cloud as the updraft. In cloud, downdrafts also follow moist adiabats, but below cloud base precipitation-laden downdrafts have temperature lapse rates somewhere between the moist and dry adiabatic values. The evaporation rate from precipitation is usually too small to keep the downdraft saturated. Upon reaching the surface, the downdrafts spread laterally as a gravity current, creating a pool of colder air beneath the cloud. The interface between the expanding rain-cooled air and the warm moist air is the gust front. In a perfectly no-shear case the cold pool will spread symmetrically around the downdraft, lifting the potentially buoyant environmental air along the

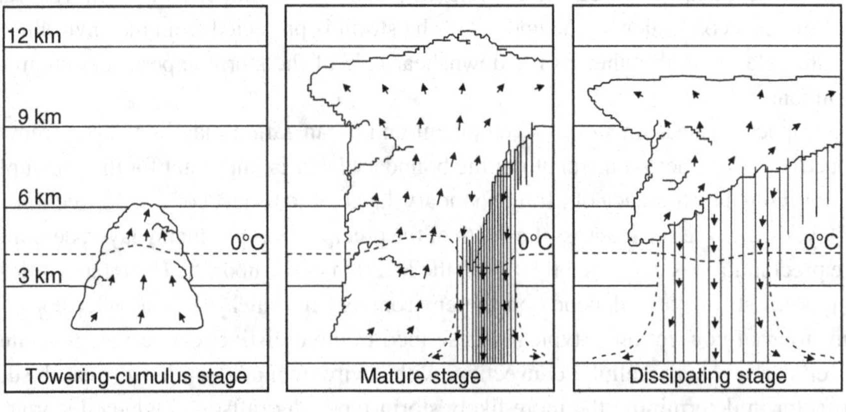

Figure 6.5 The three stages of an ordinary single-cell thunderstorm: towering cumulus, mature, and dissipating. Adapted from Byers and Braham (1949) and Doswell (1985).

gust front, sustaining the cloud during the mature stage. However, lifting in weak-sheared environments is weak and shallow, making it difficult for the gust front to initiate new convection.

As the downdraft strengthens and its root grows deeper in the cloud, vertical divergence between the up- and downdrafts forces mid-level horizontal convergence and entrainment of low-θ_e air, which further strengthens the downdraft. Downdrafts eventually dominate the cell, at which point the dissipating stage is reached. Light, but steady precipitation characterizes the dissipation stage. At this point, the cold pool has spread far from the updraft, and the forced lifting along the gust front no longer feeds into the updraft, which can now no longer be maintained. The original cloud continues precipitating until all that remains is an orphaned anvil composed of ice in the upper troposphere.

Most environments do contain some vertical shear of the horizontal wind. The effect of the interaction between the storm and its environment on the internal thermodynamic state of a cumulus congestus cloud growing in a sheared environment can be seen in Fig. 6.6, a horizontal cross section through a cell several kilometers above cloud base. We will use the conservation properties of θ_e to reveal the extent of mixing. The high-θ_e values at the back of the cloud (roughly the upshear side), are within 1.5 K of cloud base θ_e and locate the core of the updraft. A sharp-θ_e gradient exists on the upshear side of the updraft, whereas the gradient towards the downshear side is weaker. A small, protected updraft core is seen that consists of almost pure boundary-layer air, whereas a large fraction of the cloud volume consists of a mixture of boundary-layer and free-tropospheric air. The effect of mixing on the microstructure of the cloud is shown by looking at the ratio of the observed to adiabatic liquid water concentration. This ratio varies between 0.8 and 1.0 in the core of the updraft to below 0.2 in the downdraft.

Details of the mixing process may be investigated using a conserved-variable diagram, such as that shown in Fig. 6.7. The conserved variables of interest here are the total-water mass mixing ratio r_t and the liquid-water potential temperature $\theta_l \approx \theta - l_v r_L / c_p$. The atmospheric profile in r_t–θ_l parameter space is shown by the solid curve, with height indicated at the dots in hPa. The shaded oval shows the parameter space occupied by measurements taken at cloud top in a small cumulus cloud. Consistent with the picture portrayed in Fig. 6.6, we see a mixture of the conserved variables θ_l and r_t from cloud base to cloud top values. More striking, however, is the fact that most of the values fall along an approximately straight line between cloud base and cloud top. The arrangement of the observations along the mixing line suggest that entrainment in small (non-precipitating) cumulus clouds may be viewed as cloudy air mixing isobarically with environmental air near cloud top as the cloud grows.

The single-cellular storm moves approximately with the velocity of the wind averaged over its depth. The downdraft air carries the mid-level momentum down to the cold pool, such that, in a storm relative sense, the gust front advancing on the downshear flank of the storm is slowed. As a result, the low-level inflow into the cell predominantly comes from the downshear flank. At mid-levels, the horizontal pressure gradient (right-hand side Fig. 6.3) causes a P-shaped overturning of the updraft in the downshear direction close

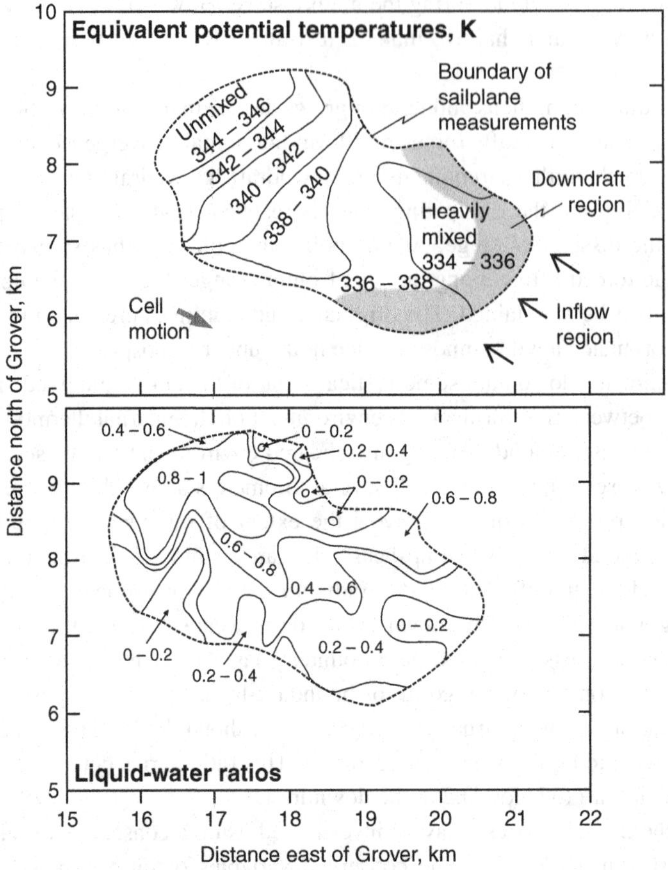

Figure 6.6 Horizontal projection of data obtained from sailplane measurements on 22 July 1976 over Grover, Colorado, in an altitude range of 3.1 km. Data have been plotted relative to the cell motion. Top: Equivalent potential temperature data (K); Bottom: Ratio of liquid water concentration to adiabatic value. Adapted from Heymsfield *et al.* (1978).

to cloud top. The downdraft then develops on the downshear side of the cloud, where the perturbation low pressure causes entrainment of environmental air into the cloud.

Multicellular convection

Multicellular convection is the most common form of convective-cloud systems. In multicellular complexes, new cells continually develop along the cold pool gust front. Individual cells cycle through the single-cell towering cumulus, mature, and dissipating stages. At any time in the system's life cycle, one may find multiple cells in the complex, all in various stages of development. As a result of the continual development of new cells, the complex survives much longer than any single cell.

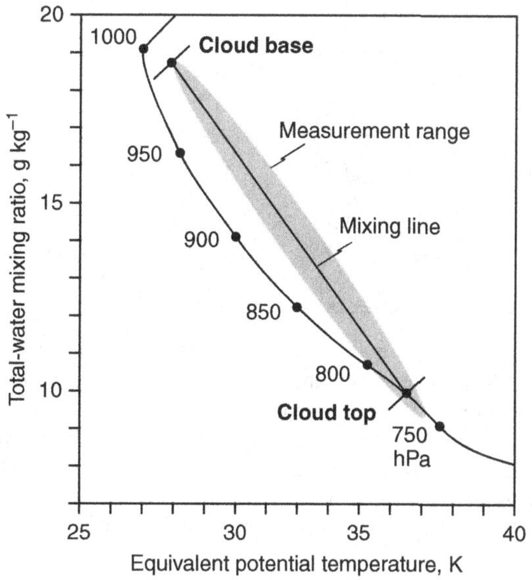

Figure 6.7 Schematic plot of total-water mixing ratio versus equivalent potential temperature (based on observations of a small cumulus). The environmental profile is shown in the solid line with heights in hPa indicated at the dots. The shaded oval shows the range of individual measurements all taken close to cloud top. The mixing line between cloud base and top is indicated by the straight line between cloud base and cloud top. Adapted from Burnet and Brenguier (2007).

The environments conducive to multicellular convection exhibit moderate shear ($\sim 10\ \mathrm{m\,s^{-1}} <$ shear $< \sim 20\ \mathrm{m\,s^{-1}}$). As in ordinary cells, individual cells *advect* with the velocity of the mean wind over their depth. However, new cells have a tendency to form on the downshear flank of the cold pool gust front where the lifting of high-θ_e environmental air is enhanced. The progress of the complex resulting from this continual development along the gust front is called the storm *propagation*. The movement of the complex is then the vector sum of the advection vector of the individual cells and the storm-propagation vector. The cause of the enhanced lifting along a preferred section of the gust front may be best understood by noting that the gust front on the downshear flank of the storm is the region where the difference in momentum between the cold pool and the environmental air is largest. The environmental shear over the depth of the cold pool is another contributing factor. The magnitude (and orientation) of the shear determines the vertical profile of the gust front, and thus the depth of lifting.

Figure 6.8 shows a conceptualization of the streamlines through a multicellular storm. This figure may be seen as an instantaneous view of four cells in different stages of development, alternatively as the development of a single cell through its different stages. A new cell (n + 1) develops out of the shelf cloud from the forced lifting by the gust front, and grows into a towering cumulus cell (n). The mature stage (n − 1) is reached when the precipitation process is well developed. The upper parts of the cell still contain an updraft,

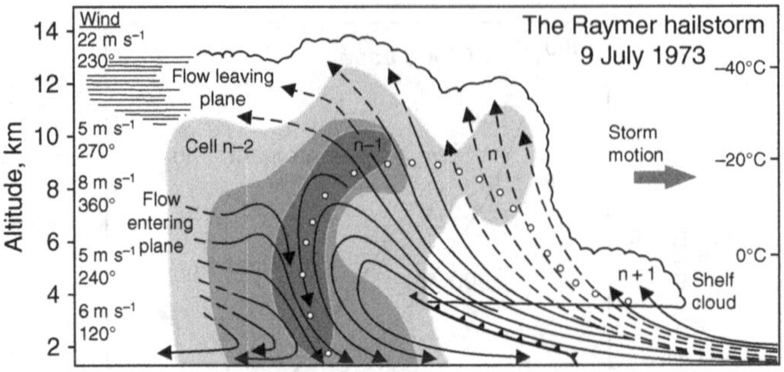

Figure 6.8 Schematic model from a northeastern Colorado multicell storm. Storm relative airflow is composited from aircraft, Doppler radar, and surface mesonet measurements. The solid lines are streamlines of flow relative to the moving system; it is shown broken on the left side of the figure to represent flow into and out of the plane and on the right side of the figure to represent flow remaining within a plane a few kilometers closer to the reader. Shading denotes radar reflectivity of 35, 45, and 50 dBZ. The empty circles represent the hypothesized trajectory of a hailstone during its growth from a small droplet at cloud base. Adapted from Browning *et al.* (1976).

but the lower parts are mostly downdraft. At the dissipation stage (n − 2) the cell contains mostly weak downdrafts, with only some vestiges of updraft remaining aloft. The open circles indicate a hypothesized trajectory of a hailstone growing from a cloud droplet at cloud base until it falls out of the cloud.

Supercellular convection

A supercell, shown schematically in Fig. 6.9, is a long-lived storm characterized by a persistent mesoscale cyclone within the updraft and by a mode of propagation distinct from that of multicellular storms. Supercells are probably the least common storm type, but they produce a disproportionate fraction of severe weather, including large hail and tornadoes. The distinctive characteristic of a supercell relates to the large shear ($> 20\,\mathrm{m\,s^{-1}}$) over the deep layer in which it forms. This shear is the source of the vertical vorticity for the mid-level mesocyclone, and it also affects the mode of propagation by inducing vertical gradients of perturbation pressure. These vertical pressure gradients may arise from horizontal momentum differences resulting from the interaction of the shear with the updraft or from the mesocyclone itself. These pressure gradients not only enhance the lifting of boundary layer air to the level of free convection, but they may also impact the vertical velocity high in the updraft, or they may aid in the generation of downdrafts. Wind shear also contributes to cell longevity by advecting hydrometeors formed in the updraft away from their source regions, thereby separating the main regions of precipitation from the updraft. The movement of the storm deviates substantially from the mean wind over the storm depth, with cyclonically (anticyclonically) storm moving to the right (left) of the mean storm wind shear.

The circulation of air in a supercell persists in a nearly steady-state pattern for periods lasting 30 minutes or more. A typical supercell storm consists of a single dominant updraft,

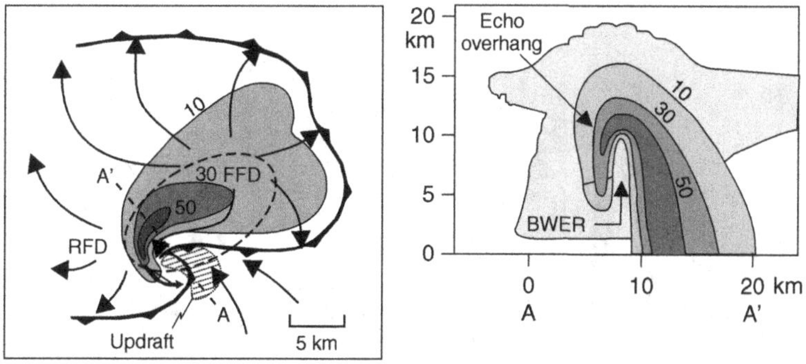

Figure 6.9 Schematic cross sections of a supercell. Left: top view. Right: side view along section A–A'. RFD = rear-flank downdraft; FFD = forward-flank downdraft; BWER = bounded weak-echo region. Radar reflectivity levels are indicated with shading, the updraft with hashing. The arrow indicates the storm motion. Adapted from Lemon and Doswell (1979) and Markowski and Richardson (2010).

although many storms have a flanking line of shallower updrafts on the right rear flank of the storm (relative to storm motion). These weaker flanking-line updrafts, with their well-developed hydrometeor populations, often merge with the main updraft, providing a source of precipitation embryos for the main storm. The vertical perturbation-pressure gradient generates a low-level updraft of sufficient strength to bring some of the negatively buoyant air from the humid, rain-cooled cold pool into the updraft, producing a spectacular wall cloud. The base of the wall cloud is thus lower than that of the main cloud, where air of high θ_e is feeding the storm. Because the air feeding the wall cloud is negatively buoyant, the cloud looks smooth, lacking the characteristic turbulent appearance of convective clouds. Radar data reveal a bounded weak-echo region (BWER) closely associated with the main updraft. This region of lower reflectivity indicates that the updraft volume has relatively small hydrometeors, a consequence of the updraft being too strong to allow the injection of precipitation-sized hydrometeors from other regions and a Lagrangian time scale too short for precipitation to form. The BWER is bounded at the top and sides by strong radar returns, suggesting that well-developed populations of precipitation particles diverge at the top of the updraft and settle along its sides. The potential role of precipitation settling next to the main updraft in the hail-formation process is discussed in Chapter 12.

The storm contains two general downdraft regions, the rear-flank downdraft (RFD) and the forward-flank downdraft (FFD). The cause of the RFD may be entrainment of mid-level, low-θ_e air, induced initially by precipitation loading and/or a downward-directed vertical pressure gradient force. The most likely location for observing large hail is around the RFD. However, the bulk of the hydrometeors are advected away from the low-level updraft region by the storm relative winds aloft toward the forward flank on the downwind side of the updraft. As these hydrometeors fall, sublimation and melting of ice and evaporation of rain produce negative buoyancy and the FFD. Each downdraft produces a cold pool with associated gust front. Collectively, these cold-pool gust fronts look similar to the

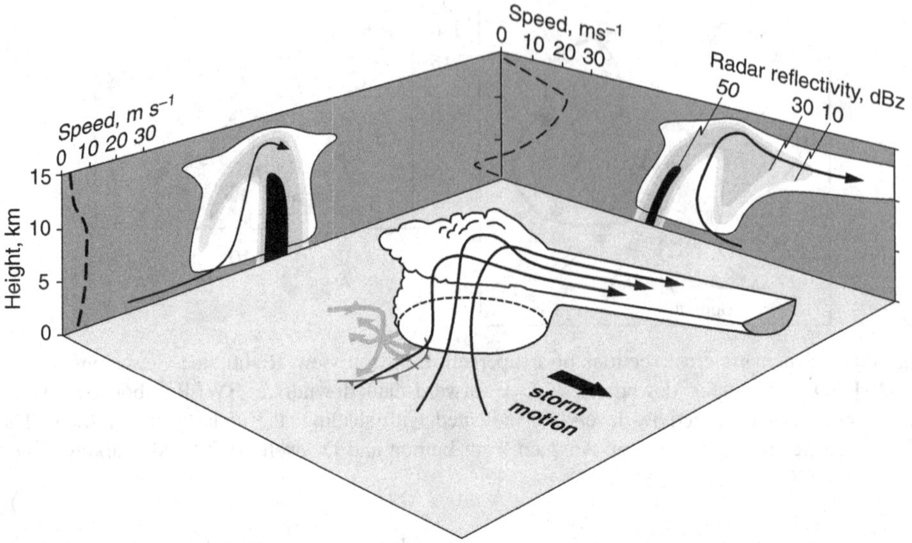

Figure 6.10 Perspective view of air flows through a supercell. Radar reflectivities and storm-relative flows are shown on side panels. The updraft is indicated with black arrows, the rear-flank downdraft with grey arrows. Adapted from Chisholm and Renick (1972).

structure of a typical mid-latitude extra-tropical cyclone. The location of the quasi-steady main updraft is typically closely associated with the intersection of the two gust fronts.

The kinematic structure of a supercell storm is complicated and dependent on the vertical profile of the horizontal wind (the wind-shear environment). The flow of high-θ_e air into a typical storm in the northern hemisphere comes from the right forward flank (Fig. 6.10; relative to storm motion). The flow ascends vertically in the main updraft, diverges and turns anticyclonically toward the left forward flank, where it forms the dominant outflow anvil. The air forming the RFD originates in the mid-troposphere, possibly from air entrained on the forward right flank that wraps around the updraft core, or from air entrained on the rear or rear right flank, which then sinks around the updraft. The FFD originates in air that, having been diverted around the updraft, passes underneath the anvil on the left forward flank of the storm, where it experiences cooling from sublimation and/or melting of ice and evaporation of rain. The complicated flow structure aids the recirculation of precipitation into the main updraft from the diverging outflow near the top of the storm.

Mesoscale convective systems

A mesoscale convective system (MCS) typically arises when the cold pools from adjoining thunderstorm cells in a region merge and form a single, contiguous cold pool. An MCS is similar to, but a larger version of multicellular convection, with new cells forming along the edge of the combined cold pool as older cells mature and decay. As with the organization of smaller-scale convective systems, shear plays an important role in the development of a mesoscale convective system. Shear establishes where the precipitation is likely to occur relative to the updraft, and shear also determines the nature of the forced lifting along the edge of the updraft. Mesoscale convective systems are relatively common at night, probably because time is required for the cold pools from isolated afternoon convection to first form and then merge.

Mesoscale convective systems are classified loosely as squall lines (propagating quasi-linear convection) or mesoscale convective complexes (large, long-lived quasi-circular convection). Quasi-linear systems prefer environments with stronger vertical shear of the horizontal wind, whereas mesoscale convective complexes occur most commonly in weak synoptic-scale features with weak wind shear.

Squall lines Squall lines are quasi-linear convective systems that occur in large-shear environments. The exact organization of the system depends on the magnitude and altitude dependence of the vertical gradient of the horizontal wind. The most prevalent conditions in which squall lines occur exhibit strong, line-normal rear-to-front, wind shear mostly confined to low levels.

The structure of linear convection developing in such an environment is illustrated by the conceptual model shown in Fig. 6.11. In this case, the squall line is moving toward the right, roughly in the direction of the low-level shear. The leading line of the system consists of convective towers producing intense precipitation, followed, after a short transition region with light precipitation, by more widespread precipitation in the trailing stratiform area. This structure is easily discernible in precipitation-radar images as a narrow high-reflectivity line along the leading edge and a larger area of enhanced reflectivity separated from the convective line by a region of lower reflectivities. The storm-relative inflow comes from the front of the system (right side of the diagram) and is lifted to the level of free convection by the propagating gust front, carrying its front-to-rear momentum into the updraft. The updraft is steep in the leading convection, where the convective instability of the environment is realized. As the cells mature, they gradually fall behind the leading edge of the gust front and weaken. The dissipating cells in effect constitute the stratiform region. The upper parts of the stratiform region, in the trailing anvil, are still

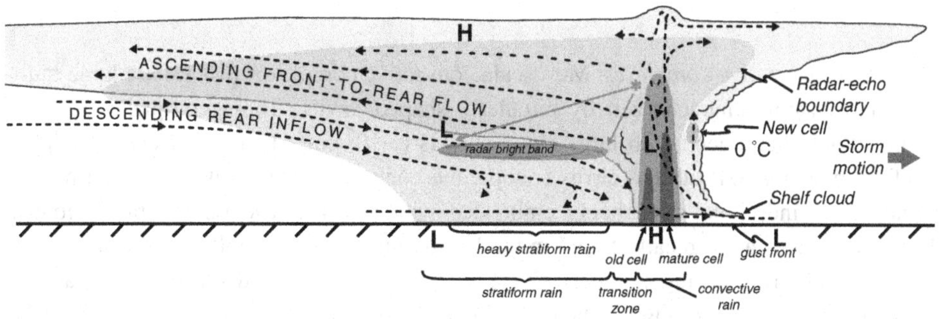

Figure 6.11 Conceptual model of a squall line with a trailing stratiform region. The vertical cross section is oriented perpendicular to the convective line. Adapted from Houze *et al.* (1989).

slightly buoyant because of the release of latent heat by the growing hydrometeors and radiative destabilization of the layer, forcing weak ($< 0.5\,\mathrm{m\,s^{-1}}$) mesoscale updrafts. As a consequence, developing hydrometeors move through the line in a general front-to-rear direction. The radar reflectivity pattern results from size-sorting of hydrometeors: the faster falling ($> 5\,\mathrm{m\,s^{-1}}$) hydrometeors fall close to the updraft along the leading edge, whereas the slower-falling ice particles are advected rearward, where they experience continued growth by vapor deposition in the mesoscale updraft region. The stratiform precipitation arises as these hydrometeors fall out and melt after the intense convective precipitation has passed.

Another characteristic feature of a squall line is the rear-to-front downdraft that develops just below the base of the trailing anvil. Evaporation and melting of the ice-phase precipitation falling from the anvil cause strong cooling of the mid-troposphere, leading to a perturbation low pressure at the anvil base. This low pressure draws environmental air in from the rear of the system towards the front. The rear-to-front movement of air is further accelerated by the baroclinically induced vorticity at the top of the cold pool and near the base of the anvil. The mesoscale downdraft descends, reaching the surface near the rear of the convective zone, where the heaviest precipitation and the largest evaporative cooling rates are found.

The persistence of squall lines arises in part from the synergy of the front-to-rear updraft and the rear-to-front downdraft. The general rearward flow in the updraft allows precipitation to fall into the mid-level inflow, the evaporative cooling and melting of hydrometeors thereby strengthening the downdraft and the gust front along the leading edge of the system. The low-level environmental shear helps keep the gust front close to the mature cells so that they last longer and propagate into the moisture-rich air ahead of the system.

Whereas many squall lines are organized according to the conceptual model discussed here, other types are also observed. In environments with strong deep-layer shear oriented rear-to-front, the squall line may have a leading (rather than trailing) anvil. Precipitation from a leading anvil does not decrease the environmental instability because evaporative cooling aloft generally exceeds that at low levels, thereby increasing the conditional instability. In environments in which the deep-layer shear has a strong along-line component the anvil is found parallel to the line, in the down-shear direction.

Mesoscale convective complexes Mesoscale convective complexes (MCC) are large multicellular systems characterized by circularly shaped anvils. These complexes tend to be nocturnal phenomena with large cloud shields ($>10^5\,\mathrm{km^2}$) that are clearly visible in satellite imagery. Beneath the cirrus canopy, the convective organization may be highly variable, the individual cells being embedded within widespread nimbostratus clouds. These complexes are typically found in the United States on the cool side of a stationary synoptic-scale front during summertime when a weak short-wave disturbance propagates through a synoptic-scale ridge in the upper troposphere.

An MCC develops its own large-scale circulation with a stable warm core that is critical to the maintenance of the system. At upper-levels, just below the tropopause, a large mesoscale high-pressure perturbation forms with anticyclonic outflow, while in mid-tropospheric levels, just below the level of maximum convective heating, a low-pressure perturbation with cyclonic circulation develops. The complex draws in free-tropospheric air from just above the stable boundary layer. The rise of this moist inflow to the level of free convection is aided by dynamics associated with the mid-level cyclone. The system maintains itself as long as it has a supply of moist, conditionally unstable air. The system as a whole outlasts the individual cells that contribute to the widespread precipitation.

Mesoscale convective systems produce prolific amounts of precipitation. Their large areal extent allows many thunderstorm cells to form and generate precipitation. Early in the life cycle of the system, precipitation rates can be locally intense and convective in nature, but as the system matures precipitation becomes lighter, but more widespread from stratiform clouds.

6.4.2 Stratiform clouds

Much of Earth is covered by extensive sheets of stratiform, or layered clouds, such as stratocumulus, altocumulus, and/or cirrus. These clouds directly impact the Earth's radiation budget, thereby playing an important role in climate. The dynamics of stratiform clouds are described by the same basic set of equations used for convective clouds. The main distinctions between these two classes of clouds is that stratiform clouds persist longer, have much smaller vertical extent, and exist in relatively thin, well-mixed layers. Layers conducive to cloud formation are often found close to the Earth's surface (yielding cloud-topped boundary layers), but they may also occur in the mid and upper troposphere (i.e., alto- and cirriform). Localized convection, driven by buoyancy forcing from below or above, may generate many small-scale cloud elements confined to the layer (e.g., altocumulus). Forcing from below requires a positive buoyancy flux (a combination of temperature and moisture flux generating relatively high θ_v in the lower part of the mixed layer). Forcing from above results predominantly from longwave radiative cooling at cloud top. Cloud layers are often capped by a statically stable layer and accompanied by a drop in vapor concentration. (In the Arctic, however, vapor concentrations frequently increase across the inversion.) Entrainment of warmer air across the inversion plays a stabilizing role by serving to off-set the effects of cloud-top radiative cooling, but the effect of entraining drier air into cloud top results in evaporative cooling, which adds to a top-down forcing of convection. The formation of precipitation within a layer, typically drizzle in boundary-layer clouds, warms the cloud layer through latent-heat release, while sedimentation of the precipitation into subsaturated air cools the subcloud layer during evaporation.

The longevity of layer clouds depends on a delicate balance of processes. Condensate forms as moist air rises or becomes cooled by a net loss of radiation. But, condensate can be lost during localized sinking motions, sedimentation out of the layer, and entrainment of drier air from above the inversion. Entrainment has the peculiar effect of reducing the concentration of condensate, while simultaneously increasing the depth of the well-mixed layer. Air circulations may also extend beyond the actual cloud layer into the subcloud layer. In the case of boundary layer clouds, these subcloud circulations constitute an important link to vapor sources at the surface. We will next briefly review two types of layered cloud systems, marine boundary clouds and elevated, mixed-phase layer clouds.

Marine cloud-topped boundary layers

Marine stratocumulus clouds cover extensive parts of most eastern sub-tropical oceans under the influence of moderate to strong free-tropospheric subsidence. The sinking motion warms the air in the free-troposphere, giving rise to a strong capping inversion that traps moisture in the boundary layer and confines cloud to the top of the atmospheric boundary layer, as shown in Fig. 6.12. The depth of the cloud layer varies in response to a complex blend of physical processes, some of which are discussed below. The liquid water concentration increases on average with altitude from cloud base to maxima close to cloud top, but entrainment and radiation both affect cloud structure and properties. The vertically integrated liquid water amounts are often large enough that the cloud behaves as a black body to longwave radiation. The areal extent and brightness of persistent cloud decks above dark ocean waters reflects sufficient amounts of solar radiation to impact the Earth's energy budget and climate.

The air circulations within marine layers are forced by a combination of radiation and entrainment. With very dry air aloft, marine clouds experience strong cooling near their tops from longwave radiative flux divergence. The negative buoyancy thus generated is the dominant driver for the circulations and turbulent eddies within the marine boundary layer. Where cloud-top cooling is especially strong, the air becomes negatively buoyant throughout the boundary layer, and the large eddies, scaled by the boundary-layer depth, create a well-mixed layer from the inversion all the way to the surface. Horizontal winds induced by divergence of these eddies at the surface enhance evaporation from the ocean surface, producing vapor-rich air that is carried back into the cloud by updrafts. When the updrafts of those same eddies encounter the capping inversion, they entrain filaments of warm dry air from the free-troposphere. The stable stratification at the inversion layer ensures that all mixing fractions are negatively buoyant and that the mixed air sinks well

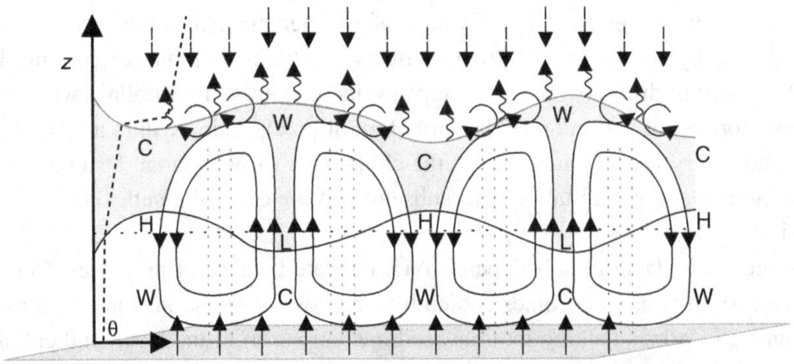

Figure 6.12 Schematic cross section of a stratiform cloud in the marine boundary layer.

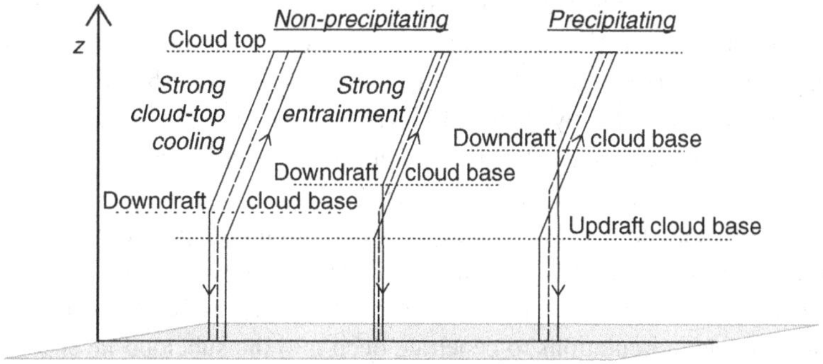

Figure 6.13 Vertical profiles of virtual potential temperature θ_v in the marine boundary layer.

into the boundary layer. It is this cross-inversion entrainment that maintains the boundary layer against the free-tropospheric subsidence. A consequence of the different physical processes at the top and bottom boundaries of the marine layer is that the updrafts are slightly warmer and moister than the downdrafts. In this sense, circulations in a cloudy marine boundary layer are thermally direct.

The thermodynamic structure of a marine boundary layer is schematically depicted in Fig. 6.13. The profiles of virtual potential temperature (θ_v) show that the boundary-layer processes are at different times dominated by strong cloud-top cooling (left), high entrainment rates of warm and dry free-troposphere air (middle), and efficient drizzle production (right). Looking at the updraft profiles, we note that in all cases θ_v in the subcloud layer is conserved and that θ_v increases in the cloud layer because of condensational warming. Radiative cooling and mixing at cloud top is represented by a θ_v decreasing at the top of the boundary layer, the magnitude of the decrease depending on the specific process that dominates. For the case (left) dominated by radiative cooling and a free troposphere having higher water vapor mole fractions (y_v), the air is cooled, but the loss of condensate needed to maintain the cloud against the drying effects of entrainment is small. As a negatively buoyant parcel descends in the downdraft, θ_v will further decrease as the parcel descends through the cloud layer to cloud base because of evaporative cooling. However, the downdraft cloud base is reached at an altitude slightly higher than that of the updraft because of the loss of condensate at cloud top. For the case of strong cloud-top cooling, the downdraft remains negatively buoyant and the circulation extends all the way to the surface, providing a continued source of water vapor to offset the drying effects of cloud top entrainment.

As the cross-inversion increase in θ_v and decrease in free-tropospheric y_v become larger, entrainment produces initial mixtures of air that are both warmer and drier than in the strong cloud-top cooling example. If entrainment rates are sufficiently high, the boundary-layer dynamics becomes more complicated. More condensate must evaporate to maintain saturation at cloud top, and though the mixed parcel cools, the decrease of θ_v at cloud top is smaller and that of liquid water concentration is larger. As a consequence, cloud base is encountered at a higher altitude in the boundary layer where θ_v is greater than in the subcloud updraft. As the downdraft descends below cloud base, it becomes positively buoyant, and decelerates. In a mean sense, the updrafts (downdrafts) in the subcloud layer then become negatively (positively) buoyant, and the cloud layer must perform work on the subcloud layer to maintain the circulations. This work is accomplished by nonhydrostatic pressure perturbations induced by the cloud-layer circulations analogous to our discussions for convective systems (Fig. 6.3). A perturbation low pressure forms at cloud base below the accelerating in-cloud updraft (shown in Fig. 6.12 with an L), which draws the negatively buoyant subcloud air into the updraft. A perturbation high pressure forms below the downdraft, accelerating subcloud air toward the updraft, thus completing the eddy circulation. The cloud will only be maintained if the induced vertical pressure perturbations are sufficient for the eddy circulations to penetrate deep into the subcloud layer to obtain the moisture necessary to off-set the effects of entrainment drying.

The microphysical processes in marine stratocumulus interact with the boundary-layer dynamics in complicated ways. Drizzle rates may be as large as the rates at which the eddy circulations bring moisture into the cloud layer, so drizzle production must be accounted for in the cloud-water budget. The effect of drizzle on the boundary-layer dynamics is similar to that of large entrainment rates at cloud top. The removal of liquid water from the cloud by drizzle effectively warms the cloud layer and cools the subcloud layer by displacing the evaporative cooling effect of the drizzle from the in-cloud downdraft to the subcloud layer. Drizzle sedimentation effectively stabilizes the cloud layer with respect to the subcloud layer. The downdrafts are drier and become buoyant above the boundary-layer mean height of cloud base. The stabilization of the boundary layer reduces the eddy circulation strength, leading to reduced entrainment at cloud top, with further impacts on the circulations within the cloud layer. The intensity of the drizzle is important. If the drizzle extends all the way to the surface, the entire subcloud layer becomes cooled, whereas if the drizzle completely evaporates in the subcloud layer the cooling is limited to a layer just below cloud base. In either scenario, the cloud base is stabilized with respect to the subcloud layer, and the subcloud layer is itself destabilized with respect to the surface. The difference between the two cases is the vertical extend of the destabilization with respect to the surface: in the heavily drizzling case, the instability occurs across the surface layer as compared to most of the subcloud layer in light drizzle. The boundary layer is dominated by fewer circulations, with updrafts more intense, but intermittent in time and space, while the downdrafts are larger in extend and weaker. The updrafts have their roots in the unstable lower subcloud layer, where they release the instability resulting from increases of θ_v in the surface layer.

The dynamics of the marine boundary layer become even more complicated when a cloudy layer advects over ocean water that becomes increasingly warmer (as in the trade winds). As the lower boundary becomes warmer, the vertical structure evolves through stages similar to those portrayed in Fig. 6.13. When the sea surface is cold, the cloud-topped boundary layer tends to be shallow and well mixed (left profile). As the sea-surface temperature increases, and with it the surface fluxes of water vapor, the circulations intensify and the cloud-top entrainment rates go up. The boundary layer deepens and the cloud bases in the updrafts and downdraft regions begin to separate (middle profile). The cloud must now work on the subcloud layer to maintain its vapor supply, and the cloud layer become progressively drier than the subcloud layer. The updrafts become progressively more intermittent in space and time, and the clouds take on the appearance of small cumulus clouds detraining into a stratocumulus layer near the top of the boundary layer. Drizzle aids in the separation of the cloud base altitudes (right profile). Eventually the updraft can no longer provide sufficient moisture, and the stratocumulus field disappears completely, being replaced by scattered trade-wind cumuli.

Elevated layer clouds

Elevated layer clouds, such as altostratus or cirrostratus, form in stable background environments, although the stability of the cloud layer itself is most often neutral as a result of in-cloud circulations. Formation of the cloud layer may result from the gradual lifting associated with large-scale cyclones or by the injection of moist or cloudy air into a stable layer from convective outflows. We confine our discussion mostly to clouds resulting from large-scale lifting.

Large-scale lifting results when air is forced to rise over a denser air mass, as occurs along synoptic-scale fronts in mid-latitudes, or from dynamical forcing in the tropics, where the lifting results from planetary-scale circulations. Gravity waves of various wavelengths, frequently observed in a stable background environment, provide an alternative lifting mechanism and contribute to the mesoscale organization of cloud systems. Once formed, circulations within a cloud layer develop from radiative cooling (near cloud top) or heating (near cloud base), or from wind shear across the layer. The circulations in such cloud layers are typically weak ($w < 0.5\,\mathrm{m\,s^{-1}}$), and the horizontal scale of the cloud elements is small (~ 1 km). In mid-latitudes, especially in the presence of significant shear, one sees precipitation fall-streaks, although the cloud layers themselves exhibit little evidence of the shear.

An elevated cloudy layer is similar in some respects to a cloud-topped boundary layer. The main differences are the lack of a solid surface along the lower boundary and the fact that cloud fills most of the elevated mixed layer. Also, the interactions between the dynamics, radiation, and microphysics on the cloud-scale circulations are stronger than in the cloud-topped boundary layer. The radiative forcing of the cloud circulations becomes increasingly more complex as the cloud optical depth decreases. Clouds with little condensate and hence smaller optical thickness experience radiative heating and cooling perturbations that penetrate deep into the cloud. Because radiative transfer is tied to the cloud particles, horizontal inhomogeneities in condensate concentration influence the circulation patterns. At nighttime, when net radiation is negative, maximum cooling occurs where condensate is being produced in the updrafts, causing the updrafts to be suppressed. Conversely, solar heating strengthens the updrafts during daytime. As a result, when the radiative forcing is comparable to the thermal perturbations, circulations within the layer are suppressed during nighttime, but enhanced during the day.

When ice is present, the microphysical details of ice-crystal nucleation, growth, and sedimentation become important determinants of supersaturation and latent heating, further contributing to non-linear interactions. Observations indicate that destabilization associated with evaporative cooling in layers below cloud may lead to a top-down development of cirrus layers as new convection is initiated below the original cloud layer. There are even hints that molecular-scale processes, such as the ability of an ice crystal to retain impinging molecules during deposition, influences whether a cirrus cloud evolves into cirrus uncinus or cirrostratus.

The picture becomes even more complicated when one considers how the large-scale environment influences elevated cloud layers. When an entire layer is lifted by a large-scale system, cooling rates can exceed those produced by the cloud-scale forcings. In such cases, the vertically integrated condensed-water concentration (i.e., the condensed water path) is determined mostly by large-scale forcing, not by the cloud-scale circulations. The separate influences of synoptic-scale systems, gravity waves, and shear-driven turbulence on the cloud dynamics and microphysics are difficult to quantify, although one can occasionally observe distinctive features in the spatial variations in the cloud cover. Dynamical processes are often thought to dominate the structure of layer clouds, but interactions among multiple weak processes are not yet well understood or appreciated.

6.5 Further reading

Cotton, W.R. and R.A. Anthes (1989). *Storm and Cloud Dynamics*. San Diego, CA: Academic Press, 880 pp. Chapters 2 (fundamental equations), 8 (cumulus clouds), 9 (severe storms), 10 (mesoscale convective systems), and 11 (middle and high clouds).

Houze, R.A. (1993). *Cloud Dynamics*. San Diego, CA: Academic Press, 570 pp. Chapters 2 (cloud dynamics), 5 (layer clouds), 7 (cumulus clouds), and 9 (mesoscale convective systems).

Rogers, R.R. and M.K. Yau (1989). *A Short Course in Cloud Physics*. New York: Pergamon Press, 293 pp. Chapter 15 (numerical cloud models).

Wallace, J.M. and P.V. Hobbs (2006). *Atmospheric Science: An Introductory Survey*. Amsterdam: Academic Press, 483 pp. Chapters 7 (atmospheric dynamics), 8 (weather systems), and 9 (atmospheric boundary layer).

6.6 Problems

1. A few years ago a town experienced a small flood when two small thunderstorms passed over the downtown area. These storms produced an approximately averaged 75 mm of rain over a 30 km^2 area. If we assume that the air entering the storm all came from the lowest 1000 m in the boundary layer, where the mean vapor mixing ratio was 10 $g\,kg^{-1}$, and that the cloud precipitation efficiency was 50% (precipitation efficiency is defined as the ratio of the rain produced to the amount of vapor entering cloud through cloud base), use the rainfall figures to estimate
   (a) the area from which the air must have come to produce this amount of precipitation,
   (b) the mass of SO_2 that ended up in the cloud if the boundary layer air contained 0.25 ppb SO_2,

(c) the cost of the energy associated with phase conversion in these storms if you had to pay for it as electricity at the rates you normally pay (approximately $0.08 per kW h).

2. A resourceful amateur meteorologist living close to a National Weather Service radiosonde release site decides to do a study on the source altitude of downdraft air from convective storms. He only has a dry- and wet-bulb thermometer at his disposal. How would he conduct this study? What assumptions must he make?

3. Consider a mixing chamber used for the nucleation of ice crystals. In this chamber two airstreams with different temperatures and vapor pressures are mixed together to produce a desired temperature and saturation ratio. Derive an expression for the temperature and saturation ratio in the mixing chamber in terms of the given variables. If the experimental setup is such that only the two inlet air mass characteristics may be controlled, how could the temperature and saturation ratio in the mixing chamber be varied? Can you produce a saturation ratio $S_i = e/e_i = 1.2$ at $-10°C$?

4. Construct a thermodynamic mixing diagram for a cloud, temperature 5°C with liquid water mixing ratio 4 $g\,kg^{-1}$, mixing with environmental air of temperature 4°C with relative humidity 90%. Repeat the calculation for an environmental relative humidity of 40%.

5. In mid-latitude winter storms, rainfall rates on the order of 20 $mm\,d^{-1}$ are not uncommon. Most of the convergence into these storms take place within the lowest 1–2 km of the atmosphere. where the water vapor mixing ratios are on the order of 5 $g\,kg^{-1}$. Estimate the magnitude of the horizontal convergence $\nabla \cdot \mathbf{v}$ into such storms.

6. In early winter, when the Great Lakes are not yet frozen over, the temperature of the open-water surface can be significantly higher than that of the air above it during a cold-air outbreak. When the warmer moist air just above the lake mixes with the colder overlaying air, a mixing cloud is formed over the lake some short distance away from the shore. With the aid of a phase diagram:
   (a) Explain why these clouds formed, using realistic values (use a computer program to plot the phase diagram!),
   (b) Assume a lake surface temperature of 4°C, and that the air temperature just above the surface reaches a relative humidity of 95% before it mixes with the air above. Show (graphically) how you would determine the minimum required relative humidity of the cold air to have at least some condensation take place when the temperature of the air is $-10°C$. Motivate your answer!
   (c) Find the temperature (of the cold air) below which condensate forms regardless of the relative humidity of that air.

7. Isobaric, adiabatic mixing may produce conditions conducive for cloud formation. Use the understanding you developed in doing problem 3 to create temperature and

liquid water mixing ratio plots as a function of mixing fraction for source air characteristics $T_1 = 25\,°C$ and relative humidity 95% mixing with air at $T_2 = 8\,°C$, relative humidity 94%.

8. Consider a stratus cloud of depth 1 km, with a mean vertical motion of $0.2\ \mathrm{m\,s^{-1}}$. Assume cloud base temperature is 15 °C at 950 hPa.
   (a) Use a thermodynamic diagram to estimate the liquid water condensed through the cloud layer.
   (b) Calculate the average rate of the wet adiabatic cooling through the cloud layer.
   (c) Compare your calculated wet adiabatic cooling rate to a cloud top radiative cooling rate of $4\,°\mathrm{C\,h^{-1}}$ and discuss your results.

Part IV

Cloud microphysics

7
Nucleation

Nucleation is the first of several microphysical processes we need to study. Cloud microphysics, by contrast with the macrophysics treated in Part III, is concerned with the liquid and solid particles that constitute clouds. We next learn how cloud particles form in the first place, how they grow, and how they interact with each other. The microphysical processes are key to determining how the atmospheric aerosol affects the optical properties and lifetimes of clouds (crucial to the Earth's climate), how water vapor is ultimately turned into precipitation, and how lightning forms. This part focuses on individual particles and interactions with their immediate environments. Effects of particle populations in the context of whole clouds are treated later.

Clouds are the places in the atmosphere where water changes phase. Water vapor enters a cloud and is converted, by one mechanism or another, into liquid and/or solid phases of water. Nucleation is the process by which these condensed phases are initiated. An overview of the possible phase transformations is shown in Fig. 7.1. The three common states of matter are identified at the apexes of the phase triangle, while the names used to define the various transformations are given in italics along each arrow.

A new phase can form only if that new phase is thermodynamically favored. As we saw in Section 3.2, the molar free energy (chemical potential) of water in the new (daughter) phase must be lower than that of the original (parent) phase for the new phase to have a chance of surviving. We say then that the parent phase is metastable with respect to the daughter phase. In other words, the molecules will be at a lower potential energy once the new phase has formed. The problem with forming a condensed phase from vapor is that the entropy of the system is lowered in the process. For instance, the molecules in solids, by contrast with gases and liquids, are highly ordered and have few ways to rearrange themselves. Entropy does not decrease easily in nature, so metastable phases (e.g., supercooled cloud water) can persist for long times. A new phase of lower entropy can form, however, if the thermodynamic driver (chemical potential difference) is sufficiently large.

Phase nucleation is categorized in several ways. We naturally want to know whether liquid droplets or ice crystals will form in any given atmospheric situation, so we treat liquids and solids separately. It is also important to know whether or not nucleation takes place spontaneously, by itself in a single parent phase, or if some external agent (a foreign substance) facilitates the process. The main categories for forming a new phase of

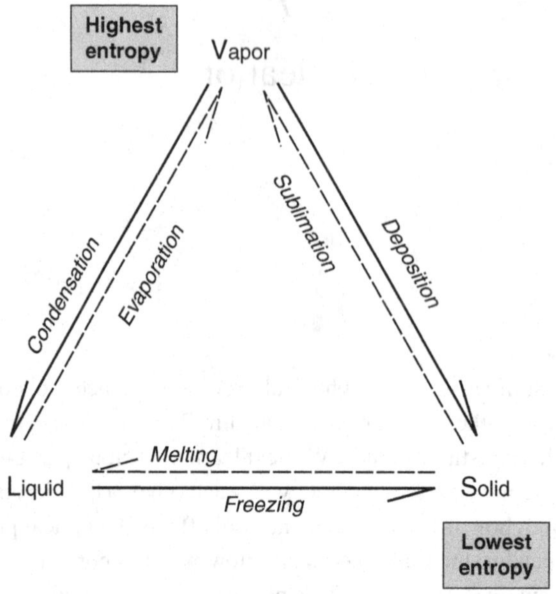

Figure 7.1 The phases in relationship to each other. The solid arrows indicate changes that lead to decreased entropy.

a single compound (e.g., water) are therefore homogeneous nucleation (uniform parent phase, no external agent) and heterogeneous nucleation (multiple phases, external agent involved). Most cloud particles form by heterogeneous nucleation because aerosol particles help the process, but homogeneous nucleation is treated here to establish the foundation of nucleation theory. When two or more chemical compounds contribute to the formation of a condensed phase, the process is termed heteromolecular nucleation. Heteromolecular nucleation may be either homogeneous or heterogeneous, but the usual atmospheric application involves chemical reactions of trace gases that lead to new aerosol particles and cloud condensation nuclei (CCN).

The procedure for calculating the conditions under which the new phase is likely to form involves four general steps:

(a) **Calculate the free energy to form embryos of the new phase.** This step uses thermodynamics to calculate the volume and surface contributions to changes in the system free energy. The result is an equation expressing the free energy change as a function of embryo size and the magnitude of the thermodynamic driver (chemical potential difference).
(b) **Determine the critical point.** Calculus is used to determine the embryo size at which the function from step (a) attains its maximum value. This particular embryo is the critical embryo, and the free energy to form it represents the barrier that must be overcome for the new phase to become stable.

(c) **Calculate the rate of nucleation.** This step couples the thermodynamics of embryo formation with kinetic theory, the discipline that enables us to estimate the frequency with which molecules are added to the critical embryo. This rate is typically found to vary greatly with the thermodynamic driver.
(d) **Estimate the threshold conditions.** The onset of nucleation is established by choosing a nucleation rate that the user deems significant, then finding the environmental conditions, expressed in terms of the applicable thermodynamic driver, that give that rate.

Watch for the use of these steps in the sections that follow.

7.1 Formation of the liquid phase

7.1.1 Homogeneous nucleation

When a pure vapor spontaneously leads to the formation of pure liquid droplets of the same substance, we say they formed by homogeneous nucleation. (Air may be present as long as it does not affect the process.) The situation can be conceptualized by considering Fig. 7.2, which illustrates the formation of a liquid droplet by the clustering of individual vapor molecules (termed monomers). The monomers behave as ordinary gas molecules in that they move around freely and collide frequently with each other. Being a vapor, however, the monomers can condense and form a liquid given the right conditions. Spontaneous condensation starts as individual monomers collide and stick together, rather than elastically bouncing off one another. The resulting dimer is not particularly stable, as the bonding is

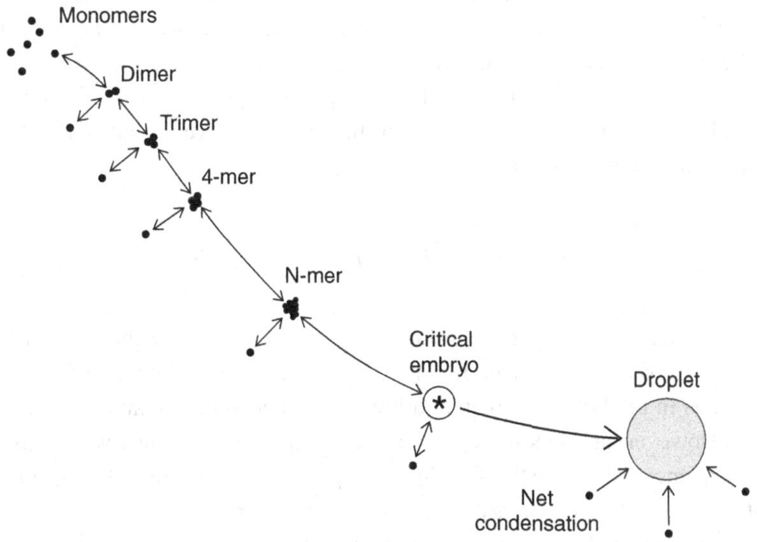

Figure 7.2 Illustration of cluster formation.

weak; even minor thermal agitation may break it apart. However, if a third monomer happens to collide (gently) and stick to the dimer during its existence, a trimer may form that is slightly more stable (lasts longer). The process of adding monomers to the existing molecular clusters gradually builds up the number of molecules in the cluster. When enough molecules have been added to a cluster to make it relatively stable, a critical embryo has been generated in the parent phase and a true water droplet can form. Molecular clustering represents an ongoing conflict between condensation (molecules entering the cluster) and evaporation (molecules leaving the cluster). Relatively few clusters reach critical size, so droplets forming by homogeneous nucleation represent rare successes in the molecular world of statistical fluctuations.

The likelihood of forming droplets by homogeneous nucleation can be made quantitative through the disciplines of thermodynamics and statistical mechanics. Our treatment here is simplified in that some of the assumptions made are not strictly valid. For instance, we base our analysis on Gibbs free energy, which is appropriate for systems maintained at constant temperature and pressure. However, as we saw in Chapter 3, the pressure inside a droplet is appreciably higher than that outside, so the system is not really at constant pressure. Strict accounting uses the Helmholtz free energy, which is appropriate when temperature and volume are fixed. Gibbs free energy is used here as it is simpler to implement and conceptually more intuitive. We also assume that the clusters are spherical and have the same surface properties as macroscopic droplets. Despite such approximations, the conclusions reached are virtually identical to those from more complete treatments.

The first step in the theory of homogeneous nucleation is to calculate the energy to form a molecular cluster in the vapor phase. We expect that this energy of formation depends on cluster size and the concentration of vapor. We assume that the cluster is in moist air with a vapor saturation ratio $S = e/e_s(T)$, where e is the water partial pressure and $e_s(T)$ is the equilibrium vapor pressure of pure, flat water at temperature T. The temperature is considered constant in time and uniform in space because the latent heat liberated each time a molecule enters a cluster is small and rapidly dissipated to the surrounding air. The chemical potential (molar free energy) of the vapor relative to that of the bulk liquid depends primarily on the saturation ratio:

$$\Delta\mu_{VL} \equiv \mu_V - \mu_L = \int_{es}^{e} v de = RT \ln S. \tag{7.1}$$

Here, we have treated the vapor as an ideal gas, so $v = RT/e$. Saturation ratios $S > 1$ serve as the thermodynamic driver for the phase change from vapor to liquid because water molecules in the liquid are then at a lower potential energy than in the vapor phase ($\mu_L < \mu_V$). However, any amount of condensed water has a surface, which also has an energy consequence. The growth of a cluster implies the generation of new surface area and a contribution to the free energy from surface tension σ_{LV}.

The overall change in the free energy of a homogeneous system, starting with vapor only and ending with vapor plus a single cluster of radius r_d, is therefore

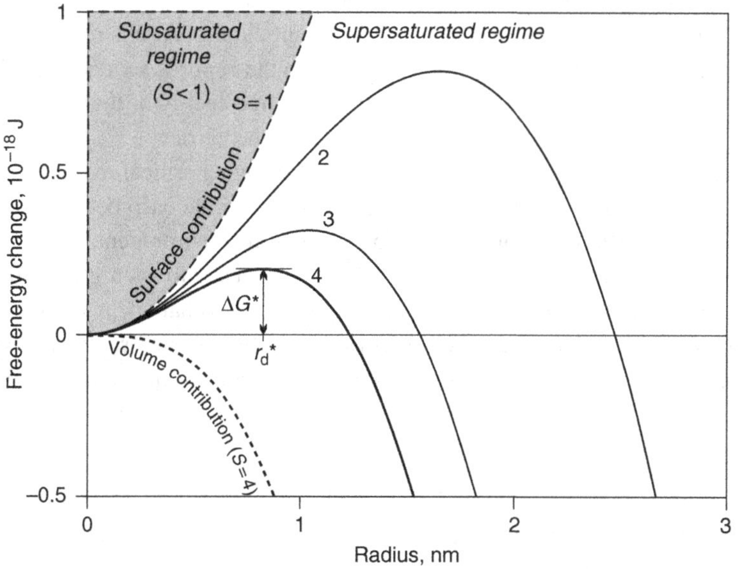

Figure 7.3 Dependence of system free energy on cluster size and saturation ratio in a homogeneous gas-phase medium.

$$\begin{aligned}\Delta G_h &= A_d \sigma_{LV} - V_d n_L \Delta\mu_{VL} \\ &= 4\pi r_d^2 \sigma_{LV} - (4/3)\pi r_d^3 n_L RT \ln S.\end{aligned} \tag{7.2}$$

The first term on the right is the surface contribution and the second term is the volume contribution to the change in system free energy. The dependence of ΔG_h on r_d is plotted in Fig. 7.3 as a family of curves, each with a specified, constant saturation ratio. For any given saturation ratio, the system free energy must initially increase with increasing cluster size, when the surface contribution $(\propto r_d^2)$ dominates the volume contribution$(\propto -r_d^3)$. As the clusters get larger, the absolute magnitude of the volume contribution increases rapidly and eventually dominates the surface contribution. The free energy of the system in the supersaturated regime must first increase before it decreases. The essence of nucleation is the presence of an energy barrier that gets in the way of forming a stable new phase.

The height of the free-energy barrier is found by calculating the mathematical critical point of each curve. The function given by Eq. (7.2) has its maximum value where the first derivative with respect to size vanishes: $\partial \Delta G_h(r_d, S)/\partial r_d = 0$. The critical radius so determined is

$$r_d^* = \frac{2\sigma_{LV}}{n_L \Delta\mu_{VL}} = \frac{2\sigma_{LV}}{n_L RT \ln S}, \tag{7.3}$$

and the maximum free energy is

$$\Delta G_h^* = \Delta G_h\left(r_d^*, S\right) = \frac{16\pi\sigma_{LV}^3}{3\left(n_L RT \ln S\right)^2}. \tag{7.4}$$

Note that the energy barrier ΔG_h^* becomes very large as $S \to 1$ and $e \to e_s(T)$. In the subsaturated regime (shaded region of Fig. 7.3), a pure liquid phase is not stable and cannot form spontaneously (because $e < e_s$). Even when the vapor is supersaturated ($S > 1$), formation of a stable new phase is assured only if the free energy of the system decreases as a result, which happens once the cluster passes over the barrier.

The critical embryo, or germ, is a molecular cluster having critical radius r_d^*, where the free energy ΔG_h has its maximum value. The germ sits on the cusp between evaporation and growth, so it exists in a state of equilibrium with its environment. This equilibrium state is unstable, for a departure from the critical point leads to even greater departures: a deviation toward smaller r_d leads to evaporation and still smaller r_d, whereas larger r_d leads to growth and even larger r_d. The equilibrium state lets us interpret Eq. (7.3) generally to calculate the ambient saturation ratio needed to maintain equilibrium with any droplet of given size r_d:

$$S_r \equiv \frac{e_r}{e_s(T)} = \exp\left(\frac{2\sigma_{LV}}{n_L R T r_d}\right). \tag{7.5}$$

Note that this relationship is the same as the Kelvin–Thomson equation derived in Section 3.4 from the point of view of mechanical stability.

The number of critical embryos in a unit volume depends strongly on the height of the free-energy barrier. The mathematical tool at our disposal to determine this number concentration is the Boltzmann factor (see Appendix C), which specifies the ratio of concentrations in terms of a difference in potential energy $\Delta V = \epsilon_j - \epsilon_0$:

$$\frac{n(\epsilon_j)}{n(\epsilon_0)} = \exp\left(-\frac{\epsilon_j - \epsilon_0}{k_B T}\right). \tag{7.6}$$

Recall the reasons that air density decreases with altitude (Chapter 2) and note that molecules are driven into higher energy states by thermal fluctuations. The thermal energy ($k_B T$) is expressed on a per-molecule basis (rather than on a molar basis, RT) when the energies represent the potential energy of the molecules. In general, the number of molecules in a high-energy state ϵ_j, compared with the number in a lower state ϵ_0, is exponentially related to the difference in energies of the two states. The greater the energy difference, the harder it is for the higher energy state to be populated. Applying the Boltzmann factor to nucleation, we interpret the generic energy difference $(\epsilon_j - \epsilon_0)$ as the barrier height ΔG_h^*, and $n(\epsilon_j)$ as the number of molecules sitting in clusters with energy ΔG_h^*. Thus, the concentration of germs is

$$n_g^* \equiv n(r_d^*) = n_V \exp\left(-\frac{\Delta G_h^*}{k_B T}\right) = n_V \exp\left(-\frac{16\pi\sigma_{LV}^3}{3k_B T (n_L R T \ln S)^2}\right), \tag{7.7}$$

where n_V is the concentration of water vapor (monomers). As S increases, ΔG_h^* decreases, which yields a higher probability of water molecules gathering together in a cluster of critical size. More advanced treatments recognize that nucleation implies a flow of embryos through the size range; as embryos pass the critical stage, new clusters of smaller size must grow to take their place. The equilibrium size distribution implied by the Boltzmann factor should be corrected for such non-equilibrium effects, in which case the concentration of critical embryos, calculated by Eq. (7.7), would be multiplied by a correction term Z called the Zeldovich factor. Typically, Z reduces $n\left(r_d^*\right)$ by about a factor of 10, which may seem significant. On the other hand, we will see that the threshold conditions are affected very little by inclusion of such non-equilibrium effects.

The rate of nucleation, the number of new droplets that form in unit time within a unit volume of vapor, depends not only on the number of embryos, but also on the rate with which they grow beyond the critical size. The concentration of clusters has been determined by considering thermodynamic equilibria, but thermodynamics does not allow us to make rate calculations. Rather, we must rely on kinetic theory. The situation is depicted conceptually in Fig. 7.4. Whereas clusters of all sizes form reversibly from the vapor by rapidly exchanging monomers, droplets form irreversibly from the critical embryos. It takes only a single new molecule added to a critical embryo, a cluster sitting at the top of the barrier, for the free energy of the system to start decreasing, thus allowing the spontaneous growth of the droplet. The nucleation rate is therefore just the rate with which clusters cross the energy barrier. An individual cluster of critical size acquires molecules at the rate J_1 [s^{-1}] given by the molecular impingement flux Φ_v [$\mathrm{m}^{-2}\,\mathrm{s}^{-1}$] times the cluster surface area A_d [m^2]:

$$J_1 = \Phi_v A_d = \left(\frac{1}{4} n_v \bar{c}_v\right)\left(4\pi r_d^{*2}\right) = \pi r_d^{*2} n_v \bar{c}_v, \tag{7.8}$$

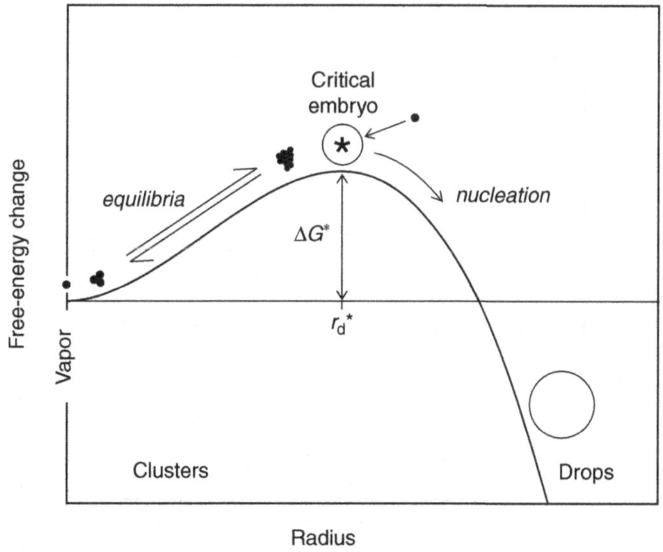

Figure 7.4 Schematic to illustrate the nature of nucleation, critical clusters passing over the free-energy barrier. Clusters maintain near equilibria with the vapor, whereas drops form irreversibly from critical clusters.

where n_v is the concentration of vapor molecules and $\overline{c}_v$ is their mean speed. Given the number concentration of such critical embryos $n\left(r_d^*\right)$ (Eq. (7.7)), the homogeneous nucleation rate is

$$J = J_1 n\left(r_d^*\right) = \pi r_d^{*2} n_v^2 \overline{c}_v Z \exp\left(-\frac{\Delta G_h^*}{k_B T}\right)$$
$$= \pi r_d^{*2} n_v^2 \overline{c}_v Z \exp\left(-\frac{16\pi\sigma_{LV}^3}{3k_B T\left(n_L RT \ln S\right)^2}\right), \tag{7.9}$$

where the Zeldovich factor Z has been included for completeness. The nucleation rate is seen to depend strongly on the saturation ratio S, as shown in Fig. 7.5. Virtually no new drops form until the saturation rate becomes several times the equilibrium value, then the rate increases dramatically. We quantify this sudden onset of droplet formation by choosing a reasonable significant rate and noting the saturation ratio that gives this rate. In this example, based on picking a significant rate of 1 cm^{-3} s^{-1} at 0 °C, the threshold saturation ratio S_{th} is about 4.4, which corresponds to a supersaturation of 340%. Because J increases so rapidly near the threshold, the precise value of J chosen as significant does not greatly change the threshold saturation ratio. For this reason, detailed considerations of higher-order corrections to the theory (as with Z) are not usually justified. Temperature affects S_{th}, as shown in Fig. 7.6, but even at reasonably high temperatures large supersaturations are needed. Homogeneous nucleation, while not physically impossible, is improbable in the atmosphere.

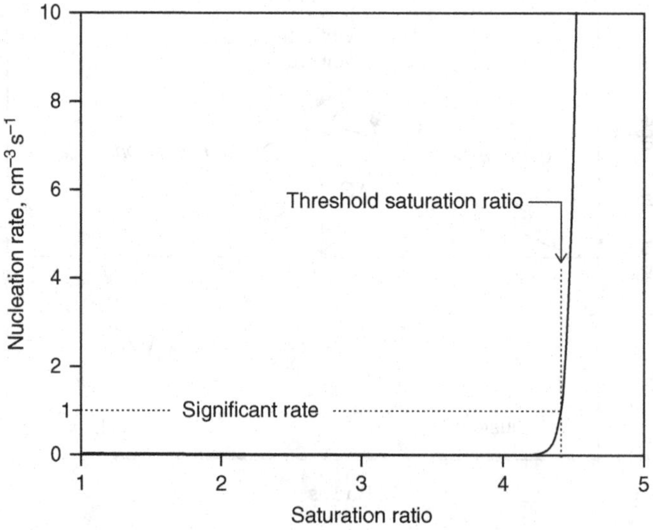

Figure 7.5 Dependence of the homogeneous nucleation rate on saturation ratio. Adapted from Fletcher (1962).

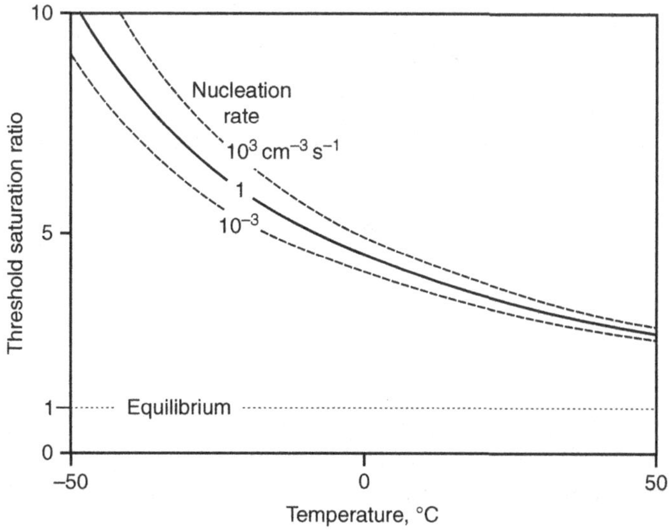

Figure 7.6 Dependence of threshold saturation ratio on temperature for selected choices of significant nucleation rate. From Fletcher (1962).

7.1.2 Heterogeneous nucleation of liquid water

Any substance that facilitates the formation of a new phase is termed a nucleating agent. Nucleating agents accomplish this role by lowering the free energy of formation below that due to homogeneous nucleation. Aerosol particles that serve this role and act as preferred sites of condensation are termed condensation nuclei. Consider a limiting case.

Flat, insoluble surface

The presence of a foreign surface, even if insoluble and unattractive toward water molecules, aids the nucleation of the liquid phase. Consider first a spherical cap of liquid on a flat surface of an insoluble solid, as illustrated in Fig. 7.7. The liquid in its resting state has a radius of curvature r and makes an angle of contact with the surface θ (because a drop of liquid placed on the surface moves relative to the surface until a force balance is established). Young's Equation expresses the balance of forces in the surface plane in terms of interfacial free energies: $\sigma_{SV} = \sigma_{LS} + \sigma_{LV} \cos\theta$.

The geometry allows the volume V_L of liquid to be calculated, as well as the surface areas of the liquid–vapor and liquid–solid interfaces. The area of the liquid–solid interface is, using the symbols defined in Fig. 7.7, $A_{LS} = \pi x_0^2 = \pi r^2 \left(1 - m^2\right)$, where $m \equiv \cos\theta$. Integration is needed to determine the area of the liquid–vapor interface and the liquid volume, so consider an infinitesimal disk (horizontal shading) of thickness dh that subtends an infinitesimal angle $d\theta'$ (radial shading). The radius of this disk is $x' = r\sin\theta'$, so the liquid surface area is

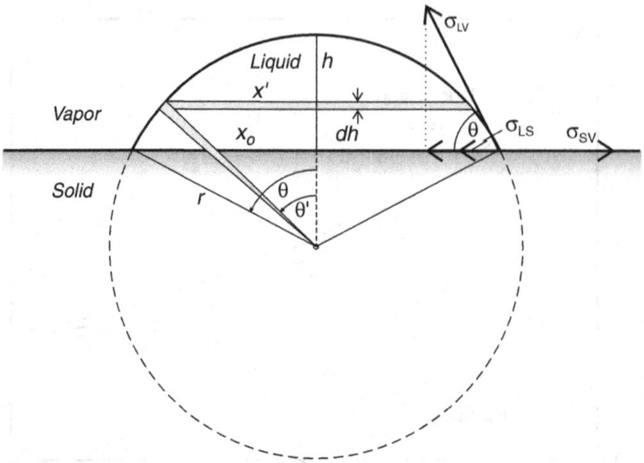

Figure 7.7 Geometry associated with forming a spherical cap on a solid substrate.

$$A_{LV} = \int_0^{\theta} 2\pi x' r d\theta'^2 (1-m) = 2\pi r^2 (1-m). \tag{7.10}$$

The differential volume of the disk inside the liquid is the area of the disk times its thickness $dh = r d\theta' \sin\theta'$: $dV_L = \pi \left(r \sin\theta'\right)^2 r \sin\theta' d\theta' = \pi r^3 \sin^3\theta' d\theta'$. The liquid volume is thus

$$V_L = \pi r^3 \int_0^{\theta} \sin^3\theta' d\theta' = \frac{\pi r^3}{3}\left(2 - 3m + m^3\right) = \frac{\pi r^3}{3}(2+m)(1-m)^2. \tag{7.11}$$

The system free energy changes in response to the formation of a spherical cap of radius r on the substrate. Considering the bulk and surface contributions, we find

$$\Delta G_s = -n_L V_L \Delta\mu_{VL} + A_{LV}\sigma_{LV} + A_{LS}\left(\sigma_{LS} - \sigma_{SV}\right), \tag{7.12}$$

where $\Delta\mu_{LV} = RT \ln S$ is the molar free energy of the vapor relative to that of the liquid when the saturation ratio S is e/e_s. Upon noting that the quantity in parentheses is $-\sigma_{LV} m$ from Young's equation and inserting the appropriate geometrical quantities, we gain the dependence of the system free energy on the radius of the spherical cap:

$$\Delta G_s = \pi\left(2 - 3m + m^3\right)\left(-\frac{n_L \Delta\mu_{VL}}{3} r^3 + \sigma_{LV} r^2\right). \tag{7.13}$$

The maximum system free energy is found by taking the derivative of this function with respect to r and setting it equal to zero. Solving for the radius at this point, we find the critical radius to be

$$r^* = \frac{2\sigma_{LV}}{n_L \Delta\mu_{VL}} = \frac{2\sigma_{LV}}{n_L RT \ln S}. \tag{7.14}$$

Note that the critical radius in the case of nucleation on a substrate is the same as that calculated for homogeneous nucleation. We should indeed expect the radii of curvature to be the same, as the equilibrium across the vapor–liquid interface is a local phenomenon; the molecules there are not affected by the presence of the solid. The height of the free-energy barrier is now

$$\Delta G_s^* = \Delta G_s\left(r^*\right) = \frac{16\pi\sigma_{LV}^3}{3\left(n_L RT \ln S\right)^2} \cdot \frac{(2+m)(1-m)^2}{4}. \tag{7.15}$$

Note that the first factor is the same as the barrier height, ΔG_h^*, in the homogeneous case, so we see the ratio $\Delta G_s^*/\Delta G_h^* = (2+m)(1-m)^2/4 \equiv f(m)$ is just a geometrical factor $f(m) < 1$. It can also be seen that $f(m)$ is the fraction of a sphere of radius r^* occupied by liquid: $f(m) = V_L/V_{sphere}$.

The presence of the insoluble surface aids the nucleation of liquid by lowering the free energy to form liquid at any given saturation ratio. As shown by the dotted curve in Fig. 7.8, the amount of lowering is the same for all radii, so the maximum occurs at the same critical radius as in the homogeneous case. In effect, a smaller volume of liquid needs to form in order to achieve the same end result. As a liquid cap forms, it covers a part of the solid surface and effectively replaces a portion of the large solid–vapor surface free energy with a smaller liquid–solid surface free energy (last term of Eq. (7.12)).

The rate of nucleation on a flat substrate depends on the number of critical embryos (spherical caps of critical size) and the rate with which those embryos acquire new vapor molecules. The number density of critical embryos on the surface is again given by the Boltzmann factor: $n\left(r^*\right) = n_{s1}\exp\left(-\Delta G_s^*/k_B T\right)$, where n_{s1} is the surface density of single adsorbed water molecules and ΔG_s^* is given by Eq. (7.15). Each critical embryo gains vapor molecules at the rate $J_1 = \Phi_V A_{LV} = \pi r^{*2} n_v \bar{c}_v$, so the surface nucleation rate [$\mathrm{m^{-2}\,s^{-1}}$] is

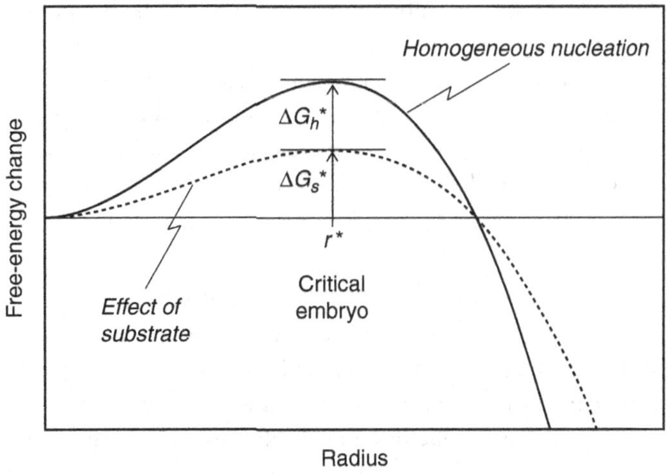

Figure 7.8 Effect of substrate on the free energy to form liquid-phase embryos from the vapor.

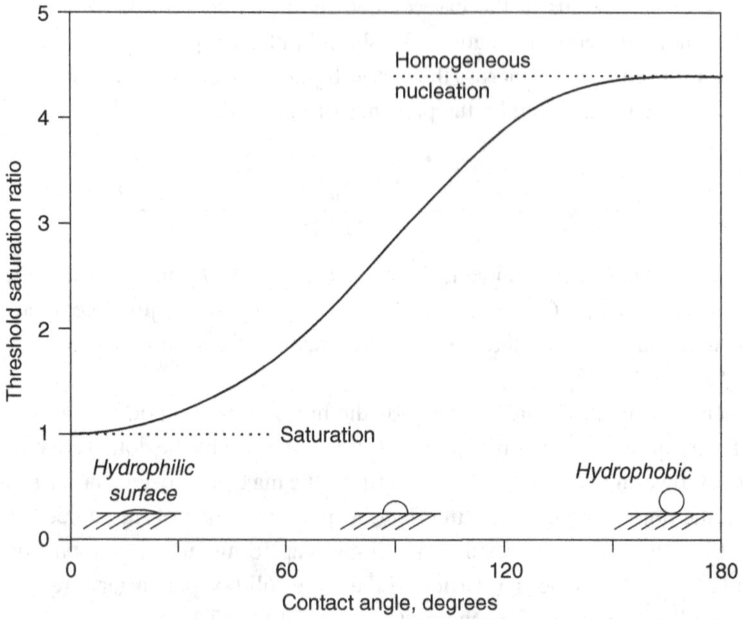

Figure 7.9 Threshold saturation ratios for nucleation on a flat, insoluble surface. Calculations based on a significant rate of 1 $\mathrm{cm^{-2}\,s^{-1}}$ at 0 °C. Adapted from Fletcher (1962).

$$J_s = J_1 n\left(r^*\right) = \pi r^{*2} n_{s1} n_V \bar{c}_V Z \exp\left(-\Delta G_s^*/k_B T\right). \tag{7.16}$$

Choosing a significant nucleation rate differs slightly from that done for homogeneous nucleation. Here, where nucleation occurs on a two-dimensional surface rather than in a volume of gas, the criterion for significant nucleation is a rate per unit surface area: 1 $\mathrm{cm^{-2}\,s^{-1}}$, which is a macroscopically sensible rate. The threshold saturation ratio S_{th} so determined is shown in Fig. 7.9 as a function of the contact angle θ. We see that surfaces that are wettable and so have small contact angles permit nucleation at saturation ratios near unity; essentially no energy barrier exists when surfaces are hydrophilic. At the other extreme, when the substrate is hydrophobic and the contact angle is large, S_{th} approaches values close to those calculated for homogeneous nucleation, as if the surface were not even present.

Nucleation on an insoluble particle

Flat surfaces do not exist in the atmosphere, so let us briefly repeat the above analysis, but for a more realistic situation, nucleation of liquid on an insoluble aerosol particle. Now, the spherical cap of liquid forms on the curved surface of the particle, which acts as a discrete nucleating agent. For simplicity, let the nucleating particle be spherical and have a radius r_N, as shown in the inset of Fig. 7.10. Although the nucleating surface is now curved and the geometry more complicated, the physical principles are identical to those used for

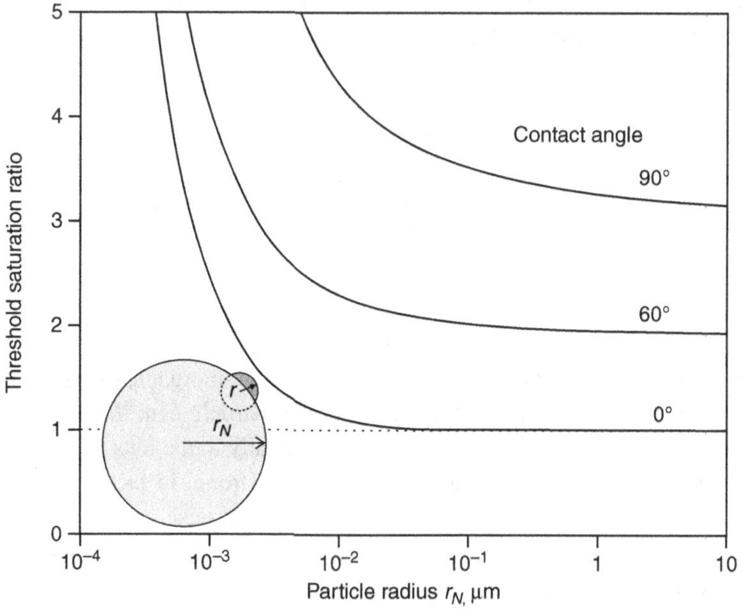

Figure 7.10 Dependence of threshold saturation ratio on contact angle and the size of the insoluble particle. Calculations based on a significant particle nucleation rate of 1 s^{-1}. Adapted from Fletcher (1962).

nucleation on a flat, insoluble surface. Note that the spherical cap again forms a portion of a hypothetical sphere of radius r. The geometrical factor $f(m, x) = V_L / V_{sphere} < 1$ is now a function of both $m = \cos\theta$ and the size ratio $x \equiv r_N / r$. The change in free energy to form the spherical cap of liquid is again calculated using Eq. (7.12), but the liquid volume and interfacial areas must now reflect the more complicated geometry:

$$f(m, x) = \frac{1}{2} \left[1 + \left(\frac{1 - mx}{g} \right)^3 + x^3 \left[2 - 3\left(\frac{x - m}{g} \right) + \left(\frac{x - m}{g} \right)^3 \right] + 3mx^2 \left(\frac{x - m}{g} - 1 \right) \right], \tag{7.17}$$

where $g \equiv \left(1 + x^2 - 2mx\right)^{1/2}$. When calculating the threshold saturation ratio, recognize that only a single cap of critical size needs to form anywhere on the particle surface in order for the particle to act as a nucleus. The criterion to use for significance is thus the product $J_s A_N = 4\pi r_N^2 J_s = 1\ \mathrm{s}^{-1}$, where J_s is the surface nucleation rate given by Eq. (7.16).

The theory of heterogeneous nucleation of liquid drops by insoluble aerosol particles yields a family of curves, such as those shown in Fig. 7.10. The threshold saturation ratio S_{th} needed for an insoluble particle to act as an effective nucleus depends both on the contact angle θ and on the size of the particle. When the particle is very large, the surface is nearly flat, so S_{th} varies with θ much as it did in the case of a flat surface (Fig. 7.9). Large particles that are easily wettable (have small θ) serve as better condensation nuclei than do particles composed of hydrophobic substances (which have large θ). Small, submicron particles may be further limited by the effect of curvature on the equilibrium vapor pressure, which raises the saturation ratio needed for effective nucleation (shown by the sharp rises in S_{th} moving toward the left in Fig. 7.10). The lowest curve, representing a particle uniformly coated with a thin layer of liquid water, reflects the same physics that yielded the Kelvin equation in Chapter 3. The curvature effect is a major impediment to the formation of small liquid droplets.

Formation of liquid on a soluble particle

An aerosol particle composed of water-soluble matter (e.g., inorganic salts) can initiate the liquid phase much more effectively than an insoluble particle can. Water, because of its polar nature (see Chapter 2), attaches itself electrostatically to the ions that typically make up soluble matter. The molecular-scale attraction is so strong, in fact, that water readily breaks the ionic bonds of salts under conditions that otherwise require temperatures of hundreds of degrees (to melt the salt). In turn, the dissolved solute ions help retain water in the condensed state by lowering the equilibrium vapor pressure. The solute ions in effect act as molecular anchors by retarding the evaporation of water molecules. Soluble matter within a particle is able to compensate for, even dominate the curvature effect of small particles. The formation of the liquid phase in an atmosphere laden with soluble aerosol particles occurs in subsaturated air, often well below cloud base.

The liquid state starts on the particle surface of an initially dry particle when water molecules accumulate faster than they leave. The process of dissolving a solid into a liquid solution is called deliquescence and occurs at relative humidities *below* 100% (equilibrium with respect to pure, flat water). Turning a solid into a liquid increases entropy, so deliquescence is an equilibrium transformation, not a nucleation process. The term nucleation is sometimes used (incorrectly) because the liquid phase first appears during deliquescence. Nevertheless, atmospheric particles, because they aid the formation of liquid from water vapor, are CN.

The overall behavior of a soluble salt particle (taken to represent a condensation nucleus) in a humid environment can be seen in Fig. 7.11. Here, a particle of $NaNO_3$ was exposed to air at different relative humidities (expressed as saturation ratio $S = e/e_s$) and allowed to come to equilibrium and its mass m measured relative to that in the dry state, m_0. At humidities below the onset of deliquescence, water vapor adsorbs onto the dry surface, but the amount of water is insufficient to disrupt the crystalline lattice. The water molecules stay on the surface for a while, then desorb with little effect on the solid. As the humidity rises, the concentration of adsorbed molecules increases, and the accumulated water is eventually sufficient to dissolve the salt. At the deliquescence point (S_{deliq}) the non-linear process of dissolution takes place: The adsorbed water breaks some of the ionic bonds holding the salt together and removes the ions from the solid. These ions that are sequestered in the liquid covering the particle hinder the evaporation of the water molecules. The lowered

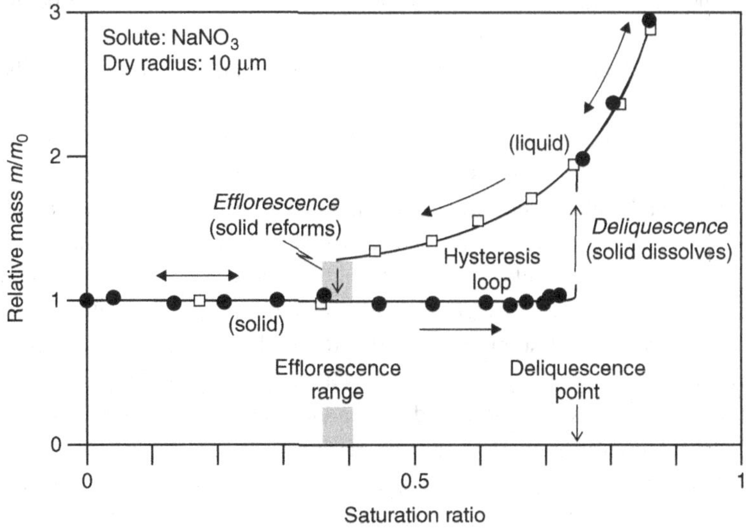

Figure 7.11 Dependence of particle mass on saturation ratio, based on data obtained from a single particle levitated electrodynamically under conditions of controlled relative humidity. The particle is large, so the effect of curvature on water uptake can be ignored and the response of the particle to changing humidity is the same as that of a bulk solution. Adapted from Lamb *et al.* (1996).

vapor pressure (from the solute effect, as described by Raoult's law) leads to further uptake of water molecules from the vapor phase and new opportunities for further dissolution of the solid. The process of vapor uptake continues until the solid completely disappears and all the ions become part of the liquid solution droplet. Beyond the deliquescence point, increasing humidity leads to further uptake of water vapor and the growth of the solution droplet.

The behavior of a liquid solution follows the same principles of equilibria discussed in Chapter 3. At high humidities ($S > S_{deliq}$), a solution droplet responds reversibly to changing humidity, as suggested by the upper two-way arrow in Fig. 7.11. However, once the humidity drops below S_{deliq}, the particle stays liquid. Decreasing ambient humidity causes the solution to lose water by evaporation, leading to progressively higher solute concentrations and a solution that becomes increasingly supersaturated with respect to solid solute. At some point the threshold solute supersaturation inside the droplet is reached, the solid phase is nucleated, and the original dry salt particle is reborn. Note that the formation of the solid solute from the liquid solution leads to a lowering of the entropy, so efflorescence is truly a nucleation event and describable by the principles of homogeneous nucleation discussed earlier. The stochastic nature of nucleation causes efflorescence to take place over a range of humidities, as suggested by the shaded bar in Fig. 7.11. Further decreases in humidity below efflorescence have little impact on the mass or size of the now-dry particle. The fact that the liquid and solid phases form at different saturation ratios is evidence of a hysteresis phenomenon, because different paths are followed with rising and falling humidities. The response of individual aerosol particles to varying humidity must be understood if one hopes to appreciate how the atmospheric aerosol affects air quality, cloud formation, the atmospheric radiation balance, and the Earth's climate.

Deliquescence is directly related to the solubility of the substance(s) making up a particle. This dependence of the deliquescence humidity on solute solubility may be quantified by recognizing that dynamic equilibria exist among the three phases, as shown in Fig. 7.12. Two interfaces separate the three phases and represent the places in the system where phase equilibria exist. Two components are involved in the system as a whole, but only the liquid phase contains both solute and water. The solid–liquid equilibrium, solute (s) $\rightleftharpoons$ solute (aq), involves the transfer of solute ions only and may be expressed generically in terms of chemical potentials: $\mu_{solute}|_S = \mu_{solute}|_L$. The aqueous solution is saturated as long as any solid solute is present. The liquid–vapor interface, on the other hand, exchanges only water, so the equilibrium, H_2O (aq) $\rightleftharpoons$ H_2O (v), can be described by the generalized form of Raoult's law: $a_w = e/e_s \equiv S$. Expressing the water activity in terms of the van't Hoff factor i (see Chapter 3), we get the deliquescence humidity for the saturated solution,

$$S_{deliq} = a_w|_{sat} = \frac{n_w}{n_w + i n_{s,sat}}, \tag{7.18}$$

where n_w is the concentration of water and $n_{s,sat}$ is the concentration of solute in the saturated solution, (i.e., the solute solubility). The solubility of various salts, as well as

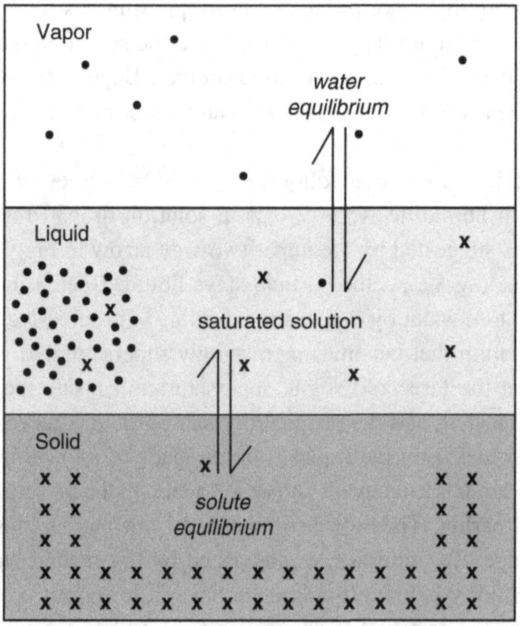

Figure 7.12 Schematic of the three phases and two components involved in deliquescence.

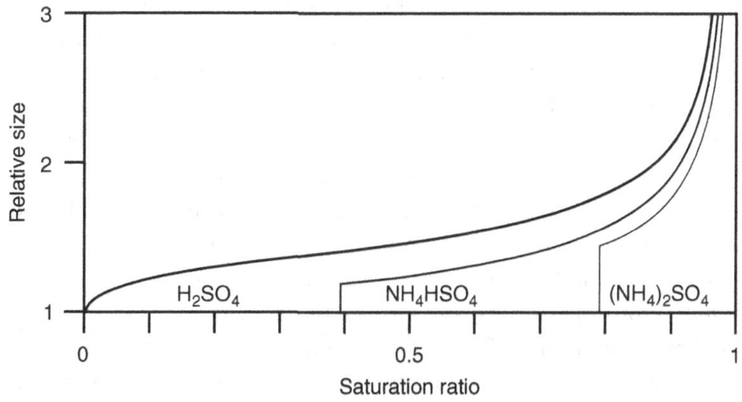

Figure 7.13 The deliquescence and growth of selected solutes containing sulfate. Supersaturated solutions and efflorescence are not displayed. Adapted from Finlayson-Pitts and Pitts (2000, p. 390).

the effective degree of dissociation (i), can be found as functions of temperature in the literature.

Deliquescence and the absorption of water from the environment are fundamental properties of any pure solute. Equation (7.18) shows that the larger the solubility, the lower the deliquescence point. The deliquescence points and hygroscopic growth curves for common solutes containing sulfate are shown in Fig. 7.13. The uptake of water is here expressed in terms of the increase in diameter of the aqueous particle relative to that of the dry particle. Note that the deliquescence point, hence solubility, depends strongly on the extent to which the acid is neutralized. Ammonium sulfate, the salt representing the fully neutralized form of sulfuric acid, is soluble, but it takes a relative humidity of 79% to dissolve the solid into a liquid. At the other extreme, sulfuric acid is infinitely soluble in water, so it has no deliquescence point at all. Pure sulfuric acid is always liquid; just the relative proportions of water and sulfate vary with the humidity. The curves above the points of deliquescence show that particle size increases strongly with increasing humidity and asymptotes to the $S = 1$ limit. Each such curve, representing the hygroscopic growth of the particle, reflects the ability of the dissolved solute to take up water. Atmospheric aerosol particles are typically composed of many compounds, each with differing physicochemical properties, so it should be no surprise that the parameters describing deliquescence and hygroscopic growth are complicated functions. Well-defined deliquescence points often do not show up with mixed particles, and hygroscopic growth is often empirically determined.

Growth of haze particles is sometimes based on the concept of a hygroscopic growth factor, $GF \equiv D_p(S)/D_p(0)$, where $D_p(0) = D_{p0}$ is the diameter of the dry particle. A suitable hygroscopic growth law for describing empirical data is the function

$$GF(S) = 1 + \frac{A_{hyg} S}{1 - S}, \tag{7.19}$$

where A_{hyg} is the fitting parameter. When the opportunities to take extensive data are limited, one often uses a single ratio as the measure of hygroscopic growth: $GF \equiv D_{p,wet}/D_{p,dry}$, where $D_{p,wet}$ may be the measured particle diameter at a pre-specified high relative humidity (e.g., 90%) and where $D_{p,dry}$ is the diameter measured at a convenient low humidity (e.g., 10%). Typical magnitudes of GF range from less than 1.1 for hydrophobic substances to over 2.0 for hydrophilic compounds. The hygroscopic growth of the atmospheric aerosol significantly affects air quality, visibility in the boundary layer, and the atmospheric radiation balance (through the direct aerosol effect).

The size of the particle must usually be taken into account, as only the largest particles exhibit bulk properties. As shown in Fig. 7.14, the curvature of submicron particles shifts the equilibria significantly. Particles still follow the same basic pattern in response to humidity variations discussed in relationship to Fig. 7.11, but we must distinguish between the humidities at which phases change in bulk and particulate systems. Because of the curvature effect, a small particle requires a larger saturation ratio for deliquescence than does a bulk solution of the same substance; the particle deliquescence point ($S_{deliq,ptcl}$) can

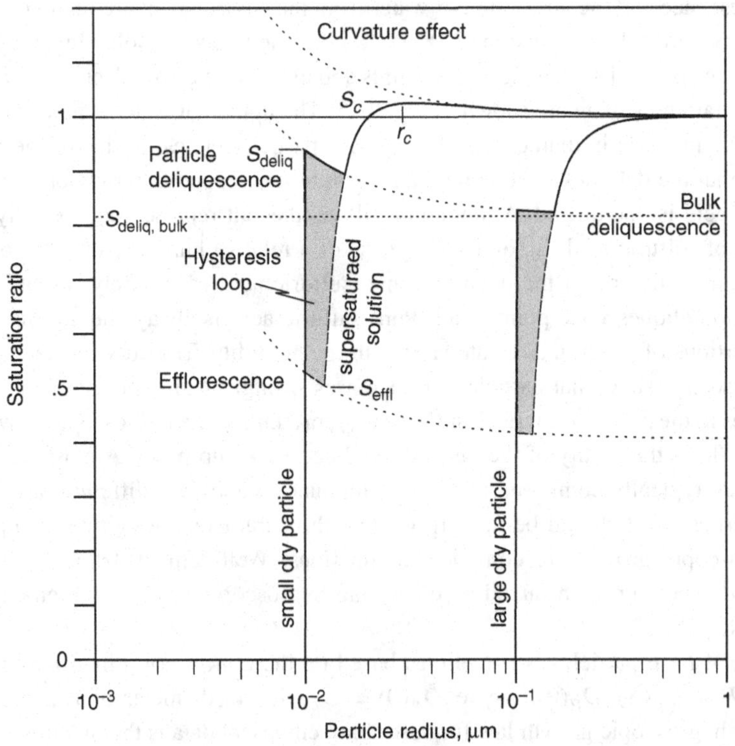

Figure 7.14 Effects of particle size and humidity on deliquescence and efflorescence. Adapted from Chen (1994).

be substantially larger than the bulk deliquescence point ($S_{deliq,bulk}$). Similar arguments apply to efflorescence.

Particle activation

The formation of cloud droplets requires that liquid solution droplets be "activated". Activation is analogous to the nucleation of water droplets in that a free-energy barrier (expressed now in terms of saturation ratio: $S = \exp(\Delta G/RT)$) must be overcome. However, the liquid phase already exists (via deliquescence), so activation is not a phase-nucleation phenomenon, rather it represents a change from stable to unstable growth in response to increasing ambient humidity. The concept of activation is crucial to our understanding of how aerosol particles act as CN and establish the initial microstructure of clouds.

The effect of humidity on the equilibrium properties of liquid solution droplets was discussed at length in Chapter 3. The mathematical analysis of the competition between the solute and curvature effects showed that the equilibrium radius r_d of the solution droplet depended in a systematic way on the ambient saturation ratio S. We found the so-called Köhler saturation ratio to be

$$S_K \doteq 1 + \frac{A_K}{r_d} - \frac{B_K i N_s}{r_d^3}, \tag{7.20}$$

where A_K and B_K are parameters that depend weakly on temperature, i is the van't Hoff factor (accounting for dissociation of the solute into ions and for other non-linear effects), and N_s is the solute content (number of moles in the particle). The Köhler function S_K, given by Eq. (7.20) and shown in Fig. 7.15, is less than unity at small radii (where the solute effect dominates), increases with increasing radius (as the solution becomes more dilute), peaks at a value slightly above unity, then gradually decreases and asymptotes to unity (as the curvature effect weakens). The maximum value of S_K, found by setting $dS_K/dr_d = 0$, is the critical saturation ratio

$$S_c = 1 + \left(\frac{4A_K^3}{27 B_K i N_s}\right)^{1/2}, \tag{7.21}$$

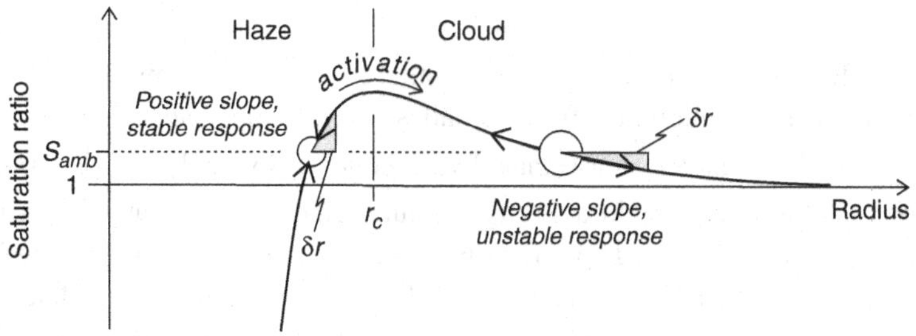

Figure 7.15 Köhler function and schematic illustration of stability analysis.

which represents the minimal difference in chemical potential ($\Delta\mu_c = RT \ln S_c$) needed for the droplet to grow beyond its critical radius

$$r_c = \left(\frac{3B_K i N_s}{A_K} \right)^{1/2}. \tag{7.22}$$

Droplets smaller than r_c are haze particles, whereas larger droplets are called cloud droplets. Activation is the process by which haze droplets grow through the peak of the Köhler function and are transformed into cloud droplets.

An important distinction between haze and cloud droplets, and the transformation of unactivated into activated particles, is the character of growth/evaporation that takes place in response to perturbations. A droplet smaller than r_c is on the rising branch of the Köhler function, meaning that the first derivative is positive: $dS_K/dr_d > 0$. On the other hand, when $r_d > r_c$, the slope is negative: $dS_K/dr_d < 0$. The importance of slope can be seen through a simple stability analysis, as suggested by the small shaded triangles in Fig. 7.15. Consider a haze droplet initially at equilibrium with its environment, characterized by saturation ratio S_{amb}. Now, imagine that its radius increases a small amount δr. The increased size of the haze droplet means its equilibrium saturation ratio is higher than S_{amb} by the amount $\delta S_K = (dS_K/dr_d)\,\delta r > 0$ (because $dS_K/dr_d > 0$). With the equilibrium value higher than the ambient value ($S_K > S_{amb}$), the excess water in the particle evaporates, causing the particle to return to its original size. A size deviation, larger or smaller, puts the haze particle out of equilibrium with its environment in such a way that the original state is restored (depicted by the arrows directed toward each other). Haze droplets grow/evaporate stably and change size only in response to changes in the ambient humidity. By contrast, a perturbation to the size of a cloud droplet leads to an unstable response: a positive increase in radius puts the droplet equilibrium state below S_{amb} by an amount $\delta S_K = (dS_K/dr_d)\,\delta r < 0$ (because $dS_K/dr_d < 0$), which encourages further deviation from the initial size. Cloud droplets can at best be in unstable equilibrium with their environment, and they grow/evaporate spontaneously away from that point (note the diverging arrows in Fig. 7.15). The distinction between cloud droplets and haze droplets, despite both being liquid solutions, is fundamental to our understanding of aerosol–cloud interactions. Atmospheric aerosol particles become cloud particles when they grow through the peak (S_c) of the Köhler function and the slope (dS_K/dr_d) changes sign from positive to negative.

The most effective condensation nuclei are those characterized by having the lowest critical saturation ratios, alternatively the smallest critical supersaturations, $s_c \equiv S_c - 1$. As seen in Eq. (7.21), small s_c is associated with large N_s $\left(s_c \propto 1/N_s^{1/2}\right)$. Particles containing more soluble matter yield a stronger solute effect and a greater lowering of the equilibrium vapor pressure than do particles with smaller solute contents (review Chapter 3 as needed). If we assume a dry particle to be spherical, we see how the solute content is related to its size: $N_s = V_p n_s = (\pi/6)\, D_p^3(\rho_s/M_s)$. Here, V_p and D_p are, respectively,

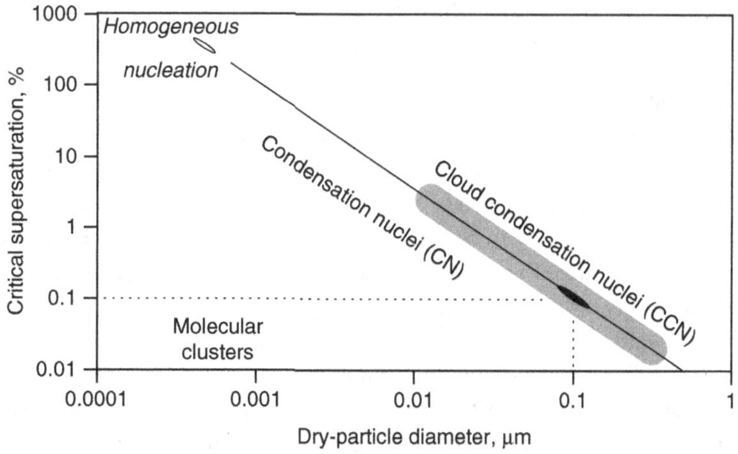

Figure 7.16 Dependence of the critical supersaturation on the dry-particle diameter. Calculation made using Köhler theory with van't Hoff factor $i = 2.5$.

the volume and diameter of the dry particle, n_s and ρ_s are the molar and mass densities of the dry solute (in its crystalline state), and M_s is the molar mass of the solute. Through Eq. (7.21), it follows that $s_c \propto D_p^{-3/2}$. The bigger the particle, the smaller is its critical supersaturation. The quantitative relationship between the critical supersaturation and the dry-particle diameter is shown in Fig. 7.16. Using logarithmic axes allows us to cover the broad range of possibilities for CN (the total aerosol) exposed to varying supersaturations. We see that large soluble particles have critical supersaturations well below 0.1%, whereas relatively small particles require critical supersaturations over 10% (saturation ratios > 1.1). In the small-particle limit, governed by molecular clusters, supersaturations of hundreds of percent are required for activation, essentially the same as the critical supersaturations found for homogeneous nucleation. The maximum supersaturations expected in atmospheric clouds are under 10%, so particles larger than about 0.01 µm are classified as CCN (shaded region). Particles with critical supersaturations less than the maximum supersaturation available from the environment activate and become cloud droplets. The limiting particle size (radius or diameter) of the dry aerosol is the so-called activation size. All particles larger than the activation size become cloud droplets; the remaining particles reside interstitially between the cloud drops. A rule of thumb to aid remembering the relationship between size and supersaturation is indicated by dotted lines aimed at the bulge on the solid line of Fig. 7.16: both diameter (in µm) and critical supersaturation (in %) have the same numerical value (0.1). Typical CCN have dry sizes of tenths of a micron and activate at supersaturations on the order of 0.1%. We will see later how the CCN are distributed in size in various parts of the atmosphere and how they grow and affect the microstructure of clouds soon after formation.

7.2 Formation of the solid phase

7.2.1 Homogeneous nucleation

The common solid phase of water, ice Ih, can in principle form from the vapor phase by either of two pathways. As suggested in Fig. 7.17, water vapor may deposit directly into the ice phase, or it may first form a liquid phase that then freezes. Whether the direct or indirect pathway is followed in nature depends on the circumstances, but we may anticipate that the direct pathway is less likely to be followed for the simple reason that it is difficult for the molecules of a disordered vapor phase (having high entropy) to gather spontaneously together into the highly ordered structure of a crystalline solid (low entropy). Rather, nature often follows Ostwald's "rule of stages", whereby an end product (here ice) arises from an intermediate, more easily formed product (liquid water), which in turn arose from other intermediates that were still easier to form. Except perhaps at uncommonly low temperatures (possibly as low as $-100\,°\mathrm{C}$), the indirect pathway is the much more likely way that ice forms from pure vapor. The formation of the liquid phase has already been treated, so we focus now on the freezing of supercooled liquid water.

Homogeneous freezing

A supercooled liquid is a metastable phase, meaning that its temperature is below the thermodynamic melting point $T_0 = 0\,°\mathrm{C}$. Ice is the more stable phase at temperatures below T_0, but the molecules of the supercooled liquid are caught in shallow wells of molecular potential energy. Transformation of the liquid into the solid becomes probable only with a sufficiently large "push", thermodynamically speaking. Supercooling $\Delta T_s \equiv T_0 - T$ serves as the thermodynamic driver for the phase change because of its direct relationship to chemical potential: $\Delta\mu \equiv \mu_L - \mu_S = RT \ln \frac{e_s}{e_i} = l_f \frac{\Delta T_s}{T_0}$. Here, we explore the theory that lets us establish the conditions needed for the liquid to freeze spontaneously.

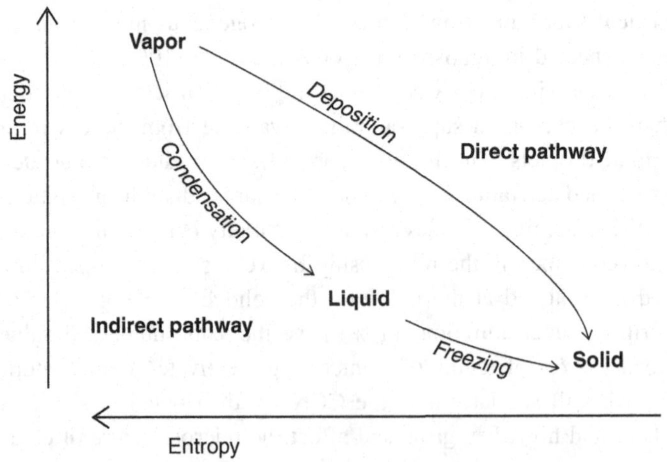

Figure 7.17 Alternative pathways for forming ice from the vapor phase.

The classical theory of homogeneous freezing starts with a specified volume of liquid water at a given supercooling and steps through the procedure outlined at the beginning of this chapter to determine the threshold temperature needed for freezing to occur. As with the homogeneous nucleation of a liquid droplet from the vapor phase, we need to calculate the change in system free energy to form a microscopic embryo of the new phase (ice) in the parent phase (a droplet of supercooled water). Now, we envision the embryo to be a hexagonal prism (a single ice crystal), rather than a sphere. The size of the ice embryo, small compared with that of the droplet, is specified to be the radius r_i of the inscribed sphere. The volume V_i of the ice embryo, being a polyhedron with sharp edges, is larger than the sphere by a geometrical factor $\alpha > 1$: $V_i = \frac{4}{3}\pi r_i^3 \alpha$. The surface area A_i of the embryo is also larger than that of the inscribed sphere: $A_i = 4\pi r_i^2 \beta$, where $\beta > 1$ is the geometrical factor for area. The first goal of the four-step procedure is to calculate the change in system free energy to form a single ice embryo:

$$
\begin{aligned}
\Delta G_i &= -V_i n_i \Delta\mu + A_i \sigma_{IL} \\
\Delta G_i &= -\frac{4}{3}\pi r_i^3 \alpha n_i l_f \frac{\Delta T_s}{T_0} + 4\pi r_i^2 \beta \sigma_{IL},
\end{aligned}
\tag{7.23}
$$

where n_i is the molar density of ice and σ_{IL} is the interfacial free energy between the solid and the liquid. The function given by Eq. (7.23) has a maximum value (calculated by finding the critical point) given by

$$
\Delta G_i^* = \frac{16\pi \sigma_{IL}^3 \xi}{3\left(n_i l_f \frac{\Delta T_s}{T_0}\right)^2},
\tag{7.24}
$$

where $\xi \equiv \beta^3/\alpha^2$ is the net geometrical factor that accounts for the non-spherical geometry of the embryo. Note that the larger the supercooling, the smaller the free-energy barrier and the easier it will be for nucleation to occur. It is noteworthy, too, that ΔG_i^* is extremely sensitive to the value of σ_{IL}, a quantity that is difficult to measure.

The rate of nucleation, the third step in the procedure, combines the thermodynamic approach that led to Eq. (7.24) with molecular kinetics to account for the rate that molecules are added to the critical embryo. The ice nucleation rate J_i [$\mathrm{m^{-3}\,s^{-1}}$] is, as with homogeneous nucleation of the liquid phase, equal to the rate J_1 that a critical embryo acquires molecules from the parent phase times the number $n\left(r_i^*\right)$ of such embryos: $J_i = J_1 n\left(r_i^*\right)$. The germ concentration is related to the concentration n_L of liquid-phase molecules by the Boltzmann factor (Eq. (7.6)): $n_g^* = n\left(r_i^*\right) = n_L \exp\left(-\frac{\Delta G_i^*}{k_B T}\right)$. Deeper consideration must, however, be given to the molecular acquisition rate (J_1).

The mechanism by which new water molecules build into the ice embryo differs from that by which molecules enter a liquid droplet. In the case of homogeneous nucleation of a droplet from the vapor phase, individual water molecules impinge onto the liquid surface with a rate given by the kinetic theory of gases (Eq. (7.8)); new molecules can be added to the liquid no faster than they arrive from the parent (gas) phase. In the case of freezing nucleation, on the other hand, the ice embryo is embedded in the liquid phase, so the parent

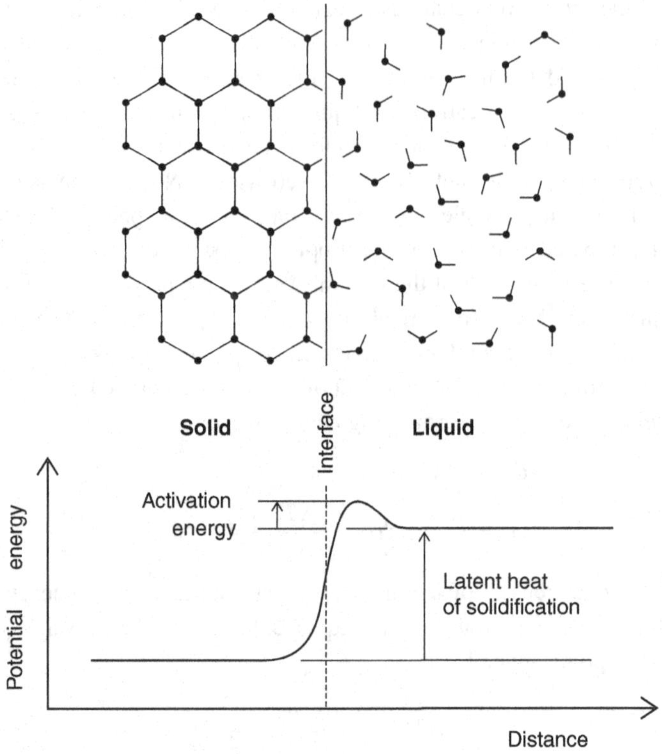

Figure 7.18 Depiction of the freezing interface. Top: molecular viewpoint. Bottom: energies associated with freezing.

water molecules are already in direct contact with the surface of the ice embryo; a supply of molecules is not the issue, rather the ability of those molecules to leave the liquid and enter the solid phase.

The situation at the freezing interface is illustrated by Fig. 7.18. As in any condensed phase, the molecules in the liquid are bonded to neighbors (bonding not shown explicitly). Those existing bonds in the liquid phase must first break before a molecule can reorient itself and adapt to the more orderly arrangement in the ice lattice. As shown in the lower part of the diagram, the potential energy of a molecule is lower in the solid phase (by the latent heat of solidification), but a liquid-phase molecule cannot get to that more favorable state without first surmounting an energy hump. This hump, the activation energy Δg_{act}, is the same energy barrier that restricts liquid-phase diffusion and accounts for the negative temperature coefficient of liquid-phase viscosity. Mobility within and freezing of a liquid involves the breaking of old bonds before new bonds can form.

The frequency of molecular incorporation into the ice, ν_{inc} [s^{-1} per molecule], the likelihood of bond breaking and molecular reorientation into the ice lattice, is obtained with the help of the Boltzmann factor: $\nu_{inc} = \nu_{molec} \exp\left(-\frac{\Delta g_{act}}{k_B T}\right)$, where $\nu_{molec} \sim k_B T/h$ is the natural (quantum mechanical) frequency of a typical molecule when the temperature is T (h is Planck's constant). The higher the activation energy and the lower the temperature, the less likely is any molecular vibration or libration (hindered rotation) able to set the molecule free of its neighbors and allow it to enter the ice lattice. We envision molecules in the liquid phase at the solid–liquid interface constantly vibrating and librating relative to neighbors in response to thermal agitations, only occasionally breaking loose and reorienting into the more favorable ice lattice. Normally, individual molecules along the interface are responsible for the freezing process, but some research points to the possibility that small molecular clusters may reorient in unison, leading to enhanced incorporation of liquid molecules into the ice at temperatures below about −40 °C. Such group responses, to the extent that they occur, serve to lower the activation energy.

The overall nucleation rate can now be estimated from the relationship $J_i = J_1 n\left(r_i^*\right)$. The molecular acquisition rate (J_1) is found by noting that the molecular incorporation frequency (ν_{inc}) applies to individual molecules across the embryo interface. Thus, we must multiply ν_{inc} by the number of molecules in the interface, yielding $J_1 = \nu_{inc} n_{s1} A_{IL}$, where $n_{s1} \sim 10^{19}\,\mathrm{m}^{-2}$ is the number density of molecules in the ice surface and A_{IL} is the area of the ice–liquid interface, that between the embryo and the liquid parent phase. Combining the acquisition rate with the number of critical embryos gives the volumetric rate of nucleation as

$$J_i(T) = J_1 n(r_i^*) = \frac{k_B T}{h} A_{IL} n_{s1} Z n_L \exp\left(-\frac{\Delta g_{act}}{k_B T} - \frac{\Delta G_i^*}{k_B T}\right), \tag{7.25}$$

where ΔG_i^* is given by Eq. (7.24) and $Z \sim 0.1$ is the Zeldovich factor that takes non-equilibrium effects into account. Note that, in contrast to the nucleation of liquid droplets from the vapor phase, the nucleation of ice in supercooled water involves two energy barriers. Despite higher temperatures favoring molecules surmounting the activation-energy barrier (Δg_{act}), the rate of freezing nucleation is dominated overall by the effect of supercooling (ΔT_s) on the height of the free-energy barrier $\left(\Delta G_i^*\right)$. The lower the temperature, the greater the supercooling, and the lower is the energy needed to form ice embryos of critical size. However, as ΔT_s increases, the magnitude of ΔG_i^* becomes comparable to Δg_{act}, so the activation energy may play a controlling role at low temperatures.

Much uncertainty surrounds the temperature dependences of the parameters associated with $J_i(T)$, including Δg_{act}. It has also been noted that the dimensionless product $A_{IL} n_{s1} Z$ in the preexponential factor of Eq. (7.25) is typically close to unity, so the rate of homogeneous freezing is often approximated as

$$J_i(T) \simeq \frac{k_B T}{h} n_L \exp\left(-\frac{\Delta g_{act}}{k_B T} - \frac{\Delta G_i^*}{k_B T}\right). \tag{7.26}$$

Despite the uncertainties and approximations, little doubt exists about the overall temperature dependence to the rate of freezing nucleation. As shown in Fig. 7.19, $J_i(T)$ increases by orders of magnitude as the temperature of the liquid is lowered by just a few degrees. It is

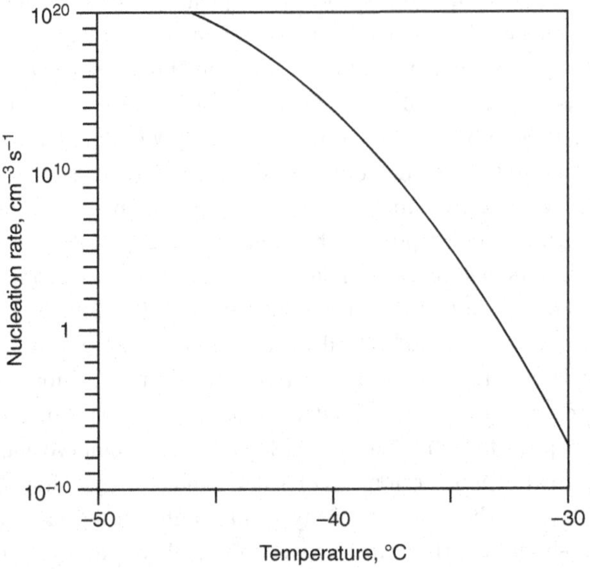

Figure 7.19 Rate of freezing nucleation in pure supercooled water. Adapted from Jeffery and Austin (1997).

the exponential function in Eq. (7.25), not the prefactor, that drives the strong temperature dependence to the nucleation rate.

Given a temperature-dependent function for the volumetric freezing rate, $J_i(T)$, how do we estimate the threshold conditions? That is, if we have a certain volume V_d of supercooled water, say that of a cloud droplet or a raindrop, at what temperature should we expect it to freeze? First, note that only a single nucleation event suffices to cause the entire volume, regardless of size, to freeze. The relevant variable is therefore the nucleation frequency $\omega_i = V_d J_i(T)$, the product of liquid volume and nucleation rate, not $J_i(T)$ alone. A larger liquid volume, V_d, implies that a smaller volumetric rate, $J_i(T)$, is needed to get the same freezing effect; equivalently, the chance of finding one critical ice embryo in a large volume is greater than that in a small volume of supercooled water. The choice of a significant nucleation rate, and an estimate of threshold conditions, thus depends on the volume of supercooled liquid, as well as the temperature. A nucleation frequency of $\omega_i = 1\ \text{s}^{-1}$ is often chosen as a reasonable criterion for homogeneous freezing.

To deal with the freezing of a large number of supercooled drops, as in a cloud, we have to recognize that all drops, even if they are all perfectly pure and the same size, are unlikely to freeze at the same time once the threshold conditions have been met. This stochastic, non-deterministic nature of nucleation can be taken into account by considering a population of N_0 drops, each of identical volume V_d, at instantaneous temperature T. If all the drops start out unfrozen at time $t = 0$, we ask what fraction of the population will have frozen by some later time t. With a nucleation frequency of ω_i, then the number of drops that freeze within some small time interval δt is $\delta N_f = N_u \omega_i \delta t$, where N_u is the number of drops still unfrozen at time t. The total number of drops remains constant, of course, so $N_u = N_0 - N_f$, where N_f is the number of drops already frozen by time t. Thus, $\delta N_f = \left(N_0 - N_f\right) \omega_i \delta t$ or, after dividing through by N_0 and taking the limit of small time intervals, we gain the differential rate equation describing the frozen fraction $F \equiv N_f / N_0$:

$$-\frac{d \ln (1-F)}{dt} = \omega_i = V_d J_i(T), \tag{7.27}$$

where the nucleation frequency ω_i is expressed in terms of its components, V_d and $J_i(T)$. This equation can be integrated to determine how the frozen fraction depends on time:

$$F(t) = 1 - \exp\left(-\int_0^t V_d J_i(T) dt'\right). \tag{7.28}$$

Both V_d and $J_i(T)$ are best left inside the integral to allow for the drop size and temperature to vary with time. However, some laboratory experiments are conducted using constant-volume drops of known size and controlled temperatures, in which case the following simpler form of Eq. (7.28) can be used to fit data and determine the nucleation rate $J_i(T)$:

$$F(t) = 1 - \exp\left(-V_d J_i(T) t\right). \tag{7.29}$$

This equation is also used to estimate the probability of freezing in atmospheric clouds. For instance, if we choose a nucleation frequency $\omega_i = 1\,\text{min}^{-1}$, then Eq. (7.29) yields $F(t = 1\,\text{min}) = 1 - \exp(-1) = 63\%$; almost two thirds of an initially unfrozen population will have frozen within 1 min. Within the next unit time interval (1 min), another 63% of the remaining unfrozen drops will have frozen, giving a total fraction of frozen drops after 2 min of 86%. The number of unfrozen drops gradually dwindles in this exponential fashion.

The threshold temperature for freezing a population of drops is determined by choosing the freezing fraction F and accounting for the time dependence, implied by Eqs. (7.27) and (7.28), by letting the population cool at a uniform rate $\gamma_c \equiv -dT/dt$. Using the chain rule of calculus, we transform Eq. (7.27) into a differential equation that depends on temperature alone:

$$-\frac{d \ln (1-F)}{dT} = \frac{V_d}{\gamma_c} J_i(T). \tag{7.30}$$

Integration of this equation from some high temperature T_0 (at which freezing in negligible) to the homogeneous freezing temperature T_F (at which the chosen fraction F of drops has frozen) yields an equation that can be solved (easily when V_d is constant) for the desired threshold temperature (T_F):

$$\int_{T_F}^{T_0} J_i(T) dT = -\ln (1-F) \frac{\gamma_c}{V_d}. \tag{7.31}$$

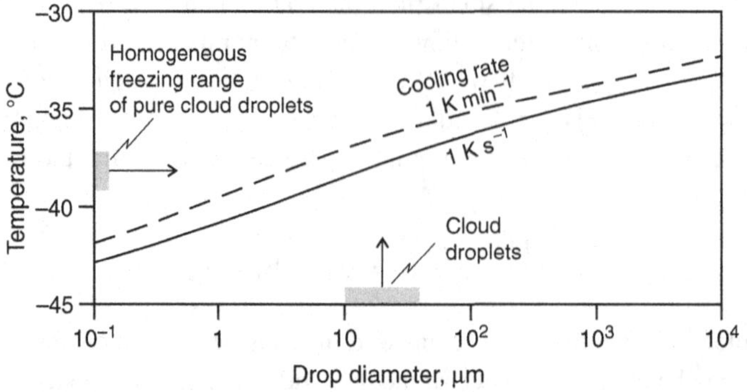

Figure 7.20 Temperature at which 99.99% of drops freeze as a function of drop size and cooling rate. Adapted from Pruppacher and Klett (1997).

Note that F is a positive quantity less than unity, so the right-hand side of Eq. (7.31) and the integral are positive. As an example, if we want to determine the median freezing temperature $T_{50\%}$, the temperature at which half of the drops freeze, then we would use the equation in the form $\int_{T_{50\%}}^{T_0} J_i(T)dT = 0.693\gamma_c/V_d$, where the value 0.693 comes from $-\ln(1-0.5)$. For an extreme case of wanting virtually all drops in a large population to freeze, such as $F = 99.99\%$, we would use $\int_{T_{99.99\%}}^{T_0} J_i(T)dT = 9.21\gamma_c/V_d$. As shown in Fig. 7.20, the so-determined freezing temperature is inversely related to the size of the drops. The dependence on drop size and cooling rate is weak because the nucleation rate $J_i(T)$ is such a strong function of temperature; even large changes in V_d or γ_c on the right-hand side of Eq. (7.31) are made up for by small compensating changes in temperature on the left-hand side. We see here that clean cloud droplets, with typical diameters of 20 μm, freeze readily once they reach the homogeneous freezing temperature for pure water, $T_{99.99\%} \equiv T_{HF} \simeq -38\,°\mathrm{C}$ (235 K).

Freezing of haze droplets

The results for pure cloud droplets need to be adjusted for submicron haze droplets, partly because they are small, but also because they contain concentrated impurities. The effect of solute on freezing may be understood by reference to Fig. 7.21, which shows how the melting point of ice and the freezing of aqueous solutions vary with water activity a_w. Recall that activity is an effective mole fraction, so high concentrations of solute imply low concentrations of water and reduced opportunities for evaporation and freezing. Higher water activities (toward the left in Fig. 7.21) imply more dilute solutions and enhanced chances that ice will form. Water activity is also a measure of the relative humidity in the air, so a point anywhere on an a_w–T graph represents the conditions of the local atmosphere in terms of composition (concentration of water vapor) and thermal energy (temperature).

Liquid haze droplets are aqueous solutions that cannot freeze, even at low temperatures, unless they first grow by condensation into a suitable activity range. As suggested by the sloping arrow in Fig. 7.21, a concentrated solution droplet must first overcome the

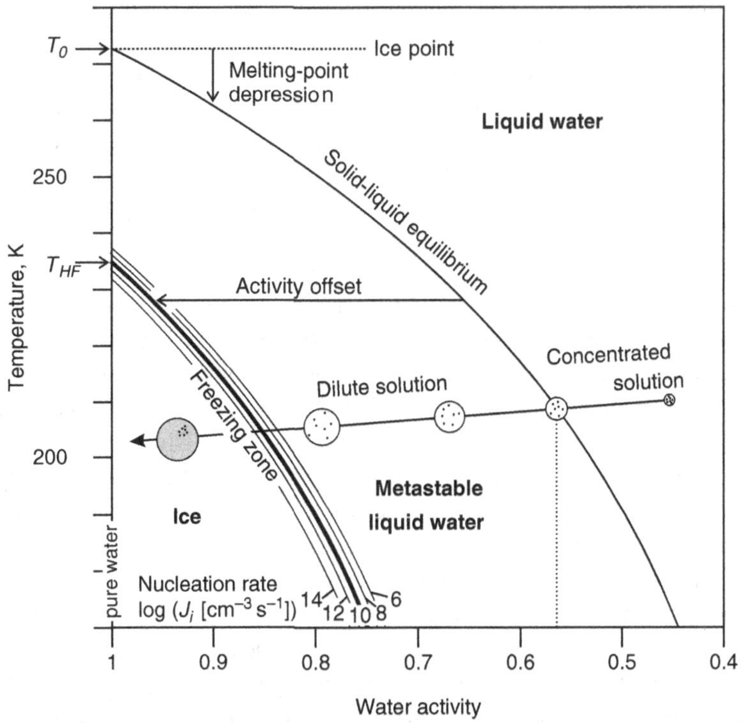

Figure 7.21 Schematic illustration of a haze droplet cooling and freezing as it rises in the atmosphere. The upper curve gives the melting point of ice in solutions of given water activity. The lower cluster of curves show how the freezing temperature varies with activity and nucleation rate. Adapted from Koop *et al.* (2000).

effect of solute on the melting point, through the addition of enough new water to move it beyond solid–liquid equilibrium and into the supercooled (metastable) region. The melting curve (upper solid curve), which represents the equilibrium boundary between the solid and liquid phases, is a well-defined thermodynamic function that can be calculated in concept from the ratio of equilibrium vapor pressures over ice and pure liquid water, giving $T_m\ (a_w)$, where $a_w = e_i(T)/e_s(T)$, as discussed in Section 3.5.3. A solution droplet in the metastable state is ripe for ice nucleation, but ice is not likely to form until the solution becomes sufficiently dilute by continued condensation.

Freezing is a rate process, not an equilibrium transformation, so deeper consideration is needed to understand the nucleation of ice in aqueous solutions. One might expect that the solute simply lowers the threshold freezing temperature below that for pure water (T_{HF}) by the magnitude of the melting-point depression, but such is not the case. Scientists have found that once diverse freezing data were adjusted to a common set of conditions and expressed in terms of water activity, freezing occurred when the water activity, not temperature, was offset by a constant amount Δa_w. The off-set that accounted for data from solution droplets several micrometers in diameter was empirically found to be $\Delta a_w = 0.305$, yielding the heavy curve in the freezing zone on Fig. 7.21. The volumetric nucleation rate corresponding to these data was calculated to be $J_i \simeq 10^{10}\,\mathrm{cm}^{-3}\,\mathrm{s}^{-1}$. (Other nucleation rates apply in other situations.) Thus, knowing how the melting temperature varies with activity, $T_m(a_w)$, the freezing temperature T_f of solutions is calculated, in concept, as

$$T_f = T_m\left(a_w - \Delta a_w\right). \tag{7.32}$$

That is, given a value for activity, as from measurements of relative humidity, one just subtracts the off-set and calculates the melting point at the new activity. A freezing curve is just the melting curve shifted toward higher activities by the amount Δa_w. The off-set in water activity at any given temperature represents the deviation from solid–liquid equilibrium, in effect the amount of dilution needed for ice to form spontaneously.

The finding that the freezing temperature is a function only of the water activity leads to important practical applications if we turn the problem around and ask what activity (rather than temperature) is needed for freezing to occur at a given temperature. The activity of water in a solution takes the non-linearities of solute–water interactions into account, so the calculations need not consider the composition of the particles, an important benefit given the uncertainty of knowing what substances constitute the atmospheric aerosol. Moreover, once we assume that the aqueous particles are in equilibrium with their local environment (a safe assumption for most atmospheric situations), the water activity of the solution equals the relative humidity of the air, giving $a_w = RH \equiv e/e_s(T)$. We may therefore establish the threshold for freezing at any given temperature as $RH^f = a_w^i(T) + \Delta a_w$, where $a_w^i(T) = e_i(T)/e_s(T)$ is the water activity in equilibrium with ice (i.e., the melting curve expressed as a function of temperature). Alternatively, one may use the saturation ratio with respect to ice, $S_i \equiv e/e_i(T)$, which is readily derived from relative humidity through their respective definitions: $S_i = RH\frac{e_s(T)}{e_i(T)}$. The threshold saturation ratio with respect to ice is then

$$S_i^f = 1 + \frac{\Delta a_w}{a_w^i(T)} = 1 + \Delta a_w \frac{e_s(T)}{e_i(T)}. \tag{7.33}$$

One could use the empirical value for the activity offset, $\Delta a_w = 0.305$, in Eq. (7.33), but we must recognize that the proper value to use depends on particle size. Particle size affects the threshold saturation ratio because of the stochastic nature of freezing. Only a single critical embryo suffices to freeze any contiguous volume of liquid, so the criterion for nucleation is the freezing frequency, $\omega_i = V_d J_i$, not the nucleation rate J_i itself. Small haze droplets will require a larger nucleation rate, and hence a larger water activity, to achieve the same nucleation frequency. The effect of size on the threshold saturation ratio S_i^f is shown in Fig. 7.22. We see that the threshold saturation ratio increases as the particle size decreases and that large ice saturation ratios are needed at low temperatures. Small haze droplets in the coldest parts of the upper troposphere (usually in the tropics) require ice saturation ratios in excess of 1.6 for ice to form.

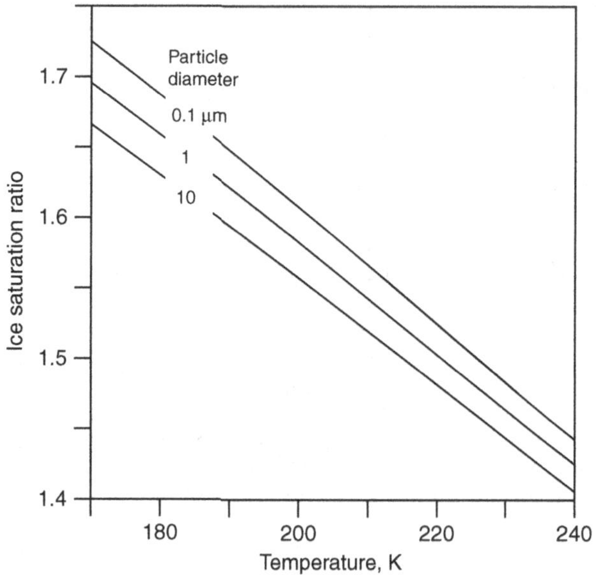

Figure 7.22 Threshold saturation ratio with respect to ice as a function of temperature for several diameters of liquid aerosol particles. Adapted from Koop *et al.* (2000).

The threshold for homogeneous freezing may be interpreted in molecular terms. The classical theory of freezing nucleation depends on knowing the properties of the ice–liquid interface and the activation energy for molecular reorientation. It has been argued, however, that the threshold temperature T_f might equally well be expressed in terms of the properties of the liquid only, because freezing is associated with the maximum in the isothermal compressibility of supercooled water (Section 2.1.3), where relatively large, thermally driven fluctuations in density occur. Where the liquid–water density becomes locally similar to that of ice, the water molecules can reorient themselves easily into ice embryos, an effect that would appear as a lowering of the activation energy (used in Eq. (7.25)). Solutes, because of their strong attachments to water molecules, most likely hinder the formation of ice embryos within their local shells of influence, but the water between such solute-influenced regions is still normal and pure in most regards; just the volume of such water has been reduced because of the solute. The volume fraction of pure water in a solution is to first approximation given by the water activity, so we can think of ice nucleation in solutions as being restricted to the smaller volume of relatively pure water in between the solute-influenced water. Interpreting observed macroscopic properties of water in terms of molecular behavior is an ongoing challenge.

7.2.2 Heterogeneous nucleation of ice

Ice that forms in a cloud under conditions not suitable for homogeneous freezing must arise via some other mechanism. Observations reveal that many clouds indeed contain ice particles at temperatures well above the threshold for homogeneous freezing of clean

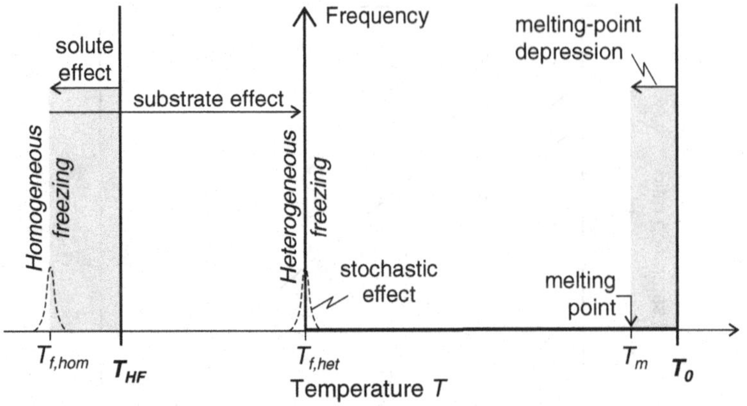

Figure 7.23 Overview of influences on freezing. Effect of solute on the melting point and on the homogeneous freezing of solutions is shown by shading. Foreign solids raise the freezing point by the substrate effect. (arrows toward right). The stochastic nature of nucleation is indicated by the dashed Gausian curves.

cloud droplets, $T_{HF} \simeq -38\,°C$. Such ice particles form heterogeneously because they require the presence of some foreign, non-aqueous phase to facilitate the nucleation process. The responsible substances are in particles contributing to a subset of the atmospheric aerosol called ice nuclei (IN). Ice nuclei initiate the ice phase in the so-called mixed-phase zone of a cloud, between the approximate temperatures of $T_{HF} \simeq -38\,°C$ and $T_0 = 0\,°C$. Figure 7.23 shows the combined effects of a substrate and any solute that may be present. An IN offers a solid substrate that facilitates ice nucleation (arrow pointing toward higher temperature), whereas soluble matter tends to lower the melting and freezing points (arrows pointing toward lower temperature). Heterogeneous nucleation is stochastic in nature (suggested by the dashed Gaussian curves), as is homogeneous nucleation, but it also has a singular, deterministic component that accounts for the shift in threshold conditions (shown by the upward arrow at the median freezing temperature $T_{f,het}$).

Nature of ice nuclei

The properties of IN differ qualitatively and quantitatively from CN. Whereas the most effective CN (i.e., CCN having critical supersaturations of at most a few percent) are composed of soluble substances, IN are generally insoluble and have crystalline structures related to the hexagonal lattice of ice Ih. Heterogeneous nucleation thus takes place on the surfaces of the IN, unlike homogeneous freezing, which is initiated randomly within the volume of liquid. Ice is more difficult to form than liquid water, so it should not be surprising to learn that the concentration of IN in the atmosphere is much smaller than that of CCN. The IN and CCN in the atmosphere may represent mutually exclusive subsets of the atmospheric aerosol, but mixed particles, those having both insoluble and soluble components, are prevalent.

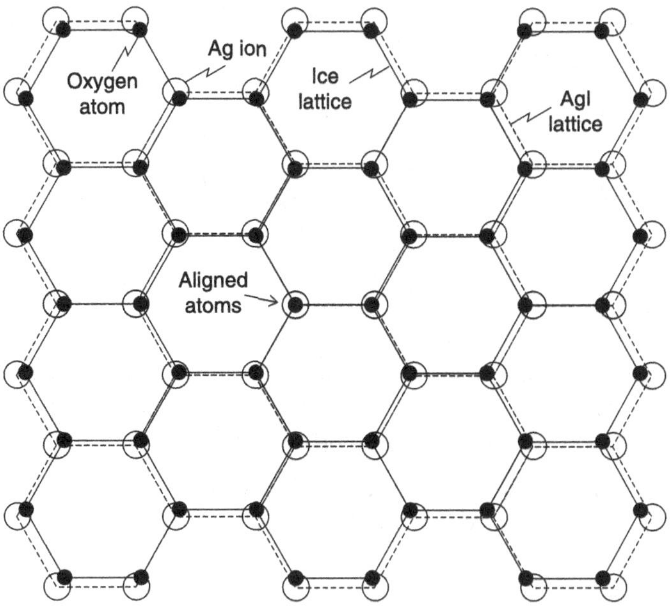

Figure 7.24 Schematic of two hexagonal lattices, those of ice Ih (closed circles joined by solid lines) and AgI (open circles joined by dashed lines). The lattices have been aligned at the arrow. The misfit is exaggerated to show the effect different lattice parameters. Adapted from Mason (1961).

An understanding of how an insoluble particle might serve to nucleate the ice phase may be gleaned from Fig. 7.24, an example of an epitaxial layer of ice deposited onto silver iodide (AgI), an insoluble salt known to be a good nucleating agent. Both ice Ih and AgI have similar crystallographic structures, so the surface of AgI can serve as a template for aligning the water molecules. The lattice parameters (dimensions between atoms) are not identical in these two crystals, however, so even if the alignment is perfect at one point (arrow), it gradually degrades with distance away from that point (note increasing separation between points). The lattice mismatch, defined as the relative difference of the interatomic distance, induces a strain in the overlying crystal, but nucleation can still be effective when the magnitude of mismatch is not too large. Experiments have shown that AgI, with a mismatch of just over 1%, can nucleate ice at temperatures as high as about −4 °C. On the other hand, detailed experimental and numerical studies of the interaction of water with various solid substrates suggest that epitaxial match, which assumes the water maintains an ice-like structure, is not the best criterion for nucleation. A good nucleating agent may indeed attract water molecules so strongly that the hydrogen bonds form preferentially with the solid, rather than with overlying water layers, giving a rather amorphous, non-ice structure near the interface. Only further research at the molecular level can reveal the true mechanism of heterogeneous ice nucleation.

Despite limited understandings of the mechanism of nucleation, the effectiveness of various crystalline substances for nucleating ice can be assessed empirically by measuring

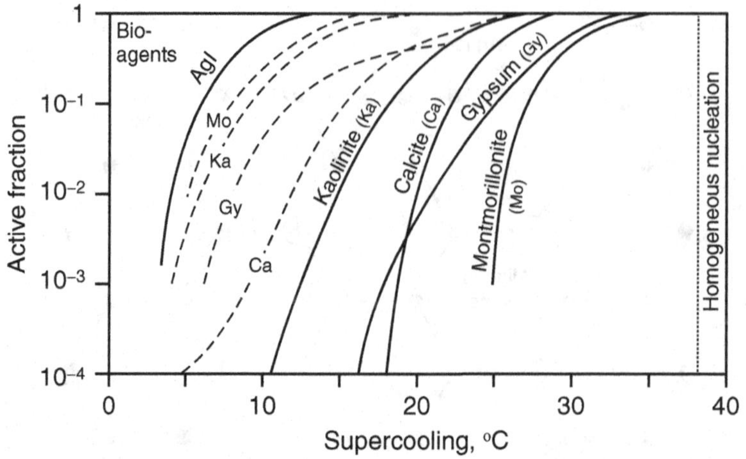

Figure 7.25 Activity curves for selected substances. Solid curves: Initial ice nucleation. Dashed curves: Effect of prior activation. Adapted from Roberts and Hallett (1968).

the fraction of a population of particles that yields ice crystals as a function of supercooling (the deviation of temperature from the ice melting point). The active fraction of several agents is seen in Fig. 7.25 to increase with supercooling. AgI forms ice at relatively high temperatures (small supercoolings) compared with clays and natural minerals, perhaps because of the smaller lattice mismatch of AgI. Only some biological agents (e.g., proteins from *Pseudomonas syringae* or *Erwinia herbicola*) serve as better ice nuclei. Ice-nucleating bacteria seem to have evolved to initiate ice formation readily, perhaps as a means for accessing nutrients from plants under cold conditions. Bioagents that can nucleate ice within a couple of degrees of 0 °C have been used to encourage freezing of dispersed water drops in the making of artificial snow at ski resorts and in weather modification. Dry ice (solid CO_2, not shown) also serves as a seeding agent, but it acts by forming tiny liquid droplets that then freeze homogeneously. Note the tendency (shown by dashed curves) for the ice-nucleating activity of some agents, especially the clay montmorillonite, to improve following prior activation. Ice that forms in cracks or in stable adsorbed layers on the IN may be preserved and stay available for subsequent ice formation. Nevertheless, factors other than temperature alone influence the ice-nucleating ability of foreign substances. It should not be surprising that supersaturation with respect to ice is necessary for ice formation, regardless of the temperature.

Measurements of ice nucleation that control both temperature and supersaturation reveal greater details of the process, specifically the modes of action. Experiments in thermal-gradient diffusion chambers, into which carefully prepared batches of aerosol are injected, show that ice nucleation occurs both when the air is subsaturated and when it is supersaturated with respect liquid water. However, the thresholds for nucleation differ depending on the foreign substance, as shown in Fig. 7.26. Each curve represents the nucleation threshold

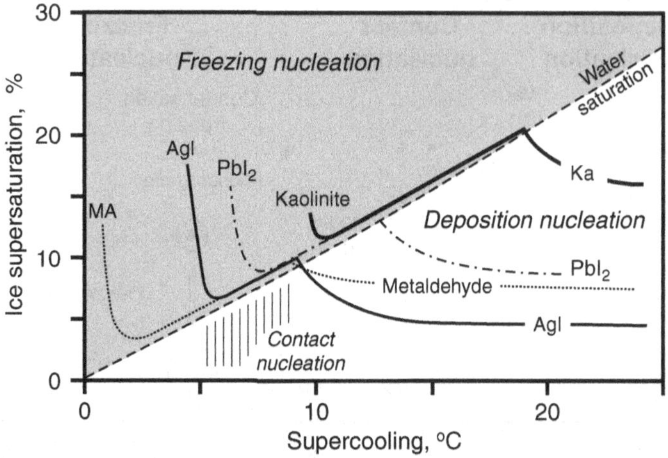

Figure 7.26 Empirically determined threshold curves for heterogeneous nucleation by selected substances. Note how the water-saturation (dashed) curve follows the numerical approximation, $s_i[\%] \simeq \Delta T_s[°C]$. Adapted from Schaller and Fukuta (1979).

for the specified substance, that is, the lower limit of ice supersaturation needed for about 1% of the given particles to form ice within one minute. Ice that appears below liquid-water saturation (identified by the dashed curve) presumably formed by direct deposition of vapor to the aerosol particles; no liquid water was needed for the process. At higher temperatures, however, each aerosol type nucleated ice only if the ice supersaturation exceeded the limit of liquid-water saturation; liquid water needed to form first, so the mechanism likely involved both condensation and freezing of supercooled water. The distinctions between the different threshold curves reflect, for the most part, differences in interfacial free energies and lattice mismatch. The experiments also revealed a range of conditions (shown by vertical bars) in which ice particles formed slowly from a haze aerosol outside the limits of deposition nucleation and condensation-freezing. This contact-nucleation mechanism is thought to arise from the slow diffusion of the fine IN to the larger liquid haze droplets, which freeze quickly once contact with the surface has been made. Contact nucleation may also act above water saturation, but its influence may be masked by other freezing modes. A fourth, immersion-freezing mode, not accessible by these experiments, is thought to account for the possibility that the freezing nucleus entered a liquid droplet earlier, perhaps while the droplet was above the nominal melting point.

Modes of action

The various ice-forming mechanisms, or modes of action, are schematically summarized in Fig. 7.27. In each case, the IN (triangle) starts as an aerosol particle exposed to water vapor. In the case of deposition nucleation, some vapor, supersaturated with respect to ice but subsaturated with respect to the liquid, presumably adsorbs onto the solid surface of the IN, where the admolecules preferentially adhere to hydrophilic sites and become aligned to

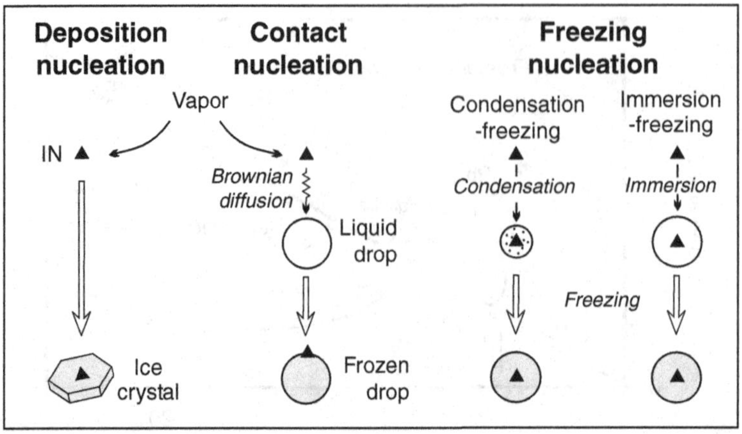

Figure 7.27 Depiction of the various modes by which ice nuclei act. Ice nuclei (IN) are shown as solid triangles, and shading represents ice. The black dots in the condensation-freezing mode represent solute, which lowers the melting point.

the underlying molecular structure of the substrate. If the lattice mismatch is not too large and the temperature is sufficiently low, ice embryos can form that lead to a high probability of nucleating ice. A deposition nucleus that is unable to form ice directly under the given conditions may nevertheless nucleate ice after being captured by a supercooled liquid drop. It is the transient contact of the aerosol particle with the liquid surface that seems to favor contact IN over deposition IN. Note, however, that the rate-limiting step for ice formation by the contact mode may not be the contact itself, rather the slow Brownian diffusion of the IN to the drop surface. This in-cloud scavenging step must be accounted for when devising measurement strategies for detecting contact nuclei or when developing numerical cloud models that include the contact mode of ice nucleation.

Effective IN are thought to be solid, insoluble particles, but they may nevertheless be coated with salts (e.g., sulfates or nitrates) that are soluble and so enhance condensation. The aqueous solution that forms on such mixed CN must become diluted (by ongoing condensation) sufficiently for supercooling to occur and ice to become stable; at least, slight supersaturation beyond liquid-water saturation would be required. By suppressing the freezing point, soluble matter operates counter to the ice-enhancing action of insoluble substances (as indicated in Fig. 7.23). Particles forming ice by this combination of processes at low temperatures are the condensation-freezing IN identified in the experiments discussed above. IN that enter the droplet early, as during condensation but before the drop became supercooled, act in the immersion-freezing mode. Immersion-freezing is not affected much by the soluble portion of mixed CN because dilution significantly reduces the effect of solute on the suppression of the freezing point. Immersion nuclei initiate the freezing of cloud droplets at temperatures characteristic of the foreign substances contained in the cloud water.

The distinctions between the modes are subtle and still not well defined. The term freezing nucleation, for instance, could also include contact nucleation (because it involves the freezing of a water drop), but tradition restricts the term to the condensation-freezing and immersion-freezing modes. Nevertheless, it is important to recognize that ice can form by different pathways and that each pathway operates on its own unique time scale. Both measurements and theory need to account for the unique properties of each nucleation mode.

Theory of heterogeneous nucleation

The theory of heterogeneous nucleation is incomplete and as yet unsatisfactory, complicated by uncertain mechanisms of action and incomplete knowledge about the surface properties of potential ice-nucleating agents. Nevertheless, a reasonable approach to follow is that used to explain the role of an insoluble surface in the heterogeneous nucleation of liquid water (see Fig. 7.7). When applying this classical approach to ice nucleation, one assumes that the spherical caps of ice that serve as the ice embryos form randomly on the surface of the IN. Good compatibility between the molecular structures of the ice and foreign substrate lowers the height of the free-energy barrier ΔG^*, as discussed in connection with Fig. 7.8, and the rate of nucleation by a particle (J_{ptcl}; units, s^{-1}) takes the general exponential form,

$$J_{ptcl} = K_X \exp\left(-\frac{\Delta G_X^*}{k_B T}\right), \tag{7.34}$$

where K_X is a kinetic coefficient adapted to parent phase X (vapor or liquid). With heterogeneous ice nucleation, one needs to recognize that an ice embryo is in simultaneous contact with the solid nucleus and with one of two parent phases, either water vapor (in the case of deposition nucleation) or liquid (with freezing nucleation). The properties of the ice-nucleating surface are describable by an interface parameter, $m_X = (\sigma_{NX} - \sigma_{IN})/\sigma_{IX}$, where the subscripts N, I, and X refer, respectively, to the nucleus, ice, and the parent phase, either water vapor (X = V) or liquid water (X = L). The lattice mismatch is accounted for by defining a misfit parameter $\delta \equiv (a_N - a_I)/a_I$, where a is the lattice spacing. The classical theory gives physical interpretation to such variables, but it is also appropriate to treat them (cautiously) as free parameters when fitting theoretical calculations to experimental data.

The theoretical approach is the same regardless of the parent phase, but the different magnitudes of the parameters lead to different effects. For instance, the critical radius of the ice embryo depends on the interfacial free energy between the ice and parent phases (σ_{IX}), as we saw earlier (Eq. (7.3)). We may write the critical radius of an ice embryo in contact with parent phase X in general terms as

$$r_X^* = \frac{2\sigma_{IX}}{n_I \Delta\mu_{XI}}, \tag{7.35}$$

where n_I is the molar density of ice and $\Delta\mu_{XI} \equiv \mu_X - \mu_I$ is the chemical potential of bulk parent phase X relative to that of ice. For a given set of environmental conditions,

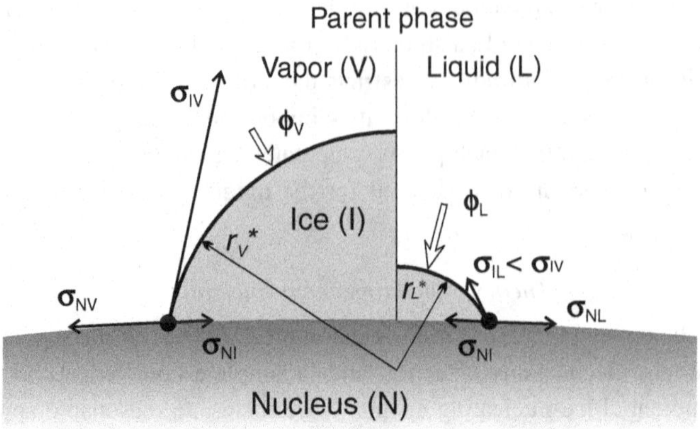

Figure 7.28 Schematic of critical ice embryos, showing the effect of the parent phase (vapor or liquid). Based on arguments of Cooper (1974).

$\Delta\mu_{XI}$ is essentially the same for an embryo in a medium of vapor or liquid. However, the surface energies differ significantly, so to good first approximation $r^*_X \propto \sigma_{IX}$. Because $\sigma_{IV} > \sigma_{IL}$, therefore $r^*_V > r^*_L$, as illustrated in Fig. 7.28. The possibility thus exists that subcritical embryos in the vapor phase would be supercritical were they to contact supercooled liquid. Perhaps it is this difference in the sizes of the critical embryos that allows a given ice-nucleating particle to act more effectively in the contact mode than in the deposition mode.

The parent phase also affects the height of the free-energy barrier and the threshold temperature for nucleation. The free energy to form a spherical ice cap on a nucleating particle can be written in general terms (a generalization of Eq. (7.12)) as

$$\Delta G_X = -n_I V_I \Delta\mu_{XI} + A_{IX}\sigma_{IX} - A_{IN}\sigma_{IX} m_X. \tag{7.36}$$

Here, V_I is the volume of the ice embryo, A_{IX} is the area of the interface between the embryo and the parent phase, and A_{IN} is the ice-nucleus interfacial area. Note that the surface free energy of the embryo in contact with the parent phase (second term on the right-hand side of Eq. (7.36)) generates the energy barrier in the first place. The ice-nucleus interfacial energy (third term), on the other hand, offsets that contribution by lowering the height of the energy barrier (when $m_X > 0$). Finding the critical point of this function yields the critical radius r^*_X (Eq. (7.35)) and the barrier height

$$\Delta G^*_X = \Delta G_X\left(r^*_X\right) = \frac{16\pi\sigma^3_{IX}}{3\left(n_I \Delta\mu_{XI}\right)^2} \cdot f(m_X, r_N), \tag{7.37}$$

where $f(m_X, r_N) < 1$ is the appropriate geometrical factor (analogous to Eq. (7.17)).

The rate of heterogeneous ice nucleation depends on the parent phase and the size of the solid nucleus, not only because of ΔG^*_X, but also because of the kinetic coefficient K_X (in Eq. (7.34)). Our goal now is to determine expressions for K_X for the two different parent media (vapor and liquid). For either medium, the particle nucleation rate is proportional to the surface nucleation rate:

$$J_{ptcl} = J_{sfc}A_N, \tag{7.38}$$

where A_N is the surface area of the nucleus. The surface nucleation rate is in turn proportional to the concentration of germs on the IN surface:

$$\begin{aligned} J_{sfc} &= J_{1,X}n_g^* \\ &= J_{1,X}n_{s1}Z\exp\left(-\frac{\Delta G_X^*}{k_BT}\right), \end{aligned} \tag{7.39}$$

where $J_{1,X}$ is the rate that new molecules are incorporation into the ice embryo from parent phase X, n_{s1} is the surface concentration of molecules at an ice interface, and where the Boltzmann and Zeldovich factors have been used. The general form for the kinetic coefficient is thus $K_X = J_{1,X}n_{s1}ZA_N$. The molecular incorporation rate ($J_{1,X}$) is the main variable that depends on the medium. As suggested by Fig. 7.28, $J_{1,X}$ depends both on the flux Φ_X of molecules across the interface and on the area A_{IX} of the ice surface in contact with parent medium X, so $J_{1,X} = \Phi_X A_{IX}$. When the parent phase is vapor, $J_{1,V} = \Phi_V A_{IV}$, where $\Phi_V = e/(2\pi m_w k_B T)^{1/2}$ is the vapor impingement flux. When the parent phase is liquid, $J_{1,L} = \Phi_L A_{IL}$, where $\Phi_L = \frac{k_BT}{h}n_{s1}\exp\left(-\frac{\Delta g_{act}}{k_BT}\right)$. Putting the diverse expressions together, we find the following ice nucleation rates for the two cases:

$$\text{Vapor:}\quad J_{ptcl}(T) = \frac{e}{(2\pi m_w k_B T)^{1/2}}A_{IV}A_N n_{s1}Z\exp\left(-\frac{\Delta G_V^*}{k_BT}\right) \tag{7.40}$$

$$\text{Liquid:}\quad J_{ptcl}(T) = \frac{k_BT}{h}A_{IL}A_N n_{s1}^2 Z\exp\left(-\frac{\Delta g_{act} + \Delta G_L^*}{k_BT}\right). \tag{7.41}$$

The kinetic coefficient thus depends on the parent medium as follows:

$$\text{Vapor:}\quad K_V = \Phi_V A_{IV}A_N n_{s1}Z = \frac{eA_{IV}A_N n_{s1}Z}{(2\pi m_w k_B T)^{1/2}} \tag{7.42}$$

$$\text{Liquid:}\quad K_L = \Phi_L A_{IV}A_N n_{s1}Z = \frac{k_BT}{h}A_{IL}A_N n_{s1}^2 Z\exp\left(-\frac{\Delta g_{act}}{k_BT}\right). \tag{7.43}$$

Note that the surface density of molecules enters the expressions linearly when the medium is vapor, but quadratically when the medium is liquid. Both the formation of the germ and the molecular flux across the interface depend on this factor when the parent phase is liquid. The areas of the respective interfaces depend on the specific geometry assumed for the embryos and IN. Uncertainties in the magnitudes of the parameters in K_X fortunately have only minor impact on the computed threshold conditions, which are controlled heavily by the exponential factor (in Eq. (7.34)).

The threshold temperature depends on the chosen criterion (typically $J_{ptcl} = 1\ \mathrm{s}^{-1}$) and becomes a function of nucleus size, as shown in Fig. 7.29 for a spherical-cap model. Note that freezing-IN activate and form ice at relatively high temperatures when the interface parameter (m) is large. The size of an IN is an important limiting factor when it is less than

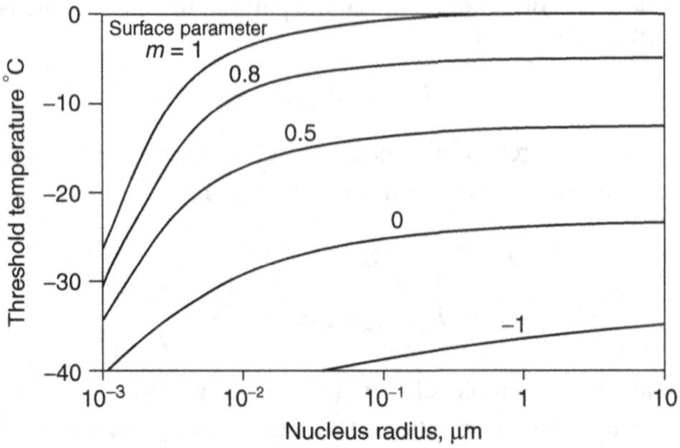

Figure 7.29 Threshold temperature for heterogeneous freezing nucleation as a function of nucleus radius and surface parameter m. The particle nucleation rate is 1 s^{-1} and the ice-liquid surface energy is 22 mJ m^{-2}. Adapted from Fletcher (1970).

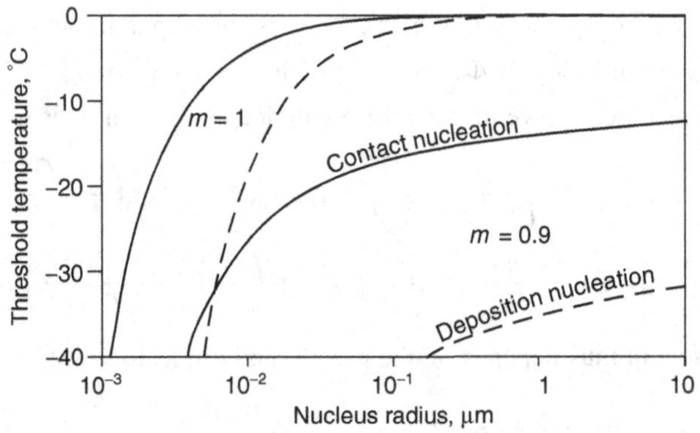

Figure 7.30 Threshold temperatures for nucleation by contact (solid curves) and deposition (dashed curves) for two values of the surface parameter m. Adapted from Cooper (1974).

the size of the critical embryo. The distinction between contact and deposition nucleation is shown in Fig. 7.30 when the ice embryos are assumed to take the form of monomolecular disks. The curves for a given interface parameter (m) reflect the difference in surface free energies for embryos in vapor and in liquid. Contact nucleation, for a given compatibility and size of nucleus, is clearly favored over deposition nucleation, especially for small IN when m is close to unity. This size effect is especially important for contact nucleation, because of the need for the IN to diffuse to the surface of a supercooled water droplet before the opportunity for nucleation arises.

Valid criticisms of the classical theory of heterogeneous nucleation can be raised. The embryos are in reality transient clusters with relatively few water molecules (~ 100), yet we give them the same properties as used to characterize stable, bulk samples of ice.

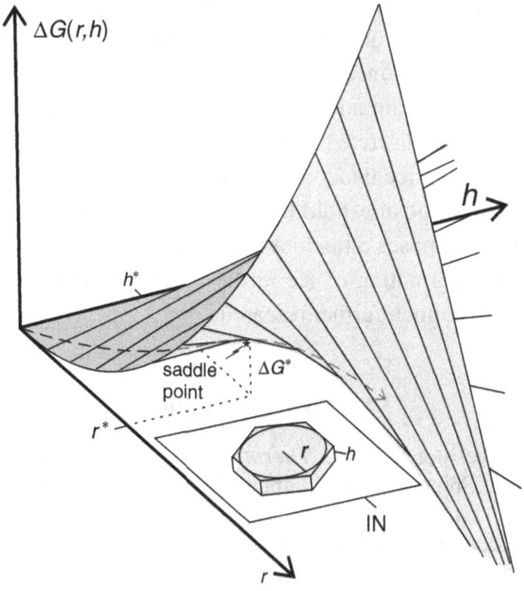

Figure 7.31 Schematic of the free energy surface for a hexagonal embryo on a substrate, where r is the radius of the inscribed circle and h is the thickness of the hexagon (see inset). The dashed curve represents the path for optimal crossing of the saddle point. Adapted from Fletcher (1965).

Almost certainly, the embryos do not assume the shape of spherical caps on a foreign substrate. The molecules in the solid phase of water must exhibit the kind of long-range order that gives ice Ih its hexagonal symmetry. Moreover, the water molecules in contact with the substrate will be forced to accommodate to the template imposed by the nucleating substate. The free energy of formation therefore depends on the geometry of the ice embryos in more complicated ways than just the radius of curvature of a spherical cap. If the embryos are hexagonal prisms, the most likely situation, then the free energy would vary in two dimensions, as depicted in Fig. 7.31 for the case of a basal face in epitaxial contact with the IN surface (see inset). Both the lateral extend, represented here by the radius r of an inscribed circle, and the thickness (h) affect the free energy because the areas exposed to the parent phase depend on the aspect ratio for a given volume. The dashed curve shows a likely path across the saddle point of the surface, which now represents the critical free energy ΔG^* to use in the calculation of nucleation rate. Unfortunately, such refined approaches, while being conceptually pleasing, require information that may not be readily obtainable. Fortunately, the threshold conditions are relatively insensitive to some of the assumptions made about crystalline embryos on foreign substrates.

Some perspective is needed to link theory with atmospheric applications. The freezing of cloud droplets is complicated by the convolution of processes that take place simultaneously in a rising parcel of air. Many aerosol particles contain sufficient soluble material to serve as effective CCN and lead to the liquid droplets that constitute a cloud in the early stages of development. Some fraction of the cloud droplets may also contain insoluble substances, gained during initial activation or later by in-cloud scavenging, that may catalyze ice formation once the droplets become sufficiently supercooled. Such a supercooled droplet may contain a number of separate solid particles or possibly a single solid particle with several discrete patches having distinct nucleating properties. Each patch is capable of nucleating ice under a range of conditions, so the supercooled droplet would exhibit a distribution of threshold temperatures. At the same time, the stochastic nature of nucleation superimposes a random time dependence to nucleation (as suggested by Fig. 7.23). The initial formation of ice in the mixed-phase zone of clouds is clearly complicated and in need of much further research.

7.3 Further reading

Curry, J.A. and P.J. Webster (1999). *Thermodynamics of Atmospheres and Oceans.* London: Academic Press, 467 pp. Chapters 4 (phase transitions) and 5 (nucleation).

Fletcher, N.H. (1962). *The Physics of Rainclouds.* Cambridge: Cambridge University Press, 386 pp. Chapters 4 (nucleation of liquid, Köhler theory) and 8 (nucleation of ice).

Mason, B.J. (1957). *The Physics of Clouds.* London: Oxford University Press, 481 pp. Chapters 1 (homogeneous nucleation of liquid), 2 (deliquescence, heterogeneous nucleation of liquid, CCN), and 4 (nucleation of ice, IN).

Pruppacher, H.R. and J.D. Klett (1997). *Microphysics of Clouds and Precipitation.* Dordrecht: Kluwer Academic Publishers, 954 pp. Chapters 6 (Köhler theory), 7 (homogeneous nucleation of liquid, ice), and 9 (heterogeneous nucleation of liquid and ice, CCN, IN).

Rogers, R.R. and M.K. Yau (1989). *A Short Course in Cloud Physics.* New York: Pergamon Press, 293 pp. Chapters 6 (nucleation of liquid, Köhler theory, CCN) and 9 (nucleation of ice, IN).

Seinfeld, J.H. and S.N. Pandis (1998). *Atmospheric Chemistry and Physics.* New York: John Wiley and Sons, 1326 pp. Chapters 9 (deliquescence), 10 (nucleation concepts), and 15 (Köhler theory).

Wallace, J.M. and P.V. Hobbs (2006). *Atmospheric Science: An Introductory Survey.* Amsterdam: Academic Press, 483 pp. Chapter 6 (nucleation theory, CCN, IN).

Young, K.C. (1993). *Microphysical Processes in Clouds.* New York: Oxford University Press, 427 pp. Chapters 3 (nucleation of liquid) and 4 (nucleation of ice).

7.4 Problems

1. Particles composed of soluble salts generally deliquesce in the atmosphere when the relative humidity exceeds a value that is characteristic for the particular salt in question. It is at this deliquescence point that a salt particle takes up enough water from the surrounding vapor phase to dissolve the solid completely. As a rule, the greater the solubility of the salt, the lower the deliquescence point.

   For this assignment, use the fundamental definition of activity in terms of the van't Hoff factor (i) and the literature value for the solubility of ammonium sulfate

($5.8\,\mathrm{mol\,l^{-1}}$ at 25 °C). Published data indicate that each molecule of ammonium sulfate dissociates on average into about 2.5 molecular components (mostly ions).

(a) Calculate the bulk deliquescence point of ammonium sulfate at 25°C.

(b) Suppose that the initially dry ammonium sulfate is contained in particles only 10 nm in radius. What is the particle deliquescence point?

(c) What is the critical supersaturation of these small aerosol particles of ammonium sulfate?

2. Explain why clouds do not contain more than a few hundred drops per cubic centimeter, considering that air contains many thousands of condensation nuclei per cubic centimeter.

3. In the "old" days, when the authors were still young (and before air conditioning was common in houses), one could hang bags of $MgCl_2$ (a salt with a particularly low deliquescence relative humidity of 33%) from the ceiling in the basement for a useful purpose during the hot, "sticky" summer months. Explain what benefit this procedure provided to the homeowners and how it worked physically.

4. Consider two aerosol particles, both with mass $m = 10^{-16}$ g, the one being sodium chloride (NaCl) and the other ammonium sulfate [$(NH_4)_2\,SO_4$]. These particles are lifted in an updraft towards a cloud base at 10 °C.

(a) At cloud base, how big will these particles be? Which is the bigger?

(b) Which of these particle will form a cloud drop first, assuming that the supersaturation keeps increasing above cloud base?

This is not a multiple choice question, please show your work.

5. We developed our homogeneous nucleation theory for pure water, and showed that the energy required to form a critically sized liquid drop from the vapor is given by ΔG^*. However, we know the atmosphere contains volatile gases that will dissolve into liquid. As a result, we rarely find pure water clusters in the atmosphere. How would you change the expression for ΔG^* of a pure water drop to account for the presence of solute in the cluster? You may assume that neither the surface tension nor the liquid bond energies are affected by the presence of the solute. To get a feeling for the possible impact of these impurities to the process, please calculate ΔG^* using water activities (a_w) of 1.0, 0.99, 0.95, and 0.9 in an environment with $S = \frac{e}{e_s(T)} = 1.001$.

6. We asserted in the text that the Zeldovich factor Z does not impact the threshold conditions for homogeneous nucleation much. Please confirm for yourself that the statement is true. Plot curves of the homogeneous nucleation rate as a function of saturation ratio S with and without the Zeldovich factor $Z \sim 0.1$. How does the difference you computed compare to an uncertainty of order 1% in surface tension close to the threshold supersaturation?

8

Growth from the vapor

8.1 Overview

At this point in our step-by-step study of cloud formation, we assume that a set of cloud particles exists via aerosol activation or nucleation. Liquid cloud droplets likely formed by condensation of water vapor onto cloud condensation nuclei, and ice crystals may be present if the temperature were sufficiently low and ice nuclei were active. We ask next how rapidly and by what mechanism the particles grow under any given set of conditions (i.e., temperature, pressure, supersaturation).

In the most general sense, growth takes place in one of two ways: a particle may grow from the vapor phase, meaning that it increases in size molecule by molecule; or, a particle may grow by collecting other particles, in which case the particle grows in size particle by particle. Both categories of growth may operate simultaneously, but we treat them separately to facilitate presentation of the underlying physics. Growth by collection is discussed in the next chapter.

A net amount of water vapor deposits onto a surface whenever the partial pressure of vapor (e) exceeds the equilibrium value (e_{eq}) of the surface. How rapidly the vapor deposits depends on the partial pressure excess ($e - e_{eq}$), the temperature and nature of the surface, and the total air pressure. If the surface is liquid, the equilibrium vapor pressure is that of a solution droplet having the given temperature, solute concentration, and size. If the surface is solid, the equilibrium vapor pressure is that of ice at the temperature of the surface. The excess of vapor partial pressure over the equilibrium value causes water molecules to enter the condensed phase, depleting vapor locally and setting up concentration gradients that drive diffusive transport of additional water vapor toward the surface. We know the process by which vapor enters the condensate either as "condensation", in the case of liquids, or "deposition", in the case of solids, but both processes represent growth of the particle by the accumulation of individual water molecules. Here, we seek the relevant growth law, the differential equation that describes the rate of growth of an individual particle under a specified set of conditions.

Throughout the following discussion, we assume that the instantaneous supersaturation (relative excess vapor pressure) is known, as perhaps calculated from consideration of vapor mass balance in a population of cloud particles within a rising air parcel. This ambient supersaturation is the average value in the interstitial air between the cloud particles. Here, we follow the meteorological convention of using a flat surface of pure liquid water as the idealized reference state and use the symbol s without a subscript to define the ambient supersaturation: $s \equiv (e_\infty - e_s(T_\infty))/e_s(T_\infty) = e_\infty/e_s - 1$, where e_∞ is the ambient, interstitial partial pressure far from the particle and $e_s(T_\infty)$ is the equilibrium vapor pressure of the hypothetical pure, flat water surface at the ambient temperature T_∞. Sometimes, we find it more convenient to use the saturation ratio, $S \equiv e_\infty/e_s(T_\infty) = s + 1$, an alternative way of specifying the relative humidity (in the US tradition).

No cloud particle ever matches the idealized reference state, however, so additional clarification is needed. Growth calculations need to use the actual cloud particle as the true reference, because, as shown in Section 3.1, flux equilibrium for liquids is affected by the solute concentration and curvature of the particle. We thus use the Köhler function $s_K = e_{eq}/e_s - 1$ (refer to Section 3.1) as the reference supersaturation, where e_{eq} is equilibrium vapor pressure of the solution droplet (see Fig. 8.1). The relative driving factor for the vapor growth of a cloud droplet is therefore the supersaturation difference $s - s_K = \left(e_\infty - e_{eq}\right)/e_s$. The difference between the ambient and droplet supersaturations continuously changes as the droplet grows or evaporates. When the particle under consideration is solid, then e_{eq} and e_s are replaced by the equilibrium vapor pressure of ice, e_i, to obtain the supersaturation with respect to ice: $s_i \equiv e_\infty/e_i - 1$.

Whether the condensate is liquid or solid, growth of a particle from the vapor proceeds via three consecutive steps, as summarized in Fig. 8.2. The first step is the transport of vapor from afar (many particle radii away) to the surface. Vapor transport occurs primarily via molecular diffusion, augmented perhaps by the mean flow of air around the

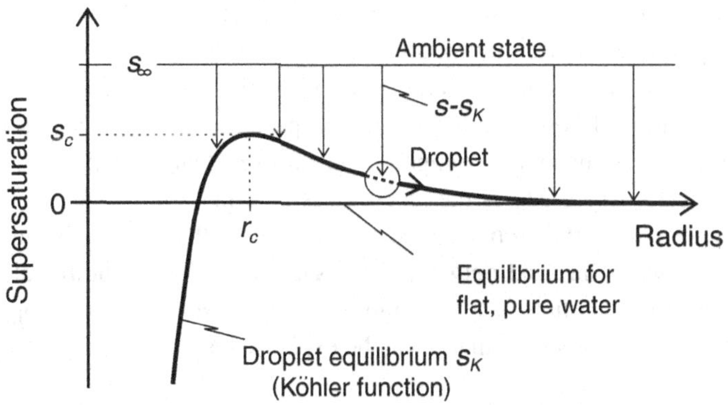

Figure 8.1 A representative Köhler curve, the supersaturation needed to maintain equilibrium with a droplet containing a fixed solute content. The down-pointing arrows show that the driver of growth is the difference between the ambient and equilibrium supersaturations. The maximum equilibrium supersaturation is the critical supersaturation s_c, which occurs at the critical radius r_c.

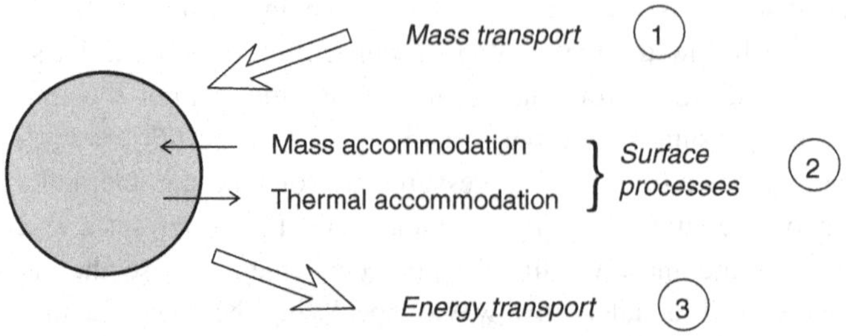

Figure 8.2 Schematic of the three major steps involved in condensation.

particle as it sediments under the influence of gravity. This diffusion of vapor, the random migration of water molecules among the more numerous non-condensing constituents of air, is described by Fick's first law of diffusion, which states that the net flux of molecules is proportional to the concentration gradient. The diffusivity (or coefficient of diffusion) D_v is the constant of proportionality, so the net flux of vapor at any point is $\Phi_\mathbf{v} = -D_v \nabla n$, where $n = e/RT$ is the molar concentration of vapor at temperature T and partial pressure e. (For simplicity, we do not use the subscript v on n, as water vapor is understood.) The second step involves molecular-scale processes on the droplet surface, where the actual phase change (and consequent warming) that contributes to the growth of the particle takes place. The removal of water vapor from the immediate vicinity of the particle establishes gradients in the vapor concentration that drive the diffusion of vapor. The latent heat released as water molecules settle into the liquid warms the particle, which in turn warms the air next to the particle and sets up temperature gradients around the particle. The third step is the transport of the released latent heat away from the particle and into the air between the cloud particles. The transport of thermal energy occurs by the molecular-scale process of conduction, which is described by Fourier's first law of conduction: $\Phi_\mathbf{T} = -k_T \nabla T$, where k_T is the thermal conductivity of the air. Any complete description of growth must account for all three steps simultaneously.

The rate at which mass is added to a particle is, in the case of growth from the vapor phase, the total flow of water molecules to the particle from the interstitial reservoir of excess vapor. Thus, by mass continuity the rate of change of particle mass m_p is $dm_p/dt = -M_w I_v$, where M_w is the molar mass of water and $-I_v$ is the total flow [in mol s^{-1}] of vapor toward the particle. (The negative sign reflects the convention of having flows be positive when directed outward in the same direction as the radial coordinate.) The vapor flow into the particle is nothing more than the integrated flux of vapor across the particle surface, so the mass growth rate can be expressed as

$$\frac{dm_p}{dt} = -M_w \oint_{sfc} \Phi_\mathbf{v.sfc} \cdot d\mathbf{A}, \tag{8.1}$$

where $\Phi_\mathbf{v.sfc}$ is the vapor flux at the surface and $d\mathbf{A}$ is an infinitesimal area of the surface. Equation (8.1) is general and applies to particles of any shape, phase, or composition. The problem with implementing Eq. (8.1) is being able to express $\Phi_{\mathrm{v,sfc}}$ in terms of the ambient conditions and the properties of the particle in question.

8.2 Vapor-growth of individual liquid drops

8.2.1 Maxwell's theory

Small liquid drops are spherical (because of surface tension), so developing a suitable growth law for condensation is relatively straightforward. Let us here follow the three steps of condensation for the simplest case of a single liquid droplet growing in stagnant air at a given temperature and supersaturation. This basic treatment, with air treated as a continuum up to the droplet surface, leads to Maxwell's growth law, named in honor of James C. Maxwell for his theory of the wet-bulb thermometer. Throughout the treatment below, we express the droplet growth rate in a variety of ways; only the last, the so-called mass growth law, is usually used.

Step 1 (Mass transport)

Apply Eq. (8.1) and recognize that integration over the area of a sphere of radius r_d is $4\pi r_d^2$ and that the flux is uniform and everywhere normal to the surface. Thus, the mass growth rate of a droplet, and hence the integrated vapor transport, is

$$\frac{dm_d}{dt} = -4\pi M_w r_d^2 \Phi_{v,sfc}, \tag{8.2}$$

where M_w is the molar mass of water. The vapor flux $\Phi_{v,sfc} = -D_v(dn/dr)_{r_d}$ (by Fick's first law), so

$$\frac{dm_d}{dt} = 4\pi M_w D_v r_d^2 \left.\frac{dn}{dr}\right|_{r_d}, \tag{8.3}$$

where the gradient of vapor concentration is evaluated at the droplet surface (at $r = r_d$). Because of the spherical symmetry, we need consider only the radial coordinate when calculating the spatial distributions of vapor and temperature about a droplet, which are needed for calculating the surface gradients.

Steady-state vapor profile The concentration of vapor surrounding a growing cloud droplet is smaller than that far away simply because the droplet is continually taking up excess vapor. The vapor concentration varies with time, too, but many problems can be solved by ignoring transients in the vapor field. We can most simply calculate the radial distribution of vapor concentration (i.e., the vapor profile) by considering mass continuity within a set of concentric shells centered about the droplet. Let two shells have arbitrary radii r_1 and r_2; the respective areas of the shells are thus $4\pi r_1^2$ and $4\pi r_2^2$. The total flow of vapor through each shell is the area of the shell times the diffusive flux (which is proportional to the local gradient of vapor concentration by Fick's law). In steady state, the total flow through one shell must be the same as that through the other, otherwise mass would accumulate

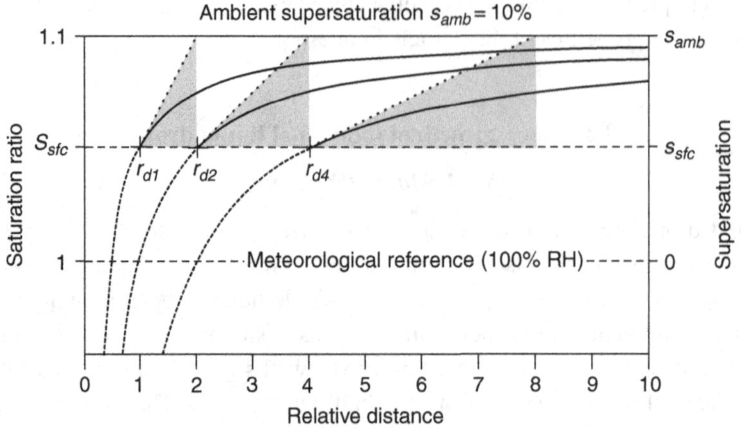

Figure 8.3 Steady-state profiles for three droplet sizes (indicated by crosses). Dashed curves represent mathematical (not physical) extrapolations of the vapor profiles.

in or be lost from the intervening volume. We thus obtain the steady-state rule for mass continuity in spherical coordinates: $r^2 \cdot dn/dr = constant$. Integration of this equation for the boundary conditions $n(r_d) = n_{eq}$ and $n(\infty) = n_\infty$ yields the steady-state vapor profile,

$$n(r) = n_\infty - \left(n_\infty - n_{eq}\right) \frac{r_d}{r}, \tag{8.4}$$

where r_d is the droplet radius and r is the radial coordinate (see Fig. 8.3). The vapor gradient that drives the diffusive flux at any r is $dn/dr = \left(n_\infty - n_{eq}\right) r_d/r^2$, which is small at large r and large at small r. The maximum gradient (shown by dotted lines in Fig. 8.3) occurs at the droplet surface and is obtained by setting $r = r_d$:

$$\left.\frac{dn}{dr}\right|_{r_d} = \frac{n_\infty - n_{eq}}{r_d}. \tag{8.5}$$

The gradient at the surface scales directly with the vapor excess and inversely with the radius of the droplet. This limiting gradient of vapor concentration determines the flux of water molecules at the droplet surface and hence the mass growth rate (from Eq. (8.3)):

$$\frac{dm_d}{dt} = 4\pi r_d M_w D_v (n_\infty - n_{eq}). \tag{8.6}$$

Notice how the r_d^2 dependence arising from surface area is partially compensated for by the $1/r_d$ dependence from the limiting gradient, giving a net linear dependence of the mass growth rate on droplet radius. Equation (8.6) yields the droplet growth rate (hence the total mass transport of vapor) in terms of the vapor concentrations at the outer and inner boundaries of the problem.

Boundary conditions play key roles in the theory of cloud-particle growth. Any process or variable that alters the excess vapor concentration $(n_\infty - n_{eq})$ changes the vapor profile, the vapor gradient at the surface, and hence the flux of vapor into the particle. The outer boundary condition (n_∞) is established by the collective influence that the cloud-particle population has on the depletion of vapor, an issue that was examined earlier. Here, we assume that the interstitial saturation ratio, $S = e_\infty / e_s\,(T_\infty) = s + 1$, is known at any instant in time. Thus, the outer boundary condition is

$$n_\infty = \frac{e_\infty}{RT_\infty} = n_s(T_\infty)S, \tag{8.7}$$

where $n_s(T_\infty) = e_s\,(T_\infty)\,/RT_\infty$ is the equilibrium concentration of vapor at the ambient temperature T_∞. The boundary condition at the surface, by contrast, depends on the nature of the particle.

Step 2 (Surface processes)

The molecular-scale processes occurring at the vapor–liquid interface affect how readily the incident vapor molecules enter the liquid, the magnitude of condensational warming, and how effectively the incident air molecules can extract the excess energy. The complexity and uncertainty of these process are discussed later, but for now, it suffices to recognize that the collective action of the surface processes influences the value of n_{eq} used in Eq. (8.6).

The surface boundary condition (n_{eq}) is highly sensitive to the temperature T_p of the particle because of the effect that temperature has on the equilibrium vapor pressure (described by the Clausius–Clapeyron equation). As with the traditional calculation of wet-bulb temperature, we first assume that the vapor concentration above the surface is that in equilibrium with the liquid. The assumption of equilibrium, made by Maxwell and many others, circumvents the complexities of the surface processes and simplifies the theory. If we let $e_s(T_p)$ be the equilibrium vapor pressure over a flat, pure water surface at the particle temperature T_p, then the equilibrium vapor concentration over such a surface would be $n_s(T_p) = e_s(T_p)/RT_p$. Cloud droplets are neither pure nor flat, however, so we must modify the equilibrium vapor pressure to account for solute and curvature, as provided by Köhler theory. Given the Köhler function expressed in terms of saturation ratio, $S_K = e_{eq}/e_s(T_p)$, we find

$$n_{eq} = \frac{e_{eq}}{RT_p} = n_s(T_p)S_K \tag{8.8}$$

to be the appropriate inner boundary condition (for now).

With expressions for the outer and inner boundary conditions, we can express the growth rate of the droplet in terms of the ambient and equilibrium saturation ratios. Combining Eqs. (8.7) and (8.8) with Eq. (8.6) yields

$$\frac{dm_d}{dt} = 4\pi M_w D_v r_d n_s(T_\infty)\left(S - \frac{n_s(T_p)}{n_s(T_\infty)} S_K\right). \tag{8.9}$$

By factoring out the saturation vapor concentration at the ambient temperature, $n_s(T_\infty)$, we see that the distinction between the ambient and particle temperatures affects only the ratio of saturation vapor concentrations. A ratio of equilibrium concentrations depends only on the temperature difference $(T_p - T_\infty)$, which we see by taking the logarithmic derivative of the ideal gas law for the saturated vapor, $n_s(T) = e_s(T)/RT$, applying the Clausius–Clapeyron equation, and integrating over the temperature range. Then, by making suitable approximations for small temperature differences, we find the ratio to be

$$\frac{n_s(T_p)}{n_s(T_\infty)} \cong 1 + \left(\frac{l_v}{RT_\infty} - 1\right)\left(\frac{T_p - T_\infty}{T_\infty}\right). \tag{8.10}$$

The use of Eq. (8.10) in Eq. (8.9) gives

$$\frac{dm_d}{dt} = 4\pi M_w D_v r_d n_s(T_\infty)\left(S - S_K + \left(\frac{l_v}{RT_\infty} - 1\right) S_K \left(\frac{T_p - T_\infty}{T_\infty}\right)\right). \tag{8.11}$$

The only unknown now is the temperature difference, for which we must consider transport of energy.

Step 3 (Energy transport)

The dissipation of condensational heating starts with the exchange of energy at the surface and continues with the conduction of thermal energy through the air. If the transport of energy is ineffective, the particle temperature and thus vapor pressure will be relatively high, causing the rate of condensation to be low. Condensation is as much dependent on energy transport as it is on vapor transport. The rate of energy transport is directly tied to the difference in temperature between the particle and the environment.

Particle temperature Calculating the particle temperature (or temperature difference) depends on knowing the rate of condensation. As water molecules condense, they bond with other molecules, lower their potential energies, and generate thermal energy that raises the temperature of the condensate and hence n_{eq}. The latent heat of the vapor is transformed into sensible heat of the particle, so condensation itself alters the subsequent rate of condensation; the increased particle temperature diminishes the concentration difference $(n_\infty - n_{eq})$ that drives vapor diffusion.

The heating due to condensation may raise the droplet temperature by only a few tenths of a degree Celsius, but it nevertheless influences n_{eq} and the distribution of vapor appreciably. Figure 8.4 illustrates the effect that latent heating has on the vapor profile of an hypothetical droplet growing in a supersaturated environment. As suggested by the inset plot, the warming raises the equilibrium vapor concentration, forcing the inner boundary condition to increase and the vapor profile to be less steep than if no heating occurred (dashed curve toward upper right of Fig. 8.4). The limiting vapor gradient is reduced with warming, as is the flux of vapor to the droplet. The rise in surface temperature depends on the rate of energy transport away from the particle, which in turn depends on the temperature gradient in the air about the particle.

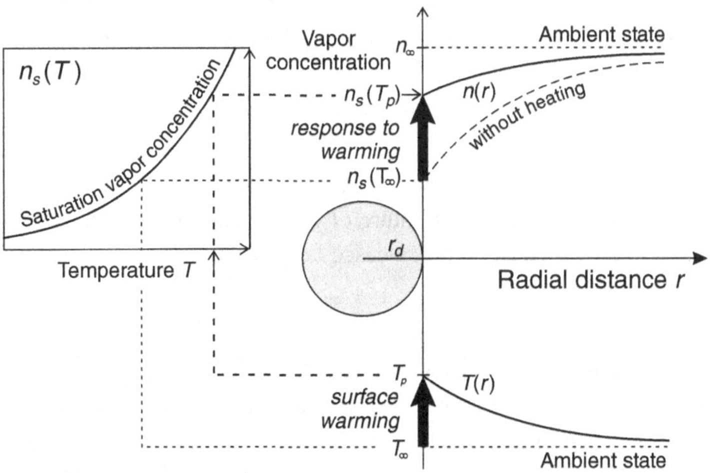

Figure 8.4 Effect of latent heating on the vapor profile of an hypothetical droplet growing by condensation. The inset is a plot of the concentration of vapor in equilibrium with a liquid surface at temperature T, $n_s(T) = e_s(T)/RT$.

The steady-state temperature profile is found, as we did for the vapor profile, by integrating the appropriate conservation relationship ($r^2 \cdot dT/dr = constant$) in spherical coordinates subject to the boundary conditions, $T(r_d) = T_p$ and $T(\infty) = T_\infty$:

$$T(r) = T_\infty - \left(T_\infty - T_p\right) \frac{r_d}{r}. \tag{8.12}$$

The largest temperature gradient, that at the droplet surface, is thus

$$\left.\frac{dT}{dr}\right|_{r_d} = \frac{T_\infty - T_p}{r_d}, \tag{8.13}$$

which establishes the flux of thermal energy away from the droplet by Fourier's law: $\Phi_{T,sfc} = -k_T(dT/dr)_{r_d} = k_T(T_p - T_\infty)/r_d$.

The magnitude of temperature rise ($T_p - T_\infty$) is determined by considering the steady-state balance of energy into and out of the particle. Under steady-state conditions, the flow of latent heat (molecular potential energy) toward the particle must be balanced by the flow of excess thermal energy (molecular kinetic energy) away from the particle. Thus, $l_v I_v + I_q = 0$, or $l_v I_v = -I_q$, where l_v is the enthalpy of vaporization and where I_v and I_q are, respectively, the flows of vapor and thermal energy. The flow of vapor toward the particle is related to the droplet mass growth rate by $I_v = -(1/M_w)dm_p/dt$. In response to the condensational warming of the growing particle, $T_p > T_\infty$, and thermal energy flows away from the particle at the rate (by analogy with the expression for mass transport, Eq. (8.6)):

$$I_q = A_d \cdot \Phi_q = 4\pi r_d k_T \cdot \left(T_p - T_\infty\right). \tag{8.14}$$

Thus, the energy balance leads to $I_q = -l_v I_v = (l_v/M_w)dm_p/dt = 4\pi r_d k_T \cdot (T_p - T_\infty)$, and the increase in particle temperature over the ambient value is

$$T_p - T_\infty = \frac{l_v}{4\pi r_d k_T M_w} \cdot \frac{dm_p}{dt}. \tag{8.15}$$

The greater the rate of condensation, the greater the temperature increase. The important point to bear in mind is that the temperature (T_p) of a growing particle is slightly higher than that of the air (T_∞) because of the released latent heat.

The growth laws – putting it all together

The rate at which water is added to a droplet growing from the vapor has been seen in Step 1 to be directly related to the concentration of excess vapor (Eq. (8.6)). The excess vapor concentration (because of how surface processes affect n_{eq}) was shown in Step 2 to depend strongly on the particle temperature (Eq. (8.8)), which in turn depends on the growth rate (Step 3, Eq. (8.15)). So, the growth rate depends on the temperature increase (from condensational heating), which in turn depends on the growth rate. The separate equations must be combined and a mathematical solution for dm_d/dt found.

The approach to solving the droplet-growth problem is now largely algebraic in nature, as we have already included the relevant physics. Start by substituting the temperature-difference Eq. (8.15) into the cumbersome mass growth rate Eq. (8.11), which yields a still-more cumbersome equation. It helps to recognize that the equation is of the form $dm_d/dt \equiv x = A[B - Cx]$, in which case the solution is $x = AB/(1 + AC)$. Upon replacing the physical variables, one gets the mass growth rate,

$$\frac{dm_d}{dt} = \frac{4\pi r_d M_w D_v n_s(T_\infty)(S - S_K)}{1 + D_v n_s(T_\infty)\left(\frac{l_v}{RT_\infty} - 1\right)\frac{l_v}{k_T T_\infty} S_K}. \tag{8.16}$$

Slight rearrangement of this equation gives the concise and much-used Maxwell's mass growth law:

$$\frac{dm_d}{dt} = 4\pi r_d \rho_L G \cdot (s - s_K), \tag{8.17}$$

where ρ_L is the mass density of liquid water, $(s - s_K)$ is equivalent to $(S - S_K)$, and where G is a slowly varying function of the physical variables:

$$G = \left[\frac{\rho_L R T_\infty}{M_w D_v e_s(T_\infty)} + \frac{\rho_L l_v}{M_w k_T T_\infty}\left(\frac{l_v}{RT_\infty} - 1\right) S_K\right]^{-1}. \tag{8.18}$$

The parameter G has SI units of $m^2\,s^{-1}$, so it acts as a diffusivity in Eq. (8.17), one that takes not only the diffusion of vapor into account, but the effect of latent heating as well. The larger the diffusivity, the more rapidly can vapor be transported to the droplet for a given amount of excess vapor (as reflected in the value of $(s - s_K)$). Alternatively, we can view diffusion as a resistance to mass transport and conduction as a resistance to energy transport, in which case we may write the mass growth law in the form of impedances:

$$\frac{dm_d}{dt} = 4\pi\rho_L r_d \cdot \frac{s - s_K}{Z_{diff} + Z_{cond}}, \tag{8.19}$$

where $Z_{diff} \equiv \rho_L R T_\infty/(M_w D_v e_s(T_\infty))$ is the mass-transfer impedance and $Z_{cond} \equiv (\rho_L l_v/M_w k_T T_\infty)\left(\frac{l_v}{RT_\infty} - 1\right) S_K$ is the heat-transfer impedance. The form of the mass growth law given in Eq. (8.19) is analogous to Ohm's law ($I = \Delta V/R$) in electrical circuitry; the driver for vapor flow is the net supersaturation, $s - s_K$, and the total resistance to that flow is proportional to the sum of impedances, $Z_{diff} + Z_{cond}$. Being able to calculate the rate at which individual droplets acquire mass by condensation is the essential first step in being able to calculate the depletion of vapor from the interstitial air inside a cloud and the evolution of a droplet size distribution.

The rate at which the size (radius or diameter) of a droplet grows follows from the same treatment of the physics. We need only recognize that the mass of a droplet of known mass density (ρ_L) is $m_d = \frac{4}{3}\pi r_d^3 \rho_L$. It follows then that the rate of change of that mass is directly related to the change in radius: $dm_d/dt = 4\pi\rho_L r_d^2 (dr_d/dt)$. Upon equating this equation and the expression given in Eq. (8.17), we obtain the "linear growth law":

$$r_d \frac{dr_d}{dt} = G \cdot (s - s_K). \tag{8.20}$$

In this form, condensation is seen to favor the smaller droplets, which thus leads to a narrowing of droplet size distributions. Total mass, however, is acquired most rapidly by the larger droplets, those having the larger surface area. Integration of Eq. (8.20) for a given net supersaturation shows that r_d increases in rough proportion to $t^{1/2}$. Condensational growth is slow compared to other mechanisms, but it is nevertheless a necessary process in virtually every cloud.

8.2.2 Refinements to the theory of condensation

The basic theory of droplet growth developed initially by Maxwell is based on several assumptions that are not strictly valid. One may apply the results derived thus far (e.g., Eq. (8.17)) with little error under some conditions, but one must know what those conditions are. In general, we must be aware of the assumptions underlying the mathematical development. For instance, the analytic solution arrived at above was made possible by assuming that the difference in temperature between the droplet and the ambient air is small. This assumption may be valid during the early stages of growth, when many droplets are growing and competing for the same excess vapor supply, but it is not likely to be valid for large (i.e., rain) drops falling into dry air (where the temperature differences can be large and constantly changing). More general solutions to the growth and evaporation of water drops are best approached by solving the mass and energy transport equations numerically. Maxwell's theory also assumes that the air is stagnant and that mass and energy are therefore transported exclusively by molecular processes (diffusion and conduction). However, motion of a drop through the air introduces a ventilation effect that should be applied to falling drops. In addition, Maxwell's theory assumes that the partial pressure of vapor at the droplet surface is the equilibrium value; growth in reality depends on slight differences between the near-surface partial pressure and the equilibrium vapor pressure of the liquid, so the molecular processes at the surface need to be considered. The basic theory also assumes that the vapor and temperature fields are constant in time about a fixed interface between the liquid and gas phases; transient effects should be considered in general and applied in a few situations (i.e., in turbulent air). Some of the refinements needed to complete the theory of condensation are discussed below.

Effects of ventilation

The condensed states of water are denser than the air in which they reside, so both droplets and ice particles sediment, that is, fall relative to the local air motions. A growing (or evaporating) particle is therefore ventilated to an extent that depends on its fallspeed (treated in detail in the next chapter). The motion of the particle through the air distorts the vapor and temperature fields surrounding it, causing the gradients to steepen on the upwind (lower) side and weaken on the downwind (upper) side. Vapor and heat are transported more rapidly where the gradients have become stronger, but less so on the other side of the particle. How much these field distortions affect the growth rate depends on how large the particle is, for fallspeed typically increases with particle size. The distortions are relatively minor for small particles because of their small terminal fallspeeds, but the distortions can be appreciable once the particles grow beyond about 60 μm in radius. The effects of ventilation are especially important on the evaporation of raindrops falling into the subsaturated air below cloud base.

The effect of ventilation on the vapor growth or evaporation of falling particles is accounted for in calculations by making suitable adjustments to Maxwell's theory. In principle, the effect of ventilation on the transport of vapor toward a growing particle, for instance, differs slightly from that on the transport of heat away from the particle. However, the distinction is slight, so one is usually satisfied with using an overall ventilation coefficient $\overline{f_{vent}}$, defined as the ratio of mass-growth rates with and without ventilation. The true, ventilated mass-growth law is thus calculated from that obtained earlier (Eq. (8.17)) under the assumption of growth in stagnant air as

$$\left.\frac{dm_d}{dt}\right|_{ventilated} = \overline{f_{vent}} \cdot \left.\frac{dm_d}{dt}\right|_{stagnant}. \tag{8.21}$$

Thus, one often sees the mass-growth law written as

$$\frac{dm_d}{dt} = 4\pi r_d \rho_L \overline{f_{vent}} G \cdot (s - s_K). \tag{8.22}$$

The ventilation coefficient has been found, through comparisons of experimental data with the theory of air flows near surfaces, to have the following functional forms:

$$\overline{f_{vent}} = 1 + 0.09N_{Re}, \qquad \text{for } N_{Re} < 2.5$$
$$= 0.78 + 0.28N_{Re}^{1/2}, \quad \text{for } N_{Re} > 2.5, \qquad (8.23)$$

where $N_{Re} \equiv 2r_d v_p / \nu_{air}$ is the Reynolds number (see next chapter) of a particle having radius r_d and falling with speed v_p through air with kinematic viscosity ν_{air}. Once the fallspeed is known as a function of radius (next chapter), N_{Re} can be calculated for any given atmospheric condition (which determines ν_{air}) and $\overline{f_{vent}}$ evaluated. The ventilation coefficient is shown as a function of particle radius in Fig. 8.5 for an altitude of about 3 km. Below a radius of about 60 μm (corresponding to the threshold $N_{Re} = 2.5$), $\overline{f_{vent}}$ varies little between 1 and 1.23. Above this threshold, however, $\overline{f_{vent}}$ increases steadily with particle size. Once the radius exceeds about 1 mm, $\overline{f_{vent}} > 8$. Ventilation effects are often ignored for the vapor growth of cloud drops, but clearly large errors would arise if Maxwell's theory were applied to the evaporation of raindrops without considering ventilation.

Effects of surface processes

The mechanisms by which water and air molecules interact with the surface of a particle are complicated and largely unknown. Still, we must account for these processes, especially because mass and energy may not be transferred between phases with 100% efficiency. For instance, some water molecules may impinge onto the surface, stay for a while in the adsorbed state, then desorb before having a chance to enter the bulk liquid. Similarly, an impinging N_2 molecule may not necessarily come to thermal equilibrium with the surface before desorbing and reentering the gas phase with more or less its original energy, thereby contributing little to the energy transfer. The effects of restricted exchanges of mass and energy with the droplet surface must be accounted for even if we do not fully understand the processes.

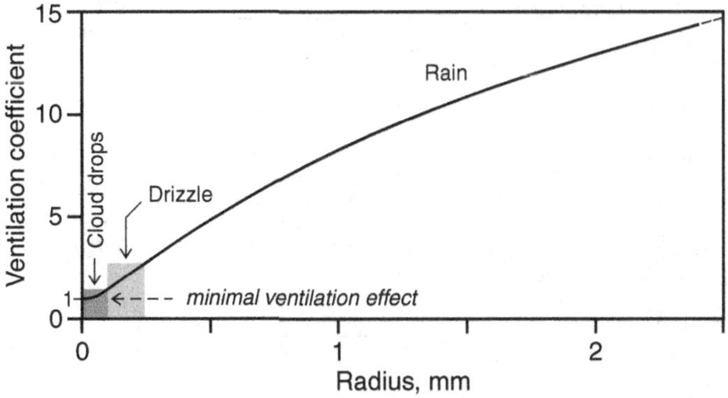

Figure 8.5 Ventilation coefficient calculated as a function of radius for a water drop falling with terminal velocity at a pressure of 700 hPa and a temperature of 0°C. Adapted from Young (1993, p. 125).

The surface processes that account for the exchanges of mass and energy across the liquid–gas interface are best represented in a way that permits the molecular-kinetic effects to be reflected easily in the theory and simultaneously allows them to be interpreted physically. The limited incorporation of impinging vapor molecules into the liquid is typically represented by a mass accommodation coefficient α_m, whereas the exchange of thermal energy is represented by a thermal accommodation coefficient α_T. Each of these kinetic coefficients range in value from 0 to 1.

The mass accommodation coefficient (α_m), also known as the condensation coefficient, is defined as the fraction of impinging water molecules that actually enter the liquid phase and contribute to the growth of the droplet: $\alpha_m \equiv \Phi_{inc}/\Phi_{imp}$, where Φ_{inc} is the incorporation flux and Φ_{imp} is the impingement flux. Note that α_m can be viewed as a measure of the efficiency with which vapor molecules condense. As suggested by Fig. 8.6, the incorporation flux is typically less (possibly much less) than the impingement flux. In contrast to the idealized view of molecular exchange portrayed in Section 3.1, the true condensation flux is

$$\Phi_{cond} = \Phi_{inc} = \alpha_m \Phi_{imp} = \frac{\alpha_m}{4} n\bar{c}_v = \frac{\alpha_m e}{(2\pi M_w RT)^{1/2}}, \tag{8.24}$$

where $n = e/RT$ is the molar concentration of water vapor and $\bar{c}_v = (8RT/\pi M_w)^{1/2}$ is the mean molecular speed at temperature T. A similar expression gives the equilibrium flux Φ_{eq}, so the *net* condensation flux is

$$\Phi_{net} = \alpha_m(\Phi_{imp} - \Phi_{eq}) = \frac{\alpha_m}{4}\bar{c}_v \delta n_i = \frac{\alpha_m(e - e_{eq})}{(2\pi M_w RT)^{1/2}}. \tag{8.25}$$

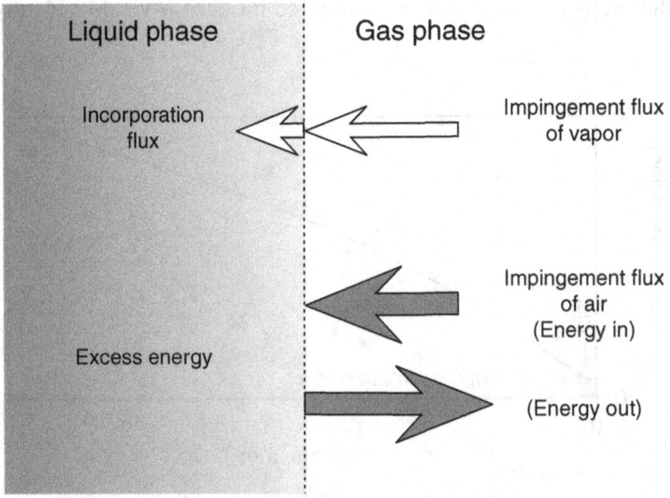

Figure 8.6 Schematic of the effects of kinetic coefficients on the mass and energy fluxes during net condensation of vapor onto a water surface.

Here, we note that α_m represents the kinetic coefficients for both condensation and evaporation, and we see that the net flux of water into the droplet is directly proportional to the product of α_m and the interfacial excess vapor concentration $\delta n_i \equiv n_i - n_{eq}$. (The subscript i refers to interfacial values, those in the vapor phase immediately over the surface.) The interfacial values n_i and δn_i will be discussed at length shortly, but we can already see (using Eq. (8.25)) how the growth rate is related to these values:

$$\frac{dm_d}{dt} = M_w \Phi_{net} A_d = \alpha_m \pi M_w \bar{c}_v r_d^2 \delta n_i. \tag{8.26}$$

The thermal accommodation coefficient (α_T) represents the efficiency with which the non-condensible molecules in air (mainly N_2 and O_2) can extract excess thermal energy from the droplet (as during condensation; see Fig. 8.6). The definition of α_T involves fluxes of energy E, but temperature T is used in practice (permitted because $E \propto T$): $\alpha_T \equiv (E_{out} - E_{in})/(E_{sfc} - E_{in}) = (T_{out} - T_{in})/(T_{sfc} - T_{in})$. Here, the subscript out refers to molecules leaving the surface, whereas in refers to the incoming molecules. We can in principle calculate the net energy exchanged as $E_{out} - E_{in} = \alpha_T(E_{sfc} - E_{in})$, but if we know the temperatures of the surface (sfc) and of the incoming molecules, then the temperature difference across the liquid–gas interface is calculated as $T_{out} - T_{in} = \alpha_T(T_{sfc} - T_{in})$. Note that if the air molecules were to adsorb and equilibrate completely with the surface temperature, they would leave with the surface temperature (i.e., $T_{out} = T_{sfc}$), in which case $\alpha_T = 1$.

Making use of the kinetic coefficients α_m and α_T in the theory assumes knowledge of their values. Someday, an adequate understanding of molecular interactions may allow for their calculation from basic principles, but for now we must rely on precise measurements of droplet growth or evaporation under well-defined laboratory conditions. Experimental data have little meaning in themselves, as they require a theory of some sort for interpretation. The process may seem circular, but actually it is just a matter of self-consistency: one determines the magnitude of the parameter (α_m or α_T) by using an equation from theory that contains the parameter sought and then fitting the calculated curve to the data. The value of the parameter that lets the theory most closely (in a least-squares sense) describe the droplet growth/evaporation data is the value used. A complication occurs in the case of water, however, in that a single set of data is insufficient to determine the two kinetic parameters independently. Various methods have been used to circumvent this difficulty and obtain independent values of α_m and α_T. Figure 8.7 gives one example of the empirical determination of α_m (horizontal axis) and α_T (vertical axis). Individual droplets of pure water were levitated electrodynamically in an environment of known temperature (near −35 °C) and known undersaturation while the droplet size and phase were determined empirically with time. The measured evaporation rate could be explained by any combination of values of α_m and α_T following the solid curve; the freezing data could similarly be explained by combinations following the dashed curve. Only one combination of α_m and α_T (that at the intersection of the two curves) describes both data sets simultaneously. This empirical method yields (with uncertainties indicated by

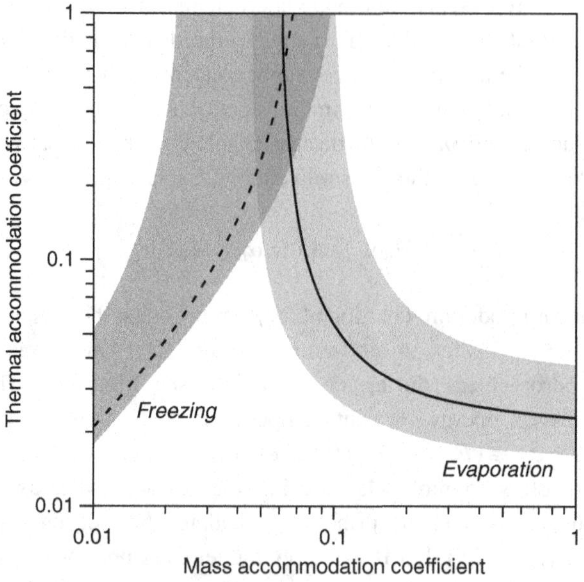

Figure 8.7 Plots of the kinetic coefficients that best fit laboratory data for individual droplets levitated in subsaturated air at −35 °C. Solid curve pertains to evaporation; dashed curve arises from the freezing of the same droplets. The best estimate of the mass and thermal accommodation coefficients is determined by the crossing of the curves. Shaded regions denote the experimental uncertainties. Reprinted with permission from Shaw and Lamb (1999), J. Chem. Phys., vol. **111**, 10659–10663. Copyright 1999, American Institute of Physics.

the overlapping shaded region) the best-estimate values $\alpha_m = 0.06$ and $\alpha_T = 0.7$. As found in other studies, the mass accommodation coefficient tends to be small (just a few percent), whereas the thermal accommodation coefficient is relatively large. We envision a mechanism whereby water molecules have difficulty entering the bulk liquid, while air molecules adsorb onto the surface and efficiently extract the excess enthalpy arising from condensation.

With attention still focused on the droplet surface and the gas phase just above it, we realize (through Eq. (8.25)) that the droplet grows only when $e > e_{eq}$. Maxwell's assumption of equilibrium at the liquid–vapor interface, if taken literally, would permit neither growth nor evaporation to occur. The partial pressure of vapor over the surface must exceed the equilibrium value by at least a small amount for the droplet to gain mass and increase in size. This imbalance in the molecular fluxes across the surface implies the existence of a discontinuity in the vapor field and a breakdown in the continuum nature of the air surrounding the droplet. We can visualize this geometry with the aid of Fig. 8.8, which shows how the profiles shown in Fig. 8.3 should be modified to reflect the realities of the molecular world. The space around each droplet is conceptually divided into two radially distinct parts, a free-molecular region (designated by Δ) and the usual quasi-continuum region

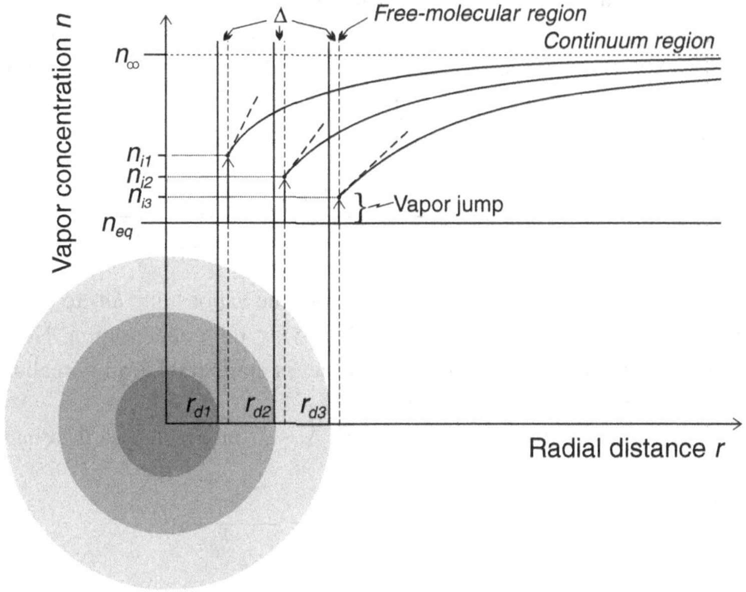

Figure 8.8 Steady-state profiles of vapor concentration outward from the free-molecular region around growing droplets of differing radius.

beyond. The free-molecular, or jump, distance Δ is in the order of the mean free path in air, $\lambda = 1/(\sqrt{2}\pi\sigma^2 n_{air})$ (approximately 0.06 μm at STP). Controversy exists over the relationship between Δ and λ, so we use the equation $\Delta = c\lambda$ to show the expected proportionality. The proportionality constant c is often taken to be the Cunningham correction for molecular slip at interfaces, in which case $c = 0.7$. In any case, the jump distance varies inversely with the density of air, which means that it increases with altitude in the atmosphere.

The magnitude of the vapor jump, $\delta n_i = n_i - n_{eq}$, can be seen in Fig. 8.8 to decrease as the droplet grows. We have already seen that the limiting vapor gradient (angled dashed lines in Fig. 8.8) gets smaller as the droplet gets larger, so the flux of vapor in the continuum region toward the droplet decreases with increasing radius. To counteract this decreasing flux, the vapor field adjusts to decrease the magnitude of the vapor jump a commensurate amount as the droplet grows. (The adjustment occurs automatically by the continual removal of vapor from the free-molecular region.) In each case, it is the vapor jump, the small deviation from equilibrium, that distinguishes the kinetically correct treatment of droplet growth from Maxwell's approach, which assumed equilibrium to prevail at the droplet surface.

We can develop the mathematical relationships between the vapor field and the vapor jump by equating the net flows of vapor in the two regions. The net flow across the free-molecular region and into the droplet is $\left|I_{fm}\right| = A_d \Phi_{net}$, whereas that entering the free-molecular region from the continuum region is $|I_{cont}| = A_i \Phi_{diff}$, where

$A_i = 4\pi(r_d + \Delta)^2$ is the area of the virtual surface at a distance Δ from the surface, and where the flux from diffusion in the continuum region is $\Phi_{diff} = -D_v(dn/dr)_{r_d+\Delta}$. Equating the two flows ($I_{fm} = I_{cont}$) and calculating the limiting gradient as $(dn/dr)_{r_d+\Delta} = (n_\infty - n_i)/(r_d + \Delta)$ lets us quantify the vapor jump in terms of the properties of the ambient air and of the droplet:

$$\delta n_i = n_i - n_{eq} = \frac{\frac{r_d+\Delta}{r_d}(n_\infty - n_{eq})}{\frac{r_d+\Delta}{r_d} + \frac{\alpha_m \bar{c}_v r_d}{4D_v}} = \frac{(n_\infty - n_{eq})}{1 + \frac{\alpha_m \bar{c}_v}{4D_v}\frac{r_d^2}{r_d+\Delta}}. \tag{8.27}$$

The transport limitations imposed by diffusion cause the vapor jump δn_i to be just a fraction of the concentration difference, $n_\infty - n_{eq}$, driving the condensation. Note that the magnitude of Δ is important only when the droplet is comparable to or smaller than the mean free path λ.

For most applications in cloud physics, $r_d >> \Delta \sim \lambda$, in which case the concentration ratio is

$$\frac{\delta n_i}{n_\infty - n_{eq}} \simeq \frac{1}{1 + \frac{\alpha_m \bar{c}_v r_d}{4D_v}} = \frac{1}{1 + \frac{R_{gas}}{R_{sfc}}}, \tag{8.28}$$

where $R_{gas} \equiv r_d/D_v$ and $R_{sfc} \equiv 4/(\alpha_m \bar{c}_v)$ are the resistances (reciprocals of effective speeds) due to gas transport and to molecular incorporation at the surface, respectively. The form of Eq. (8.28) is identical to that which one calculates for the ratio of voltages $((V_i - V_{sfc})/(V_\infty - V_{sfc}))$ across two electrical resistances in series, as shown in Fig. 8.9. We can therefore view the transport and uptake of water vapor by a droplet in analogous fashion, as restricted by the sum of two resistances in series. Condensation of water is indeed sequential in nature, as the molecules must first get to the surface of a droplet, then those molecules must fight through the surface processes to become part of the liquid phase. The transport of water molecules through the air and across the droplet surface is

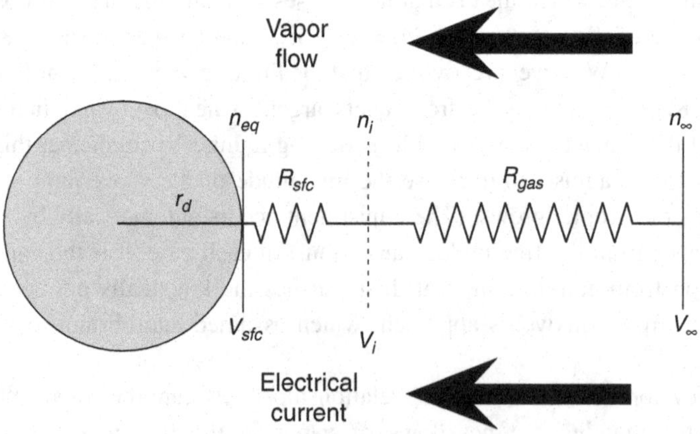

Figure 8.9 Schematic illustrating the analogy between vapor transport and electrical current. Sawtooth symbols represent resistances to the flow of vapor/electrical charge.

restricted by two separate processes, just as the transport of electrons is restricted by passage through two electrical resistances in series. The mathematics in both situations are identical.

Two limiting cases can be defined depending on the relative magnitudes of the transport resistances. When the surface processes are inefficient (large R_{sfc} because of small α_m), they impose the primary limitation to growth (i.e., $R_{sfc} >> R_{gas}$) and the second term in the denominators of Eqs. (8.27) and (8.28) can be neglected, yielding $\delta n_i \approx n_\infty - n_{eq}$. In this kinetically limited case, diffusion is relatively fast and allows the ambient concentration of vapor to extend all the way to the interface. By contrast, when diffusion poses the dominant limit to growth (which tends to occur at large droplet sizes), then $R_{gas} >> R_{sfc}$ and $\delta n_i \approx 4D_v(n_\infty - n_{eq})/(\alpha_m \bar{c}_v r_d)$. In this diffusion-limited case, the vapor jump varies inversely with the accommodation coefficient: $\delta n_i \propto 1/\alpha_m$. This inverse relationship between the excess vapor concentration over the surface and the molecular efficiency of growth proves important for understanding the growth of ice and snow from the vapor.

The mass growth rate can now be calculated in a way that takes the surface kinetics into account. We already have an equation for mass growth in terms of δn_i (Eq. (8.26)), so we need only combine that equation with Eq. (8.27):

$$\frac{dm_d}{dt} = 4\pi r_d M_w D_v' (n_\infty - n_{eq}), \tag{8.29}$$

where

$$D_v' \equiv \frac{D_v}{\frac{r_d}{r_d+\Delta} + \frac{4D_v}{\alpha_m \bar{c}_v r_d}} \tag{8.30}$$

is termed the modified diffusion coefficient because it adjusts the physical diffusion coefficient (D_v) to account for mean free path effects and restrictive surface processes. Note that Eq. (8.29) is identical in form to Eq. (8.6), which was developed earlier in connection with Maxwell's theory. Occasionally, one combines the molar mass of water (M_w) with the molar concentration of vapor (n) to obtain the mass density of water vapor, $\rho = M_w n$, thus gaining a succinct form, $dm_d/dt = 4\pi r_d D_v' (\rho_\infty - \rho_{eq})$.

The problem with the mass growth rate expressed here is that it does not yet account for the warming due to the release of latent heat during condensation. Fortunately, the same three-step procedure used to develop the Maxwellian growth law can be used here. One first expresses the outer and inner boundary conditions in terms of the ambient and equilibrium saturation ratios (Eqs. (8.7) and (8.8)), respectively. Next, the temperature rise and its dependence on the mass growth rate are determined (Eq. (8.15)) from consideration of the energy balance. The kinetic limitations to energy exchange, represented by the thermal accommodation coefficient α_T, are accounted for in much the same way that the limitation to mass accommodation is accounted for. As shown in Fig. 8.10, a growing droplet is warmer than its surroundings, so thermal energy (heat) first passes through the

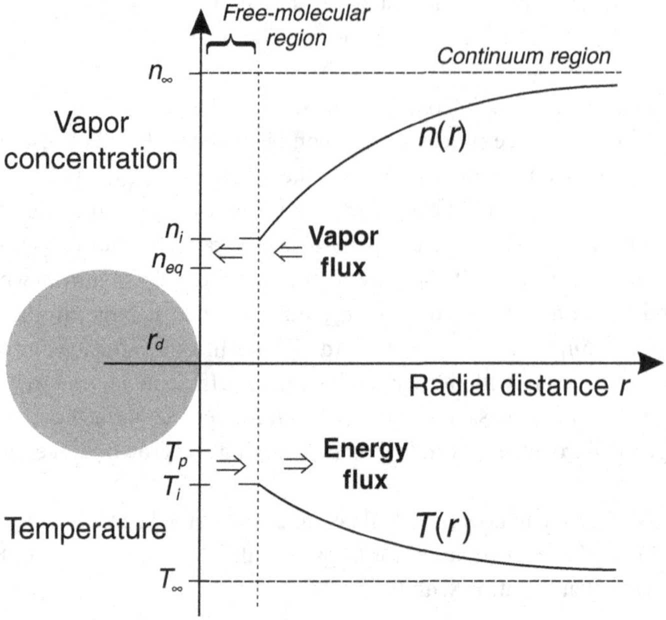

Figure 8.10 Schematic showing the profiles of vapor concentration and temperature influenced by surface processes.

free-molecular region at the rate $I_{q,fm} = A_d \Phi_q$, where the heat flux is the impingement flux of air times the net energy carried by those air molecules:

$$\Phi_q = \left(\frac{1}{4}\bar{c}_{air} n_{air}\right)(c_{p,air}\alpha_T(T_p - T_i)) = \frac{\alpha_T}{4}\bar{c}_{air} n_{air} c_{p,air}(T_p - T_i). \tag{8.31}$$

Here, n_{air} is the molar concentration of air, $c_{p,air}$ is the molar heat capacity of air at constant pressure, T_p is the particle (surface) temperature, and T_i is the air temperature a distance Δ from the surface. Once the energy flow across the free-molecular region ($I_{q,fm}$) is balanced by the energy leaving the interface and entering the continuum region ($I_{q,cont} = A_i k_T (T_i - T_\infty)/(r_d + \Delta)$), one can solve for the interfacial temperature (T_i) and then calculate the temperature jump $\delta T = T_p - T_i$. One finds an equation identical in form to Eq. (8.15), but with

$$k_T' = \frac{k_T}{\frac{r_d}{r_d+\Delta} + \frac{4k_T}{\alpha_T \bar{c}_{air} n_{air} c_{p,air} r_d}} \tag{8.32}$$

replacing k_T. It helps to remember that energy is transferred via air molecules, unlike the transport of mass, which is based on water vapor. In principle, the magnitude of the jump distance (Δ) for vapor transport may differ from that for energy transport, but the distinction is small and ignored here.

The final mass growth rate is obtained, as before, by solving the resulting algebraic equation for dm_d/dt and rearranging terms. One thus gains the steady-state, kinetically corrected mass growth law:

$$\frac{dm_d}{dt} = 4\pi r_d \rho_L G'(s - s_K), \tag{8.33}$$

where

$$G' = \left[\frac{\rho_L R T_\infty}{M_w D_v' e_s(T_\infty)} + \frac{\rho_L l_v}{M_w k_T' T_\infty}\left(\frac{l_v}{RT_\infty} - 1\right) S_K\right]^{-1} \tag{8.34}$$

has the same form as Eq. (8.18) except for the primes on G, D_v, and k_T. The corresponding linear growth law also takes on the Maxwellian form (Eq. (8.20)); just use G' instead of G.

Evaluation of the theoretical growth laws lets us appreciate the evolution of droplets in real clouds. Using realistic values for the parameters and variables, one can integrate Eq. (8.33) and plot the droplet mass as a function of time, as shown in Fig. 8.11. The uppermost curve shows how a droplet would grow at a supersaturation of 1% if the assumptions behind Maxwell's theory were valid. The effect that the surface processes have on the growth, shown by the other curves, is always to lower the mass of the droplet at any given time t. Notice that the surface processes have a large influence on the mass growth when α_m is small, but they have little effect when large. This non-linear behavior is understandable by considering the analogy with electrical circuitry, discussed earlier in connection with Eq. (8.28) and Fig. 8.9. When α_m is large, R_{sfc} is relatively small relative to R_{gas}, and the rate of condensation is restricted mainly by diffusion of vapor through the air.

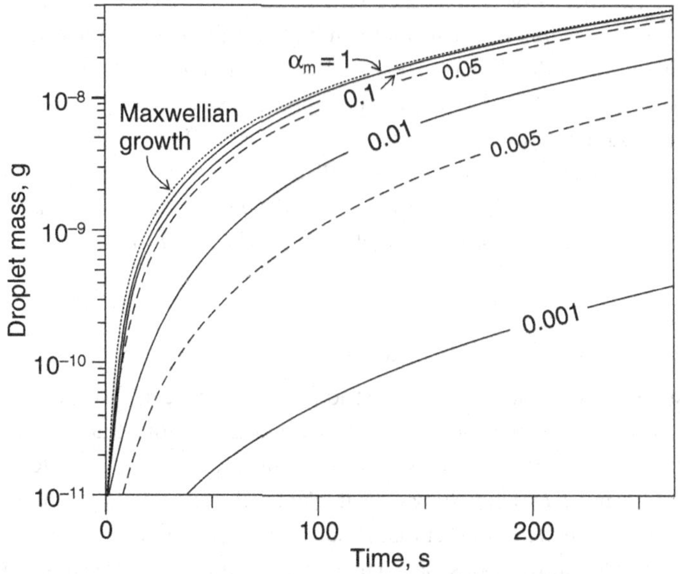

Figure 8.11 Integrated mass of droplets during growth by condensation with different mass accommodation coefficients α_m. Adapted from Fukuta and Walter (1970).

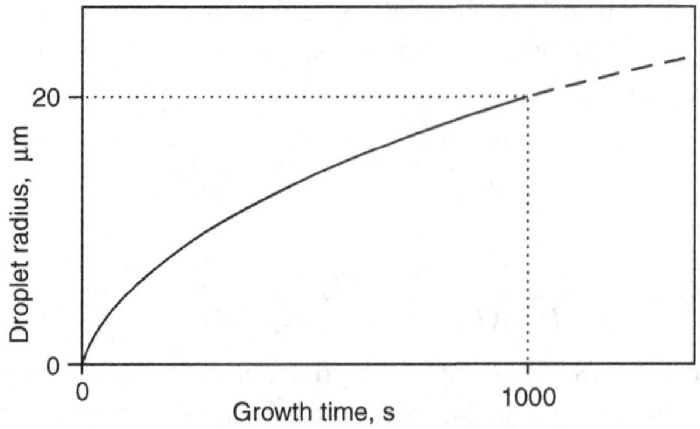

Figure 8.12 Dependence of droplet radius on time during condensational growth, showing square-root dependence on time.

The equation for the linear growth rate of a droplet (Eq. (8.20), modified for surface-kinetic effects) can also be integrated to determine how the radius increases with time. At this point, however, it suffices to simplify the situation and see the general properties of the solutions. Consider a droplet that has grown well beyond its critical radius; the value of s_K is therefore close to zero, and the initial radius taken for integration may be neglected. Furthermore, the ambient supersaturation then varies relatively slowly, as do the parameters constituting G'. The right-hand side of Eq. (8.20) is then approximately constant and the integration yields a solution for the radius that increases approximately as the square root of time: $r_d \sim \sqrt{t}$. This behavior is illustrated schematically in Fig. 8.12, which also gives the approximate limiting radius (20 μm) that can be expected from condensation within the typical lifetime of cloud droplets ($\sim$ 1000 s). We need to appreciate that condensation by itself is normally slow and unable to yield precipitation-size drops ($r_d \sim$ 1 mm). The collective mass of liquid cloud water formed by condensation is nevertheless very important for the subsequent evolution of the drop size distribution. The tendency of small droplets in a population to grow in radius faster than large droplets leads to an initial narrowing of the size distribution; the small droplets catch up to the larger droplets, but they do not pass them in size.

Transient behavior

When the assumption of steady state is not strictly valid, transient effects need to be treated. The gradients in vapor concentration are established in the first place around a water droplet because of the continual removal of water vapor by a condensing droplet. How does a steady-state profile arise, and how long does it take for the vapor field to achieve this steady state? Imagine a hypothetical droplet that is suddenly injected into moist air having properties that are initially uniform in space. The concentration n of vapor in any infinitesimal volume element changes only when the fluxes into and out of the volume do not balance. This principle of mass conservation is expressed by Fick's second law of diffusion:

$$\frac{\partial n}{\partial t} = -\nabla \cdot \Phi_{\mathbf{v}} = D_v \nabla^2 n, \tag{8.35}$$

where $\Phi_{\mathbf{v}} = -D_v \nabla n$ comes from Fick's first law and where the diffusivity (D_v) is assumed to be independent of position. The mathematical analysis is simplified when the geometry is spherically symmetric, for then the variables depend only on the radial coordinate r. The concentration of vapor thus varies in space and time as $n = n(r, t)$. The radial component of the Laplacian is $\nabla_r^2 = \partial^2/\partial r^2 + (2/r)(\partial/\partial r)$, so we need to solve

$$\frac{\partial n}{\partial t} = D_v \left(\frac{\partial^2 n}{\partial r^2} + \frac{2}{r}\frac{\partial n}{\partial r} \right) \tag{8.36}$$

subject to the following initial and boundary conditions for a droplet of radius r_d:

$$\begin{aligned} n(r, 0) &= n_\infty\,, \quad r > r_d \\ n(r_d, t) &= n_{eq}\,, \quad t > 0 \\ n(\infty, t) &= n_\infty\,, \quad t > 0. \end{aligned} \tag{8.37}$$

The mathematical solution of Eq. (8.36) involves the complementary error function and takes the form

$$n(r, t) = n_\infty - \left(n_\infty - n_{eq}\right)\frac{r_d}{r} erfc\,(X_v)\,, \tag{8.38}$$

where $erfc\,(X_v) = 1 - erf\,(X_v)$, $erf\,(X_v) = \frac{2}{\sqrt{\pi}}\int_0^{X_v} e^{-\xi^2} d\xi$, and the argument $X_v \equiv (r - r_d)/\left(2\sqrt{D_v t}\right)$. Plots of $erf\,(X_v)$ and $erfc\,(X_v)$ are shown in Fig. 8.13. Note the opposing dependences of these functions on X_v and the fact that $erf\,(X_v)$ and $erfc\,(X_v)$ have equal values of about 0.5 when $X_v = 0.5$.

The general behavior of the solution is readily seen, independent of specific boundary conditions, in the non-dimensional form

$$Y_v \equiv \frac{n(r, t) - n_{eq}}{n_\infty - n_{eq}} = 1 - \frac{r_d}{r} erfc\,(X_v)\,. \tag{8.39}$$

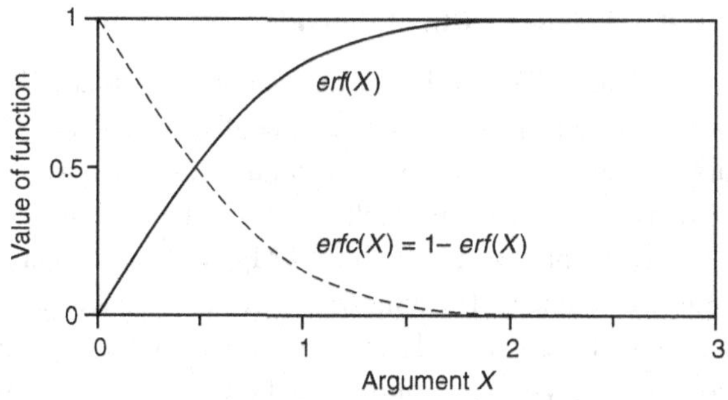

Figure 8.13 Dependence of the error function and the complementary error function on the argument X.

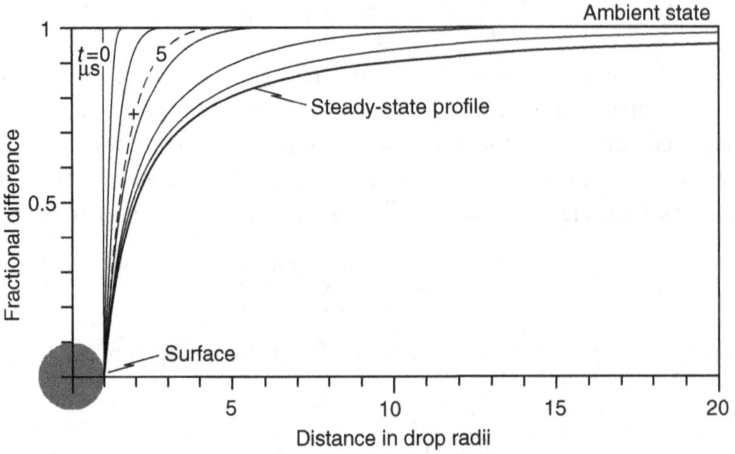

Figure 8.14 Profiles of vapor concentration about a water droplet in non-dimensional coordinates. Time is in µs. Lowest curve: the steady-state profile. Dashed curve: the "half-relaxed" profile.

All of the transient behavior is contained in $erfc\,(X_v)$, which gradually relaxes toward unity as $t \to \infty$ and $X_v \to 0$. Figure 8.14 shows a set of non-dimensional vapor profiles for selected times in non-dimensional coordinates. We see that the vapor concentration close to the droplet changes rapidly with time because of the strong gradients established by the condensation of vapor onto the droplet. The vapor field at a distance of one radius away from the droplet surface ($r/r_d = 2$) will have changed halfway to the final state when $erfc\,(X_v) = 1/2$, so we use $X_v = 1/2$ as the criterion for determining a relevant time scale for the vapor field to relax to its steady-state distribution: $\tau_{1/2} = r_d^2/D_v$. For a 10-µm droplet $\tau_{1/2} \doteq 5\,\mu\text{s}$, and the vapor concentration at $r/r_d = 2$ is midway between n_∞ and n_{eq}. This "half-relaxed" vapor profile is shown as the dashed curve in Fig. 8.14, the midway point as the cross. Because of the rapid adjustment of the vapor field, we often ignore the transients and use only the steady-state profile for the growth of cloud droplets.

8.3 Vapor-growth of individual ice crystals

The growth of ice crystals in a cloud is similar in many ways to that of liquid droplets, but distinctions exist that must be dealt with. For instance, the mass density of solid water, as we may remind ourselves every time we drink from a glass of ice water, is less than that of liquid water. The density difference is small (only about 8%), but it must be taken into account in the growth theory; thus, we use $\rho_I = 0.92\,\text{Mg m}^{-3}$ rather than $\rho_L = 1.0\,\text{Mg m}^{-3}$ in the growth law (Eqs. (8.17) and (8.18)). Snow and ice particles tend to be relatively clean and large, so we can ignore the solute and curvature effects on the equilibrium vapor pressure of ice. The theory of ice crystal growth is simplified by not having to deal with Köhler theory; we merely set $s_K = 0$ in Eq. (8.17) and $S_K = 1$ in Eq. (8.18). A significant distinction, however, is the relatively low equilibrium vapor pressure of ice compared with that of liquid water. The hydrogen bonds in a solid are more orderly and complete (hence stronger on average) than those in a liquid, so water molecules in ice are unable to escape as readily as those in the liquid at any given temperature. We therefore need to replace the supersaturation with respect to liquid water (s) by that with respect to ice (s_i) in the growth law (Eq. (8.17)). The stronger bonds also cause the latent heat of sublimation (l_s) to be larger than that of vaporization (l_v). Such differences between ice particles and liquid droplets are easy to implement compared with the fundamental distinction, that of shape.

The shapes of solid particles are seldom spherical. The diversity of snow and ice found in the atmosphere attests to the complex shapes that are possible in nature. Moreover, those shapes tend to be preserved once the particles are formed and conditions remain constant. The rigidity of ice, arising from the relatively strong molecular bonding in the solid, prevents the kind of adjustments in shape that causes liquid droplets to remain spherical throughout much of their growth histories. An ice particle in air has a larger surface tension than does a liquid droplet, but the surface forces of the solid are nevertheless too weak to alter the shape of the particle.

How one adjusts the growth laws to account for non-spherical shapes is still controversial and the subject of further discussion here. In principle, one could apply Eq. (8.1) to calculate the rate of mass accumulation to a crystal of arbitrary shape. However, without further information, one cannot readily determine the surface flux ($\boldsymbol{\Phi}_{\mathbf{v.sfc}}$) at every point, making it impossible to perform the integration. Attempts to apply this method to simple hexagonal prisms have met with some success, but relatively few ice particles in nature have such simple geometries. A more general and convenient approach uses Maxwell's growth law for droplet growth with the simple replacement of the droplet radius (r_d) by a particle capacitance C. Then, using the other substitutions, one gets the compact form

$$\frac{dm_p}{dt} = 4\pi C \rho_I G'_I s_i, \tag{8.40}$$

where m_p is the mass of the growing ice particle and

$$G'_I = \left[\frac{\rho_I R T_\infty}{M_w D'_v e_i(T_\infty)} + \frac{\rho_I l_s}{M_w k'_T T_\infty}\left(\frac{l_s}{RT_\infty} - 1\right)\right]^{-1} \tag{8.41}$$

is the growth parameter appropriate for ice. The particle capacitance acts as a size, but one influenced by shape. The idea behind such a substitution arises from an analogy to electrostatics.

8.3.1 Electrostatic analogy

Problems in electrostatics have mathematical solutions that are similar in form to those in other branches of physics, so it is often appropriate to use analogies. Typically, only the physical interpretations of the variables differ, the mathematical manipulations remaining identical from one application to another. Here, we show the analogy between electrostatics and the depositional growth (or evaporation) of non-spherical particles (i.e., ice crystals).

Electric charge responds to spatial variations of electric potential in much the same way that water vapor responds to variations of its concentration. In both cases, the rate of transport is directly proportional to the gradient of a field property. The flux of electric charge is a current density $\mathbf{j_e} = -\sigma_e \nabla V_e$, where σ_e is the electrical conductivity and V_e is the electric potential (the potential energy per unit charge). The flux of vapor is given in similar form by Fick's first law as $\Phi_\mathbf{v} = -D_v \nabla n$, where D_v is the diffusivity of vapor in air and n is the concentration of vapor. The negative signs in these expressions indicate that the fluxes are directed down the respective gradients, from high to low values of either electric potential or vapor concentration. As needed, we can replace V_e with n.

Solutions to problems in electrostatics are often expressed through relationships between electric charges and the fields they cause. A single point charge q_e generates an electric field (the force on a unit test charge) that radiates outward with a strength that diminishes inversely with the square of distance r:

$$\mathbf{E} = \frac{q_e}{4\pi \varepsilon_0 r^2}\hat{\mathbf{r}}, \tag{8.42}$$

where ε_0 is the permittivity of free space and $\hat{\mathbf{r}}$ is the outward-pointing unit vector. The electric field is just the negative gradient of electric potential, $\mathbf{E} = -\nabla V_e$, so the potential due to a single point charge varies radially as

$$V_e = -\int_\infty^r \mathbf{E} \cdot d\hat{\mathbf{r}} = \frac{q_e}{4\pi \varepsilon_0 r}. \tag{8.43}$$

A collection of charges affects the electric field in more complicated ways. The divergence theorem of mathematics is then useful because we can take advantage of appropriate conservation principles. In electrostatics, the divergence theorem is called Gauss' law and takes the form

$$\oint_{sfc} \mathbf{E} \cdot d\mathbf{A} = \frac{1}{\varepsilon_0}\int_{V_p} \rho_e dv = \frac{Q_e}{\varepsilon_0}, \tag{8.44}$$

where the integration is taken over the surface area of the particle, which bounds a volume V_p. The volume, which we may take to be that of the particle, contains variable charge density ρ_e and total charge $Q_e = \int_{V_p} \rho_e dv$. In effect, Eq. (8.44) states that the electric field lines passing through the surface of a particle arise from the collection of charges on or within the particle. From some distance away the electric field resembles that of a single point charge of magnitude Q_e located at the center of the particle.

The electric potential of the particle relative to its distant surroundings increases in direct proportion to its total net charge, so the ratio of charge to potential is constant and defined as the electrostatic capacitance $C_e \equiv Q_e/\Delta V_e$, where $\Delta V_e = V_{e,p} - V_{e,\infty}$ is the electric potential of the particle with respect to that at infinity. Upon replacing $Q_e = C_e \Delta V_e$ and $\mathbf{E} = -\nabla V_e$ in Eq. (8.44), we obtain an expression involving only the electric potential as the unknown variable:

$$\oint_{sfc} \nabla V_e \cdot d\mathbf{A} = -\frac{C_e}{\varepsilon_0}\Delta V_e. \tag{8.45}$$

Such an equation allows us to solve for the spatial distribution of potential surrounding a particle having an electrostatic capacitance C_e. In a medium having an electrical conductivity σ_e, an electrical current of density $\mathbf{j}_e$ would stream from the particle at the rate

$$I_e = \oint_{sfc} \mathbf{j}_e \cdot d\mathbf{A} = -\oint_{sfc} \sigma_e \nabla V_e \cdot d\mathbf{A} = \sigma_e \frac{C_e}{\varepsilon_0} \Delta V_e. \tag{8.46}$$

This current would discharge the particle at the rate $dQ_e/dt = -I_e$.

We now draw on the analogy between electric potential and vapor concentration. By replacing V_e with ρ_v (in Eq. (8.46)), and the transport coefficient σ_e with D_v, as well as recognizing that the electric current (I_e) is analogous to the molar flow of vapor (I_v), we obtain the mass growth rate of the particle:

$$\frac{dm_p}{dt} = -M_w I_v = M_w \oint_{sfc} D_v \nabla n \cdot d\mathbf{A} = -4\pi C D_v M_w \Delta n. \tag{8.47}$$

Note that we have let the electrostatic capacitance C_e become the "particle capacitance" $C \equiv C_e/4\pi\varepsilon_0$. (The particle capacitance is identical to the electrostatic capacitance in Gaussian units and has dimensions of distance.) For the common situation of growth by vapor deposition, $\Delta n \equiv n_p - n_\infty < 0$, in which case

$$\frac{dm_p}{dt} = 4\pi C D_v M_w \left(n_\infty - n_p\right). \tag{8.48}$$

Continued development of the theory to include heat transfer leads to the Maxwell-like growth law for ice particles, Eq. (8.40). Such equations for the mass growth of an ice particle allow us to treat the particle as a point sink of vapor, without consideration for how the vapor is distributed across the surface of the particle, spherical or not. Only near the particle surface (within distances comparable to the particle size) does the shape of the particle affect the vapor gradients and hence pattern of deposition. We will need to distinguish far-field from near-field effects.

The extension of Maxwell's theory of droplet growth to ice is commonly used in cloud models because of its simplicity; a single parameter (C) captures both the size and the shape of the ice particle. At the same time, however, the value of C needed for the model requires independent knowledge of how particle shape evolves with time, one of the very features one would like the theory to accomplish. We will also see that any model based on capacitance is based on assumptions that cannot hold in general. More advanced theories of ice-particle growth depend on a deeper understanding of the physics involved.

8.3.2 Refinements to the theory of deposition

The non-spherical nature of ice particles greatly complicates the theory of growth. We must now treat the other, non-radial coordinates, for the shapes of real crystals distort the distributions of vapor and temperature in the near-field around the ice particle. Shape alone distorts the vapor field, but the non-uniform uptake of vapor across the surface of the ice particle further influences the field. Vapor gradients can be as significant laterally, (i.e., along the surface) as they are radially outward from the center of the particle, at least in the near-field.

Growth of ice from the vapor is highly non-linear and affected by several feedback mechanisms of consequence. When the linear growth rate in the c-axis direction of a crystal, for instance, is greater than that along any of the other crystallographic axes, vapor is removed more rapidly from the gas phase over the basal faces than over other parts of the crystal. The same surface-kinetic processes that cause the crystal to elongate along the c axis (become more columnar) also draw down the local vapor supply and establish lateral gradients of vapor concentration that channel additional vapor toward the rapidly growing basal faces. In effect, an inherently fast-growing face conditions its own environment in a way that encourages even faster growth of that face. Both the macroscopic form of the crystal and the local conditions over a face affect the growth of that face and the evolution of crystal shape. Such non-linear interactions between the molecular-scale processes on the surface and the particle as a whole introduce a complexity into crystal-growth theory that challenges our ability to make predictive calculations.

The following treatment proceeds in the order of increasing complexity. First, we discuss the properties of a single, flat surface that permit incorporation of vapor into the crystal. Then, we develop the theory that lets us calculate the linear growth rate of a crystallographic face in the absence of air. We then consider in turn a spherical ice particle and a single hexagonal ice crystal growing in air. The complexity introduced by particles of arbitrary shape compels us to end with a parameterization that captures only part of the physics, yet permits practical growth calculations of atmospheric ice particles. The treatment here is facilitated by simplifying the notation for supersaturation to minimize double subscripting: We drop the subscript i on s_i, as the reference state is understood to be ice and because latent heating of the surface is not considered until near the end.

Surface physics and the linear growth of facets in pure vapor

The molecular processes occurring on a given face of an ice crystal are diverse and ill understood, even in the simplest case of growth in pure vapor. Nevertheless, more is understood about the incorporation of water molecules into ice than into liquid, and much of what we know about the growth of ice comes from many past studies of crystal growth in general.

The structure of a solid surface depends on its orientation relative to the underlying crystal. The flat surfaces, or facets, commonly seen on snow crystals are the basal and prism faces, which form where these low-index planes, (0 0 0 1) and (1 0 $\bar{1}$ 0), respectively, meet the gas phase (refer to Section 2.1). Facets are also affected by temperature, tending to be molecularly flat at low temperatures and rough at high temperatures (within about 10 K of the melting point).

Each facet of a crystal growing in the absence of air is exposed to water vapor at a given, uniform supersaturation. Concentration gradients are essentially non-existent in a pure-vapor environment, so the local supersaturation over each facet is essentially the same as the ambient supersaturation. The presence of air changes the situation dramatically, but the principles explained here remain valid as long as one restricts attention to the local environment, that within about a mean free path of the surface.

Facets grow relatively slowly because of the difficulty with which vapor molecules can enter the lattice; the molecules making up the low-index planes of a crystal are already strongly bonded together in a regular geometrical pattern. In the basal (0 0 0 1) plane of ice Ih, for instance, the water molecules are arranged in a puckered hexagonal pattern. New molecules cannot be added to this layer without disrupting the established order, and adsorbed molecules sitting on a facet are only weakly attached. Specific mechanisms are required for molecules impinging onto a facet to incorporate into the lattice and contribute to the growth of the crystal.

Vapor deposition onto a facet is a multistage process, as shown in Fig. 8.15. Vapor molecules first impinge onto the surface at random locations and get attached weakly to the underlying crystal, a process called adsorption (Stage 1). The adsorbed molecules (called admolecules) may have enough thermal energy to break free of the initial point of attachment, if not completely, then at least enough to migrate randomly across the surface (Stage 2; a two-dimensional version of diffusion). If an admolecule happens to encounter one of the kink sites along a step, where the opportunity for bonding to neighbors is strong, the molecule incorporates into the lattice and becomes a permanent member of the crystal (Stage 3). If an opportunity for incorporation is not encountered within its residence time on the surface, the admolecule desorbs and reenters the gas phase without contributing to the growth of the crystal. As the processes of vapor adsorption, surface diffusion, and incorporation are repeated over and over, molecule by molecule, the step

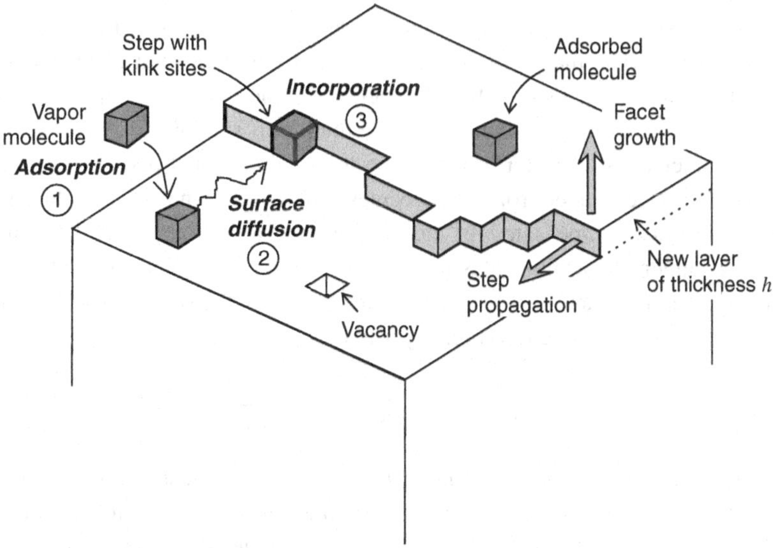

Figure 8.15 Schematic view of a crystal facet, identifying the three main stages of vapor deposition.

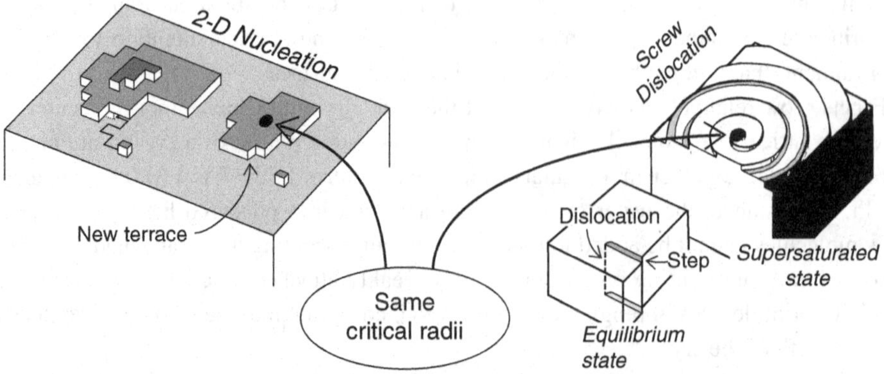

Figure 8.16 Schematic of the two origins of steps on a facet. The grey areas on the right side represent the catchment area associated with the spiral step.

advances across the facet and adds a new molecular layer to the crystal. Faceted crystals grow by layers. However, once a step advances completely across a facet, a new layer has been added to the crystal, but the surface is again left molecularly smooth and unable to accommodate new molecules into the lattice, at least until a new step appears.

Steps can originate on a facet in either of two ways: by the two-dimensional (2-D) nucleation of a new layer or from crystalline defects (screw dislocations). Figure 8.16 contrasts these two possibilities. Both mechanisms generate steps that have kink sites (not shown) and allow the facet to advance normal to itself, but the origins of the step differ significantly.

In the absence of defects, new layers must form by the process of nucleation, as discussed in Chapter 7. The main difference now is that the nucleation of layers occurs in two, rather than three dimensions. In 2-D nucleation, the embryo is envisioned to be a cylindrical island (a disc) just a single molecule thick (left side of Fig. 8.16). As with the homogeneous nucleation of water droplets in a three-dimensional parent phase of vapor, the nucleation of a new layer from admolecules requires the embryo to achieve a critical size, one that just compensates for the effect of curvature on the equilibrium "vapor pressure", the tendency for molecules to leave the cluster and reenter the adlayer. Upon considering the geometry and the surface free energy σ_{SV} of the solid–vapor interface, one finds the critical radius of the embryonic layer to be

$$r_c = \frac{v_{ice}\sigma_{SV}}{\Delta\mu_{SV}} = \frac{v_{ice}\sigma_{SV}}{RT\ln S} \simeq \frac{v_{ice}\sigma_{SV}}{RTs}, \tag{8.49}$$

where v_{ice} is the molar volume of ice and $\Delta\mu_{SV} \simeq RTs$ is the chemical potential difference between the two phases. As the local supersaturation with respect to ice, s, increases, r_c decreases, making it more probable that clusters of admolecules can grow to critical size before dissipating. Once larger than the critical size, islands are thermodynamically stable and can grow rapidly to cover the facet with a new monomolecular layer. For all practical purposes, layers nucleate and facets advance only when the supersaturation with respect to ice exceeds the critical value, s_0. When steps have to be generated by 2-D nucleation, crystal growth is effectively either on ($s > s_0$) or off ($s < s_0$). It is still uncertain whether or how often ice crystals grow in this dichotomous fashion in the atmosphere.

Many real crystals contain various kinds of crystallographic defects, including screw dislocations that intersect the surface. The theory of crystal growth based on screw dislocations was developed in 1951 by Burton, Cabrera, and Frank as an alternative explanation for the growth of crystals at low supersaturations, when layer nucleation was not expected. The starting point of the BCF theory is the presence of at least one screw dislocation, a line defect about which the lattice planes spiral, thus generating a step in the surface (see smaller image on the right side of Fig. 8.16). A step arising from a screw dislocation is permanent and pinned at the point of emergence. At equilibrium, the step is straight, but under supersaturated conditions (larger image on right), the step winds into a spiral that rotates about the dislocation. Spiral growth of a facet is characterized by widely spaced steps and slow growth at low supersaturations and by closely spaced steps and increasingly rapid growth as the supersaturation (s) increases. Spiral (or pyramidal) growth of a facet does not require a threshold or critical supersaturation, but the applied supersaturation does establish a minimum radius of curvature of the spiral that is the same as the critical size (r_c) of an embryo formed by 2-D nucleation.

The rate with which a facet advances normal to itself, termed the linear growth rate, can be described both kinematically and mechanistically. Refer to Fig. 8.17, then imagine observing a specific spot on the surface and counting the number of steps that pass by each second. If the steps all have height h and pass the spot with frequency f_{step}, then the linear growth rate (typical units, $\mu m\, s^{-1}$) is

$$R = f_{step} h. \tag{8.50}$$

The step height is generally taken to be equivalent to the dimension of a single molecule packed in the crystal and so depends on the molar volume of ice, $v_{ice} = M_w/\rho_{ice} \simeq 19.6$ $cm^3\, mol^{-1}$. The step frequency depends on the mechanism of step origin and on the local supersaturation immediately over the facet. With both mechanisms, not many steps pass by the observation spot when the supersaturation is low, but the step frequency and linear growth rate naturally increase as the supersaturation increases and more excess molecules are available for surface diffusion and incorporation into the lattice.

Quantifying step speed (v_{step}) needs special consideration because of possible interactions between neighboring steps. Refer again to Fig. 8.17 and consider the behavior of admolecules. Adsorbed molecules are able to migrate only so far, a mean migration distance x_s, before random thermal motions kick them back into the gas phase. Vapor molecules that impact the surface within the distance x_s away from any step adsorb inside the catchment zones (shaded areas of Fig. 8.17) and can be incorporated into the crystal at kink sites along the step. By contrast, molecules landing outside the catchment areas have little chance of incorporating before desorbing. Such a rigid distinction is artificial,

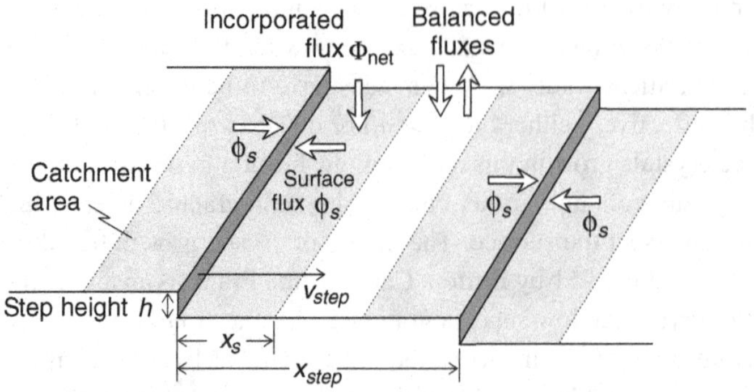

Figure 8.17 Perspective view of two steps progagating across a facet by incorporating those vapor molecules that impinge onto the catchment areas.

as surface diffusion, as with volume diffusion in a gas, involves many random encounters with other molecules. The boundaries of the catchment areas are truly diffuse, but defining the regions with sharp boundaries gives valid average results and helps us understand the principles of crystal growth.

Mass-balance arguments let us see how the step speed is related to the admolecule flux ϕ_s [mol m^{-1} s^{-1}] across the surface and to the net impingement flux Φ_{net} [mol m^{-2} s^{-1}] from the vapor phase. After considering the number of molecules that build into the lattice along a unit length of step within a small time interval, one finds the speed of a straight step to be proportional to the surface flux ϕ_s:

$$v_{step} = \frac{2v_{ice}}{h}\phi_s. \tag{8.51}$$

The factor 2 arises from the assumption that a step can gather admolecules equally from both the top and bottom terraces.

The surface flux is determined by assuming that all of the molecules entering the step come from vapor impinging onto the catchment area. When steps are spaced far apart, so that the catchment area of one step does not overlap that of a neighboring step (the situation depicted in Fig. 8.17), the surface flux is simply $\phi_s = x_s \Phi_{net}$. When, as occurs at high supersaturations, the step spacing becomes small and the catchment areas overlap, one must take the competition for admolecules into account. After solving the second-order differential equation arising from mass continuity on the surface and calculating the admolecule gradient at the step, one finds the general equation for the surface flux to be

$$\phi_s = x_s \Phi_{net} \tanh\left(\frac{x_{step}}{2x_s}\right). \tag{8.52}$$

Two limiting cases can be identified, which helps shed light on the mechanism of step advancement. The nature of the mathematical function $\tanh(u)$ is such that at small u, $\tanh(u) \to u$, whereas at large u, $\tanh(u) \to 1$. Thus, in the small-u limit, when $x_{step} << x_s$, $\phi_s \to x_{step}\Phi_{net}/2$, each step must share the available admolecules equally with its neighbors. In this case, the surface flux depends on the step spacing (x_{step}), not on the mean migration distance (x_s). At the other extreme, when steps are far apart ($x_{step} >> x_s$), the surface flux to the step is the maximum possible and proportional to x_s: $\phi_s = x_s \Phi_{net}$, as noted above.

The step speed can now be calculated by combining the relationships already developed. The step speed is found by inserting Eq. (8.52) into Eq. (8.51):

$$v_{step} = \frac{2x_s v_{ice}}{h} \Phi_{net} \tanh\left(\frac{x_{step}}{2x_s}\right). \tag{8.53}$$

We now see the influence on step speed of possible competition for admolecules through the tanh function.

Steps originating by 2-D nucleation pass any spot on the facet at a rate that depends indirectly on step speed. At low supersaturations, the limiting factor for growth of the facet is the rate of forming embryonic islands of critical size, so it is the frequency of layer nucleation that dictates the growth rate of the facet. Following the procedure outlined in Chapter 7, but using two-dimensional geometry, we find the free energy to form a cylindrical island (a new embryonic layer) of critical radius r_c to be

$$\Delta G^* = \Delta G(r_c) \simeq \frac{\pi v_{ice} h \sigma_{SV}^2}{RTs}. \tag{8.54}$$

The number density of such critical islands on the facet is, if we may assume a Boltzmann distribution, $n(r_c) = n_0 \exp(-\Delta G^*/RT)$, where n_0 is the number density of unclustered admolecules. Each critical island becomes stable and grows across the facet with the addition of a single admolecule. The rate J_{2D} [$\mathrm{m^{-2}\,s^{-1}}$] of nucleating new layers is therefore proportional to the frequency of adding new admolecules to a critical island, $2\pi r_c v_{step} h/v_{ice} \simeq 2\pi r_c x_s \Phi_{net}$, times the number density of critical islands, $n(r_c)$:

$$J_{2D} = 2\pi r_c x_s \Phi_{net} n_0 \exp\left(-\frac{\Delta G^*}{RT}\right). \tag{8.55}$$

Note that only a single new layer needs to be nucleated anywhere on a facet of area A_{facet}, so the step frequency becomes $f_{step,2D} = A_{facet} J_{2D}$. The larger the area of a facet, the smaller the layer nucleation rate needs to be to contribute to any specified growth rate. Through Eqs. (8.54) and (8.55), we see that the growth rate of a facet normal to itself depends strongly on the local supersaturation s and can be expressed in the form

$$R_{2D} = c_0 \exp\left(-\frac{s_0}{s}\right). \tag{8.56}$$

Here, c_0 is a pre-exponential factor that depends on the mean migration distance x_s, and s_0 is the critical supersaturation, which depends on surface free energy and is a characteristic of the surface in question. As the supersaturation increases, new layers form on top of previous layers even before those layers are complete. The catchment areas of the steps overlap, and the dependence of R_{2D} on s eventually approaches a linear relationship that is independent of facet size.

As we have learned from BCF theory, the emergence of a screw dislocation on a facet generates a permanent spiral step that advances even at low supersaturations. Recall that a step passes a given spot each time the spiral makes one complete rotation and that the spiral rotates in the first place only because the step acquires admolecules at kink sites and propagates normal to itself. The greater the step speed, the faster the rotation of the spiral. Growth pyramids rotate as a solid body, maintaining the shape of an Archimedean spiral (constant spacing between steps, $x_{step} = 4\pi r_c$). The step speed is inversely related to curvature and so decreases systematically toward zero at the center of the spiral (where equilibrium with the critical embryo is maintained). The step frequency for spiral growth is, according to the BCF theory,

$$f_{step,spiral} = \frac{\omega}{2\pi} = \frac{v_{step}}{4\pi r_c}, \tag{8.57}$$

where ω is the angular rotation rate of the spiral, v_{step} is the speed of a straight step (far from the center), and r_c is the critical radius given by Eq. (8.49).

The step frequency for spiral growth is obtained by substituting Eqs. (8.53) and (8.49) into Eq. (8.57):

$$f_{step,spiral} = \frac{x_s}{2\pi h} \frac{RTs}{\sigma_{SV}} \Phi_{net} \tanh\left(\frac{x_{step}}{2x_s}\right). \tag{8.58}$$

Unfortunately, this equation contains two variables, x_{step} and x_s, that are not easily derivable from experimental data. However, recall that the spacing between steps (x_{step}) in an Archimedean spiral is directly proportional to the critical radius (r_c) and inversely proportional to supersaturation (through Eq. 8.49):

$$x_{step} = 4\pi r_c \simeq \frac{4\pi v_{ice}\sigma_{SV}}{RTs}. \tag{8.59}$$

Thus, the distance ratio $x_{step}/2x_s$ is equivalent to a ratio of supersaturations s_1/s, if we let $s_1 \equiv 2\pi v_{ice}\sigma_{SV}/(x_s RT)$. This primary surface parameter, s_1, is alternatively termed the transitional or threshold supersaturation, which can be diagnosed from experimental data. Thus, Eq. (8.58) can be written in the more practical form

$$f_{step,spiral} = \frac{v_{ice}}{h} \Phi_{net} \left(\frac{s}{s_1}\right) \tanh\left(\frac{s_1}{s}\right). \tag{8.60}$$

The linear growth rate of a facet containing a screw dislocation may now be calculated. Note that the net impingement flux can be expressed in terms of the supersaturation as $\Phi_{net} = \Phi_{eq}(T_{sfc})s$, where $\Phi_{eq}(T_{sfc}) = e_i(T_{sfc})/(2\pi M_w RT_{sfc})^{1/2}$ is the impingement flux at growth equilibrium. Thus, by combining Eqs. (8.60) and (8.50), we gain the linear growth law in the succinct form

$$R_{spiral} = v_{ice}\Phi_{eq}(T_{sfc}) \left(\frac{s^2}{s_1}\right) \tanh\left(\frac{s_1}{s}\right). \tag{8.61}$$

Note the limiting cases, now relative to s_1. At low supersaturations, when $s << s_1$, $R_{spiral} \sim v_{ice}\Phi_{eq}(T_{sfc})(s^2/s_1) \propto s^2$, whereas at high supersaturations, $s >> s_1$, and

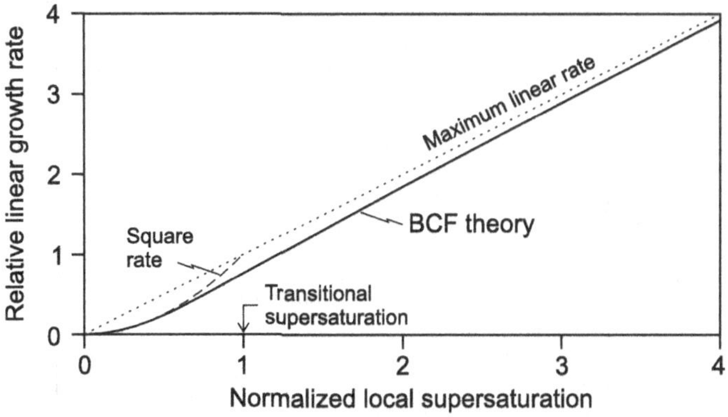

Figure 8.18 Relative growth rate of a facet in pure vapor as a function of the normalized supersaturation, s_i/s_1. The growth rate has been normalized to the maximum rate at the transitional supersaturation s_1.

$R_{spiral} \sim v_{ice}\Phi_{eq}(T_{sfc})s \propto s$. The surface parameter s_1 thus defines the supersaturation at which the growth law shifts from one following a dependence on s^2 to one following a linear dependence on supersaturation. This behavior is plotted in Fig. 8.18. With increasing local supersaturation, the linear growth rate starts out slowly, but it increases rapidly as the square of the supersaturation, before eventually becoming linearly proportional to supersaturation (shown as a dashed line).

With any theory of growth, it is useful to compare the actual rate of growth with the theoretical maximum. Such a comparison provides a measure of the efficiency with which a given surface takes up molecules from the vapor phase. As we saw earlier, the growth efficiency for liquid droplets was the mass accommodation coefficient for condensation (the condensation coefficient). The analogous measure of growth efficiency for ice is the mass accommodation coefficient for deposition (the deposition coefficient). We should anticipate here that the BCF growth rate will be low relative to the maximum at low supersaturations, when sizeable regions exist between the catchment areas that are not effective at incorporating vapor molecules (refer again to Fig. 8.17). But, as the local supersaturation increases and the critical radius decreases, the step spacing decreases proportionally and a greater fraction of the total surface area becomes available for capturing impinging molecules. (Thus, α_m can be interpreted as the fraction of the total surface covered by catchment areas.) If every impinging molecule were to enter the lattice, the maximum linear growth rate would be $R_{\max} = v_{ice}\Phi_{net} = v_{ice}\Phi_{eq}(T_{sfc})s$. The actual growth rate, however, is determined by the incorporation flux, $R = v_{ice}\Phi_{inc} = v_{ice}\alpha_m\Phi_{eq}(T_{sfc})s$, so the deposition coefficient, determined by the ratio of growth rates, is $\alpha_m = R/R_{\max}$. The deposition coefficient from BCF theory is thus

$$\alpha_m = \left(\frac{s}{s_1}\right)\tanh\left(\frac{s_1}{s}\right). \tag{8.62}$$

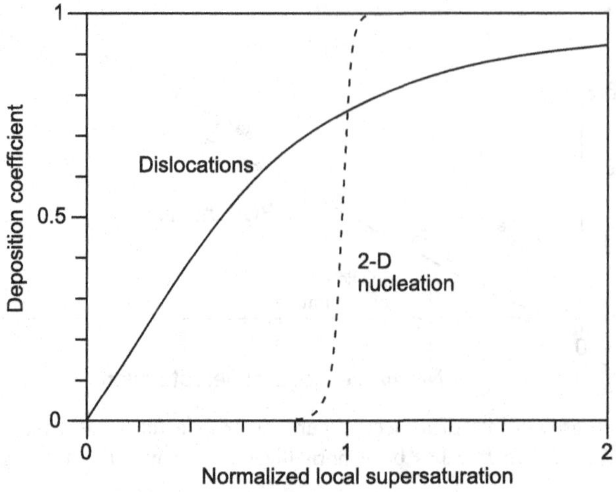

Figure 8.19 Dependence of the deposition coefficient on the local supersaturation with respect to ice.

This function is shown in Fig. 8.19, along with the corresponding deposition coefficient for growth by 2-D nucleation. Recall that the radius of curvature of a two-dimensional nucleus and the minimum radius of curvature of a growth spiral are the same, so these two curves for deposition efficiency always maintain the same relationship to each other. The growth of every facet of a crystal can in principle be described in terms of a deposition coefficient that varies with the interfacial supersaturation in the way depicted here. A significant problem is, however, that we seldom know the conditions in the gas phase immediately over a facet. And, the theory is predicated on knowing the transitional or critical supersaturations of the various faces on a crystal as a function of temperature.

The growth of individual facets represents only a part of the overall problem of crystal growth. Thus far, we have assumed that the surface supersaturation, that within a mean free path of the surface is given. However, the presence of air decouples the surface and ambient supersaturations. Moreover, knowing how rapidly one facet grows tells us little about how the other facets and the crystal as a whole is growing. To gain additional perspective on the problem, consider an ice particle growing in air.

8.3.3 Growth of a spherical ice particle in air

An ice particle gathers excess water vapor from the surrounding air by gas-phase diffusion. Diffusion is slow, and the air readily supports gradients of vapor concentration, so the surface supersaturation will be much smaller than the ambient supersaturation, the one based on conditions far from the surface. We now temporarily consider the hypothetical case in which the deposition coefficient is uniform over the entire surface. The particle thus grows as a sphere, and we can make use of principles developed for condensational growth. Giving the particle spherical symmetry lets us focus on the radial gradients.

However, in contrast to the growth of droplets, the need for layered growth on faceted crystals causes nonlinear interactions between the surface supersaturation and the surface kinetics.

As discussed earlier in conjunction with Fig. 8.8, we expect a vapor jump to occur immediately over the particle surface, because it is this finite difference between the surface and equilibrium concentrations of vapor that drives the growth of the particle. The presence of air and the consequent gradient of vapor concentration, causes the magnitude of this vapor jump, hence the interfacial supersaturation (that immediately over the surface), to vary with the size of the particle. The concentration gradient in the continuum region near the particle surface decreases with particle size, requiring in turn a smaller vapor jump to accommodate the smaller flux of molecules approaching the particle from afar. The fundamental difference between the growth of a liquid droplet and that of a spherical ice particle is that the mass accommodation coefficient for liquid is usually assumed to be constant (a single value), whereas that for ice is known to vary with the interfacial supersaturation. It is the non-constant nature of the deposition coefficient that adds complexity to the growth of ice.

The fact that the deposition coefficient depends on the supersaturation gives the impression that there is also a size dependence to the deposition coefficient. An apparent size dependence arises only indirectly, a consequence of the effect that size has on the interfacial supersaturation. We can understand the implicit nature of the size dependence by referring to Fig. 8.20. Each of the three particle sizes depicted is associated with a unique interfacial supersaturation, which we here term the surface supersaturation s_{sfc} (identified by up arrows on the right in this figure and analogous to the vapor jump of Fig. 8.8). Each surface supersaturation can be mapped onto the supersaturation-dependent deposition coefficient (left side of Fig. 8.20). As the size of a particle increases and the surface supersaturation decreases, the magnitude of the deposition coefficient decreases. Only the

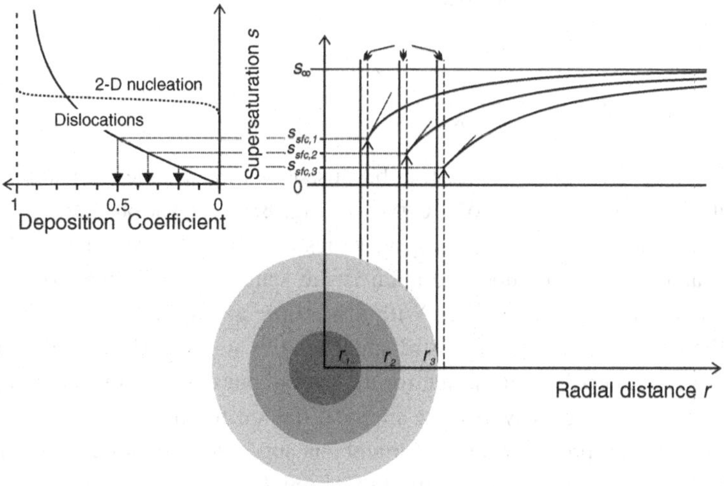

Figure 8.20 Vapor jumps and deposition coefficients for three different sizes of spherical ice particle.

supersaturation dependence $\alpha_m(s_{sfc})$ is explicit (i.e., real); the size dependence is implicit (i.e., apparent): $\alpha_m(s_{sfc}(r_p))$. This indirect effect arises because of the presence of air. The non-condensible components of the gas phase impede the transport of vapor from the distant environment to the particle surface and cause the surface supersaturation to vary with both the ambient conditions and particle size.

A further consequence of growth in air is a change in the apparent dependence of the deposition coefficient on supersaturation. The theory of crystal growth discussed in conjunction with BCF theory presupposes a static, independent interfacial supersaturation, not one that changes as the particle grows or encounters different ambient supersaturations. Air, by causing large gradients in vapor concentration and temperature to exist around the particle, forces us to infer the interfacial conditions and hence to expect the deposition coefficient to vary with the ambient supersaturation in a different way. We can see the influence of air by recalling the treatment of liquid droplets, in which the matching of flows at the interface between the free-molecular and continuum regions leads to a significant reduction in the excess vapor concentration over the surface compared with that of the ambient air (Eq. (8.28)). The corresponding relationship for a spherical ice particle growing in air with an ambient supersaturation with respect to ice, s_∞, is

$$\frac{s_{sfc}}{s_\infty} \simeq \frac{1}{1+\frac{\alpha_m \bar{c}_v r_p}{4D_v}} = \frac{1}{1+K\alpha_m} = \frac{1}{1+\frac{R_{gas}}{R_{sfc}}}. \tag{8.63}$$

where the dimensionless group $K \equiv \bar{c}_v r_p/(4D_v)$. Note that $K\alpha_m = R_{gas}/R_{sfc}$ is the ratio of transport resistances due to molecular diffusion in the gas phase and on the surface. The larger the gas-phase resistance ($R_{gas} = r_p/D_v$) compared with the surface resistance ($R_{sfc} = 4/\bar{c}_v\alpha_m$), the smaller is the surface supersaturation and the driver of growth. With the understanding that the supersaturation used in Eq. (8.62) is the surface value, we obtain an implicit equation for the deposition coefficient in terms of the ambient supersaturation appropriate to BCF theory:

$$\alpha_m = \left[\frac{s_\infty}{(1+K\alpha_m)s_1}\right] \tanh\left[\frac{(1+K\alpha_m)s_1}{s_\infty}\right]. \tag{8.64}$$

This equation must be solved by iteration because of its transcendental nature. The form of the solution is shown by the solid curves in Fig. 8.21 for various values of K. With $K = 0$ (implying pure vapor, $D_v \to \infty$), α_m varies with supersaturation as in BCF theory (Fig. 8.19), and the surface supersaturation is the same as the ambient supersaturation (gas-phase transport is infinitely fast). With just a little air in the system (represented by $K = 0.1$), the gas-phase resistance is still small relative to the surface-kinetic resistance, and the α_m-curve differs little from that for the pure-vapor case. As the ice particle grows in environments with progressively more air, the diffusion coefficient D_v becomes smaller, and K becomes larger; progressively larger ambient supersaturations are needed to achieve the same deposition coefficient. In the limit of extremely large gas-phase resistance (large K because of small D_v), solution of Eq. (8.64) yields

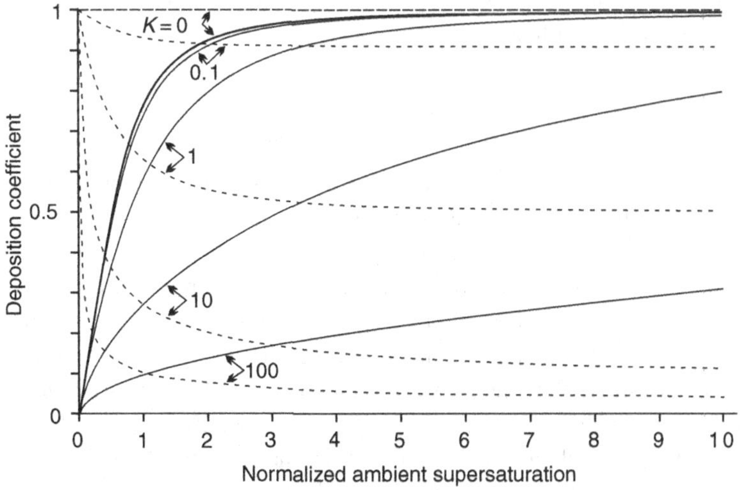

Figure 8.21 Dependence of deposition coefficient (solid curves) and supersaturation ratio (dashed curves) on ambient supersaturation for selected values of the dimensionless group K.

$$\alpha_m \sim \left(\frac{s_\infty}{K s_1} \right)^{1/2}. \tag{8.65}$$

Of particular importance is the fact that the form of this limiting relationship between α_m and s_∞ differs qualitatively from that given by BCF theory ($\alpha_m \sim s/s_1$ in the low-s limit of Eq. (8.62)). Whereas the deposition coefficient at low supersaturations varies linearly with supersaturation in the absence of air (BCF theory), it varies approximately as the square root of ambient supersaturation in the presence of air. If one were to measure the growth rate of a spherical ice particle as a function of the applied (i.e., ambient) supersaturation and extract the apparent deposition coefficient, one might misinterpret the data and draw invalid conclusions about the surface processes. The effects of air must be accounted for.

Air not only retards the growth of the particle, it also changes the character of the growth process. Beyond the fact that the deposition coefficient varies nonlinearly at low ambient supersaturation $\left(\alpha_m \sim s_\infty^{1/2}\right)$, the limiting case considered above (when $K\alpha_m >> 1$ in Eq. (8.63)) also implies that the ratio of supersaturations is

$$\frac{s_{sfc}}{s_\infty} \sim \frac{1}{K\alpha_m}. \tag{8.66}$$

When the limiting function for α_m (Eq. (8.65)) is used in this relationship, we find the surface supersaturation to vary as the square root of the ambient supersaturation: $s_{sfc} \sim (K s_1 s_\infty)^{1/2} \propto s_\infty^{1/2}$. The particle surface experiences a vastly different supersaturation than does the air far from the particle. Finally, note in Eq. (8.66) how the surface supersaturation varies with the deposition coefficient: $s_{sfc} \sim 1/\alpha_m$. This inverse relationship affects the development of crystal habits, for the more efficient the surface process are on a given part of a crystal, the more effectively is the vapor removed from the vapor phase in the immediate vicinity of that facet.

8.3.4 Habit formation

An ice crystal develops into non-spherical forms when the linear growth rate is not uniform across the surface. An ice particle may start out spherical, as when a cloud droplet freezes, but different growth rates on the different crystallographic faces soon lead to preferential advancement of the surface in one direction over another. Non-uniform growth rates and a non-spherical shape arise in the first place from non-uniform deposition coefficients, but the particle shape subsequently alters the spatial distributions of vapor and temperature, field properties that then affect the later evolution of shape. As we learned in Section 1.2, the diverse forms of realistic ice crystals growing in air are characterized by distinct primary habits (defined by a dimensional ratio that depends on temperature) and by a myriad of secondary habits (growth features superimposed onto a primary habit and driven mainly by supersaturation). Our goal here is to identify the key processes responsible for the complex forms of snow found in nature.

Consider the idealized case of a single hexagonal crystal of ice bounded by molecularly smooth facets, two basal faces and six prism faces (see Section 2.1.3 and Fig. 8.22). The basic geometry of ice Ih (hexagonal prism) arises from the way in which the water molecules bond to each of four neighbors in the ice lattice, as discussed in Section 2.1.3. The primary habit (plate or column) of such a crystal at any given time is characterized by the aspect ratio c/a, but it evolved to this point because of the differential growth of the

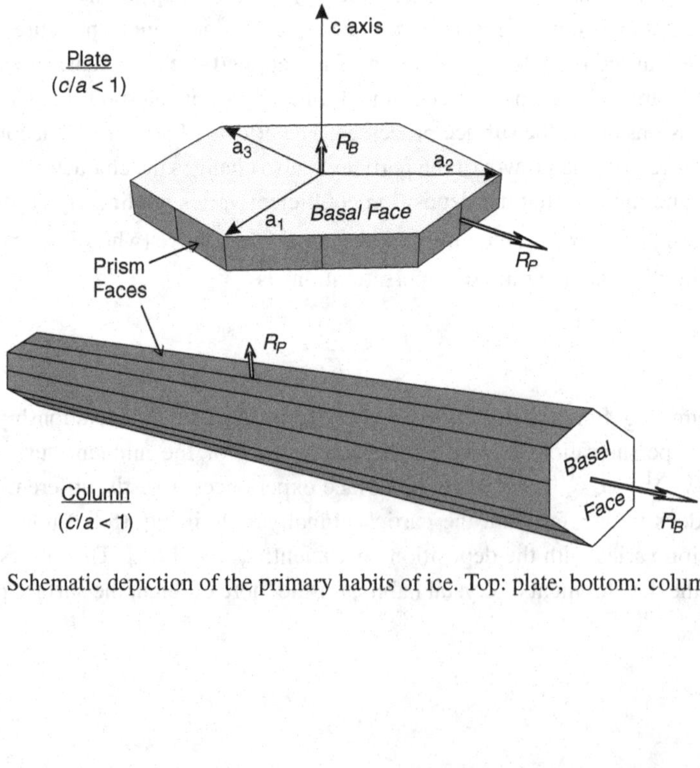

Figure 8.22 Schematic depiction of the primary habits of ice. Top: plate; bottom: column.

facets. If we let R_B and R_P be the linear growth rates of the basal and prism faces, respectively, then the primary habit being formed at that instant could equally well be described by the growth ratio R_B/R_P. When the growth of the basal face is greater than that of the prism faces ($R_B > R_P$), the crystal develops a columnar shape, whereas preferential growth of the prism faces ($R_P > R_B$) leads to a plate-like form. The primary habit arises incrementally from the growth ratio:

$$\frac{R_B}{R_P} = \frac{dc/dt}{da/dt} = \frac{dc}{da}. \tag{8.67}$$

(Note: The linear growth rate of a prism face, R_P, is traditionally taken to be the rate of change along the a axis, da/dt, which is not strictly correct because the a axes point toward the corners of the hexagon. However, R_P and da/dt are always related to each other in constant proportion by the hexagonal geometry, so we continue the tradition.)

The growth rates along the c and a directions depend on both the respective deposition coefficients and the local supersaturations. As shown previously, the growth rate of a generic facet is $R = \alpha_m v_{ice} \Phi_{eq}(T_{sfc})s$, a relationship used to measure α_m in laboratory experiments. By holding $\Phi_{eq}(T_{sfc})s$ constant as the surface temperature T_{sfc} is varied and measuring R of individual facets in the absence of air, one finds the deposition coefficient to depend on temperature in the manner shown in Fig. 8.23. We clearly see the alternation of the primary habit found in nature (e.g., plates where $\alpha_{m,prism} \equiv \alpha_P > \alpha_B \equiv \alpha_{m,basal}$), but the reason for such temperature dependences is still unclear. Some past studies have suggested that the surface free energy varies with temperature, others argue that the cause lies with how the rate of admolecule diffusion depends on temperature. Future research will likely find that the mechanistic basis for the variation of primary habit with temperature depends on the details of the surface structure of ice and its variation with temperature.

If the deposition coefficient were simply a function of the crystallographic axis and temperature, but not dependent on the local supersaturation s, we would expect the growth ratio

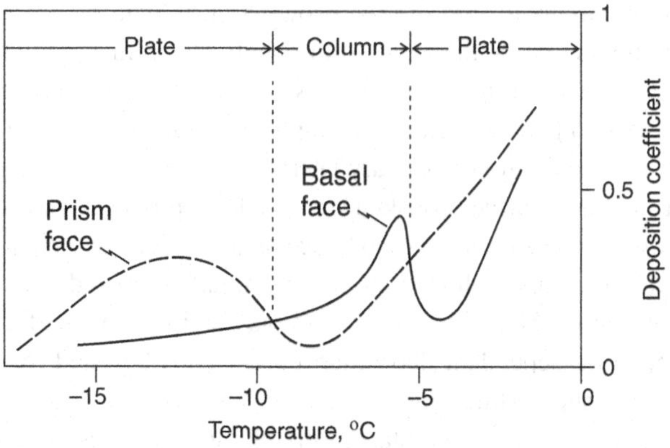

Figure 8.23 The deposition coefficients of the basal face (solid curve) and prism face (dashed curve) as derived from the data of Lamb and Scott (1972).

to be $R_B/R_P = \alpha_B/\alpha_P$. However, the deposition coefficients also depend on the respective local supersaturations, which differ over the basal and prism faces whenever crystals grow in air. Thus, the growth ratio is more accurately written as

$$\frac{R_B}{R_P} = \frac{\alpha_B(s_B)s_B}{\alpha_P(s_P)s_P}, \tag{8.68}$$

where s_B is the surface supersaturation with respect to ice over the basal face and s_P is that over the prism face. Note the non-linearity implied by Eq. (8.68); the supersaturation enters the problem both as the fundamental driver of growth (a measure of the excess vapor available) and as an influence on the efficiency with which excess vapor is able to be incorporated into the crystal. The faster-growing faces, those with larger deposition coefficients, remove vapor more rapidly from their local environments than do the slower-growing faces and so reduce the local supersaturation. It may seem initially counterintuitive that the faster a face grows, the lower the local supersaturation over that face. The competing effects do not off-set each other completely, and the dilemma is resolved by realizing that the faster growth of a face arises because the vapor molecules incorporate into the lattice more efficiently on that face; the lower supersaturation over that face is a consequence of the rapid growth, not a cause. By the same reasoning, a slow-growing face will experience a relatively large local supersaturation.

The supersaturations vary greatly in the vicinity of a growing crystal, as can be appreciated by considering Fig. 8.24. In this computational example, a thin plate-like crystal is growing in air characterized by a uniform and constant ambient supersaturation (specified far from the crystal). Note that the field has been divided into two regions, called the far field and the near field. In the far field (many crystal dimensions away), the gradients tend to be uniform, with vapor fluxes directed radially toward the crystal; at a sufficient distance, a growing crystal acts as a point sink of vapor (as treated in the electrostatic analogy, Eq. (8.44)). In the near field (inset), the gradients have components both normal and parallel to the facets; the shape of the crystal dictates the distribution of vapor, and the vapor is redistributed preferentially toward the fast-growing faces. The largest gradients are normal to the fastest-growing faces, the smallest gradients are normal to the slowest-growing faces. The vapor gradients near the sharp corners can be large and cause substantial fluxes of vapor to be transported from the slow-growing faces to the fast-growing faces. Note also that the interfacial supersaturation must vary laterally across each facet in a way that holds the vapor flux normal to the surface constant (i.e., the local product $\alpha_m s$ is fixed). However, the lateral variations in supersaturation (and hence in deposition coefficient) across individual facets of simple plates and columns are usually small compared with the face-to-face differences. The primary habits of single ice crystals are driven by the face-to-face differences in the temperature-dependent deposition coefficients, adjusted for the face-to-face differences in the local supersaturations.

The secondary habits of an ice crystal, morphological features superimposed on the primary habit, are caused primarily by interactions of the crystal with excess vapor in the near field (refer again to Fig. 8.24). In effect the growing crystal distorts the vapor field

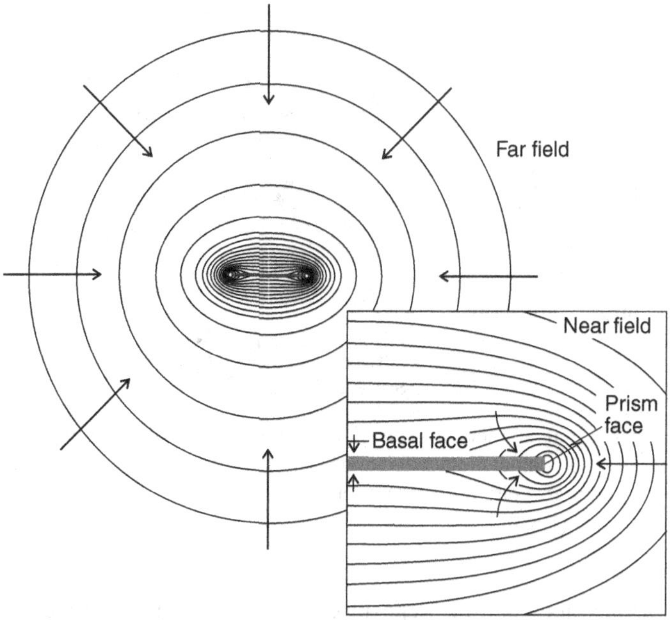

Figure 8.24 Surfaces of constant supersaturation around a thin plate growing in a uniform ambient field. Inset is an expanded view of the right half of the crystal (shaded rectangle). The arrows identify the fluxes of vapor down the gradients. The gradients are predominantly radial in the far field, but in the near field they respond to the shape of the crystal. Adapted from Libbrecht (2005).

around it, causing gradients in the vapor concentration that redirect the vapor fluxes toward the crystal. The plate-like crystal depicted in Fig. 8.24, for instance, has caused the vapor to be preferentially depleted over the prism faces, which leads to enhanced vapor fluxes toward those same prism faces. The crystal remains a plate ($c/a < 1$), but the c/a ratio increasingly deviates from unity (gets smaller in the case of plates); under appropriate conditions, exceedingly thin plates can form, with aspect ratios $c/a \sim 1/100$. Similarly, in temperature regimes that favor columnar growth ($c/a > 1$), the basal face can advance relatively rapidly and produce long columns, even needles. In such cases, the secondary habits are simply exaggerations of the primary habit.

At sufficiently high ambient supersaturations, the shape of the crystal, even for a fixed primary habit, undergoes morphological (i.e., shape) instability: a change in shape changes the vapor field in a way that reinforces the original shape change. A good way to think of morphological instability is to visualize a small bump that forms on the surface and thereby experiences a higher local supersaturation (because of the vapor gradients). As the bump grows further out into the vapor field, it gets ever closer to the high supersaturations in the ambient air. We can see this tendency applied to hexagonal ice crystals by referring to Fig. 8.25, a representation of a plate-like ice crystal. At low supersaturations, the vapor gradients (not shown) are relatively weak, so the crystal shape remains dictated by the

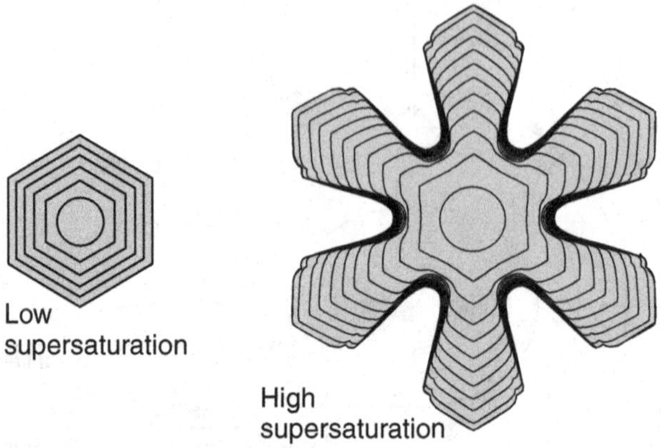

Figure 8.25 Schematic diagram showing the effect of morphological instability on the formation of complex patterns. The computations were applied in two dimensions, so the starting form was an infinite cylinder (seen as a circle here). The curves depict the external forms at discrete times. Adapted from Yokoyama (1993).

underlying crystallography and the deposition coefficients. At high supersaturations, the strong radial gradients of vapor concentration lead to the preferential growth of any feature that sticks farthest out into the vapor-rich air. The corners of the hexagon experience higher local supersaturations than do the flat parts, so they grow faster. Continued enhancement of the corners is reinforced by the progressively higher local supersaturations they see. Meanwhile, the flat parts are robbed of vapor and so advance very slowly. Carried to an extreme, such action leads to the dendrites so common in natural snow (see Fig. 8.26). Not only do the corners of the hexagon grow rapidly into the vapor-rich air, but side branches also form by slight perturbations in the ambient conditions and follow the same rules of instability.

8.3.5 Mass growth of ice particles

The mass of a crystal represents the accumulated effects of vapor deposition across the various surfaces, in effect an integration of the linear growth rates over the surface, as suggested by Eq. (8.1). With simple hexagons (e.g., left side of Fig. 8.25), one could compute the mass growth rate of the crystal geometrically from the linear growth rates. Even then, the problem is non-trivial, because growth on one face affects the growth of neighboring faces (as discussed above). When the shapes are complex (as exhibited by the dendrite in Fig. 8.26), simple computational methods are definitely not feasible. Not only is the geometry itself complicated, but that geometry changes in uncertain ways throughout the growth time. The situation is even worse for the many irregular and polycrystals found in atmospheric clouds. Calculation of the mass growth rates of ice crystals is crucial for estimating

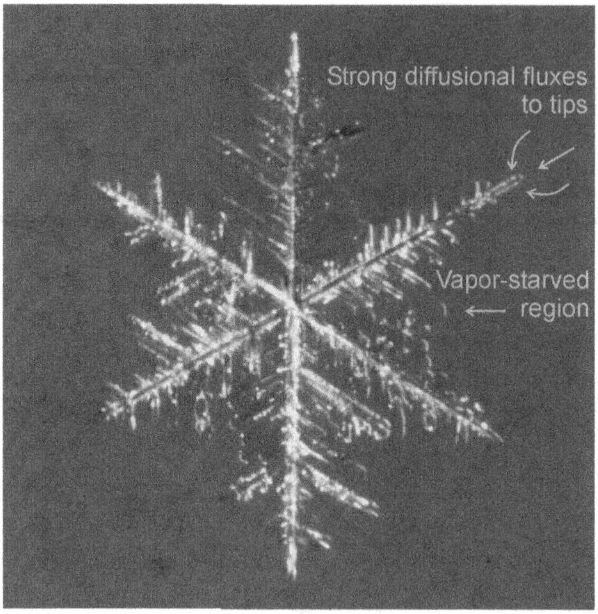

Figure 8.26 A photo of a natural dendrite, annotated to emphasize the role of vapor gradients in the formation of complex secondary habits. Photo by D. Lamb.

the rate of vapor depletion in cloud models, so we must depend on efficient, approximate methods.

Suitable extension of the classical capacitance model (Maxwell's growth law applied to ice, Eq. 8.40) offers a way to calculate the growth in mass and the evolution of the primary habit of a representative ice particle simultaneously. The key to success here is using a shape that resembles real ice particles in some regard and yet allows the capacitance to be calculated analytically. The hexagonal prism, even though it is the true shape of simple ice crystals, is not appropriate; the flat faces and the need for flux boundary conditions make it difficult to calculate an equivalent capacitance. Moreover, the shapes of most atmospheric ice crystals are not at all prism-like, so even with a convenient way to calculate the capacitance of an hexagonal prism, it would be valid for only a small fraction of the ice particles in a cloud anyway. A geometrical form is needed that is general and simple.

Spheroids are suitable mathematical shapes with properties that can be exploited. Formed by rotating an ellipse about one of its axes, a spheroid is a three-dimensional shape that allows the area, volume, even the capacitance to be computed easily. Two classes of spheroids can be defined, depending on which axis of the ellipse is used as the rotation axis. A prolate spheroid is generated when the ellipse is rotated about the major axis, whereas an oblate spheroid arises when the ellipse is rotated about its minor axis. As shown in Fig. 8.27, a prolate spheroid reasonably well reflects the form of a columnar ice particle. Oblate spheroids represents plates. Suitable adjustment of the aspect ratio allows spheroids

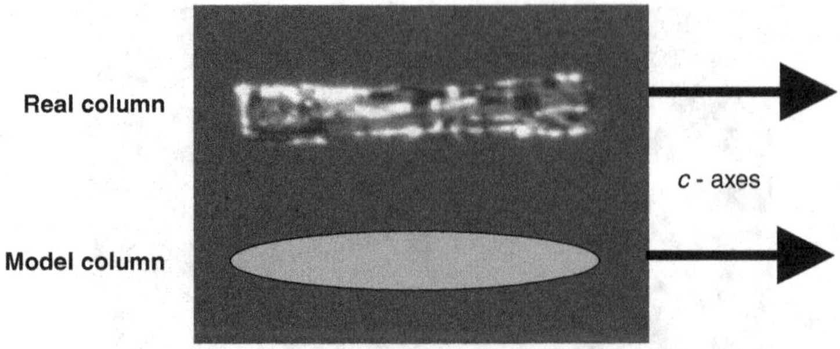

Figure 8.27 The representation of a column by a prolate spheroid.

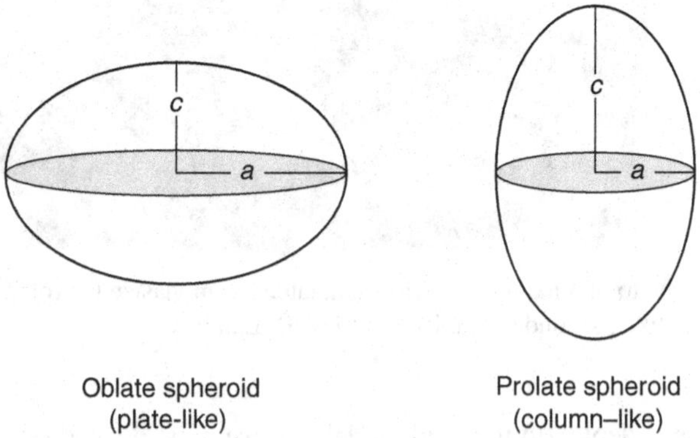

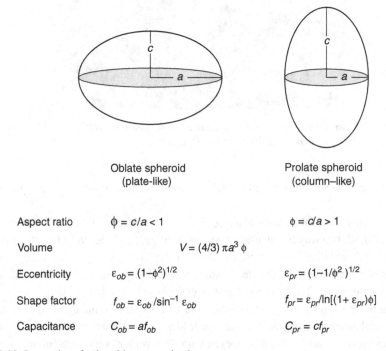

	Oblate spheroid (plate-like)	Prolate spheroid (column–like)
Aspect ratio	$\phi = c/a < 1$	$\phi = c/a > 1$
Volume	$V = (4/3)\,\pi a^3\,\phi$	
Eccentricity	$\varepsilon_{ob} = (1-\phi^2)^{1/2}$	$\varepsilon_{pr} = (1-1/\phi^2)^{1/2}$
Shape factor	$f_{ob} = \varepsilon_{ob}/\sin^{-1}\varepsilon_{ob}$	$f_{pr} = \varepsilon_{pr}/\ln[(1+\varepsilon_{pr})\phi]$
Capacitance	$C_{ob} = af_{ob}$	$C_{pr} = cf_{pr}$

Figure 8.28 Properties of spheroids summarized.

to represent a wide variety of ice forms, including many polycrystals. In order to maintain consistency with the usual terminology for single ice crystals, we identify the rotation axis of the spheroid as the principal crystallographic (c-axis) of ice Ih in both cases; the lateral axis is designated the a-axis. Distances along the respective axes are the half-maximum dimensions c and a. Prolate spheroids (with aspect ratios $\varphi \equiv c/a > 1$) thus represent columns; oblate spheroids ($\varphi < 1$) represent plates. Figure 8.28 summarizes the distinctions between these two main categories of spheroids. A sphere results when $c = a$ and $\varphi = 1$.

The capacitance C accounts for both the size and the shape of the particle. We can separate these two components and define the shape factor as the contribution to particle capacitance remaining after size is factored out. For the shape factor to make mathematical sense, size must be considered carefully. One could use the volume-equivalent radius of a sphere, $r_{sp} = (3V/4\pi)^{1/3}$, but we prefer to define the size of a spheroid to be the semi-major length $r_{\max}$ (i.e., half the maximum dimension). Think radius rather than diameter, and use $r_{\max} = c$ for the size of a prolate spheroid and $r_{\max} = a$ for the size of an oblate spheroid. The general shape factor is thus $f_{sh} \equiv C/r_{\max}$, so for a prolate spheroid $f_{sh,pr} \equiv C/c$ and for an oblate spheroid $f_{sh,ob} \equiv C/a$. The top part of Fig. 8.29 shows how the computed shape factors depend on the aspect ratio φ. The shape factor depends only on φ, directly and through the eccentricity, which also depends only on φ. The shape factor for a sphere is unity, but it decreases as the shape deviates from a sphere. The limiting case for an oblate spheroid ($\varphi \to 0$) is $f_{sh,ob}\big|_{\substack{\text{thin}\\\text{plate}}} \to 2/\pi$, whereas that for a prolate spheroid ($\varphi \to \infty$) is $f_{sh,pr}\big|_{\substack{\text{long}\\\text{column}}} = 1/\ln(2\varphi) \to 0.$

The lower part of Fig. 8.29 shows the particle capacitance for selected volumes. Even though the shape factor decreases as the shape deviates from a sphere, the capacitance increases markedly as the maximum dimension increases for a given particle volume. Note a subtle, but important asymmetry: the capacitance of a prolate spheroid is greater than that of an oblate spheroid at extreme deviations. This geometrical effect arises because of the different number of spatial dimensions represented by c and a. Think of a cylinder with a cross-sectional area A ($\sim a^2$) and length L ($\sim c$); for a given volume ($V = AL$), $L \propto 1/A$, so $c \propto 1/a^2$. Changes in distance along the c axis impact only one dimension, whereas changes along the a axis impact two, lateral dimensions simultaneously. We will

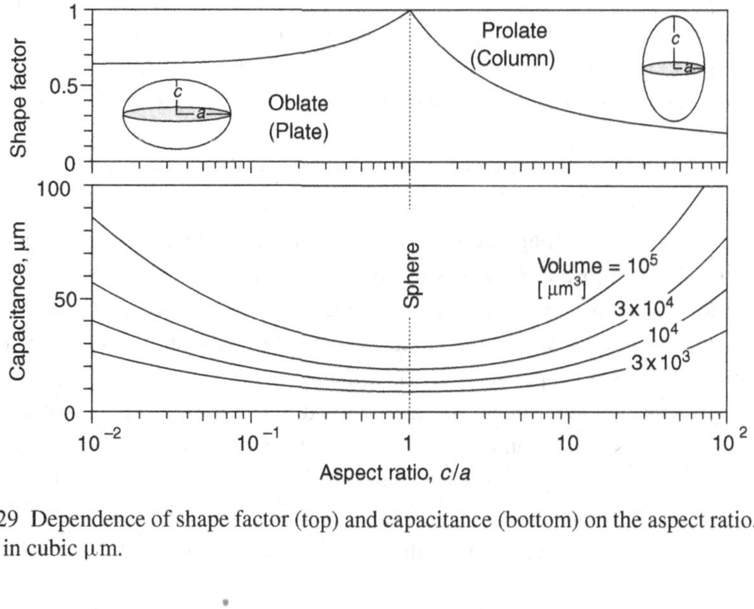

Figure 8.29 Dependence of shape factor (top) and capacitance (bottom) on the aspect ratio. Volumes are given in cubic µm.

see that preferred deposition on the basal face (as with columns) leads to more rapid mass accumulation (at any given ambient supersaturation) than does deposition on the basal face because of this geometrical effect.

Application of spheroidal capacitance in an ice-particle growth model is convenient for numerical cloud models, but it is realistic only if both the particle mass and the aspect ratio can be calculated simultaneously. Such a mathematical model was formulated by J.-P. Chen in 1994. Chen's model is said to be adaptive in the sense that the mass growth rate and shape of the particle both evolve with time in a self-consistent way in response to the prevailing environmental conditions. This adaptive growth model represents a distinct advance over the classical capacitance model, which required the capacitance to be specified from information outside of the growth model. The adaptive model, by contrast, computes the capacitance internally at each instant from the evolving size and shape of the particle. The model ice particle may start as a sphere (as from the freezing of a supercooled cloud droplet), but its shape (aspect ratio) changes appropriately as mass is added. The instantaneous rate with which the particle gains mass is assumed to be given by the capacitance model (Eq. (8.40)): $dm_p/dt = 4\pi C\rho_I G_I' s_i$. In line with the reasoning behind the electrostatic analogy, this mass growth law represents the overall flow of vapor toward the particle as if the particle were a point sink. Chen's ice-particle growth model couples this classical growth law (to determine the total flow in the far field) with a relationship for estimating how excess vapor is distributed in the near field to the different parts of the particle.

The near-field distribution of vapor is based on the physical principles discussed earlier. The molecular-scale surface physics determines the growth efficiencies (deposition coefficients), and the consequent gradients of vapor and temperature established in the near field drive the transport of mass and energy to and from the particle. The fundamental driver of habit evolution is the "inherent growth ratio" $\Gamma \equiv \alpha_B/\alpha_P$, which varies primarily with temperature. The shape of the particle evolves in response to the relative magnitudes of α_B and α_P, but it also conditions the vapor field. Vapor diffusing toward an ice particle from the far field is thus hypothesized to be distributed in the near field according to

$$\frac{dc/dt}{da/dt} = \frac{dc}{da} = \Gamma \cdot \varphi, \text{ alternatively } \frac{d\ln c}{d\ln a} = \Gamma. \tag{8.69}$$

The relative growth rates along the c and a axes at any instant clearly depend on the growth efficiencies in those directions (hence Γ), but they also reflect the particle shape that evolved to this point in time (hence φ). (Recall that the surface supersaturations and vapor gradients are greatest where the crystal extends farthest into the vapor field, so the ratio of vapor gradients is proportional to the dimensional ratio φ.)

The mass growth rate and aspect ratio are coupled through the geometry of a spheroid. First, recognize that a small amount of mass, dm_p, added to a particle increases its volume by $dV = dm_p/\rho_{dep}$, where ρ_{dep} is the ice density of the deposit. (If the particle is solid ice, then $\rho_{dep} = \rho_I$, the bulk density of ice.) This change in volume is related to the changes in the size and aspect ratio of the particle through the total derivative of spheroidal volume (given in Fig. 8.28): $d \ln V = 3d \ln a + d \ln \varphi$. Changes in the aspect ratio are similarly related to changes in the particle dimensions through the total derivative of its definition: $d \ln \varphi = d \ln c - d \ln a = (d \ln c / d \ln a - 1) d \ln a$. The mass distribution hypothesis (Eq. (8.69)) lets us replace $d \ln c / d \ln a$ with the inherent growth ratio $\Gamma = \alpha_B / \alpha_P$, so $d \ln \varphi = (\Gamma - 1) d \ln a$. Any change in a affects both the aspect ratio and the volume at the same time, so simultaneous solution of the two equations gives

$$d \ln \varphi = \frac{\Gamma - 1}{\Gamma + 2} d \ln V. \tag{8.70}$$

As an ice particle grows in volume, its aspect ratio changes in proportion to the deviation of the inherent growth ratio from unity. In temperature ranges conducive to the growth of plates, for instance, $\Gamma < 1$ and φ decreases with time. In columnar growth regimes ($\Gamma > 1$), φ increases with time. Equations (8.69) and (8.70) form the basis of Chen's adaptive growth model and let the classical capacitance model take on new importance. The changes in φ given by Eq. (8.70) cause the shape factor and capacitance to change (Fig. 8.28), which affects the subsequent rate of mass uptake (Eq. (8.40)) and the near-field distribution of vapor (Eq. (8.69)). The continual feedback of information makes the model versatile and adaptive to changing conditions.

The main input to Chen's adaptive growth model is the inherent growth ratio Γ, prescribed as a function of temperature and supersaturation. Due to the complexity of ice formation and growth, values of Γ are best derived from laboratory experiments or field observations. The top of Fig. 8.30 shows how a wide range of data can be used to construct a function $\Gamma(T)$. Considerable uncertainty exists in the accuracy of Γ at any one temperature T, but at least the transitions ($\Gamma = 1$) occur at the correct temperatures to define the observed primary-habit regimes. At lower temperatures, where appropriate growth data do not exist, one can employ observed relationships between the mass and maximum dimension of ice particles.

Running Chen's ice-particle growth model at discrete temperatures gives the results shown in the lower two panels of Fig. 8.30. The ice particle, assumed to be spherical initially with a radius of 1 μm, grew at liquid water saturation in air at a pressure of 1000 hPa. As shown for selected growth times in the middle panel, the aspect ratio of the particle gradually evolves away from that of a sphere, depending on the $\Gamma(T)$ function given by the solid curve in the top panel. The bold solid curve gives the results after 10 min of growth and is seen to agree well with data from wind-tunnel experiments. The mass of the particle similarly shows good agreement with the experimental data obtained after 10 min of growth. Note that mass accumulates on the particle preferentially when the aspect ratio deviates most from a sphere, in line with the capacitance curves shown in Fig. 8.29. Chen's ice growth model, despite the numerous simplifying assumptions and mathematical approximations, captures enough of the growth physics to yield reasonably accurate results. It is computationally efficient and appropriate for use in numerical cloud models.

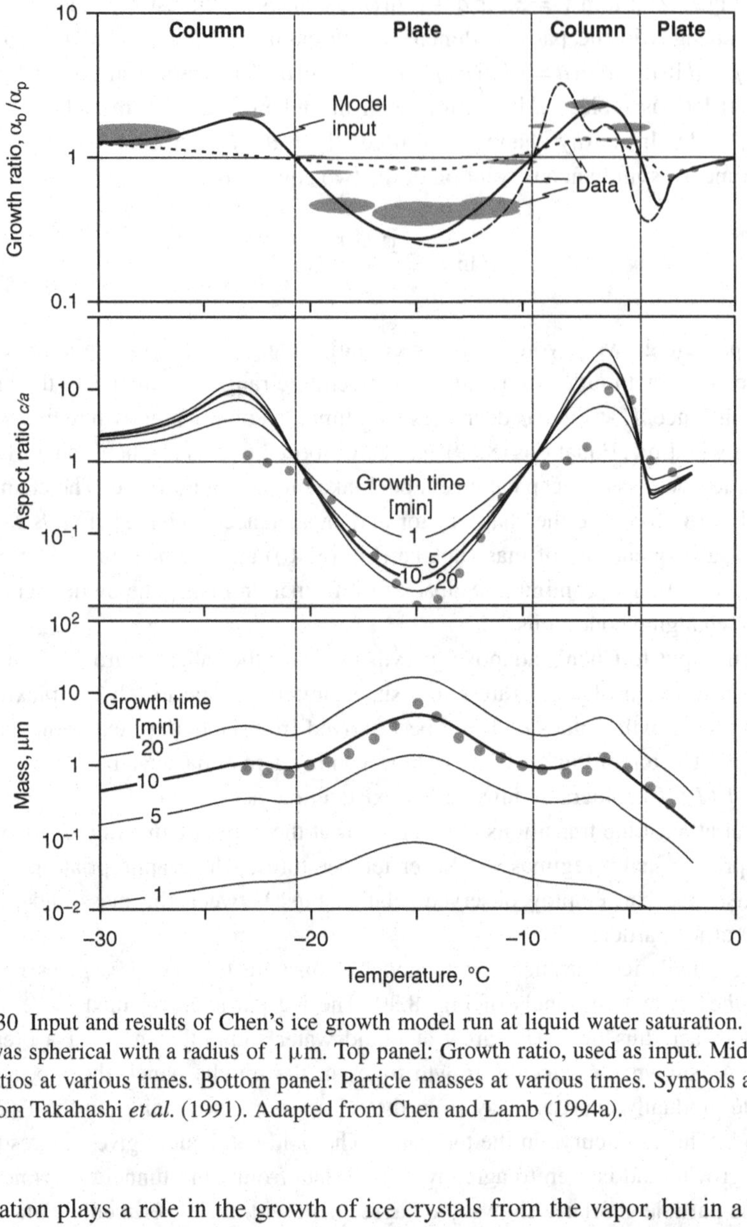

Figure 8.30 Input and results of Chen's ice growth model run at liquid water saturation. Initial ice particle was spherical with a radius of 1 μm. Top panel: Growth ratio, used as input. Middle panel: Aspect ratios at various times. Bottom panel: Particle masses at various times. Symbols are data at 10 min from Takahashi *et al.* (1991). Adapted from Chen and Lamb (1994a).

Ventilation plays a role in the growth of ice crystals from the vapor, but in a way that differs from that involved with liquid drops. We learned earlier that sedimentation of a particle accelerates the exchanges of mass and energy between the particle and its gaseous environment. The effect of ventilation is small for the growth of cloud droplets, but not for the evaporation of raindrops, mainly because of the large difference in fallspeeds. The overall effect of ventilation on the growth of an ice crystal is similarly minimal because of the low terminal fallspeeds. However, the non-spherical shape of a single ice crystal induces a distinctly non-uniform ventilation effect over the individual facets. An hexagonal plate, for instance, falls on average with the basal (slow-growing) face perpendicular to the airflow. For a given sedimentation speed, the air flows much faster across the prism (fast-growing) faces than it does across the basal face. Once a crystal grows beyond several hundred micrometers in maximum dimension, the non-uniform ventilation causes significantly accelerated growth of the already fast-growing faces. This asymmetric ventilation effect may contribute to the extreme aspect ratios observed for larger snow crystals.

8.4 Melting

The melting of ice particles is an important mechanism by which snow and graupel are transformed into rain in cold clouds. The process occurs commonly in winter warm-frontal bands and in summer convective storms as relative large solid-phase precipitation particles fall into warmer atmospheric layers. By cooling the air, melting also affects the thermodynamic stability of the lower atmosphere and thereby the dynamics of convective systems. The effect of melting is often observed as the so-called bright band displayed by weather radar, because melting changes the dielectric properties of the particles from those of the solid to those of liquid water. Melting also increases the densities and fallspeeds of the particles. The vertical depth of the melt layer (and bright band) below the 0 °C isotherm depends on how rapidly the melting occurs and the particles fall, processes that depend strongly on particle size.

The rate that an ice particle melts is limited by the transfer of energy to the liquid–ice interface, where the latent heat of fusion is absorbed. At the interface, thermal energy is added to the network of bonds holding the solid together. The molecular kinetic energy gets converted into molecular potential energy (latent heat of fusion). It is not necessary to treat the details of the molecular-scale processes here, but we do need to consider the conservation of energy on the scale of individual particles.

The concept of an energy balance, used earlier in the growth of cloud droplets and ice crystals, plays a key role in calculating rates of melting. The various processes that can contribute to the flow of thermal energy to the ice–liquid interface include heat conduction through the liquid in the particle and the air surrounding the particle, condensation of vapor, and the collection of cloud water. Radiation could be added to this list, but its contribution is usually insignificant on typical melting time scales. Note that conduction through the liquid occurs inside the particle, whereas the other mechanisms of heat transfer are external to the particle. We can therefore set up a qualitative relationship that serves as an energy budget:

$$\textit{Energy to melt ice} = \textit{internal conduction} = \textit{external conduction} \\ + \textit{condensation/evaporation}.$$

When the ice is contained inside its existing melt water, all of the energy to melt additional ice passes through the liquid, in which case the internal conduction of heat through the

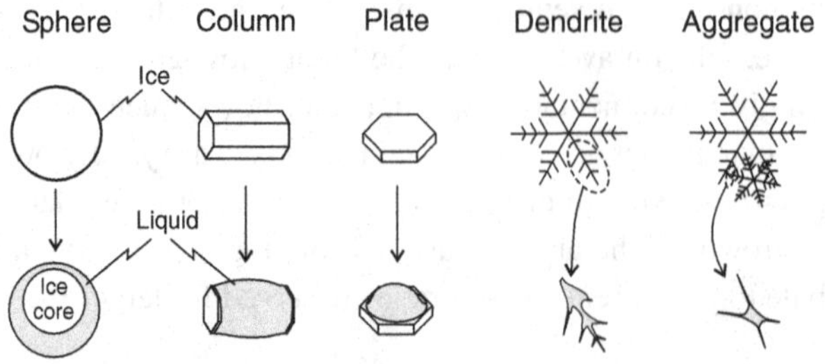

Figure 8.31 Geometries of selected ice particles and their patterns of melting. Shading represents melt water during an intermediate stage of melting. Adapted from various sources, including Rasmussen *et al.* (1984), Knight (1979), and Oralty and Hallett (2005).

water stands alone in the equation and is balanced by the three external transfer mechanisms. The rate of energy transferred internally depends on the growing thickness of the liquid and on the temperature difference between the inner and outer interfaces of the liquid. The temperature of the inner, ice–liquid boundary always stays close to $T_0 = 0\,°\text{C}$, but that of the outer, air–liquid boundary depends on the external heat-transfer mechanisms and the particle geometry.

A set of particle geometries is shown in Fig. 8.31 to illustrate the range of complexities that commonly arise in the atmosphere. The simplest case to consider is an initially spherical ice particle (approximation of a hailstone), but even here the ice core seldom remains in the center of the particle throughout the melting period. The energy for melting any ice particle invariably comes from the air, so newly formed liquid (the melt water) often resides between the remaining ice and the air. Most melting occurs by stages, with the amount of liquid increasing at the expense of the ice and surface tension affecting the distribution of liquid. Observations show that columns start melting uniformly over the surface, but later non-uniformly. As melting proceeds, the liquid gradually accumulates into one or more bulges (one is shown in the figure). Toward the end of melting, an initially columnar crystal takes the form of a spherical drop with only a tiny fraction of the original ice embedded. Plates melt similarly, but without any tendency to form multiple liquid bulges. During intermediate stages (see figure), a melting plate takes on the shape of a double-convex lens, with the edges defined by the original hexagon. Dendrites and aggregates are complex forms with multiple branches in close proximity. Melt water tends to fill interior corners, and surface tension further acts to keep aggregates together, but in progressively more compact shapes. Melting often occurs when the surrounding air is subsaturated, in which case parts of large aggregates may be evaporating while other parts melt. Arms of dendrites weakened by evaporation may detach, along with some melt water, and contribute in complicated ways to the population of cloud particles.

Despite the complexity of melting geometries, it is instructive to treat the idealized case of concentric spheres to understand the method of calculating a melting rate. The energy to melt a unit amount of ice is the enthalpy of fusion l_f, so melt water is formed at the rate

$dm_l/dt = I_q/l_f$, where m_l is the mass of liquid, I_q [J s^{-1}] is the flow of enthalpy into the ice–liquid interface, and l_f [J g^{-1}] is here expressed in mass units. By mass conservation, the ice disappears at the mass rate $dm_i/dt = -dm_l/dt$. A spherical ice core of radius r_i will therefore melt back at the linear rate

$$\frac{dr_i}{dt} = -\frac{I_q}{4\pi\rho_i l_f r_i^2}, \tag{8.71}$$

where ρ_i is the density of ice. The problem can be solved once the enthalpy flow (I_q) is determined as a function of time for each geometrical situation.

Calculation of enthalpy flow I_q and the surface temperature of the particle, as well as quantification of the energy budget in general, requires a mathematical description for each heat-transfer process appropriate to the geometry of the particle. Heat-transfer problems are typically solved using the convective-conduction equation ($\partial T/\partial t + \mathbf{u} \cdot \nabla T = \kappa_T \nabla^2 T$), which for our steady-state situation ($\partial T/\partial t = 0$) is

$$\mathbf{u} \cdot \nabla T = \kappa_T \nabla^2 T, \tag{8.72}$$

where $\mathbf{u}$ is the local velocity of the fluid and $\kappa_T = k_T/\rho c_p$ is the thermal diffusivity of the medium having thermal conductivity k_T, density ρ, and specific heat c_p. For the simplest case of concentric spheres and no fluid motion ($\mathbf{u} = 0$), the Laplace equation applies: $\nabla^2 T = 0$. The solution of this equation yields the temperatures of the inner and outer boundaries, from which the enthalpy flow can be calculated. Let the radius of the inner (ice) sphere be r_i and the radius of the particle (ice plus liquid) be r_p. Then, the temperature in the liquid portion of the particle varies with radius r as

$$T(r)|_{Liquid} = T_0 + \frac{T_{sfc} - T_0}{r_p - r_i} r_p \left(1 - \frac{r_i}{r}\right), \tag{8.73}$$

where T_{sfc} is the temperature of the particle surface (the air–liquid interface). The heat flow through the liquid to the ice is the area of any concentric sphere within the liquid times the inward energy flux, which by Fourier's first law is $\Phi_T = -k_{T,L}(dT/dr)$:

$$I_{q,melt} = 4\pi k_{T,L} \frac{r_p r_i}{r_p - r_i} \left(T_{sfc} - T_o\right). \tag{8.74}$$

Note that the liquid depth ($r_p - r_i$) is very small in the early stages of melting, but then so is the temperature difference $\left(T_{sfc} - T_o\right)$. The main issue now is determining the surface temperature $\left(T_{sfc}\right)$.

The surface temperature can be calculated only by treating the energy budget in some detail. We assume steady state and expect that the flow of energy passing through the liquid (given by Eq. (8.74)) is the same as that provided by the external heat-transfer mechanisms. Conduction through the air is described by the same mathematics as is conduction through the liquid; just the constants need to be made appropriate for air in Eq. (8.72). The temperature profile outside the drop can be obtained by repeating the mathematical procedure, or one can use Eq. (8.73) with the following replacements: $r_p \to \infty$, $r_i \to r_p$, $T_{sfc} \to T_\infty$, and $T_0 \to T_{sfc}$:

$$T(r)|_{air} = T_\infty - \left(T_\infty - T_{sfc}\right)\frac{r_p}{r}. \tag{8.75}$$

The overall temperature profiles inside and outside the particle are shown schematically in Fig. 8.32 for the case of concentric spheres in stagnant air. The energy flow into the drop by conduction through the air is (Eq. (8.74) with the same replacements)

$$I_{q,air} = 4\pi k_{T,air} r_p \left(T_\infty - T_{sfc}\right). \tag{8.76}$$

Note that the energy flow must be the same into the surface from the air side as it is away from the surface on the liquid side (by energy conservation), but the temperature gradients at the particle surface differ in proportion to the conductivities:

$$(dT/dr)_{air} \,/\, (dT/dr)_{liquid} = k_{T,L}/k_{T,air} \sim 24. \tag{8.77}$$

The surface temperature, calculated by equating Eqs. (8.74) and (8.76), is

$$T_{sfc} = \frac{T_\infty + C_{melt} T_0}{1 + C_{melt}}, \tag{8.78}$$

where

$$C_{melt} = \frac{k_{T,L}}{k_{T,air}} \frac{r_i/r_p}{1 - r_i/r_p}. \tag{8.79}$$

Note the limiting cases: When melting just starts, the radius ratio, r_i/r_p, is close to unity, which means the melting coefficient C_{melt} is very large, and $T_{sfc} \to T_0$. Toward the end of the melting process, $r_i/r_p \to 0$, giving a limiting surface temperature $T_{sfc} \to T_\infty$. Over the course of melting, the magnitudes of the surface temperature, the energy flow, and the size of the ice core all change continuously, making the solution look complicated.

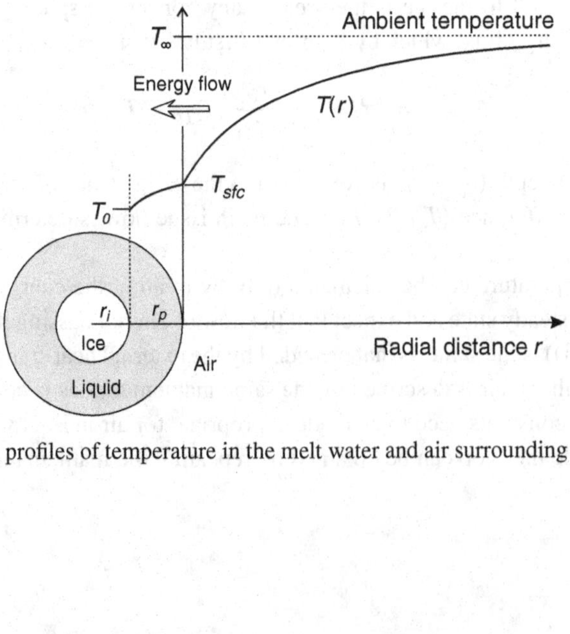

Figure 8.32 Radial profiles of temperature in the melt water and air surrounding an ice core.

But, the problem is straightforward in concept: One substitutes Eqs. (8.78) and (8.79) into Eq. (8.76) to determine the heat flow. This expression for I_q is then used in Eq. (8.71) to determine the linear melting rate. Integration of Eq. (8.71) then yields the time-dependent radius of the ice core. The idealized situation of concentric spheres in a stagnant environment with constant ambient temperature allows the surface temperature, the heat flow, and the melting rate to be calculated readily.

More realistic situations involve greater complexity. As an ice particle begins to melt and a layer of liquid water develops, the surface temperature gradually rises, causing an increase in the equilibrium vapor pressure of the water. Net evaporation may result if the ambient humidity is sufficiently low, thereby cooling and offsetting the warming tendency. The effect of evaporation on the energy budget is calculated as it was for the growth from the vapor (Section 5.2). The rate of energy lost through evaporation is similar in form to Eq. (8.76):

$$I_{q,evap} = 4\pi l_v D_v r_p \left(\rho_{v,\infty} - \rho_{v,sfc}\right), \tag{8.80}$$

where l_v is the enthalpy of vaporization, D_v is the diffusion coefficient, and where $\rho_{v,\infty}$ and $\rho_{v,sfc}$ are, respectively, the vapor densities far from the particle and in equilibrium with the surface. Net evaporation occurs when $\rho_{v,\infty} < \rho_{v,sfc}$. Particles falling into dry layers outside of cloud may in fact cool enough to delay the onset of melting by several degrees (see Fig. 8.33). The onset condition for melting can be estimated by setting the sum of Eqs. (8.76) and (8.80) to zero and solving for the ambient temperature that yields $T_{sfc} = T_0$:

$$T_{\infty,onset} = T_0 - \frac{l_v D_v}{k_{T,air}} \left(S_\infty \rho_v (T_\infty) - \rho_v (T_0)\right). \tag{8.81}$$

Here, S_∞ is the ambient saturation ratio (relative humidity), and the function $\rho_v(T)$ is the vapor density in equilibrium with a surface at temperature T. An ice particle inside a cloud

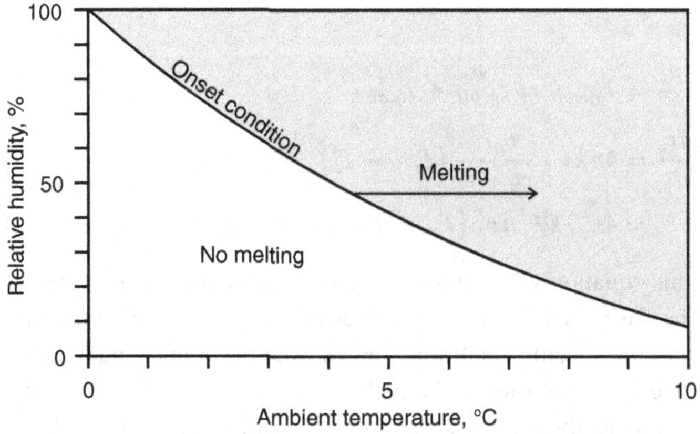

Figure 8.33 Ambient relative humidity and temperature needed for melting to begin. Adapted from Rasmussen and Pruppacher (1982).

may well begin melting near the nominal melting point (T_0), but outside a cloud, melting may not start until it has fallen several hundred meters below the 0 °C isotherm.

Melting typically occurs when ice particles are large enough to fall into warmer parts of the atmosphere, so effects of advection on the transport of energy and mass must also be taken into account. For energy transport, the steady-state convective conduction equation (8.72) is used, and for mass transport, the steady-state convective diffusion equation,

$$\mathbf{u} \cdot \nabla \rho_v = D_v \nabla^2 \rho_v. \tag{8.82}$$

The flow of air around a falling particle contracts the boundary layer and has the effect of enhancing the exchange of heat and vapor between the particle and its environment. The thermal conductivity and the diffusivity each appear to have larger values when comparing experiments and theory, so they are multiplied by appropriate factors called ventilation coefficients. A variety of research has yielded the following semi-empirical expressions:

$$\text{For energy}: f_h = 0.78 + 0.308 N_{\text{Pr}}^{1/3} N_{\text{Re}}^{1/2}, \tag{8.83}$$

where $N_{\text{Pr}} = \upsilon/\kappa_T$ is the Prandtl number and N_{Re} is the Reynolds number.

$$\text{For vapor}: f_v = 0.78 + 0.308 N_{\text{Sc}}^{1/3} N_{\text{Re}}^{1/2}, \tag{8.84}$$

where $N_{\text{Sc}} = \upsilon/D_v$ is the Schmidt number. These coefficients yield mass-average values that vary with the size of the particle (through the Reynolds number), and they differ from each other only because the thermal diffusivity κ_T is less than the vapor diffusivity D_v by less than 30%. The ventilation coefficients are multiplied by the respective physical coefficients to give effective parameters: $f_h k_{T,air}$ for energy, and $f_v D_v$ for vapor.

The energy budget can now be written for melting particles and used to calculate the rate of melting. Including the effects of ventilated conduction and ventilated condensation, we have

$$\begin{aligned}
-l_f \frac{dm_i}{dt} &= I_{q,melt} = I_{q,air} + I_{q,evap} \\
-4\pi \rho_i l_f r_i^2 \frac{dr_i}{dt} &= 4\pi k_{T,L} \frac{r_p r_i}{r_p - r_i} \left(T_{sfc} - T_o\right) \\
&= 4\pi f_h k_{T,air} r_p \left(T_\infty - T_{sfc}\right) + 4\pi l_v f_v D_v r_p \left(\rho_{v,\infty} - \rho_{v,sfc}\right).
\end{aligned} \tag{8.85}$$

Integration of this equation yields a melting curve, the dependence of the ice-core radius (r_i) on time. The dashed curve in Fig. 8.34 results when r_p is held constant and the particle is exposed to the specific conditions shown. The ice core changes slowly at first, but the rate of change increases toward the end, when the surface-to-volume ratio of the ice core increases dramatically. The time for complete melting computed with this model does not agree well with measurements (shown near lower axis) taken in wind-tunnel experiments, so other factors must come into play. Observations during the experiments showed

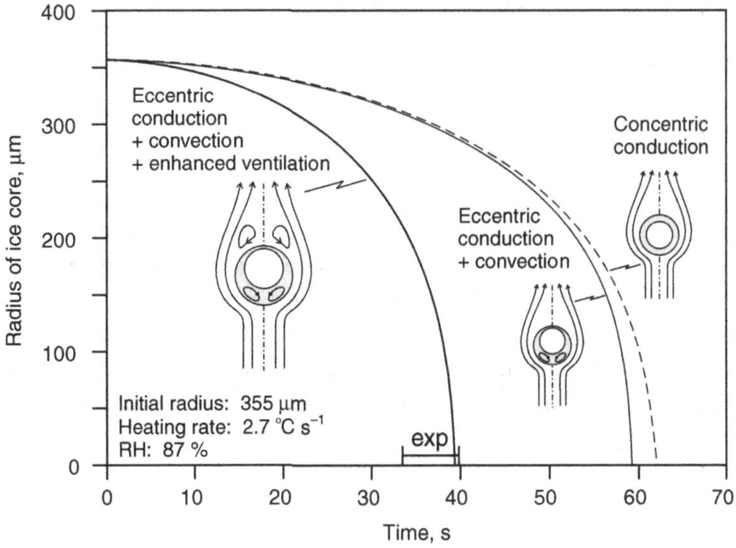

Figure 8.34 Melting curves for three scenarios under the conditions indicated in the lower right corrner. The range of observed end points is indicated by "exp" near the bottom. Adapted from Rasmussen *et al.* (1984).

that the concentric feature inherent in the model is not valid. The ice core tends to rise toward the top of the particle, partly because ice is less dense than liquid water. The eccentric geometry must be accounted for, as well as the likely convection within the liquid (shown by the ovals in the middle and left insets). When such additional features are introduced into the mathematical model, melting is slightly more rapid, as shown by the light solid curve in Fig. 8.34, but the improvement is minor. The only way the model could be made to agree with the data was to enhance the ventilation coefficients by a factor of 2.4. The most realistic melting curve, resulting with eccentric geometry, internal circulations, and enhanced ventilation, is shown by the heavy solid curve. The enhancement in ventilation of large ice particles is thought to arise physically from a combination of surface irregularities, the shedding of eddies in the wake of the falling particle, and non-steady motions.

Aggregates melt in very different ways from frozen drops, graupel, and hailstones. Their complicated and time-varying geometries preclude the development of any convenient mathematical model. Moreover, the densities, fallspeeds, and ventilation all increase during melting. Experiments with wind tunnels have led researchers to identify several stages of melting:

Stage 1: Exposed parts of crystals melt, especially on lower surface, leaving small drops on tips.

Stage 2: Melt water seeps by capillary action into the interior of the aggregate, while the outer parts remain exposed to the warm air.

Stage 3: The mechanical structure becomes altered, the weakest inter-crystal bonds giving way first.

Stage 4: The main frame collapses into the shape of a drop, with the melt water engulfing the remaining ice.

The driving force for the structural changes arises predominantly from surface tension, giving the tendency for minimization of the total surface area with time. Aerodynamic forces play at best minor roles in the structural rearrangements.

In contrast to the situation with graupel and hailstones, the energy for melting does not necessarily pass through liquid during the melting of aggregates. Nevertheless, the two mechanisms of heat transfer external to the particle, conduction and evaporation/condensation, still dominate and must be considered. The appropriate energy-balance equation for aggregates is thus

$$\begin{aligned} l_f \frac{dm_l}{dt} &= I_{q,air} + I_{q,evap} \\ &= 4\pi C \overline{f} k_{T,air} (T_\infty - T_0) + 4\pi C \overline{f} l_v D_v (S_\infty \rho_v (T_\infty) - \rho_v (T_0)), \end{aligned} \tag{8.86}$$

where it is assumed that the particle is non-spherical and best represented as a spheroid having a particle capacitance C. Note that conduction is driven physically by the difference in temperature between the environment and the particle, which is at the melting point. Condensational warming is driven by the ambient saturation ratio and the equilibrium vapor densities at the same two temperatures. The ventilation coefficients for heat

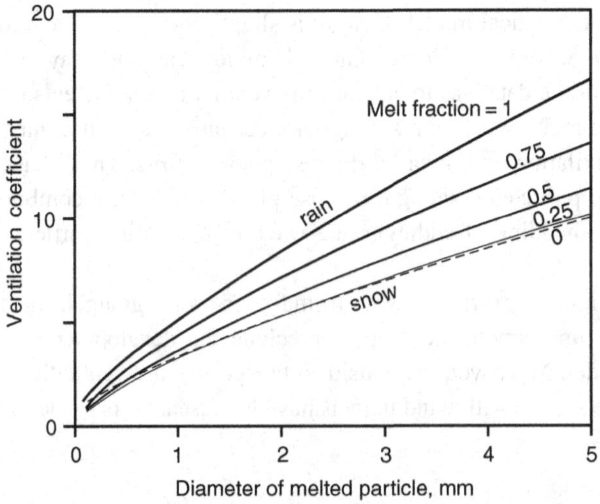

Figure 8.35 Variation of the ventilation coefficient on size and melt fraction of aggregates with density 0.1 g cm^{-1}. Adapted from Szyrmer and Zawadzki (1999).

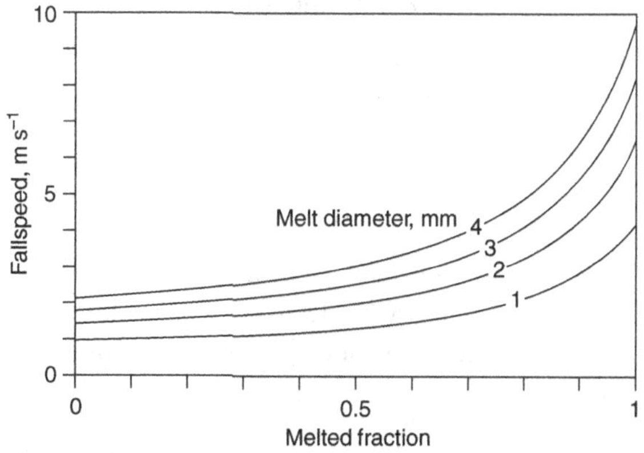

Figure 8.36 Dependence of fallspeed on the melt fraction and size of aggregates. Adapted from Szyrmer and Zawadzki (1999).

and vapor transfer are often assumed to differ little from each other, so a mean ventilation coefficient $\overline{f}$ is used here. The ventilation coefficient depends on the fallspeed, therefore on the size of the particle and the extent of melting, in a manner suggested by Fig. 8.35. One should expect $\overline{f}$ to increase with the size of the particle, here specified as the diameter of the melted particle, because ventilation, hence the rate of melting, also increases as the particle becomes more compact and the fallspeed increases during melting (see Fig. 8.36). The fallspeed changes little during the early stages of melt, but it increases rapidly once structural rearrangement of the aggregate begins, when the melt fraction exceeds about 0.7. Detailed calculations show that aggregates can fall several hundred meters before melting completely.

8.5 Further reading

Fletcher, N.H. (1962). *The Physics of Rainclouds*. Cambridge: Cambridge University Press, 386 pp. Chapters 6 (growth of cloud droplets) and 10 (growth of ice crystals).

Hobbs, P.V. (1974). *Ice Physics*. Oxford: Oxford University Press, 837 pp. Chapter 8 (vapor growth), Appendix B (dislocations in crystals).

Pruppacher, H.R. and J.D. Klett (1997). *Microphysics of Clouds and Precipitation*. Dordrecht: Kluwer Academic Publ., 954 pp. Chapters 13 (diffusional growth of liquid and ice) and 16 (melting of ice particles).

Rogers, R.R. and M.K. Yau (1989). *A Short Course in Cloud Physics*. New York: Pergamon Press, 293 pp. Chapters 7 (droplet growth) and 9 (ice growth).

Seinfeld, J.H. and S.N. Pandis (1998). *Atmospheric Chemistry and Physics*. New York: John Wiley and Sons, 1326 pp. Chapter 15 (growth of droplets).

Wang, P.K. (2002). *Ice Microdynamics*. San Diego, CA: Academic Press, 273 pp. Chapter 4 (ventilation effects on ice growth).

Young, K.C. (1993). *Microphysical Processes in Clouds*. New York: Oxford University Press, 427 pp. Chapters 5 (diffusional growth of water drops, ventilation effects) and 6 (diffusional growth of ice crystals).

8.6 Problems

1. Ignoring curvature and solution effects, we have shown that the rate of growth of a droplet by vapor deposition can be expressed by

$$r_d \frac{dr_d}{dt} = G \cdot s.$$

If a cloud sustains a supersaturation of $s = 0.01$ at a temperature of 4°C and pressure of 1000 hPa, calculate the time it takes a droplet to grow from 10 μm to 1000 μm. Discuss your results with respect to the formation of precipitation in cumulus and stratiform clouds.

2. Consider a raindrop falling from a tilted updraft through the side of a small Colorado cumulus cloud. Right before the drop exits the cloud, it is in steady-state growth in an environment with supersaturation $s = 0.0001$ and temperature 5°C. Assume that environmental conditions change instantaneously to that of the cooler (3°C) and drier ($s = -0.5$) environmental air as the drop exits the cloud. Discuss what happens to the drop's vapor and thermal fluxes as it makes this transition from supersaturated to subsaturated conditions.

3. We derived an expression for the growth of cloud drop by vapor deposition, given by Eq. (8.17). Suppose this cloud droplet is situated near the top of the cloud where it experiences a net radiative cooling at the same time that it is growing by condensation. Derive the mass growth rate equation for this cloud droplet. Be sure to include the effect of the radiative cooling as well as the effect of latent heating.

4. Suppose we have a 500 μm ice crystal (maximum dimension) in a water-saturated environment with a temperature of −12°C. You may assume that the crystal is well approximated by an oblate spheroid with aspect ratio 0.2. Calculate the crystal mass growth rate with and without ventilation included. How long will it take the crystal to grow to a size of 1 mm under these conditions?

5. Consider a cloud drop ($r_d = 20$ μm) growing in a supersaturated environment. The environmental temperature $T = 5\,°\mathrm{C}$, pressure $p = 850$ hPa, and supersaturation $s = 0.01$. Calculate the drop surface temperature with and without ventilation. If the drop is detrained from a cloud, it may suddenly find itself in a dry environment. Repeat your previous calculation, but now for $s = -0.3$. Note the relative magnitude differences. Discuss your results in the context of our assumptions in derivation of the vapor depositional growth equation.

6. The onset of melting depends not only on the environmental temperature, but also the relative humidity. The temperature 0 °C level is therefore not the best indicator of when melting will begin. What single variable determined from atmospheric temperature and dewpoint profiles is a better indicator for the onset height of melting? Given similar

temperature profiles in Colorado and Florida, where would one have a higher expectation for the survival of hail to the surface if the hail exits the cloud at the 0 °C level? Explain.

7. In the trailing stratiform precipitation behind a squall line, the rain results from aggregates melting as they fall to the ground. Where, relative to the height of the 0 °C level will the aggregate start to melt if
   (a) the anvil cloud base is warmer than 0°C,
   (b) the anvil cloud base is colder than 0°C?
   Please explain.

9

Growth by collection

9.1 Overview

The particles constituting a cloud, once they have grown large enough, often collide with other particles. Initial growth from the vapor causes cloud particles to acquire mass and increasingly come under the influence of the Earth's gravitational attraction. With new mass, the particles fall ever faster relative to the air in their immediate environments. At some point, as faster particles overtake slower particles, the inertia suffices to cause them to collide and possibly stick together. The growth of one particle by the collection of others gives that particle favored status (by virtue of its new mass and increased fallspeed) as a collector of still other particles. Growth by gravitational collection is an important class of mechanisms responsible for the development of rain and snow.

Cloud particles can be either liquid or solid, and they can be small or large. We therefore find it useful to devise a set of categories by which particles grow, as shown in Fig. 9.1. Particles growing from the vapor do so as individual members of the cloud, but particles growing by collection always involve pairs. When both members of an interacting pair are liquid, the process is called collision-coalescence. The larger ("collector") drop of a liquid–liquid pair collects one or more smaller drops, because of the difference in fallspeeds, and leads to the growth of the collector at the expense of the collected drops. Collision-coalescence is the basic mechanism by which rain develops in warm clouds, those in which ice plays no significant role.

When at least one of the interacting pairs is solid (ice), then the situation is complicated by the range of possibilities (refer again to Fig. 9.1). When the temperature is below the melting point (nominally $T_0 = 273.15\,\text{K}, 0\,°\text{C}$), a relatively large ice crystal (which likely grew by vapor deposition) may collide with supercooled cloud droplets that freeze rapidly on contact with the ice surface. This riming process leads initially to lightly rimed ice crystals, then to moderately and heavily rimed crystals, eventually to graupel and hail when conditions are appropriate. If the situation is reversed, such that a relatively large supercooled drop (which may have grown by collision-coalescence) collects a small ice crystal, a frozen drop results, with little evidence of the original ice crystal remaining. This process is appropriately called capture nucleation because the liquid drop is suddenly converted to an ice particle, much as it would were it to have collected an ice nucleus

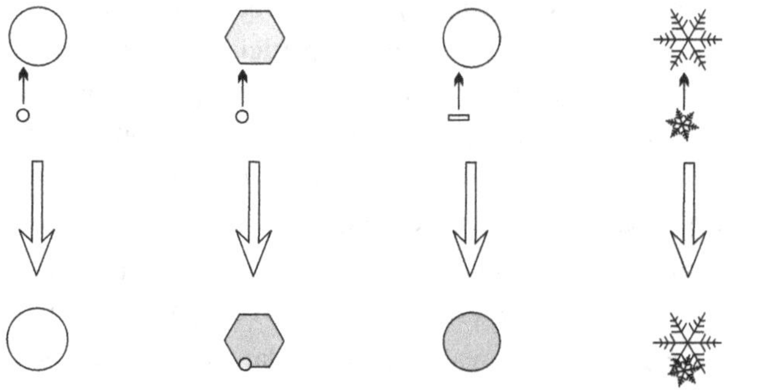

Figure 9.1 Categories of growth involving collection.

operating in the contact mode. Capture nucleation, because it results from the collection of a preexisting ice particle, is neither a homogeneous nor a heterogeneous nucleation event. By itself, capture nucleation does not lead to any new ice particles. Finally, the collection of small ice crystals by a larger ice crystal is termed aggregation, the process responsible for many of the snowflakes from winter storms. Collection growth involving ice is responsible for much precipitation from cold clouds. Ice–ice collisions that do not lead to growth may nevertheless play important roles in the electrification of clouds. In every case, the outcome of collisions depends on the relative motions of the particles involved, so we digress to consider the microdynamics of the various types of cloud particles.

9.2 Particle fallspeeds

Collisions between two neighboring cloud particles occur when their paths through the air cross a given point in space at the same time. It is the difference in the three-dimensional, vector velocities of the two particles that causes them to collide. The requisite velocity difference may be established by small-scale turbulent air motions, but most commonly it arises from differences in particle fallspeed, that downward component of velocity that causes particles denser than air to sediment because of gravity. Here, we develop the microdynamics of single particles in air and show the effects of size and shape on the fallspeeds of individual cloud particles before collision.

9.2.1 Aerosol particles and liquid drops

The settling behavior of a cloud particle depends on how it interacts with the air in its immediate environment. All particles, being bits of condensed matter, are pulled downward by gravitational attraction to a greater extent than is the air they displace. So, even with no motion relative to the air, a particle of volume V_p experiences a slight buoyancy force that is directed upward: $F_B = \rho_{air} V_p g$, where ρ_{air} is the local air density and g is the acceleration due to gravity. However, the weight of the particle, $F_g = m_p g = \rho_p V_p g$, is about a factor $F_g/F_B = \rho_p/\rho_{air} \sim 10^3$ larger, so we ignore the buoyancy effect. The net force on a particle at rest in stagnant air, $F_{net}|_{rest} = F_B - F_g \simeq -F_g$, is thus downward and imparts a downward acceleration to the particle. As the particle gains speed, it pushes air out of the way and so experiences air resistance, expressed as the drag force $F_D = F_D\left(v'_p\right)$, which increases with the instantaneous (time-dependent) speed v'_p of the particle relative to the air (see Fig. 9.2). The magnitude of the net force on a moving particle is therefore $F_{net}\big|_{moving} = F_g - F_D\left(v'_p\right)$, and an appropriate scalar equation of motion is

$$m_p \frac{dv'_p}{dt} = F_g - F_D\left(v'_p\right). \tag{9.1}$$

Being scalar in form, this equation describes only the magnitude of the fall velocity, which itself is negative (points downward). A particle allowed to fall from an initial state of rest (when $F_D = 0$) accelerates, its speed v'_p increasing with time until $F_D\left(v'_p\right) = F_g$. Once these counteracting forces balance, the particle speed no longer increases. This maximum downward speed is called the terminal fallspeed v_p (without a superscript), or just fallspeed. The time to achieve this end state depends on the mass and size of the particle, but for most atmospheric applications, we can ignore the transient acceleration and assume that cloud particles always fall at their terminal speeds. We must recognize, however, that the fallspeed (a measure of motion relative to the air in which it resides) is distinct from its ascent rate $dz_p/dt = w - v_p$, which is the rate of vertical motion relative to the Earth when the air is rising with speed w. Note that the concept of fallspeed rests on the existence of a drag force, which in turn rests on the presence of air. Particles falling in a vacuum experience no drag force and so would continue to accelerate; particles falling in vacuum have no terminal fallspeed.

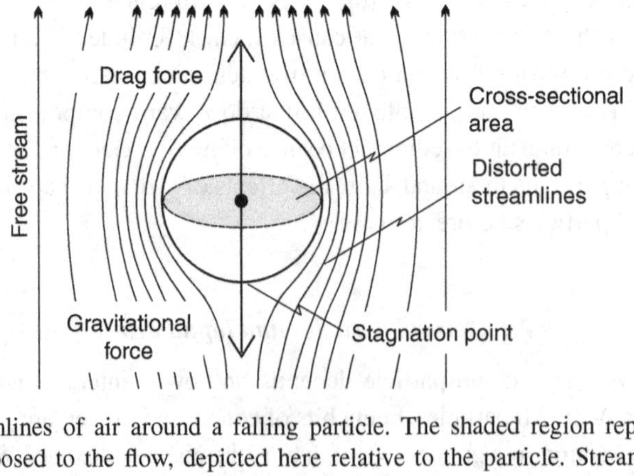

Figure 9.2 Streamlines of air around a falling particle. The shaded region represents the cross-sectional area exposed to the flow, depicted here relative to the particle. Streamlines courtesy of P. Bannon.

Calculations of particle fallspeed rely on knowledge of how the drag force varies with the properties of the particle and the air through which it falls. As we shall see, the drag force depends on the size and shape of the particle in rather complicated ways. However, all calculations follow the same general procedure. Refer again to Fig. 9.2 and note that the falling particle exposes a cross-sectional area A_c to the air, which must flow around the particle. The point on the leading edge of the particle where the streamlines divide is called the stagnation point because the air there does not move relative to the particle. The pressure at this point is the stagnation pressure, $p_{stag} = p_{atm} + p_{dyn}$, where p_{atm} is the ambient pressure at this altitude and p_{dyn} is the dynamic pressure, the additional pressure that arises from the relative motion of the particle (v_p). Pressure is a force per unit area, but it can be interpreted as an energy density $[\mathrm{J\,m^{-3}}]$: $p_{dyn} = \frac{1}{2}\rho_{air} v_p^2$, which is the change in kinetic energy of the air from its value at the stagnation point relative to that in the free stream (away from the influence of the particle). The drag force is proportional to the dynamic pressure and to the cross-sectional area of the particle, so we can write

$$F_D\left(v_p\right) = p_{dyn} A_c C_D = \frac{1}{2}\rho_{air} v_p^2 A_c C_D, \tag{9.2}$$

where the proportionality constant C_D is known as the the drag coefficient, an empirical factor that accounts for all non-geometrical components to the drag force. Equation (9.2) is a defining relationship for the drag coefficient, which itself depends on fallspeed. The fallspeed, defined as that speed (v_p) which yields the force balance, $F_D\left(v_p\right) = F_g = \rho_p V_p g$, can in principle be calculated in terms of the particle volume V_p and cross-sectional area A_c with the help of Eq. (9.2):

$$v_p = \left(\frac{2g}{C_D}\frac{\rho_p}{\rho_{air}}\frac{V_p}{A_c}\right)^{1/2}. \tag{9.3}$$

This formula is impractical without knowledge of the function, $C_D = C_D\left(v_p\right)$, and the particle geometry. Nevertheless, we see from Eq. (9.3) that fallspeed ultimately depends on the ratio of particle mass ($\rho_p V_p$, a measure of how strongly gravity pulls on the particle) to the effective cross-sectional area ($C_D A_c$, how much air is forced aside because of the particle motion). Large particles that are of compact shape (e.g., raindrops or hailstones) will have large fallspeeds, in contrast to small particles that expose large areas to the flow (e.g., pristine ice crystals).

The generic calculation of fallspeed can be made specific to cloud drops by assuming a sphere of diameter D_p. Then, the particle volume $V_p = (\pi/6)\,D_p^3$ and cross-sectional area $A_c = (\pi/4)\,D_p^2$, yielding the volume-to-area ratio $V_p/A_c = (2/3)\,D_p$. When the different terms in the (Navier–Stokes) equation of motion, which describes the air flow around the drop, are compared with one another, one finds that the flow is characterized rather well by the dimensionless Reynolds number, the ratio of the inertial and viscous forces:

$$N_{\mathrm{Re}} \equiv D_p v_p / \nu_{air}, \tag{9.4}$$

where $\nu_{air} = \mu_{air}/\rho_{air}$ is the kinematic viscosity of air having dynamic viscosity μ_{air}. The drag force is thus conventionally expressed (using Eq. (9.2)) in the form

$$F_D(v_p) = 3\pi \mu_{air} D_p v_p \left(\frac{C_D N_{\mathrm{Re}}}{24}\right) = F_{D,Stokes}\left(\frac{C_D N_{\mathrm{Re}}}{24}\right), \tag{9.5}$$

so that the general drag force can be compared with the Stokes drag force $F_{D,Stokes} = 3\pi \mu_{air} D_p v_p$, which is the only drag force that can be calculated analytically. The factor in parentheses is a dimensionless number that corrects the drag force when the flow is not Stokesian (creeping). One gains the fallspeed by equating the gravitational and drag forces or using Eq. (9.3):

$$v_p = \frac{1}{18}\frac{\rho_p g D_p^2}{\mu_{air}\left(\frac{C_D N_{\mathrm{Re}}}{24}\right)} = \frac{4}{3}\frac{\rho_p g}{\mu_{air}}\left(\frac{D_p^2}{C_D N_{\mathrm{Re}}}\right). \tag{9.6}$$

All of the size dependence is contained in the final factor, $D_p^2/(C_D N_{\mathrm{Re}})$. This equation is general for spherical particles, but the dependence of C_D on N_{Re} (and therefore on D_p) must first be established.

The drag coefficient (C_D) and the Reynolds number (N_{Re}) vary greatly with particle size and the type of flow around the particle. Whereas the mass, hence gravitational force, of a spherical particle depends simply on particle diameter and density: $F_g = m_p g = \frac{\pi}{6}\rho_p D_p^3 g$, the drag force (Eq. (9.5)) and thus fallspeed (Eq. (9.6)) depend on particle size in indirect and complicated ways. The problem of calculating fallspeed hinges on being able to calculate the drag force. The problem is facilitated by dividing the full range of possible sizes into several regimes, as shown in Fig. 9.3. Particle sizes range over seven decades in the atmosphere, so it should not be surprising to learn that the smallest (aerosol) particles behave very differently from the largest (rain or hail) particles. The small end (~ 1 nm) is defined by particles that may contain only a few molecules, whereas the large end (~ 1 cm) of liquid particles is limited by rupture (breakup of raindrops). Small (submicron) particles tend to stay suspended in the air, whereas large particles fall out and precipitate. The regimes are dictated by how particle diameter compares with the mean free path

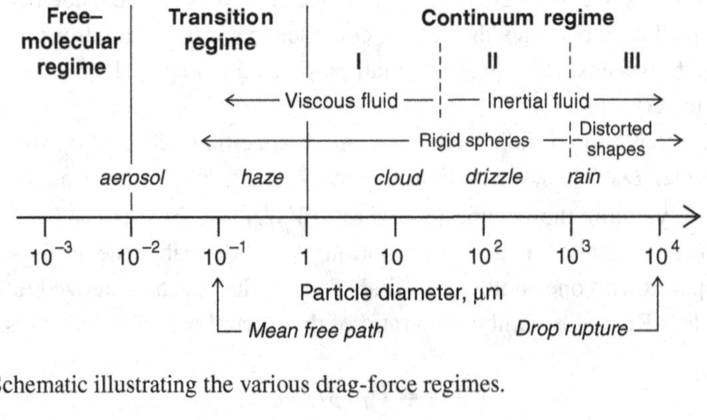

Figure 9.3 Schematic illustrating the various drag-force regimes.

λ_{air} ($\sim 0.1\mu m$) of air molecules, whether the air flow around the particle is viscous or turbulent, and whether the particle shape stays spherical or not. Each of the major drag-force regimes are treated separately before developing the overall dependence of fallspeed on size.

Free-molecular regime ($D_p << \lambda_{air}$)

When a particle is small compared with the mean free path, air must be treated as a collection of molecules, not as a continuum. The drag force in the free-molecular regime is dominated by collisions of individual air molecules with the particle. Each molecular collision transfers a bit of momentum to the particle, which causes the particle to experience a small force. It is these discrete molecular impacts that cause small aerosol particles to experience Brownian motion and diffuse through the air. The collective action of molecular collisions on larger particles cancels out when the particle is stationary, but a net force results when the particle moves through the air. The problem is complicated in its details, but a conceptual derivation suffices.

A reasonable theoretical treatment begins by estimating the momentum transferred by collisions of air molecules with the particle. Recall from Section 3.1 that the molecular impingement flux onto a surface is $\Phi_{air} = \frac{1}{4} n_{air} \overline{c}_{air}$, where n_{air} is the concentration of air and $\overline{c}_{air} = \left(\frac{8k_B T}{\pi m_{air}}\right)^{1/2}$ is the mean speed of the air molecules (having mass m_{air}) when the temperature is T. Air molecules thus collide with the surface of a stationary spherical particle with a frequency $\omega_{air} = \Phi_{air} A_p = \left(\frac{\pi}{4}\right) D_p^2 n_{air} \overline{c}_{air}$. The momentum transferred to the particle during each collision depends upon the nature of the collision (whether the molecule sticks or rebounds), but it is nevertheless proportional to the mass of the molecule and its impact speed. The impact speed varies around the surface, being larger on the lower (upstream) surface and weaker on the upper (downstream) surface compared with those on the lateral surfaces, and it is this asymmetry that gives rise to the drag force proportional to the particle speed v_p. Integration of the forces due to all collisions around the surface yields the free-molecular drag force

$$F_{D,FM} = C_{FM} m_{air} n_{air} \overline{c}_{air} D_p^2 v_p, \tag{9.7}$$

where C_{FM} is a constant on the order of unity. Note that $m_{air} n_{air} = \rho_{air}$, the local density of the air. Equating the drag force with the weight of the particle allows one to calculate the fallspeed as

$$v_{p,FM} = \frac{\pi}{6C_{FM}} \frac{\rho_p}{\rho_{air}} \frac{g}{\overline{c}_{air}} D_p. \tag{9.8}$$

The fallspeed in the free-molecular regime is proportional to the diameter of the (aerosol) particle.

Transition regime ($D_p \sim \lambda_{air}$)

When the particle diameter is comparable to the mean free path of air, the physics becomes complicated. Complete molecular-scale interactions are difficult to calculate, and the assumptions upon which continuum microdynamics rests are not strictly valid. The common approach is to calculate the drag coefficient as an empirical modification to the Stokes drag $F_{D,Stokes}$ (which will be developed in the next subsection). The assumption of no flow at the particle surface, used to calculate $F_{D,Stokes}$, breaks down when the particle size is comparable to the spacing between gas molecules, so the transition-regime drag is smaller than the Stokes drag by the so-called Cunningham slip-correction factor, $C_c = 1 + (1.26)\frac{2\lambda_{air}}{D_p}$. The drag force in the transition regime is based on the following interpolation formula:

$$F_{D,trans} = \frac{F_{D,Stokes}}{C_c} = \frac{3\pi\mu_{air} D_p v_p}{1 + (1.26)\frac{2\lambda_{air}}{D_p}}. \tag{9.9}$$

The transition-regime fallspeed, calculated using this drag force, is therefore

$$v_{p,trans} = C_c v_{p,Stokes} = \frac{\rho_p g}{18\mu_{air}}\left(1 + (1.26)\frac{2\lambda_{air}}{D_p}\right) D_p^2. \tag{9.10}$$

Note the limiting conditions based on the relative magnitudes of D_p and λ_{air}. When $D_p << \lambda_{air}$, the drag force and fallspeed revert to the respective expressions appropriate to the free-molecular regime (recognizing the kinetic-theory relationship between viscosity and mean free path: $\mu_{air}/\lambda_{air} = \frac{1}{3} m_{air} n_{air} \overline{c}_{air}$). At the other extreme, when $D_p >> \lambda_{air}$, Eqs. (9.9) and (9.10) give expressions appropriate to the Stokes regime, which is the first subrange in the continuum regime.

Continuum regime ($D_p >> \lambda_{air}$)

When the diameter of a particle is much larger than the mean free path, the air can be treated as if it were a continuous medium. For the purposes of calculating drag forces and fallspeeds, we ignore the true molecular nature of air and envision the particle to be embedded in a continuous fluid having the same macroscopic properties as air. Nevertheless, we must realize that some properties, such as diffusivity, thermal conductivity, and the viscosity of this hypothetical fluid, owe their existence to molecular interactions. Treating air as a continuum is only an approximation, albeit a convenient one that allows us to simplify the calculations. Figure 9.4 summarizes the three subregimes of the continuum regime.

Continuum regime I ($1\ \mu m < D_p \leqslant 30\ \mu m$; $N_{Re} << 1$) Continuum regime I (also known as the Stokes regime) is characterized by slow, creeping flow around the particle. When the Reynolds number is very small, the inertia of the fluid can be ignored, and the flow is symmetric around the particle, as shown on the left-most side of Fig. 9.4. The forced flow of air around the droplet introduces gradients in air velocity, which cause momentum to be transferred from the higher-speed air to the lower-speed air. The resulting viscous stress, arising from the fluid viscosity μ_{air}, is transferred to the droplet and appears as a force retarding the downward motion of the droplet. At the same time, the dynamic pressure of

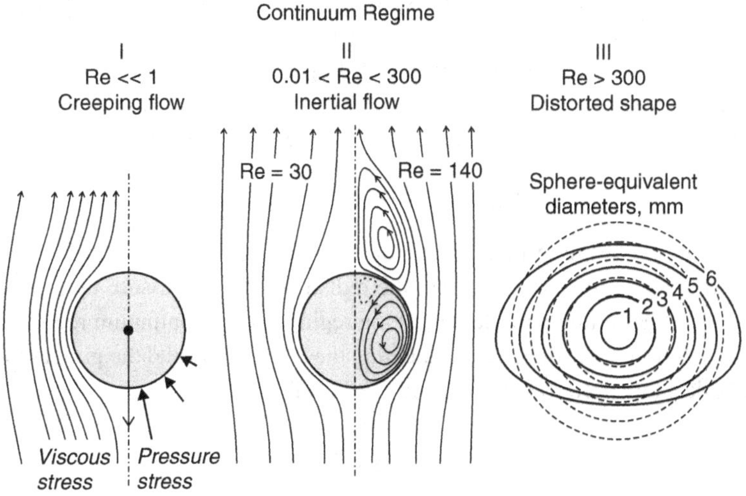

Figure 9.4 Schematic of the subregimes of the continuum regime, as determined by the indicated range of Reynolds number (Re). The downward-pointing arrow in I represents the fall velocity, which applies to each regime though not shown. The dashed curves in III represent the volume-equivalent spheres for drops of the indicated diameters. Composited after Pruppacher and Klett (1997, p. 387, 396).

the air acting on the droplet surface (arrows on right half) induces a pressure stress, which also retards the fall of the droplet.

The two components of the drag force can be derived analytically by considering the differential equations for momentum balance and mass continuity. Under the conditions of creeping flow, the flow does not change with time, and the air can be treated as an incompressible fluid. The steady, incompressible equation of motion thus reduces to

$$\nabla p = \mu_{air} \nabla^2 \mathbf{u}, \tag{9.11}$$

which states that the pressure gradient force is balanced by viscous dissipation. Solution of the equations of motion (Eq. (9.11)) and continuity ($\nabla \cdot \mathbf{u} = \mathbf{0}$), subject to the no-slip boundary condition, yields the distribution of pressure and shearing stress around the droplet. In polar coordinates (r, θ) referenced to the droplet center, the pressure field in the air surrounding a droplet of diameter D_p is calculated to be

$$p = p_\infty - \frac{3}{4}\mu_{air}\frac{u_\infty D_p}{r^2}\cos\theta, \tag{9.12}$$

where p_∞ and u_∞ are the pressure and relative air speed far from the droplet, respectively. Notice that downstream of the particle ($\cos\theta > 0$) the pressure is less than the ambient value, whereas on the upstream (lower) side of the particle ($\cos\theta < 0$) the pressure is greater. Notice that the dynamic pressure at the stagnation point, calculated using $\theta = \pi$ and $r = D_p/2$, is $p_{dyn} = 3\mu_{air}u_\infty/D_p$. The generic expression $\left(p_{dyn} = \frac{1}{2}\rho_{air}u_\infty^2\right)$, used to define the drag coefficient (Eq. (9.2)), is appropriate only when air viscosity is insignificant. Integrating the surface pressure over the particle surface yields the pressure, or form drag, $F_{D,p} = \pi \mu_{air} D_p u_\infty$. The component of drag contributed by the shearing stress is the viscous drag, or skin friction, $F_{D,v} = 2\pi \mu_{air} D_p u_\infty$. The total drag force acting on the particle in the Stokes regime is therefore $F_{D,Stokes} = F_{D,p} + F_{D,v}$ or

$$F_{D,Stokes} = 3\pi \mu_{air} D_p u_\infty. \tag{9.13}$$

Two-thirds of the Stokes drag force is due to the shearing of the viscous fluid as it flows around the particle, one-third arises from the redistribution of pressure that forces the air out of the way of the falling particle. Of all the regimes, only continuum regime I permits an analytical solution to the drag force. Equating the Stokes drag and the gravitational force allows the Stokes fallspeed to be calculated. The result is

$$v_{p,Stokes} = \frac{\rho_p g}{18\mu_{air}} D_p^2. \tag{9.14}$$

The Stokes fallspeed is proportional to the square of particle diameter and is used in Eq. (9.10) to calculate the fallspeeds of aerosol particles and small cloud droplets.

Continuum regime II ($30\,\mu\text{m} < D_p \leqslant 1\,\text{mm}$; $0.01 < N_{\text{Re}} \leqslant 300$) The inertia of the air can no longer be ignored once drops exceed about 30 μm in diameter. However, liquid water drops can still be considered rigid spheres in this regime (refer again to Fig. (9.3)). The inertia of the air causes the streamlines to be asymmetric, with those along the upstream side hugging the drop and those on the downstream side extending well beyond the surface (see left side of the center panel in Fig. 9.4). This overshooting of the air on the upper part of the drop lowers the downstream pressure and results in enhanced drag (compared with creeping flow). The drag coefficient in continuum regime II is larger than that in the Stokes regime: $C_{D,II} > C_{D,Stokes} = 24/N_{\text{Re}}$. We should therefore expect to find a weaker dependence of fallspeed on diameter (C_D appears in the denominator of Eq. (9.6)).

Calculation of the drag force in continuum regime II is complicated by the non-linearity of the inertial effect. With steady (time-independent) flow, the appropriate momentum-balance equation is

$$\mathbf{u} \cdot \nabla \mathbf{u} = -\frac{1}{\rho_{air}} \nabla p + \nu_{air} \nabla^2 \mathbf{u}, \tag{9.15}$$

where $\mathbf{u}$ is the air flow velocity and $\nu_{air} = \mu_{air}/\rho_{air}$ is the kinematic viscosity. It is the term on the left-hand side, the advection of velocity, that is non-linear and makes direct computation difficult. However, Oseen showed (about 1910) that this term could be linearized with little loss of generality in the small-drop part of regime II by replacing $\mathbf{u} \cdot \nabla \mathbf{u}$ with $\mathbf{u}_\infty \cdot \nabla \mathbf{u}$, where $\mathbf{u}_\infty$ is the fixed free-stream velocity far from the drop. Solution of the simplified equation, again subject to a no-slip boundary condition on the drop surface, yields the Oseen drag force

$$F_{D,Oseen} = F_{D,Stokes}\left(1 + \frac{3}{16}N_{\mathrm{Re}}\right) = 3\pi\mu_{air}D_p u_\infty\left(1 + \frac{3}{16}\frac{\rho_{air}D_p u_\infty}{\mu_{air}}\right). \quad (9.16)$$

The Oseen drag coefficient is thus

$$C_{D,Oseen} = C_{D,Stokes}\left(1 + \frac{3}{16}N_{\mathrm{Re}}\right) = \frac{24}{N_{\mathrm{Re}}}\left(1 + \frac{3}{16}N_{\mathrm{Re}}\right), \quad (9.17)$$

which gradually becomes independent of Reynolds number above $N_{\mathrm{Re}} \simeq 5$. The Oseen approximations are helpful over a narrow range of conditions for rigid spheres, but they do not work as well for liquid water drops. The liquid–air interface is not rigid, so the shearing stresses at the surface set up circulations inside the drop (see right side of the central panel in Fig. 9.4). At the same time, especially toward the large end of regime II, standing eddies form in the wake of the drop that partially counter the internal circulations. Such complexities suggest a need to resort to empiricism.

Drop fallspeeds in continuum regime II can be computed semi-empirically. It is helpful to realize that some variables (e.g., particle diameter and density) are physical by nature, whereas others (e.g., Reynolds number) are grouping of variables that characterize the flow and aid computations. For instance, if we know the magnitude of the Reynolds number, we can easily calculate the fallspeed from its definition, $N_{\mathrm{Re}} \equiv \rho_{air}D_p v_p/\mu_{air}$. The problem of knowing how the Reynolds number independently varies with particle size has been determined through measurements of drop fallspeeds in a vertical wind tunnel over a range of conditions. Knowing that the gravitational and drag forces balance each other when the drop is falling at its terminal speed allows one to calculate the drag force, but then the drag coefficient is still unknown. This problem is resolved by computing the so-called Davies number $N_{\mathrm{Da}} \equiv C_D N_{\mathrm{Re}}^2$, which is a function of physical variables only and derived in the following way. First, equate the drag and gravitational forces:

$$F_D = F_g$$

$$3\pi\mu_{air}D_p v_p\left(\frac{C_D N_{\mathrm{Re}}}{24}\right) = m_p g = \frac{\pi}{6}\rho_p g D_p^3 \quad (9.18)$$

and note that part of the Reynolds-number definition is contained in the product $D_p v_p (= \mu_{air}N_{\mathrm{Re}}/\rho_{air})$. Making this substitution and multiplying Eq. (9.18) through by $(8\rho_{air}/\mu_{air}^2)$ yields

$$N_{\mathrm{Da}} \equiv C_D N_{\mathrm{Re}}^2 = \frac{4}{3}\frac{g\rho_{air}\rho_p}{\mu_{air}^2}D_p^3. \quad (9.19)$$

Note that the Davies number, even though defined in terms of the drag coefficient and Reynolds number, actually contains only physical variables. The Davies number can always be calculated from the properties of the particle (size and density) and the air (viscosity and density). The Davies number is also called the Best number, but it is best thought of as a measure of particle size; small particles have small Davies numbers, large particles have large Davies numbers.

The semi-empirical procedure for calculating the fallspeed of a liquid water drop in continuum regime II involves three steps:

(a) Calculate the Davies number N_{Da} (Eq. (9.19)).
(b) Calculate the particle Reynolds number according to an empirical formula of Beard

$$N_{\mathrm{Re},II} = \exp\left(\sum_{j=0}^{6} A_j X^j\right), \tag{9.20}$$

where $X \equiv \ln N_{Da}$ and the coefficients are $A_0 = -3.18657$, $A_1 = 0.992696$, $A_2 = -1.53193 \times 10^{-3}$, $A_3 = -9.87059 \times 10^{-4}$, $A_4 = -5.78878 \times 10^{-4}$, $A_5 = 8.55176 \times 10^{-5}$, and $A_6 = -3.27815 \times 10^{-6}$. Note that all quantities and variables here are non-dimensional.
(c) Calculate the fallspeed from the definition of the Reynolds number:

$$v_{p,II} = \frac{\mu_{air}}{\rho_{air}} \frac{N_{\mathrm{Re},II}}{D_p} = \nu_{air} \frac{N_{\mathrm{Re},II}}{D_p}. \tag{9.21}$$

This procedure is valid for all water drops in continuum regime II, but it is noteworthy that the fallspeeds of drops in the size range $200\,\mu\mathrm{m} < D_p < 1\,\mathrm{mm}$ vary approximately linearly with diameter. A reasonable approximation for fallspeed in this drizzle size range is given by

$$v_{p,drizzle}[\mathrm{m\,s^{-1}}] \simeq k_{drizzle} D_p[\mathrm{mm}], \tag{9.22}$$

where $k_{drizzle} = 4\,\mathrm{s^{-1}}$ when diameter is given in mm and fallspeed is expressed in $\mathrm{m\,s^{-1}}$. Proper evaluation of fallspeed at any specific altitude requires using the appropriate dependences of the physical variables (mainly ρ_{air} and μ_{air}) on pressure and temperature.

Continuum regime III ($1\,\mathrm{mm} < D_p$; $300 < N_{\mathrm{Re}}$) Air flow around large water drops is complicated by the changing shapes of the drops. As a liquid drop grows (via ongoing collection of smaller drops), the surface tension that keeps small drops spherical becomes less able to contain the growing mass of liquid. The larger fallspeed increases the dynamic pressure, which flattens the lower side, making the drop look more like a hamburger bun than a ball, as suggested by the right-most panel in Fig. 9.4. As the cross-sectional area for a given volume increases, the drag eventually becomes so large that its increase with drop size almost matches the increase in weight with size. The fallspeed thus exhibits a very weak dependence on size toward the large end of continuum regime III. Complicating the situation further are enhanced circulations of the liquid inside the drop, the shedding of turbulent vortices in the wake of the falling drop, and oscillations of shape excited by vortex shedding. No theory exists that can take all such effects into account in a meaningful way.

Calculations of fallspeed in continuum regime III depend mainly on experimental data taken in wind tunnels. The approach follows that used for continuum regime II, except that two new dimensionless variables are introduced, the Bond number N_{Bo} and the physical-property number N_{pp}. The Bond number is a relative measure of the gravitational force acting on the drop compared with the surface-tension force holding the drop together:

$$N_{\text{Bo}} \equiv \frac{g\rho_p D_p^2}{\sigma_{LV}} \sim \frac{F_g}{F_{sfc}}. \tag{9.23}$$

A large Bond number indicates that surface tension is relatively weak and of limited ability to contain the mass of water in the drop; think of a water balloon that has been filled a bit too much. The physical-property number is formed from the Bond number and the Davies number (Eq. 9.19) in a way that eliminates the dependence on drop size:

$$N_{\text{pp}} \equiv \frac{\rho_{air}^2 \sigma_{LV}^3}{g\rho_p \mu_{air}^4} \sim \frac{N_{\text{Da}}^2}{N_{\text{Bo}}^3}. \tag{9.24}$$

The physical-property number depends only on gravity and the intensive properties of the particle and air. Careful analysis of the experimental data showed that the Reynolds number of the falling drops depended on the Bond and physical-property numbers according to the empirical formula

$$N_{\text{Re},III} = N_{pp}^{1/6} \exp\left(\sum_{j=0}^{5} B_j X^j\right), \tag{9.25}$$

where

$$X^j = \ln\left(\frac{16}{3} N_{\text{Bo}} N_{\text{pp}}^{1/6}\right). \tag{9.26}$$

The fitting coefficients are $B_0 = -5.00015$, $B_1 = 5.23778$, $B_2 = -2.04914$, $B_3 = 0.475294$, $B_4 = -5.42819 \times 10^{-2}$, and $B_5 = 2.38449 \times 10^{-3}$. The fallspeed is then obtained through the definition of Reynolds number

$$v_{p,III} = \frac{\mu_{air}}{\rho_{air}} \frac{N_{\text{Re},III}}{D_p}. \tag{9.27}$$

Note that the volume-equivalent diameter of the drop enters the formula through the Bond number (Eq. (9.23)). Altitude dependence is introduced implicitly by specifying the temperature and pressure dependence of the physical variables ρ_{air}, ρ_p, μ_{air}, and σ_{LV}.

Summary of particle fallspeeds

The fallspeeds of individual aerosol particles and liquid water droplets are the terminal velocities relative to the air in their immediate surroundings. Calculations make use of the fact that this terminal velocity occurs when the drag force on a particle balances its weight (the gravitational attraction). The force-balance equation,

$$F_D\left(D_p, v_p\right) = F_g\left(D_p\right), \tag{9.28}$$

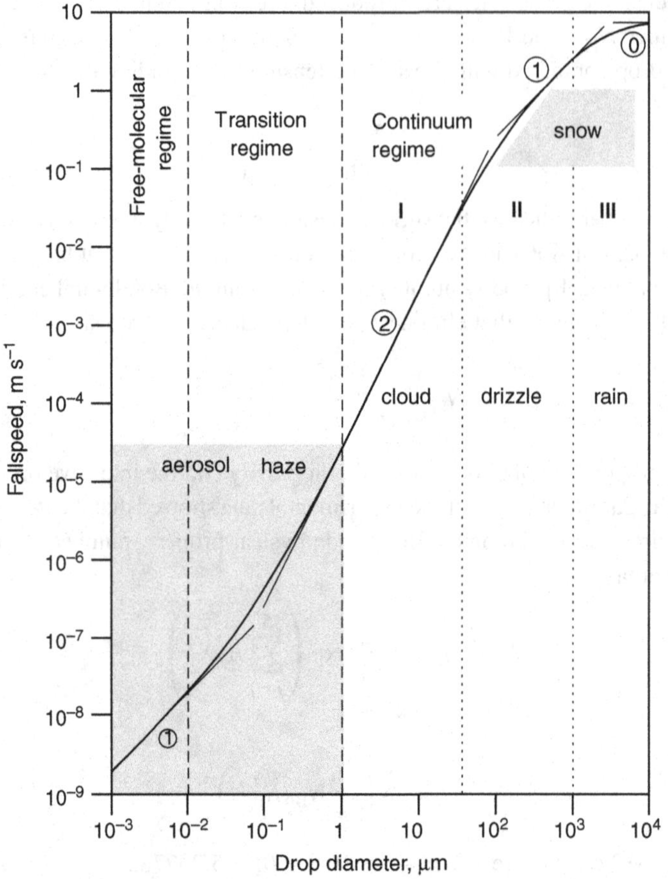

Figure 9.5 Summary of fallspeed as a function of particle diameter for average conditions in the lower troposphere. The light lines with associated circled number n identify power-law dependences, $v_p \propto D_p^n$. The lower shaded region suggests the types of particles suspended by atmospheric turbulence.

is solved in concept for the fallspeed v_p. In effect, what value of v_p yields the requisite force balance (Eq. (9.28))? The drag force can be a complicated function of particle diameter D_p, so it was found convenient to break the full size range into regimes. Figure 9.5 provides a graphical summary, with the regimes identified. Some uncertainty exists near the transitions between regimes, as no master formula has been given. Nevertheless, a robust feature of every formulation of drop fallspeeds is its monotonic nature; large drops always fall faster than small drops. One must also recognize that the very small fallspeeds of small aerosol particles are largely academic for atmospheric purposes, as the speeds of turbulent air motions in excess of the fallspeeds serve to keep the particles suspended in the atmosphere (shaded region of graph). Sedimentation becomes an important removal mechanism only for supermicron particles (D_p>1 μm). Cloud droplets are such supermicron particles, but they stay aloft by having fallspeeds less than the organized vertical motions that formed the clouds ($\sim 1\,\mathrm{cm\,s^{-1}}$). Cloud droplets tend to follow the average air motions in cloud. By contrast, rain consists of water drops that have fallspeeds exceeding the average updraft speeds in cumuliform clouds ($\sim 5\,\mathrm{m\,s^{-1}}$). Drizzle drops are of intermediate size and represent the transition from cloud droplets to raindrops.

9.2.2 Snow and ice particles

By contrast with liquid cloud drops, snow and ice particles exhibit distinct flow patterns and fallspeeds that vary greatly with particle size, shape, and density. Given the vast variety of crystalline shapes in falling snow, we can expect difficulties when attempting to determine fallspeeds of individual particles. Even for relatively simple shapes, the flow around a falling crystal can be complicated, as shown in Fig. 9.6. Objects tend to fall with an orientation that maximizes drag, so a plate-like crystal exposes its large basal face to the flow. The large cross-sectional area forces air around the sharp edges where the basal and prism faces meet. The air overshoots the back side of the crystal by wide margins, generating a low-pressure zone that encourages wake air to recycle downward along the flow axis. The three-dimensional flow can be especially complicated with branched crystals

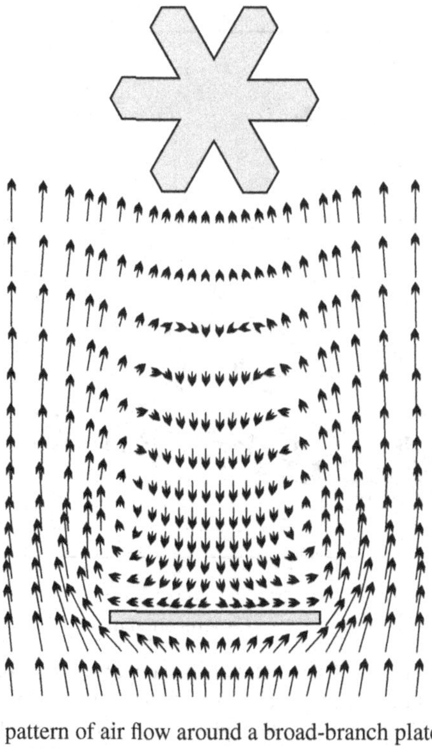

Figure 9.6 Example of the pattern of air flow around a broad-branch plate, Reynolds number ~ 80. Adapted from Ji and Wang (1990).

and aggregates. The wake flow may become turbulent and unsteady with large crystals, leading to side-slipping and tumbling. Eddies affect the drag force and hence terminal fallspeed in complicated and uncertain ways.

Measurements in the atmosphere confirm the complexity and diversity of fallspeeds. Therefore, it proves helpful to classify the shapes when presenting the data, as shown in Fig. 9.7. Despite the large differences in fallspeeds between crystal habits and in the scatter with such data, distinct patterns emerge. Compact crystals (e.g., short solid columns, left-most curve) fall faster than do plate-like forms of a given size. Moreover, the fallspeeds of compact crystals increase most rapidly with size. More complicated crystalline forms offer greater cross-sectional area to the flow for a given mass and hence generate greater drag. The fallspeeds of delicate dendrites, for instance, increase very little with maximum dimension, because new mass is added preferentially to surfaces near the edges (e.g., prism faces). The drag force increases more or less as rapidly as the mass increases for such thin plates. In general, crystals that expose a large cross-sectional area to the air flow fall slower than do compact crystals of the same mass. Such a tendency is similar in concept to the effect a parachute has on the fallspeed of a person jumping from an airplane; the mass stays the same as the parachute is deployed, but the rate of descent is dramatically decreased by the large increase in area from the parachute.

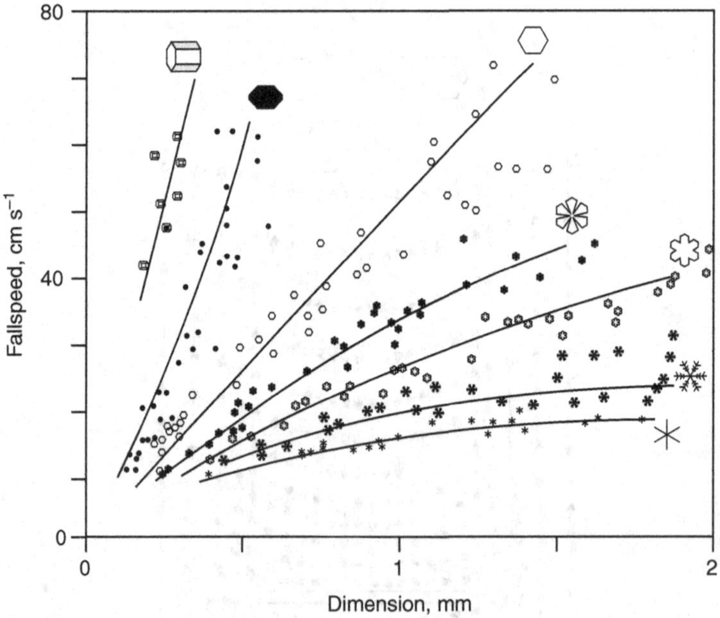

Figure 9.7 Fallspeeds of short columns and plate-like snow crystals as measured on a mountain top at a temperature of −7 °C. From left to right: solid columns, thick plates with some hollowing of facets, simple plates, sector plates, broad-branch plates, dendrites, and stellar crystals. Adapted from Kajikawa (1972).

The theory available for calculating drag forces and fallspeeds of snow and ice particles is limited to semi-empirical treatments. A reasonable approach for calculating the fallspeeds of particles having irregular forms is to follow that for water drops falling in continuum regime II, but using appropriately modified parameters. We saw that the Davies number ($N_{\mathrm{Da}} \equiv C_D N_{\mathrm{Re}}^2$, Eq. (9.19)) is a non-dimensional parameter that characterizes the air (through viscosity and density) and the particle (its density and size) without the need to specify the flow itself. The Davies number may be made general and applicable to non-spherical particles by explicitly retaining the particle mass (m_p in Eq. (9.18)) and introducing the cross-sectional area $A_c = \frac{\pi}{4} D_c^2$, where D_c is a characteristic (e.g., maximum) dimension of the particle. We thus gain a more general form of the Davies number,

$$N_{\mathrm{Da}} \equiv C_D N_{\mathrm{Re}}^2 = \frac{2g\rho_{air}}{\mu_{air}^2}\left(\frac{m_p}{A_c}\right) D_c^2 \propto \left(\frac{m_p}{A_c}\right) D_c^2. \tag{9.29}$$

The dependence of the Davies number on the characteristic size of the particle is emphasized through the proportionality, but we also see that N_{Da} depends explicitly on the mass-to-area ratio (quantity in parentheses), which we intuitively expect to determine the fallspeed (through the parachute effect). Then, following the procedure for drops, we attempt to learn the functional dependence of the Reynolds number on the Davies number, $N_{\mathrm{Re}} = f(N_{\mathrm{Da}})$, a function that depends on the drag coefficient C_D. Once we know N_{Re}, we calculate the fallspeed from its defining relationship as

$$v_p = \frac{\mu_{air}}{\rho_{air}} \frac{N_{\mathrm{Re}}}{D_c}. \tag{9.30}$$

The drag coefficient of a non-spherical particle is notoriously difficult to determine accurately, but its variation with Reynolds number can be estimated nevertheless. As a particle falls, it carries with it a thin layer of air called the flow boundary layer. This boundary layer is viscous and varies in thickness with the fallspeed. With Stokes (creeping) flow, the air appears viscous everywhere, and we saw (Eq. (9.13)) that two-thirds of the drag arises from the viscous shear stresses, one-third from pressure effects. At the higher Reynolds numbers typical of rain and snow, the boundary layer becomes thin and restricted to the immediate vicinity of the particle. As a reasonable approximation, we can think of all viscous effects being confined to the thin boundary layer. We thus view the particle plus its boundary layer falling as an ensemble through the air, which is represented by an idealized, inviscid fluid describable by potential-flow theory. The drag on a particle in potential flow is due entirely to form, or pressure effects. The viscous boundary layer surrounding a falling particle makes the particle seem fatter, exposing a larger cross-sectional area to the flow, but at least most of the complicated shape effects and viscous-shear drag are isolated from the larger flow field. The theory of potential flow shows that the thickness δ of the boundary layer, relative to the characteristic radius r_c of the particle, is inversely related to the square root of Reynolds number, so

$$\frac{\delta}{r_c} = \frac{\delta_0}{N_{\text{Re}}^{1/2}}, \tag{9.31}$$

where δ_0 is a dimensionless constant. Note that the relative thickness of the boundary layer decreases as the Reynolds number increases. The faster a particle falls, the thinner is its boundary layer and the smaller is the viscous drag.

The effect of the viscous boundary layer on the drag is due to the modified cross-sectional area of the ensemble, which is geometrically determined as $A_{ens} = \pi\,(r_c + \delta)^2 = \pi r_c^2\,(1 + \delta/r_c)^2 = A_c\left(1 + \delta_0/N_{\text{Re}}^{1/2}\right)^2$. This modified cross-sectional area affects the drag coefficient in the following way. In general, the drag force is $F_D = \frac{1}{2}\rho_{air} v^2 A_c C_D$, but here, because the particle ensemble is in an inviscid environment and characterized by the potential-flow drag coefficient $C_{D,0}$, the drag force is $F_D = \frac{1}{2}\rho_{air} v^2 A_{ens} C_{D,0}$. Thus, $A_{ens} C_{D,0} = A_c C_D$, and

$$C_D = C_{D,0}\,(A_{ens}/A_c) = C_{D,0}\left(1 + \delta_0/N_{\text{Re}}^{1/2}\right)^2. \tag{9.32}$$

The drag coefficient of a particle experiencing some viscous drag is affected by the presence of the boundary layer, but the thickness and effect of the boundary layer diminish with increasing N_{Re}. The overall dependence of C_D on N_{Re} is shown in Fig. 9.8. Note the limiting cases: At very low N_{Re} (viscous limit), Eq. (9.32) becomes $C_D \simeq C_{D,0}\delta_0^2/N_{\text{Re}}$. But, in this viscous limit, we know (from Eq. (9.18)) that $C_D = 24/N_{\text{Re}}$, implying that $C_{D,0}\delta_0^2 \simeq 24$. At the other extreme, when viscous effects can be ignored, the dependence on N_{Re} vanishes, leaving $C_D \simeq C_{D,0}$. Fits to empirical data, suggested by the broken curves in Fig. 9.8, yield the best estimates, $C_{D,0} = 0.6$ and $\delta_0 = 5.83$.

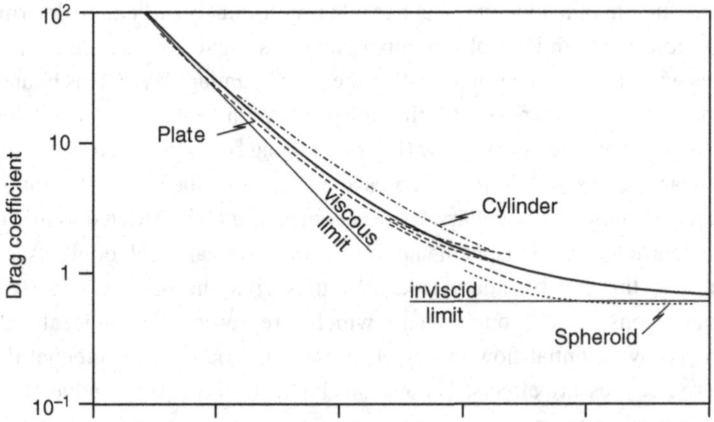

Figure 9.8 Drag coefficient versus Reynolds number. Solid curve: computed; lighter curves: various objects for comparison. Adapted from Böhm (1989).

Now that we know how the drag coefficient varies with Reynolds number for the particle-boundary layer ensemble, we can determine the Reynolds number itself. Substituting Eq. (9.32) into the defining relationship for Davies number $\left(N_{Da} \equiv C_D N_{\mathrm{Re}}^2\right)$, we get

$$N_{\mathrm{Re}} = \left(\frac{N_{Da}}{C_D}\right)^{1/2} = \left(\frac{N_{Da}}{C_{D,0}}\right)^{1/2} \frac{1}{\delta_0\left(1+\delta_0/N_{\mathrm{Re}}^{1/2}\right)}. \tag{9.33}$$

When this equation is solved algebraically for N_{Re}, we gain the sought-after function, $N_{\mathrm{Re}} = f(N_{Da})$:

$$N_{\mathrm{Re}} = \frac{\delta_0^2}{4}\left[\left(1+\frac{4N_{Da}^{1/2}}{\delta_0^2 C_{D,0}^{1/2}}\right)^{1/2} - 1\right]^2. \tag{9.34}$$

With the estimated values for the coefficients, $C_{D,0}$ and δ_0, we now have a continuous relationship that spans several orders of magnitude in Reynolds number, as shown in Fig. 9.9. Comparisons with data from a variety of studies gives confidence that the parameterization given by Eq. (9.34) is reliable.

The fallspeeds of snow and ice particles may be calculated by Eq. (9.30) once the Reynolds number has been determined (Eq. (9.34)). One needs to know the characteristic

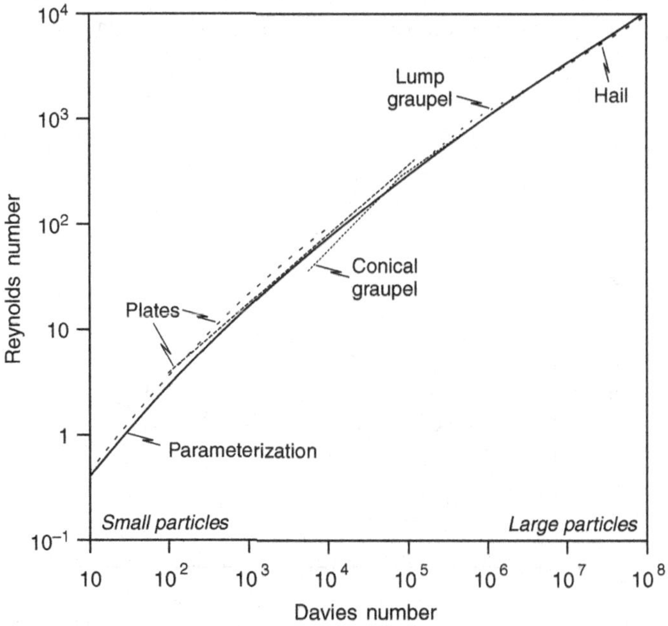

Figure 9.9 Empirical relationships between the Reynolds and Davies numbers for various forms of ice particles. Adapted from Mitchell (1996).

dimension (D_c) of the particle, as well as its mass-to-area ratio in order to calculate the Davies number (Eq. (9.29)). However, such specific information varies widely even within a given habit class (e.g., planar dendrites). Many researchers have therefore tediously measured the masses and areas of ice particles as a function of their sizes and have provided sets of mass-dimensional and area-dimensional relationships for many habit classes. At least within restricted size ranges, all mass- and area-dimensional relationships take the form of a power law:

$$m_p = aD_c^b$$
$$A_c = cD_c^d, \tag{9.35}$$

where a, b, c, and d are parameters unique to each class of particle. The mass-to-area ratio is therefore

$$\frac{m_p}{A_c} = \frac{a}{c}D_c^{b-d}. \tag{9.36}$$

Insertion of this equation into Eq. (9.29) yields

$$N_{Da} = \frac{2g\rho_{air}a}{\mu_{air}^2 c}D_c^{2+b-d}. \tag{9.37}$$

Now, one needs to know only the class of particle (values of a, b, c, d) and its characteristic diameter D_c. Examples of fallspeeds calculated with this approach are shown in Fig. 9.10. Note that riming adds mass with only minor increases in area, so rimed particles fall faster than the corresponding pristine (unrimed) crystals. The combination of theory and careful

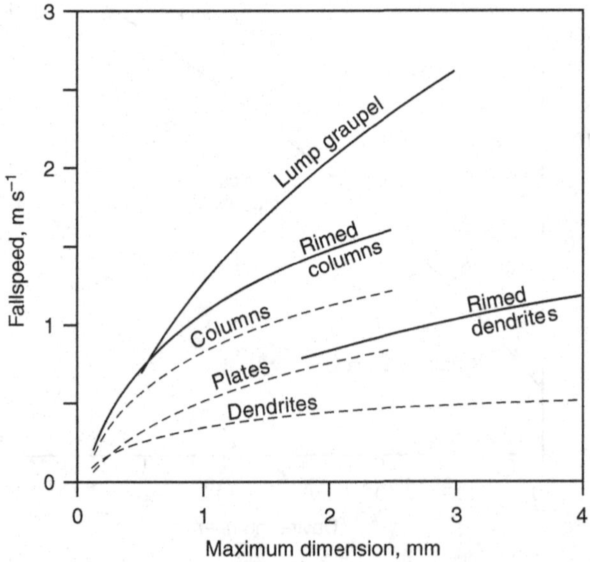

Figure 9.10 Fallspeeds of snow particles, as parameterized by Mitchell (1996).

observations proves to be a powerful way of gaining the ability to compute the fallspeeds of diverse types and sizes of cloud particles. Fallspeeds are needed in numerical cloud models for estimating precipitation rates, as well as for calculating growth of cloud particles by collection.

9.3 Collision-coalescence

The collision of one liquid drop with another is a necessary first step in the development of rain in warm clouds. Given a collision, coalescence of the pair may occur and so contribute to the growth of the larger (collector) drop at the expense of the smaller (collected) drop. Gravitational collection requires that the one (faster) drop overtake the other (slower) one. Now that we know how to calculate particle fallspeeds (previous section), we can estimate the rates of collision, coalescence, and collection.

9.3.1 Collisions

Rates of collisions are calculated by recognizing that a falling drop sweeps out a volume of air that scales with the size of the drop. The volume swept out enters directly into the calculation of collision rates because any other drops in that swept volume have a chance of colliding with the falling drop and getting caught. A large drop offers a large cross-sectional area, and it falls a greater distance in unit time than does a smaller drop. The size of the swept volume can be calculated by considering the situation illustrated in Fig. 9.11. The collector drop has a well-defined cross-sectional area, but the cross section for collision must take the size of the collected drop into account as well (contact occurs at the edges of objects, not at their centers). The geometrical cross-sectional area for collision between a large drop of radius r_L and a smaller drop of radius r_S is thus $A_{geom} = \pi(r_L + r_S)^2$. The geometrical volume swept out in unit time is a vertically aligned cylinder calculated as the geometrical area times the relative, or impact speed Δv of the drop pair:

$$\dot{v}_{geom} = A_{geom}\Delta v = \pi(r_L + r_S)^2 (v_L - v_S), \tag{9.38}$$

where v_L and v_S are the fallspeeds of the large and small drop, respectively. However, the geometrical volume ($\dot{v}_{geom}$) by itself is insufficient for calculating collision rates.

The flow of air around each of the drops in a colliding pair complicates the calculation of collision rates. We therefore have to account for the microdynamics of interacting flow fields. Only under a restricted set of conditions can we expect any and all drops in the geometrical swept volume to collide with the collector drop. Drops tend to follow air streamlines, so collisions occur only when a drop crosses streamlines, which it can do only with sufficient inertia. The net effect is that only a fraction of the drops in the path of a falling drop actually collide with the collector.

We account for the limited success of potential collisions by introducing an efficiency factor. The collision efficiency E is defined as the ratio of successful events (actual collisions) to all possible collisions (based on drops in the geometrical volume swept out by

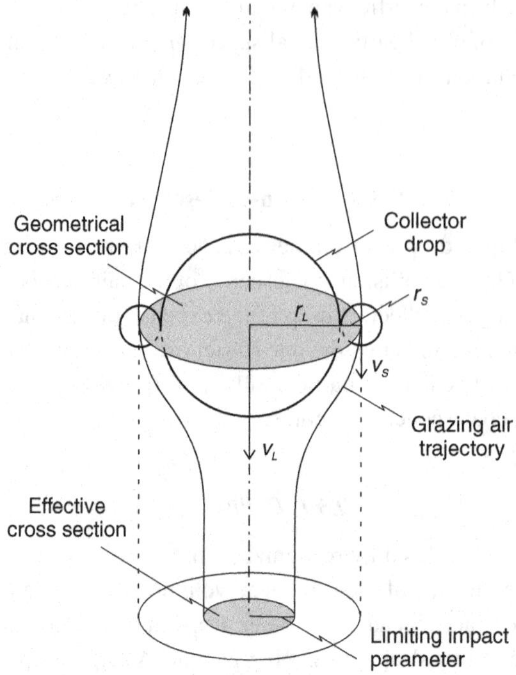

Figure 9.11 Schematic of the geometry associated with the collision of a small drop with a large drop. Air flow is shown relative to the large drop, which actually falls with speed v_L.

the collector). The lower shaded disk shown in Fig. 9.11, inside the grazing trajectories, identifies the effective area A_{eff} conducive to collision; drops that pass through the disk collide with the collector, those that pass outside it do not. Thus, in terms of the limiting value y_c of the impact parameter, the collision efficiency

$$E = \frac{A_{eff}}{A_{geom}} = \frac{\pi y_c^2}{\pi (r_L + r_S)^2}. \tag{9.39}$$

The collision efficiency, once it is known for any given drop pair (r_L, r_S), is used to calculate the effective swept volume (for collision purposes) as $\dot{v}_{eff} = E\dot{v}_{geom} = \pi(r_L + r_S)^2 E (v_L - v_S)$.

The problem with evaluating the collision efficiency E is being able to determine the grazing trajectory and hence the limiting impact parameter y_c. Various empirical and theoretical approaches for arriving at the function $E(r_L, r_S)$ have been attempted in the past with reasonable success. A graphical display of representative results is presented in Fig. 9.12 as a family of constant-r_L curves. Several features of these curves are worth noting. Perhaps the most notable feature of any collision efficiency function is the low values of E for small-drop radii below about 5 μm, regardless of the size of the collector drop. The collision efficiency never equals zero, but for practical purposes a small-drop

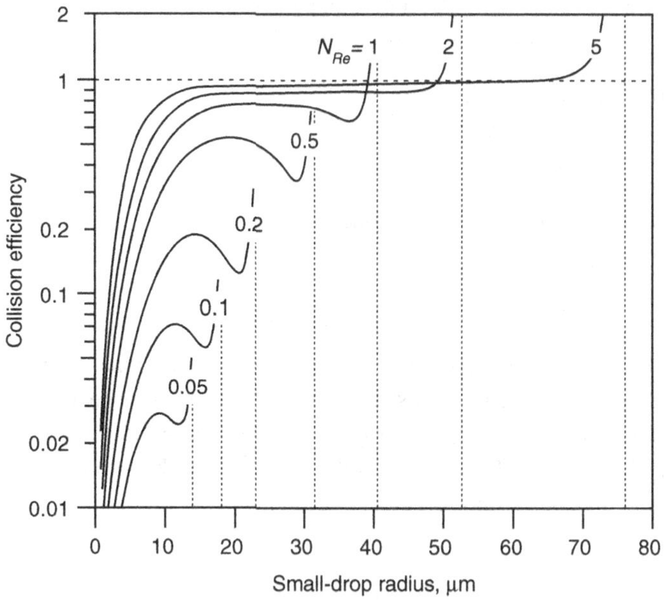

Figure 9.12 Collision efficiencies between two liquid drops. Adapted from Böhm (1992).

threshold exists for meaningful values of $E > 0.1$: $r_S > 5\,\mu\text{m}$. The reason small drops do not collide readily with other drops is their small inertia and tendency to follow the air motions; small drops are simply dragged around the larger drop by the air as it is forced out of the way of the falling collector drop.

Once the small-drop size exceeds the threshold ($r_S > 5\,\mu\text{m}$), the collision efficiency increases rapidly. The mass of a drop increases as the cube of its radius, so small drops of increasing size become less able to respond to the sudden sideward acceleration of the air as the collector drop approaches; they tend to get stuck in the path of the collector. But note that the maximum values of E are low for small collector drops, so a practical large-drop threshold also exists: $r_L > 20\,\mu\text{m}$. When both the collector and collected drops are of modest size, they tend to move out of the way of each other and avoid each other. Such is the effect of interacting flows on low-inertia particles. The decrease in E beyond the maximum is due to the similarity of fallspeeds and hence greater time available for the drops to respond to each others approach. Not until the large drop exceeds a radius of about $50\,\mu\text{m}$ and the small-drop radius is greater than about $10\,\mu\text{m}$ does the collection efficiency approach unity (meaning that all drops in the geometrical swept volume collide with the collector). The upturn in E when the collector and collected drops are nearly identical in size is attributed to a wake effect: the lower drop distorts the flow field in its wake, allowing the trailing drop to fall faster than its free-stream fallspeed and so catch up to the lower drop. The interactions of liquid drops are complicated even in stagnant air, but the basic physics underlying collision efficiencies is understood.

Liquid drops that collide with each other may or may not coalesce. The collision of two liquid drops actually has three possible outcomes. Given a collision, the drops may

(a) coalesce (merge into a single drop),
(b) rebound (bounce off one another), or
(c) disrupt (break up into smaller drops).

Each outcome class will be treated individually.

9.3.2 Coalescence

Coalescence, the merging of one drop into another, is the only outcome that leads directly to the growth of a collector drop. Collisions that result in coalescence are termed collision-coalescence events and form the basis of growth by collection. Coalescence cannot occur without collision, so the efficiency of collection, E_c, is the product of the collision efficiency, E, and the coalescence efficiency ϵ:

$$E_c = E \cdot \epsilon. \tag{9.40}$$

The coalescence efficiency is the fraction of all collisions that result in permanent attachment.

The determination and interpretation of coalescence efficiencies are based on past laboratory experimentation. Collector drops of controlled size were allowed to fall and collide with other (collected) drops of various sizes, and photographic methods were used to determine if a collision resulted in coalescence or rebound. It was found that small collected drops tend to coalesce, whereas the larger collected drops tend to rebound. Such findings were hypothesized to result from drop distortion and the trapping of an air layer during impact, so the data were analyzed in terms of the Weber number (N_{We}), a non-dimensional measure of distortion used in fluid mechanics. In general terms, the Weber number is the ratio of the dynamic pressure force (an external force acting to distort the drop) to the surface tension force (the cohesive force resisting distortion). The pressure acts over an area, whereas surface tension acts over a characteristic length (L), so a general definition is

$$N_{We} \equiv \frac{\rho L v^2}{\sigma}, \tag{9.41}$$

where ρ is a relevant density, v is a relative speed, and σ is the surface tension of the relevant drop. Small values of the Weber number ($N_{We} << 1$) imply that surface tension dominates and a drop stays spherical. Large values ($N_{We} >> 1$), on the other hand, indicate that the external disturbance dominates the surface force and causes the drop to distort from its normal spherical shape.

Application of the Weber number to drop coalescence requires careful definitions of the parameters in Eq. (9.41). The external disturbance arises from the collision of another liquid drop, so the appropriate density $\rho = \rho_L$, the density of liquid water. The characteristic length of the offending object is logically taken to be the radius of the small drop, so $L = r_S$. The relative speed is the impact speed, the difference in fallspeeds of the large and small drops; thus, $v = v_L - v_S$. Finally, the surface tension $\sigma = \sigma_{LV}$, that of the liquid–vapor interface. With these definitions, the so-called collisional Weber number used to analyze the laboratory results is

$$N_{We} \equiv \frac{\rho_{LV} r_L (v_L - v_S)^2}{\sigma_{LV}}. \tag{9.42}$$

Each drop pair in the experiments yields a unique value for N_{We}, so the coalescence efficiency can be plotted against N_{We}, as in Fig. 9.13. As expected, the coalescence efficiency decreases as the Weber number approaches, then exceeds unity, giving strong evidence that the lack of coalescence is due to the distortion of the drop surface at the point of collision. When the inertia of the impacting drop is sufficiently large, the surface of the large drop develops a cavity that traps air for the brief time of collision. The air apparently acts as a cushion that prevents molecular contact of the two liquids. The trapped air may also help in recoiling the drops apart during rebound. At the other extreme ($N_{We} << 1$), the drops stays spherical during contact, which prevents air from getting trapped and allows the surfaces of the drops to make contact.

The collection efficiency represents a convolution of the collision and coalescence efficiencies (Eq. (9.40)). Computations of collection efficiency over the complete parameter space (all combinations of large- and small-drop sizes) is complicated and not fully developed. However, a general understanding of the effect of limited coalescence on the collection process can be obtained from Fig. 9.14, a plot of the three efficiencies against the small-drop radius for a single large-drop radius $r_L = 100\,\mu\text{m}$. We saw earlier (Fig. 9.12) that the collision efficiency increases rapidly once the small-drop radius exceeds about $5\,\mu\text{m}$, but it is in just this range of sizes that the coalescence efficiency ϵ is largest;

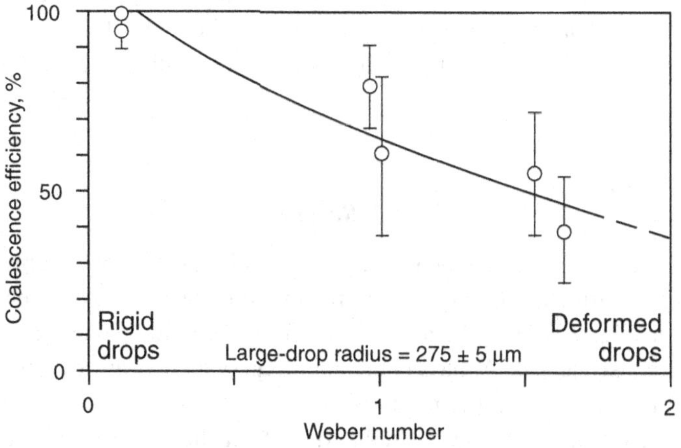

Figure 9.13 Dependence of the coalescence efficiency on the Weber number. Adapted from Ochs *et al.* (1986).

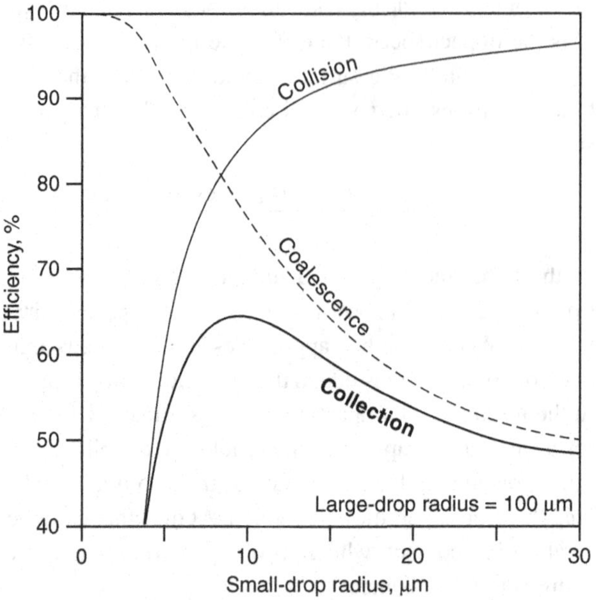

Figure 9.14 Efficiencies of collision (fine curve), coalescence (dashed curve), and collection (bold curve). Adapted from Beard and Ochs (1984).

ϵ decreases significantly as the small-drop size increases. The product of these two curves (Eq. (9.40)) gives the net effect of collision and coalescence on the collection efficiency (shown as a bold curve in Fig. 9.14). The collection efficiency E_c is very small initially, maximizes at intermediate small-drop sizes, then drifts gradually downward because of limited coalescence. Beyond the de facto threshold for small drops ($r_L > 5\,\mu\text{m}$), the overall collection efficiency is close to 50% over a broad range; E_c is nowhere close to the unity value assumed in many computer simulations of clouds. The growth of a small rain drop inside a cloud is limited by its ability to collide with and collect smaller cloud droplets.

9.3.3 Rebound

A rebound event, which occurs when a small drop collides with but simply bounces off the larger drop, contributes nothing to the growths of either drop. Drop rebound, shown schematically in Fig. 9.15, may be a non-event from the point of view of growth, but it can have other consequences. A mass bouncing off an object always transfers momentum to the object. In the case of colliding drops, the force of impact can set the larger drop into oscillations, for instance. These oscillations affect the fall characteristics, and they can initiate disruption of the larger, less stable drop. The close proximity of the two drop surfaces during collision generates gradients of various properties that can be large, which

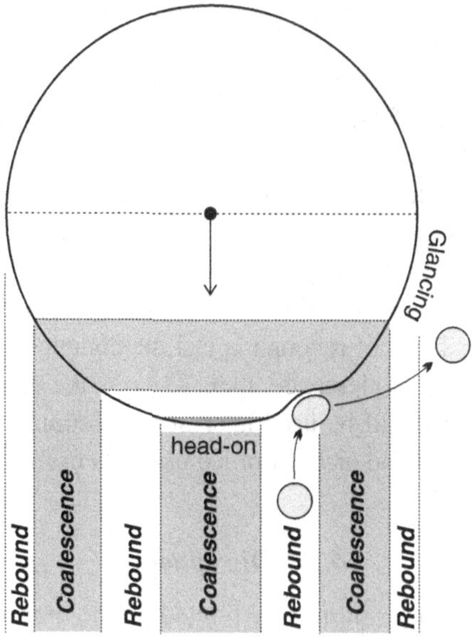

Figure 9.15 Schematic depiction of the contact zones that define whether rebound or coalescence is the outcome of the collision between two liquid drops. Adapted from Jayaratne and Mason (1964).

in turn may lead to the exchanges of energy, electric charge, and trace gases between the drops.

As suggested by the detent at the point of impact in Fig. 9.15, the inertia of the colliding drop distorts the surface and traps a thin layer of air. It is this viscous layer of air that must be drained away before contact of the two liquid surfaces can take place. If the thickness of the air layer never becomes less than about 0.1 μm, the drops may well rebound. A gap less than this critical thickness permits molecular attractions to be large enough to cause coalescence. How rapidly the air drains away determines whether coalescence or rebound is likely to occur.

The conditions under which rebound takes place are difficult to determine with precision, but the variables that need to be considered include the sizes of the colliding drops, their relative (or impact) speed, and the angle of impact. These parameters collectively determine the time of contact and hence the time the trapped air has to escape. Detailed measurements with high-speed photography show that typical contact times are on the order of 1 ms or less, but the contact time is sensitive to the component of the impact speed normal to the surface of the larger drop. The laboratory studies place great importance on the impact angle, and they show that rebound and coalescence occur in alternating zones symmetrically oriented around the vertical fall axis of the larger drop (see Fig. 9.15). In the extreme situations, nearly head-on collisions (large impact angle) readily lead to coalescence, whereas grazing collisions (small contact angle) favor rebound. The boundaries of these impact zones are imprecise and vary in complicated ways with the sizes of the colliding drops, but studies of this kind shed important light on the mechanisms that determine the outcome of collisions.

Drops approach each other randomly in a cloud, so the impact angle cannot be predicted. Moreover, less detailed types of experimental investigations, those that cannot resolve the exact points of impact, obtain only overall or net efficiencies resulting from many collisions. A numerical cloud model, too, seldom needs to deal with the details of the collisions and their outcomes. One usually needs just an overall measure of rebound, one that can be applied in a statistical sense. Fortunately, when the drops have not yet grown to the point of rupturing, the outcome of any given collision is either coalescence or rebound. Given only two possibilities, the probability of rebound is just the complement of coalescence. So, if ϵ represents the efficiency of coalescence, then $1 - \epsilon$ is the efficiency of rebound. This principle of complementarity will be used later in connection with cloud electrification, which depends crucially on the bounce-off of ice particles from one another.

9.3.4 Disruption

Raindrop growth by collection is ultimately limited by the breakup of one or both of the drops in a collision. Liquid water drops are held intact by surface tension, so at some point the mass in the drop (which increases as the cube of size) becomes larger than the surface-tension forces (which increase as the square of size) can reasonably contain. The situation with a large raindrop is somewhat analogous to that of a water balloon about to rupture. The surface membrane (rubber in the case of a balloon, asymmetrically bonded molecules in the surface of a water drop) eventually gives way and the interior water breaks free. Raindrops can break up spontaneously (due to self-induced turbulence) once they exceed diameters about 1 mm, but more commonly they break up as a result of collisions with smaller drops well before aerodynamic disruption occurs.

The disruption of raindrops is important in cloud physics because it redistributes water mass into a number of smaller fragments, each of which may grow by collection anew and stimulate a broadening of the drop-size distribution. Some studies also hint at the possibility that the breakup of supercooled raindrops may initiate ice, perhaps by the redistribution of energy during the creation of new surface area. Here, we limit discussion to findings from experiments of collisional breakup.

The past experimental studies with wind tunnels point to three distinct types of breakup. Each type is identified schematically along the left-hand side of Fig. 9.16; representative examples of the fragment size distributions are shown on the right. Filament, or neck breakup is the most common type, as it occurs at all impact parameters (off-set between the fall lines of each colliding drop) when the collision kinetic energy is small, as well as with glancing collisions at high collision kinetic energies. The identities of the parent drops (arrows in Fig. 9.16) is remembered, but small satellite drops split off from the neck during separation. Both sheet and disk breakup occur only at high collision kinetic energies when the impact parameter is smaller (less than that from glancing, more toward head-on collisions). Sheet breakup occurs when the impact of the small drop sets the larger drop into

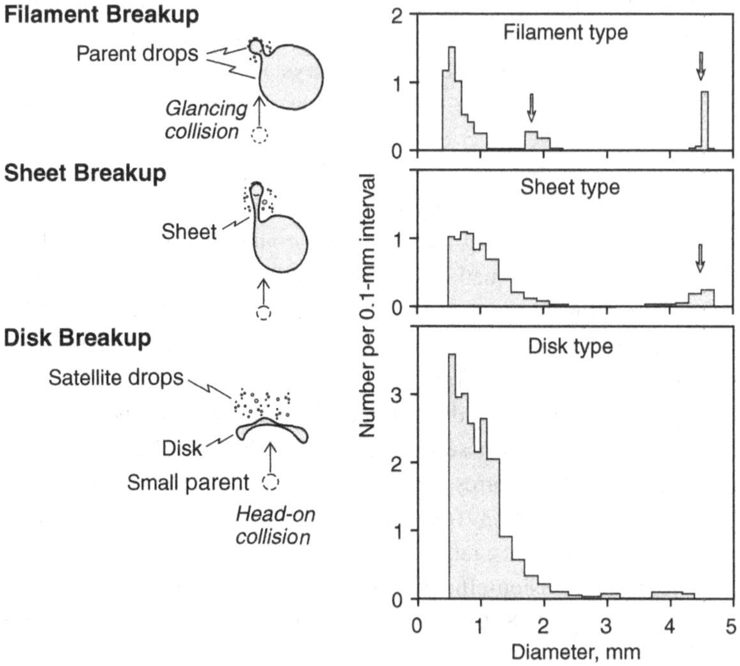

Figure 9.16 Types of collision breakup. Left side: schematics. Right side: histograms of fragment sizes. Broad arrow indicates the parent drop. Adapted from Low and List (1982).

rapid rotation because of the off-center collision. The largest parent drop is remembered, but the small drop contributes to the array of fragments associated with the sheet. Disk, or bag breakup occurs during head-on collisions at high kinetic energies. The identities of the parent drops are lost as the disk breaks up (sometimes explosively) into many drops of intermediate size. Disk breakup is especially effective in limiting the sizes of raindrops in clouds.

9.4 Riming

Riming is the process by which ice particles collect supercooled cloud water. Riming is sometimes also called the accretional growth of ice. Cloud drops at temperatures below the melting point readily freeze as soon as they contact an ice surface. Therefore, it is normally assumed that a supercooled drop colliding with an ice particle sticks to the surface; collection is a matter of collision only, not coalescence. We will see later that the riming process may sometimes throw off tiny splinters of ice and so contribute to the secondary production of ice in mixed-phase clouds, but the growth of the collecting ice particle is not appreciably affected by such minor losses. The accretion of cloud water onto an ice particle adds to its mass with only minor extension of its surface area, so the fallspeed of the ice particle can increase rapidly in clouds with high concentrations of supercooled liquid water. The growth of ice particles by collection is important for generating significant rain in cold stratiform clouds and for the formation of graupel and hail in convective storms.

Collisions of small liquid drops with falling ice crystals are treated theoretically in much the same way as was done for drop–drop collisions. The main distinctions are the differences in fallspeeds and the effect of shape on the collision efficiency. In the simplest limiting case, riming starts after single crystals (plates and columns) have grown by vapor deposition to some critical size. Determination of the critical size, as well as the efficiencies of collisions, is made by computing the flow field around the falling crystal (solving the incompressible Navier–Stokes equations, as done for the crystal in Fig. 9.6), then superimposing individual drop trajectories to see if a collision results or not. As shown in Fig. 9.17, the starting positions (horizontal offsets) of several trajectories are varied to identify the set of grazing collisions. These critical trajectories outline the effective collision cross section A_{eff} (darkened plate) that enters the definition of the collision efficiency: $E = A_{eff}/A_{geom}$, where A_{geom} is the geometric cross section, the area of the crystal expanded by the radius of the cloud drop. This method corresponds well with that used to calculate the efficiencies of liquid–liquid collisions (Eq. (9.39)), except that the initial offsets of liquid–ice collisions must be varied in two dimensions, rather than just one.

The results from the collision-efficiency calculations for initially unrimed plates and columns are shown in Fig. 9.18. Each curve corresponds to the specified type of crystal

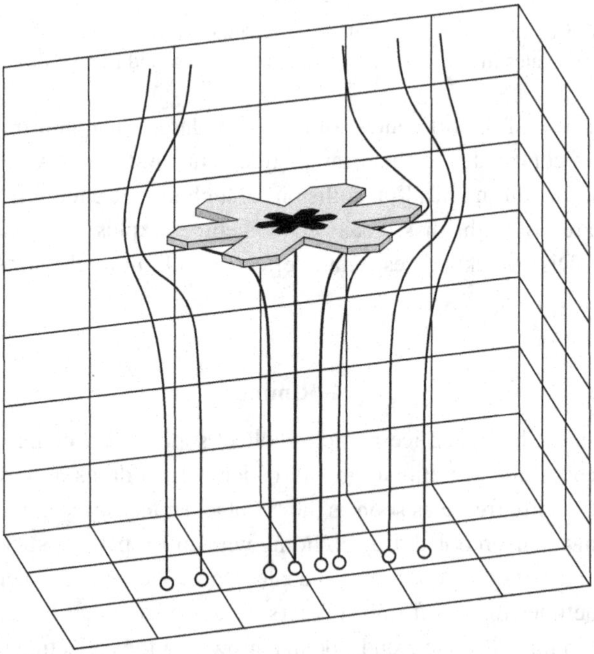

Figure 9.17 Schematic of trajectories of 2 μm drops around a broad-branch plate having Reynolds number of 10. The dark inner plate represents the effective collision cross section. Adapted from Wang and Ji (2000).

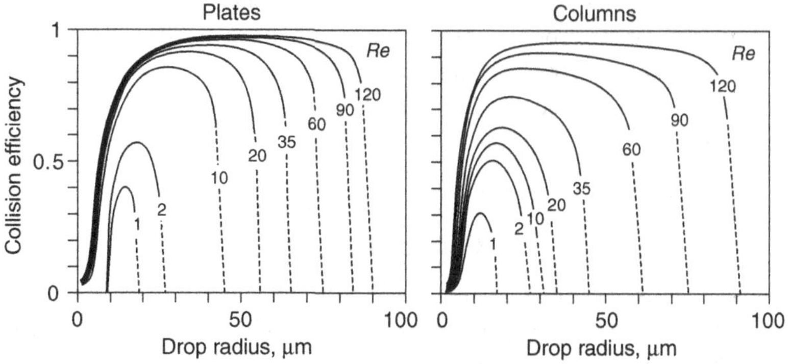

Figure 9.18 Collision efficiencies for plates (left) and columns (right) during riming of supercooled drops. Adapted from Wang and Ji (2000).

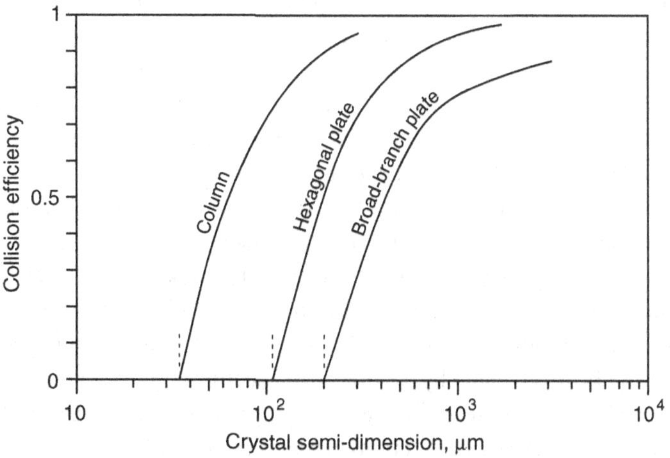

Figure 9.19 Critical semi-dimension (dashed lines) for the onset of riming by crystal habit. Adapted from Wang and Ji (2000).

falling at its terminal speed with the attitude that maximizes drag (c-axis vertical for plates, c-axis horizontal for columns). As occurs with liquid–liquid collisions, the collision efficiency is small for small drops (radii $< 5\,\mu\text{m}$), but it increases rapidly as the drop size increases for any given crystal until the maximum in E is reached. The falloff in E at large-drop sizes occurs as the fallspeeds of the two colliding particles become similar.

The maximum values of E for small crystals (those with small Reynolds numbers, N_{Re}) are well below unity, suggesting that a minimum crystal size is needed for riming. Extrapolating the maximum values of E to zero yields the cutoff crystal size, the lateral dimension (perpendicular to the c-axis) below which riming does not take place, as shown in Fig. 9.19. Riming of a column sets in only after the width exceeds about $35\,\mu\text{m}$, whereas the onset of riming for a plate occurs when its diameter is greater than about 110 μm. Broad-branch plates must be even larger, around 200 μm, before riming starts. The collection of small cloud drops by relatively larger ice crystals can be an important mechanism for adding mass to falling crystals, but only once they have grown sufficiently from the vapor to exceed the critical dimensions.

9.5 Capture nucleation

Capture nucleation is the other mixed-phase collection process, the one involving a relatively large drop and a smaller ice crystal. The significance of capture nucleation is, however, not the directly added mass, rather the phase change that results. Once the ice crystal contacts a supercooled drop, the liquid freezes rapidly, and the latent heat of fusion is suddenly released. This rapid infusion of thermal energy into a cloud can stimulate cloud development. Also, turning the initially liquid drop into an ice particle prevents it from disrupting. The newly formed ice particle can become an instant riming particle, one that can grow virtually without bounds as a graupel particle or even a hailstone.

The efficiency of collection depends on the sizes of the particles, as usual, but it is also affected significantly by the shape of the ice particle. Again, we assume that the coalescence efficiency is unity and that each collision results in a single freezing event. The collection efficiency is thus the same as the collision efficiency. Here, we limit discussion to the capture of columnar ice crystals by spherical water drops because of the likely importance of these two types of particles in cloud glaciation.

Calculations of collision efficiency are made using the superposition technique, as done for riming. Once the flow around the falling drop is calculated, the crystal trajectory is superimposed onto that flow. It is assumed that the crystal has no influence on the flow field and that it can be represented by a cylinder of specified length along the c-axis (c) and aspect ratio $\phi = c/a$. Note, by reference to Fig. 9.20, that a cylindrical form can approach the drop in any one of several orientations. Examples a and b have the c-axis horizontal, whereas example c has it vertical. The normal orientation of columnar ice particles

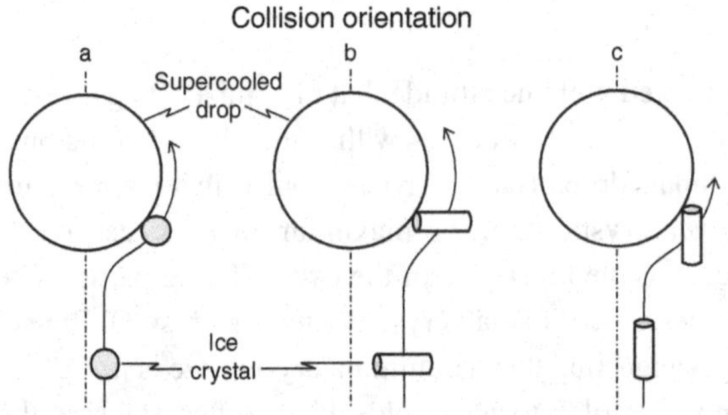

Figure 9.20 Schematic views of possible orientations a cylindrical form can take during collision with a spherical drop. Adapted from Lew and Pruppacher (1983).

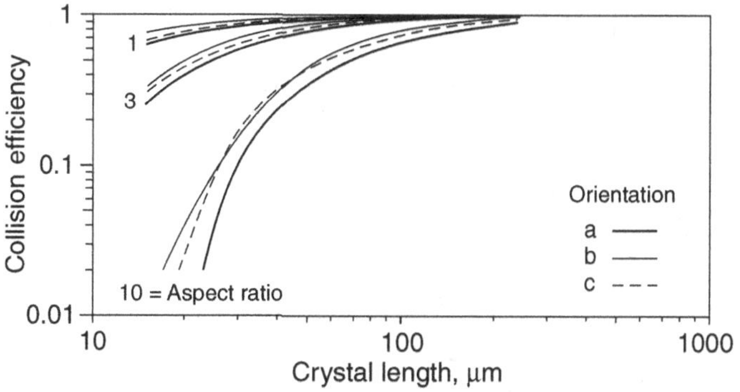

Figure 9.21 Computed collision efficiencies of cylindrical ice crystal with a spherical water drop. Adapted from Lew and Pruppacher (1983).

is with the c-axis in a horizontal plane, so collision orientations a and b are likely to be common. However, many other azimuthal variations are possible beyond the two shown. Short columns, or other uniaxial ice particles, tend to lack any preferred fall orientation, so collision orientation c needs consideration. We also have little information about how the orientation of the crystal changes as it approaches the crystal. Despite such uncertainties, the calculated collision efficiencies, shown in Fig. 9.21, yield interesting insights. The three clusters of curves are distinguished by the differing aspect ratios (1, 3, and 10), whereas each curve within a cluster corresponds to one of the three orientations identified in Fig. 9.20. Perhaps the most apparent feature of these results is the relative importance of aspect ratio over orientation. Orientation b yields the largest efficiency for any given aspect ratio, probably because the tip of the crystal points directly toward the drop surface, but shorter, more compact crystals (aspect ratio ~ 1) are clearly favored for capture over long columns. Compact forms, as do spheres, offer the smallest cross-sectional areas for a given mass and so have the least drag. In every case, size matters, as collisions depend on the mass of the smaller particle to have enough momentum to cross the flow streamlines.

9.6 Aggregation

Aggregation is the joining of two or more individual ice crystals together to form an aggregate, more commonly known as a snowflake. Aggregates are commonly observed in winter snowfalls, but they are also found in the upper parts of many summertime convective clouds, as well as in cirriform clouds. Casual observations during snowfalls onto the ground demonstrate that aggregation is especially effective at relatively high temperatures (within a few degrees of 0 °C), as suggested by Fig. 9.22. The largest aggregates are found within a few degrees of 0 °C. The types of crystals constituting a given aggregate can vary widely, but dendrites are perhaps most common and account for the large sizes between −15 and −12 °C. On occasion, snowfalls are observed to be dominated by

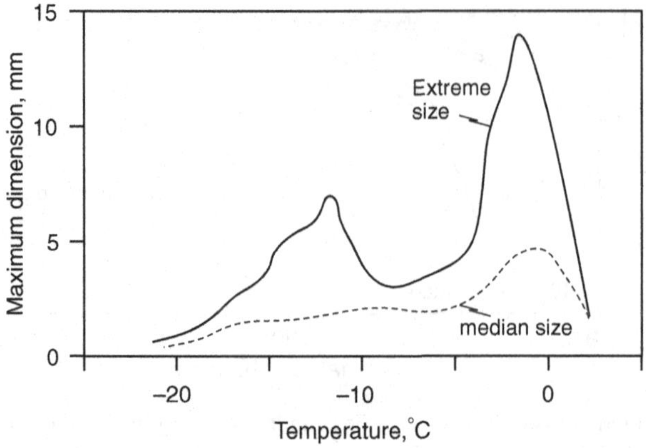

Figure 9.22 Maximum dimensions of aggregates measured at the given temperature. Solid curve: extremes of the sampled set. Dashed curve: medians of the sampled set. Adapted from Young (1993), based on data from Hobbs *et al.* (1974).

aggregates of needles, which form around −5 °C. The study of aggregation in the upper troposphere requires research aircraft and specialized equipment. Aggregation, by consolidating mass and reducing exposed surface area, augments precipitation rates by increasing fallspeeds and enhancing riming efficiencies. However, aggregates are low-density ensembles with much pore space, so the fallspeeds do not increase as dramatically with size as do drops resulting from collision-coalescence.

The process of aggregation is complicated and still not well understood. The diverse habits of snow crystals make calculating the flow field challenging, a problem made all the more difficult once dozens, perhaps hundreds of crystals have combined to form a geometrically complex mass of porous ice. Solid particles tend to bounce off each other, so adhesion (the more appropriate term for coalescence) cannot be guaranteed for any collision; specific mechanisms of adhesion must be identified before calculations can be trusted.

The large sizes of aggregates in snowfalls near 0 °C (shown in Fig. 9.22 to exceed 1 cm on occasion) are thought to be caused by sintering, a surface phenomenon common to many substances near their respective melting points. Molecules near the surface of a solid have fewer neighbors and so are less tightly bound. Near the melting point, these molecules vibrate extensively and become arranged in very disordered ways. Surface molecules can be mobile and act more like a liquid that a solid. When two such layers contact each other, the molecules in these quasi-liquid layers suddenly find other like molecules with which to bond. Where the liquid–gas interface disappears, relatively strong, solid-like bonds reform, and the colliding particles adhere to one another. Sintering is a strong mechanism of adhesion, but it is effective only within a few degrees of the melting point.

The secondary set of large aggregates at lower temperature requires an alternative explanation for adhesion. The common observation that such aggregates are composed of dendrites has led to the suggestion that the intricate structural features let one crystal hang onto another by mechanical interlocking. Each arm of a dendrite acts as a set of hooks that get intertwined with those of a neighboring crystal. One may argue that dendrites form only within a degree or so of $-15\,^{\circ}\mathrm{C}$, but the temperatures in Fig. 9.22 refer to those recorded at the point of observation, which are not likely to be the temperatures at which growth occurred. Mechanical interlocking is still the most likely explanation for adhesion at these lower temperatures. Crystal habits, both primary and secondary, thus affect the sizes of aggregates and hence their fallspeeds, riming efficiencies, and precipitation rates.

9.7 Further reading

Fletcher, N.H. (1962). *The Physics of Rainclouds.* Cambridge: Cambridge University Press, 386 pp. Chapters 6 (collision-coalescence) and 11 (hail formation).

Mason, B.J. (1957). *The Physics of Clouds.* London: Oxford University Press, 481 pp. Chapter 6 (hail formation).

Pruppacher, H.R. and J.D. Klett (1997). *Microphysics of Clouds and Precipitation.* Dordrecht: Kluwer Academic Publishers, 954 pp. Chapter 15 (collision-coalescence of liquid drops, breakup) and 16 (aggregation of ice particles, melting).

Rogers, R.R. and M.K. Yau (1989). *A Short Course in Cloud Physics.* New York: Pergamon Press, 293 pp. Chapters 8 (collision-coalescence, effects of turbulence), 9 (accretion), and 10 (aggregation).

Seinfeld, J.H. and S.N. Pandis (1998). *Atmospheric Chemistry and Physics.* New York: John Wiley and Sons, 1326 pp. Chapters 8 (aerosol fallspeeds) and 15 (rain formation).

Wallace, J.M. and P.V. Hobbs (2006). *Atmospheric Science: An Introductory Survey.* Amsterdam: Academic Press, 483 pp. Chapter 6 (drop collisions, coalescence, fallspeeds, riming, hailstone growth).

Wang, P.K. (2002). *Ice Microdynamics.* San Diego, CA: Academic Press, 273 pp. Chapter 4 (riming efficiencies).

Young, K.C. (1993). *Microphysical Processes in Clouds.* New York: Oxford University Press, 427 pp. Chapters 7 (collision-coalescence, particle fallspeeds) and 8 (riming, hail growth, aggregation).

9.8 Problems

1. Consider a small liquid drop of size $D = 20\,\mu\mathrm{m}$ falling from rest in the Earth's atmosphere, the density and viscosity of which can be taken as $\rho_{air} = 1\,\mathrm{kg\,m^{-3}}$ and $\mu_{air} = 2 \times 10^{-5}\,\mathrm{Pa\,s}$, respectively. Start with the fundamental physical principles and develop the mathematics that would allow you to calculate the speed v of the drop as a function of time t. In terms of the physical variables, what is the steady-state fall velocity and the time constant τ to achieve this terminal velocity. Are we justified in assuming that the cloud droplet is always falling at its terminal fall speed? Explain.

2. Create graphs of liquid drop fallspeeds for drops ranging from $0.1\,\mu\mathrm{m}$ to $1000\,\mu\mathrm{m}$ at the surface ($T = 290\,\mathrm{K}$, $p = 1000\mathrm{h\,Pa}$) and aloft ($T = 255\,\mathrm{K}$, $p = 500\mathrm{h\,Pa}$). This may best

be done by writing a small program. Be sure to include the effects of drop deformations. These graphs may also be used to get an estimate of the drop size given an observation of its fallspeed. On a separate graph, plot the Reynolds number N_{Re} as a function of diameter for these drops. These are useful graphs to keep around.

3. Create graphs for the fallspeeds of hexagonal crystals and stellar crystals with broad arms as a function of diameter. You may use the mass- and area-dimensional power laws for each crystal type as given below:

	Hexagonal plates	a	b	c	d
(a)	$15\,\mu\mathrm{m} \le D \le 100\,\mu\mathrm{m}$	0.00739	2.45	0.24	1.85
	$100\,\mu\mathrm{m} < D \le 3000\,\mu\mathrm{m}$	0.00739	2.45	0.65	2.00

	Stellar crystals	a	b	c	d
(b)	$10\,\mu\mathrm{m} \le D \le 90\,\mu\mathrm{m}$	0.00583	2.42	0.24	1.85
	$90\,\mu\mathrm{m} < D \le 1500\,\mu\mathrm{m}$	0.00027	1.67	0.11	1.63

The mass- and area-dimensional relationships are specified in cgs units.

4. Colloidal stability in clouds may be defined as a measure of the probability of droplets in a cloud experiencing collisional grow. A colloidally stable cloud will have few collisions.
   (a) Compare and contrast the colloidal stability (with respect to collisions between particles in warm clouds) of a warm-based maritime cumulus cloud and a cold-based continental cumulus for clouds having the same vertical depth, both having life-times of about 30 min.
   (b) If the cold-based continental cumulus forms in a polluted air mass, what can you say about the likelihood that the cloud would produce precipitation? Explain you answer.

5. Processes in a cloud do not work in isolation. We derived an expression for the vapor depositional growth of an ice crystal in Chapter 8 as

$$\left.\frac{dm_p}{dt}\right|_{VD} = 4\pi C \rho_I G_I s_i,$$

where C is the electrostatic capacitance of the crystal and G_I is a slowly varying function of the temperature T and the pressure p. However, it frequently happens that an ice crystal will grow by vapor deposition and riming at the same time. How will riming impact the growth by vapor? If we assume a riming rate of growth given as $\left(\left.\frac{dm_p}{dt}\right|_{\mathrm{riming}}\right)$, please modify the above equation to include the effect of riming on the vapor depositional growth rate.

Part V

Cloud-scale and population effects

10
Evolution of supersaturation

10.1 Overview

Supersaturation, the thermodynamic driver of phase changes, is linked intimately to cloud dynamics and microphysics. Upward motions of an air parcel always lead to lower pressures and densities, so the work done by the rising parcel on the environment causes the internal energy and temperature of the air to decrease during uplift. This cooling reduces the equilibrium vapor pressure of water, so the relative humidity increases. Above cloud base, the level of 100% relative humidity and the cusp between subsaturation and supersaturation, condensation gradually depletes the excess vapor. The delicate balance that exists between the generation and depletion of excess vapor depends on the updraft speed and the concentration of cloud-active aerosol particles in the rising parcel. The thermodynamic development of supersaturation was described earlier, so this chapter focuses on the influences that the aerosol population and cloud microphysical processes have on the evolution of supersaturation in clouds.

Thermodynamic considerations permitted us to derive the adiabatic supersaturation development equation (see Chapter 6). We now extend the theory to include the explicit effects of aerosol particles on the evolution of supersaturation. If we let w represent the speed that a moist air parcel rises adiabatically inside a cloud, then supersaturation, $s \equiv e/e_s(T) - 1$, increases at the rate $Q_1 w$, where $Q_1 \simeq \left(l_v/\left(c_p T\right) - 1\right)\left(M_{air} g/RT\right)$. Condensation, by contrast, removes some of the vapor at the rate $Q_2 d\omega_L/dt$, where $Q_2 \simeq l_v^2/(M_w p c_p T) + RT/(M_w e_s(T))$ and ω_L is the instantaneous concentration of liquid water. Supersaturation changes with time at a rate equal to the difference between the production and loss of excess vapor:

$$\frac{ds}{dt} = Q_1 w - Q_2 \frac{d\omega_L}{dt}. \tag{10.1}$$

This differential equation must be solved (numerically) to gain the evolution of supersaturation, $s(t)$.

The form of the solution to the adiabatic supersaturation development equation (Eq. (10.1)) can be understood qualitatively without further study. As shown by Fig. 10.1, several stages in the evolution of supersaturation can be identified. Initially, as air rises through cloud base, supersaturation s increases more or less linearly and causes aerosol

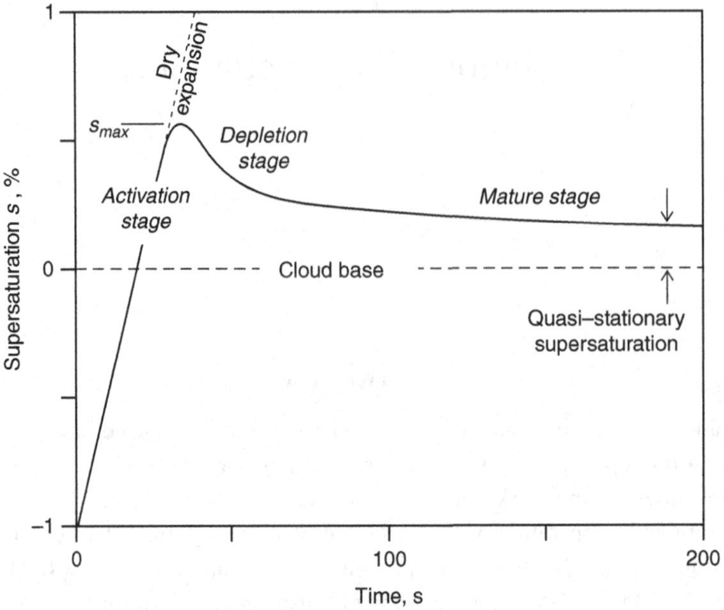

Figure 10.1 Evolution of supersaturation during the adiabatic rise of an air parcel through cloud base.

particles to become activated, transformed into cloud droplets that act as sinks of excess vapor. This early, activation stage is driven mainly by the cooling of the rising parcel, and it would continue in the absence of depletion by condensation. With active condensation, however, the rate of vapor depletion increases and the supersaturation reaches a maximum, $s_{\max}$. Beyond the time of $s_{\max}$, the supersaturation remains positive, but it decreases rapidly in the vapor-depletion stage; excess vapor is being lost by condensation faster than it is being produced by parcel cooling. Eventually, in the mature stage, a quasi-stationary supersaturation is reached, when the rates of production and loss are in approximate balance. This general pattern is realized in many clouds near their bases.

10.2 Extended theory

More exact solutions of Eq. (10.1) require knowing how $d\omega_L/dt$ changes with time. Note at the onset that the principle of mass conservation dictates that the rate at which liquid increases in a parcel is just the negative of the rate at which vapor is depleted: $d\omega_V/dt = -\,d\omega_L/dt$. Also recognize that each aerosol particle that activates forms a single new cloud droplet, so the increase in the droplet concentration is

$$dN_d = n_{CCN}(s)ds, \qquad (10.2)$$

where $n_{CCN}(s)$ is the supersaturation spectrum of CCN, the number of particles that activate in the supersaturation range s to $s + ds$. Thus, the cumulative concentration of droplets (N_d) is related to the differential CCN spectrum by

$$N_d(t) = \int_0^{s(t)} n_{CCN}(s_c)ds_c, \tag{10.3}$$

where s_c is the critical supersaturation of a given particle (review Köhler theory as needed) and $s(t)$ is the instantaneous supersaturation. New droplets are added to the population only as long as s increases with time and viable CCN remain available in the parcel. Each new droplet grows by condensation and augments the rate of vapor depletion:

$$-\frac{d\omega_V}{dt} = \frac{d\omega_L}{dt} = \int_0^{N_d} \frac{dm_d}{dt} dN_d = \int_0^{s(t)} \frac{dm_d}{dt} dn_{CCN}(s_c)ds_c. \tag{10.4}$$

The mass growth rate of each droplet (dm_d/dt) is related kinematically to the linear growth rate by

$$\frac{dm_d}{dt} = 4\pi\rho_L r_d^2 \frac{dr_d}{dt} = 4\pi\rho_L r_d \left(r_d \frac{dr_d}{dt} \right). \tag{10.5}$$

Factoring the square of droplet radius (r_d) facilitates introducing the physical growth law for condensation,

$$r_d \frac{dr_d}{dt} = Gs(t), \tag{10.6}$$

where G is the growth parameter derived earlier (Section 8.2). Integration of Eq. (10.6) yields

$$r_d(t) \simeq (2G)^{1/2} \left[\int_{t_{act}(s_c)}^{t} s(t')dt' \right]^{1/2}, \tag{10.7}$$

where $t_{act}(s_c)$ is the instant at which a particle having critical supersaturation s_c activates. (The time interval $t - t_{act}$ is the time during which a droplet grows and actively takes up vapor.) Inserting Eqs. (10.6) and (10.7) into Eq. (10.5) gives us the dependence of the mass growth rate on the ever-changing supersaturation, $s(t)$:

$$\frac{dm_d}{dt} = 2\pi\rho_L (2G)^{3/2} s(t) \left[\int_{t_{act}(s_c)}^{t} s(t')dt' \right]^{1/2}. \tag{10.8}$$

Insertion of this equation into Eq. (10.4) yields the rate that liquid water builds up in the cloud:

$$\frac{d\omega_L}{dt} = 2\pi\rho_L (2G)^{3/2} s(t) \int_0^{s(t)} \left[\int_{t_{act}(s_c)}^{t} s(t')dt' \right]^{1/2} n_{CCN}(s_c)\, ds_c. \tag{10.9}$$

Note that supersaturation itself does not directly depend on s_c, so $s(t)$ is brought outside the integration in Eq. (10.9). Integration of this equation still needs knowledge of $s(t)$, but nevertheless one can recognize that the integration over s_c reaches a fixed value once the supersaturation reaches its maximum value (s_{max}). Thereafter, the concentration of liquid water continues to increase at a rate proportional to $s(t)$.

The evolution of supersaturation can now be calculated by integrating Eq. (10.1) while coupled to Eq. (10.9). Mathematical analysis is facilitated by recognizing that the differential equation to be solved can be written in a form that isolates the supersaturation dependence:

$$\frac{ds}{dt} = Q_1 w - A_2 s I(s), \tag{10.10}$$

where $A_2 \equiv 2\pi Q_2 \rho_L (2G)^{3/2}$ is a factor independent of s. The integral,

$$I(s) = \int_0^s \left[\int_{t_{act}(s_c)}^{t} s(t')dt' \right]^{1/2} n_{CCN}(s_c)\, ds_c, \tag{10.11}$$

contains all the information about the formation of new droplets through the aerosol activation. The inner integral can be estimated in terms of s_c by recognizing that vapor loss is minimal until near the time of $s_{\max}$. Thus, the slope $ds/dt \simeq Q_1 w$ is approximately constant, so t and t_{act} in the integral limits are $s/Q_1 w$ and $s_c/Q_1 w$, respectively. The inner integral is therefore

$$\int_{t_{act}(s_c)}^{t} s(t')dt' = \frac{s^2 - s_c^2}{2Q_1 w}. \tag{10.12}$$

Most of the remaining mathematical complications arise from the dependence of liquid water concentration on the number of cloud droplets, which is continually changing throughout the activation stage. One must of course measure or assume a CCN spectrum, $n_{CCN}(s_c)$, and know how the updraft speed w varies with time.

To gain a qualitative feel for the activation of an aerosol population, the results from a simple numerical model of an adiabatic parcel rising at a constant updraft speed are shown in Fig. 10.2. Plotted here are the sizes of selected droplets having the solute contents indicated along the lower axis. Supersaturation resulting from the collective action of aerosol activation and growth is shown by the dashed curve (with axis along the top). Note that the ordinate is logarithmic relative to cloud base, so new droplet formation takes place rapidly within the first 10 to 20 m. Very small particles (toward left-hand side) grow slightly as the supersaturation increases toward and beyond the maximum supersaturation, but their small solute contents prevent them from becoming activated. Unactivated particles remain interstitially as haze droplets within the cloud. Particles having solute contents greater than about 10^{-18} mol (in this example) readily activate and grow by condensation. However, those particles with the largest solute contents (curves with hooked arrows), even though they have the smallest critical supersaturations, may become so massive that they sediment out of the parcel before reaching critical size. As we learned earlier, the linear growth rate of a droplet varies inversely with its size, so the large droplets grow relatively more slowly than do the smaller droplets. The combined effects of aerosol activation and growth kinetics lead to a narrowing of the droplet spectrum well separated in size from the interstitial haze population.

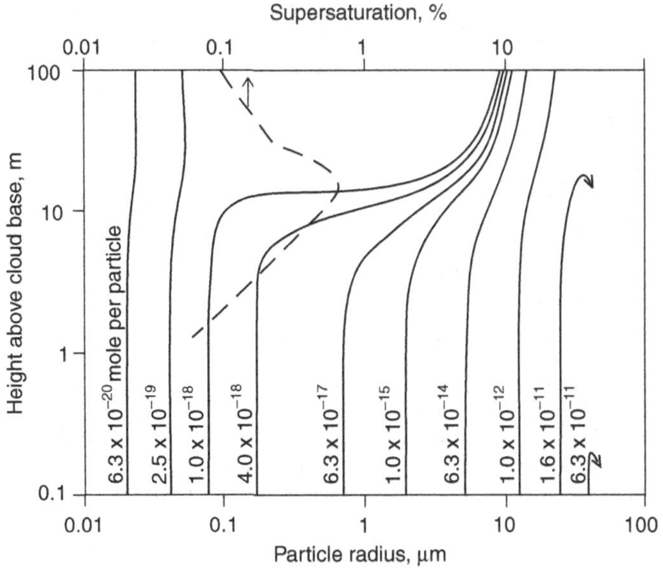

Figure 10.2 Results from an adiabatic-parcel model with a sparse aerosol population (several hundred cm^{-3}) rising at the rate of $15\,cm\,s^{-1}$. Radii of particles having the indicated solute contents are shown on the lower axis. Arrows on larger particles suggest sedimentation should also be considered. Supersaturation (dashed curve) is given on the upper axis. Adapted from Mordy (1959).

10.3 Aerosol influences

Aerosol particles serve as the sites of condensation in a cloud, so it is natural to question how the aerosol population in the pre-cloud air influences supersaturation development. Inspection of Eq. (10.9) shows that the liquid water concentration builds up at a rate roughly proportional to the integrated concentration of CCN. A low concentration of CCN thus yields a lower concentration of cloud droplets and a larger maximum supersaturation than does a high concentration of CCN. The growth rates of the droplets that do form are proportional to the supersaturations, so we are left with the qualitative expectation that clouds forming in clean air (having small [CCN]) have a small number of relative large cloud droplets compared with clouds forming in dirty air (having large [CCN] and therefore many small droplets). In effect, the aerosol concentration controls the concentration of cloud droplets and their early distribution with size.

The general influences of the aerosol concentration on the development of supersaturation and cloud droplets are shown in Fig. 10.3. The two sets of curves pertain to cases of low and high concentrations of condensation nuclei (CN, representing the total aerosol); each set shows how supersaturation evolves (solid curves) and how the population of cloud droplets (dashed curves) increases until the maximum supersaturation is reached. The shaded regions identify the respective stages of aerosol activation. Beyond the time of maximum supersaturation, no new aerosol particles can become activated and

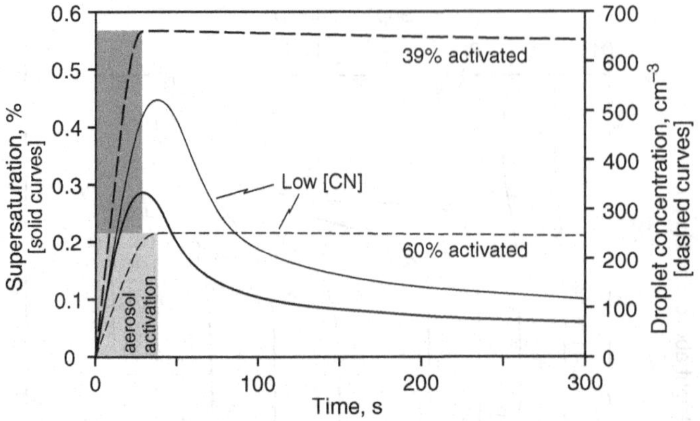

Figure 10.3 The evolution of supersaturation (solid curves, left axis) and droplet concentration (dashed curves, right axis) in clouds of low (light curves) and high (heavy curves) aerosol concentrations. For the low case, [CN] = 425 cm^{-3}; for the high case, [CN] = 1700 cm^{-3}. Results from a parcel model of J.Y. Harrington.

so the concentration of cloud droplets remains roughly constant; the gradual decrease with time is due to the expansion of the parcel during uplift. These results clearly show that an increased concentration of aerosol particles leads to a weaker development of supersaturation; the maximum value is lower, as are the magnitudes in the mature stage. The concentration of particles in the high-[CN] case is larger than that in the low-[CN] case, but the fraction of particles activated is smaller.

The supersaturation represents the balance between production and loss of excess vapor (Eq. (10.1)), so it responds to the aerosol spectrum in predictable ways. Using the power-law CCN parameterization, $N_{CCN}(s_c) \simeq cs_c^k$ (considered in Chapter 13), we note a fundamental distinction between maritime and continental clouds, that of aerosol abundance. Maritime air masses, having had long times to cleanse the air of particulate matter without the introduction of new aerosol sources, are typically clean, characterized by low values of c (hundreds per cm^{-3}). By contrast, continental air, because of the diverse sources of particulate matter available, is aerosol rich and characterized by c values more than ten times greater. Figure 10.4 shows the impact of these aerosol differences on the evolution of the supersaturation in a simple parcel model. Note that, for each updraft speed, the maximum supersaturation is significantly greater in the representative maritime cloud than it is in the continental cloud. The maximum supersaturation in the maritime case occurs later, simply because supersaturation can develop longer before the sparse aerosol is able to deplete the excess vapor. In the continental case, the high aerosol concentration quickly draws down the excess vapor, retarding the supersaturation at an early stage. The initial rates of rise are independent of aerosol concentration (as shown by the overlapping parts of the curves), but they do scale directly with the updraft speed (in accordance with the

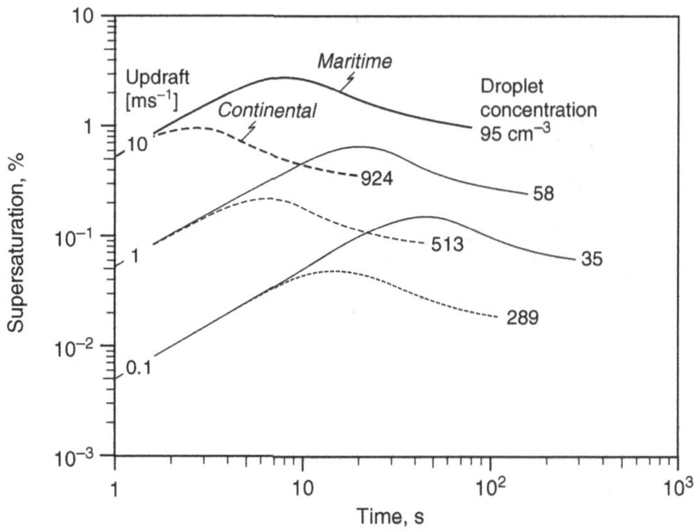

Figure 10.4 Evolution of supersaturation in maritime and continental clouds for three different updraft speeds. Calculations based on an adiabatic parcel model using the CCN parameters $c = 310\,\text{cm}^{-3}$ and $k = 0.33$ for the maritime case; $c = 6000\ \text{cm}^{-3}$ and $k = 0.40$ for the continental case. The numbers at the end of each curve give the maximum concentration of cloud droplets. Adapted from Fletcher (1962).

first term in Eq. (10.1)). Note, too, that the droplet concentrations are always larger in the continental case than in the maritime case for a given updraft speed, but that higher updraft speeds in a given air mass yield larger droplet concentrations, because larger supersaturations permit greater opportunities to activate aerosol particles. Maritime clouds have low concentrations of droplets because of low concentrations of CCN, in contrast to continental clouds that have high concentrations of both aerosol particles and cloud droplets.

Variations in aerosol abundance, such as occurs because of pollution on local, regional, and even intercontinental scales, can impact the concentrations and size distributions of cloud droplets early in their development. In general, the higher the concentration of aerosol particles present at the time of cloud formation, the greater is the competition for the available excess vapor and the smaller are the resulting droplets. This pattern can be seen in Fig. 10.5, where the data have been selected from several observed cases in which the total liquid water concentrations were comparable. The smallest droplets tend to be found in clouds when their concentrations are the largest. By contrast, sparser clouds have relatively few, but larger droplets. Such distinctions reflect the quality of the air mass through the CCN concentration, and they impact the ability of clouds to scatter radiation and form precipitation. Alterations in cloud microstructure in turn affect precipitation efficiencies (higher in maritime clouds) and radiative energy balances (higher albedos in continental clouds).

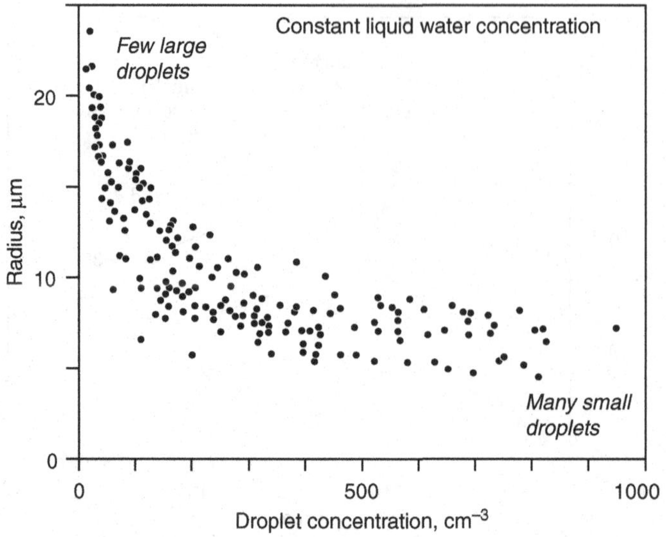

Figure 10.5 Sizes of droplets versus concentration in cloud of comparable liquid water concentration. Adapted from Dong *et al.* (1997).

10.4 Quasi-stationary supersaturation

The supersaturation in an adiabatic parcel tends to vary little in the mature stage, at times well beyond that of supersaturation maximum. When the rate that excess vapor is depleted roughly compensates for its generation (by continued uplift of the parcel), the supersaturation maintains a quasi-stationary value. The magnitude of this quasi-stationary supersaturation s_{QS} can be calculated by setting the rate in Eq. (10.1) equal to zero.

The mathematical procedure for finding s_{QS} is similar to that for calculating the maximum supersaturation $s_{\max}$, but we no longer need to account for new activation and droplet formation. Once past the time of $s_{\max}$, the droplet concentration is fixed at $N_d = \int_0^{s_{\max}} n_{CCN}(s_c)ds_c$. Thus, the liquid water concentration in Eq. (10.1) depends in a relatively simple way on the instantaneous droplet size distribution:

$$\frac{d\omega_L}{dt} = \int_0^\infty \frac{dm_d}{dt} n_d\,(r_d)\,dr_d = 4\pi\rho_L G s \int_0^\infty r_d n_d\,(r_d)\,dr_d, \tag{10.13}$$

where G is the condensational growth parameter, and where the mass-growth law for an individual droplet of radius r_d has been used: $dm_d/dt = 4\pi\rho_L G r_d s$. The liquid water concentration thus builds up in the mature stage in direct proportion to the supersaturation, so we can solve for the quasi-stationary value after setting the production and depletion terms in Eq. (10.1) equal to each other:

$$s_{QS} = \frac{Q_1}{4\pi\rho_L G Q_2} \cdot \frac{w}{\int_0^\infty r_d n_d\,(r_d)\,dr_d}. \tag{10.14}$$

At this point, it proves helpful to simplify the equation by calling the first factor $A \equiv Q_1/(4\pi\rho_L G Q_2)$ and recognizing how the first moment of a size distribution, sometimes called the integral radius, is related to the mean radius: $\overline{r_d} = (1/N_d) \int_0^\infty r_d n_d (r_d)\, dr_d$. Thus,

$$s_{QS} = \frac{Aw}{\overline{r_d} N_d}. \tag{10.15}$$

The quasi-stationary value during the mature stage of supersaturation evolution depends directly on the updraft speed and inversely on the mean size and total number concentration of the droplet population. Of particular importance to later analysis, any process that alters the magnitudes of w, $\overline{r_d}$, or N_d affects s_{QS} and the subsequent development of the cloud.

Typical values of s_{QS} can be determined by considering common atmospheric situations. Use of reasonable values for the variables in Q_1 and Q_2 give $A \simeq 2$ s m^{-3}, so for an updraft speed $w = 5$ m s^{-1} and a mean droplet radius $\overline{r_d} = 10$ μm $= 10 \times 10^{-6}$ m, we find to good approximation $s_{QS} \simeq 1/N_d$. In fact, this relationship gives the numerical value of s_{QS}, expressed as a fraction, when the droplet concentration (N_d) is given in units of cm^{-3}. Thus, in maritime clouds ($N_d \simeq 100$ cm^{-3}), the quasi-stationary supersaturation $s_{QS} \simeq 0.01 = 1\%$, whereas in continental clouds ($N_d \simeq 1000$ cm^{-3}), the quasi-stationary supersaturation $s_{QS} \simeq 0.001 = 0.1\%$. Again, such distinctions, which arise from differences in aerosol abundances, give rise to important differences in the properties of clouds forming in maritime and continental air masses.

The time scale to achieve steady state must be understood if one is to apply the concepts embodied in Eq. (10.15) to actual cloud situations. Variations in the pertinent variables ($w, \overline{r_d}, N_d$) will always lead to variations in s_{QS}, but if the changes are too rapid, the supersaturation will not have time to respond and Eq. (10.15) will be invalid. We start a time-scale analysis with the original differential equation (Eq. (10.1)). When no new droplets are formed, we can write the equation in the simple form,

$$\frac{ds}{dt} + c_1 s = c_0, \tag{10.16}$$

where $c_1 \equiv 4\pi \rho_L G Q_2 \overline{r_d} N_d$ and $c_0 \equiv Q_1 w$. The first term on the left-hand side is the response term, the second term is the first-order restoring term, and the right-hand term is the source term. Analysis of such first-order differential equations shows that the exponential time-constant τ_{QS} is determined by the coefficient of the restoring term. Thus,

$$\tau_{QS} = \frac{1}{c_1} = \frac{1}{4\pi \rho_L G Q_2 \overline{r_d} N_d}. \tag{10.17}$$

Note that the time constant is inversely related to the first moment of the size distribution $\left(\overline{r_d} N_d = \int_0^\infty r_d n_d (r_d)\, dr_d\right)$. Large mean sizes and high droplet concentrations offer a large surface area for taking up excess vapor, so the supersaturation responds quickly in such cases. Comparison of Eqs. (10.14) and (10.17) shows that the time constant can be related to the quasi-stationary supersaturation by $\tau_{QS} = s_{QS}/(Q_1 w)$. Conditions favoring large s_{QS} also imply longer times to achieve that quasi-stationary value, but time scales for phase relaxation seldom exceed 10 s, even in clean maritime clouds.

10.5 Microphysical and dynamical influences

Supersaturation evolves in response to the continuously changing microphysical and dynamical environment of a cloud. Supersaturation develops initially as moist air rises above cloud base, but the positive supersaturation activates aerosol particles, which then grow as liquid or solid cloud particles. Analyses leading to idealized patterns (e.g., Fig. 10.1) assume that the updraft speed is constant. In reality, the magnitude of w often fluctuates about some mean value, especially in convective clouds where turbulent motions are common. Fluctuations in w lead to fluctuations in the rate at which s varies (via Eq. (10.1)). At the same time, the depletion of excess water vapor by condensation and deposition, expressed by the right-most term in Eq. (10.1), depends on the ever-changing size distributions and phases of the particles. If the values of all pertinent variables were known at each instant in time, one could in principle integrate Eq. (10.1) numerically to obtain a reasonable estimate of how the supersaturation truly evolves in a cloud. Lacking the requisite information, we resort to generalities and merely identify the types of influence that small-scale dynamical and microphysical processes have on supersaturation.

For the moment, consider how the supersaturation might evolve in a constant updraft later in the lifecycle of a cloud, as when the warm-cloud microphysics becomes active. Collision-coalescence causes some of the drops to merge, generating larger drops at the expense of smaller drops. The mean size ($\overline{r_d}$) of the population will increase slightly, but the main impact is on the number concentration (N_d). (Mass is conserved, so when two similar size drops coalesce, the size increases only by the cube root of two ($\sim$ 1.26); however, the number of drops decreases by a factor of two.) As anticipated by Eq. (10.15) and shown in Fig. 10.6, coalescence leads to a significant decrease in N_d and a corresponding increase in s. The initially large droplet concentration, established during initial activation,

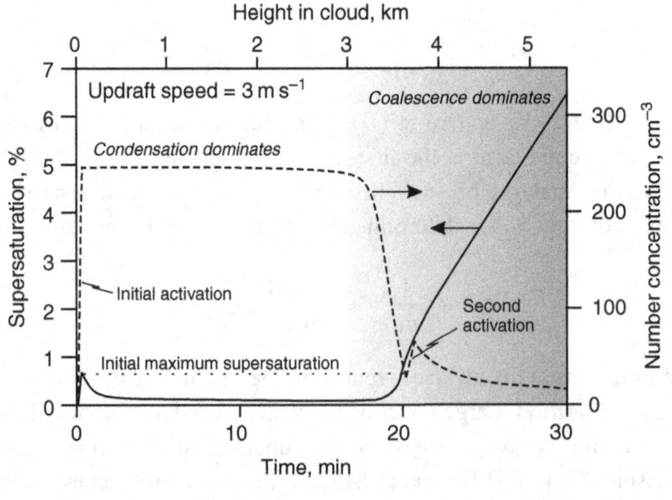

Figure 10.6 Evolution of supersaturation (left axis) and droplet concentration (right axis) in a cloud with a constant updraft speed of 3 m s^{-1}. Adapted from Young (1993).

retains its magnitude following the initial supersaturation maximum when condensation alone is the mechanism of droplet growth. Once the droplets grow sufficiently large to initiate collision-coalescence ($\sim$ 17 min), the droplet concentration decreases rapidly, resulting in less competition for the vapor and an imbalance in the two terms on the right-hand side of Eq. (10.1). The supersaturation increases and soon exceeds the initial maximum value. Some of the interstitial haze particles that were not activated during the initial rise in supersaturation are now activated, forming a new set of small droplets among the larger drops. These new droplets grow rapidly by condensation and then quickly get collected by the large drops. As the droplet concentration decreases again, little prevents the supersaturation from rising sharply and well above the initial supersaturation maximum. This scenario nicely illustrates the strong control that the droplet population has on supersaturation, but it is idealized in that no account has been taken of other processes, such as entrainment mixing and sedimentation.

Entrainment of dry air into a cloud by turbulent mixing alters the thermodynamic and microphysical development of a cloud. As droplets evaporate by exposure to dry air, the air cools further and changes the local stability of the rising cloudy air. Buoyancy and the rate of rise may be reduced in convective clouds, leading to a reduction in the rate of production of excess vapor. Very near cloud edge, the air may cool enough to become negatively buoyant and sink. However, deeper within the cloud, where the air continues to rise, the lack of condensate by entrainment leads to a slower uptake of excess vapor. The net rate of supersaturation development (according to Eq. (10.1)) is no doubt affected by entrainment, but it is not immediately clear which term (production or loss) is most affected. The dynamical forces responsible for air motions in a convective cloud are complicated and affected by events elsewhere in the cloud, not just those directly affected by a particular entrainment event. If one accepts that the air motions are in the first instance decoupled from a given entrainment event, then the dominant effect of entrainment is to reduce the competition for excess vapor among the surviving droplets (expressed by a smaller integral radius in Eq. 10.14). The counterintuitive effect may well be higher supersaturations, new droplet formation, or enhanced condensational growth rates of droplets in entrainment-affected parcels.

Evidence for large supersaturations and new droplet formation inside real clouds affected by entrainment is shown by some aircraft data (see Fig. 10.7). The various panels display the time series of measured and computed variables acquired during passage through a vigorous continental cumulus cloud. The inset shows how the droplets were distributed in size at the time indicated by the arrow. The top panel shows the measured updraft speed w, which generates positive supersaturation (third panel down) when $w > 0$. The total concentration of droplets (across all sizes) varies greatly across the width of the cloud, from near zero at cloud edges to over 200 cm^{-3} in a couple of places. Note (from inset) the minimum in droplets having radii near 6 μm, which divides the total population into two modes. The concentration of droplets larger than about 6 μm in radius, considered to be mature, well-established cloud droplets in this study, was found to decrease suddenly in the region identified by shading in Fig. 10.7. The quasi-stationary supersaturation,

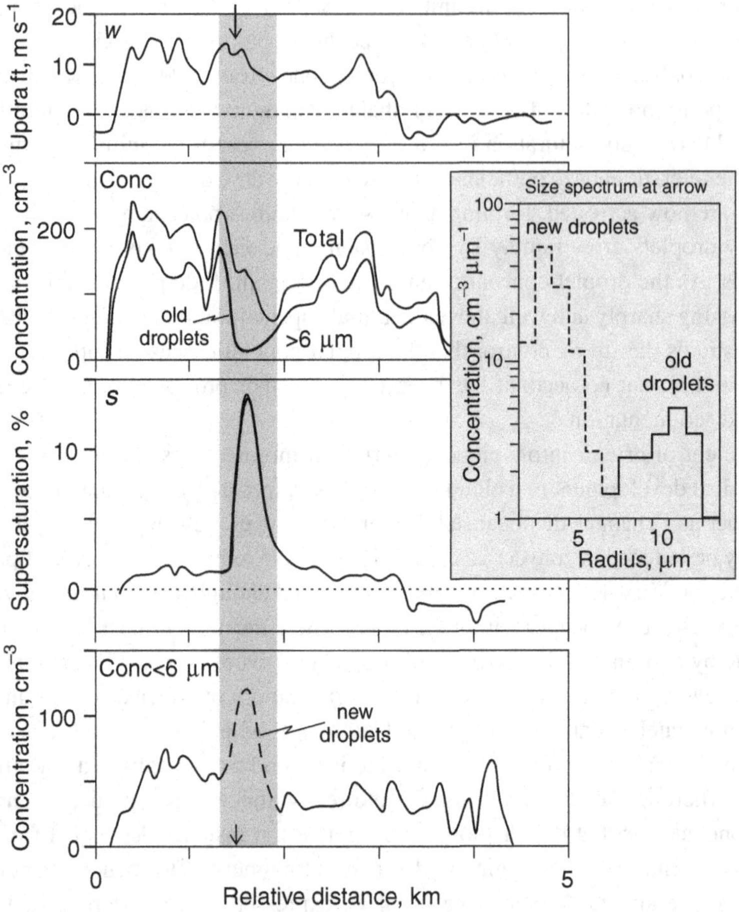

Figure 10.7 Aircraft data suggesting the formation of new droplets. Arrow identifies point at which the size spectrum (inset) was acquired. Adapted from Paluch and Knight (1984).

computed from Eq. (10.14), but with the lower limit set to 6 μm rather than 0, gives the expected supersaturation in the absence of the very smallest droplets (dashed part of inset). This conditional supersaturation is shown in the third panel to have increased greatly in the shaded region as a result of the decrease in the large-droplet population (bold curve in second panel). The sudden increase in supersaturation is thought to have triggered the activation of haze aerosol, which manifested itself as the sudden appearance of small droplets (dashed curve in bottom panel). These data offer strong support for the extended theory of supersaturation development summarized in this chapter.

The dynamical and microphysical influences on supersaturation in real clouds is made especially complicated by the simultaneous fluctuations in air motions and droplet concentrations. Small-scale turbulence at the high Reynold numbers found in convective clouds

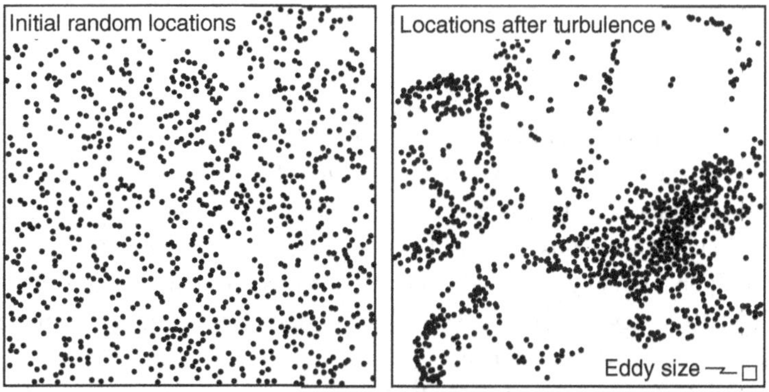

Figure 10.8 Particles in turbulent flow. Left panel, random locations of particles at start of DNS simulation. Right panel, pattern after several eddy-turnover times. The size of a Kolmogorov eddy is shown by the box in the lower right corner. Adapted from Shaw *et al.* (1998).

is thought to generate regions of high and low vorticity that are superimposed on the mean updraft. In particular, vortex tubes some centimeters in diameter and perhaps ten times as long are hypothesized to form frequently. The rapid rotational motions cause droplets to centrifuge out of high-vorticity regions and into low-vorticity regions. Such a mechanism of preferential concentration would give rise to pockets of alternately low and high concentrations of cloud droplets, as shown in Fig. 10.8. The traditional theory leading to quasi-stationary supersaturation (Eq. (10.14)) suggests that supersaturation should increase in regions of low droplet concentration (high vorticity) and decrease in regions of high concentration (low vorticity). The high supersaturations generated by such turbulent effects could lead to the formation of new droplets by aerosol activation and/or the enhanced condensational growth of the few lucky droplets able to survive the turbulence. By contrast, droplets in the highly concentrated regions, even though they may not grow effectively by condensation, may find nearby neighbors with which to collide and coalesce. In any case, turbulence may offer opportunities for enhanced growth of cloud particles and a broadening of the droplet size distributions. The effects of turbulence on supersaturation development are still not well understood, partly because other processes (e.g., particle sedimentation) add compounding complexities.

The formation of ice exerts a strong influence on the partial pressure of water vapor in a supercooled cloud. The very terms excess vapor and supersaturation in a mixed-phase cloud become ambiguous until one specifies the reference state (liquid or ice). As soon as ice particles appear, supersaturation with respect to ice, s_i, becomes relevant, and the supersaturation with respect to liquid, s, develops more slowly because of the lower equilibrium vapor pressure of the solid and the more numerous opportunities for vapor to be removed. A second term must therefore be added to Eq. (10.1) to account for the simultaneous phase transformations from vapor to liquid and from vapor to solid. In the mixed-phase regions of a cloud, the supersaturation with respect to liquid develops at a rate given by

$$\frac{ds}{dt} = Q_1 w - Q_2 \frac{d\omega_L}{dt} - Q_3 \frac{d\omega_I}{dt}, \tag{10.18}$$

where $Q_3 \simeq l_v l_s/(M_w p c_p T) + RT/(M_w e_s(T))$. The new term on the right-hand side represents the loss of excess vapor to the ice particles, whose mass is collectively given by the ice–water concentration ω_I. The complete solution to Eq. (10.18) is challenging to obtain when new droplets and ice crystals are allowed to form simultaneously. Perhaps the most realistic situation in cold clouds involves the freezing of cloud droplets to form ice crystals. Then, the total concentration of cloud particles remains constant, because the concentration of droplets decreases at exactly the same rate as does the concentration of ice particles; only the phases of the particles change.

In many mixed-phase clouds, relatively few ice crystals grow from the vapor while surrounded by many supercooled liquid droplets, which slowly evaporate because of the difference in equilibrium vapor pressures of ice and liquid water. This thermodynamically driven process, whereby the ice crystals grow at the expense of the liquid droplets, is known as the Wegener–Bergeron–Findeisen (or just Bergeron) process. The Bergeron process is responsible for the initiation of much cold-cloud precipitation and can be modeled relatively easily. For example, consider a parcel of cloudy air containing given concentrations of droplets and ice particles. As the mixed-phase parcel rises adiabatically at the rate $w = 0.8$ m, the quasi-stationary supersaturations (s, s_i) follow the trends shown in Fig. 10.9. Initially, while the ice particles are still very small, their impact on s is minimal and both supersaturations are positive; both the liquid and solid particles grow. However, the ice particles grow rapidly in an environment near water saturation, which causes very

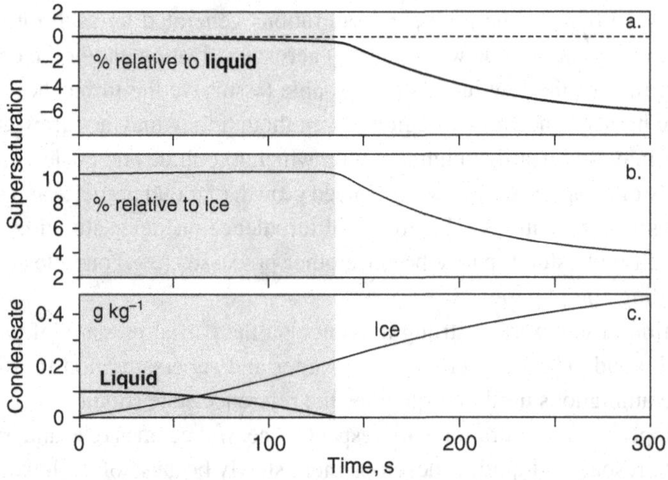

Figure 10.9 Changes of supersaturation and condensate with time during the glaciation of an initially mixed-phase cloud. a. Supersaturation relative to liquid water. b. Supersaturation relative to ice. c. Condensate mixing ratios of liquid (heavy curve) and solid (light curve) phases. Shading: mixed-phase period. Adapted from Korolev and Mazin (2003).

large supersaturations with respect to ice. Vapor is taken up more rapidly by the relatively few ice particles than can be supplied by uplift, so the liquid water supersaturation falls below zero, and the droplets evaporate from then on. Note that s remains close to liquid water saturation until glaciation is complete (at about 140 s when the liquid water concentration falls to zero). Thereafter, in the all-ice parcel, vapor is depleted by the continued growth of the ice particles and s_i gradually relaxes toward the quasi-stationary value

$$s_{QS,i} = \frac{A_i w}{\overline{r_i} N_i}, \tag{10.19}$$

where $A_i \equiv Q_1/(4\pi \rho_I G_I Q_3)$ is the parameter used in Eq. (10.15) made applicable to ice, $\overline{r_i}$ is the mean radius of the volume-equivalent ice particles, and N_i is the number concentration of ice particles. The supersaturations in mixed-phase clouds can be complicated functions of time, especially when phase changes are coupled with particle sedimentation and other microphysical and dynamical influences.

10.6 Further reading

Curry, J.A. and P.J. Webster (1999). *Thermodynamics of Atmospheres and Oceans.* London: Academic Press, 467 pp. Chapter 5 (supersaturation development).
Fletcher, N.H. (1962). *The Physics of Rainclouds.* Cambridge: Cambridge University Press, 386 pp. Chapter 6 (supersaturation development).
Pruppacher, H.R. and J.D. Klett (1997). *Microphysics of Clouds and Precipitation.* Dordrecht: Kluwer Academic Publishers, 954 pp. Chapter 13 (supersaturation development).
Rogers, R.R. and M.K. Yau (1989). *A Short Course in Cloud Physics.* New York: Pergamon Press, 293 pp. Chapter 7 (supersaturation development).
Young, K.C. (1993). *Microphysical Processes in Clouds.* New York: Oxford University Press, 427 pp. Chapters 9 (supersaturation development in warm clouds) and 6 (supersaturations in cold clouds).

10.7 Problems

1. We will work our way through the derivation of the development of supersaturation close to cloud base while drops are still activating. The governing equation is

$$\frac{ds}{dt} = Q_1 w - 2\pi Q_2 \rho_L (2G)^{3/2} s I(s),$$

where

$$I(s) = \int_0^s \left[\int_{t_{act}(s_c)}^t s(t')dt' \right]^{1/2} n_{CCN}(s_c)\, ds_c.$$

Use Eq. (10.12) (convince yourself that this is a reasonable approximation!) and assume that the aerosol spectrum is well represented by a power law of the form $n_{CCN}(s_c) = cks_c^{k-1}$ to write the integral in a form that may be integrated. Rewrite the integral in terms of $x = s_c^2/s^2$. Remember the integration limits! Use the definition of the beta function $B(a, b) \equiv \int_0^1 y^{a-1}(1-y)^{b-1}\,dy = \Gamma(a)\Gamma(b)/\Gamma(a+b)$ to solve the integral. Rewrite the governing equation in the form

$$\frac{ds}{dt} = nA - Bs^{k+2}.$$

Solve for the expression of the maximum supersaturation. Please calculate the maximum supersaturation attained close to cloud base for updraft velocities of $w = 1$ m s^{-1} and $w = 5$ m s^{-1} if
(a) $k = 0.33$ and $c = 310$ cm^{-3} (maritime air)
(b) $k = 0.4$ and $c = 6000$ cm^{-3} (continental air).
Assume a constant temperature 280 K and pressure 850 hPa in your calculations.

2. Calculate the quasi-steady supersaturation and the exponential time scale to achieve steady state for updraft velocities of $w = 1$ m s^{-1} and $w = 5$ m s^{-1} if
(a) $r_d = 20$ μm and $N_d = 30$ cm^{-3} (maritime air)
(b) $r_d = 10$ μm and $N_d = 500$ cm^{-3} (continental air).
Assume a constant temperature 273 K and pressure 750 hPa in your calculations.

11
Warm clouds

11.1 Overview

Warm clouds typically form in the lower troposphere when ice is not important to the microphysics. The development of warm clouds depends on condensation to activate aerosol particles and grow liquid droplets initially. Subsequent processes include drop–drop collisions, coalescence, and disruption during the formation of rain in mature clouds. Such processes may also occur in colder, mixed-phase clouds as long as the presence of ice does not interfere significantly with any of the warm-cloud microphysical processes.

The focus of this chapter is the spectral evolution of drop populations, how small ($\sim 10\,\mu$m radius) cloud droplets grow to large (~ 1 mm radius) raindrops. Many of the individual processes responsible for particle formation and growth have already been treated, so we now look at how these processes work collectively to form rain.

A perspective of how liquid-phase microphysics leads to rain in a warm cloud can be obtained by viewing the different processes as discrete boxes operating simultaneously inside the cloud. As shown in Fig. 11.1, one process leads to another, from initial activation of the aerosol entering the cloud in its updraft to rain falling out through its base. Overall, rain formation requires several broad categories of spectral evolution: condensation, coalescence initiation, and continuous collection. As supersaturation develops by adiabatic cooling of the rising air, the aerosol particles grow first as haze droplets, then as cloud droplets by condensation. Adiabatic condensation, the idealized kind treated earlier, is slow because of competition for the available excess vapor, and it moreover leads to a narrow drop size distribution. Adiabatic condensation is not conducive to collection because few if any droplets attain the sizes required for collisions to occur. Rain formation requires relatively large drops and an active collision-coalescence process, often termed continuous collection because the growth of a raindrop can be treated mathematically by approximating the population of collected drops as a continuum. The interaction labeled autoconversion represents a type of parameterization suitable for some cloud models in order to circumvent the need to deal with the complicated physical processes (e.g., diabatic condensation and stochastic collection) that broaden the drop spectrum and enable collection to be initiated. Mechanisms for broadening the spectrum of drop sizes will be treated later in this chapter.

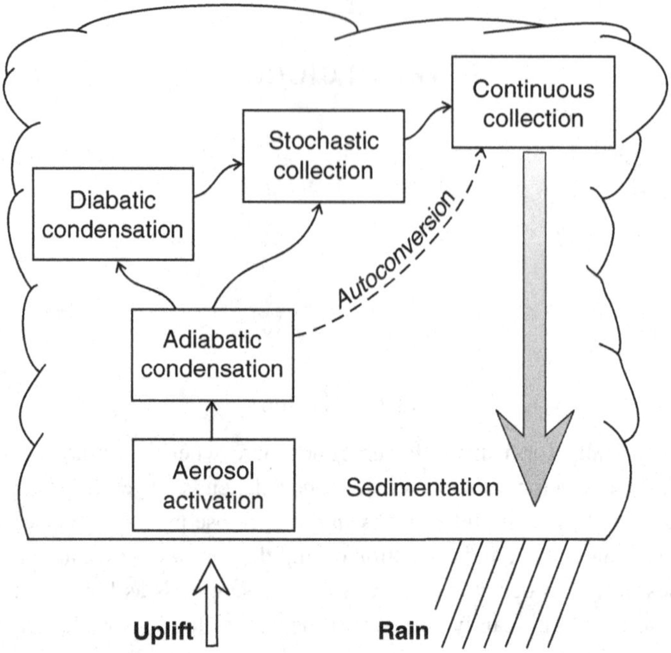

Figure 11.1 Schematic diagram of the types of microphysical processes involved in the formation of rain in a warm cloud.

A typical warm cloud first develops as moist air rises, either by slow ascent in a synoptic-scale weather system or by rapid ascent during convection, beyond the thermodynamic cloud base (100% relative humidity). The early stages of cloud formation follow a common pattern regardless of the cause of uplift. Consider the simplified scenario in which a parcel of moist air starts from near ground level and rises adiabatically at a constant rate. The consequences of this uplift are illustrated in Fig. 11.2. The air is initially subsaturated and contains a population of dry aerosol particles distributed in size according to the curve shown near the bottom. Most of the particles have diameters close to 0.02 μm, but a significant number have larger diameters. Three representative sizes are shown by vertical lines that extend upward into the growth diagram above. As uplift proceeds, the relative humidity in the parcel increases, as shown on the left by the supersaturation curve becoming less negative. In response, the aerosol particles (assumed to be soluble) deliquesce and become liquid droplets. With continued uplift, the droplets grow by condensation and follow the Köhler-theory equilibrium curves (shown by dashed curves). As the parcel passes through cloud base and the supersaturation becomes positive, large particles having small critical supersaturations activate first and grow rapidly beyond their respective critical sizes because the ambient supersaturation is larger than the respective equilibria. This growth behavior is exemplified by the two bold curves that separate away from the Köhler curves at their peaks. The uptake of vapor by the growth of the cloud droplets gradually depletes the

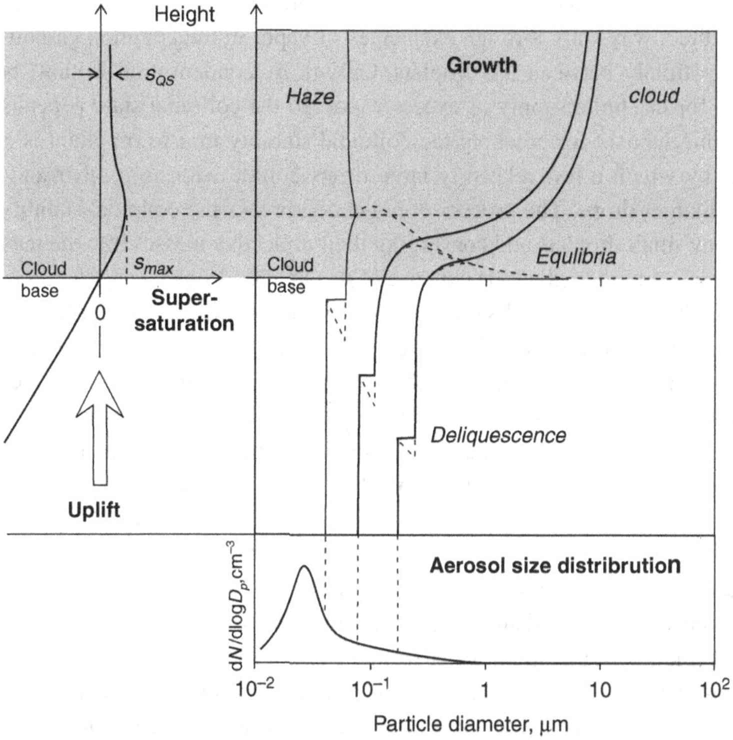

Figure 11.2 Evolution of the microphysical properties of a warm cloud during the adiabatic uplift of a moist air parcel. Left: Evolution of supersaturation as the parcel passes through cloud base. Right: Diameters of three representative particles as they grow from dry sizes (distribution shown at bottom) to aqueous solution droplets in response to the changing supersaturation. Heavy curves follow droplets that activate, light curve follows a haze droplet, an aerosol particle that did not grow beyond its critical size. Dashed curves identify the Köhler-theory equilibria.

excess vapor, which causes the supersaturation to reach a maximum value (s_{max}) and begin decreasing toward quasi-stationary values (s_{QS}). Smaller aerosol particles, those having critical supersaturations larger than the maximum ambient supersaturation (exemplified by the light curve), grow slightly but are unable to activate (grow beyond the equilibrium maximum). These unactivated particles remain as small haze droplets. The cloudy parcel is thus left with a population of cloud droplets that continue to grow (as long as the supersaturation remains positive) and a separate population of haze droplets that reside interstitially among the cloud droplets. The haze particles often outnumber the cloud droplets, but it is the more effective scattering of solar radiation off the much larger cloud droplets that lets us see the cloud (represented by the shaded region of Fig. 11.2).

A warm cloud is initially a mixture of small liquid droplets in air, a stable colloidal system that can persist for long times. A suitable microphysical mechanism is needed to break the colloidal stability, because early in their formation the droplets are all small and well separated from one another. They follow the air motions faithfully and interact with each other in no direct way, only through exchanges of vapor in the common gaseous medium existing interstitially between the droplets. Growth by condensation is slow, because of competition for the finite supply of excess vapor, so the colloidal state is persistent. The only viable mechanism for breaking the colloidal stability in a warm cloud is collection, the process by which a few relatively large drops collide with each other and coalesce, forming still larger drops. The process of collection breaks the colloidal stability by gathering up many small droplets and converging their collective masses into the mass of large drops. The big get bigger, all at the expense of the smaller droplets, and the drop spectrum broadens rapidly. However, collisions require significant relative motions between drops, either through turbulence or sedimentation, so collision-coalescence starts only with difficulty. In effect, a mechanistic gap exists between growth by condensation (which supplies only small droplets with radii less than 20 μm) and by collection (which requires drop radii well in excess of 20 μm). The situation is shown schematically in Fig. 11.3. The radii of droplets growing by condensation tend to follow a square-root law, whereas the radii of drops growing by collection tend to increase exponentially with time. Rain, representing the release of liquid from the cloud, forms only after some of the drops grow by collection to a size sufficient to let them fall against the updraft. At that point, the colloidal stability of the cloud has been broken and an active warm-rain process takes over, one that can remove a significant fraction of the liquid water from the atmosphere.

The initiation of collection in a warm cloud is still uncertain, but it likely depends in one way or another on the aerosol population and the early droplet spectra. The aerosol particles entering a cloud in the updraft serve as the sites of condensation, so a natural correlation exists between the initial concentrations of aerosol particles and cloud droplets. Under similar thermodynamic settings (given profiles of temperature and moisture), a cloud forming

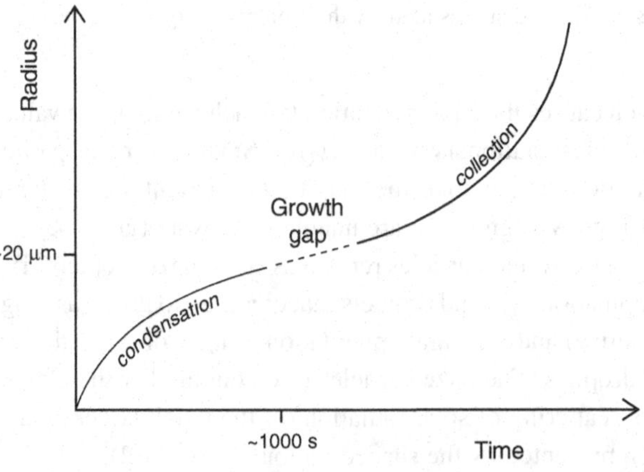

Figure 11.3 Relative growth of drops by condensation and collection. The growth gap refers to the uncertain mechanisms by which drops grow beyond adiabatic condensation to sizes required for the initiation of collision-coalescence.

in a clean air mass (one having low aerosol concentrations) will have fewer droplets than one forming in a dirty air mass. Fewer droplets competing for the excess vapor means higher supersaturations and larger condensational growth rates. Larger drops in a population increase the opportunities for differential fallspeeds and early growth by collection. The colloidal stability of clean (e.g., maritime) clouds can therefore be expected to break earlier than that of dirty (continental) clouds. Indeed, cumuli in maritime air are observed to precipitate more readily than similarly sized cumuli in continental air. At extremely large aerosol concentrations, one might expect rain to be inhibited almost completely. Such a tendency toward pluvial constipation does seem to exist, especially in highly polluted air. The situation is more complicated than simply comparing mean concentrations, however, as a relatively low concentration of giant aerosol particles, if present, may serve to initiate the collection process when it would otherwise be inactive. We have already considered CCN activation and adiabatic condensation in detail, so we next develop the mathematical basis for continuous collection before focusing on how the growth gap may be bridged. The few drops that serve to bridge the growth gap are sometimes referred to as coalescence embryos because they get the coalescence process started.

11.2 Continuous collection

Continuous collection assumes that at least a few large drops already exist among the population of much more numerous and smaller cloud droplets. As suggested by Fig. 11.4, individual collision-coalescence events do not need to be considered. Rather, the cloud droplet population may be treated as a continuum having a collectable liquid water concentration $\omega_L = \frac{4}{3}\pi\rho_L \int_{r_{\min}}^{r_L} r_d^3 n(r_d) dr_d$, where r_L is the radius of the large, collector drop and $r_{\min} \simeq 5\ \mu\text{m}$ is the minimum size for collection. Similarly, the mean radius of the small, collected droplets is $r_S = \frac{1}{N_d}\int_{r_{\min}}^{r_L} r_d n(r_d) dr_d$, where $N_d = \int_{r_{\min}}^{r_L} n(r_d) dr_d$ is the concentration of collectable drops. A representative large drop falls with speed v_L through the cloud of small droplets (having average fallspeed v_S) and collects them with efficiency E_c. Consideration of the geometry (Fig. 11.4) allows us to write the rate at which mass is accreted by the large drop:

$$\begin{aligned}\frac{dm_d}{dt} &= A_c E_c \Delta v \omega_L \\ &= \pi \left(r_L + r_S\right)^2 E_c \left(v_L - v_S\right) \omega_L \\ \frac{dm_d}{dt} &= K\left(r_L, r_S\right) \omega_L.\end{aligned} \tag{11.1}$$

The collection kernel $K(r_L, r_S) \equiv \pi (r_L + r_S)^2 E_c (v_L - v_S)$ is the effective volume of cloud swept out by the collector drop in unit time, so multiplying this volume by the mass concentration (mass per unit volume) of liquid water gives the mass growth rate of the collector. Recall that the collection efficiency $E_c = E_c(r_L, r_S)$ is itself dependent on the drop size distribution, so the collection kernel is in general a very complicated function of the drop size distribution.

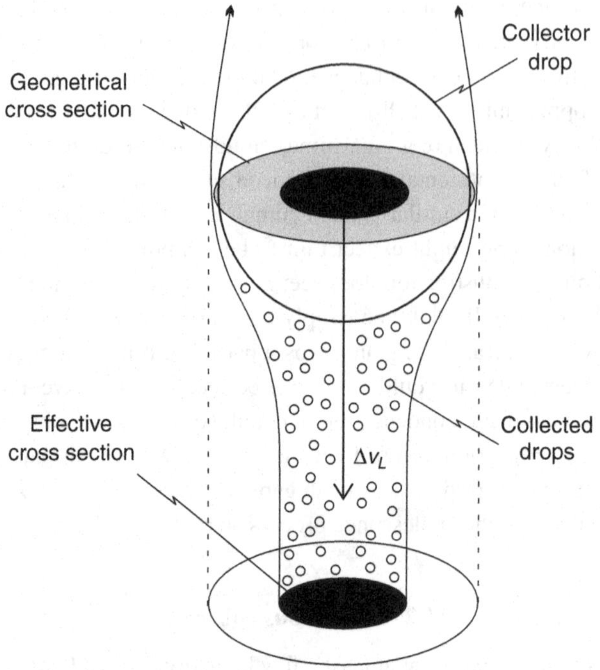

Figure 11.4 Geometry for continuous collection. Light shading: geometrical cross section. Dark shading: effective cross section, within which all droplets are collected. (Concentration of collected droplets is exaggerated for emphasis.)

Despite the complicated dependence of the collection kernel on drop sizes, some simplification is possible in the limit of continuous collection. When the difference in size of the collector and collected drops is large (i.e., when $r_L >> r_S$), then $v_L >> v_S$ and $\Delta v \simeq v_L$. The collection efficiency may then be approximately constant. For the sake of simplicity, set $E_c = 1$. If the collector drop remains spherical, its fallspeed increases roughly in proportion to its size (continuum regime II), and we may use the linear approximation, $v_L \simeq k_1 r_L$. Thus, the simplified kernel becomes $K \simeq \pi r_L^2 \cdot k_1 r_L = \pi k_1 r_L^3$. Note how rapidly the volume swept out by a large drop increases as it grows, as the cube of its radius. We thus see that the mass growth rate also increases dramatically with size: $dm_d/dt \simeq \pi k_1 r_L^3 \omega_L \propto r_L^3 \propto m_d$. The mass growth rate is proportional to the mass itself, giving exponential behavior and the general expectation that the big get bigger when the growth mechanism is collection.

Drop growth by collection differs fundamentally from that for condensation. Whereas the mass growth law for collection is $dm_d/dt \propto r_d^3$, that for condensation yields $dm_d/dt = 4\pi\rho_L G r_d s \propto r_d$. We may convert from mass to linear growth rates by employing the geometrical relationship between mass and radius: $m_d = \frac{4}{3}\pi\rho_L r_d^3$. The general relationship between mass and linear rates of growth is thus $dm_d/dt = 4\pi\rho_L r_d^2 dr_d/dt$. Integration of these relationships and solving for radius give the types of curves shown in

Fig. 11.3. The radius of a drop growing by continuous collection increases exponentially with time: $dr_d/dt \sim \exp(t)$, whereas that of a drop growing by condensation increases with the square root of time: $dr_d/dt \sim r_d^{1/2}$. Continuous collection is an effective growth mechanism, once it gets started. Condensation cannot easily provide droplets of the size needed to start the collection process, but it is a necessary process for supplying the liquid water upon which the collector drops feed.

Incipient raindrops evolve out of the huge population of cloud droplets as exceptions to the law of averages. Typical concentrations of millimeter-size raindrops are only about $1000\,\mathrm{m}^{-3}$, compared with about $10^9\,\mathrm{m}^{-3}$ of cloud droplets (each about $10\,\mu\mathrm{m}$ in radius). Comparison of their relative sizes (a factor of about 100) suggests that something like a million cloud droplets must be collected by each raindrop. Computations based on such mean values provide important perspectives of the problem, perhaps, but they hide most of the mechanistic details about how rain arises in warm clouds from rare events.

11.3 Diabatic condensation

Diabatic processes occur when the system of interest (a cloudy air parcel in our case) is influenced by its surroundings. The term diabatic is used when the assumption of adiabaticity no longer applies. Energy is not confined within a system once the system becomes open, able to exchange mass with its surroundings. An air parcel may rise adiabatically for a while in the core of a convective cloud, for instance, but turbulent motions eventually mix dry environmental air into the parcel, causing its thermodynamic and microphysical properties to change. Turbulent mixing is a unifying process, which causes the cloudy air to get drier and the entrained blob to get moistened. The microphysical properties, in particular the drop size distribution, are affected in complicated ways.

To understand how entrainment mixing affects the microphysics, one must appreciate the differences in the time scales of the various processes. Here, we need to deal only with the representative times for droplets to evaporate (τ_{evap}) and for environmental air to mix into a cloud (τ_{mix}). We care less about the actual magnitudes of the times, only their relative values. Past studies have shown that $\tau_{evap} << \tau_{mix}$, meaning that cloud droplets respond relatively rapidly to the changes in supersaturation brought about by mixing. Thus, when a blob of dry environmental air enters a cloud under such inhomogeneous-mixing conditions, the droplets that first encounter the dry air all evaporate rapidly, but in so-doing they add water vapor to the blob. Enough droplets evaporate completely to resaturate the blob, leaving the sizes of the remaining droplets largely unaffected. The subsequent mixing of the moistened blob with the cloud serves mainly to dilute the cloudy air. As shown in Fig. 11.5, the size distribution of the droplets after an inhomogeneous mixing event has the same shape as that before mixing; only the number concentration has changed. Note in particular that the mean size of the droplets is largely unchanged. Entrainment mixing simply reduces the concentrations of droplets and liquid-water in proportion to the mixing fraction.

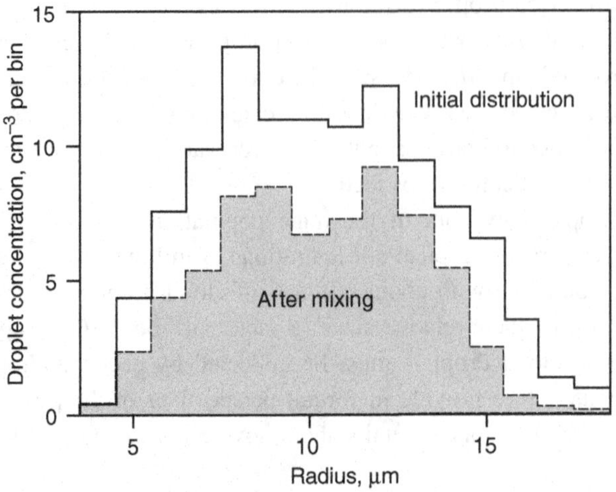

Figure 11.5 Distributions of droplet sizes before and after an inhomogeneous entrainment event. From Baker *et al.* (1980), *Quart. J. Roy. Meteorol. Soc.*, vol. **106**, pp. 581–598 Fig. 5, used with permission. Copyright John Wiley and Sons, Inc.

Diabatic condensation brought on by entrainment is an important process, one that can enhance droplet growth, broaden the drop size distribution, and promote the onset of collection. Recall that any process that reduces the number concentration N_d of droplets permits the supersaturation to rise faster than it would were many droplets to be present. We saw earlier that the quasi-stationary supersaturation is given by $s_{QS} = Aw/(\overline{r_d} N_d)$, so for a given updraft and mean droplet size $s_{QS} \propto 1/N_d$. Smaller drop concentrations (here due to entrainment) lead to larger supersaturations and thus faster growth rates (because $dm_d/dt \propto s_{QS}$). Subsequent mixing inside the cloud merges droplet populations having diverse histories. The entrainment of dry air into a cloud leads to faster growth (of the surviving droplets) and a broader distribution of sizes than is possible by adiabatic condensation. Droplets whose condensational growth have been enhanced may act as important seeds for initiating the collection process.

11.4 Stochastic collection

A stochastic process is one that is not deterministic and so occurs with some degree of chance. The stochastic process of interest for our study of warm rain is the initiation of coalescence. At the early stages of collision-coalescence, the collection of other drops by a given drop is inherently indeterminate because we cannot predict with certainty when a collection event will occur. Cloud droplets are distributed randomly in space, so a collision may be imminent when a partner is nearby, or it may take a while if potential partners are not in the immediate vicinity. Moreover, each coalescence event is discrete, not continuous. Whenever a collection event occurs, the mass of the collector drop increases incrementally

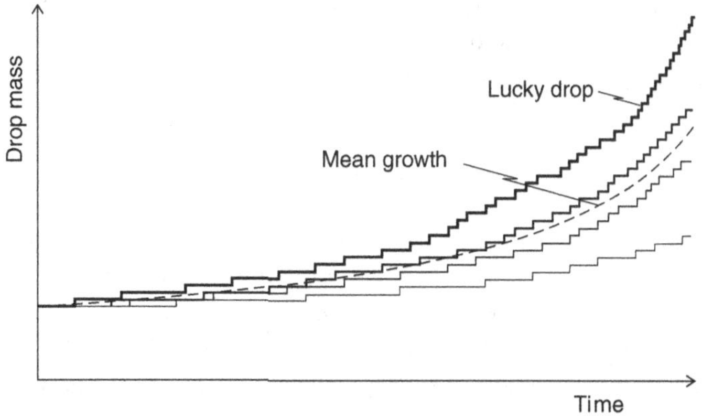

Figure 11.6 Schematic representation of the increases of mass due to stochastic collection for four independent drops. This figure was published in *A Short Course in Cloud Physics* by Rogers and Yau (1989), page 135. Copyright Elsevier 1989.

by the complete mass of the newly collected drop, not by some fraction of its mass. Our earlier attempt to describe collection as a continuous process is an approximation, valid only after the difference in sizes between the collector and collected droplets has become large. The mass change by each event in continuous collection is tiny, so the growth of the collector can be approximated by an integration rather than a summation. Stochastic collection accounts for both the discreteness and randomness of coalescence events.

The discrete and random nature of early coalescence is illustrated in Fig. 11.6. In each of the four examples shown, the times between collection events vary randomly, and the masses increase discretely (shown as sudden jumps). Drops that happen to encounter a collection partner quickly (upper case) grow more rapidly than do drops that missed early encounters (lower cases). Clearly, statistical fluctuations (deviations from mean quantities) must be considered in addition to mean values themselves (dashed curve). The times for rain to form, when calculated from mean quantities only, are usually long compared with observed times, so it is thought that a few cloud drops on the tail of the initial size distribution find coalescence partners by chance and so go on in favored status to collect other drops. Once a drop becomes even slightly larger, it gains a larger cross-sectional area and it falls faster, giving it a larger collection kernel. Only a few such lucky drops (~ 1 in 10^6) are needed to account for rain formation.

The mathematical formulation of an equation for stochastic collection starts with the realization that drops grow by accumulating the water from drops of all smaller sizes, not just those that are much smaller (as is assumed in the continuous-collection model). The rare collection of just a few drops close in size (because of their relatively large masses) is more important than the more common collection of many much-smaller drops. The likelihood that a given collector drop collides with and coalesces with another drop is given by the collection probability, a function related to the collection kernel discussed earlier in conjunction with Eq. (11.1). Consider a time interval δt that is so small that the collector drop can collect at most one other drop. Then, in general, the probability $P(m_L, m_S)$ that a collector drop of mass m_L collects a smaller drop of mass m_S within interval δt is given by

$$P(m_L, m_S) = K(m_L, m_S)n\,(m_S)\,\delta t < 1, \tag{11.2}$$

where $K(m_L, m_S)$ is the collection kernel and $n\,(m_S)$ is the number concentration of smaller drops having mass between m_S and $m_S + dm_S$. (Recall that the collection kernel (K) represents the effective volume swept out by the collector drop (m_L) in unit time, so multiplying it by a concentration of m_S-drops yields a number per unit time.) We are interested in calculating how many collection events occur in a unit volume of cloud, however, so we need to multiply the probability that a single drop collects another drop (Eq. (11.2)) by the concentration of collector drops. Thus, the number of collection events per unit volume of cloud per unit time is $K(m_L, m_S)n\,(m_S)\,n\,(m_L)$. These microdynamical interactions between drops form the basis for calculating the evolution of drop spectra.

The spectral distribution of drops in a cloud changes dramatically with time due to collection. Large drops grow ever larger by collecting smaller drops, causing the number of large drops to increase at the same time that the number of small drops decreases with time. The drop spectrum shifts from small toward larger sizes. The rate with which the distribution changes is calculated by considering both the gains and losses of drops within any given mass interval. Consider Fig. 11.7, which represents an arbitrary distribution $n\,(m)$ at a given time t. Expressing the distribution in terms of drop mass (rather than radius) is preferred because mass (not radius) is the conserved quantity. Thus, if two drops of masses m_1 and m_2 collide, the new mass is $m_{new} = m_1 + m_2$. Now, pick an arbitrary mass m and the interval $m + dm$, which we call the m-bin (dark shading). Drops are added to the m-bin when various smaller drops (light shading) collide such that their masses add to m. That is, if we arbitrarily choose a smaller mass m_x, then the collection partner must be drops in bin $m - m_x$. By definition, collisions of m_x and $m - m_x$ drops, if they coalesce, result in new drops of mass m. Drops in the m-bin can also collide and coalesce with other drops, of course, which invariably results in the loss of drops from the m-bin. Calculating the net change in the concentration of drops in the m-bin, the difference between the rates of gain and loss from the m-bin, yields the stochastic collection equation (SCE):

$$\begin{aligned}\frac{\partial n\,(m,t)}{\partial t} = &\frac{1}{2}\int_0^m K\,(m_x, m - m_x)\cdot n\,(m_x)\,n\,(m - m_x)\,dm_x - n\,(m)\\ &\int_0^\infty K\,(m, m_x)\cdot n\,(m_x)\,dm_x.\end{aligned} \tag{11.3}$$

The factor 1/2 prevents the counting of each collision twice. This integro-differential equation, the heart of the theory of stochastic collection, describes the evolution of the drop spectrum due to collision-coalescence.

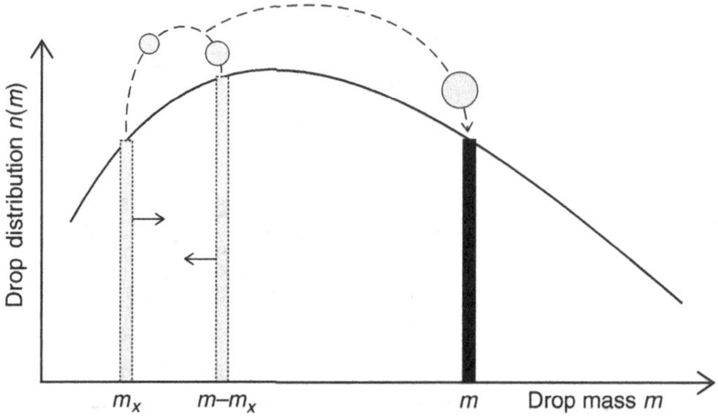

Figure 11.7 Distribution of drops by mass. Curve: Continuous approximation to the discrete process of coagulation (columns).

Solutions of the stochastic collection equation are usually performed numerically because of the complicated dependences of the kernels on the masses of the drops. An example is shown in Fig. 11.8. The initial distribution (solid curve) contains 135 drops and is relatively narrow, as one expects from adiabatic condensation. After 3 min of growth by collection, one discerns little difference on the linear plot between growth calculated by the continuous model and by the stochastic model (Eq. (11.3)). However, an expansion of scales (inset) reveals that a subset of large drops has indeed emerged after just 20 s. It will be these relatively few large drops that go on to produce rain. When drop growth is calculated according to the stochastic model over a longer time, the mass distribution evolves as shown in Fig. 11.9. The initial distribution again starts out narrow, but eventually develops a second mode after about 15 min. The growth of the drops in the large-drop mode could in principle be calculated by the continuous model, but it takes the stochastic model to develop that mode in the first place.

Careful scrutiny of the stochastic model discloses the fact that calculations from Eq. (11.3) are repeatable; that is, each execution of the model yields the same result for the same starting conditions. One way of quantifying the true stochastic nature of collection is to consider the time between collection events as a random variable and use the statistics of improbable (i.e., unlikely) events. Collisions occur among drops distributed randomly in space and time, so the times between collisions vary statistically about some mean value (as suggested by Fig. 11.6). If we let the mean time to collection be $\tau = 1/(N_d A_c E_c v)$, where N_d is the drop concentration, E_c is the collection efficiency, and A_c and v are the geometrical cross-sectional area and fallspeed of the collector, respectively, then the probability density function of actual time t to collection is given (according to Poisson statistics) by

$$p(t) = \frac{1}{\tau} \exp\left(-\frac{t}{\tau}\right). \tag{11.4}$$

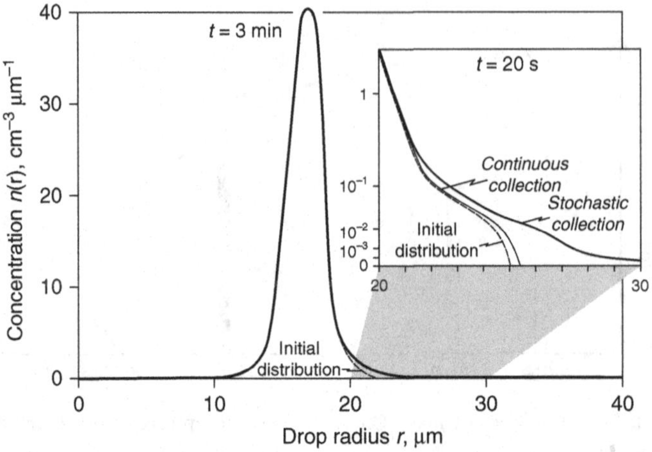

Figure 11.8 Drop size distribution after growth by continuous and stochastic collection. Adapted from Twomey (1964).

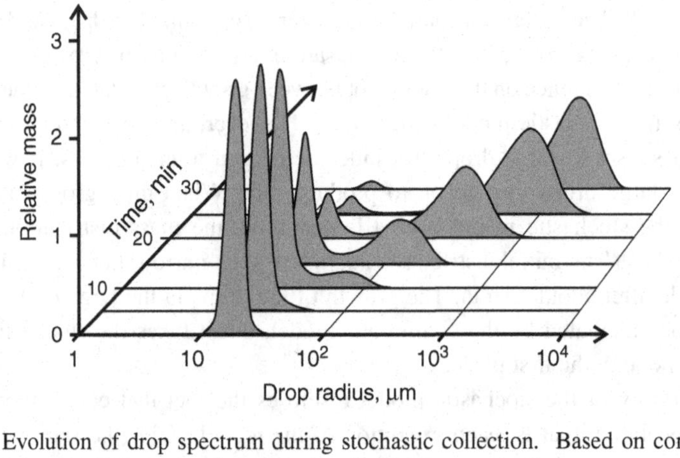

Figure 11.9 Evolution of drop spectrum during stochastic collection. Based on computations by Berry and Reinhardt (1974). Figure here from Lamb (2001).

The important rare collisions occur early, when $t << \tau$, in which case the cumulative probability $P(t) = \int_0^t p(t')\,dt' \simeq t/\tau << 1$. We ultimately care how much the early exceptional events beat the average growth time, so we must also take into account how the mean time between collisions changes as the drop grows. The average cumulative time to the nth collection event is just the sum of the individual times: $\tau_{tot} = \sum_1^n \tau_n$, but the times between collection events decrease as the size of the collector r_L, the relative speed Δv, and the collection efficiency E_c increase. For drops in the radius range 10 to 50 μm, each of these variables increases roughly as the square of radius, so the compound effect on the time between the $(n-1)$th and the nth collision varies as $\tau_n \sim r_L^{-6}$. It turns out that about

125 collection events are needed to grow cloud drops to drizzle drops, but the accumulated time to reach such sizes is driven strongly by the first few coalescence events. Early collection events by just a few ($\sim 10^{-6}$) lucky drops may, even in the absence of giant nuclei that act as coalescence embryos, be sufficient to stimulate active collection and ultimately rain formation.

11.5 Warm rain

Rain represents the culmination of all the diverse microphysical processes that make liquid drops larger. With reference again to Fig. 11.1, it can be seen that rain formation can be traced back to the initial upward motions of moist air that generates vapor concentrations in excess of the equilibrium values. That excess vapor is taken up by the aerosol particles that are carried along in the updraft, allowing them to grow by condensation. The cloud droplets, although much larger than the original aerosol particles, are too small to sediment significantly; they faithfully follow the air motions. Nevertheless, cloud water is necessary for rain development because it becomes the feedstock for the larger cloud drops and incipient drizzle drops that grow by collision-coalescence. Subsequent growth of the drizzle drops, which may themselves be considered rain from stratiform clouds with weak updrafts, leads to mature rain drops. Raindrops, with typical diameters in excess of a millimeter, have fallspeeds that exceed the updraft speeds of most clouds. Rain results from the complicated interplay between the microphysics and dynamics of a cloud.

Rain events can be as varied as light drizzle from a fog or stratus on an overcast day to heavy downpours from severe thunderstorms and hurricanes. In the fogs and stratus of clean maritime air masses (as off the west coast of the United States) rain rates may be only a fraction of a millimeter per day, yet the removal of condensed water by the drizzle often controls the water mass balance and the albedo of the clouds. Such clouds shield vast areas of the world's oceans from solar insolation to the extent that condensed water remains aloft. Arguably, the microphysics responsible for precipitation formation in stratus is an important control over the Earth's radiation balance and climate.

Convective clouds produce rain in more patchy patterns than do stratiform clouds. Unstable air causes air to rise and descend rapidly, so rich supplies of condensate are produced, but they are consumed quickly by the rain or by evaporation in the downdrafts. Trade-wind cumuli are noted for the ease with which they produce rain, in large part because of the low aerosol concentrations in maritime air masses. Rain production is enhanced when such clouds intercept mountainous islands (e.g., Hawaii) and the updrafts are strengthened by the orography. Extreme cases of heavy rainfall occur when tropical systems, already rich in moisture, impact large mountain ranges, such as when the Asian monsoons are forced to rise by the Himalayas. Flooding can occur, especially when the rainfall is channeled into narrow canyons, such as occurred with the Big Thompson Canyon Storm of 1976, when moist air was forced against the eastern slopes of the Rocky Mountains of Colorado. That storm is technically classified as a cold cloud, because ice was prevalent in the upper levels, but the rain was produced primarily by the warm-rain microphysical processes active in the lower levels. Rain from convective clouds is complicated by the strong interplay between cloud dynamics and microphysics.

A general perspective of rain generation can be gained by implementing the relatively simple model often ascribed to Bowen in 1950. Our version reflects new understandings of fallspeeds and growth rates. One envisions air rising at a specified rate w while a representative raindrop grows by continuous collection and falls against the updraft. The fallspeed v of a water drop during the intermediate stages of rain formation is proportional to its radius r_d, so let $v = kr_d$, where k is the constant of proportionality. The drop falls through the cloud with a liquid water concentration ω_L and collects the much smaller cloud droplets with collection efficiency E_c. The mass m_d of the collector drop thus increases at the rate $dm_d/dt = K\omega_L$, where the collection kernel $K \simeq E_c\pi r_d^2 v = E_c\pi k r_d^3$. The radius of the drop changes at the rate

$$\frac{dr_d}{dt} = \frac{1}{4\pi\rho_L r_d^2}\frac{dm_d}{dt} = \frac{E_c k\omega_L}{4\rho_L} r_d, \tag{11.5}$$

where ρ_L is the density of liquid water. Solving this equation for r_d gives

$$r_d = r_0 \exp\left(\frac{E_c k\omega_L}{4\rho_L} t\right), \tag{11.6}$$

where r_0 is the drop radius at time $t = 0$. The tendency for the drop to grow exponentially by collection means that it starts slowly, but quickly grows to sizes that let it fall against the updraft. The trajectory can be calculated by noting that the altitude z_d of the drop varies at the rate $dz_d/dt = w - v_d = w - kr_d$ and using Eq. (11.6) for r_d. Integration of the resulting differential equation yields the altitude of the drop versus time. As seen in Fig. 11.10, the drop first rises with the air, then lags ever more behind the air. Once the fallspeed of the drop exceeds the updraft speed, it has reached its maximum altitude. Continued growth makes the drop fall ever faster, until it falls out of the cloud altogether.

An explicit dependence of the drop radius on its altitude above cloud base can be obtained by transposing coordinates from time to altitude. Use the chain rule from calculus to modify the drop growth rate: $dr_d/dz_d = (dr_d/dt) \cdot (dt/dz_d) = (dr_d/dt)\,/\,(w - v_d)$. We thus gain an approximate rate of change of radius with change in altitude,

$$\frac{dr_d}{dz_d} = \frac{E_c k r_d \omega_L}{4\rho_L\,(w - kr_d)}. \tag{11.7}$$

This equation may be solved explicitly for radius as a function of altitude when w and the other parameters are constant. Thus, if a drop starts out with radius r_0 at altitude z_0, then integration of Eq. (11.7) gives the following height-radius relationship:

$$z - z_0 = \frac{4\rho_L w}{Ek\omega_L}\ln\left(\frac{r_d}{r_0}\right) - \frac{4\rho_L}{E\omega_L}(r_d - r_0). \tag{11.8}$$

More complete treatments allow for more general dependences of fallspeed and collection efficiency on drop size, in which case Eq. (11.7) may have to be integrated numerically.

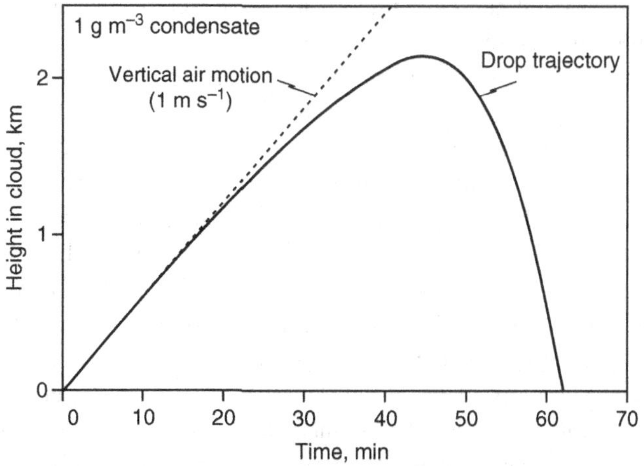

Figure 11.10 Motions of air (dashed line) and of a representative drop (solid curve) growing by collection of cloud water with condensate concentration 1 g m^{-3}. From Bowen (1950), available at <http://www.publish.csiro.au/nid/51/paper/CH9500193.htm>. Used with permission of CSIRO.

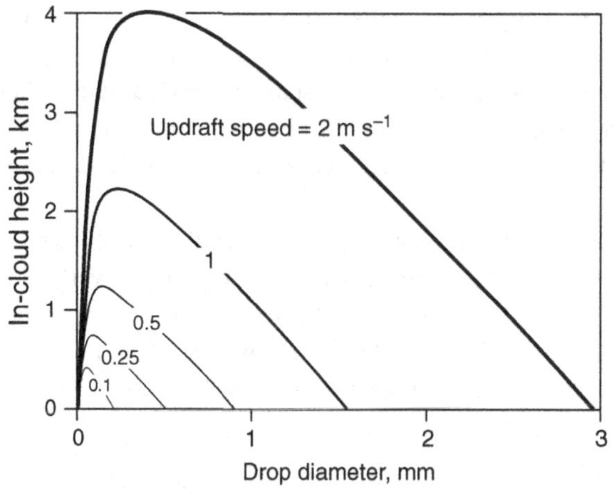

Figure 11.11 Change of drop size with height inside cloud due to collision-coalescence while traveling upward at the indicated speeds. From Bowen (1950), available at <http://www.publish.csiro.au/nid/51/paper/CH9500193.htm>. Used with permission of CSIRO.

Nevertheless, the general pattern always follows that shown in Fig. 11.11. Each curve, representing a distinct, but constant updraft speed, shows that the drop first travels upward with the airflow while it is small, then it deviates progressively away from the air motions as it gains mass and falls faster. At some point (near the apogee) the drop fallspeed equals and then exceeds the updraft speed, causing the drop to fall rapidly out of the cloud (taken as close to or below z_0). Of particular importance is the fact that the larger the updraft, the larger are the drops that fall through cloud base.

Rain is composed of drops of many sizes, reflecting the broad range of microphysical histories, so calculating rainfall rates needs special consideration. Let the concentration of drops be related to diameter D by the function $n(D)$ and the individual fallspeeds by $v(D)$. The rate that drops in the size range D to $D+dD$ pass through a unit horizontal area fixed at a given level in a cloud is therefore the product $(w - v(D))\,n(D)dD$. Note that we must take the vertical speed w of the air into account, as fallspeed is referenced to the air, not to the Earth. Drops that are small, such that $v(D) < w$, tend to move upward in an updraft. Large drops, with $v(D) > w$, fall downward through the horizontal plane. With the convention that quantities are positive when directed upward, the net vertical flux of drops $[\mathrm{m^{-2}\,s^{-1}}]$ is $\Phi = \int_0^\infty (w - v(D))\,n(D)dD$. We may break the overall drop distribution into two parts based on the drop size that represents the transition between small drops going up and large drops going down. If we define this threshold diameter D_{th} by the relationship $v(D_{th}) = w$, then the upward flux is $\Phi_{up}(D_{th}) = \int_0^{D_{th}} (w - v(D))\,n(D)dD > 0$ and the downward flux is $\Phi_{dwn}(D_{th}) = \int_{D_{th}}^\infty (w - v(D))\,n(D)dD < 0$. Rain in the ordinary sense represents the downward part of the overall flux. The larger that w is, the larger D_{th} must be. This seemingly arbitrary distinction takes on importance when calculating rain rates within cloud, and it helps us realize that it is the updraft speed that defines whether drops fall out of a cloud or not. Visible clouds typically represent the scattering of sunlight off small droplets, those having fallspeeds less than the updrafts spawning condensation. At the other extreme, as we saw from the Bowen model (Fig. 11.11), larger updraft speeds support the larger precipitation-size drops for longer times, letting them grow larger before their inevitable fall to the ground.

The rate of rainfall onto the ground is closely related to the concept of drop fluxes just considered. For rain at the Earth's surface, the area defining the vertical flux is a horizontal section of the ground itself, where the vertical component of air motion vanishes (i.e., $w = 0$). Each falling drop of diameter D has a unique mass $m(D) = \frac{\pi}{6}\rho_L D^3$ and downward speed $v(D)$, so the absolute magnitude of the mass flux $[\mathrm{kg\,m^{-2}\,s^{-1}}]$ of rain hitting the ground is

$$\Phi_{mass} = \int_0^\infty m(D)v(D)n(D)dD. \tag{11.9}$$

However, mass is not usually measured at a recording station. Rather, it is the depth Δh_L of liquid that accumulates over a time interval Δt that is most easily measured. The instantaneous rain rate $R = \lim_{\Delta t \to 0}(\Delta h_L/\Delta t)$ is therefore

$$R = \frac{1}{\rho_L}\Phi_{mass} = \frac{\pi}{6}\int_0^\infty D^3 v(D)n(D)dD. \tag{11.10}$$

Rain rate is related directly to the third-moment of the number flux of drops hitting the ground. Large rain rates tend to have large drops, such as those produced in convective storms. The units of the quantities in Eq. (11.10) must be treated carefully, as diameters are often measured in mm, while rain rate is measured in mm h^{-1}, fallspeed in m s^{-1}, and size spectra in m^{-3} mm^{-1}. Computations using Eq. (11.10) are best made after converting all values to SI units. Care must also be taken to use appropriate values for the drop speed v. Lacking direct measurements of drop speed, one is tempted to use the unique terminal fall-speed $v(D)$. However, many intermediate size drops may actually be fragments of recent break-up events, in which case the momenta of the large parent drops are preserved for a time and give the smaller drops super-terminal speeds. Without due concern for such non-equilibrium phenomena, significant errors can occur in measured and calculated rain rates, as well as in the determination of size spectra from disdrometers and vertically pointing radars.

The size distribution $n(D)$ of raindrops to be used with Eq. (11.10) are usually represented by one of several analytical functions. As shown in Fig. 11.12, not all functions (shown as curves) work well for a given set of data (symbols). The data shown here are best fitted with a log-normal distribution, which can be related to the total number concentration $N_{tot} = \int_0^\infty n(D)dD$ and the geometrical standard deviation σ_g (standard deviation of the log of drop diameters) by

$$n(D) = \left(\frac{(N_{tot})}{((2\pi)^{1/2} D ln\sigma_g)}\right) \exp\left(-\left(\frac{((lnD - lnD_m)^2)}{(2ln^2\sigma_g)}\right)\right), \tag{11.11}$$

where D_m is the mean diameter of the drops. Note the skewness of the log-normal distribution when plotted with a linear D-axis. The log-normal distribution is a standard Gaussian in logarithmic coordinates, so it would look symmetric about D if D were plotted logarithmically. Other distributions in common use include the modified gamma distribution and its special case, the purely exponential distribution. The gamma distribution takes the general form

$$n(D) = N_{tot} f_g(D) = \left(\frac{(N_{tot})}{(\Gamma(\nu))}\right)\left(\left(\frac{D}{(D_n)}\right)\right)^{\nu-1}\left(\frac{1}{(D_n)}\right)\exp\left[-\left(\frac{D}{(D_n)}\right)\right], \tag{11.12}$$

which can be expressed first in terms of the total number concentration, N_{tot}, multiplied by the gamma probability distribution function, $f_g\,(D)$. Parameters of the distribution include the shape parameter, ν, which determines the peakedness of the distribution, and the scaling diameter, D_n, which is closely related to the mean diameter of the distribution by

$$D = \left(\frac{(\Gamma(\nu+1))}{(\Gamma\,(\nu))}\right) D_n. \tag{11.13}$$

The factor $\Gamma\,(\nu)$, the gamma function, ensures that $f_g\,(D)$ integrates to unity when all drop sizes have been accounted for. The exponential distribution can be obtained by setting $\nu = 1$ in Eq. (11.12), in which case $D = D_n$. The gamma distribution allows for a fall-off in concentration at small sizes, as does the log-normal distribution, and both have long tails toward large sizes.

Physical interpretations of some analytical functions have been obscure in the past. The gamma distribution, for instance, has often been written in the form

$$n(D) = n_0 D^{\beta} \exp(-\Lambda D), \qquad (11.14)$$

where the parameters here are related to those in Eq. (11.12) as follows: $\Lambda = 1/D_n$, $n_0 = N_{tot}/\left(\Gamma(\nu)D_n^{\nu}\right)$, and $\beta = \nu - 1$. This particular modification of the gamma function is much used in cloud physics, but it may not always give the best representation of data, as seen in Fig. 11.12. The exponential function is derived from Eq. (11.14) with $\beta = 0$:

$$n(D) = n_0 \exp(-\Lambda D). \qquad (11.15)$$

This exponential function is known as the Marshall–Palmer distribution when the parameters are specified as $n_0 = 8 \times 10^3\ \mathrm{m^{-3}\ mm^{-1}}$ and $\Lambda = 41 R^{-0.21}$ [R expressed in $\mathrm{mm\,h^{-1}}$, D in mm]. An exponential distribution, as shown in Fig. 11.12, captures the large-drop tail of the observed spectrum, but it has no capability of representing falloffs at small

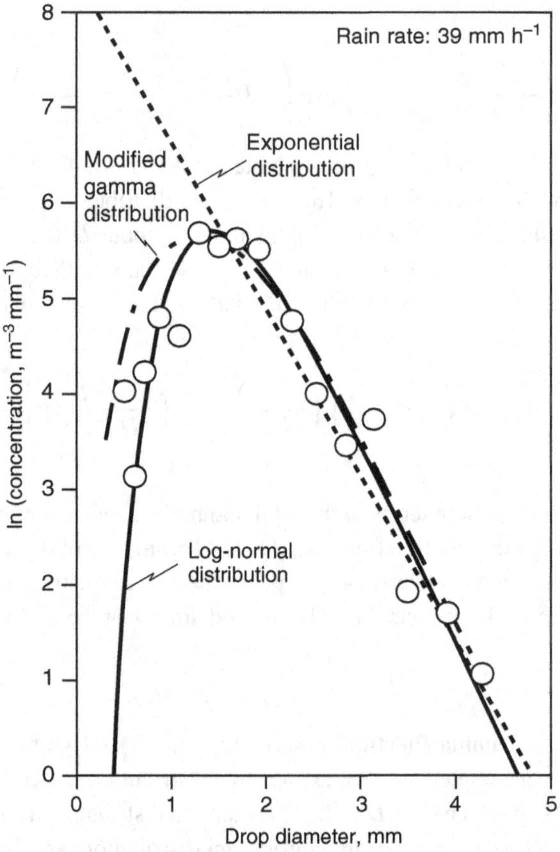

Figure 11.12 Measured (symbols) and approximated (curves) distributions of raindrops for a rain rate of 39 mm h^{-1}. Adapted from Feingold and Levin (1986).

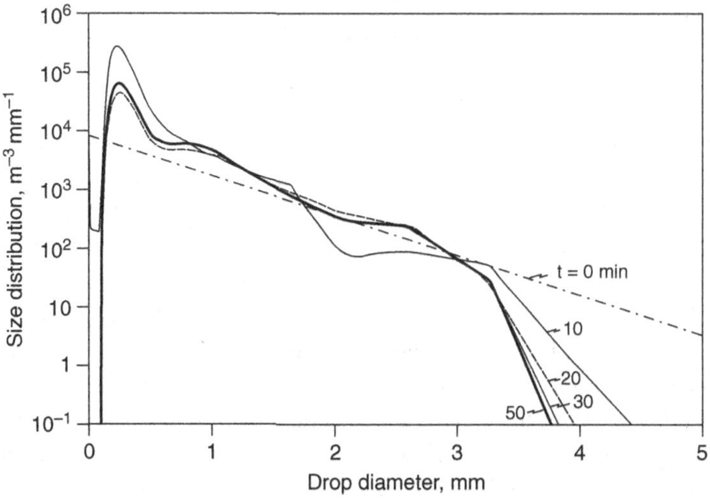

Figure 11.13 Evolution of the drop-size distribution from Marshall–Palmer (dot-dashed) at time t = 0 to trimodal at t = 50 min. Adapted from Chen and Lamb (1994b).

sizes. Nevertheless, the Marshall–Palmer distribution is still used because of its simplicity (having a free parameter Λ that is conveniently related to rain rate R).

The mechanisms leading to various raindrop size distributions are complicated by interactions between processes that lead to the formation of large drops and those that lead to their destruction. The proposed mechanisms are mainly microphysical in nature, but dynamical influences should be included when possible. In general, collision-coalescence tends to make drops larger, whereas drop breakup limits the upper end of the spectrum and produces smaller drops. Figure 11.13 shows how a spectrum that starts out as a Marshall–Palmer distribution (dot-dashed curve) gradually evolves into one with three distinct peaks. The smallest drops are consumed rapidly by collection and may be replenished by condensation onto still smaller drops. At the same time, the unrealistically large drops, initially allowed by the exponential distribution, readily break up, causing the concentration of the largest drops to decrease monotonically with time and that of submillimeter drops to increase above that specified by the Marshall–Palmer distribution. The concentrations of intermediate size drops oscillate up and down as the collection and disruptions processes interact with each other, but after some 30 min the drop spectrum becomes relatively stable and trimodal. The intermediate mode reflects the distribution of drops formed during the breakup of the large drops. The Marshall–Palmer distribution represents at best an empirical fit to drop spectra over a limited size range, possibly over a broader range after temporal averaging. Truly large drops, allowed by any exponential function, cannot exist for physical reasons. Rare drops up to about 8 mm in diameter have been found in maritime cumuli and ascribed to the early growth by collection of drops that started as giant condensation nuclei (possibly from salt spray). When giant nuclei are available, rain forms readily in warm convective clouds. Otherwise, some combination of microphysical and dynamical mechanisms may be needed to explain rain formation from collision-coalescence alone.

11.6 Aerosol effects

The impact that aerosol particles have on the micro- and macrostructure of clouds is of great interest because of the role clouds play in the radiant energy balance of the Earth and climate. Figure 11.14 illustrates how warm clouds differ in clean (top row) and polluted (bottom row) environments. Aerosol effects are particularly evident in shallow sub-tropical marine cloud systems, where changes in the microstructure or spatial extent of clouds near the surface alter the shortwave radiation budget. Low-level clouds have little impact on the longwave budget, however, because their temperatures are similar to those of the underlying ocean surface. The shortwave radiation budget is sensitive mainly to the albedo (broad-band reflectance) of the clouds, which is increased by increased concentrations of cloud-active aerosol particles. This impact of aerosol concentration on cloud reflectance is generally referred to as the albedo or Twomey effect and stems from the strong relationship that exists between the total aerosol concentration in the updraft feeding the cloud and the resulting cloud droplet concentration. As shown in Section 10.3, when the available condensed liquid must be distributed over a larger number of droplets, as in a polluted case, the droplets are smaller on average, but the collective cross-sectional area of the dropet population is larger than that in the clean case. In-coming solar radiation is scattered

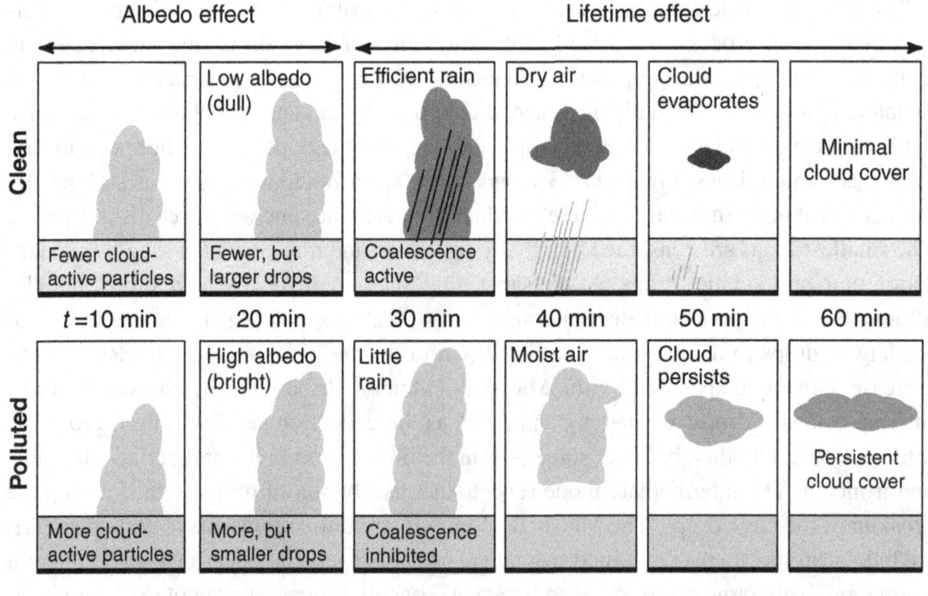

Figure 11.14 Hypothesized effects of aerosol populations on cloud albedo (left two columns) and cloud lifetime (right four columns). Top panel: Effects of low aerosol concentrations. Bottom panel: Effects of high aerosol concentrations. Air may become polluted by emissions from either natural or anthropogenic sources. Adapted from Stevens and Feingold (2009) by permission from Macmillan Publishers Ltd: *Nature*, copyright 2009.

more in aerosol-rich clouds, which thus appear brighter from space than do clouds forming in clean air masses. Such albedo effects are readily seen in satellite images as bright tracks in an otherwise dull field of stratocumuli over an ocean. The visible tracks are caused in part by the particluate pollution emitted into clean maritime air by ships. More reflective clouds reduce the net solar energy incident onto the surface, and thus have a cooling effect on climate.

The physical arguments underpinning the albedo effect can easily be extended to include aerosol impacts on precipitation, and through them cloud lifetime and cloud cover. These second-order impacts of aerosols, loosely referred to as cloud-lifetime effects, assume that precipitation efficiency decreases monotonically with increasing drop concentration. The basis for this assumption rests on our common experience that clouds forming in clean marine air masses precipitate more readily than similar clouds forming in aerosol-rich continental air masses. Such a relationship, to the extent that it really exists, may arise because the larger droplets in clean clouds can initiate collision-coallescence more readily. Lower precipitation efficiency in continental clouds means that more condensate remains in the cloud, other factors remaining constant. A diet rich in aerosol tends to give a cloud pluvial constipation, but greater longevity in the atmosphere. Individual clouds in a polluted air mass may thus persist longer and contribute to an increase in total cloud cover and less solar energy reaching the surface.

Whereas each hypothesized link may rest on reasonable observational evidence, quantifying the overall effects of aerosol burden on realistic cloud fields remains elusive. Clouds are highly non-linear systems, and cloud-environment interactions can lead to many unanticipated results, confounding attempts to interpret data. The relative importance of the diverse processes active in an individual cloud depends on the conditions encountered by that cloud. For example, strong winds over the ocean surface produce sea spray that thus increases the aerosol burden in the environment. We may expect precipitation to be inhibited, but sea spray also introduces especially large cloud condensation nuclei (CCN), which by activating faster may reduce the maximum supersaturation attained at cloud base and therefore the number of small aerosol particles activating. The albedo effect is then less than expected. Moreover, the large CCN seed the coalescence process, which may help to maintain or even enhance the precipitation processes, altering the expected effect on cloud lifetime. The magnitudes of aerosol impacts on precipitation efficiency and cloud lifetime are complicated and depend on the regimes. For instance, the precipitation efficiency in clouds that generate little precipitation (low liquid-water concentrations) or clouds that precipitate very effectively (high liquid-water concentrations) are less likely to be susceptable to changes in aerosol particle concentrations than clouds with intermediate liquid water concentrations.

Given the complexity of aerosol–cloud interactions, one must view clouds as more than a suspension of water drops in air. We tend to think of how a cloud is affected by its immediate environment, but the interactions are two-way. For instance, a less efficient precipitation process implies that more condensate remains aloft, where it evaporates, moistens, and cools the environment aloft. The now-modified environment becomes more conducive to the development of deeper clouds with higher liquid water concentrations. These deeper clouds may then produce more precipitation. The net result may be that the reduction in precipitation efficiency, because of an increased aerosol burden, is less than anticipated. Similarly, the greater tendency for precipitation to form at night in stratocumulus clouds leads to weaker entrainment across the inversion and therefore shallower and moister boundary layers. Reduced precipitation during the day, by contrast, helps maintain stronger boundary-layer circulations. A lower aerosol concentration, which produces more precipitation, actually promotes cloud formation, which is counter to what one might have anticipated from simpler arguments.

The processes in warm clouds, despite involving only liquid drops, are deceptively simple on the microscale, but they involve complexities on the macroscale that pose a stern warning to take care drawing conclusions, but also offer enticing challenges to prospective cloud physicists. You are entering an exiting field in which much is yet to be learned. However, the subject of cloud physics becomes even more complicated as new interactions emerge involving ice particles in cold clouds, the subject considered next.

11.7 Further reading

Fletcher, N.H. (1962). *The Physics of Rainclouds*. Cambridge: Cambridge University Press, 386 pp. Chapters 6 (coalescence, colloidal stability) and 7 (rain formation, Bowen model, rain measurements).

Lamb, D. (2003). Cloud microphysics. In *Encyclopedia of Atmospheric Sciences*, ed. J.R. Holton, J. Pyle, and J.A. Curry. London: Academic Press, pp. 459–464.

Levin, Z. and W.R. Cotton (eds.) (2009). *Aerosol Pollution Impact on Precipitation*. New York: Springer, 386 pp. Chapters 6 (observed effects of pollution on precipitation) and 7 (calculated effects of pollution on precipitation).

Pruppacher, H.R. and J.D. Klett (1997). *Microphysics of Clouds and Precipitation*. Dordrecht: Kluwer Academic Publishers, 954 pp. Chapter 15 (continuous and stochastic collection).

Rogers, R.R. and M.K. Yau (1989). *A Short Course in Cloud Physics*. New York: Pergamon Press, 293 pp. Chapters 8 (collection, Bowen model, turbulence effects), 10 (Marshall–Palmer distribution), and 12 (precipitation processes).

Wallace, J.M. and P.V. Hobbs (2006). *Atmospheric Science: An Introductory Survey*. Amsterdam: Academic Press, 483 pp. Chapter 6 (continuous and stochastic collection).

Young, K.C. (1993). *Microphysical Processes in Clouds*. New York: Oxford University Press, 427 pp. Chapters 7 (collection models) and 9 (development of precipitation, drop spectra).

11.8 Problems

1. Modeling and observations of cloud droplets indicate that turbulence in clouds creates clustering of droplets, with some areas having much higher number density than the cloud on average. Discuss potential implications of such clustering of droplets on the initiation of rain through the warm-rain process.

2. The warm-rain process in a cloud involves collisions between liquid water drops that may or may not coalesce to form a larger drop. As larger drops fall through a cloud of smaller drops, the small-drop population is gradually reduced in number density, while the larger drops grow in size. Consider a simple, idealized cloud composed initially of two monodisperse drop populations: the small-drop population, containing drops of radius 10 μm and fallspeed 1 cm s^{-1}, has a number density $n_1 = 10^9$ m^{-3}, whereas the large-drop population (drop radius 500 μm and fallspeed 5 m s^{-1}) has a number density $n_2 = 1000$ m^{-3}. Assume the collision efficiency to be 0.9 and the coalescence efficiency to be 0.7. Further assume that the large drops never collide with other large drops.

   Calculate the exponential time constant for depletion of the small drop population under assumption that the initial magnitude of the collection kernel is constant throughout the process. What is the final size of the large drops after all the small drops have been collected? What would you have to do differently to make the problem more realistic?

3. The vertical motion of a droplet in an updraft is determined by its net motion, such that the rate of altitude change is described by

$$\frac{dz_d}{dt} = w - v_d.$$

   Assume two drops, radii $r_d = 50$ μm and $r_d = 150$ μm, recycles from cloud base into the updraft. These drops grow by continuous collection as they ascend in the updraft. Derive an expression for the drop altitude as a function of time in the cloud. Assume the fallspeed [m s^{-1}] of a drop is well approximated by

$$v_d = 80 r_d$$

   for r_d specified in cm. You may also assume a collection efficiency of 1. Plot the trajectory of the drops for
   (a) constant vertical motion speeds of 0.5 m s^{-1}, 1.5 m s^{-1}, and 2.5 m s^{-1}, and liquid water concentration of 1.5 g m^3.
      (i) How long do the drops stay in cloud for each case?
      (ii) How big are the drops when they leave cloud?
   (b) constant vertical motion of 1.5 m s^{-1}, and liquid water concentrations of 0.5 g m^3 and 2.5 g m^3.

      Discuss the relative importance of the initial size, the vertical velocity and the liquid water concentration to the final size of the drop.

4. Calculate the ratio of the rate of continuous collection to vapor depositional growth in a cloud with a temperature of 4 °C at a pressure of 1000 hPa, supersaturation $s = 0.01$ and liquid water concentrations $\omega_L = 0.2$ g m^{-3} and $\omega_L = 1$ g m^{-3} for
   (a) a 60 μm diameter drop,
   (b) a 140 μm diameter drop.
   Discuss your results.

5. Calculate the expression for the instantaneous rain rate at the surface for a gamma distribution. The fallspeeds of the raindrops are well approximated by

$$v_d = -0.193 + 4960D - 9.04 \times 10^5 D^2 + 5.66 \times 10^7 D^3.$$

Use this expression to determine the rainfall rate for a rain distribution with total number concentration $N_{tot} = 5600\,\mathrm{m}^{-3}$, shape factor $\nu = 1.5$ and scaling parameter $D_n = 4 \times 10^{-5}\,\mathrm{m}$. If this rain is embedded in a uniform downdraft of $2\,\mathrm{m\,s}^{-1}$, now what will be the rainfall rate?

12
Cold clouds

12.1 Overview

Cold clouds, because they often involve both ice and liquid water, are more complicated than warm clouds. But, cold clouds also offer greater opportunities for precipitation formation, and they enhance interactions with radiation. Early in the history of cloud physics, ice formation was thought to be a prerequisite for precipitation in all clouds, but we now realize that the warm-cloud processes of condensation and collision-coalescence are sufficient to produce rain in some clouds (e.g., tropical maritime cumuli), and they can coexist with, even augment, the cold-cloud processes that produce snow, graupel, and hail. The so-called ice-crystal mechanism is required for rain from many continental clouds, even in summertime. The high concentrations of aerosol particles in a continental air mass form many cloud droplets, which leads to competition for the finite supply of excess vapor and limits droplet growth rates. The formation of ice in such situations invariably breaks the colloidal stability of the cloud once it supercools. Cold clouds, of course, are always responsible for wintertime snow. High cirrus clouds are an extreme type of cold cloud, dominated by ice crystals that readily absorb and reradiate infrared radiation.

Both cold and warm clouds produce rain by transforming excess water vapor into condensate that eventually results in large liquid drops falling to the ground. Some of the diverse types of interactions that can take place inside clouds to produce rain are summarized in Fig. 12.1. The left-hand portion identifies the categories of liquid particles and the processes by which they interact, whereas the right-hand side emphasizes the solid particles and their interactions. Small particles, those that fall slowly and so follow the air motions, are shown in the upper part of the diagram; large particles, those that readily sediment, are near the bottom. The small liquid particles form by activation of cloud condensation nuclei (CCN) and grow by condensation. The small ice particles form most commonly from the freezing of these droplets when supercooled, but they can also form and grow by direct deposition onto suitable ice nuclei (IN).

Large particles form from the small particles by diverse interactions. Large liquid drops are generated initially from coalescence embryos, which in turn form from giant CCN (dashed arrow), or by stochastic-collection events involving the small drops. The large drops then grow in mass by continuous collection of the small cloud drops until they fallout

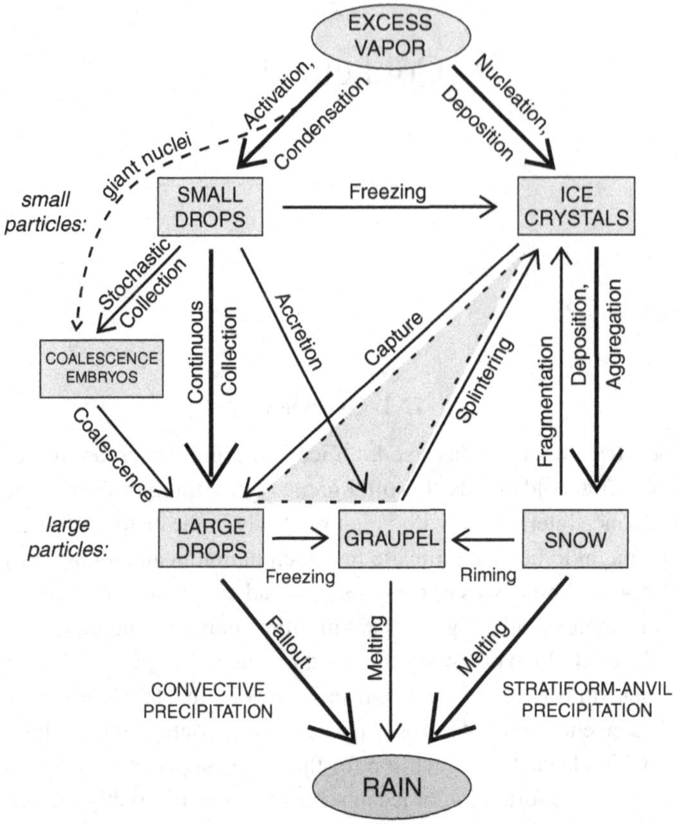

Figure 12.1 Schematic diagram of particles and processes involved in the formation of rain in cold clouds. Particle categories are shown in boxes, processes of interactions as arrows. The shaded triangle identifies a feedback loop important for the glaciation of some clouds. Adapted from Lamb (2001).

as rain. This warm-rain process is common in convective clouds having low (warm) bases, whether or not the ice phase is present. In addition, the processes that generate a broad size range of supercooled drops aid ice formation and cloud glaciation. Ice particles are common forms of precipitation in stratiform clouds and in the upper parts of deep convective systems (e.g., thunderstorms). Snow forms by continued deposition onto the ice crystals, or by aggregation. Some of the more delicate snow crystals might fragment during collisions and resupply the small-ice population, while some other snow crystals may rime and gradually become graupel particles. Graupel particles may also form directly from the freezing of supercooled raindrops. Graupel particles, whether arising from rimed ice crystals or frozen rain drops, may continue to grow by accretion of supercooled cloud water until they fall out and melt (depending on temperature) along with the larger snow aggregates. Graupel particles also serve as the embryos for hail production, sites of electrical charge separation, and as a catalyst for ice multiplication. Details are discussed in the following sections.

12.2 Ice initiation

Ice is a fundamental attribute of cold clouds, so we need to consider the processes by which ice particles originate. Our understanding of ice formation and propagation in cold clouds is not well understood, but we can identify broad classes of mechanisms that may operate at one time or another in every cold cloud. In the broadest sense, we classify ice origins as primary or secondary. Primary ice is the first ice formed, either from supersaturated vapor (as with deposition nucleation) or from supercooled liquid (homogeneous or heterogeneous freezing nucleation). By contrast, secondary ice forms only once some ice is already present.

Evidence for the distinction between primary and secondary ice came initially from field observations. When the concentrations of ice particles in the mixed-phase region of cold clouds ($-38° < T < 0\,°C$) agree with the measured concentrations of ice nuclei (IN), we suspect that the ice particles had primary origins, meaning that each ice particle likely formed by the activity of one ice nucleus. By contrast, when the ice concentrations exceed the IN concentrations by several orders of magnitude, we are left with the conclusion that secondary ice processes must have been active. Measurements of IN concentrations are notoriously difficult to make accurately, but comparisons are nevertheless feasible because of the wide range of concentrations found, as shown in Fig. 12.2. Concentrations of IN (hence of primary ice particles) vary with temperature roughly as indicated by the shaded region; a typical concentration is $1\,L^{-1}$ at $-20\,°C$ (cross), but it increases

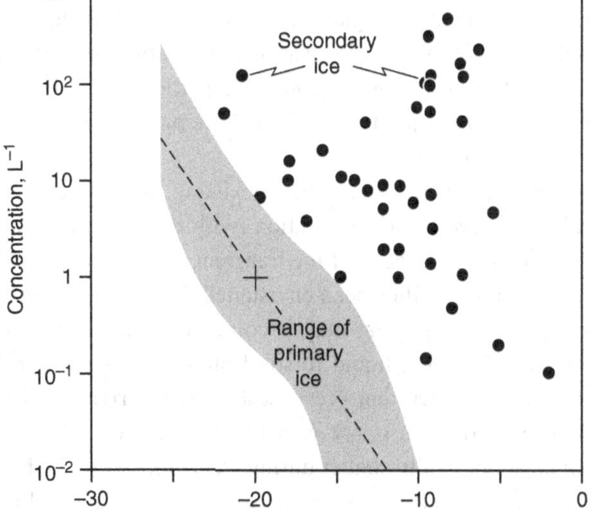

Figure 12.2 Measured concentrations of ice particles versus temperature, separated into primary (shaded region) and secondary (points) categories. Adapted from Mossop (1985).

by a factor of 10 for each 4 °C decrease in temperature (dashed line). A classic representation of this temperature trend is given by Fletcher's global-average formula for the IN concentration,

$$[\mathrm{IN}] = A \exp(\beta \Delta T_s), \tag{12.1}$$

where $A = 10^{-5}\,\mathrm{L}^{-1}$, $\beta = 0.6\,°\mathrm{C}^{-1}$, and $\Delta T_s = T_0 - T$ is the supercooling. However, use of this parameterization in cloud models is now discouraged because of its inability to predict ice concentrations accurately. Ice concentrations at really low temperatures (large ΔT_s) are often well below those predicted by Eq. (12.1), while those at high temperatures (small ΔT_s) often contain much more ice than predicted. Clouds with ice concentrations consistent with Fletcher's parameterization have been found throughout the world, but such clouds usually contain very small cloud droplets (typical of continental clouds). On the other hand, clouds found to have enhanced concentrations of ice (dots), often greatly in excess of the IN concentrations for a given temperature (usually taken to be that at cloud top), are typically maritime in nature or of the deep-convective type. Clouds giving evidence of secondary ice production generally have broad drop-size distributions in the mixed-phase region, especially near the −5 °C level.

Primary ice particles arise in the mixed-phase region of a cold cloud by heterogeneous nucleation. The ice nuclei that serve as the sites for initiating ice act in one of several modes, as described in Chapter 7. Supersaturated vapor may deposit directly onto an IN, in which case the crystal subsequently grows first by vapor deposition, then possibly by riming. Alternatively, a potentially active IN may diffuse to a supercooled cloud droplet and nucleate the ice phase upon contacting the liquid surface. Droplets may also freeze when suitable IN are immersed inside the liquid. The frozen droplets may subsequently grow by vapor deposition as either single or poly crystals, depending on the temperature, then by riming. Each nucleation mode (deposition, contact, or freezing) acts under a unique set of circumstances, but in every case one new ice particle is born from one nucleation event. When conditions are appropriate, the primary ice particles may serve to catalyze the production of secondary ice particles.

Secondary ice particles are generated from preexisting ice in several conceivable ways. Ice nucleation can be sensitive to supersaturation (with respect to ice), which is enhanced in the wake of a freezing drop because of the transient evaporation of vapor into the cold air from the warmed drop. It has thus been envisioned that IN, which are not active under the prevailing average supersaturation, may become active as they pass close to a drop during its transformation from the liquid to solid phases. This mechanism is dependent on transient events and the concurrent presence of just the right kind of nuclei, so this mechanism of enhanced activation is not very likely. Large drops (greater than 400 μm diameter) have also been thought to shatter during freezing, due to the build-up of internal pressure as a symmetric ice shell forms around the freezing liquid. However, laboratory studies have shown drop shattering to be minimal under normal atmospheric conditions. Ice crystals growing from the vapor often exhibit delicate morphological features, such as the arms of dendrites, and field observations indeed show many examples of broken crystals, which may have resulted from collisions with other crystals. Each fragment of an ice crystal, even when that crystal arose from a primary nucleation event, represents a new, secondary ice particle. Once an ice particle, regardless of its origin, is large enough to collide with supercooled cloud droplets, it accretes rime ice, some of which may shed small ice splinters. The laboratory experiments of Hallett and Mossop have confirmed that approximately one splinter is released from a graupel particle for every 250 drops accreted at a temperature of $-5\,^{\circ}\mathrm{C}$. It is important to note that the graupel particles must be at least 0.5 mm in diameter, the accreted drops must be larger than about 25 μm in diameter, and the temperature for splinter production ranges from -8° to $-3\,^{\circ}\mathrm{C}$. The mechanism by which ice splinters form during riming in such a narrow temperature range is still uncertain. Of the diverse mechanisms proposed, only mechanical fragmentation and rime-splintering (Hallett–Mossop mechanism) have been shown to be viable for producing significant amounts of secondary ice.

12.3 Glaciation

Glaciation, in the context of cloud physics, is the conversion of supercooled cloud water into ice. The mechanisms of glaciation invariably involve primary nucleation, but secondary ice formation, when active, can rapidly propagate the ice phase throughout a cloud. An overview of the various processes within the context of a convective cloud is shown in Fig. 12.3. We are again reminded of the three main regions of a deep cloud based on the phase of condensed water and the approximate temperature T: all liquid ($T > 0\,^{\circ}\mathrm{C}$), mixed phase ($-38\,^{\circ}\mathrm{C} < T < 0\,^{\circ}\mathrm{C}$), and all ice ($T < -38\,^{\circ}\mathrm{C}$). The warm cloud is dominated by liquid drops, including any pockets of supercooled water in the mixed-phase region. The mixed-phase region also includes the melting zone that extends somewhat below the 0 °C-level. The left-hand side of Fig. 12.3 emphasizes the primary mechanisms of ice formation, whereas the right-hand side identifies the two main mechanisms for generating secondary ice. Capture nucleation, whereby a small ice particle collides with and freezes a large supercooled drop, is placed in the center to emphasize its role in both primary and secondary mechanisms of ice formation.

Overall, glaciation takes place via two classes of mechanisms, mass conversion and ice multiplication. Mass conversion of liquid to ice occurs when relatively few primary ice crystals form in the midst of a cloud of supercooled cloud drops. Ice multiplication results when secondary ice-production mechanisms lead to significant increases in the concentrations of ice particles.

12.3.1 Mass conversion

Mass conversion of liquid to ice is stimulated by the inherent difference in the equilibrium vapor pressures of liquid and ice at any temperature below 0 °C. The phase difference alone is sufficient to drive the growth of ice in mixed-phase clouds having weak updrafts

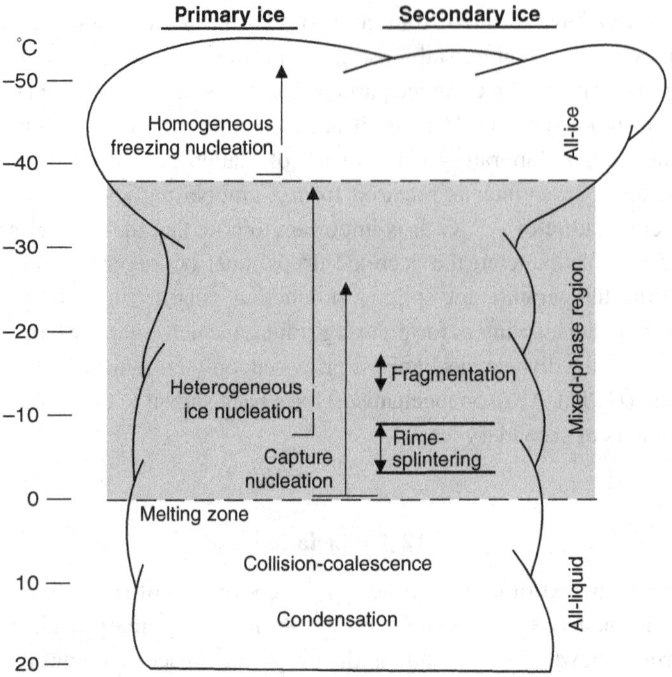

Figure 12.3 Schematic diagram of the primary (left half) and secondary (right half) mechanisms responsible for glaciation in the mixed-phase region (shaded). Adapted from Lamb (1999).

or downdrafts. Under such conditions vapor deposits preferentially onto ice particles, leading to the idea, suggested by Wegener in 1911 and developed further by Bergeron and Findeisen in the 1930s, that precipitation in supercooled clouds is initiated by ice formation. As suggested by Fig. 12.4, water vapor is transferred via gas-phase diffusion from liquid drops (which evaporate) to ice crystals (which grow). The Bergeron process implies that ice crystals grow at the expense of the liquid drops, which does occur when external forcings are weak. As shown in Fig. 12.5, the partial pressure of water vapor must therefore lie between the equilibrium vapor pressures of the solid and liquid phases: $e_i(T) < e < e_s(T)$. Figure 12.6 shows how the equilibrium vapor pressure difference (solid curve) varies with temperature and how the ice crystals respond (dashed curves) when the vapor phase is near liquid-water saturation, that is, when $e \simeq e_s(T)$. Note that the maximum vapor-pressure difference occurs at a temperature close to $-12\,°\mathrm{C}$, whereas the growth rates at liquid-water saturation peak at lower temperatures. This lowering of the temperature of maximum growth results from the warming that occurs because of deposition. This effect of latent heating is stronger the greater the growth rate, so the upper dashed curve is shifted more toward lower temperatures than is the lower dashed curve. The depositional growth of ice is limited by the diffusion of vapor through the air, so crystals grow faster (at a given temperature) at lower pressures, where the diffusion coefficient is larger.

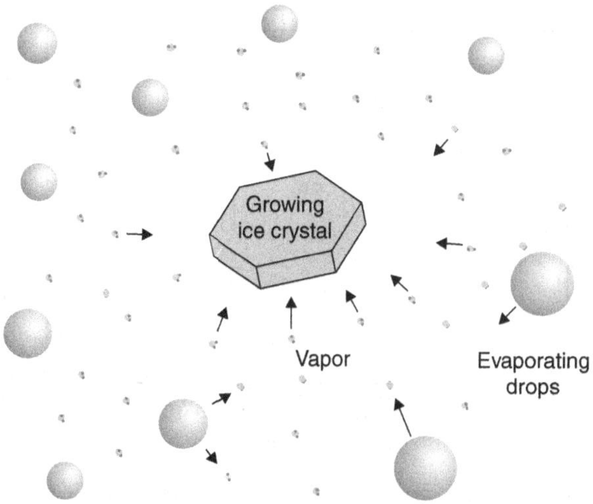

Figure 12.4 Cartoon illustrating the growth of an ice crystal at the expense of evaporating liquid drops.

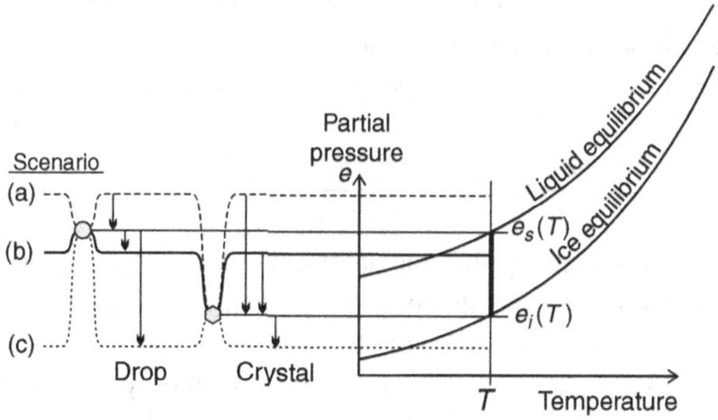

Figure 12.5 Schematic showing three scenarios for mixed-phase clouds. Adapted from Korolev (2007).

Under steady-state and quiescent conditions, the rate at which the collective mass of ice increases is the same as the rate at which the liquid mass decreases. The Bergeron process implies a more or less commensurate balance in the conversion of water mass from liquid to ice.

The vertical motions in a mixed-phase cloud can, however, alter the mass balance of the phase transformations. Intuitively, we expect that air rising rapidly generates excess vapor (because of rapid cooling) faster than either the liquid drops or the ice crystals can take it up, in which case both the liquid and ice particles grow, as the upper dashed curve in Fig. 12.5 indicates. At the other extreme, when air descends rapidly, the strong warming causes all particles to evaporate (lower dashed curve). Following theoretical arguments

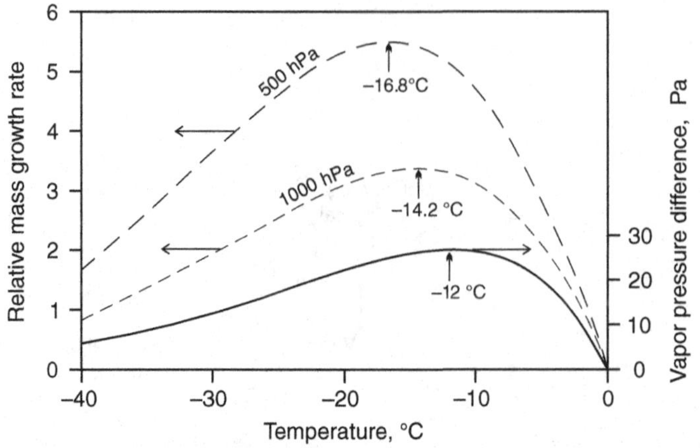

Figure 12.6 Dependences of vapor pressure difference (solid curve, right axis) and crystal growth rates (dashed curves, left axis) on temperature near liquid saturation. Adapted from Pruppacher and Klett (1997).

by Korolev, we can calculate the threshold updraft speed for phase equilibrium by again considering the development of supersaturation in an adiabatic parcel. We saw earlier that the supersaturation with respect to liquid water, s, develops in air ascending with speed w at the rate

$$\frac{ds}{dt} = Q_1 w - Q_2 \frac{dy_L}{dt} - Q_3 \frac{dy_I}{dt}, \tag{12.2}$$

where y_L and y_I are respectively the (molar) mixing ratios of liquid and solid condensate, and where the coefficients are slowly varying functions of temperature and pressure. By assuming no formation or loss of particles, we can simplify the analysis and illustrate the main point that the Bergeron process is limited to only a small range of vertical velocities. First, note that the liquid condensate grows in proportion to the instantaneous supersaturation, at the rate

$$\frac{dy_L}{dt} = 4\pi G s N_d \overline{r_d}, \tag{12.3}$$

where G is the condensational growth parameter and $N_d \overline{r_d} = \int_0^\infty r_d n_d (r_d)\, dr_d$ is the integral radius of the droplet population. Similarly, the growth of the solid condensate (ice) is given by

$$\frac{dy_I}{dt} = 4\pi G_i s_i N_i \overline{r_i}, \tag{12.4}$$

where G_i is the growth parameter for deposition onto spherical ice particles at supersaturation s_i and $N_i \overline{r_i} = \int_0^\infty r_i n_i (r_i)\, dr_i$ is the integral radius of the ice-particle population. Before combining equations, we must express the supersaturation with respect to ice (s_i) in terms of the supersaturation with respect to liquid (s), which we do by using their respective definitions and letting the vapor-pressure ratio $e_s(T)/e_i(T) \equiv \xi(T) = (s_i + 1) / (s + 1) > 1$. Thus, $s_i = \xi s + \xi - 1$, alternatively $s = (s_i + 1) / \xi - 1$. The supersaturation development equation thus takes the form

$$\frac{ds}{dt} = Q_1 w - (A_2 + A_3 \xi) s - A_3 (\xi - 1), \tag{12.5}$$

where $A_2 \equiv 4\pi G Q_2 N_d \overline{r_d}$ and $A_3 \equiv 4\pi G_i Q_3 N_i \overline{r_i}$. The quasi-steady supersaturation is therefore

$$s_{QS} = \frac{Q_1 w - A_3 (\xi - 1)}{A_2 + A_3 \xi}. \tag{12.6}$$

Liquid droplets cannot grow unless the supersaturation $s_{QS} > 0$, so we obtain the threshold updraft speed that yields liquid-phase equilibrium by setting $s_{QS} = 0$ and solving Eq. (12.6) for w:

$$w_L^* = \frac{A_3}{Q_1} (\xi - 1) = 4\pi G_i \frac{Q_3}{Q_1} (\xi - 1) N_i \overline{r_i} > 0. \tag{12.7}$$

Note that this threshold updraft for liquid equilibrium depends only on the properties of the ice population, in particular the ice-particle integral radius. In similar fashion, after replacing s_{QS} with its equivalent with respect to ice, $s_{QS} = \left(s_{i,QS} + 1\right) / \xi - 1$, we obtain the threshold speed for ice-phase equilibrium by setting $s_{i,QS} = 0$:

$$w_I^* = \frac{A_2}{Q_1} \left(\frac{1}{\xi} - 1\right) = -4\pi G \frac{Q_2}{Q_1} \left(\frac{\xi - 1}{\xi}\right) N_d \overline{r_d} < 0. \tag{12.8}$$

Now, the threshold speed is a downdraft, for which ice equilibrium depends only on the liquid-drop population. Such an analysis shows the power of theory in quantifying intuition and for revealing the limitations of traditional ideas. The Bergeron process is not the universal feature of mixed-phase clouds it was once thought, rather it operates in just one of three possible updraft regimes.

The dependence of particle-growth behavior on vertical velocity and integral radius is shown in Fig. 12.7 for adiabatic ascent (top) and descent (bottom). In rapidly rising air, the large amount of excess vapor produced condenses onto the liquid drops, as well as onto the ice crystals. In this regime, the liquid drops compete with the ice crystals and hinder the growth of the ice compared with a situation in which the liquid were not present. At the other extreme, when the air is sinking rapidly, both the ice and the liquid evaporate. Only in the regime where liquid evaporates and the ice grows (shaded regions in Fig. 12.7) can one technically say that the Bergeron process is operating. A proper accounting of the mass transfer between the vapor phase and both types of condensate is needed for accurate cloud modeling.

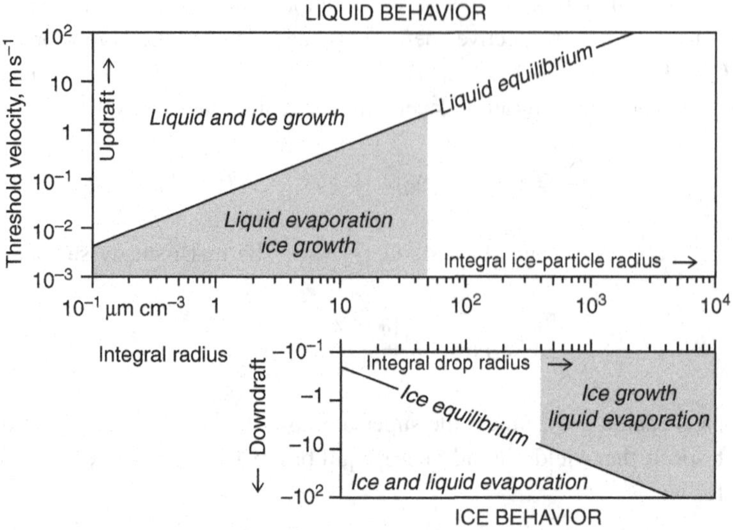

Figure 12.7 Behavior of liquid water and ice in ascending (top) and descending (bottom) air. Shaded regions identify ranges of atmospherically realistic integral radii over which ice grows at the expense of the liquid. Adapted from Korolev (2007).

12.3.2 Ice multiplication

Ice multiplication refers to the propagation of the ice phase throughout a cloud by increased numbers of ice particles. Primary ice particles are in effect multiplied by the average number of secondary ice particles spawned per primary ice particle. Of course, the processes are non-linear and seldom give rise to constant multiplication factors. Some situations will yield no secondary particles, some many. We can appreciate the problem of glaciation by recalling that primary ice nucleation is rare in the mixed-phase region of a cloud and that the liquid water is highly dispersed, divided into many droplets, each of which freezes independently. How can nearly a billion droplets freeze within a short period of time in each cubic meter of cloud? Nevertheless, observations suggest that much of the supercooled liquid is somehow able to freeze rapidly in some clouds. Our goal here is to illustrate one possibility, when conditions are right, by which an all-liquid cloud transforms itself rapidly into an all ice cloud.

Consider the rime-splintering mechanism operating within a deep convective cloud, such as can be found in tropical continental air masses. We idealize the setting, mainly by ignoring the cloud-scale dynamics, in order to keep the mathematics tractable and the principles clear. Figure 12.8 shows the collisional interactions possible in a cold cloud with a deep warm-cloud zone (i.e., a cloud whose base temperature exceeds about 20 °C). The warm-cloud process of collision-coalescence serves to produce a broad drop size distribution, so supercooled raindrops, as well as a rich supply of small droplets, are available during glaciation near the −5 °C level. The process starts as some ice nuclei (shown as a solid

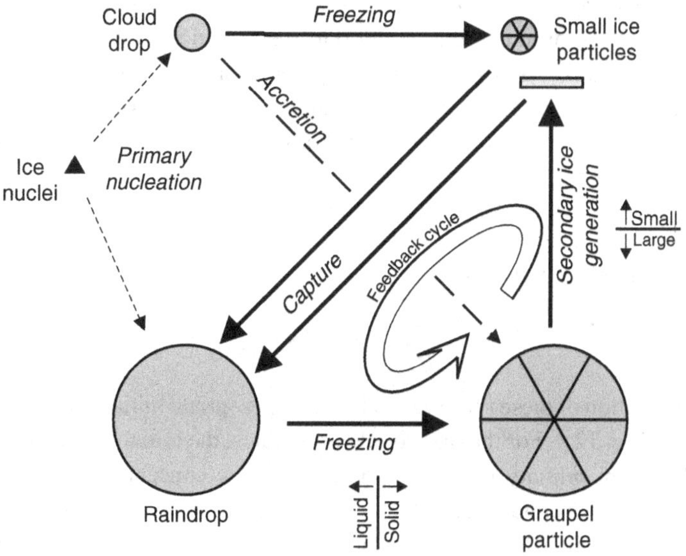

Figure 12.8 Schematic diagram of the mechanistic pathways by which glaciation takes place in a cold cloud when rime-splintering is active. From Lamb (1999), in *Encyclopedia of Applied Physics*, Update 1 (Trigg, G. ed.), pp. 1–25. Copyright Wiley-VCH Verlag GmbH & Co. KGaA. Reproduced with permission.

triangle) cause a few of the cloud drops to freeze and form an equal number of small ice particles. Some of the raindrops may also freeze by heterogeneous nucleation, but the greater likelihood is that they will freeze by capturing small ice particles. Frozen raindrops make excellent graupel particles that accrete supercooled cloud drops and generate secondary ice splinters. These splinters add to the population of small-ice particles, many of which are in turn captured by other supercooled raindrops. A regenerative feedback cycle is thus set up that rapidly increases the populations of both large and small ice particles. The process of ice multiplication continues until the supply of raindrops is exhausted. But, by then, the concentration of ice particles will be orders of magnitude larger than that of the ice nuclei. Primary nucleation serves mainly as the trigger for explosive growth of the secondary ice population.

A simple mathematical model of ice multiplication by rime splintering can be set up to explore the time scale for glaciation. Let the concentrations of the cloud particles be represented by the symbol n subscripted with the letters c, s, R, and G for the cloud drops, splinters (small-ice particles), raindrops, and graupel, respectively. The volumetric rates at which each species is produced is specified by J subscripted with the species name. For example, the rate of splinter production is designated J_s [$\mathrm{m}^{-3}\,\mathrm{s}^{-1}$]. The splinters are produced at a rate proportional to the concentration of graupel particles, so $J_s = S_s n_G$, where $S_s \simeq \beta_s \pi r_G^2 E_c v_G n_c$ is the proportionality constant, which serves as a first-order rate coefficient. Here, β_s is the probability of producing splinters ($\beta_s \sim 1/200$ at -5°C) and E_c is the efficiency with which graupel collects cloud water. In this simple model, we expect each capture event to cause a raindrop to freeze and instantly form a new graupel particle; thus, $J_G = S_R n_s$, where $S_R \simeq \pi r_R^2 E_{cap} v_R n_R$ and E_{cap} is the capture efficiency. The concentration of a species changes due to any imbalance in its rate of production and loss, so we use a material-balance relationship for the small ice particles:

$$\frac{dn_s}{dt} = J_s - J_G = S_s n_G - S_R n_s. \tag{12.9}$$

Similarly for the graupel particles:

$$\frac{dn_G}{dt} = J_G = S_R n_s. \tag{12.10}$$

Simultaneous solution of these two equations, with appropriate initial conditions, yields the trends shown in Fig. 12.9. For the first couple of minutes, the initial frozen drop population had to develop from primary nucleation. Thereafter, the concentrations of splinters and graupel increased exponentially until the raindrops were consumed. The entire glaciation process took only about 10 min in this example and resulted in a total ice concentration more than five orders of magnitude larger than that of the primary ice (indicated by the dotted curve). The conditions in real clouds may vary significantly from those assumed in this model, but when the setting is appropriate, glaciation can be efficient and rapid.

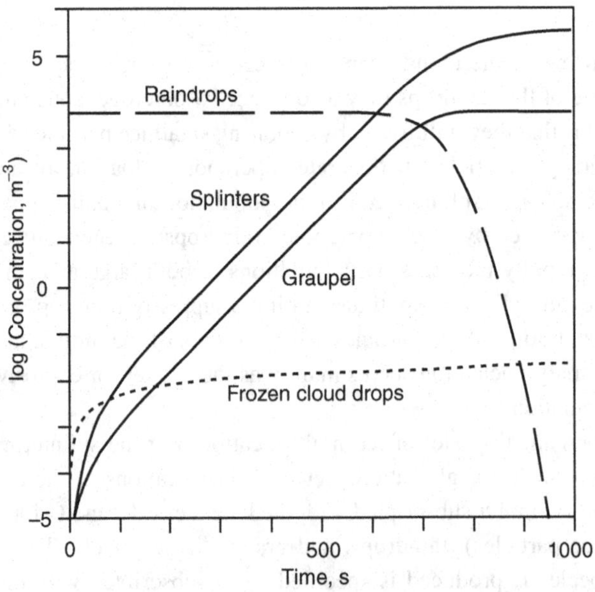

Figure 12.9 Concentrations of various cloud particles during the rapid glaciation of a cold cloud. Calculations based on the model of Lamb *et al.* (1981). From Lamb (1999), in *Encyclopedia of Applied Physics*, Update 1 (Trigg, G. ed.), pp. 1–25. Copyright Wiley-VCH Verlag GmbH & Co. KGaA. Reproduced with permission.

Rapid glaciation almost always requires an active warm-cloud process to provide the large supercooled raindrops that capture the small secondary ice particles generated by rime splintering.

12.4 Snow and cold rain

A significant fraction of rain that falls in summer in the mid-latitudes originates from the melting of some type of ice hydrometeor. We refer to rain that originates via the ice phase as cold rain. As shown in Fig. 12.10, the pathways through which ice grows into precipitation differ by the nature of the cloud system, whether stratiform or convective. Consistent with the naming convention, the pathways are closely linked to the dynamical environment, represented by the vertical velocity (w) of the air motion in which the ice grows. Stratiform precipitation results in environments in which the fallspeeds of the pristine ice crystals and aggregated ice particles (V_i) are greater than the magnitude of the updraft speed (i.e., $|w| < V_i$). Convective precipitation, by contrast, forms in environments with updraft speed, $w > V_i$. The bifurcation results from slow, vapor deposition dominated growth processes in stratiform precipitation versus fast, accretion-dominated growth processes in convective precipitation.

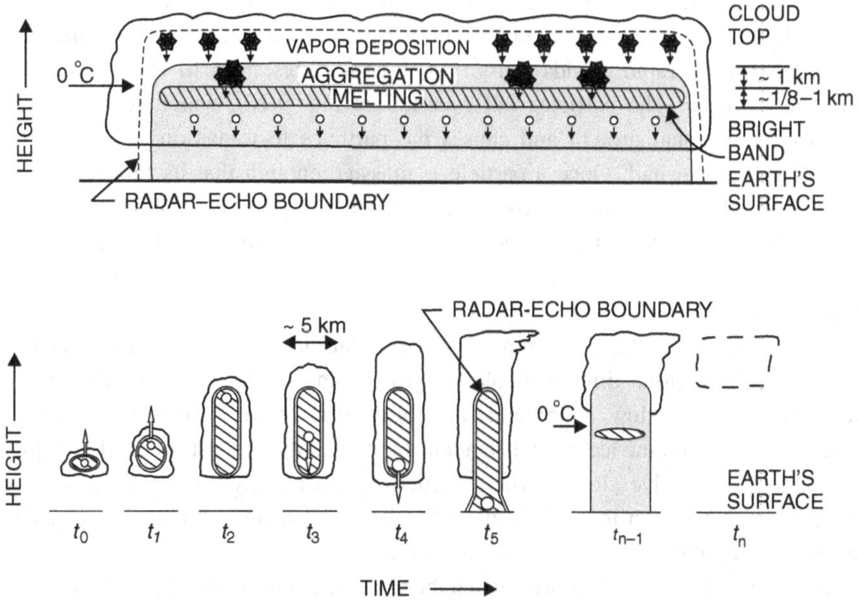

Figure 12.10 Cross sections of precipitation forming in stratiform (top) and convective (bottom) clouds. From Houze (1981), Structures of atmospheric precipitation systems: a global survey. *Radio Sci.*, vol. **16**(5), pp. 671–689. Copyright 1981 American Geophysical Union. Reproduced by permission of American Geophysical Union.

In a stratiform cloud, which often forms ahead of a warm front or in the upper outflow from deep convection, ice crystals near the upper reaches of the cloud grow by vapor deposition and fall slowly against the weak updraft. As we have already seen in association with Fig. 12.7, small vertical velocities in mixed-phase clouds are conducive to ice growth from the vapor at the expense of the liquid phase. Hence, with smaller drops and lower liquid water concentrations, growth by accretion is relatively ineffective. The slow net vertical motion of the ice crystal, coupled with the fact that depositional growth of non-spherical particles increases their fallspeeds only modestly, means that the ice crystals can experience prolonged periods of depositional growth (perhaps as much as 1 to 3 h in thick nimbostratus clouds). As these growing ice particles gradually fall into warmer parts of the cloud, sintering becomes possible and they begin to aggregate together. Growth by riming also becomes increasingly likely as the ice particles fall closer to the center of the warm-cloud microphysics. Both aggregation and riming concentrate the condensate into larger particles, which eventually become fast-falling raindrops upon passing through the melting layer. When the temperature stays below 0 °C all the way to the ground, as in winter or at high altitudes in the mountains, the surface precipitation appears as snow.

In a convective situation, when vertical wind speeds are appreciable and larger than the particle fallspeeds, ice crystals are carried aloft into colder parts of the cloud. As shown in Fig. 12.7, updraft speeds in excess of a given integral radius cause both the liquid drops and the ice crystals to grow simultaneously. As the sizes of both drops and crystals increase, conditions become increasingly favorable for accretional growth (Fig. 9.19). Riming rapidly adds mass, but it contributes little to the cross-sectional area of the particles. The particles fall ever faster (Fig. 9.10), which further increases the rates of accretional growth and causes the particles to transition from pristine ice crystals to graupel to hail. Once a particle is massive enough that its fallspeed exceeds the updraft speed, the particle falls from the cloud, possibly melting if it falls through the melting layer. Rain at the surface, even in summertime, often owes its existence to the diverse ice-phase microphysical processes of vapor deposition, aggregation, riming, and melting.

A critical element in the development of precipitation from both stratiform and convective clouds is the total time available for growth. In stratiform clouds, where growth by vapor deposition is slow, appreciable precipitation rates are realized by the relatively small fallspeeds of pristine ice crystals that closely match the modest speed of the upward-moving air. In convective clouds, where particles grow rapidly by accretion, significant precipitation results when the fallspeeds of the more compact and massive particles are large and also comparable to the large updraft speed.

What happens in convective storms when the time available is not sufficient for ice particles to grow to precipitation sizes in the convective core? Two situations may be envisioned. When the ice phase initiates late (i.e., well into the supercooled zone), any convective precipitation must come mostly from the warm-rain process of collision-coalescence. Or, if the concentration of ice crystals is excessively large (perhaps because of rapid nucleation), the concentration of liquid water may become depleted through the Bergeron process, in which case the individual ice crystals grow slowly by vapor deposition in a competitive environment. Either situation results in the ice particles being carried upward through the $-40\,^{\circ}$C level without attaining the fallspeeds needed to overcome the strong updraft in the convective core of the storm. These ice particles are therefore blown upward and horizontally away from the main updraft by the divergent air flows near the level of neutral buoyancy. The decreasing updraft speeds in the anvil gradually let the particles fall downward to serve as seeds for the formation of stratiform precipitation downshear of the core. Particles advected far into the anvil may simply fall into the subsaturated air beneath the anvil and evaporate before reaching the ground. The microphysics of precipitation formation in convective storms is complicated by the three-dimensional character of the air flows.

The arguments put forth above suggest a potential link between convective and stratiform precipitation in general. In a deep convective cloud, the updraft often supplies the seed ice that subsequently initiates the stratiform rain. However, convection does not have to be deep to produce the seeds for stratiform precipitation. For instance, radiative loss of energy from the upper levels of a stratiform cloud can produce a shallow layer of potentially unstable air within which smaller-scale convection develops. The updraft speeds attained in these convective cells may be modest compared with those in deep convective clouds, but they may nevertheless be sufficient to produce precipitation-sized particles. As the ice particles are advected laterally away from the convective updrafts, they settle into the stratiform region in which the convection is embedded. The particles continue to grow because of the supersaturation generated by the gentle mesoscale uplift. This process in which embedded convection in the upper layers provides the seed ice for the stratiform precipitation is called a seeder–feeder mechanism. Seeder–feeder precipitation is typically showery in nature, the intermittency coming from the discrete and stochastic nature of the seed-generating cells.

The formation of snow and rain in cold clouds is complicated both by the diverse microphysics involving different phases of water and by the interactions that exist between the microphysical processes and complicated flow patterns. As shown earlier in Fig. 12.1, snow can fall directly from stratiform clouds, or it can, in more convective environments, accrete supercooled drops and initiate graupel and hail formation (discussed below). Cold rain appears if the ice particles melt before they reach the ground. Convective precipitation is important in its own regard for producing large rainfall rates. But, we have also seen two ways in which convection contributes to stratiform precipitation. Deep convection produces modest-sized ice particles that get carried aloft and ejected into the adjacent stratiform region, good examples being squall lines (Fig. 6.11) and mesoscale convective complexes. Convection can also be shallower and of smaller scale, embedded in the upper layers of a stratiform cloud where it seeds widespread stratiform precipitation. An example of such a relationship is seen in warm fronts. A complete picture of precipitation formation in cold clouds requires simultaneous consideration of the microphysics and the mesoscale organization of the air flows.

12.5 Hail formation and growth

Hail is the end product of riming, occurring when the growth of large ice particles is dominated by the rapid collection of supercooled cloud water. By convention, the individual hailstones must exceed 5 mm in diameter and exhibit alternating layers of opaque and clear ice, as shown in Fig. 12.11. Hailstone diameters can occasionally exceed 15 cm, about the size of a grapefruit. The shape of a large hailstone may become highly irregular, as lobes jutting into the airstream offer preferred sites for collection of cloud droplets (i.e., the local collection efficiency is higher than the particle-average efficiency). The internal structure of such a hailstone only partially reflects the external morphology because the lobes grow in a regime distinct from that of the overall hailstone. Smaller hailstones tend to be spheroidal or conical and have more symmetric layering. The opaque layers contain many air bubbles trapped in the ice during so-called dry growth; opaque layers are porous and have low densities of ice. The clear, dense layers represent wet growth, which occurs when the collection of supercooled droplets is so rapid that not all of the water has a chance to freeze. Excess liquid water is often shed from the particle, or it may fill in the air spaces of the underlying dry-growth layer, thus forming a so-called spongy hailstone.

A hailstone can originate via one of two possible pathways. As shown in Fig. 12.12, the precursor ice particle (the hail embryo) can be either a graupel particle or a frozen drop. Graupel embryos are bubbly throughout and often conical in shape, whereas frozen-drop embryos are clear and spheroidal. The frequency with which each type of embryo initiates hail correlates with the temperature of cloud base, as shown in Fig. 12.13. Embryos of the graupel type are favored in cool-base clouds, in which a relatively shallow region exists where the warm-cloud process of collision-coalescence can operate. Frozen-drop

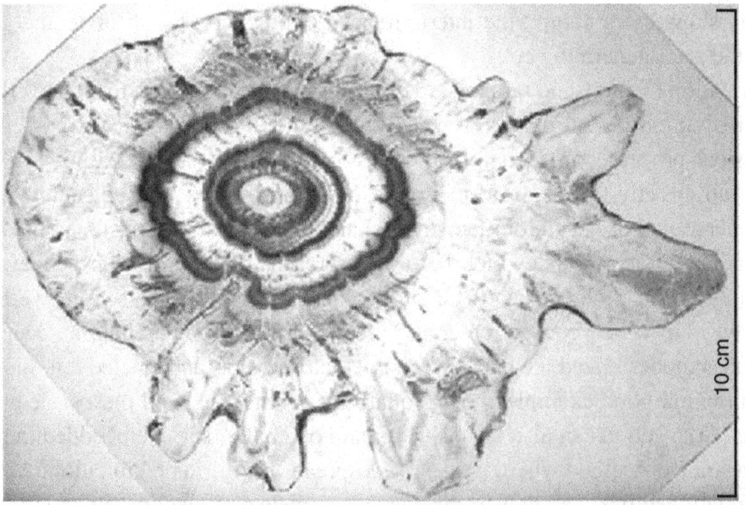

Figure 12.11 Photograph of a large hailstone that fell near Aurora, Nebraska. Wet growth results in relatively clear ice, dry growth in opaque layers. Adapted from Knight and Knight (2005).

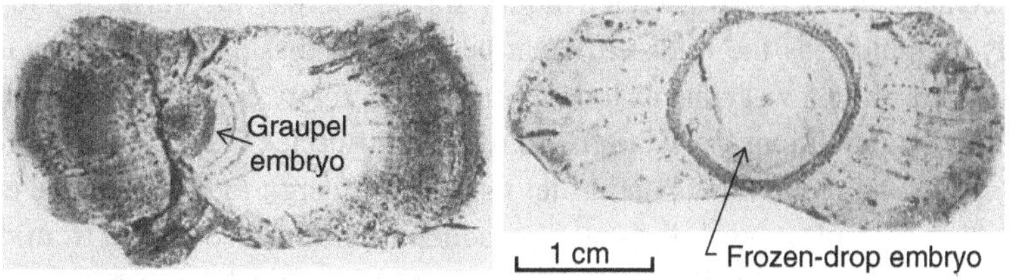

Figure 12.12 Photographs of hailstones originating from different embryo types. Left, hailstone with graupel embryo. Right, hailstone with frozen-drop embryo. Adapted from Knight (1981).

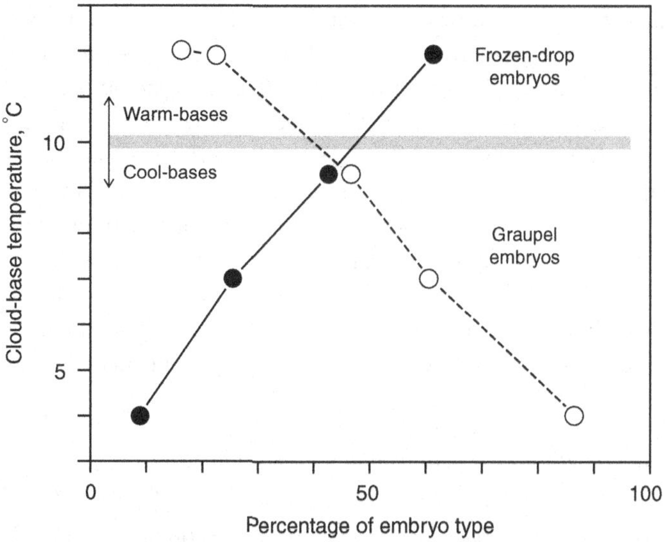

Figure 12.13 Frequency of embryo type versus the average cloud-base temperature. Adapted from Knight (1981).

embryos require large, millimeter-size drops to exist prior to freezing, which can arise only in clouds with warm bases (i.e., $\geq 10\,^\circ$C). Hailstone origins thus reflect a regional climatology because cool-base clouds tend to be found in relatively dry continental zones, whereas warm-base clouds are found in humid coastal and maritime zones. In every situation, only a small fraction (< 1 in 10^4) of embryos become hailstones. Successful embryos start growth early and happen to grow in regions where the local updraft speed more or less matches the particle fallspeed.

Continued growth of a hailstone beyond the embryo stage depends on the accretional sweepout of supercooled water. Accretion, the physical process by which an ice particle collects cloud drops when the air temperature $T_{air} < T_0$ (the nominal melting point of ice), has important energy consequences: The freezing of the accreted droplets releases the latent heat of fusion and warms the hailstone. With increasing rates of accretion come higher temperatures. Once the surface temperature T_{sfc} gets to $0\,^\circ$C, the droplets no longer freeze on contact. Rather, the accreted liquid spreads across the surface, and freezing occurs only as rapidly as the released latent heat can be dissipated to the environment. The hailstone grows in a wet state as long as the accretion rate exceeds a threshold value.

Calculating the conditions for wet growth requires a surface energy balance. It suffices to assume steady-state conditions and to let the rate of energy into the surface equal that leaving it. At the simplest level, the enthalpy added to the surface comes from the freezing of the accreted supercooled water. The hailstone is typically much larger than the droplets it collects, so the continuous collection model is appropriate (review Chapter 11). The thermal power into the surface is therefore $l_f K \omega_L$, where l_f is the latent heat of fusion, $K \sim (\pi/4)\, E_c D_p^2 v_p$ is the collection kernel, and ω_L is the liquid water concentration. This added enthalpy warms the surface, which causes some energy to leave the surface by conduction in proportion to the difference in temperatures between the surface and the air: $2\pi D_p f_h k_T \left(T_{sfc} - T_{air}\right)$, where f_h is the ventilation coefficient for heat transfer and k_T is the thermal conductivity of the air. The surface warming also causes the hailstone to sublimate, so the power leaving the surface by sublimation is proportional to the differences in vapor densities at the surface, ρ_{sfc}, and in the free-stream air, ρ_{air} : $2\pi D_p f_v D_v \left(\rho_{sfc} - \rho_{air}\right) l_s$, where f_v is the ventilation coefficient for vapor transport, D_v is the diffusivity of vapor in air, and l_s is the mass-based enthalpy of sublimation. The accreted water also gets warmed, from its original, air temperature to the surface temperature, so we include the term $K\omega_L \left(T_{sfc} - T_{air}\right) c_L$, where c_L is the specific heat of liquid water. The first-order surface energy balance (ignoring radiation and non-steady effects) is thus

$$\begin{aligned} \textit{freezing of accreted water} = \textit{heat conduction} &+ \textit{sublimation} \\ &+ \textit{warming of accreted water} \end{aligned} \tag{12.11}$$

$$\begin{aligned} l_f K \omega_L = 2\pi D_p f_h k_T \left(T_{sfc} - T_{air}\right) &+ 2\pi D_p f_v D_v \left(\rho_{sfc} - \rho_{air}\right) l_s \\ &+ K\omega_L \left(T_{sfc} - T_{air}\right) c_L. \end{aligned} \tag{12.12}$$

In water-rich regions of cloud (large ω_L), the enhanced enthalpy input is balanced by increased values of T_{sfc} and $\rho_{sfc} = \rho_s \left(T_{sfc}\right)$, which lead to increased losses of energy by conduction and sublimation. The wet-growth limit is reached when the surface temperature $T_{sfc} = T_0 \equiv 273.15\,\mathrm{K}\,(0\,°\mathrm{C})$, so the so-called Schumann–Ludlam threshold for wet growth is

$$\omega_{L,wet} \geq \frac{2\pi D_p}{K} \frac{\left[f_h k_T \left(T_0 - T_{air}\right) + f_v D_v \left(\rho_{s.0} - \rho_{air}\right) l_s\right]}{l_f + c_L \left(T_0 - T_{air}\right)}, \tag{12.13}$$

where $\rho_{s.0} \equiv \rho_s\left(T_0\right)$ is the saturation vapor density at 0 °C. Appropriate simplifications can be made by evaluating the collection kernel in terms of particle size. The collection kernel is an effective sweepout volume per unit time, so for the continuous collection model, $K \sim (\pi/4)\, E_c D_p^2 v_p$. Because the hailstone is relatively large and not deformable, we may assume the collection efficiency $E_c \simeq 1$, and we can usually ignore the warming of the accreted water. We may also use the estimation for hailstone fallspeed, $v_p = 9D_p^{0.8}$, thus finding $K \sim (9\pi/4)\, D_p^{2.8}$ and

$$\omega_{L,wet} \geq \frac{8}{9 l_f D_p^{1.8}} \left[f_h k_T \left(T_0 - T_{air} \right) + f_v D_v \left(\rho_{s.0} - \rho_{air} \right) l_s \right]. \tag{12.14}$$

The Schumann–Ludlam limiting conditions can thus be plotted as shown in Fig. 12.14. Note that the criteria for wet growth depend on air temperature, liquid water concentration, and particle size. Wet growth is favored, for any given hailstone diameter, when the temperature is high and the concentration of supercooled water is large (upper right portion of Fig. 12.14).

Hail forms effectively only when the dynamic structure of the storm is appropriate. We see in Fig. 12.15 that three zones may be defined within the cloud. The embryo-formation zone (EFZ) is characterized by weak updrafts, so graupel particles have time to grow to diameters of 5 mm or more. The embryos that can grow sufficiently and also enter the stronger updraft regions of the cloud go on to develop clear layers of ice (glaze) in the hail-growth zone (HGZ). Strong updrafts are needed to hold the hailstones in water-rich regions by compensating for their increasing fallspeeds. If the updraft is too weak, the hailstones fall out before they grow very large; if it is too strong, the hailstones are pushed up into the anvil where little liquid water is available. The terminal fallspeeds of hailstones have been estimated to increase with particle diameter D_p [in cm] according to $v_p[\text{m s}^{-1}] = 9D_p^{0.8}$, which implies updrafts varying approximately between about 10 and 60 m s^{-1}. One should appreciate the delicate balance between particle growth and the dynamic structure of the cloud that is needed to let hailstones grow. Conditions must be just right for large hailstones to develop and fall out in the fallout zone (FOZ). A suggested pathway for the formation and growth of hail in this cloud is shown in Fig. 12.16.

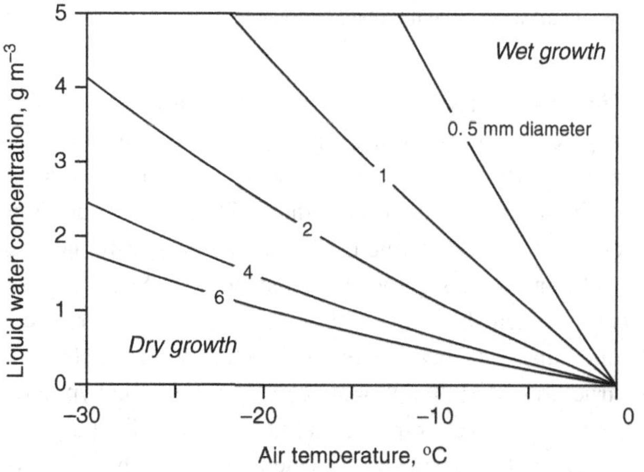

Figure 12.14 Threshold liquid water concentrations for wet growth as a function of ambient air temperature and hailstone diameter. Adapted from Young (1993).

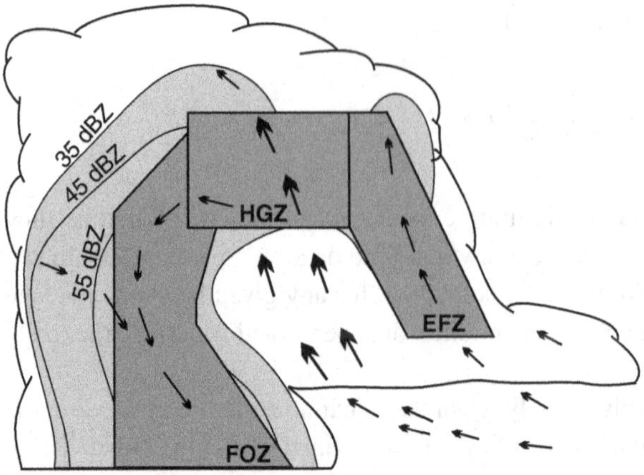

Figure 12.15 Zones in which hail develops within a convective storm. Arrows indicate wind velocities; heavy arrows suggest stronger winds. Adapted from Young (1993).

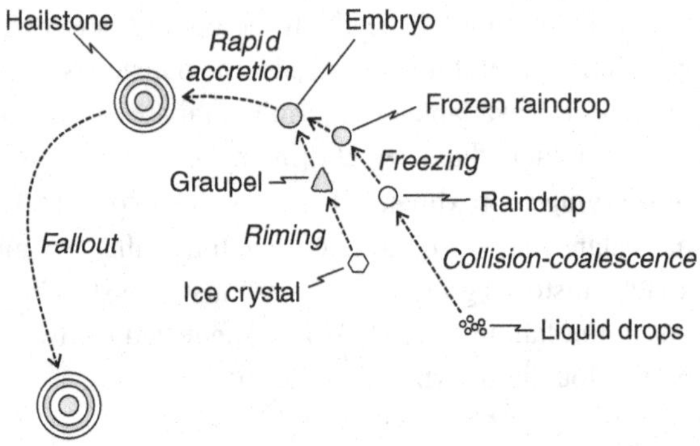

Figure 12.16 Suggested pathways by which hail forms and grows in the cloud shown in Fig. 12.15.

Hail can be produced in all types of deep convective storms, from single-cell ordinary thunderstorms to supercells, but most large hail forms in multicellar or supercell convection. The criterion that must be met in a hail-producing cloud is that the updraft speed must closely match the hailstone fallspeed in the HGZ region, where high concentrations of supercooled liquid water have formed. The hailstone then remains suspended under optimal-growth conditions for a significant period of time. Such conditions may be met in ordinary convection when the updraft strength increases at a rate comparable to the increases in hailstone fallspeeds, but they are more likely to be satisfied in the complicated dynamical cloud structures of multicell or supercell storms. The hypothesized pathway of a hailstone growing from a cloud droplet at cloud base to its final arrival at the surface is shown in Fig. 6.8 as the path of open circles, where the strengthening updraft carrying the hailstone is protected from the drying effects of entrainment by the adjacent cells.

The largest hailstones, those having diameters of 5 cm or larger, seem to be produced in supercells. It is thought that the optimal production of large hail occurs when the embryos find themselves above and slowly falling into the slanted updraft of the storm as they are carried aloft. Embryos may enter this region after forming in cells along the rear-flank gust front that subsequently merge with the main updraft. However, it is more likely that embryos form in the main updraft and are then ejected into the embryo curtain around the bounded weak-echo region. Some, favored, embryos fall back into the updraft at lower levels, where their increasing fallspeeds stay matched to the updraft speed by moving progressively closer to the core of the tilted updraft, where the vertical velocity is largest. When the hailstone fallspeed exceeds the updraft velocity, the hailstone falls to the surface, growing through continued accretion of supercooled water above the melting layer. Once it passes into the layer in which the wet-bulb temperature is 0 °C, the hailstone starts to melt and shed liquid drops. The height above the surface at which melting begins plays a large role in determining the size of the hailstone when it reaches the ground. Hail in storms near Alberta, Canada, has a much greater chance of surviving to the surface than does hail near Denver, Colorado, or that near Phoenix, Arizona.

12.6 Further reading

Fletcher, N.H. (1962). *The Physics of Rainclouds*. Cambridge: Cambridge University Press, 386 pp. Chapter 11 (rain from cold clouds).

Houze, R.A. (1993). *Cloud Dynamics*. San Diego, CA: Academic Press, 573 pp. Chapter 6 (snow and cold rain precipitation processes).

Lamb, D. (2003). Cloud Microphysics. In *Encyclopedia of Atmospheric Sciences*, ed. J.R. Holton, J. Pyle, and J.A. Curry. London: Academic Press, pp. 464–467.

Mason, B.J. (1957). *The Physics of Clouds*. London: Oxford University Press, 481 pp. Chapter 6 (precipitation processes).

Rogers, R.R. and M.K. Yau (1989). *A Short Course in Cloud Physics*. New York: Pergamon Press, 293 pp. Chapters 10 (snow size distribution), 12 (mesoscale precipitation, precipitation efficiency), and 13 (thunderstorm precipitation).

Wallace, J.M. and P.V. Hobbs (2006). *Atmospheric Science: An Introductory Survey*. Amsterdam: Academic Press, 483 pp. Chapter 6 (precipitation from cold clouds).

Young, K.C. (1993). *Microphysical Processes in Clouds*. New York: Oxford University Press, 427 pp. Chapters 9 (warm-rain process) and 10 (ice-crystal mechanism for precipitation).

12.7 Problems

1. A hypothetical cloud that is cylindrical in shape has a cross-sectional area of $10\,\text{km}^2$ and a height of 3 km. The entire volume of the cloud is initially supercooled and the liquid water concentration is $2\,\text{g m}^{-3}$. All of the water in the cloud is transferred onto ice nuclei present in a uniform concentration of one per liter.

(a) How much latent heat is released as a result of the glaciation?
(b) Assuming all crystals are the same size, calculate the mass of each ice crystal produced and the total number of ice crystals in the cloud.
(c) If all the ice crystals precipitate and melt before they reach the ground, what will be the total rainfall (in cm) produced?
(d) How does the number obtained in (c) compare with what you would expect from a heavy thunderstorm that lasts for 3 h?

2. Why is it that raindrops are never found to be as large as big hailstones?

3. On a snowy day a real cloud physicist will always walk with arm parallel to the ground, catching and analyzing ice crystals.
(a) What can you conclude about the physical processes in the cloud overhead when you observe large numbers of needle aggregates (aggregates formed exclusively from needles)?
(b) What can you conclude about the physical processes in the cloud overhead when you observe large numbers of rimed dendrites?
(c) The majority of pristine crystals collected are delicate crystals such as needles, dendrites, or stellar crystals: rarely does one see the more isometric crystals such as thick plates or solid columns. Can you explain why?

4. In a recent press release, it was reported that *basketball*-sized hail fell from a convective storm over a remote region of China. Such reports appear from time to time in the news media. Please calculate, to a first order approximation, the storm characteristics (vertical motion, w, and liquid water concentration, ω_L) that would be necessary for the formation of such a large hailstone. In your calculation you may assume that the liquid-containing layer of the cloud is 10 km deep, that the hailstone grows by continuous collection, and that when it exits the cloud it has a diameter of 30 cm. Furthermore, assume a collection efficiency of 0.5. Please discuss your results in the context of what you know about convective storms.

5. Consider a population of ice crystals growing in cloud environment with temperature $-15\,°\mathrm{C}$ and supersaturation $s = 0.0005$ with respect to liquid. Discuss the various mass growth paths available to these ice crystals, with particular focus on how the various pathways will impact the fallspeed of the particles.

6. What changes might occur to a typical hurricane during its mature (steady state) stage of development if the hydrogen bonds in liquid water and ice were to double suddenly in strength throughout the world?

7. When a hailstone falls through liquid layers in the cloud, hailstone growth by collection of liquid drops contributes to its melting rate. Please include the contribution of the energy by these drops to the hailstone surface energy budget. Calculate the surface

temperature of hailstones with sizes $D_p = 0.5$ mm, 1 mm, 2 mm, 4 mm, and 6 mm, liquid water concentration $\omega_l = 0\,\mathrm{g\,m^3}$, $0.5\,\mathrm{g\,m^3}$, $1\,\mathrm{g\,m^3}$, and $2\,\mathrm{g\,m^3}$, and environmental temperatures $T = -20\,°\mathrm{C}$, $-10\,°\mathrm{C}$, and $10\,°\mathrm{C}$.

(a) At what point does the ventilation effect become important?

(b) At what point can we safely ignore the effect of the rimed water?

8. This exercise is designed give you experience in putting together an ice microphysical model. Consider a cloud in which the instantaneous concentration of cloud droplets is n_c and that of raindrops is n_R. Ice forms initially in this hypothetical cloud by primary contact nucleation (concentration of contact nuclei n_N), which freezes either cloud droplets (producing frozen cloud droplets, n_{fc}) or raindrops (instantly producing graupel, concentration n_G). Each nucleation event results in the loss of a nucleus. Now, once present, graupel may collect cloud droplets (via continuous growth), and produce rime splinters (n_s) at a rate $\beta = 1/200\,\mathrm{s^{-1}}$. Rain may also collect cloud droplets, thus reducing their numbers.

   (a) Write down the complete set of differential equations for the time rate of change of the concentrations of contact nuclei, cloud droplets, rain drops, frozen cloud drops, splinters, and graupel. You may assume that contact nucleation of cloud drops and raindrops proceeds at the respective rates of

$$J_{nC} = 4\pi D_N n_N n_c r_c$$

   and

$$J_{nR} = 4\pi D_N n_N n_R r_R$$

   where $D_N = 5 \times 10^{-5}\,\mathrm{cm^{-2}\,s^{-1}}$ is the diffusivity of aerosol particles, and r_c and r_R are the radii of the cloud and raindrops, respectively. You may assume that all collection processes proceed via continuous grow. Make any other assumptions you need to close the problem.

   (b) Write a simple computer model that integrates the coupled set of equations. Using the known concentrations of constituents at time t, calculate the rates of change of all constituents at that time. Using a time step of $\delta t = 1$ s, calculate the concentrations at time $t + \delta t$. Use $r_c = 7.5\,\mu\mathrm{m}$, $r_R = r_G = 1$ mm, fallspeeds $v_R = 500\,\mathrm{cm\,s^{-1}}$ and $v_G = 140\,\mathrm{cm\,s^{-1}}$. The collection efficiencies for graupel and rain collecting cloud drops is $E_{cc} = 0.5$, while that for rain collecting ice is $E_{cR} = 0.1$. Do several runs, playing with different assumed values for the parameters, but use the same initial concentrations of contact nuclei ($n_N = 10^{-4}\,\mathrm{L^{-1}}$), cloud droplets ($1100\,\mathrm{cm^{-3}}$), and rain drops ($5.7\,\mathrm{L^{-1}}$). Start the calculation with all liquid water (i.e., no frozen drops, splinters, or graupel) and let it run until all the rain is depleted.

13
Cloud chemistry

13.1 Overview

Clouds play important roles in the composition of the atmosphere and in the chemical quality of precipitation. Cloud particles form in the first place by condensation onto aerosol particles composed of diverse compounds. Then, they take up additional chemicals from the air, change their chemical properties, and eventually release modified compounds back into the air or transfer them to large, sedimenting particles. Clouds effectively cleanse the air through precipitation, which serves as the carrier of atmospheric chemicals to terrestrial and aquatic ecosystems. Clouds simultaneously depend on the chemicals in the air and influence the composition of the atmosphere through a variety of microchemical processes.

The term microchemistry in cloud physics parallels that of microphysics. Both disciplines deal with the particles making up clouds, but the emphasis in cloud microchemistry is on the chemicals contained in the particles, not on the particles themselves. Atmospheric trace chemicals influence cloud properties in important ways, and the cloud microphysics also determine the fates of atmospheric chemicals. Important goals of microchemical research include understanding source–receptor relationships, the chemical quality of precipitation, and the influence trace chemicals have on clouds and climate.

As soon as pollutants enter a cloud, they become intertwined with the cloud processes at both the macro- and microscales. The active dynamics of a large convective cloud, for instance, often serves to vent the planetary boundary layer, pulling pollutants into the free troposphere along with the moisture that serves as the fuel for cloud formation (see Fig. 13.1). Venting simultaneously cleanses the boundary layer (though not the atmosphere as a whole) and transports chemicals vertically. At the microchemical level, the compounds contained in the aerosol particles carried into a cloud may enter the cloud drops directly, if their solute contents allow the particles to serve as cloud condensation nuclei (CCN). The other particles remain interstitially in the air between the drops, possibly only later interacting with the liquid and solid particles through a variety of scavenging mechanisms (discussed later). Trace gases brought into the cloud may also be scavenged by the cloud particles, in the process leaving the interstitial air and adding non-aqueous matter to the cloud particles. Particles and gases not scavenged by the time the air passes through the cloud are simply transported away with the outflow. Clouds act as species-selective filters during the vertical transport of pollutants.

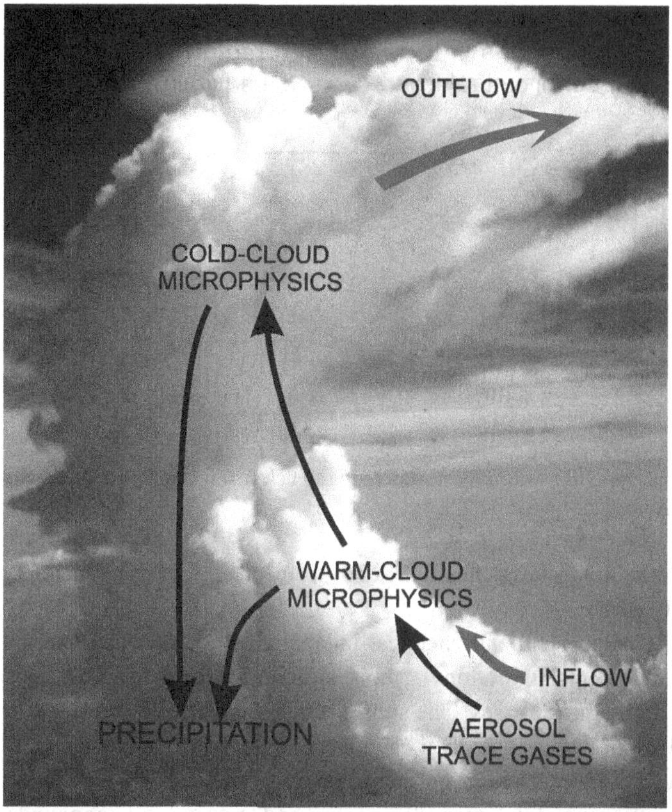

Figure 13.1 Schematic depiction of flows through a small cumulonimbus cloud. Background photo by D. Lamb.

The microchemical processing of pollutants by clouds is intimately linked to the microphysics of precipitation formation. Through precipitation, clouds not only provide fresh water to the Earth's surface, they also cleanse the atmosphere of pollutants, both natural and anthropogenic. In the process of cleaning the air, clouds necessarily increase the chemical content of the rain and snow they release, a manifestation of the mass-balance principle. A summertime example of the cleansing power of precipitating clouds is shown in Fig. 13.2. The warm, moist air that rides over a cooler air mass forms cloud that is often convective near the surface front and stratiform farther downwind. Particulate matter and trace gases brought into the cloud from the warm sector are first processed by the convective cloud, later by the stratiform cloud. The microphysical processes responsible for precipitation formation allow some of the chemicals to be removed from the most polluted air and end up in the rain closest to the surface front. As verified by field measurements, this first rain is relatively dirty compared with the rains that fall farther from the surface front. Precipitation, with its chemical burden, simultaneously closes the hydrologic cycle and the atmospheric cycles of many trace

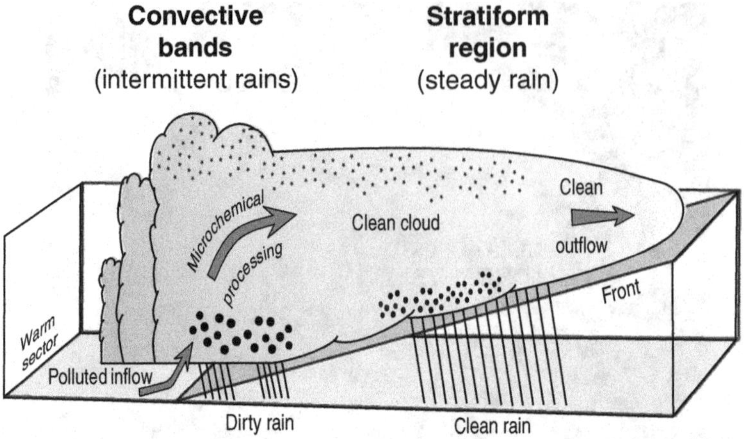

Figure 13.2 Schematic of a warm front illustrating the cleansing capability of precipitating clouds. Adapted from Ahrens (1994).

chemicals. Determining how chemicals end up in precipitation is an important goal of cloud microchemistry.

The diversity of physical and chemical interactions inside a cloud can be appreciated from Fig. 13.3. The upward motions of moist air required for cloud formation also bring in aerosol particles and trace gases. The subset of the atmospheric aerosol that provides the sites for condensation contribute their chemical compounds to the cloud water, the process known as nucleation scavenging. Further uptake of chemicals by the cloud water occurs via in-cloud scavenging of the interstitial aerosol and trace gases. Compounds contained in the cloud droplets are directly transferred to larger raindrops in the warm cloud (the lower part not containing ice), as collision-coalescence involves no additional phase change, and the chemicals are simply carried along with the collected water.

Collection of cloud water by precipitation in the cold cloud, by contrast, is significantly more complicated. The riming growth of graupel, for instance, entails a phase change when the supercooled droplets freeze on contact with the ice surface. The large difference in chemical solubilities in liquid water (high) and ice (low) means that most of the solute is forced into small brine pockets of liquid trapped within the polycrystalline ice matrix. A further distinction must be made on the basis of the volatility of the solute: Non-volatile solutes (e.g., sulfate) tend to stay with the growing graupel, although their distribution within the particle is far from uniform. But, as the concentration of solute in the brine pockets increases, the equilibria for volatile solutes (e.g., nitric and sulfurous acids) shift in favor of the gas phase; volatile solutes are only partially trapped during riming. Another barrier to the transfer of solute from small to large particles exists during the Bergeron process: The evaporation of droplets, the source of water vapor for the growing ice crystals, similarly leads, albeit more slowly, to higher concentrations of solute in the cloud water. Whether the solute stays with the liquid particle or reenters the gas phase again depends on the solute volatility. In either case, only water, and virtually none of the solute, is transferred

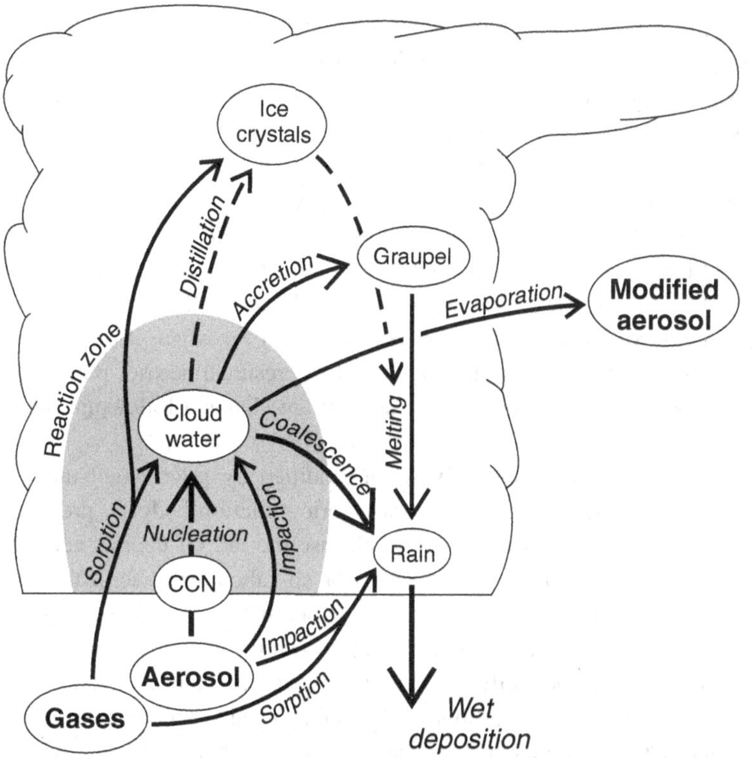

Figure 13.3 Schematic view of a cloud showing the dominant pollutant pathways (solid curves, widths suggesting importance). Dashed curves identify processes with minimal chemical impact. Impaction scavenging of aerosol particles to ice crystals and graupel may be important, but is not shown for clarity.

to the ice crystals. The depositional growth of ice in supercooled clouds acts as a low-temperature distillation process that yields relatively clean snow. The microchemistry in the mixed-phase portion of clouds is complicated by the diversity of physicochemical processes that are possible. The transfer of chemicals from the small cloud particles to the large, sedimenting particles depends on the specific mechanisms of precipitation formation prevailing at the time, and it is often the precipitation microphysics that limits the removal of atmospheric trace chemicals.

Even non-precipitating clouds have important microchemical consequences. Such clouds affect the chemical quality of the air in which they form partly because the many small cloud droplets, often as many as 10^9 m^{-3}, offer opportunities for chemical interactions not available in the gas phase. The large surface-to-volume ratio of the droplets in the reaction zone (see Fig. 13.3) readily lets trace gases enter the aqueous phase, where they may react and form new compounds. For instance, sulfur dioxide (SO_2) reacts little if at all in the gas phase with ozone (O_3) and hydrogen peroxide (H_2O_2). Once dissolved in liquid water, however, the SO_2 can be oxidized by these strong oxidants to sulfuric acid

(H_2SO_4), a highly soluble compound. This additional, non-volatile solute stays with the particles as sulfate $\left(SO_4^{2-}\right)$ even after the cloud droplets evaporate along cloud edges (refer to Fig. 13.3). Through such in-cloud chemistry, the atmospheric aerosol becomes acidified, contributes to fine-particle health effects, and degrades visibility. Also, as we have learned from Köhler theory (Section 3.5), higher solute contents mean lower critical supersaturations, so sulfate-enriched cloud water often leads to increased concentrations of CCN active at any given supersaturation. The sulfate readily reenters subsequent, possibly precipitating clouds, where it may contribute to the acidity of rain. Clouds may also enhance the solute contents of individual droplets through collision-coalescence: The merging of water mass ensures a merging of solute. Drop–drop coalescence, even when it does not contribute to precipitation, increases the solute contents of the residual aerosol particles following evaporation. Both the trace gases and the aerosol are changed following microchemical processing by clouds.

The atmospheric aerosol, especially when modified by prior cloud microchemistry, influences cloud microstructure, hence atmospheric radiation budgets, precipitation efficiencies, and the effectiveness of clouds to cleanse the air. Of course, aerosol particles outside of clouds scatter and absorb radiation (the so-called direct aerosol effect on radiation), but influences of aerosol particles on radiation budgets are perhaps even more pronounced when the aerosol interacts with clouds (leading to a couple of indirect effects). The first indirect, or Twomey, effect is the increase in reflectance of solar radiation due to an increase in droplet concentration, which in turn is due to an increase in the number concentration of CCN. The increase in reflectance is especially important when pollution impacts low, relatively clean clouds. The second indirect effect of aerosol particles on the planetary radiation balance is due to increases in cloud lifetimes and cover because of higher aerosol concentrations. Higher concentrations of CCN lead to higher concentrations of cloud droplets, but those droplets are then also smaller and less likely to be removed by precipitation. The solutes contained in those droplets, once they evaporate, are similarly left in the air and able to influence subsequent cycles of cloud formation. The increase in planetary albedo arising from higher aerosol concentrations interacting with low clouds has a cooling effect and so may compensate to some extent for a warming of the climate by IR-active gases. High concentrations of aerosol particles may also decrease the efficiency of precipitation, but the linkages are not fully established. Aerosol–cloud interactions are complicated and dependent on still inadequately understood scavenging mechanisms.

13.2 Scavenging of aerosol particles

Scavenging in the atmospheric sciences refers to the removal of trace substances from the air by cloud and precipitation particles. The removal may take place inside the cloud, in which case the process is called in-cloud scavenging or rainout. Below-cloud scavenging, sometimes referred to as washout, occurs when the trace substances interact with precipitation during its fall to the ground. Although the trace substances may be either gaseous or particulate in nature, we focus here on the scavenging of aerosol particles. The uptake of trace gases is discussed in the next section.

Classification of aerosol scavenging reflects the physical processes responsible for the removal of the particles from the air. A fundamental distinction is therefore made between nucleation scavenging and impaction scavenging. As suggested by Fig. 13.3, nucleation scavenging is intimately tied to the activation of cloud-active aerosol particles (CCN), which occurs just above cloud base during the initial formation of cloud droplets. Impaction scavenging, by contrast, refers to contact of the interstitial (non-CCN) particles with cloud drops or ice particles. Contact at the surfaces of the colliding particles is usually deemed sufficient to ensure capture. Impaction scavenging removes aerosol particles either inside the cloud or below cloud base.

13.2.1 Nucleation scavenging

Incorporation of particulate matter into cloud water during aerosol activation is the dominant means by which trace substances are removed from the air by a cloud (heavy arrow pointing upward in Fig. 13.3). Any substances, soluble or not, contained in the CCN automatically enter the cloud droplets during activation. Condensation adds only water to the particles, so the compounds in the CCN become progressively more diluted as the droplets grow by condensation. Throughout the activation process, the chemical contents of the individual particles remain fixed.

The effectiveness of nucleation scavenging can be appreciated by considering how the various processes are linked. As shown in Fig. 13.4a, supersaturation develops near the base of a cloud in ascending air, the adiabatic expansion causing the air to cool and the equilibrium vapor pressure of water to decrease (review Chapter 6 as needed). The small supersaturations early in the uplift initially activate only the largest particles (Fig. 13.4b, from Köhler theory, Chapter 3), those with the smallest critical supersaturations. But, with continued uplift, the supersaturation rises and activates progressively smaller and smaller particles as their higher critical supersaturations are gradually exceeded. Once the supersaturation has reached its maximum value (s_{max}, 0.5% in this example), particle activation ceases. The limiting diameter of the dry aerosol capable of being activated is the activation diameter D_{act}. Note that D_{act} is an inverse measure of CCN activity for a population of particles: higher activated fractions of the total aerosol are associated with smaller activation sizes. The part of the aerosol incorporated into the cloud water is shown in Fig. 13.4c by shading. The number (light shading) of particles activated may be small (often less than 50% of the total) when nucleation-mode particles are present, but the aerosol volume (dark shading) typically exceeds 90% of the total because most of the aerosol mass is contained in the largest particles.

Further insight into the interaction of an aerosol with a new cloud can be gained through analytical, albeit approximate, means. The method ascribed to Twomey assumes that the cloud-active part of the atmospheric aerosol can be represented by a power law, which may be valid over a suitable range of the aerosol spectrum. The atmospheric aerosol is

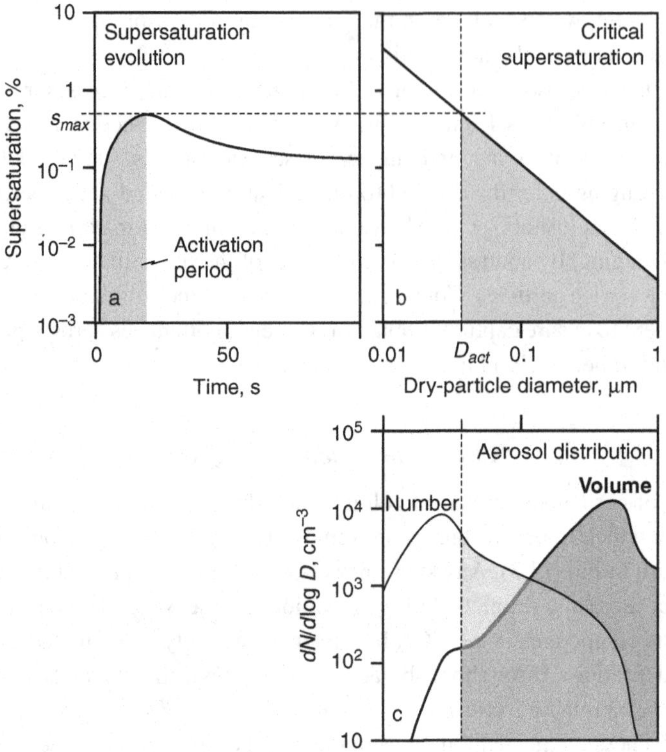

Figure 13.4 Relationships between the evolution of supersaturation in an adiabatic parcel (a), the critical supersaturation (b), and the activated portions of the aerosol (c).

typically characterized in size by three spectral modes: (1) the nucleation mode, (2) the accumulation mode, and (3) the coarse mode, as Fig. 13.5 illustrates. Thus, the total aerosol concentration (solid curve) is simply the sum of the concentrations of the individual modes (dashed curves, each assumed here to be described by a log-normal function). Recall that the critical supersaturations (s_c) of the particles are inversely related to their sizes (see Köhler theory in Chapter 3), and note that the axes in Fig. 13.5 are logarithmic. Over a narrow range of sizes (and solute contents), the total concentration can be approximated by the power-law relationship suggested by the heavy straight line on the log–log plot. The differential distribution function, $n_{CCN}(s_c)$, implied by this line yields the more commonly used cumulative distribution function, which has the power-law form

$$N_{CCN}(s_c) \equiv \int_0^{s_c} n_{CCN}(s_c)ds \simeq cs_c^k, \tag{13.1}$$

where c and k are parameters characteristic of the air mass in which the cloud forms. Clean air masses are characterized by small values of c, dirty air masses by large values. The value of k varies slightly with air mass quality (from about 0.3 in aged, clean air to 0.4 in recently polluted air), but it is mainly the value of c that links the aerosol population to the initial

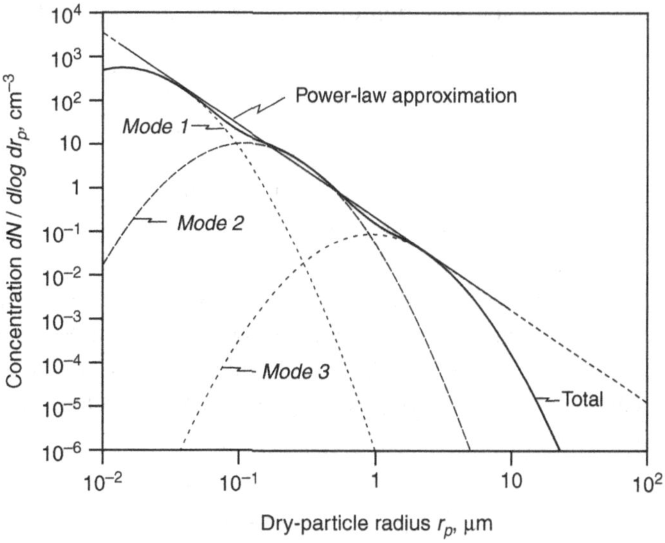

Figure 13.5 Differential size distributribution of the atmospheric aerosol, assumed to be composed of three distinct log-normal modes (dashed curves). Mode 1 is the nucleation mode, Mode 2 is the accumulation mode, and Mode 3 is the coarse mode. Solid curve: Total concentration (sum of all modes). Straight line: Power-law approximation. Adapted from Zahn (1993).

cloud microstructure and scavenging effectiveness. Supersaturations are usually expressed as a percent, so the value of c represents the cumulative concentration [usually as number per cm^{-3}] of aerosol particles that would be active up to a supersaturation of 1%.

Given a power-law representation of the aerosol (Eq. (13.1)), Twomey could develop a quantitative aerosol–drop relationship. Activation of aerosol particles is tantamount to the formation of cloud droplets, but the relationship is complicated by the competition the activated particles have for the excess vapor generated in ascending air. The larger the concentration of particles, the greater the competition and the less rapidly supersaturation can develop during the activation period. The number of cloud droplets formed is determined by the maximum supersaturation achieved, which will be small when the aerosol concentration is large. As shown in Chapter 6, the supersaturation develops at the rate $ds/dt = Q_1 w - Q_2\,(d\omega_L/dt)$, where w is the ascent rate. After taking the growth rates of individual droplets into account ($(dr_d/dt = Gs/r_d)$, see Chapter 8), we find the total liquid water concentration ω_L to increase at the rate

$$\frac{d\omega_L}{dt} = 2\pi\rho_L\,(2G)^{3/2}\,sI(s), \tag{13.2}$$

where ρ_L is the density of liquid water and $I(s) = \int_0^s \left[\int_{t_{act}(s_c)}^{t} s(t')dt'\right]^{1/2} n_{CCN}(s_c)\,ds_c$ is an integrated measure of water vapor taken up up by the time the supersaturation is s. When the supersaturation spectrum of the aerosol is represented by a power law, $n_{CCN}(s_c) = cks_c^{k-1}$, we find

$$I(s) = \frac{kc}{2(2Q_1 w)^{1/2}} s^{k+1} B\left(\frac{k}{2}, \frac{3}{2}\right), \tag{13.3}$$

where $B(a,b) \equiv \int_0^1 y^{a-1}(1-y)^{b-1}\,dy$ is the beta function. The beta function can be evaluated by noting its relationship to the gamma function: $B(a,b) \equiv \Gamma(a)\,\Gamma(b)/\Gamma(a+b)$. The beta function in Eq. (13.3) varies only about 30% for the range of k values normally found in the atmosphere (0.3 to 0.4). Most of the variation in $I(s)$ comes from the variable quality of the air mass (through c) and from the magnitude of the updraft speed (through w). Once Eq. (13.3) is substituted into Eq. (13.2), we can obtain the differential equation that describes the evolution of supersaturation with a power-law CCN distribution:

$$\frac{ds}{dt} = Q_1 w - 2\pi Q_2 \rho_L (2G)^{3/2} \frac{ckB\left(\frac{k}{2}, \frac{3}{2}\right)}{(Q_1 w)^{1/2}} s^{k+2}. \tag{13.4}$$

The maximum supersaturation is determined by setting $ds/dt = 0$ and solving for $s_{\max}$:

$$s_{\max} = c^{\frac{-1}{k+2}} \left[\frac{(Q_1 w)^{3/2}}{2\pi Q_2 \rho_L G^{3/2} kB\left(\frac{k}{2}, \frac{3}{2}\right)}\right]^{\frac{1}{k+2}}. \tag{13.5}$$

The prefactor shows that the maximum supersaturation decreases as the CCN concentration increases. Finally, the maximum concentration of droplets that can form up to this maximum supersaturation is given by the CCN parameterization itself (Eq. (13.1), or equivalently $N_{d,\max} \equiv N_d(s_{\max}) = cs_{\max}^k$):

$$N_{d,\max} = c^{\frac{2}{k+2}} \left[\frac{(Q_1 w)^{3/2}}{2\pi Q_2 \rho_L G^{3/2} kB\left(\frac{k}{2}, \frac{3}{2}\right)}\right]^{\frac{k}{k+2}}. \tag{13.6}$$

Therefore, for given values of w and k, the concentration of droplets in a parcel rising adiabatically increases. The concentration of droplets, for a given aerosol, increases weakly with the updraft speed: $N_{d,\max} \propto w^{\frac{k}{2k+4}} \propto w^{0.22}$ and, for a given updraft, not quite linearly with the concentration of aerosol particles: $N_{d,\max} \propto c^{\frac{2}{k+2}} \sim c^{0.85}$. The fraction of the total aerosol that activates and contributes particulate chemicals to the cloud water may be estimated simply if one assumes the aerosol concentration and shape of the distribution are independent. Thus, for a given value of k, the activated fraction is $F = N_{d,\max}/N_p \simeq c^{\frac{2}{k+2}}/c = c^{\frac{-k}{k+2}} \simeq c^{-0.15}$. The greater the aerosol concentration entering a given cloud, the smaller is the fraction of particles activated. Thus, clouds in clean (e.g., maritime) air masses efficiently remove what little particulate matter is present and keep the air clean. Clouds in polluted regions often have greater difficulty cleansing the atmosphere.

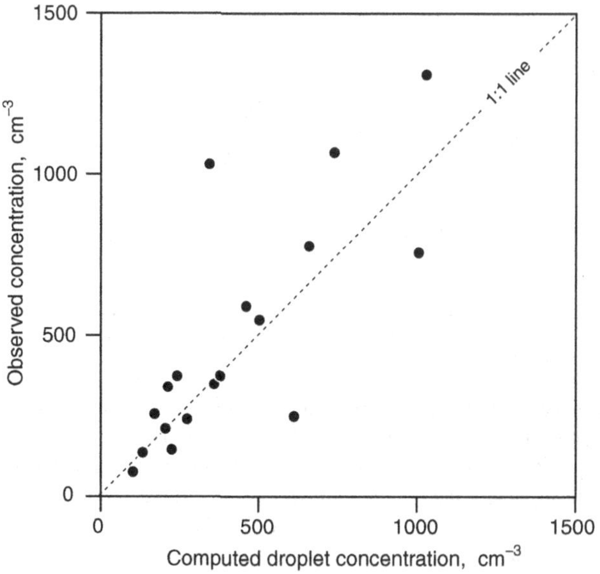

Figure 13.6 Comparison of droplet concentrations observed and computed. The observations were taken in warm, non-precipitating cumuli within 300 m of cloud base; updraft speeds ranged from 2 to 6 m s^{-1}. The computed concentrations were based on measured CCN spectra taken in the air below cloud bases; calculations used an updraft speed of 3 m s^{-1}. Adapted from Twomey and Warner (1967).

The essential correctness of Twomey's theory is demonstrated by comparing the concentrations of droplets observed near cloud base and those calculated by Eq. (13.6). As seen in Fig. 13.6, the agreement is quite good considering the scatter in the data and the uncertainties inherent with measuring the updrafts and CCN spectra. Analytical approaches help us visualize physical linkages involved in aerosol activation and nucleation scavenging, but they are limited by the numerous approximations that must be made. Measured aerosol spectra must often be described by functions more complicated than power laws, and the composition and properties of the aerosol are not properly accounted for in the Twomey model.

Precise quantification of nucleation scavenging requires detailed cloud models. Such models show that the activated fraction depends non-linearly on the aerosol abundance. A simple example serves to illustrate the main points: The adiabatic ascent of a parcel of moist, aerosol-laden air is simulated through numerical integration of equations that account for the development of supersaturation (see Chapter 6) and the growth of the haze and cloud droplets. As shown in Fig. 13.7, the concentration of activated particles increases as the total aerosol concentration increases, but the fraction of the total decreases. Relatively few particles are activated at high aerosol loadings mainly because the maximum supersaturation is kept low by the condensation of vapor onto the many particles. In the case here of a purely soluble and hygroscopic aerosol, significant amounts of vapor are

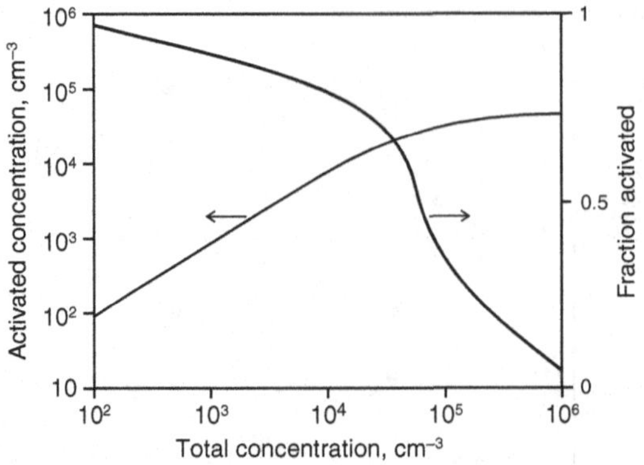

Figure 13.7 Dependences of activated aerosol concentration (light curve, left axis) and number fraction (heavy curve, right axis) on the total aerosol concentration, as calculated from an adiabatic parcel model. The updraft was fixed at 10 m s^{-1}, and parcel rise started with a relative humidity of 75% at the 900 hPa-level, where the temperature was 15 °C. The cloud base was at 1.5 km, where the pressure was 850 hPa and the temperature was 10 °C. The aerosol was soluble and distributed log-normally into 30 critical supersaturation classes. The modal critical supersaturation was 0.5%, and the geometrical standard deviation of the distribution was 3. Droplet growth was calculated by a kinetic-diffusion model using unity condensation coefficient.

removed by the haze particles below cloud base, which gives rise to the sharp decrease in the activated fraction beyond an aerosol concentration of about 10^4 cm^{-3}. Except in such extreme cases, nucleation scavenging efficiently transfers particulate matter from the aerosol to the cloud.

Understanding the dependence of nucleation scavenging efficiency on chemical composition is complicated by the diversity of compounds and particle sizes found in the atmosphere. The largest particles, those greater than a couple of micrometers in diameter, often enter the atmosphere by mechanical means (e.g., resuspension of dust by strong surface winds). Such particles tend to be insoluble, but they may nevertheless be scavenged efficiently once they have acquired a coating of some soluble material, such as ammonium sulfate. Combustion by-products represent another primary source of particulate matter, but the particles are smaller and contribute mostly to the accumulation mode. Many other accumulation-mode particles are derived from secondary sources, mainly through gas-phase reactions. The first aerosol particles formed by gas-to-particle conversion appear in the nucleation mode and so may escape being scavenged by droplet nucleation. However, as such particles grow by subsequent chemical deposition or by coagulation, they enter the accumulation mode and then become vulnerable to nucleation scavenging. Inorganic compounds containing sulfates and nitrates usually reside in the accumulation mode and are very soluble, so they are readily scavenged during activation.

Organic compounds impose special challenges toward our understanding of nucleation scavenging. Many volatile organic compounds (VOC) are released into the atmosphere naturally, others by industrialized societies. Some of these organic vapors undergo photochemical reactions involving ozone (O_3) or reactive radicals (e.g., OH, HO_2, NO_3) and yield products that often have low volatility. These products either form new particles in the nucleation mode or condense onto preexisting particles in the accumulation mode. Such gas-to-particle reactions contribute to the secondary organic aerosol (SOA), which can constitute anywhere from 10 to 90% of the dry fine-particle mass. One problem with learning how SOA interacts with clouds stems from the fact that the particulate organic matter is composed of hundreds of separate compounds, many of which have yet to be identified. Each class of compounds has its own aqueous-phase solubility, which in turn depends on the mix of chemicals residing in the particle. Some organic compounds also change the surface tension of the haze droplets, making the assumptions underlying Köhler theory questionable and requiring empirical methods to measure nucleation effectiveness. Other compounds may coat the particles with a thin film that hinders the rate of water uptake during growth.

The ability of the SOA to serve as CCN is usually determined empirically. Interpretations of the data in terms of Köhler theory allow comparisons with better understood inorganic compounds. Once an aerosol of known composition is prepared in the laboratory, it is size selected and passed into a CCN counter (a device that exposes the particles to a known supersaturation and counts the particles that become droplets). The fraction of the particles that activate and form cloud droplets is then plotted as a function of the dry-particle diameter for each setting of the CCN counter, as shown in Fig. 13.8 for norpinic acid (a compound formed during the oxidation of alpha-pinene, a VOC released by trees). At a given supersaturation, the smallest particles yield only a small number of droplets, while larger droplets of the same compound yield a larger number of droplets. The diameter at which the sigmoidal curve used to fit the data crosses the 50th percentile line is the activation diameter for that compound at that applied supersaturation. Repetition of the experiments at other settings of the CCN counter yield other sigmoidal curves and other activation diameters. Again, we see that higher supersaturations cause smaller particles to activate. A larger fraction of any given aerosol is incorporated into cloud droplets during activation at larger supersaturations.

When compared with expectations from Köhler theory, many organic compounds are found to deviate significantly from those of comparably sized inorganic salts. The unique properties (mainly solubility and surface tension) of each compound are accounted for by suitable selection of the parameters used in the theory (see Chapter 3). The theoretically expected activation size is then plotted against the measured size, as shown in Fig. 13.9 for single-component aerosols exposed to a supersaturation of 0.3%. The solid points to the right of the 1:1 line show that low-solubility species are expected by theory to have large activation diameters, whereas they are measured to have low activation diameters. These compounds serve as better CCN than one should expect for compounds that are only weakly soluble in water. On the other hand, these same compounds are readily wetted by water, a property that lets them activate readily despite their lack of solubility. Better agreement between measurements and theory is achieved by assuming that wettable compounds act as if they were completely soluble, as shown by the arrows in Fig. 13.9. Such

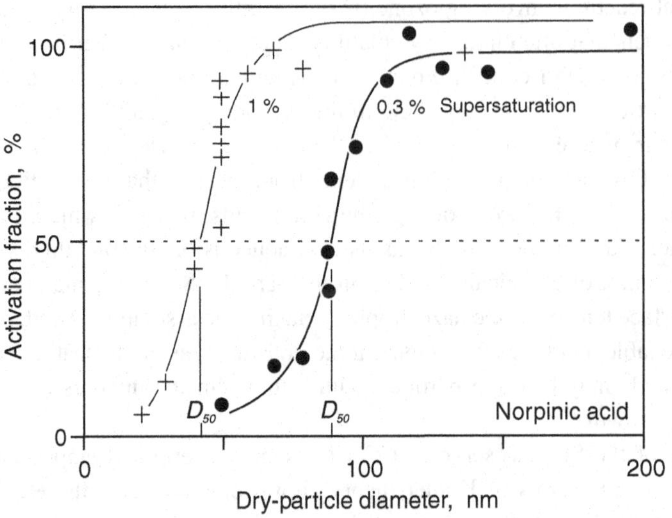

Figure 13.8 Activity of norpinic acid particles at two supersaturations. Adapted from Raymond and Pandis (2002).

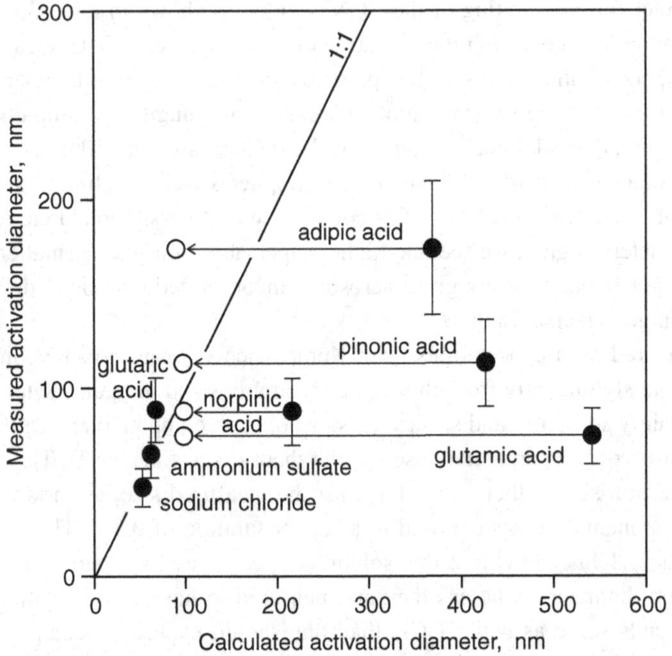

Figure 13.9 Comparison of measured and calculated activation diameters of selected organic and inorganic compounds. Arrows indicate the adjustment made when wettable compounds are assumed to be completely soluble. Based on data from Raymond and Pandis (2002).

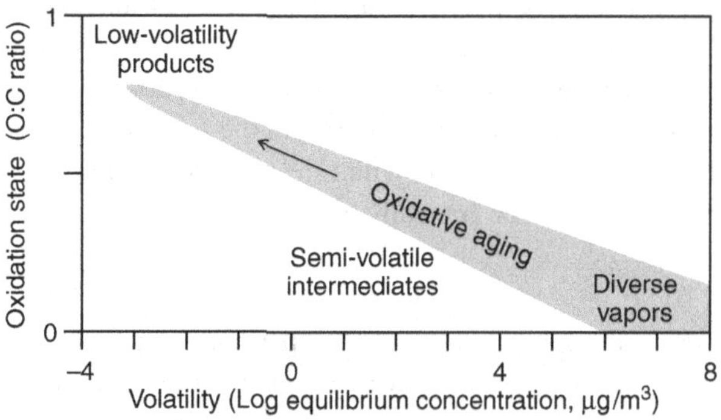

Figure 13.10 Effect of aging on the oxidation state and volatility of organic compounds. Aging includes oxidation of semi-volatile compounds that leads to a convergence of properties with time. Adapted from Jimenez *et al.* (2009).

findings are consistent with particles containing an insoluble (e.g., dust) core and coated with a soluble salt. In both cases, the dry particle provides a starting size to overcome, at least partially, the curvature effect, while the wettable surface lets the particle grow readily in supersaturated air.

The CCN activity is complicated when many organic compounds are internally mixed, that is, blended within the same particles. Each chemical compound has its own physical properties, so the activation of a particle containing a complex and uncertain mixture of organic and inorganic compounds is almost impossible to predict. Fortunately, some studies find that CCN activity depends primarily on the fraction of the aerosol that is water soluble, not the exact composition. Many organic aerosol particles contain semi-volatile compounds that exchange rapidly between the particulate and gaseous phases. Such compounds become exposed to strong oxidants (e.g., O_3 and OH) while in the gas phase and form new products. Oxidation usually leads to polar groups on organic molecules, which provides opportunities for attachment of water molecules. The products formed from oxidation reactions are therefore less volatile in aqueous solutions. As shown in Fig. 13.10, the aging of an aerosol in an oxidizing medium (air) gradually lowers the volatility as the oxidation state increases. With time, the properties of the organic components in the aerosol particles converge to a common, sulfate-like set of properties, and the origins of the individual components gradually become irrelevant. Problems still remain with knowing the chemical age of an aerosol, but findings that enable scientists to group properties help them predict the ability of clouds to scavenge the particulate matter during activation.

13.2.2 Impaction scavenging

That subset of the atmospheric aerosol that did not contribute directly to the formation of cloud droplets resides interstitially between the cloud particles until removed by one of several impaction scavenging mechanisms. All but one of the scavenging mechanisms require the aerosol particle to deviate from the streamlines of air flowing around a sedimenting cloud particle, the scavenging hydrometeor. The hydrometeor is typically a liquid drop in the warm part of a cloud or an ice particle in the cold part. Regardless of the specific capture mechanism, the chemical matter contained in the scavenged aerosol particles is of course added to that of the hydrometeor. With time, the cloud particles become enriched chemically, while the interstitial air gradually becomes cleaner.

Interception

Interception is the one scavenging mechanism that permits capture even when the trajectories of the aerosol particles align perfectly with the air streamlines around the hydrometeor. Interception follows from the geometrical argument that contact between an aerosol particle and the scavenging agent occurs at their surfaces, not at the centers of mass. It is the finite size of the aerosol particle that permits contact with and retention by the scavenging agent when the particle follows streamlines. Thus, the efficiency of scavenging scales with the size ratio r_p/r_d, where r_p is the radius of the aerosol particle and r_d is the radius of the drop. Interception is most important when the sizes of the aerosol and cloud particles are comparable, but then inertial scavenging may dominate. A schematic view of interception in relationship to other impaction-scavenging mechanisms is given in Fig. 13.11. The principles are illustrated adequately for the situation of a spherical water drop scavenging a spherical aerosol particle, but realize that the physical geometries may be more complicated.

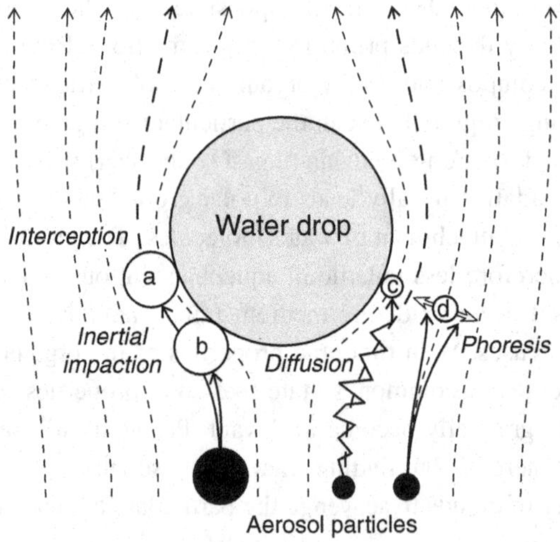

Figure 13.11 Schematic of impaction scavenging mechanisms. Gray circle represents a water drop acting as the scavenging agent. Dark circles identify aerosol particles about to be scavenged. White circles show particles near point of scavenging. a, Interception: particle follows streamline. b, Inertial impaction: particle crosses streamline due to inertia. c, Diffusion: particle deviates from streamline due to Brownian motion. d, Phoresis: particle drifts either toward or away from particle depending on gradients of temperature and vapor concentration.

Inertial collection

Inertial collection arises when an aerosol particle crosses the streamlines because of its inertia. Particle inertia increases in proportion to mass, which varies as the cube of the radius, so inertial collection increases rapidly with size and relative impact velocity. Analogous to the problem of growth of small raindrops by continuous collection (Section 11.2), a single water drop of radius r_d collects aerosol particles at a rate proportional to the effective volume $K_I\left(r_d, r_p\right)$ swept out by the falling drop and to the number concentration $n_p\left(r_p\right)$ of aerosol particles inside that volume having radii in the interval r_p to $r_p + dr_p$:

$$-\frac{dN_p}{dt\,dr_p} = K_I(r_d, r_p) n_p\left(r_p\right), \tag{13.7}$$

where N_p is the number of particles scavenged by the drop and equal to the number removed from the air. Note that the rate must be specified as number scavenged per unit time and per unit size interval of the distribution function. The kernel K_I for inertial collection is usually expressed in terms of a collection efficiency E_c, defined as the ratio of the actual collision cross section A_c to the geometrical cross section $A_{geom} = \pi\left(r_d + r_p\right)^2$. The geometrical sweepout volume $\dot{V}_{geom} = A_{geom}\Delta v = \pi\left(r_d + r_p\right)^2 \Delta v$, and so

$$K_I = E_c \dot{V}_{geom} = E_c(r_d, r_p)\pi\left(r_d + r_p\right)^2 \Delta v, \tag{13.8}$$

where Δv is the relative speed.

Inertial collection is often controlled by gravitational sedimentation, in which case the relative speed is simply the difference in fallspeeds, $\Delta v = v_d - v_p$. When the aerosol particles are small compared with the collecting drop, $A_{geom} \simeq \pi r_d^2$. Then, the fallspeeds of the aerosol particles are negligible, so $\Delta v \simeq v_d$, and the collection kernel becomes $K_I(r_d, r_p) = E_c(r_d, r_p)\pi r_d^2 v_d$. A single drop (of radius r_d) thus scavenges interstitial aerosol particles of all sizes at the rate

$$-\left.\frac{dN_p}{dt}\right|_{\text{drop}} = \int_o^{\infty} K_I(r_d, r_p) n_p\left(r_p\right) dr_p. \tag{13.9}$$

All of the drops in a unit volume, the population being characterized by a distribution function $n_d\left(r_d\right)$, collectively deplete the aerosol at the rate

$$-\frac{dN_p}{dt} = \int_0^{\infty} n_d\left(r_d\right)\left[\int_o^{\infty} K_I(r_d, r_p) n_p\left(r_p\right) dr_p\right] dr_d. \tag{13.10}$$

If we now normalize this equation by the total concentration of aerosol particles, $N_p = \int_0^{\infty} n_p\left(r_p\right) dr_p$, we get the fractional rate of change in the aerosol concentration:

$$-\frac{1}{N_p}\frac{dN_p}{dt} = \int_0^{\infty} n_d\left(r_d\right)\left[\int_o^{\infty} K_I(r_d, r_p) f_p\left(r_p\right) dr_p\right] dr_d, \tag{13.11}$$

where $f_p\left(r_p\right) = n_p\left(r_p\right)/N_p$ is the probability distribution function of the aerosol. The fractional rate of change is sometimes called the scavenging coefficient, $\Lambda \equiv -\frac{1}{N_p}\frac{dN_p}{dt}$, because it characterizes the time dependence of the aerosol population. A characteristic time τ for depletion is the inverse of the scavenging rate: $\tau = 1/\Lambda$. If the drop population changes little over this time, the aerosol concentration decays exponentially with time:

$$N_p(t) = N_{p,0} \exp(-\Lambda t)\,. \tag{13.12}$$

The depletion time scale, the time for the population to be reduced by the factor $1/e$, depends strongly on the collection efficiency E_c, so the assumption of a stable drop population should be scrutinized in applications.

Brownian diffusion

Small aerosol particles, those with radii less than about 0.1 μm, have too little mass to be caught effectively by inertial forces, but they do nevertheless deviate from streamlines due to Brownian diffusion. As suggested by Fig. 13.11 (mechanism c), a small particle wanders from its average trajectory randomly in response to impacts by individual air molecules. Particles suspended in a gaseous medium diffuse much the same way that specific gases diffuse, except that particles diffuse more slowly because of their larger masses. Each time an air molecule bumps a particle, momentum is transferred to the particle, and the particle is forced to move suddenly in a new direction. The molecular impacts affect small particles more than they do large particles because of the differences in mass; small aerosol particles diffuse more rapidly than do large particles.

The dependence of particle diffusivity on particle size can be calculated with the help of thermodynamics. As with gases, particles diffuse in accord with Fick's first law. The magnitude of the flux Φ_p of particles is related to the concentration gradient ∇n_p through

$$\Phi_p = -D\nabla n_p, \tag{13.13}$$

where D is the coefficient of diffusion, or simply the diffusivity. The gradient of concentration (n_p) in effect drives particles down gradient with a mean drift speed $v \equiv \Phi_p/n_p$. The diffusivity can be expressed in terms of this drift speed by solving Eq. (13.13) for D and making the substitution $\Phi_p = v n_p$:

$$D = -\frac{v n_p}{\nabla n_p} = \frac{v}{-\frac{1}{n_p}\nabla n_p}. \tag{13.14}$$

The denominator in Eq. (13.14) is related to the thermodynamic driving force F_{TD} through its definition in terms of chemical potential μ:

$$F_{TD} \equiv -\nabla\mu = -\nabla(\mu_0 + k_B T \ln n) = -k_B T \frac{1}{n_p}\nabla n_p. \tag{13.15}$$

Combining Eqs. (13.14) and (13.15) yields

$$D = k_B T \left(\frac{v}{F_{TD}}\right). \tag{13.16}$$

The quantity in parentheses is the particle mobility $B \equiv v/F_{TD}$, a measure of how rapidly the particle moves under the influence of a given external force. We thus gain the famous Einstein relationship between diffusivity and mobility:

$$D = k_B T B. \tag{13.17}$$

Both D and B provide measures of how readily particles respond to external driving forces. The dependence of B (hence of D) on particle radius r_p can be determined by letting the force in the definition of B be the generalized Stokes drag force F_D in air. Thus, setting

$$F_{TD} = F_D = \frac{F_{D,Stokes}}{C_c} = \frac{6\pi \mu_{air} r_p v}{1 + (1.26)\frac{\lambda_{air}}{r_p}}, \tag{13.18}$$

where v is the terminal fallspeed of the particle and λ_{air} is the molecular mean free path in air, we find the particle mobility to be

$$B = \frac{1 + (1.26)\frac{\lambda_{air}}{r_p}}{6\pi \mu_{air} r_p}. \tag{13.19}$$

The particle diffusivity is therefore, from Eq. (13.17),

$$D = \frac{k_B T}{6\pi \mu_{air} r_p}\left(1 + (1.26)\frac{\lambda_{air}}{r_p}\right). \tag{13.20}$$

This equation is shown graphically in Fig. 13.12. Note how the dependence differs on either side of the mean free path, assumed here to be 0.1 μm. When the particles are much smaller than the mean free path ($r_p << \lambda_{air}$), their diffusivities vary inversely as the square of their radii: $D \sim 1/r_p^2$. At the other extreme, when $r_p >> \lambda_{air}$, a simple inverse

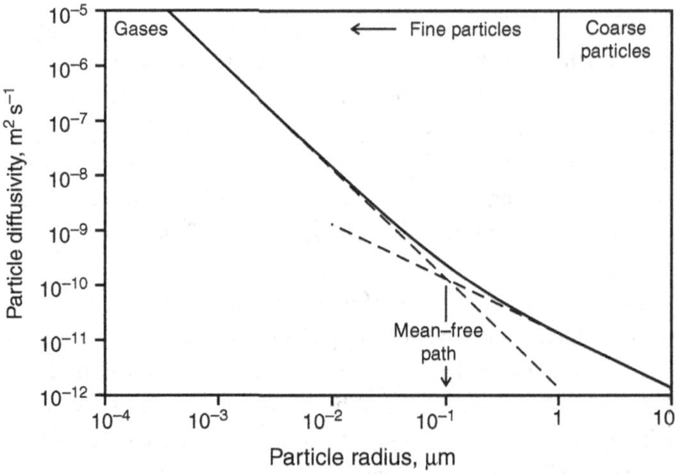

Figure 13.12 Dependence of Brownian diffusivity on particle size. Adapted from Slinn (1984).

proportionality applies: $D \sim 1/r_p$. Small aerosol particles are highly mobile and can be easily scavenged by the much larger cloud particles.

The rate of scavenging by Brownian diffusion is determined in much the same way that condensation is on a sedimenting drop. As a first approximation, think of the aerosol particles surrounding a drop as being analogous to the excess vapor surrounding a growing drop. The only real difference is the magnitude of the diffusion coefficient and the assumption that all particles contacting the surface of the drop are retained. We account for the simultaneous motion of the air around the drop and the diffusion of particles relative to that mean motion by assuming that the mean and random motions can be superimposed. The particle flux is therefore

$$\mathbf{\Phi_p} = n_p\mathbf{v} - D\nabla n_p, \tag{13.21}$$

where $\mathbf{v}$ is the local velocity of the air and the particle concentration is n_p. When no particles are produced or destroyed in the local volume, the usual continuity relationship applies, $\partial n_p/\partial t = -\nabla \cdot \mathbf{\Phi_p}$. The time-dependent convective-diffusion equation in an incompressible fluid ($\nabla \cdot \mathbf{v} = 0$) thus becomes

$$\frac{\partial n_p}{\partial t} = \mathbf{v}\cdot\nabla n_p - D\nabla^2 n_p. \tag{13.22}$$

Solution of this equation in spherical coordinates (for a spherical hydrometeor), subject to the boundary conditions, $n_p = 0$ at the surface and $n_p = n_{p\infty}$ far away, yields the spatial variation of particles around the drop. The rate that particles are scavenged by a drop of radius r_d is then calculated by integrating the particle flux over the drop surface (**A**):

$$\frac{dN_p}{dt} = f_p \oint_{sfc} \mathbf{\Phi}_{p,sfc} \cdot d\mathbf{A} \equiv K_B n_{p\infty}, \tag{13.23}$$

where f_p is a ventilation coefficient that effectively accounts for the convective enhancement due to the mean motion of the air around the drop and $K_B = 4\pi r_d D f_p$ is the collection kernel for Brownian diffusion to a drop of radius r_d. If one compares the Brownian collection kernel with the usual kernel for inertial collection (Eq. (13.8)), one can define the Brownian collection efficiency, $E_B \equiv K_B/\dot{V}_{geom} \simeq 4\pi r_d D f_p/\pi r_d^2 v_d = 4Df_p/(r_d v_d)$. The Brownian collection efficiency varies strongly with size through $D = D(r_p)$, as shown in Eq. (13.20). Small particles efficiently deviate from the air streamlines and are able to contact the drop surface. However, small particles carry relatively little mass, so their uptake by the hydrometeor contributes little to the chemical load of the drop.

Overall effects

The efficiency with which interstitial aerosol particles are scavenged depends strongly on the mechanism of scavenging and is generally a complicated function of the sizes of the hydrometeors (drops) and of the particles. When the effects of Brownian and inertial scavenging are combined, one obtains the filter function shown by the solid curve in Fig. 13.13. As the analyses above indicate, the scavenging efficiency is large both for very small particles (because of their high Brownian mobility) and for large particles (because of their

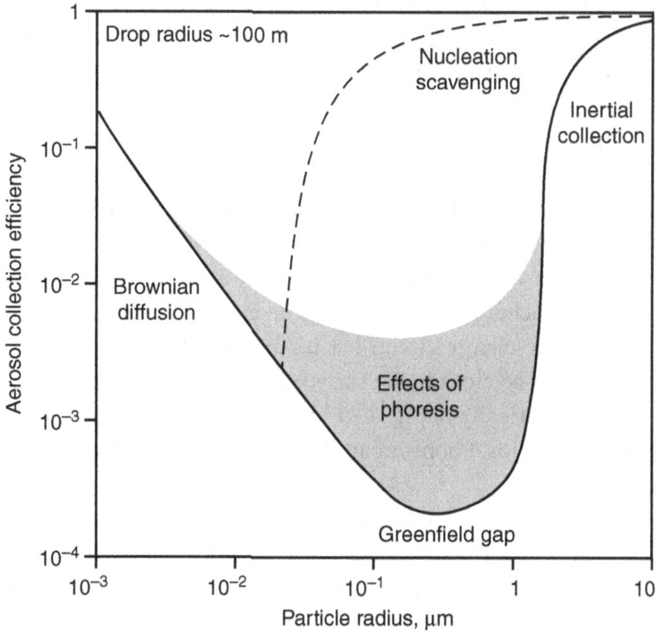

Figure 13.13 Combined effects of various scavenging mechanisms on the efficiency of aerosol scavenging by drizzle drops. Adapted from Wang et al. (1978).

inertia). Fine particles, those with radii near 0.1 μm, are only weakly scavenged by these two mechanisms, so a minimum exists in the filter function. The size range near the minimum in scavenging efficiency is called the Greenfield gap, named for a pioneer in the theory of precipitation scavenging. A broad minimum in collection efficiency at intermediate sizes is characteristic of all filter functions, in both nature and industry. It is no coincidence that the accumulation mode of the atmospheric aerosol prevails in the Greenfield gap.

Other scavenging mechanisms exist in the atmosphere that help fill the Greenfield gap, at least partially. Soluble or wettable particles in the accumulation mode have already been identified as good CCN, so nucleation scavenging (dashed curve in Fig. 13.13) is most effective for removing aerosol particles larger than the activation size. Lacking opportunities for activation, particles in the Greenfield gap may also be scavenged by phoresis, a class of transport mechanisms that depend on gradients in fluid properties. As suggested by path (d) in Fig. 13.11, phoretic forces can cause systematic drifts away from the air streamlines. A phoretic force can operate in either sense, either toward or away from the hydrometeor, depending on the specific gradient in question. A gradient in temperature leads to thermophoresis, which pushes the particle down gradient toward lower temperature. A growing drop is warmed by the latent heat of condensation, so scavenging by thermophoresis is inhibited inside of cloud, but it is enhanced outside of cloud during evaporation. A gradient in vapor concentration leads to diffusiophoresis and a tendency for particles to move in the direction of lower partial pressure (i.e., with the flow of vapor). The lower partial pressure lowers the total pressure ever so slightly, causing a convective flow, known as Stefan flow. Thus, scavenging is enhanced by diffusiophoresis during condensation, inhibited during evaporation. Diffusiophoresis and thermophoresis operate in opposing senses, but the large latent heat of vaporization for water causes thermophoresis to dominate diffusiophoresis under most circumstances. A gradient in electric potential causes a positively charged particle to move down the potential gradient, the phenomenon known as electrophoresis. Electrical forces are attractive when a charged aerosol particle approaches a hydrometeor of opposite charge, but they are repulsive when the charges are of the same sign. Even uncharged particles may be electrically attracted to an uncharged hydrometeor by the image-charge effect, but the hydrometeor must be suitably polarizable, and the particle must be close to the surface for this mechanism to work. The effects of phoresis are shown generically in Fig. 13.13 by shading, as exact efficiencies depend on the details of specific scenarios. Phoresis can augment scavenging, but it cannot completely fill the Greenfield gap.

The interactions of aerosol particles with clouds are complicated and in need of further research. The scavenging of aerosol particles by cloud drops and ice crystals is the first step in their removal from the atmosphere. At the same time, the scavenged aerosol particles alter the microphysical properties of the cloud, and they themselves become modified in the process. Large aerosol abundances lead to numerous, smaller drops that coalesce less readily. Cloud water is then retained longer, leading to lowered precipitation efficiencies. Coalescence, when it does occur, combines the chemical contents of the colliding partners, enhancing the sizes of particles released back into the atmosphere during evaporation along cloud edges. Aerosol-impacted clouds appear brighter and so return more solar radiation back into space and alter the radiation balance of the Earth. Predicting the precise nature and extent of the mutual changes to cloud and aerosol require detailed mathematical models, which in turn depend on parameters obtained by laboratory experimentation and on verification through focused field observations.

13.3 Uptake of trace gases

Trace gases, as do aerosol particles, interact in important ways with clouds. Both atmospheric gases and particles are drawn into clouds as part of the natural air circulations before being exposed to the cloud particles. However, being components of the air itself, gases have no distinguishing inertia, so the only viable capture mechanism is diffusion. Once contact is made with the surface of the hydrometeor, whether it be a liquid water drop or an ice particle, the gas molecules may be sorbed or not. That is, the molecules may adhere to the hydrometeor and add to its solute content, or the molecules may desorb back into the gas phase and leave no net effect on the hydrometeor.

The uptake of a trace gas by a hydrometeor is a multi-step process involving two distinct phases, the gas and the condensate of the hydrometeor, either liquid or solid water. An overview of the various processes involved in trace-gas scavenging is shown in Fig. 13.14

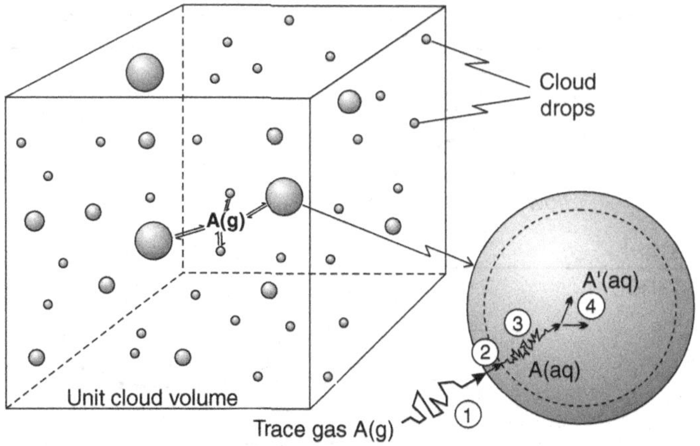

Figure 13.14 Schematic of the interaction of a generic trace gas A with cloud drops. Left: Perspective of a unit volume of cloudy air. Right: Expanded view of one drop with individual steps identified. 1: Gas-phase diffusion; 2: Interfacial exchange; 3: Aqueous-phase transport; 4: Chemical reaction leading to new substance A′.

for liquid hydrometeors. The left-hand view illustrates the multi-phase nature of warm clouds, a collection of disconnected drops suspended in a continuous gaseous medium (air). Most of the volume of a cloud is gaseous, only a tiny fraction ($\sim 1/10^6$) being the aqueous phase (the collective liquid water of the drops). Still, it is this small amount of liquid that accounts for removal of the trace gas and acts as a medium for possible reaction with other chemicals in the cloud water. A generic trace gas, represented by A(g), interacts with all drops in the vicinity, as suggested by the double arrows. Some molecules of A are removed from the interstitial air by sorption into the drops, while others escape the aqueous phase and return to the gas phase. Net uptake of A requires that the gas-phase concentration over the surface be greater than the value needed for solubility equilibrium (review Henry's law in Section 3.5).

13.3.1 Mechanisms

The various steps involved in the uptake of trace gas A by a representative drop are shown on the lower right-hand part of Fig. 13.14. Trace gas A(g) must first get to the drop (Step 1) before it can cross the gas–liquid interface (Step 2) and become a part of the aqueous solution. The aqueous species, A(aq), can then be transported internally within the drop (Step 3) and possibly react with other components to become a new species A′(aq) with different properties (Step 4). Net transport and uptake of a gas require gradients of concentrations, so concentration varies both outside and inside the drop, as illustrated in Fig. 13.15. The concentration profiles shown here have been normalized to the final, equilibrium values in each phase ($[\widehat{A}] = [A(r,t)]/[A]_{eq}$) and presented at three different times (earliest time, t_1, shown by the lower, bold curve). It is appropriate, in the absence of significant reactions, to

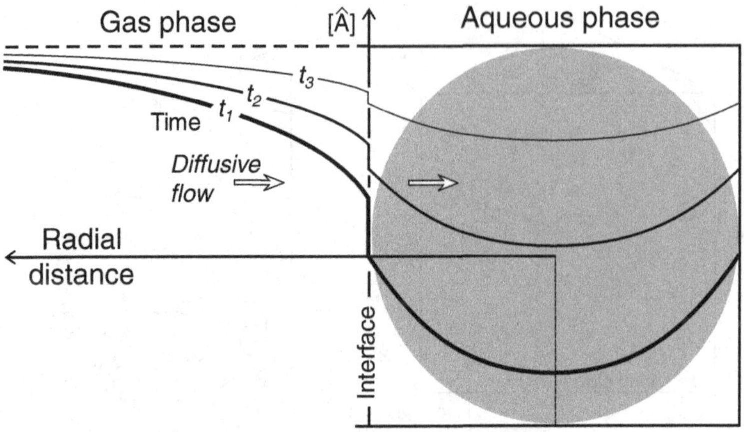

Figure 13.15 Radial variation of normalized concentration ([Â], vertical axis) outside and inside a drop during the scavenging of a trace gas A. Each profile pertains to a specific time $t_1 < t_2 < t_3$. Adapted from Schwartz (1986).

think of the drop as a reservoir of dissolved gas that gradually fills with time as molecules of the trace gas leave the interstitial air and enter the drop. The discontinuities at the interface represent the deviations from solubility (Henry's law) equilibrium at the air–liquid interface. Details of the pertinent processes follow.

External transport

Step 1 is gas-phase transport by local convection and molecular diffusion from the ambient interstitial air to the drop surface. The principles of gas-phase transport to a sedimenting drop are exactly the same as those for aerosol scavenging by Brownian diffusion, except that the diffusivity (D) used in Eq. (13.22) must be appropriate for the specific gas in question (D_A). Solution and analysis of the convective-diffusion equation allow estimation of the time scale for steady-state conditions to be attained in the gas phase, $\tau_A \simeq r_d^2/D_A$ (see Section 8.2). Gas diffusivities are always relatively large ($\sim 10^{-5}$ m^2 s^{-1}), so steady state is reached in the gas phase in small fractions of a second, even for large drops. The steady-state form of the convective-diffusion equation,

$$\mathbf{v} \cdot \nabla n_A - D_A \nabla^2 n_A = 0, \tag{13.24}$$

may therefore be used to calculate the local concentrations of A and the flow of A to the drop surface. The real concern we have is not the time to reach diffusive steady state, rather the time for molecules to move along the concentration gradient and fill the drop reservoir. The diffusive flux is, according to Fick's first law, always down the concentration gradient:

$$\mathbf{\Phi}_A = -D_A \nabla n_A. \tag{13.25}$$

As we found for Brownian diffusion, the diffusivity may be augmented by the air motions around a sedimenting drop, so we may replace D_A with $f_v D_A$ as appropriate, where f_v is the mean ventilation coefficient for vapor transport.

Interfacial exchange

Step 2 is the interfacial exchange across the surface of the drop. It is the molecular-scale interaction of A with H_2O molecules in the liquid phase that causes the trace gas to become solvated (loosely bound to water molecules) and a part of the aqueous solution. A trace gas in solution is designated A(aq), and the interfacial exchange can be written

$$\mathrm{A(g)} \overset{H_A}{\rightleftarrows} \mathrm{A(aq)}, \tag{13.26}$$

where H_A is the Henry's law coefficient for physically dissolved A. The double arrow indicates that the exchange can operate in both directions, although not necessarily at equal rates. As soon as these new molecules contact the water molecules in the solution, they may be dissociated into molecular ions, as discussed in Section 3.5. Such ionic reactions, exemplified generically by

$$\mathrm{A(aq)} + \mathrm{H_2O} \overset{K_{1A}}{\rightleftarrows} \mathrm{H^+} + \mathrm{AOH^-}, \tag{13.27}$$

are typically very fast and let local equilibrium be achieved rapidly. Here, K_{1A} is the equilibrium constant for the first dissociation of A; further dissociation is possible, as in the formation of sulfite from bisulfite $\left(\mathrm{HSO_3^-} \overset{K_{2S}}{\rightleftarrows} \mathrm{H^+} + \mathrm{SO_3^{2-}}\right)$. When both the first and second ionizations are rapid, the overall Henry's law coefficient is most applicable at the interface:

$$H_A^* = \left(1 + \frac{K_{1\mathrm{A}}}{[\mathrm{H^+}]}\left(1 + \frac{K_{2\mathrm{A}}}{[\mathrm{H^+}]}\right)\right) H_A, \tag{13.28}$$

so the total concentration of A in solution near the interface is, for a given local partial pressure p_A^0,

$$[\mathrm{A_{tot}(aq)}] = H_A^* p_A^0. \tag{13.29}$$

Note that the overall solubility is dependent on the solution pH ($\equiv -\log[\mathrm{H^+}]$) and that Eq. (13.29) is based on equilibrium considerations. At equilibrium, A molecules enter and leave the drop at equal rates. The partial pressure p_A^0 is sometimes termed a "potential" partial pressure because it prevails in reality only at equilibrium. Only deviations from the equilibrium state allow net amounts of the trace gas to exchange between the two phases.

The rate that gas molecules cross the air–liquid interface is limited by their impingement onto the surface. The maximum flux of A molecules hitting a surface can be calculated from kinetic-theory impingement:

$$\Phi_{imp,A} = \frac{1}{4}\bar{v} n_A = \frac{p_A}{(2\pi m_A k_B T)^{1/2}}, \tag{13.30}$$

where n_A and p_A are the concentration and partial pressure of A over the surface, m_A is the mass of A, k_B is the Boltzmann constant, and T is the temperature. The mean thermal speed of the gas molecules is $\overline{v} = (8k_BT/\pi m_A)$. However, only a fraction α of the impinging molecules actually accommodate with the water surface, so the actual flux absorbed into the liquid is $\Phi_{abs,A} = \alpha\Phi_{imp,A}$. At the same time, and quite independently, A molecules desorb from the drop at the rate

$$\Phi_{des,A} = \frac{\alpha p_A^0}{(2\pi m_A k_B T)^{1/2}}. \tag{13.31}$$

The net flux of A molecules across the interface is just the difference between the absorption and desorption fluxes,

$$\Phi_{net,A} = \Phi_{abs,A} - \Phi_{des,A} = \frac{\alpha\left(p_A - p_A^0\right)}{(2\pi m_A k_B T)^{1/2}}. \tag{13.32}$$

Because the desorption flux is proportional to the aqueous-phase concentration of A, we may replace the potential partial pressure, p_A^0, by its equivalent in terms the overall Henry's law coefficient:

$$\Phi_{net,A} = \frac{\alpha}{H_A^* (2\pi m_A k_B T)^{1/2}} \left(H_A^* p_A - \left[\mathrm{A_{tot}(aq)}\right]\right), \tag{13.33}$$

where it is understood that $\left[\mathrm{A_{tot}(aq)}\right]$ represents the total dissolved gas on the liquid side of the interface. The quantity $H_A^* p_A$ represents the capacity of the reservoir, the concentration of solute in equilibrium with the gas partial pressure p_A. The rate of molecular exchange across the interface depends both on the departure from equilibrium, $\left(H_A^* p_A - \left[\mathrm{A_{tot}(aq)}\right]\right)$, and on the kinetic coefficient α for the trace gas in question.

Estimating the time for the drop to fill with dissolved trace gas provides insight into the scavenging process. The meaning of a time scale to fill a reservoir (the drop) can be appreciated by thinking of the drop as a tiny bucket into which a substance (A) is introduced until its capacity is reached. Consider first a conventional bucket into which one pours a fluid. If the input rate I_0 [mol s^{-1}] were constant and the bucket capacity $N_{\max}$ [mol] were to be specified, the amount in the bucket would increase linearly with time, from $N = 0$ to $N_{\max}$, as seen from the material-balance equation,

$$\frac{dN}{dt} = I_{in} - I_{out}, \tag{13.34}$$

with $I_{in} = I_0$ and $I_{out} = 0$. Integration of Eq. (13.34) shows that the time [s] to fill the bucket is $\tau_0 = N_{\max}/I_0$. The filling of a drop with volatile solute differs from the bucket analogy in that the input rate varies with time and the reservoir capacity is dictated by the ambient conditions.

A water drop takes up a trace gas from the surrounding air, but it also discharges some of that same gas previously taken up. A certain fraction of the gas molecules striking the liquid surface become absorbed, while some of the molecules near the surface in the aqueous phase have a chance to escape back into the gas phase. This two-way exchange of molecules across the interface has been accounted for in the calculation of the net flux, as given by Eq. (13.32) or (13.33). Thus, an appropriate version of the material-balance equation for a drop is

$$\frac{dN}{dt} = V_d \frac{dn}{dt} = A_d \Phi_{net}. \tag{13.35}$$

Here, we drop the subscript A (for notational simplicity) and let the radius of a spherical drop be r_d. The drop volume is thus $V_d = \frac{4}{3}\pi r_d^3$, and its surface area is $A_d = 4\pi r_d^2$, giving the surface-to-volume ratio $A_d/V_d = 3/r_d$. We assume that the drop size remains constant, and we express the solute content (N) in terms of the aqueous-phase concentration ($n \equiv [A(aq)]$). The maximum rate with which the concentration in solution increases occurs at the start of the process, when $n = 0$:

$$\frac{dn}{dt} = \frac{A_d}{V_d}\Phi_{net,\max} = \frac{3}{r_d}\frac{\alpha p_\infty}{(2\pi m_A k_B T)^{1/2}} = \frac{3\alpha\overline{v}}{4r_d RT} p_\infty, \tag{13.36}$$

where Eq. (13.33) has been used and we assume that the partial pressure of gas at the surface is the ambient value p_∞. Note that $(2\pi m_A k_B T)^{-1/2} = \overline{v}/4RT$ from the kinetic-theory impingement flux (Eq. (13.30)). This maximum rate cannot be sustained, however, because the partial pressure over the surface decreases and desorption begins as soon as any gas becomes dissolved in the liquid. The trace gas removed from the air near the drop surface must be replenished by diffusion from afar, and the desorption flux must be accounted for.

The coupling of gas-phase diffusion (Step 1) and interfacial exchange (Step 2) is achieved mathematically by balancing the respective fluxes at the drop surface. In steady state, any gas diffusing toward the surface must cross the interface and become part of the aqueous solution: $\Phi_{diff} = \Phi_{sfc}$. The diffusive flux Φ_{diff} is found by applying Fick's first law to the gas profile, as we did for the flux of water vapor in the calculation of droplet growth by condensation (Section 8.2):

$$\Phi_{diff} = \frac{D}{r_d RT}\left(p_\infty - p_{sfc}\right) \equiv c_{diff}\left(p_\infty - p_{sfc}\right), \tag{13.37}$$

where p_∞ is the ambient partial pressure and p_{sfc} is the partial pressure at the drop surface. The driving factor for diffusion is the difference in partial pressures $\left(p_\infty - p_{sfc}\right)$, which is emphasized by defining the proportionality constant $c_{diff} \equiv D/(r_d RT)$. The net flux across the surface, Φ_{sfc}, comes from Eq. (13.32):

$$\Phi_{sfc} = \frac{\alpha\overline{v}}{4RT}\left(p_{sfc} - p^0\right) \equiv c_{sfc}\left(p_{sfc} - p^0\right). \tag{13.38}$$

Recall that the potential pressure p^0 is a measure of the aqueous-phase concentration and the rate of desorption at any given instant. The driving factor for interfacial exchange is the partial-pressure difference $\left(p_{sfc} - p^0\right)$, and its proportionality constant is $c_{sfc} \equiv \alpha\overline{v}/(4RT)$. The partial pressure at the drop surface, p_{sfc}, varies with time and plays a pivotal role in the transport of gas both through the air and across the interface. Were p_{sfc}

to be too large, the diffusive flux (Eq. (13.37)) would be too small and the surface flux (Eq. (13.38)) would be too large. The larger flux into the drop would deplete the gas phase and cause p_{sfc} to decrease. Only when the two fluxes, acting in series, are equal does p_{sfc} have the correct value. We thus determine the magnitude of p_{sfc} mathematically by equating the two fluxes ($\Phi_{diff} = \Phi_{sfc}$) and solving for p_{sfc}:

$$p_{sfc} = \frac{c_{diff}\, p_{\infty} + c_{sfc}\, p^0}{c_{diff} + c_{sfc}} = \frac{4Dp_{\infty} + \alpha\bar{v}r_d\, p^0}{4D + \alpha\bar{v}r_d}. \qquad (13.39)$$

This intermediate variable is then used in Eq. (13.38) to determine the net flux into the drop:

$$\Phi_{diff} = \Phi_{net} = \frac{c_{diff}\, c_{sfc}}{c_{diff} + c_{sfc}} \left(p_{\infty} - p^0\right) = \frac{\alpha\bar{v}D}{RT\,(\alpha\bar{v}r_d + 4D)} \left(p_{\infty} - p^0\right). \qquad (13.40)$$

This flux causes the solute concentration in the drop to increase at the rate (from Eq. (13.33))

$$\frac{dn}{dt} = \frac{3}{r_d}\Phi_{net} = \frac{3\alpha\bar{v}D}{H_A^* RTr_d\,(\alpha\bar{v}r_d + 4D)} \left(H_A^* p_{\infty} - n\right). \qquad (13.41)$$

This differential equation is of the form $dn/dt = K\,(n_{\max} - n)$, where $K \equiv 3\alpha\bar{v}D/\left(H_A^* RTr_d\,(\alpha\bar{v}r_d + 4D)\right)$ and $n_{\max} \equiv H_A^* p_{\infty}$, which has the solution $n(t) = n_{\max}\left[1 - \exp(-Kt)\right]$, or

$$n(t) = \left[1 - \exp\left(-\frac{3\alpha\bar{v}Dt}{H_A^* RTr_d\,(\alpha\bar{v}r_d + 4D)}\right)\right], \qquad (13.42)$$

for the initial condition $n(0) = 0$. As Fig. 13.16 shows, the rate of uptake is most rapid at first, but it then it gradually declines as equilibrium is approached. We see that the aqueous-phase concentration (n) increases toward the equilibrium value $\left(H_A^* p_{\infty}\right)$ with an exponential time constant $\tau = 1/K = H_A^* RTr_d\,(\alpha\bar{v}r_d + 4D)\,/\,(3\alpha\bar{v}D)$.

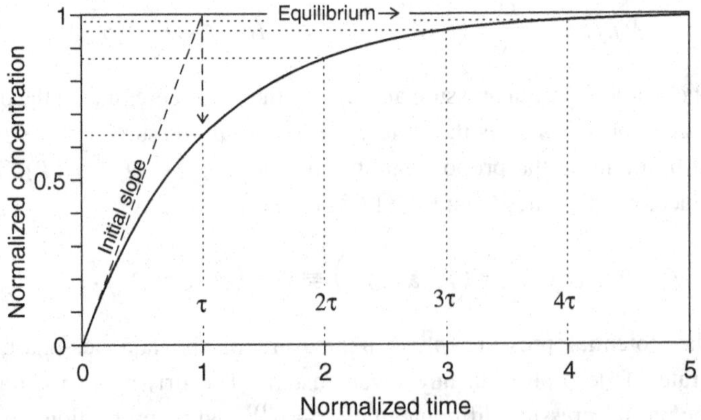

Figure 13.16 Time dependence of the normalized concentration. The exponential time constant is indicated as τ.

Analysis of the equation (13.42) describing the uptake of trace gas by a water drop lets us see the individual contributions to uptake. The total time scale may be resolved into two components by rewriting it as follows:

$$\tau = \frac{1}{K} = \frac{H_A^* RT r_d\,(\alpha\overline{v} r_d + 4D)}{3\alpha\overline{v}D} = \frac{H_A^* RT r_d^2}{3D} + \frac{4H_A^* RT r_d}{3\alpha\overline{v}} \equiv \tau_{diff} + \tau_{sfc}. \quad (13.43)$$

We see two distinct contributions, $\tau_{diff} = H_A^* RT r_d^2/(3D)$ and $\tau_{sfc} = 4H_A^* RT r_d/(3\alpha\overline{v})$, which align with the separate, but coupled processes of gas-phase diffusion and interfacial exchange, respectively. As we could have anticipated, the time constant for diffusion depends inversely on the diffusivity D of the trace gas in air, and the time constant for interfacial exchange depends inversely on the gas accommodation coefficient. Both time constants depend on the size of the drop, but with different dependences, and both increase linearly with the solubility of the gas (through the dimensionless Henry's law coefficient, $\widehat{H}_A^* \equiv H_A^* RT$). We may divide each time constant by the solubility and so define normalized times that are insensitive to the equilibrium state: $\tau'_{diff} \equiv \tau_{diff}/\left(H_A^* RT\right)$ and $\tau'_{sfc} \equiv \tau_{sfc}/\left(H_A^* RT\right)$. One need only multiply the normalized times by the dimensionless overall solubility $\left(\widehat{H}_A^*\right)$ in order to obtain the times appropriate to a particular gas. The dependence of these characteristic time constants on the gas accommodation coefficient α are plotted in Fig. 13.17 for several representative drop sizes. Note how the time scale for gas-phase diffusion is independent of α, but it increases rapidly as the drop radius r_d increases. The time scale for interfacial exchange, by contrast, depends in an inverse sense on both α and r_d. The composite time constants show breaks in slope near where the component time scales are equal.

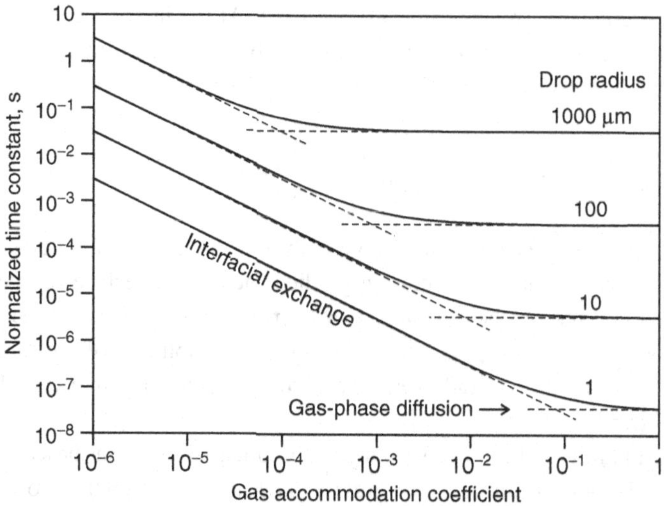

Figure 13.17 Dependence of the time to fill a well-mixed drop on the trace-gas accommodation coefficient. The time constant has been normalized to the dimensionless overall Henry's law constant, and a representative gas diffusivity of 10^{-5} m^2 s has been used. Adapted from Schwartz (1986).

A key purpose of such an analysis is to identify regimes in which a single rate-limiting process dominates the uptake. If the rapid process can be ignored in such a regime, then the mathematical analysis becomes simpler. For instance, Fig. 13.17 shows that the characteristic time for large drops to take up a trace gas is independent of surface-kinetic processes when α is larger than a certain value; in this regime, only gas-phase diffusion need be considered. Unfortunately, the specific values of α for different gases is still the subject of laboratory research. At the other extreme, gas uptake by small cloud drops is relatively insensitive to gas-phase diffusion whenever the gas accommodation coefficient is less than about 0.01; in this regime, only surface kinetics needs to be treated. The mathematical analysis here shows that a single process often limits gas uptake, even when both are involved mechanistically. A complete mathematical treatment, one involving all processes, is needed whenever the time scales of the individual processes are comparable.

Internal transport

Step 3 in the uptake of a trace gas is the transport of solute within the drop. Internal transport occurs partly by molecular diffusion through the liquid water, but the dissolved gas is also carried along with any liquid-phase circulations inside the drop. Small drops, such as those found in hazes, fogs, and the early stages of cloud formation, are carried by the wind with little separate motion relative to the air; they may be treated as stagnant and motionless within the viscous medium that supports them. Such an extreme view is only partially valid, but it does let us appreciate the importance of molecular-scale processes, both outside and inside the drop. Transport of a trace gas to the surface of a small haze or cloud droplet requires molecular diffusion of the gas molecules through the air. Once inside the drop, the dissolved gas molecules diffuse away from the surface in response to concentration gradients (refer again to Fig. 13.15). At the interface, mass conservation requires that the fluxes match, which in turn means that the radial gradients on either side of the interface scale inversely with the respective diffusivities:

$$\frac{\left.\frac{\partial n}{\partial r}\right|_{aq}}{\left.\frac{\partial n}{\partial r}\right|_{g}} = \frac{D_g}{D_{aq}}. \tag{13.44}$$

The diffusivity of a substance in liquid is typically four orders of magnitude smaller than that in the gas phase, so the concentration gradient near the interface on the liquid side is much larger than that on the gas side. However, the gradients vanish near the center of a spherically symmetric drop, as shown in Fig. 13.15. Solution of the time-dependent diffusion equation for a drop of radius r_d reveals the characteristic time scale for diffusion to be $\tau_{aq} \simeq r_d^2/\left(\pi^2 D_{aq}\right)$.

Transport inside a large liquid drop is more complicated than that based on molecular diffusion alone. Under the influence of gravity, such drops fall relative to the surrounding air. The air movement against the surface generates a drag force that causes the liquid surface to move upward along the sides (recall discussion in Section 9.2). Complicated internal circulations, sometimes laminar, sometimes turbulent, serve to augment transport

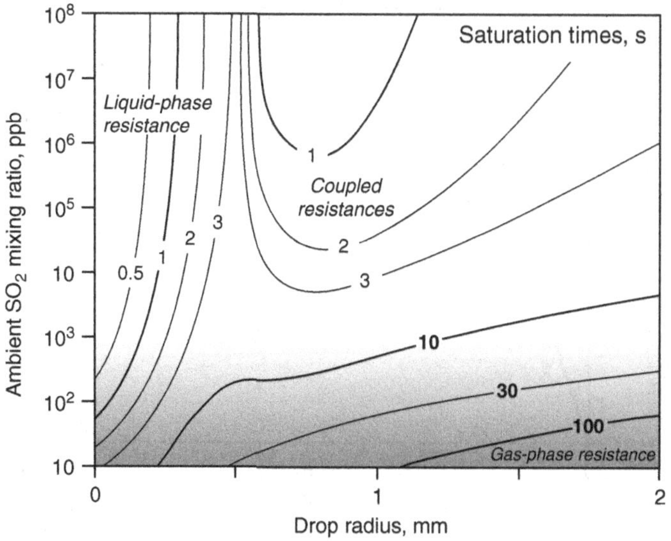

Figure 13.18 Times (isopleths, s) to reach 63% of the saturation value inside drops of specified radius (horizontal axis) within a gaseous environment of the specified mixing ratio (vertical axis). Ambient range of SO_2 mixing ratios is shown by shading near the bottom. Adapted from Walcek and Pruppacher (1984).

relative to molecular diffusion, but transport in the liquid is nevertheless slower than if the drop is assumed to be well mixed (implying infinitely fast transport). The approach to equilibrium inside the drop follows the pattern illustrated in Fig. 13.16, except that the rate is multiplied by a factor less than unity. A complete model of transport by convective-diffusion in the two phases, including ionic equilibria in the aqueous phase, is needed to calculate the uptake of a reactive gas accurately. An example of the saturation time scales needed for water drops exposed to specified mixing ratios of SO_2 is shown in Fig. 13.18. Note the complex nature of the uptake in the different regimes: Uptake of a gas at low ambient concentrations is limited by gas-phase transport to large drops (because internal circulations permit rapid transport in the liquid); uptake at high ambient concentrations is limited by liquid-phase transport to small drops (because of the limitation of molecular diffusion in a stagnant liquid). The resistances to transport are intimately coupled for intermediate concentrations and drop sizes. The magnitudes of the saturation time constants vary greatly, from a fraction of a second to a couple of minutes. Large drops require long times to saturate, and they fall large distances in those times. One must question how uniform the atmosphere is through which such drops fall and whether the mathematical analysis given here is valid.

Desorption occurs when a drop already laden with volatile solute falls into a region depleted in the trace gas. Desorption is essentially the reverse of absorption, except when the gas species taken up has dissociated into ionic fragments in the aqueous solution. When

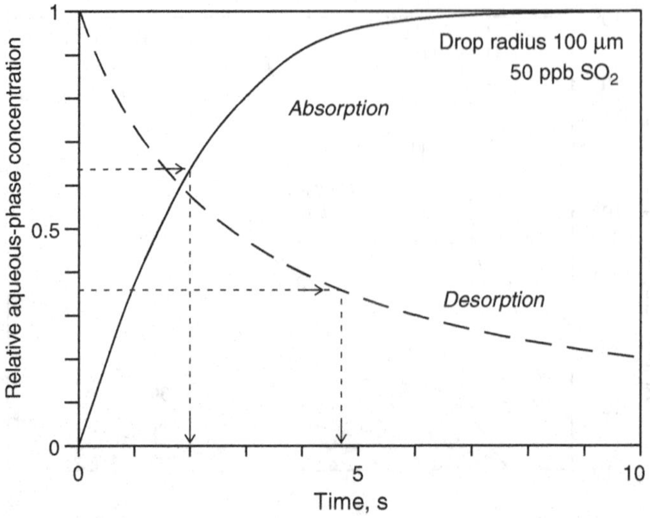

Figure 13.19 Time dependence of aqueous-phase concentration relative to that in equilibrium with 50 ppb SO_2 in a drop of 100 μm radius during absorption (solid curve) and desorption (dashed curve). Arrows identify times to come to within $1/e$ of the final value. The aqueous phase was assumed to be well mixed, and its pH was determined solely by SO_2 dissociation. Adapted from Mitra and Hannemann (1993).

ions are produced from a neutral species, such as SO_2 hydrating and dissociating into bisulfite and sulfite ions during uptake, these same ions must again come together to form the molecular species before desorption from the liquid can take place. The processes of dissociation and re-association are pH dependent (shown by Eq. (13.28), where A refers to SO_2), so the rate of desorption from an otherwise pure water drop is then not equal to the rate of uptake. The asymmetric nature of gas–liquid interactions is shown for SO_2 in Fig. 13.19. Note that this time scale for desorption is more than twice that for absorption. The distinction is not so great when the solution pH is dominated by other components in solution.

13.3.2 Aqueous-phase reactions

Step 4 in the scavenging of a trace gas is its possible reaction to a new chemical species. One of the more common reactions is the aqueous-phase oxidation of SO_2 (sulfur in oxidation state IV) to sulfuric acid (H_2SO_4, sulfur in oxidation state VI). The liquid water of the drops in a cloud offers an alternative medium to the gas phase for chemical reactions. Ozone (O_3), for instance, can coexist indefinitely with SO_2 in the gas phase, but in the aqueous phase dissolved O_3 acts as a strong oxidant for several of the dissociation components of SO_2. Aqueous-phase chemical reactions are often irreversible and lead to new compounds that permanently alter the solute content and properties of the drop.

The rate with which two compounds react in liquid water must be distinguished from the rate that a cloud converts one species to another. That is, we need to distinguish homogeneous, aqueous-phase reactions from heterogeneous, in-cloud conversions. The reason for being precise here can be appreciated by considering the multiphase nature of clouds (Fig. 13.14). A unit volume of cloudy air contains only a small volume of liquid, and it is only in the liquid that the reactions take place (for the sake of this analysis). Trace gas A dissolves into the cloud water, where it reacts to compound A′, which is later released back to the cloud-free air in the particulate phase. Cloud-mediated reactions represent a form of gas-to-particle conversion. From the point of view of atmospheric chemistry, we may want to know how rapidly a cloud converts A to A′, from an initial trace gas to a component in the aerosol left behind after cloud dissipation. The heterogeneous conversion may be summarized as

$$\mathrm{A(g)} \underset{\text{cloud}}{\rightarrow} \mathrm{A'(p)}. \tag{13.45}$$

Trace gas A got converted to new compound A′, with the cloud acting as the medium. If we let the initial trace gas concentration be n_A [mol m^{-3}], the fractional heterogeneous conversion rate may be written [s^{-1}, or % h^{-1}].

$$r_A \equiv -\frac{1}{n_A}\frac{dn_A}{dt} = \frac{1}{n_A}\frac{dn_{A'}}{dt} \equiv \frac{RT}{p_A} R_{A'} \upsilon_L, \tag{13.46}$$

where $R_{A'} \equiv dn_{A'}/dt$ is the homogeneous aqueous-phase rate of reaction [units, mol m^{-3} s^{-1}] and υ_L is the volume fraction of liquid water. Note that the conversion rate scales directly with the volume concentration of liquid available for reaction, $\upsilon_L \sim 10^{-6}$. Clouds or hazes with low liquid water concentrations would convert less of the reactant gas A to the product A′ than they would in more water-rich environments for a given aqueous-phase reaction rate.

Reactions within a cloud drop impose new complications on our analysis of trace-gas uptake. Once the gas enters the aqueous phase, it simultaneously diffuses and reacts, destroying the original component and replacing it with the reaction product. Reactions deplete the dissolved gas at the same time that it diffuses through the drop. The reactions thus alter the concentration gradients and hence the rates of mass transport. To avoid undue complications arising from time-dependent phenomena, we consider simultaneous transport and reaction only under steady-state conditions. We also limit the aqueous-phase transport to molecular diffusion within small drops, a restriction that is not all that limiting in reality; most reactions in clouds occur in the small end of the drop spectrum where the surface-to-volume ratio is relatively large and the time scales for transport are small (see Fig. 13.17). The appropriate mass-balance equation for a system in which a component A is simultaneously diffusing with aqueous-phase diffusivity $D_{A,aq}$ and reacting with rate $R_{A,aq}$ is

$$\frac{dn_A}{dt} = D_{A,aq}\nabla^2 n_A - R_{A,aq}, \tag{13.47}$$

where $n_A \equiv [A(aq)]$ is the aqueous-phase concentration of A. Most reactions are approximately linear in the reactant concentration, so we assume a rate $R_{A,aq} = k_A n_A$, where k_A is the first-order rate coefficient. The equation to solve is therefore

$$\frac{dn_A}{dt} = D_{A,aq} \nabla^2 n_A - k_A n_A. \tag{13.48}$$

The solution to this equation for a spherical drop of radius r_d exposed to a constant gas-phase concentration is

$$n_A(r) = n_{A,0} \frac{r_d}{r} \frac{\sinh\left(\left(\frac{k_A}{D_{A,aq}}\right)^{1/2} r\right)}{\sinh\left(\left(\frac{k_A}{D_{A,aq}}\right)^{1/2} r_d\right)}, \tag{13.49}$$

where $n_{A,0}$ is the aqueous-phase concentration at the air–liquid interface. This equation may be put into more generic form by normalizing the concentration to the surface value, $\widehat{n}(\widehat{r}) \equiv n_A(\widehat{r})/n_{A,0}$, where the relative radial distance $\widehat{r} \equiv r/r_d$. Then, Eq. (13.49) becomes

$$\widehat{n}(\widehat{r}) = \frac{1}{\widehat{r}} \frac{\sinh(q\widehat{r})}{\sinh(q)}, \tag{13.50}$$

where the dimensionless reaction parameter $q \equiv r_d (k_A/D_{A,aq})^{1/2}$ is a measure of the rate of reaction compared with transport by diffusion. The pattern exhibited by Eq. (13.50) is shown in Fig. 13.20 for several selected values of q. The concentration of reactant decreases away from the surface (left side) at rates that vary systematically with the value of q. We may give physical meaning to q as follows: The rate of a

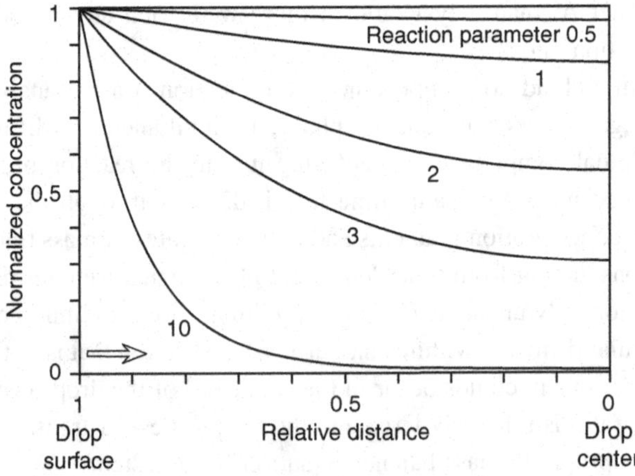

Figure 13.20 Profiles of relative concentration inside a water drop during reaction characterized by the indicated reaction parameter. The arrow shows the direction of reactant flow into the drop from the surface. Adapted from Schwartz and Freiberg (1981).

first-order reaction is closely related to chemical lifetime: $\tau = 1/k$, and the distance that the substance diffuses in that time is (by Einstein's relation) $x_\tau = (D\tau)^{1/2} = (D/k)^{1/2}$. Thus, $q \equiv r_d / \left(D_{A,aq}/k_A\right)^{1/2} = r_d/x_\tau$ can also be viewed as a dimensionless distance, the ratio of the drop radius (r_d) to the average distance (x_τ) the reactant penetrates beneath the surface before reaction destroys it. When reaction is slow, $q << 1$, the reactant can diffuse throughout most of the drop before being significantly depleted by reaction; the reactant concentration remains near the gas-equilibrium value. At the other extreme, when the reaction is fast, $q >> 1$, the reactant gets depleted rapidly before it can diffuse very far into the drop.

The rate that any specific substance reacts depends on the details of the reaction mechanism. The chemical principles can be understood by considering the aqueous-phase oxidation of sulfur dioxide (SO_2). Even though SO_2 is itself not very soluble, it and its ionization products can become oxidized to form the sulfate ion $\left(SO_4^{2-}\right)$, a stable compound of sulfur that is non-volatile for all practical purposes. The reactants are the various species having sulfur in oxidation state IV, namely the physically dissolved form, $SO_2{\cdot}H_2O$; the first ionization product, the bisulfite ion, HSO_3^-; and the second ionization product, the sulfite ion, SO_3^{2-}. Collectively, these aqueous species are designated S(IV). The oxidation product $\left(SO_4^{2-}\right)$ contains sulfur in oxidation state VI. Thus, we may write the oxidation reaction generically as

$$\mathrm{S(IV)} + \text{oxidant} \rightarrow \mathrm{S(VI)}, \tag{13.51}$$

where "oxidant" refers to one of several possible oxidizing agents. Oxidation is an irreversible reaction, so the arrow points one way. From an atmospheric perspective, oxidation of S(IV) to S(VI) represents a permanent fixation of sulfur.

The oxidants of sulfur are diverse and yield widely varying rates of oxidation. Some organic peroxides and nitrogen dioxide (NO_2) serve to oxidize S(IV), but the low solubilities of these compounds in water limit their importance. Ordinary oxygen (O_2) may also oxidize S(IV) compounds in aqueous solution, but a catalyst, such as the manganese ion Mn^{2+} or the iron ion Fe^{3+}, is thought to be needed. The strongest and most common oxidants in the atmosphere are ozone (O_3) and hydrogen peroxide (H_2O_2). The rates of S(IV) oxidation by some of these oxidants is shown in Fig. 13.21. Oxidation rates tend to be pH dependent, largely because of the pH dependence of the ionization products of S(IV) in solution.

We can understand the pH dependencies of the S(IV) oxidation rates, particularly those by O_3 and H_2O_2, by considering the specific mechanisms. Dissolved ozone reacts with all forms of S(IV), but at different rates. The individual reaction steps are

$$\begin{aligned} SO_2 \cdot H_2O + O_3 &\overset{k_0}{\rightarrow} SO_4^{2-} \\ HSO_3^- + O_3 &\overset{k_1}{\rightarrow} SO_4^{2-} \\ SO_3^{2-} + O_3 &\overset{k_2}{\rightarrow} SO_4^{2-}. \end{aligned} \tag{13.52}$$

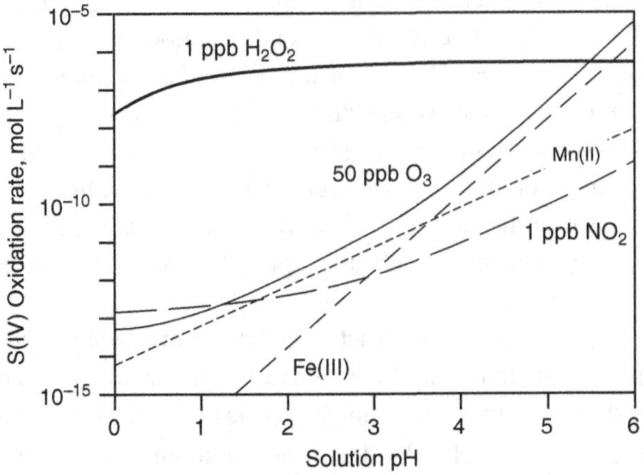

Figure 13.21 The rates of S(IV) oxidation by various agents as a function of solution pH. Adapted from Seinfeld and Pandis (1998).

The common oxidant here is O_3, so the overall rate law may be written as

$$R_{H_2SO_4}\big|_{O_3} \equiv \frac{d\left[SO_4^{2-}\right]}{dt} = \left(k_0\left[SO_2 \cdot H_2O\right] + k_1\left[HSO_3^-\right] + k_2\left[SO_3^{2-}\right]\right)\left[O_3\right], \tag{13.53}$$

with the individual bimolecular rate coefficients having the measured magnitudes (at 298 K) $k_0 = 2.4 \times 10^4$ L mol^{-1} s^{-1}, $k_1 = 3.7 \times 10^5$ L mol^{-1} s^{-1}, and $k_2 = 1.5 \times 10^9$ L mol^{-1} s^{-1}. Each term is weighted by the concentration of the respective S(IV) species; the last two are strongly dependent on pH and give the trend shown in Fig. 13.21 for a gas-phase concentration of O_3 of 50 ppb. The curve increases steeply with pH as more and more of the dissolved SO_2 is able to ionize.

The significantly different pH dependence shown by H_2O_2 arises from its more complicated reaction with the bisulfite ion. Oxidation of HSO_3^- by H_2O_2 proceeds in the aqueous phase via two mechanistic steps. The first step is rapid and achieves near equilibrium with an acidic intermediate, called peroxymonosulfurous acid (SO_2OOH^-):

$$HSO_3^- + H_2O_2 \underset{k_{1b}}{\overset{k_{1f}}{\rightleftarrows}} SO_2OOH^- + H_2O, \tag{13.54}$$

where k_{1f} is the forward rate coefficient and k_{1b} is the coefficient for the back reaction. As with any balanced reaction, this reaction is characterized by an equilibrium constant $K_1 \equiv k_{1f}/k_{1b}$. The second reaction step is relatively slow and irreversible:

$$SO_2OOH^- + H_3O^+ \overset{k_2}{\rightarrow} H_2SO_4. \tag{13.55}$$

The sulfuric acid (H_2SO_4) so formed dissociates rapidly in the aqueous solution to form the sulfate ion $\left(SO_4^{2-}\right)$ and hydronium ions (H_3O^+):

$$H_2SO_4 \rightarrow 2H_3O^+ + SO_4^{2-}. \tag{13.56}$$

The overall rate law can be found by assuming that the concentration of the acid intermediate changes little with time. Its steady-state concentration is

$$\left[SO_2OOH^-\right] = \frac{k_{1f}\left[HSO_3^-\right][H_2O_2]}{k_{1b} + k_2\left[H_3O^+\right]}. \tag{13.57}$$

The rate of sulfuric acid formation is therefore

$$R_{H_2SO_4}\big|_{H_2O_2} = k_2\left[H_3O^+\right]\left[SO_2OOH^-\right] = k_2\left[H_3O^+\right]\frac{k_{1f}\left[HSO_3^-\right][H_2O_2]}{k_{1b} + k_2\left[H_3O^+\right]}. \tag{13.58}$$

In cloud water having pH > 2, this rate expression simplifies to

$$R_{H_2SO_4}\big|_{H_2O_2} = K_1 k_2\left[H_3O^+\right]\left[HSO_3^-\right][H_2O_2]. \tag{13.59}$$

Note that this rate may be put into the standard form for a second-order reaction:

$$R_{H_2SO_4}\big|_{H_2O_2} = k_2^{II}\left[HSO_3^-\right][H_2O_2], \tag{13.60}$$

where we see that the pseudo-second-order rate coefficient $k_2^{II} \equiv K_1 k_2\left[H_3O^+\right]$ is acid catalyzed. It is this acid catalysis (in Step 2) that lets hydrogen peroxide be such a strong oxidant in low-pH solutions. As seen in Fig. 13.21, the rate of oxidation is essentially independent of pH over much of the range of atmospheric interest. The inability of SO_2 to dissociate readily in acidic solutions is almost exactly compensated for by the acid catalysis of the oxidation rate. Hydrogen peroxide is of great importance in the atmosphere, not only because of its strong oxidizing power in acidic solutions, but also because it is very soluble. Oxidation of SO_2 in clouds is limited primarily by the availability of this important trace gas.

Chemical reactions leading to a low-volatility compound, such as sulfuric acid, play important roles in cloud and atmospheric chemistry. Through such aqueous-phase reactions, a gas (e.g., SO_2) of low solubility is transformed into a non-volatile solute $\left(SO_4^{2-}\right)$ in cloud water. As long as the cloud water maintains contact with a strong oxidant (H_2O_2, O_3), the trace gas continues to be absorbed and consumed, being transformed into new, permanent solute in the cloud water. The added solute alters the properties of the residual aerosol upon evaporation of the cloud drops. Such solute-enriched particles are favored for activation in subsequent cloud-formation cycles, in turn giving further opportunities to acquire additional solute. The positive feedback lets clouds act as agents in the conversion of reactive trace gases into aerosol particles.

Microchemical processes occurring in non-precipitating clouds are partly responsible for observed transitions in aerosol characteristics. Data taken during research cruises show how the aerosol size distribution changes from one characteristic of a continental air mass to one typically found in the remote parts of the oceanic boundary layer. The curves in Fig. 13.22 show averaged size distributions at the four locations identified in the inset map

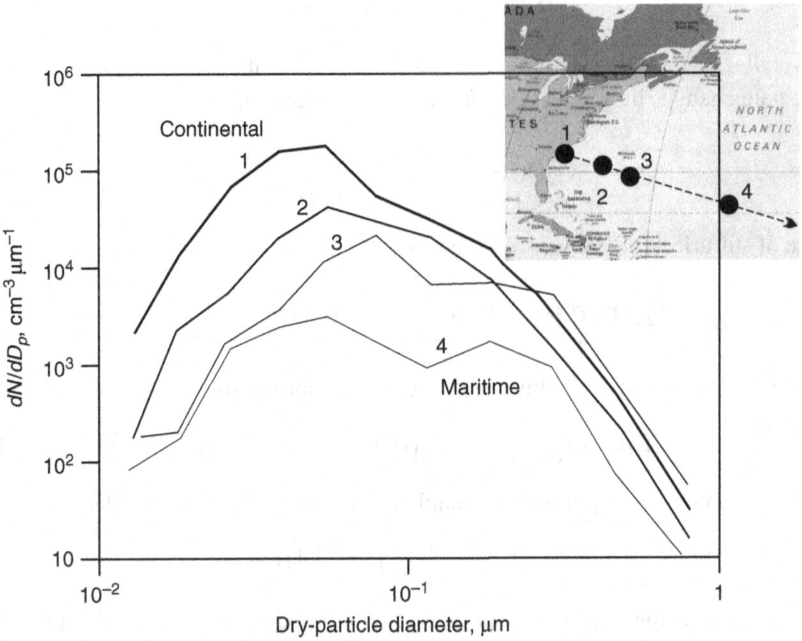

Figure 13.22 Averaged aerosol size spectra obtained from data taken aboard ship at the four locations in the North Atlantic Ocean shown on the inset map. Adapted from Hoppel *et al.* (1990).

of the western Atlantic Ocean. The westerly airflow off North America brings the continental aerosol into the oceanic region, where the sources of aerosol particles are sparse. Even in the absence of precipitation (a mechanism for removing aerosol from the atmosphere), the transition from a continental aerosol (curve 1; characterized by high concentrations of relatively small particles) to a maritime aerosol (curve 4; low concentrations, bimodal distribution) can be brought about by several microchemical processes acting in concert. As we saw above, the conversion of reactive trace gases to particulate matter causes a general increase in the dry sizes of the particles. The in-cloud scavenging of unactivated aerosol particles reduces the small end of the spectrum, while simultaneously adding additional solute to the drops. As the aerosol concentration is gradually lowered, the average drop sizes increase, which facilitates collision-coalescence, a process that merges both water mass and solute. Coalescence is favored by the largest drops in a population, so their evaporation could well contribute to the development of the larger mode in the maritime aerosol. The microchemical processes mostly lead to fewer, but larger particles, and to enhanced efficiencies with which the particles are removed from the atmosphere by precipitation.

13.4 Precipitation chemistry

Some of the gases and particulate matter scavenged by clouds end up in precipitation and are removed from the atmosphere. We have seen that soluble aerosol particles are scavenged effectively during activation near cloud base, and trace gases are taken up most rapidly when the drops are still small (while the surface-to-volume ratio is large). Aqueous-phase reactions scale with the concentration of liquid water, but they require the rapid uptake of the reacting gases, so again the small drops are favored. In general, many of the most important microchemical processes take place within drops toward the small end of the size distribution. The question thus becomes how the chemicals in small cloud droplets migrate up the size scale and enter precipitation particles at the large end of the size spectrum. We will see that the fates of chemicals in a cloud are intimately tied to the microphysics of precipitation formation. Throughout the following discussion, recall the tendencies for stratiform clouds to initiate precipitation by vapor deposition during the ice-crystal process and convective clouds to depend more on collection, of liquid drops in warm clouds or of supercooled drops in cold clouds.

13.4.1 Spectral migration of chemicals

Chemicals contained in small cloud drops migrate upscale toward larger particles in the size spectrum by some of the same mechanisms that generate precipitation. The tendency of chemicals to follow the water through the cloud is not simple, however. The couple of precipitation-formation mechanisms that involve phase changes, shown in Fig. 13.23, limit chemical migration. We will see shortly that spectral migration in mixed-phase clouds is more complicated than that in either all-liquid or all-ice clouds. This subsection focuses on the relevant processes of precipitation formation that impact the interphase partitioning of chemicals and the efficiency with which chemicals migrate into precipitation.

Precipitation formed primarily from smaller cloud particles by collection always contains chemicals once borne by smaller cloud elements. In warm clouds, collision-coalescence brings the liquid water of a small drop (radius r_1) and that of a larger drop (radius r_2) together and at the same time lets the solutes merge. The resultant drop contains a mass of water equal to the sum of the individual masses: $m_{tot} = m_1 + m_2 = \rho_L V_1 + \rho_L V_2 = (4/3)\,\pi\rho_L\left(r_1^3 + r_2^3\right)$, where ρ_L is the density of liquid water, and V_1 and V_2

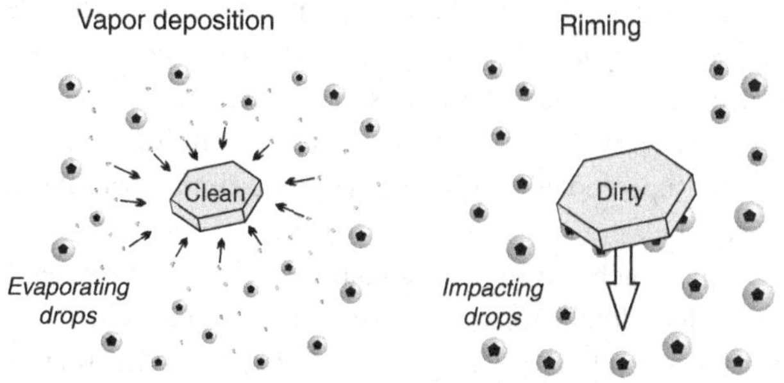

Figure 13.23 Cartoon distinguishing the chemical implications of two precipitation-formation processes. Solid pentagons represent solute. Left: Vapor deposition, the transfer of water vapor from liquid drops to ice crystals. Right: Riming, the direct collection of supercooled drops by a falling ice crystal.

are the respective drop volumes. If the molar solute contents of the individual drops are N_1 and N_2, respectively, then the solute content of the coalesced drop is $N_{tot} = N_1 + N_2$. The final concentration of the merged solution is thus $n_f \equiv N_{tot}/V_f = (N_1 + N_2)/V_f = (n_1 V_1 + n_2 V_2)/V_f = n_1 f_1 + n_2 f_2$, where $f_1 \equiv V_1/V_f$ and $f_2 \equiv V_2/V_f$ are the volume fractions of the respective colliding partners. The final concentration is just the volume-weighted fraction of the starting concentrations. Such simple calculations work well for non-volatile, chemically stable compounds. When the solute is volatile, as results from the uptake of trace gases, a shift in chemical equilibria and a possible expulsion of some gas is possible, especially if the drops had different temperatures or pH prior to coalescing. Still, collision-coalescence is a process that causes both water and solute to migrate upscale at more or less the same rate. It is worth noting, however, that coalescence of liquid drops always reduces the surface-to-volume ratio, and so it increases the time scales for trace-gas equilibration. Also realize that the enhanced solute contents of coalesced drops means that they may serve as better CCN should they evaporate before contributing to precipitation. The aggregation of ice crystals to form snow in cold clouds similarly leads to enhanced solute contents, but the chemical burden of pristine ice crystals is typically small, so this process is often neglected altogether in microchemical calculations. In both of these cases, the phases of the colliding partners are the same, either liquid in the case of warm clouds or solid in the case of cold clouds.

Cold-cloud precipitation processes that involve a phase change impose unique challenges to understanding the upscale migration of chemicals. For non-volatile solutes, riming, the accretional sweepout of supercooled cloud drops by a falling ice particle (right side of Fig. 13.23), is a direct collection process not unlike that of collision coalescence. A solute like sulfate will be rejected from the ice lattice and end up in tiny liquid inclusions after the collected drop has frozen, but it is nevertheless caught by the riming particle. Particle-average concentrations follow the same mathematical principle outlined above for collision-coalescence. Riming is an important precipitation growth process that is also effective in migrating non-volatile solutes from small to larger particles.

The chemical implication of riming differs considerably when the solute in question is volatile. Recall that a volatile solute is essentially a dissolved trace gas. The gas is taken up by a liquid cloud drop when the gas-phase concentration exceeds the Henry's law equilibrium value in the aqueous phase. Should conditions change and shift the aqueous-phase concentration to the other side of equilibrium, some of the solute will desorb and return to the gas phase. Riming is a process that causes just such a shift in equilibrium. Retention of solute contained in the supercooled drops collected by a riming particle is not guaranteed when the solute is volatile.

Consider the case in which the volatile solute is derived from gaseous SO_2 in equilibrium with supercooled cloud drops at temperature $T < T_m$, where T_m is the melting point of the solution. The melting point typically differs but a fraction of a degree from the ice point $T_0 = 273.15\,\text{K}$, so the drops freeze readily on contact with the surface of the collecting ice particle. The overall Henry's-law equilibrium coefficient H_S^* varies slightly with the absolute temperature, but the aqueous-phase concentration of S(IV),

$[S(IV)] \simeq [SO_2 \cdot H_2O] + [HSO_3^-]$, changes greatly and suddenly upon freezing of the solution. Before a drop contacts the riming surface, the solute is uniformly distributed throughout the liquid and in chemical equilibrium with SO_2 in the air. As soon as freezing starts, the solute concentration increases as the liquid volume diminishes by conversion to ice. The components of S(IV) in solution cannot enter the ice lattice because of their large sizes, so the solute is confined to small liquid inclusions within the dendritic mesh of ice that forms during the rapid adiabatic stage of freezing. Freezing, by reducing the liquid volume, puts the dissolved components well beyond the equilibrium concentration and forces partial desorption of the dissolved gas. Some of the solute is retained with the solution, because the time scale for diffusion ($\sim$10 ms) is large compared with the freezing time scale ($\sim$1 ms). One might correctly expect that the amount of volatile solute retained would increase with the supercooling $\Delta T_s \equiv T_0 - T$, because the fraction of the drop frozen during the adiabatic phase increases with supercooling: $f_{ice} \simeq (c_L / l_f)\,\Delta T_s$, where c_L is the specific heat of liquid water and l_f is the latent heat of fusion. Indeed, laboratory measurements show such a dependence (Fig. 13.24). The data suggest that the fraction of the initial S(IV) retained during riming takes the linear form $f_{S(IV)} = a\Delta T_s + b$. A small, but nevertheless significant fraction of the original dissolved gas stays with the frozen drop during riming and is carried on to larger particle sizes.

Complications arise with this scenario when other compounds are also present. A volatile solute with ionic components readjusts rapidly to maintain chemical equilibrium with any other ionic components in solution. As freezing reduces the liquid volume, the concentrations of all species increase, but not necessarily at the same rate. Any oxidation reactions that are pH dependent may thus speed up. Measurements have suggested that

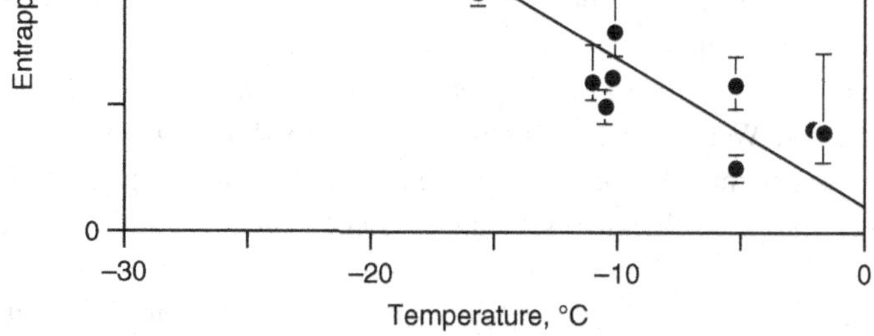

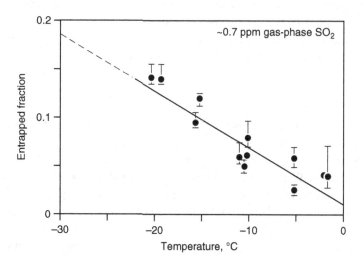

Figure 13.24 Measured fraction of aqueous-phase S(IV) retained in rime ice at the indicated temperature. The ambient (gas-phase) SO_2 mixing ratio $\sim$0.7 ppm. Reproduced with permission of Elsevier Science and Technology Journals, from Lamb and Blumenstein (1987), *Atmospheric Environment*, vol. **21**, pp. 1765–1772; permission conveyed through Copyright Clearance Center, Inc.

the presence of ammonium hydroxide (NH_4OH, derived from dissolved NH_3) in sufficient concentrations to raise the pH above 5 causes a substantial fraction of S(IV) to become oxidized to S(VI). A volatile solute can thus be transformed into a non-volatile solute during the riming process, increasing the fraction of total S retained and passed on to larger particles. The empirical and theoretical understanding of solute reaction and retention during riming is still an ongoing pursuit.

Precipitation growth by vapor deposition to ice crystals in supercooled drops (Fig. 13.23) imposes its own unique restrictions to the upscale migration of chemicals. This process, the heart of the Bergeron mechanism for initiating precipitation in mid-latitude stratiform clouds, depends on a steady flow of water vapor to the crystal for its growth. The driver for this transfer of water from the liquid drops to the ice is the equilibrium vapor pressure difference that results from the slightly different molecular bonding in the two distinct phases. If the crystal is large enough to sediment, drops may pass close to the droplet surface, increasing the local vapor gradient and contributing extra vapor to the crystal, but direct contact of the liquid with the ice particle is not part of the story. Given sufficient time and a relatively high ice crystal number concentration, the drops will completely evaporate and the interstitial vapor concentration gradually relaxes to that in equilibrium with ice, as discussed in Section 12.3. As the drops evaporate, the concentrations of solute in the aqueous phase increase. Volatile solute will be expelled, releasing dissolved trace gases back into the interstitial air. Non-volatile solutes cannot desorb, so they will stay with the evaporating drop, eventually becoming part of the interstitial aerosol when evaporation is complete. None of the solute originally contained in the small cloud drops is directly transferred to the larger particle. Vapor deposition serves as a low-temperature distillation process, one that effectively halts the upscale migration of all chemicals. The interstitial air may become enriched by the evaporation of the drop population, perhaps, but the growing ice crystals emerge relatively clean from the process. Field studies indeed find pristine snow to be much cleaner than snow that shows evidence of significant riming.

13.4.2 Wet removal

The removal of atmospheric chemicals by precipitation is termed wet deposition. Chemical wet deposition from the atmosphere must be distinguished from vapor deposition, the term used to describe the growth of ice crystals from the vapor phase; the context makes the distinction clear. Wet deposition focuses on the cloud as a whole, more so than on the individual microchemical processes, and it is responsible for supplying surface ecosystems with a wide range of chemical substances, some beneficial, some detrimental to the environment.

Precipitation, whether rain, snow, or hail, is key to the wet removal of any substance from the atmosphere. When viewed as a whole (see Fig. 13.25), a cloud represents a physicochemical agent that processes chemicals imported near its base. The fates of those chemicals depend on the properties of the compounds (e.g., their solubilities) and on the cloud's ability to produce precipitation. The microphysical processes responsible for

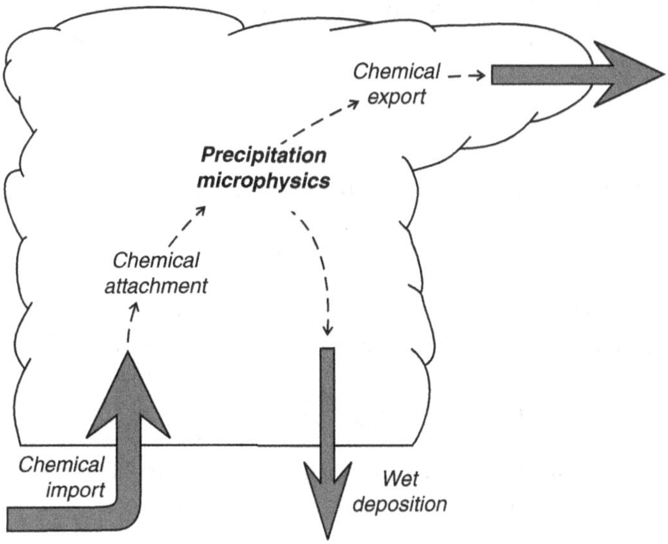

Figure 13.25 Cartoon of a cloud emphasizing the central role that precipitation microphysics plays in the wet removal of imported chemicals.

precipitation formation act as a regulator of the chemical flows, channelling them either into the precipitation itself and down the wet-deposition pathway or releasing them into the upper-level outflow. The disposition of imported chemicals depends crucially on the precipitation microphysics.

Still viewing the cloud as a processing agent, we can treat the system as a chemical flow reactor, one with two sequential steps (attachment and removal). A chemical imported into the reactor must first attach itself to the cloud particles, otherwise it simply passes through the cloud. For instance, an insoluble gas, or an aerosol particle that does not activate as a CCN and also fails to be scavenged inside the cloud, would stay in the interstitial air and be exported. Only through attachment to cloud particles can a chemical have a chance to be caught up in the precipitation microphysics and be removed from the atmosphere, the second step in the processing sequence.

A first level of quantification may be achieved by assuming steady-state conditions and applying simple, first-order kinetics to the overall process. Let the rate coefficient for the first step (attachment) be k_1, and let the rate coefficient for the second step (wet deposition) be k_2. Let the volumetric (i.e., molar) mixing ratios of the chemical in the air and in the condensate be y_A and y_C, respectively. Then, as the chemical is taken up by the cloud particles, its mixing ratio in the air decreases at the rate

$$\frac{dy_A}{dt} = -k_1 y_A. \tag{13.61}$$

As with any simple first-order process, the mixing ratio decays exponentially with time, the time scale for attachment being $\tau_1 = 1/k_1$. The chemical mixing ratio of the chemical in the condensate, however, involves two processes, one that adds solute (because of in-cloud scavenging) and one that removes it from the air (as precipitation forms and falls to the ground):

$$\frac{dy_C}{dt} = k_1 y_A - k_2 y_C. \tag{13.62}$$

When combined with Eq. (13.61), this equation may be rewritten in the form

$$\frac{dy_C}{dt} + k_2 y_C = k_1 y_A = k_1 y_{A0} \exp(-k_1 t), \tag{13.63}$$

where y_{A0} is the initial mixing ratio in the air. The solution of Eq. (13.63), for the case of initially clean condensate, gives the time dependence of the chemical in the condensate:

$$y_C(t) = \frac{k_1 y_{A0}}{k_2 - k_1} (\exp(-k_1 t) - \exp(-k_2 t)). \tag{13.64}$$

We naturally care most about how well the cloud, and the precipitation it yielded, was able to remove the chemical, so we calculate the total mixing ratio:

$$y_{tot} \equiv y_A + y_C = \frac{y_{A0}}{k_2 - k_1} (k_2 \exp(-k_1 t) - k_1 \exp(k_2 t)). \tag{13.65}$$

Now, two exponential time scales are involved, and we need to compare the two. When attachment of the chemical to the condensate is the rate-limiting step, meaning $k_1 << k_2$ (or $\tau_1 >> \tau_2$), then $y_{tot} \simeq y_{A0} \exp(-k_1 t)$. At the other extreme, when removal by precipitation is the slowest step, $k_2 << k_1$ (or $\tau_2 >> \tau_1$), we find $y_{tot} \simeq y_{A0} \exp(-k_2 t)$. The slowest step in a sequence always imposes the greatest limit to the overall removal rate and so shows up in the exponential expression. Only when the two time scales are comparable do we need the full equation (13.65).

Determining the rate coefficients to be used in Eq. (13.65) in terms of physical scenarios is challenging, but fortunately we need to make only order-of-magnitude comparisons. Our earlier analyses of both aerosol and gas scavenging suggested time scales on the order of seconds, at most minutes. Precipitation forms and removes condensate over time frames closer to an hour, so it is generally accepted that removal by precipitation is the rate-limiting step in chemical wet deposition. One can qualitatively appreciate such an assertion by thinking of the many clouds that form but never yield precipitation at all. Chemicals may well be taken up by the condensate in any cloud, but it takes precipitation to remove the substances from the atmosphere (at least via wet deposition). When condensate removal by precipitation is the rate-limiting step, we may simplify our view of clouds and use a single measure of wet deposition.

A single-parameter approach to quantifying the effects of clouds on the removal of chemicals from the atmosphere has been employed for many years. Historically, all the details of the scavenging processes and precipitation formation were ignored, but even now the use of a single parameter to describe wet deposition is all one can hope for in field programs when mechanistic details are not available, or when large-scale models of atmospheric chemistry need to take at least the first-order effects of clouds into account.

The traditional terminology for wet deposition differs slightly from that used above, but the concept follows from Eq. (13.65) when the second step (precipitation microphysics) is rate limiting ($k_2 << k_1$). The single parameter used for wet deposition has traditionally been called the scavenging coefficient (although a more accurate term might be the wet-deposition coefficient) and given the symbol Λ. The scavenging coefficient is the first-order rate coefficient for the removal of atmospheric chemicals from the air, so following Eq. (13.61) we see that Λ is the relative depletion rate of a chemical in the air (due to the precipitating cloud):

$$\Lambda \equiv -\frac{1}{n}\frac{\partial n}{\partial t}. \tag{13.66}$$

(The traditional measure of chemical abundance in the air is concentration n, but we could also use the equally valid molar mixing ratio for greater generality.) The dimensions of Λ are always reciprocal time [s^{-1}], and its magnitude ranges from about 10^{-5} s^{-1} to 10^{-3} s^{-1}. The times scales for removal by wet deposition thus vary from 10^5 s ($\sim$1 d) to 10^3 s ($\sim$ 1/3 h).

Integration of Eq. (13.66) over the time period that a cloud processes chemicals and removes them from the air connects first-order removal to an alternative measure of cloud-scavenging effectiveness, the scavenging ratio. Let a typical time scale for precipitation be τ_{precp}, then integration of the first-order rate equation (13.66) yields the before and after abundance of chemical j in the air:

$$\frac{n_j(\tau_{precp})}{n_{j,0}} = \exp\left(-\Lambda\tau_{precp}\right). \tag{13.67}$$

Any substance removed from the air must have gone into the precipitation itself, so we use the mass continuity principle to calculate its concentration in the precipitation. In a unit volume of air, the concentration difference, $\Delta n_j = n_{j,0} - n_j(\tau_{precp}) \simeq n_{j,0}\Lambda\tau_{precp}$, is related to the concentration of chemical j in the precipitation by $c_j f_L = \Delta n_j$, where $f_L \sim 10^{-6}$ is the volume fraction of liquid-equivalent precipitation. So,

$$c_j = \frac{n_{j,0}\Lambda\tau_{precp}}{f_L}. \tag{13.68}$$

The two variables relevant to the overall process are this concentration in the precipitation and the original concentration in the air. The ratio of these two concentrations is the so-called scavenging ratio

$$W_j \equiv \frac{c_j}{n_{j,0}} = \frac{\Lambda\tau_{precp}}{f_L} \propto \Lambda. \tag{13.69}$$

The scavenging ratio is thus seen to be directly related to the scavenging coefficient. Both W_j and Λ are measures of a cloud's ability to remove trace chemicals from the air.

Practical application of scavenging ratios must take the units of measure into consideration. A ratio of concentrations, as used in Eq. (13.69), is technically dimensionless, but the scavenging ratio represents the amount of chemical in two different phases. Ideally, the

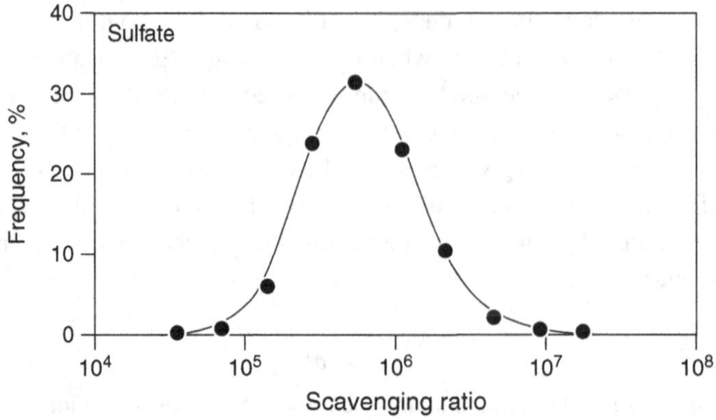

Figure 13.26 Frequency of occurrence of the specified scavenging ratio. Based on data from stations in the eastern United States. Adapted from Hicks (2005).

concentrations in each phase are based on the same volume [e.g., m^3]. Sometimes, however, the liquid-phase concentration is expressed as an amount per liter of liquid, while the gas-phase concentration is based on amounts per m^3. Such scavenging ratios will then be off by a factor of a thousand. Scavenging ratios derived from atmospheric data and based on a common volume are shown in Fig. 13.26, where we see that scavenging ratios tend to be distributed log-normally, with magnitudes of W_j ranging from about 10^5 to 10^7. Because of the small fraction of a given volume occupied by precipitation (the factor f_L), a substance removed from the air becomes confined to a much smaller volume in the condensed phase. It is sometimes helpful to keep the concentrations as those in air, in which case the concentration of aqueous-phase j per unit volume of air is $c_j f_L$, and a new dimensionless parameter is defined as the scavenging factor,

$$F_j \equiv \frac{c_j f_L}{n_{j,0}} = W_j f_L. \tag{13.70}$$

Scavenging factors require knowledge of the liquid fraction (f_L), but then they offer a truly dimensionless measure of scavenging effectiveness.

Scavenging ratios, while serving as transfer functions relating output to input, are strictly empirical in nature. They can be determined from field programs, if the measurements are appropriately made, and they can be calculated from detailed numerical models. However, interpreting scavenging ratios in terms of specific microchemical processes or meteorological scenarios is difficult. The magnitudes of W_j often depend on the properties of both the chemical species in question and the cloud doing the scavenging. Improvements have been sought, for instance, by taking the solubility of a trace gas into account, but additional efforts are needed.

The concept of fractionation may lead to improved measures of cloud-scavenging effectiveness by helping to reduce the variability introduced by differing meteorologies and precipitation efficiencies. Fractionation often occurs in nature when a process leads to a separation of constituents; one constituent in a mixture becomes enriched relative to another as a result of the process. Fractionation occurs in clouds because the chemicals and water originally in the air are processed differently. During initial condensation, for instance, some of the particles are activated and enter the cloud water, while the others remain in the interstitial air. Both the water and the particulate matter enter the cloud drops, but at different rates. With continued condensation onto a non-volatile solute, for instance, the drops become enriched in water relative to the solute. Gas uptake, too, is a fractionation process. Some gas is absorbed as new water is condensed onto a drop, but the gas and the water enter the drop at different rates. In general, clouds process water (from vapor to precipitation) at the same time that they process chemicals (from an atmospheric constituent to a component in precipitation). We seek a general measure of scavenging, one that emphasizes the fates of the chemicals within a given set of cloud processes.

Within the context of cloud scavenging and wet deposition, fractionation may be quantified by forming a factor that is a measure of the enrichment of the chemical in question relative to the condensate. If a cloud processes a particular chemical more efficiently than it does the water, the chemical will be enriched relative to the water. The concept is easiest to understand by expressing the concentrations of substances as mole fractions, for then we are simply counting molecules. Values in air use the symbol y, whereas values in liquid (precipitation) use x. The mole fraction of substance j in the air flowing into a cloud is y_j and that of water vapor is y_{H_2O}. The cloud processes each of these compounds in its own way, giving rise to the liquid-based mole fractions in the precipitation, x_j and x_{H_2O}. If the cloud were (surely only by coincidence) to remove substance j with the same efficiency that it removed water from the atmosphere, the relative abundances of j and H_2O in the air and in the precipitation would be the same; the ratio y_j/y_{H_2O} would then equal x_j/x_{H_2O}. In general, these two ratios would not be equal, meaning that the cloud processed the one substance more effectively than it did the other. With this motivation, we can see that a good measure of the enrichment of j relative to water is given by the following ratio of ratios, which we call a relative removal efficiency:

$$\epsilon_j \equiv \frac{x_j/x_{H_2O}}{y_j/y_{H_2O}}. \tag{13.71}$$

Through this particular type of enrichment factor, the relative proportion of j to water in the precipitation (numerator) is compared directly with its relative proportion to water in the air entering the cloud (denominator). The relative removal efficiency is insensitive to the dynamical structure of the cloud because both substance j and the water vapor are drawn into the cloud at the same time and in the same way. The relative removal efficiency responds mostly to the microchemical mechanisms responsible for scavenging.

Practical application of the relative removal efficiency can be made by rearranging Eq. (13.71). Even heavily polluted precipitation is still essentially all water, so the mole fraction of water in precipitation, $x_{H_2O} \simeq 1$. The mole fraction of j in a dilute solution is therefore calculated from measured concentrations, c_j, by $x_j \simeq c_j/c_{H_2O}$, where

$c_{H_2O} = 55.6$ mol L^{-1} is a constant, the molar concentration of liquid water. At this point, Eq. (13.71) becomes

$$\epsilon_j \equiv \frac{y_{H_2O}}{y_j} \frac{c_j}{c_{H_2O}}. \tag{13.72}$$

The gas-phase mole fractions may be similarly converted to more convenient measures. A mole fraction is by definition a ratio of molar concentrations, so in air $y_j \equiv n_j / n_{air} = n_j RT/p$, where R is the gas constant, T is the temperature, and p is the barometric pressure at the measurement site. The mole fraction of water vapor in air at temperature T and pressure p is related to a measured dew-point temperature T_d by $y_{H_2O} \equiv n_{H_2O}/n_{air} = e_s(T_d)/p$. Thus, the ratio $y_{H_2O}/y_j = e_s(T_d)/\left(n_j RT\right)$. The molar concentration of liquid water can be expressed in mass units, giving $c_{H_2O} = \rho_L / M_{H_2O}$, and so we get

$$\epsilon_j \equiv \frac{\rho_V}{\rho_L} \frac{c_j}{n_j} = \frac{\rho_V}{\rho_L} W_j, \tag{13.73}$$

where we have used the defining relationship for the scavenging ratio (Eq. (13.69)) to show the tight connection that exists between the relative removal efficiency and the traditional scavenging ratio. The ratio of vapor density to liquid density $\rho_V/\rho_L \sim 10^{-5}$, so magnitudes of ϵ_j are much smaller and vary around unity. It is primarily the vapor density that varies from situation to situation that distinguishes the two measures of scavenging effectiveness.

Now, if the concentration of j in air is calculated from data based on a mass concentration m_j [e.g., g m^{-3}], then $n_j = m_j / M_j$, where M_j [g mol^{-1}] is the molar mass of j. So, a reasonable operational formula for calculating the relative removal efficiency is

$$\epsilon_j = \frac{M_{H_2O}}{\rho_L} \frac{e_s(T_d)}{RT} \frac{c_j^m}{m_j}, \tag{13.74}$$

when the aqueous-phase concentration of j is also measured in mass units (so $c_j^m = c_j M_j$). The main restriction to the determination of the relative removal efficiency, as it is for scavenging ratios in general, is that the gas-phase measurements be made in air that is representative of that entering the cloud in question.

Precipitating clouds have profound effects on the abundance of chemicals in the atmosphere. By effectively removing soluble compounds from the air by the various in-cloud and below-cloud scavenging mechanisms, such clouds provide the balance needed to counteract the emissions of chemicals into the atmosphere. Between rain or snow events, especially during prolonged droughts, dry deposition of course plays an important role in this chemical balance. Even then, the compounds most susceptible to deposition tend to be soluble gases and hygroscopic aerosol particles. Clouds, through aqueous-phase reactions contribute, along with gas-phase photochemistry, to transform less soluble gases into soluble ones that are then able to be scavenged efficiently. Clouds do not just act on the trace gases and the atmospheric aerosol along a one-way path; rather they themselves are influenced by the non-aqueous constituents ingested. Aerosol-rich air imposes important restrictions on the growth of the cloud particles and the ability of the clouds to precipitate and remove water and the trace substances. Pollutant-impacted clouds tend to be brighter and more reflective of solar radiation, giving them opportunities to influence climate. Cloud-chemical interactions are highly non-linear, and only through ongoing research can we hope to understand the full impact of clouds on our regional and global environments.

13.5 Further reading

Levin, Z. and W.R. Cotton (eds.) (2009). *Aerosol Pollution Impact on Precipitation.* New York: Springer, 386 pp. Chapters 6 (observed aerosol effects on clouds) and 7 (modeling of aerosol-cloud interactions).

Pruppacher, H.R. and J.D. Klett (1997). *Microphysics of Clouds and Precipitation.* Dordrecht: Kluwer Academic Publishers, 954 pp. Chapter 17 (cloud chemistry, scavenging).

Seinfeld, J.H. and S.N. Pandis (1998). *Atmospheric Chemistry and Physics.* New York: John Wiley and Sons, 1326 pp. Chapters 6 (aqueous-phase chemistry), 8 (particle diffusion, phoresis), 9 (aerosol thermodynamics), 11 (mass transfer of gases and aerosols), and 15 (scavenging processes).

Wallace, J.M. and P.V. Hobbs (2006). *Atmospheric Science: An Introductory Survey.* Amsterdam: Academic Press, 483 pp. Chapter 6 (cloud and precipitation chemistry).

13.6 Problems

1. Several different sets of units are commonly used in atmospheric and cloud chemistry, so we need to know how to convert from one set to another. Standard temperature and pressure (STP) are $T_0 = 273.15$ K (0 °C) and $p_0 = 101325$ Pa.
   (a) Derive a general formula for converting mole fraction y_j into mass concentration m_j of trace gas j (molar mass M_j) at any temperature T and pressure p.
   (b) What is the mass concentration m_{SO_2} (units, $\mu g\, m^{-3}$) of $y_{SO_2} = 1$ ppb of SO_2 ($M_{SO_2} = 64$ g mol^{-1}) under STP conditions?
   (c) What is the molar concentration of air n_{air} at STP?
   (d) Calculate the mole fraction y_v of water vapor in air when the specific humidity is 10 g mol^{-1}.

2. The trace gas SO_2 dissociates when it dissolves in cloud water to form a weak acid, very much as CO_2 does. Refer back to Chapter 2 as needed and calculate the following quantities under the assumption that no other non-aqueous substances are present. For this problem, the Henry's law coefficient $H_{SO_2} = 1.2 \times 10^{-2}$ mol L^{-1} atm^{-1}, the first dissociation constant $K_{1S} = 1.7 \times 10^{-2}$ mol L^{-1}, the second dissociation constant $K_{2S} = 6.6 \times 10^{-8}$ mol L^{-1}, and the water dissociation constant is $K_w = 1 \times 10^{-14}$ mol^2 L^{-2}. Use approximations when appropriate.
   (a) The solution pH when the gas-phase mole fraction $y_{SO_2} = 1$ ppb;
   (b) the total concentration of dissolved S(IV); and

(c) the partitioning of the various S(IV) species in solution. That is, what fraction (by moles) of the total dissolved gas is in the various forms: undissociated molecules ($SO_2 \cdot H_2O$), bisulfite ions $\left(HSO_3^-\right)$, and sulfite ions $\left(SO_3^{2-}\right)$?

3. Approximately how much would the pH of otherwise pure water change when the mole fraction of atmospheric CO_2 doubles from its current value of approximately $y_{CO_2} = 400$ ppm?

4. Rain is never pure water. Discuss the various microphysical processes by which sulfur compounds are scavenged by a warm stratiform cloud and removed from the atmosphere in its precipitation.

5. What compounds are typically found in rain forming in industrial regions of North America? How does the composition vary by season (summer, winter)?

14
Cloud electrification

14.1 Overview

The electrification of large cumulonimbus clouds often leads to lightning, an exciting, yet sometimes frightening phenomenon of nature. The generation of electric fields of sufficient strength to cause electrical breakdown of the air involves a broad range of scales, all the way from the size of individual units of electrical charge (electrons and protons) to that of the cloud itself. Both the microphysical processes and the macroscale motions of air within thunderstorms must act in coordinated ways for charges to separate and large electric fields to develop. The subjects of charging mechanisms, electric-field evolution, and discharge events are each extensive in their own rights. The treatment here offers a basic overview to show the interconnectivity of the various scales and processes.

14.1.1 Electrical structure of thunderstorms and the atmosphere

Thunderstorms, by definition, produce thunder, the audible consequence of lightning. Ever since the pioneering work of Franklin and d'Alibard in the mid-1700s, it has been recognized that lightning is an electrical phenomenon resulting from excess charges in various parts of the parent cloud. Subsequent investigations by Wilson in the 1920s and many others have enabled us to develop a valid conceptual model of thunderstorm electrification.

The simplest charge structure of a thunderstorm is envisioned to be similar to that shown in Fig. 14.1. A convective storm during its mature stage of development exhibits an anvil that represents the upper outflow of cloudy air from the storm's interior. The anvil of an electrically active storm typically contains excess positive charge, which resides on the relatively small ice particles thrown upward by the strong vertical motions in the central part of the storm. The body of the storm contains the main negative charge region, where the temperature is between about -10 and -20°C, and a smaller region of positive charge in or just below the melting layer. The diffuse upper region of positive (+) charge and the concentrated region of negative (−) charge together constitute the principal dipole, which resides in a positive orientation (+ above −). The poles often contain some tens of Coulombs each and are separated vertically by several kilometers. The lower positive charge region, when it exists, contains just a few Coulombs and gives simple storms

Figure 14.1 Cartoon overlay of the observed electrical structure of a thunderstorm. Photo by D. Lamb.

a tripolar structure. The distribution of charged particles in many large or multicellular storms can be much more complicated, involving multiple, horizontally off-set charge centers, as well as screening layers due to interactions with ions and electric fields in the larger environment.

Spatially separated charges generate electric fields and cause currents to flow. Recall that an electric field ($\mathbf{E_e}$) is a vector, defined as the Coulombic force ($\mathbf{F_e}$) per unit charge (q_1): $\mathbf{E_e} = \mathbf{F_e}/q_1$. The polarity of the field is defined as the direction a positive charge would move if it were otherwise unconstrained (i.e., the field points in the same direction as the force acting on the positive charge). So, any general charge $\pm q$ caught in an electric field $\mathbf{E_e}$ experiences a force: $\mathbf{F_e} = \pm q\mathbf{E_e}$. Electric field is an alternative description for the gradient of electric potential Φ_e, but with the sign reversed: $\mathbf{E_e} = -\nabla\Phi_e$. The SI unit of potential is the volt [V], so the strength of an electric field is typically specified as the number of volts per unit distance [V m^{-1}]. Electric potential represents the work to move a unit charge against the field, so the volt is an energy per unit charge [J C^{-1}]. Note that potentials are scalar quantities, but gradients of potentials are vectors. Positive charges exposed to a gradient of electric potential experience forces that try to move them down the gradient (i.e., toward lower potential). How readily the charges move depends on the medium in which they are embedded and is characterized by an electrical conductivity σ or resistivity $\rho \equiv 1/\sigma$. The current density $\mathbf{j}$ [A m^{-2}] is proportional to the conductivity and the electric field: $\mathbf{j} = \sigma\mathbf{E}$. A highly conductive medium (e.g., a metal or ionized air) lets a lot of charge flow for a given potential gradient, whereas a highly resistive medium (e.g., a neutral gas) lets only a small amount of charge flow. Note that current density represents a flux of charge, as it is equivalent to the product of the charge density qn_q and the average drift velocity $\mathbf{v}_q$ of the charges in the field: $\mathbf{j} = qn_q\mathbf{v}_q$, where q is the charge on particles moving with velocity $\mathbf{v}_q$, and n_q is the number density of the charges.

The specific mechanisms by which electrical charges separate from neutral matter may be uncertain, but we do know that charge (as with mass and energy) must be conserved. For instance, when two dissimilar solids (e.g., rubber and wool) are rubbed together to generate a static electric field, electrons (the units of negative charge) move from one substance (the one holding electrons weakly) to the other (the one that attracts electrons more strongly). The one substance gains electrons and a net negative charge, while the other loses an equal number of electrons and acquires a net positive charge; the electrons have simply moved from the one substance to the other. In fluids, the exchange of electrons between one component and another may be more complicated and involve freely moving ions (molecules with unequal numbers of electrons and protons), but the principle of charge conservation persists. The generation of static electric fields is always caused by the separation of charges, while the total (or net) charge of the system remains unchanged.

Throughout the atmosphere, air molecules are occasionally ionized (converted to ions) in ways that affect the electrical conductivity of the air. Cosmic rays (mostly high-energy protons) routinely enter the atmosphere from outer space and collide with the nitrogen and oxygen molecules, liberating some of their electrons and many sub-atomic particles. The primary event causes a cascade of secondary collisions that also ionize the air. Some of the molecular, or "small" ions are highly mobile; they may recombine rapidly with ions of opposite charge and become neutral again, or they may react with atmospheric gases and form more exotic ions. Other ions may attach themselves to aerosol particles to become "large", sluggish ions. The rates that ion pairs are generated vary greatly with altitude and latitude, but they can exceed $1\,\mathrm{cm}^{-3}\,\mathrm{s}^{-1}$ in the upper troposphere. Cosmic-ray induced ionization gives the atmosphere a more or less uniform background conductivity. Natural radioactive decay (especially from radon) also ionizes the air, but its effects are generally limited to continental air masses. Solar ultraviolet (UV) radiation is less energetic than nuclear radiation, but it is the primary cause of ionization in the upper atmosphere above about 60 km. This upper-atmospheric region, called the electrosphere, is especially conductive and so tends to have a uniform electric potential (about 300 kV relative to the ground). The electrosphere strongly affects the long-range propagation of radio waves, and it permits thunderstorm activity to have global impacts.

The remote regions between thunderstorms exhibit weak electric fields that vary only slightly throughout the day. This fair-weather field amounts to slightly over $100\,\mathrm{V\,m}^{-1}$ near the surface and a few $\mathrm{V\,m}^{-1}$ in the middle atmosphere. The fair-weather field arises from a perpetual excess of positive charge in the electrosphere. As suggested by Fig. 14.2, the electrosphere and earth together act as a spherical capacitor. The fair-weather electric field between the upper and lower electrodes thus points downward (while the potential gradient points upward). This sign convention, as used in physics, is not always adhered to in the atmospheric sciences, so care is warranted when reading the literature. The atmosphere acts as non-uniform resistance filling the space between the capacitor electrodes, so the fair-weather field causes a small (positive) electric current to flow downward. The magnitude of the conduction current density is about $j \simeq 2 \times 10^{-12}\,\mathrm{A\,m}^{-2}$, which amounts to a global current of about 1 kA. This leakage current continuously drains positive charge from the electrosphere and is sufficient to discharge the capacitor in just a few minutes were it not for mechanisms that replenish the charges on the capacitor.

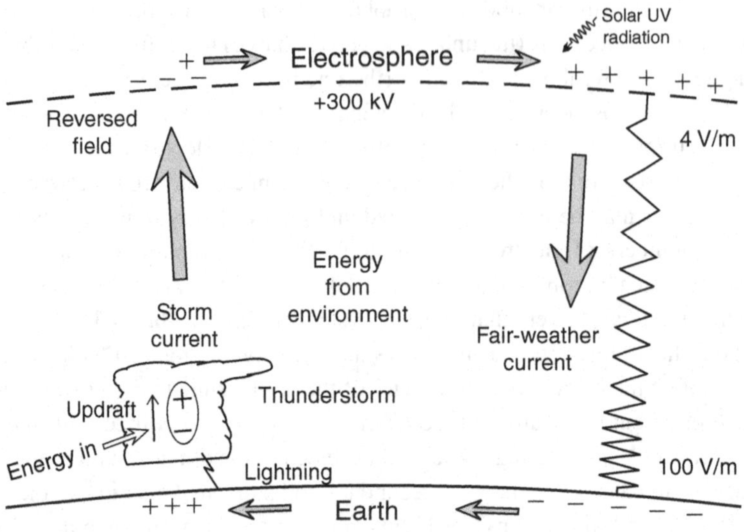

Figure 14.2 Schematic of the global electric circuit. Shaded arrows depict the direction of positive current; the open arrow suggests that the source of energy to drive the updraft and thus the thunderstorm generator (ellipse) is the atmosphere. The electric potential of the electrosphere, derived physically from the storm current, is relative to that of the ground. The return flow of charge through the non-uniform resistance of the atmosphere, shown on the right-hand side, causes the indicated potential gradients in fair weather. Adapted from MacGorman and Rust (1998).

The excess positive charge in the electrosphere is maintained through electrical discharges from thunderstorms. Wilson, in 1920, suggested that the estimated 1800 thunderstorms that are always active around the globe serve as electrical generators that collectively supply positive charge to the electrosphere (the storm current identified in Fig. 14.2) and negative charge to the Earth (mainly via lightning). The principal dipole within each thunderstorm, formed by processes discussed later, contributes its share of current to the global electric circuit by locally reversing the electric fields below and above cloud. The energy to drive the global electric circuit comes from the mesoscale environments that generate the cloud motions. Solar UV radiation gives the electrosphere its high conductivity (by causing pairs of ions to form from neutral air molecules), but it takes thunderstorms to give the electrosphere its net positive charge.

Any mechanism of cloud electrification must account for several commonly observed features of thunderstorms. Past research suggests that the principal dipole in a thunderstorm is generated within the core of deep convection. The processes of charge separation are still debated, but they appear to be associated with strong radar echoes, hence precipitation-size particles. The convection is clearly driven by buoyancy and mesoscale pressure gradients, but the precipitation is a product of the cloud microphysics. Consistent with the observed electrical structure, new charge appears to be generated within an active thunderstorm cell at a rate of about $1\,\mathrm{C\,km^{-3}\,min^{-1}}$, and that charge is separated on the cloud scale at a rate sufficient to produce breakdown electric fields of 100 to 300 kV m^{-1} within about 20 min. The carriers and possibly also the generators of charge appear to be graupel particles (which grow by riming) and pristine ice crystals (which grow by vapor deposition). The concentration of the charged particles is in the order of 100 m^{-3}, with each particle carrying an average charge of 10 to 100 pC. These diverse features serve as a basis for testing any hypothesized mechanism of electrification.

14.1.2 Schools of thought

Our understanding of thunderstorm electrification follows two schools of thought. The relationships of these two schools to each other and to cloud electrification are shown in Fig. 14.3. The so-called convective-charging school emphasizes the dynamic nature of the clouds, whereas the collisional-charging school focuses on the microphysics of precipitation formation. Collisional charging is further categorized into inductive and noninductive mechanisms. Inductive-charging mechanisms require the presence of an electric field, whereas the noninductive mechanisms do not. Various charging mechanisms may contribute simultaneously to cloud electrification, but each is treated separately for the sake of clarity.

The convective-charging (or Grenet–Vonnegut) mechanism starts with the upward motion of air in the presence of ions within a preexisting electric field. At first, the ions arise from cosmic radiation, later from corona discharge near sharp objects, where the electric field can become locally large. The ambient electric field is initially the fair-weather field, but it later increases rapidly as the convective cloud matures. The subsequent stages of charge separation by convection are shown in Fig. 14.4. The positive charge that resides on aerosol particles in the boundary layer, later on the cloud particles, is advected through cloud base and into the free troposphere. The overall positive charge of the cloud preferentially attracts negative ions in the environmental air, forming a negative screening layer

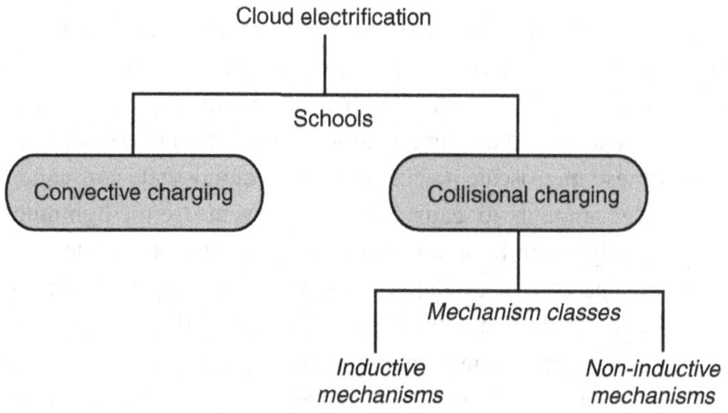

Figure 14.3 Heirarchical depiction of the alternative schools of thought regarding the electrification of clouds.

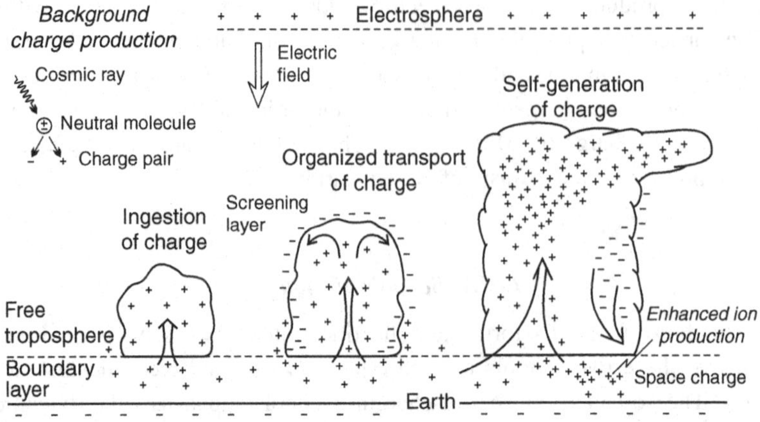

Figure 14.4 Cartoon illustrating the main stages of convective charging. Adapted from MacGorman and Rust (1998).

that is carried downward along cloud edges (where drop evaporation cools the air and provides negative buoyancy). The redistribution of charge by the convection enhances the ambient electric field, which eventually causes corona discharge in the boundary layer and a new supply of positive space charge that is advected into the cloud. The process is self-reinforcing in principle and could lead to lightning under ideal conditions. Although critics of convective charging point to problems with the amount of space charge available and to the long time scales for field development, little doubt surrounds the importance of convection in moving electric charge within a storm. Indeed, the lower positive charge found in many storms has been thought by some to result from the advection of positive space charge into the cloud by the updraft.

Collisional-charging is based on an exchange of charge between colliding cloud particles. Collisional interactions are of course necessary for precipitation formation, but for collisions to transfer charge they must be followed by rebound, not coalescence or aggregation. Collisional charging takes place during precipitation formation, but it involves those collisions that do not contribute to particle growth. Charging by collisions may be thought of as the complement of growth by collisions, requiring both collision and rebound. How charge might be exchanged during collision-rebound events is discussed later, but for now recognize that these microscale mechanisms have cloud-scale consequences.

Electric fields large enough to cause discharge events (e.g., lightning) arise from collisional charging only when both the small and large scales operate together, as suggested by Fig. 14.5. The charge exchanged during microscale collision-rebound events in the generation zone yields two sets of particles with differing net charge, relatively large negatively charged particles and smaller positively charged particles. The organized motions within the cloud then carry each set of particles to different parts of the cloud. The microphysical collisions may separate the charge from initially neutral cloud particles, but such microscale charge exchange would not lead to significant cloud electrification without

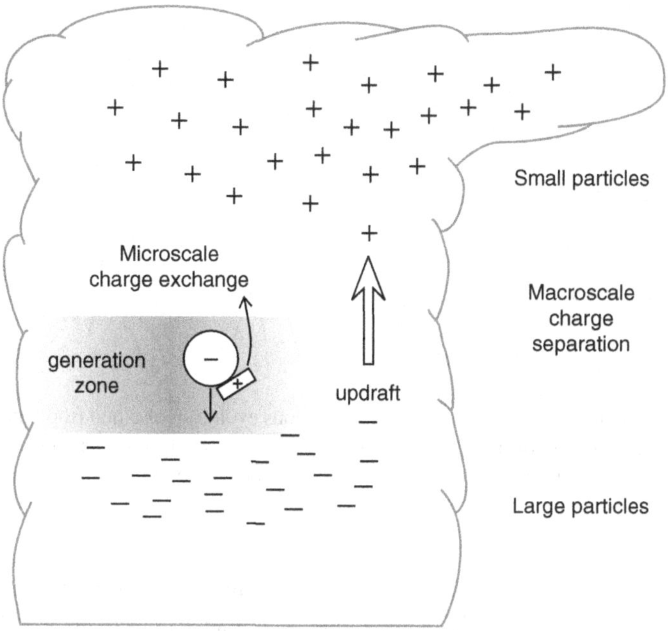

Figure 14.5 Schematic illustration of the two scales involved in collisional charging.

gravitational separation of the charged particles on the cloud macroscale. It is the differential sedimentation of the particles that lets the charge separated on the microscale develop into cloud-scale electric dipoles.

14.2 Macroscale charge separation

The cloud as a whole gets electrified by collisional charging mechanisms because the particles involved are different in size and therefore have different fallspeeds. Collisions occur only when one particle falls faster than another, as during precipitation formation, and large particles tend to settle lower in a given updraft than do small particles. Under typical situations, the smaller particles gain positive charge and ascent with the updraft toward cloud top, while the larger particles gain negative charge and lag behind or settle toward lower altitudes.

The coevolution of a cloud and its electric field has been studied observationally on several occasions. The findings from a research mission into a small thunderstorm involving several aircraft and radar are summarized in Fig. 14.6. The cloud started with predominantly upward motions (broad arrows), when the electric field was weak and probably dominated by the fair-weather field. Precipitation, presumably in the form of small graupel several millimeters in diameter, began to develop at altitudes between about 7 and 8 km within about 5 min after the onset of organized convection. The electric field developed

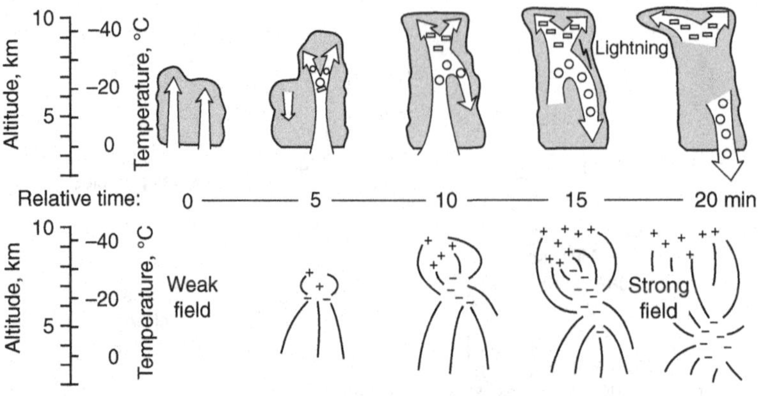

Figure 14.6 Schematic representation of the simultaneous evolution of cloud (top) and electric fields (bottom) in a small thunderstorm. Circles in top panel represent graupel, rectangles represent pristine ice crystals. Time is given relative to the onset of organized convection (left-most figure). Adapted from Dye *et al.* (1986).

appreciably only after precipitation-size particles sedimented (after relative time 10 min) in a weak downdraft on one side of the cloud. It is likely that graupel particles carried the negative charges downward, while the smaller ice crystals carried the positive charges upward. The resulting separation of charge caused a cloud-scale positive dipole to form and the electric field to increase rapidly over the next 7 min. A single in-cloud lightning strike was thought to occur at a relative time of about 17 min. Thereafter, both the cloud and the electric field decayed rapidly. Such observations confirm our view that cloud electrification is tied closely to the development of precipitation within convective clouds.

The formation of a cloud-scale electric dipole may also be visualized by considering a simple semi-quantitative model of a cloud containing a bimodal population of particles. Let the smaller particles have a mean radius r_1 and fallspeed v_1. The larger particles have a mean radius r_2 and fallspeed v_2. The concentrations of the two sub-populations are n_1 and n_2, respectively. These two sub-populations collide with each other at a volumetric rate [SI units, $m^{-3}\,s^{-1}$] given by

$$J_{12} = K_{12} n_1 n_2, \tag{14.1}$$

where $K_{12} = \pi\,(r_2 + r_1)^2\,(v_2 - v_1)\,E$ is the gravitational collision kernel and E is the collision efficiency. A fraction ϵ of these collisions causes growth of the larger particle by coalescence, but the remaining fraction $(1 - \epsilon)$ results in rebound and a small separation of charge δq per collision-rebound event. The rate of collision-rebound events is therefore $J'_{12} = (1 - \epsilon) J_{12}$, and the rate that charge develops in a unit volume of cloud within the generation zone is

$$dQ/dt = J'_{12}\delta q = \pi\,(r_2 + r_1)^2\,E(1 - \epsilon) n_1 n_2 \Delta v \delta q. \tag{14.2}$$

Here, Q represents the volumetric concentration of positive charge [$\mathrm{C\,m^{-3}}$] residing on the smaller particles (which will soon be pushed upward out of the generation zone). The same difference in fallspeed ($\Delta v = v_2 - v_1$) that lets collisions take place also causes differential sedimentation of the particles on the cloud scale. In a cloud of constant and uniform updraft speed, the two populations separate inside the cloud at the rate $d\Delta z/dt = \Delta v$, where Δz is the vertical distance between charge centers. After time interval Δt, the separation distance becomes $\Delta z = \Delta v \Delta t$. The electric dipole strength $P = Q\Delta z$ thus develops at a rate having two components:

$$\frac{dP}{dt} = Q\Delta v + \frac{dQ}{dt}\Delta z. \tag{14.3}$$

The first term represents the macroscale separation of charge, the vertical displacement of charge centers of magnitude Q away from each other at the differential sedimentation rate Δv. The second term represents the contribution due to the microscale charge exchange, the development of new charge within centers already separated by Δz. Combination of Eqs. (14.2) and (14.3) show that both components are proportional to the difference in particle fallspeeds, and that the microscale generation of charge (by some combination of inductive and noninductive processes) is additionally proportional to the amount of charge exchanged during each collision-rebound event. The updraft speed does not enter directly into such a simple model, but a substantial updraft is required to keep the large particles aloft over the time period needed for the dipole strength to build to breakdown potential. Only more complete models, those that include appropriate air motions and detailed microphysical interactions, can yield a realistic picture of electric field evolution in a thunderstorm.

Numerical cloud models are valuable tools for investigating the development of the electric field in a growing thunderstorm. Not all models necessarily need to include all the dynamical and microphysical processes known to contribute to cloud development; rather, a model can be useful simply by providing a means for comparing the effects of contrasting charging mechanisms. The relative importance of inductive versus noninductive charge exchange as alternative collisional-charging mechanisms, for instance, has been shown in a model representing the cloud circulation simply as a steady-state, two-dimensional vortex. The evolution of electric field inside such a model cloud is shown in Fig. 14.7. Inductive charge transfer between colliding particles requires the presence of an ambient electric field, so when the field strength is small, the charge transferred between particles is small. As shown in Fig. 14.7, the field caused by inductive charging increases slowly at first, but it can eventually grow exponentially with time. By contrast, a noninductive mechanism, one not dependent on the ambient field, causes the field to develop from the start of convection. However, noninductive mechanisms become less effective as charge builds up on the graupel particles: the transferred charge must move against an ever-increasing potential difference between the colliding particles, and a balance is eventually reached between the rates of up-gradient and down-gradient charge transfer. The electric field due to noninductive mechanisms acting alone tends to saturate well below the value needed

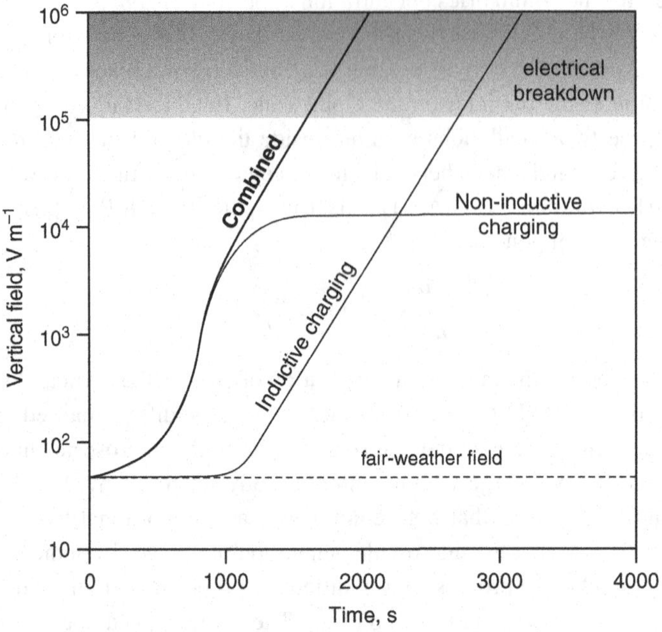

Figure 14.7 Evolution of the electric field in a model thunderstorm. Adapted from Kuettner *et al.* (1981).

for the air to breakdown electrically. The best opportunity for effective development of the electric field seems to exist when both the noninductive and the inductive processes operate simultaneously (bold curve in Fig. 14.7). The inductive exchange of charge acts to amplify the charge transferred by the noninductive process. One of the biggest uncertainties in any model of cloud electrification is the mechanism and magnitude of charge transfer between colliding particles, so further consideration of microscale charge separation is warranted.

14.3 Microscale charge separation

How electric charge is exchanged between ice particles experiencing rebounding collisions is still hotly debated. Theories abound, but little agreement has been reached on the mechanism by which charge is transferred from one particle to the other. As we have seen, most ideas fall into either of two classes, inductive (field-dependent) and noninductive (field-independent) mechanisms. Some sense of the diversity of ideas is provided below, but firm conclusions should not be expected. Some ideas are based on the physical properties of pure systems, others depend on the influence that atmospheric chemicals have on the surface properties of ice.

The essential features of inductive charging were developed initially in 1885 by Elster and Geitel. Any conductive object in an external electric field will become polarized, as

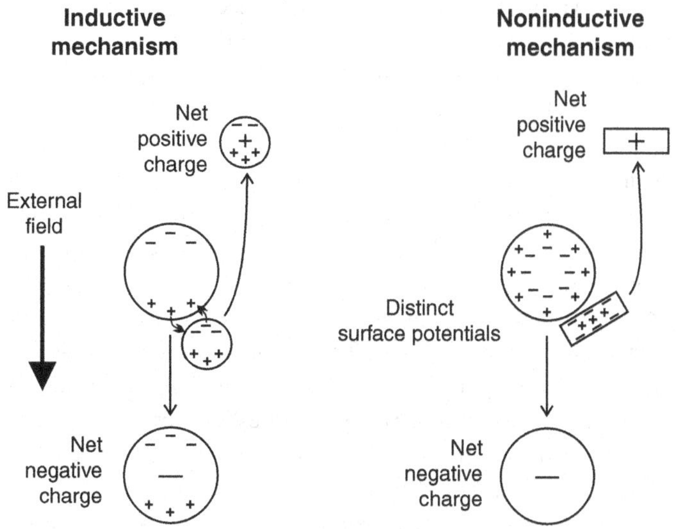

Figure 14.8 Schematic illustrating the two classes of particle charging mechanisms. Left panel: Inductive charging (circles represent any types of colliding particles). Right panel: Noninductive charging (circles represent a graupel particle growing by riming, the rectangles an unrimed, vapor-grown ice crystal).

shown in the left panel of Fig. 14.8. In the normal fair-weather field, for instance, the electric field points downward (direction a positive charge would move), so mobile positive charges migrate toward the bottoms of the particles, negative charges toward the tops. Contact of the smaller particle anywhere on the lower hemisphere of the larger particle would let positive charge be transferred to the smaller particle. The result, following particle rebound, is a net positive charge on the small particle (which continues ascending in the cloud) and negative charge on the larger, faster-falling particle. The degree of polarization and thus amount of charge exchanged is proportional to the field strength. Weak fields, as at the start of electrification, allow only minimal charges to be exchanged. But, inductive-charging mechanisms lead to positive (reinforcing) feedbacks and rapid growth of the electric field. The stronger the field becomes, the greater the amount of charge separated per collision and hence the greater the rate of field development. The field eventually develops exponentially with time (and appears as a straight line on the log-linear plot in Fig. 14.7), at least until dielectric breakdown and discharge occurs.

Noninductive charging results from an inherent difference in the electrical properties of the colliding bodies, as suggested by the right panel in Fig. 14.8. By one means or another, one of the collision partners hangs onto its electric charge more tenaciously than does the other. Otherwise, the charge transferred to one particle would be the same as that transferred to the other, and no net exchange would take place. One sometimes talks about a difference in contact or surface potential as a measure of the ease with which charge transfers during collision. The time scales for charge equilibration and mass exchange must be considered, and one needs to account for the differences in the molecular structure, ionic charge density, and electrical conductivity of the respective surfaces.

Laboratory studies help us identify the physical factors responsible for charge separation in ways that field observations or theory cannot. Empirical data let us quantify charge exchange, and they constrain our thinking about possible mechanisms of action. Experiments by numerous research groups over many decades have been focused specifically on the charge transferred to graupel particles during collisions with unrimed ice crystals in the absence of external electric fields. Several physical factors have been found to affect the magnitude or sign of noninductive charge transfer: temperature, the concentration of supercooled liquid water as it affects riming rate (hence, the effective liquid water concentration), and impact velocity. The findings from the diverse experiments are not all self-consistent, but they do show that significant charging of the graupel particle occurs only during active riming (but below the threshold for wet growth). The graupel particle normally gains the negative charge, the ice crystal the positive charge, unless the effective liquid water concentration exceeds a threshold value (about $1\,\mathrm{g\,m^{-3}}$) or the temperature is greater than about $-10\,°\mathrm{C}$; then, the sign of charge exchange reverses. The precise conditions for sign reversal and the magnitudes of charge transfer may be uncertain, but the empirical evidence shows that electrification by noninductive mechanisms requires collisions between rimed and unrimed ice particles.

Given the fact that all cloud particles (including rimed and unrimed ice) are composed of water, one has to wonder what distinguishes the electrical properties of colliding particles in a thunderstorm. The situation in clouds clearly differs significantly from that of rubbing two completely different substances together to generate a static charge in the laboratory. The need for distinctive electrical properties of the collision partners in clouds is emphasized on the right side of Fig. 14.8 by giving the particles different shapes. These two classes of particles, while both are ice, do differ in their growth mechanisms, as well as in their life histories within the cloud. Ice crystals (shown as rectangles) grow by vapor deposition, whereas graupel particles (circles) grow by riming. Moreover, each type of particle usually arrives at the collision point from different regions of the cloud: the small vapor-grown ice crystals tend to follow the updraft and thus arrive from below the point of collision, whereas graupel particles are larger and arrive from higher up. Each particle type grows at distinct rates and with different temperatures.

Numerous explanations have been put forth to explain particle charging in pure water systems. Temperature differences alone, as caused by the release of latent heat during riming, can cause potential differences by the thermoelectric effect, which arises in ice by the greater mobility of positive protons (H^+) than of the negative hydroxyl ions (OH^-). However, the magnitudes of charge transferred are too small to account for thunderstorm electrification. Some laboratory studies have shown that the rate of growth of ice from the vapor can affect particle charge, so one expects local saturation ratios and surface temperatures to play roles in charge exchange. Experimental data involving colliding ice particles consistently show that those particles growing most rapidly by vapor deposition acquire the most positive charge. Explanations for such an effect vary, but they typically depend on the relative volatilities and mobilities of the H_3O^+ and OH^- ions in the ice lattice. The defect structure of ice and the thickness of the quasi-liquid layer have also been invoked to explain how the surface of one particle differs electrically from that of another.

Atmospheric clouds never form in pure environments, so the non-aqueous components must be considered in any complete explanation of charge separation. Laboratory researchers often treat ambient trace chemicals as contaminants, rather than as reagents in the electrification process. However, the charge transferred from one particle to another is invariably borne by one chemical compound or another. In truly pure systems, the carrier is a fragment of water, but in the natural world it may well be one or more chemicals derived from the air in which the cloud formed.

Trace chemicals can possibly influence the electrical state of ice surfaces because of the way they interact with ice. Before being accreted by a graupel particle, supercooled cloud drops invariably contain the substances in the original CCN, as well as acidic trace gases (e.g., HNO_3, SO_2, CO_2) taken up during condensational growth. The surfaces of pristine ice crystals, too, take up trace gases by adsorption during vapor deposition. The solutes typically dissociate and form ions (positive cations and negative anions). We can represent the chemical processes by letting an acidic gas be designated generically as HA and showing dissociation by the reaction $HA \rightarrow H^+ + A^-$, as shown in Fig. 14.9. The resulting solution (cloud water, left side; or the quasi-liquid layer, right side) is often acidic, meaning that the concentration of hydrogen ions (H^+) exceeds that of the hydroxyl ions (OH^-). The subsequent distribution of the various ions in the ice particles affects the electrical state of their respective surfaces. Freezing of supercooled cloud water (as occurs during contact with a graupel particle) excludes the anions from the ice phase because of their relatively large sizes. As shown in Fig. 14.9, anions, such as sulfate, nitrate, and carbonate, cannot enter the ice lattice itself and so are relegated to liquid inclusions within the polycrystalline ice when freezing is rapid and latent heat is transferred to the air. The protons (H^+), on the other hand, can permeate the ice lattice because their high mobilities (note that ice is a protonic conductor). Rapid freezing of a supercooled cloud drop during contact with a graupel particle retains its initial electrical neutrality, but the positive and negative charges are distributed differently. The surface of the frozen particle (hence of rime ice) is likely to have a positive charge, whereas the negative charges are restricted to the interior. By contrast, the anions of an acid acquired by adsorption during the vapor-growth of ice (right side of Fig. 14.9) are restricted to the surface, again because their large sizes make them energetically unfavorable for inclusion in the ice lattice. As with frozen drops, the H^+ ions may again permeate the ice lattice, leaving the surface of vapor-grown ice with a negative charge. The fate and distribution of ions differ significantly in rime ice and in vapor-grown ice.

The chemical stage is thus set for contact of pristine ice crystals with graupel particles to yield a net transfer of negative charge to the graupel and positive charge to the vapor-grown ice. It matters little whether the transfer occurs by mass exchange or electrical conductivity; only the concentrations of anions and cations on the respective surfaces are important. The

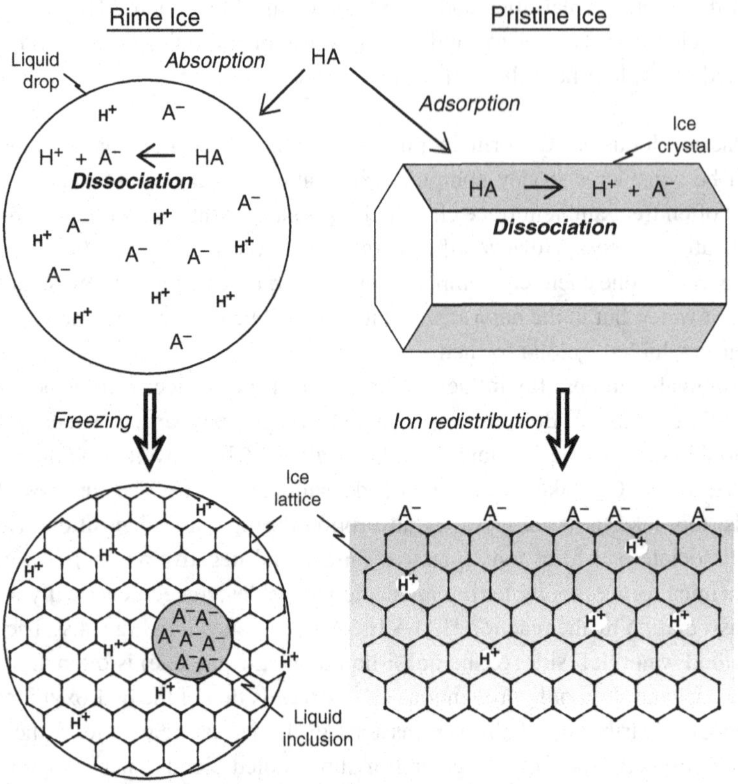

Figure 14.9 Hypothesized distribution of a generic acid (HA) in rime ice (left side) and in a vapor-grown ice crystal (right side). The anion is designated by A^-, whereas the hydrogen ion is shown as H^+.

electrical effect may actually reverse at higher temperatures or near the onset of wet growth, for then the direction of heat transfer during the slow freezing of accreted water reverses and favors the accumulation of anions on the surface of the graupel.

The difference in growth mechanism (riming or deposition) within a common chemical environment may help account for noninductive charging. However, a complete picture will emerge only after due consideration is made for all physical and chemical factors, including the size distribution of collected drops, gas solubility in supercooled water, solute retention during freezing, drop spreading on impact, relative heat transfers to the substrate and air, and gas uptake by ice surfaces. In any case, two steps seem to be required for significant charge transfer: (1) establishment of appropriate electrical states of the respective surfaces before collision, and (2) sufficiently rapid exchange of charge during contact (either by conduction or mass transfer). Despite much speculation as to the cause of particle charging, some combination of physicochemical effects most likely cooperate with each other to separate charge on the microscale.

14.4 Discharge events

As charges separate within a cloud, the electric field continues to builds up. At some point, however, the air that had been so effective in resisting the flow of electricity by conduction breaks down and becomes highly conductive. The process of breakdown is non-linear and leads to rapid changes in the properties of the air. The charges of opposite sign that were initially pushed slowly apart by the cloud-scale circulations suddenly find conductive paths toward recombination. The electrical breakdown of the air leads to the discrete discharge events we recognize as lightning within or below a thunderstorm and as transient luminescent events (TLE) above the storm.

14.4.1 Breakdown

The electrical breakdown of air starts with the acceleration of free electrons by the electric field. Any electric field of course causes electrons to accelerate and collide with the neutral molecules in its path. When the field strength is low, however, the amount of kinetic energy contributed by the field to the collisions is also low, perhaps little more than that caused by ordinary thermal collisions. As the field strength increases, so does the acceleration of electrons in the space between molecules; the kinetic energy at impact increases and causes increased agitation of the neutral molecule. Once the field is strong enough to give the accelerating electron enough kinetic energy at impact to strip a new electron from the air molecule, then the process of breakdown has begun. Ionization leads to a new population of electrons, each one of which feels the effect of the electric field and becomes a possible new source of ionization. The process leads to an avalanche of electrons and the catastrophic failure of the air to serve as an insulator.

An early stage of breakdown is corona. Corona, sometimes seen as a faint glow near sharp objects, is the precursor of sparks, not the main electrical discharge itself. A corona, or point discharge, is the mechanism by which new space charge is generated in the boundary layer, as indicated toward the right-hand side of Fig. 14.4. It is the high curvature near the corners of conductors that causes the electric field to reach the breakdown threshold before it is reached elsewhere. As suggested by Fig. 14.10, geometry plays a fundamental role in distorting the local field, generating space charge, and initiating sparks. The conductivity of an object, even if not metallic, allows electrons to flow along its surface in response to the applied electric field. But, the object itself distorts the field in its immediate vicinity in a way that ensures that the surface of the conductor stays at a constant potential. Gradients of electric potential are then zero parallel to the surface, and non-zero only normal to the surface. Free electrons in the vicinity of a tip, where the potential gradient is greatly enhanced, may well achieve the large kinetic energies needed to ionize air molecules. Note that the ions so produced move in opposite directions in the local field. The negative electrons are drawn into a positive object (the example shown) and so are lost from the air. The positive ions, on the other hand, are forced away from the object and into regions of weaker field. Were the object to have negative polarity, it would remove the

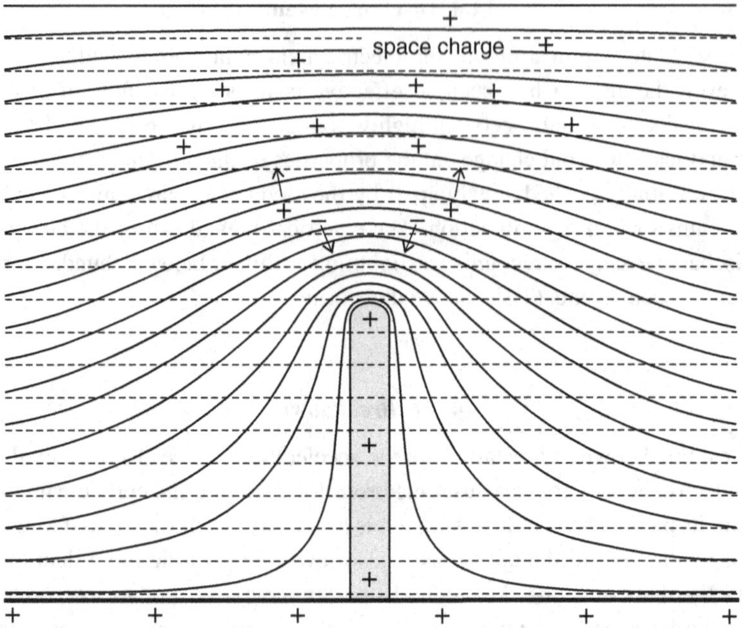

Figure 14.10 Distortion of the electric field in the vicinity of a conductive object. Solid curves represent equipotential surfaces in the presence of the object; dashed curves, show the equipotentials in the absence of the object. Close spacing of equipotentials indicate high potential gradients and strong electric fields that drive space charge toward or away from the object.

positive ions and release electrons into the air. The ions having the same sign as the object become the new space charge, first as small, molecular ions, later as large, aerosol-based ions. The presence of the space charge reduces the local field strength and hence the production of new space charge. The process is ultimately limited by how rapidly the ions are advected away by air motions (wind). Corona, through its ability to generate space charge and increase the conductivity of the air, leads to favored current paths during lightning discharges.

14.4.2 Lightning

As the electric field increases within and below a cloud, and the air becomes progressively more conductive, a continuous path to another object becomes possible. Lightning may be viewed as a large spark between one charge center and another, a visible manifestation of an electrical connection having been made. However, parallels between lightning in the atmosphere and sparks in the laboratory must be drawn carefully; the charge centers in clouds are diffuse and contain finite charge, quite unlike metal electrodes connected to power supplies. Caution is also needed when learning about lightning. The processes that lead to the initiation and evolution of lightning are still not well understood, so explanations range widely and must be interpreted critically.

Electrical phenomena have been classified in many ways. The terminology often stems from common observations, sometimes from consideration of the physical processes involved. Common designators of lightning include heat (glows from distant lightning on a hot day), forked (highly branched), sheet (diffuse-looking), and dry (from clouds yielding little rain). Discharges above the thunderstorm are generally called transient luminescent events (TLE). TLE are sub-classified into sprites (vertical carrot-like tendrils in the mesosphere that occur in association with strong positive lightning), jets (narrow bluish columns emanating from the anvil into the stratosphere), and elves (broad reddish disks near the base of the electrosphere that may be associated with lightning-induced electromagnetic pulses). TLE may help transport charge through the middle atmosphere and into the electrosphere, whereas lightning transports charge within the troposphere.

Perhaps the most fundamental distinction between lightning types is based on the locations of the charge centers, whether from one part of a cloud to another (cloud-to-cloud lightning) or between the cloud and the ground (cloud-to-ground lightning). Most cloud-to-cloud discharges occur within the parent cloud, and is then designated intra-cloud lightning. Lightning between charge centers in different clouds or cells of a thunderstorm is called inter-cloud lightning. Cloud-to-ground lightning, even though not most common, is the most spectacular because it is readily visible from the ground and a threat to life and structures on the surface. Cloud-to-ground lightning usually helps discharge the main negative charge center in a cloud and so bring negative charge to Earth. Positive lightning occurs infrequently, but it is then very powerful because it travels a great distance through the air, from the upper positive charge center in the anvil to the ground (often well away from the cloud).

Lightning arises from a complex and still little-understood sequence of events. At least for cloud-to-ground discharges, some researchers have suggested that the growing electric field increasingly polarizes and distorts raindrops in the lower positive charge region, thereby forming many streamers. The electrons liberated from such water-point coronas make the air in the lower cloud highly conductive, which in turn allows negative charge to propagate downward. As the distance between the main negative charge region and the ground decreases, the magnitude of the potential gradient increases. A number of new streamers, called stepped leaders, form and advance, by distorting the local field and generating new corona, some tens of meters downward, increasing the conductivity of the air farther down. As new stepped leaders form and approach the ground, the field strength increases near the ground, which causes new corona and space charge to form in the lower boundary layer. Eventually, the potential gradient between one of the stepped leaders and an upward-propagating streamer becomes large enough to forge a direct electrical connection. A large current then flows along the entire length of the highly conductive channel, partially depleting the in-cloud charge and transporting it to the ground via a so-called return stroke. Although the stepped leaders can be seen as numerous dead-end branches, it is the return stroke that is so highly visible and the cause of the thunder we hear.

The subsequent evolution of a lightning flash may involve several strokes. The first return stroke may have transported much charge to Earth, but it is unlikely to have drained all of the charge from the negative charge center in the cloud. But, the air in the channel is now highly ionized, a plasma that remains conductive for nearly a second after the first stroke. Subsequent return strokes, each initiated by dart leaders from the cloud and ground, continue the process of transporting negative charge to Earth. These subsequent strokes, because they follow a well-established conduit, look smooth compared with the forked appearance of the first return stroke amidst the many stepped leaders. A single overall flash of lightning is thus often composed of numerous strokes, each of which contributes to the discharge of the cloud.

The electrification of thunderstorms, as with precipitation development in general, arises only after a long series of microscale processes operates within a suitable dynamical, macroscale environments. Initially, cloud forms when upward motions, forced by mesoscale pressure gradients, cause moist air to cool and become supersaturated. The excess vapor condenses onto the aerosol particles also brought in with the air motions. The cloud droplets, chemically enriched by those same particles, may grow actively by condensation, but they themselves remain too small to precipitate or cause significant electrification. It is their collection by a few large drops that contributes to warm-rain development, alternatively by ice particles to yield cold rain or snow. One consequence of precipitation formation and its requirement for collection is the transfer of trace chemicals and electrical charge from one population of particles to another, from small to large. The difference in fallspeeds of these populations lets the collisions happen in the first place, and it allows the populations to separate gravitationally inside the cloud; the small particles rise to the top of the cloud, while the large particles hover near mid-levels when the updraft is suitably strong. The cloud becomes electrically polarized because the microphysics of collision and charge exchange operate within a macrophysical environment characterized by organized air motions. The processes continue until the air can no longer accommodate the electrical stress; the breakdown in resistance lets massive currents flow that discharge the charge centers. The effects of precipitation formation and cloud electrification are felt not only locally in the form of rain and lightning, but also broadly as contributions to the hydrological cycle and the global electric circuit.

14.5 Further reading

Cotton, W.R. and R.A. Anthes (1989). *Storm and Cloud Dynamics*. San Diego, CA: Academic Press, 880 pp. Chapter 9 (thunderstorm dynamics).

Doswell, C.A. (ed.) (2001). *Severe Convective Storms*. Boston, MA: American Meteorological Society, 561 pp. Chapter 13 (electrification in severe storms).

MacGorman, D.R. and W.D. Rust (1998). *The Electrical Nature of Storms*. New York: Oxford University Press, 422 pp. Tutorial (E and M basics), Chapters 1 (overview of atmospheric electricity), 3 (thunderstorms), and 5 (lightning).

Pruppacher, H.R. and J.D. Klett (1997). *Microphysics of Clouds and Precipitation*. Dordrecht: Kluwer Academic Publishers, 954 pp. Chapter 18 (microscale cloud electricity).

Uman, M.A. (1987). *The Lightning Discharge*. Orlando, FL: Academic Press, 377 pp. Appendix A (review of electrostatics) and Chapters 1 (history, categorization), 2 (phenomena), and 3 (cloud and lightning charges).

Wallace, J.M. and P.V. Hobbs (2006). *Atmospheric Science: An Introductory Survey*. Amsterdam: Academic Press, 483 pp. Chapter 6 (charge generation, lightning, global circuit).

Williams, E.R. (2009). The global electric circuit: a review. *Atmos. Res.*, **91**, 140–152.

14.6 Problems

1. In your own words, what distinguishes the two main schools of thought regarding the separation of electric charge in a convective cloud. In what ways are the two schools similar?

2. To the extent that collisions between cloud particles are needed to separate electric charge in a thunderstorm, what role is played by the particles sticking together? With the efficiency of sticking together following collision being represented by the coalescence efficiency ϵ, what would you expect of a thunderstorm in the extreme cases of (a) $\epsilon = 0$ and (b) $\epsilon = 1$? (Assume that the air motions are the same in both cases.)

3. How much larger in magnitude is the electric field below a typical thunderstorm compared with the fair-weather electric field?

4. What causes the reversal of the electric field near the surface in the vicinity of an active thunderstorm?

5. Estimate the total downward flow of positive electric charge due to the fair-weather electric field around the entire Earth. (Ignore the area of active thunderstorm activity.) How does this downward flow of charge compare with the upward flow due to all ($\sim$ 1800) thunderstorms around the globe?

6. Global (e.g., satellite) observations of lightning show the frequency to be much higher over land than over oceans. Offer compelling arguments why thunderstorm activity tends to be so much more pronounced over continental landmasses than over the oceans.

7. Thunderstorms with relatively cool cloud bases ($T_{base} \sim 5\,°C$) tend to cause more wild fires than do than do those with warm bases ($T_{base} \sim 20\,°C$). Explain.

8. The chemical-based hypothesis put forth in conjunction with Fig. 14.9 offers one explanation for the separation of electric charge on the microscale. Offer an alternative explanation for charge separation that depends only on water (in any phase).

Appendix A
Cloud classification

The naming of cloud types began in the early 1800s. Jean Babtiste Lamarck, a French naturalist, proposed a naming scheme in about 1802, but his nomenclature did not catch on, perhaps because of the cumbersome French terms he used. Lamarck is, however, credited with proposing that clouds be identified with the level (high, middle, or low) at which they appear in the atmosphere. Then, within a year or so, Luke Howard, an English chemist/pharmacist, suggested that clouds could be grouped into four main categories: Cirrus (Latin for hair curl), Cumulus (heap), Stratus (layer), and Nimbus (rain). The simple Latin terminology likely helped Howard's categorization to be accepted more readily.

The cloud classification scheme used internationally today thus traces its roots back to the suggestions of Lamarck (for organization into three levels) and Howard (for use of Latin names). The names in Latin follow the convention used in biology: *Genus species*. Subspecies or varieties are often appended to distinguish one species of cloud from another. The names given to various cloud types, now as then, are all based on human observations from the surface. The modern methods of observing clouds (aircraft, radar, satellites, etc.) were not available to the early observers. The determination of cloud type is based on visual appearance only, mainly the shape/form and size of the cloud. The altitude of cloud base is difficult to judge precisely, but the form usually helps us distinguish high clouds from mid-level and low clouds.

A summary of the various cloud types is shown graphically in Fig. A.1. Clouds are classified into ten genera (plural of genus), each of which can have several species and varieties. The internationally recognized two-letter abbreviations for genera are shown in parentheses. The genera are the principal cloud types and are given in italics with initial capital letters, while species names indicate a sub-class given here in italics without capitalization.

Clouds in layers are shown on the left side of Fig. A.1, whereas clouds that develop vertically, possibly spanning several levels, are shown on the right. Low clouds, with bases up to about 2 km, have a prefix of strato- and include *Stratocumulus* (Sc; continuous mass of cloud with gray and whitish patches) and *Stratus* (St; uniform gray cloud). Fog, often classified separately as visibility degradation, may also be considered a stratus cloud touching the ground. Mid-level clouds (bases between about 2 and 7 km altitude) have the prefix alto-, except for *Nimbostratus* (Ns; rain- or snow-producing), which tends to hover near

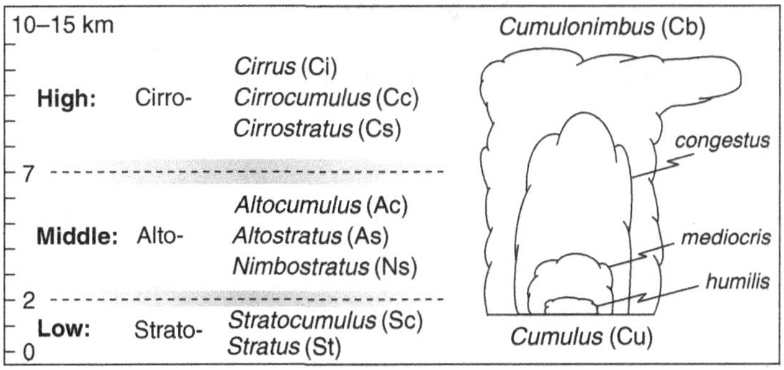

Figure A.1 Schematic summary of the major cloud types. Italicized words are the latin names for the genera (capitalized) and species (lower case). Clouds in layers are shown on the left side. The indicated altitude ranges for the low, middle, and high clouds are approximate (suggested by the shading). Clouds with vertical development are shown on right, with three examples of Cu species.

the boundary between low and middle clouds. Middle clouds also include *Altocumulus* (Ac; well-defined white or gray clouds in patches or rolls) and *Altostratus* (As; striated or uniform gray mass of cloud). High clouds have the prefix cirr- or cirro- and include *Cirrus* (Ci; detached filaments of delicate white cloud), *Cirrocumulus* (Cc; small and thin white patches of cloud without shadows), and *Cirrostratus* (Cs; smooth or fibrous veil of whitish cloud, sometimes producing halos).

Cumuliform clouds, shown on the right side of Fig. A.1, represent convective motions that produce detached clouds having vertical dimensions comparable to or larger than the horizontal dimensions. The largest clouds are *Cumulonimbus* (Cb; dense rain-producing clouds, often spreading out into anvils near the tops). Cb clouds often produce lightning or severe weather when mature, but they develop from lesser *Cumulus* clouds (Cu; detached clouds with well-defined edges). The term cumulus covers many sub-classes, including Cu humilis (fair-weather clouds having uniform bases and minimal vertical extend), Cu mediocris (dense clouds with moderate vertical development), and Cu congestus (large, dense clouds with great vertical development and sharp edges).

Appendix B
Overview of thermodynamics

Thermodynamics is the study of energy and its transformations. Traditionally, thermodynamics has been used to understand the transfer of "heat" and the mechanical work that can be realized from it. Mechanical heat engines became important to society during the Industrial Revolution, but natural heat engines exist, too, although we do not often speak of thunderstorms and hurricanes in such terms. All such "engines" ultimately derive their organized, macroscopic motions from the random motions of the molecules making up the systems.

Thermodynamics is useful because it applies to many phenomena in the Universe. At the same time, the discipline can become abstract, especially when one is not sure exactly what part of the Universe is being considered. It is therefore important to define the components and variables of the system carefully. Traditionally, the "system" is the part of the universe we are interested in for a particular application. Any system is separated from the rest of the Universe by a "control surface" situated between it and its "surroundings" or "environment". The "system" plus "surroundings" together make up the "universe". The "state" of the system at any given time is specified by the magnitudes of all relevant macroscopic variables, such as temperature, pressure, and volume of the system. In atmospheric physics, a commonly used system is a parcel of air, an amount of gas that is small enough to have uniform properties, but large enough to contain many molecules. A convenient parcel size is about $1\,m^3$, which contains about 1.3 kg of air at standard temperature and pressure (STP).

"Energy" is the primary concern of thermodynamics. However, beyond being defined generally as the capacity to do work, energy is not an easy concept to grasp. Nevertheless, the concept of energy, especially when suitably qualified, is incredibly useful. In simple mechanical systems, such as a ball falling toward the ground, we find that the macroscopic kinetic energy of the ball increases in exact proportion to the decrease in gravitational potential energy (when friction can be neglected, that is). Energy also resides in the molecules making up all matter, either as energy of motion (i.e., molecular kinetic energy) or as energy of position (molecular potential energy). It helps to distinguish microscopic (molecular) and macroscopic (observable) scales, and potential (arising from position in a force field) and kinetic (arising from motions) forms of energy.

Thermodynamics embodies several useful principles that help us understand the behavior of systems undergoing change. These principles are typically called "laws", but we must recognize that natural laws represent descriptions of macroscopic phenomena in terms of measurable quantities; they are not legal prescriptions for how systems should behave for fear of prosecution. The laws are summarized here in a logical way, not in the order in which they were developed historically. The explicit statements given below for each law should be accepted as practical expressions of diverse ideas developed over many years.

Zeroth law. If two systems are independently in thermal equilibrium with a third system, then these two systems are also in thermal equilibrium with each other. The zeroth law implies the existence of the intensive variable that we call "temperature", a measure of the hotness or coldness of a system. "Thermal equilibrium" implies that no net exchange of molecular kinetic energy (sometimes incorrectly called "heat") occurs between the systems. By counter argument, when two systems are not in thermal equilibrium, then the two systems are at different temperatures and may transfer energy from the warmer system (the one with the higher temperature) to the cooler one. The process of transferring energy by virtue of a temperature difference is called "heating". The rate that energy is exchanged between two systems depends not only on the temperature difference, but also on the degree of thermal contact (i.e., the thermal conductivity of the medium between the two systems). The principles of thermometry stem from the zeroth law and allow us to construct devices ("thermometers") with which to measure and compare the average molecular kinetic energies of various systems.

First law. The internal energy of a system changes only to the extent that energy is exchanged with its environment. The first law is basically a statement of the conservation of energy. If energy enters a system from its surroundings faster than it leaves, then the "internal energy" of the system increases by exactly the net difference. The energy transferred between a system and its surroundings can be categorized as (1) heating and (2) working. "Heating" is energy transferred as a consequence of a temperature difference and so involves the random motions of molecules (which we call "thermal energy") or electromagnetic radiation. We may think of heating as transferring a disorganized form of energy into (or out of) the system. By contrast, "working" is an organized form of energy transfer, because it results from a net external force acting in some coherent way on the system over a distance along the direction of force.

We may express the first law mathematically as

$$\frac{dU}{dt} = Q + W, \tag{B.1}$$

where U is the internal energy of the system, Q is the heating rate, and W is the working rate. Note that the sign convention used here means that positive values of heating and working contribute to an increase in system energy with time. Positive heating occurs when the temperature of the surroundings is greater than that of the system. Positive working occurs when the surroundings push on the control surface and cause the system (e.g., an air parcel) to be compressed (i.e., made smaller in volume). Normally, we need consider only work of expansion/contraction in atmospheric problems. We can therefore express the working rate as

$$W = -p\frac{dV}{dt}, \tag{B.2}$$

where p is the pressure and V is the volume of the system. The first law may now be put in the so-called "energy" form,

$$Q = \frac{dU}{dt} + p\frac{dV}{dt}. \tag{B.3}$$

This equation says that heating of a system (positive left-hand side) results in an increase in the internal energy ($dU/dt > 0$) and/or an increase in volume ($dV/dt > 0$).

Many atmospheric applications involve processes in which the pressure remains constant (at least approximately). In this case, Eq. (B.3) can be readily integrated over the time interval Δt during which the system changes from one arbitrary state "1" (at time t_1) to another arbitrary state "2" (at time $t_2 = t_1 + \Delta t$):

$$\begin{aligned} E_{q,p} &\equiv \int_{t_1}^{t_1+\Delta t} Q dt = \int_{U_1}^{U_2} dU + p\int_{V_1}^{V_2} dV \\ &= U_2 - U_1 + p(V_2 - V_1). \end{aligned} \tag{B.4}$$

This relationship may be rearranged algebraically to give

$$E_{q,p} = (U_2 + pV_2) - (U_1 + pV_1), \tag{B.5}$$

from which we see that the particular group of state variables, $U + pV$, is uniquely associated with each state. This combination of variables defines the state function called "enthalpy",

$$H \equiv U + pV. \tag{B.6}$$

We thus see (from Eqs. (B.5) and (B.6)) that a change in system enthalpy, $\Delta H \equiv H_2 - H_1$, is equivalent to the amount $E_{q,p}$ of thermal energy transferred (as a result of temperature differences) into the system at constant pressure (i.e., $\Delta H = E_{q,p}$). The enthalpy of a system is greater than its internal energy U by the amount pV, so enthalpy may be thought of as the "total energy" of the system, the internal energy of the matter contained in the system plus the work of expansion expended in bringing that matter to the finite volume it has at the given environmental pressure p. Differentiation of Eq. (B.6) with respect to time, followed by subtraction of the result from Eq. (B.3), yields the "enthalpy" form of the first law:

$$Q = \frac{dH}{dt} - V\frac{dp}{dt}. \tag{B.7}$$

Here, we again see that changes in the enthalpy of a system are equivalent to the heating rate (Q) at constant pressure ($dp/dt = 0$).

When applied to an ideal gas, the first law becomes especially useful for atmospheric applications. In this case, the internal energy depends only on the temperature, so the rate of change of U is given simply by

$$\frac{dU}{dt} = C_V \frac{dT}{dt}, \tag{B.8}$$

where C_V is the heat capacity at constant volume. Equation (B.3) then becomes

$$Q = C_V \frac{dT}{dt} + p\frac{dV}{dt}. \tag{B.9}$$

The work of expansion in an ideal gas is obtained by including the equation of state

$$pV = NRT, \tag{B.10}$$

where N is the (constant) number of moles of gas in the system (think air parcel) and R is the universal (i.e., molar) gas constant. In general, we allow the pressure, volume, and temperature to vary arbitrarily, so long as the relationship between these variables does not violate the equation of state. Mathematically, this constraint is expressed by taking the total derivative of Eq. (B.10):

$$p\frac{dV}{dt} + V\frac{dp}{dt} = NR\frac{dT}{dt}. \tag{B.11}$$

Solving Eq. (B.11) for $p\,dV/dt$ and substituting the resulting expression into the working term of Eq. (B.9), we obtain

$$\begin{aligned} Q &= C_V \frac{dT}{dt} + NR\frac{dT}{dt} - V\frac{dp}{dt} \\ &= (C_V + NR)\frac{dT}{dt} - V\frac{dp}{dt}. \end{aligned} \tag{B.12}$$

At this point, it is useful to change from extensive to intensive variables, so we divide Eq. (B.12) through by N to obtain

$$q = (c_v + R)\frac{dT}{dt} - v\frac{dp}{dt}. \tag{B.13}$$

The intensive heating rate q is thus expressed as a rate of energy transfer per unit amount of material (e.g., $\mathrm{W\,mol^{-1}}$), while the volume is the molar volume v ($\mathrm{m^3\,mol^{-1}}$), and the heat capacity becomes the "specific heat" at constant volume, c_v ($\mathrm{J\,K^{-1}\,mol^{-1}}$). The factor modifying the dT/dt term in Eq. (B.13) can be seen to be the specific heat when the pressure is unchanging (i.e., when $dp/dt = 0$), hence we find the expression for the specific heat at constant pressure (for an ideal gas) to be

$$c_p = c_v + R. \tag{B.14}$$

The first law of thermodynamics, constrained by the ideal gas law, can therefore be written in a form convenient for atmospheric applications:

$$q = c_p\frac{dT}{dt} - v\frac{dp}{dt}. \tag{B.15}$$

Equation (B.15) provides a succinct relationship between the changes in temperature and pressure that occur in the atmosphere. The specific enthalpy can now be seen, by comparing Eqs. (B.7) and (B.15), to change in direct proportion to changes in temperature according to

$$\frac{dh}{dt} = c_p \frac{dT}{dt}, \tag{B.16}$$

the proportionality constant being the specific heat at constant pressure, c_p.

The first law of thermodynamics is truly universal in its applicability, but it is sometimes limited in its usefulness. Energy must of course be conserved, preserved in all isolated systems, including the universe itself. The first law thus gives us a quantitative tool for solving many practical problems. At the same time, however, we commonly observe phenomena that cannot be predicted by the first law alone, even though the outcome is clear to us in everyday life. For example, we expect a hot beverage sitting on a tabletop to cool off until it reaches the temperature of its surroundings. Such a "spontaneous" or natural change is irreversible, for we also know that the "reverse" process will never occur by itself; the liquid will never spontaneously extract energy from the room and warm back up. Similarly, a bouncing ball will assuredly come to rest as its initial macroscopic potential energy is lost and converted into molecular kinetic energy. Energy in macroscopic systems becomes "degraded" in quality (through friction), although never lost. All such reverse processes are "allowed" by the first law, but the first law cannot predict the natural course of events. A second law is needed.

Second law. Kelvin's version: No process is possible for which the sole result is the absorption of heat from a reservoir and its complete conversion to work. This statement says in effect that not all of the thermal energy transferring into a system can perform work. Some fraction of the energy entering from a warm reservoir can indeed be extracted for useful purposes, but the remaining energy must be discarded to a colder reservoir. Thus, all heat "engines" (e.g., locomotives or hurricanes) are intrinsically inefficient, as if nature extracts a "tax" (the discarded energy) from every operation.

Clausius' version: No process can be devised that transfers heat from a cold reservoir to a warm reservoir without also producing other effects. This statement claims that the natural tendency for thermal energy to transfer from warm to cold objects can be reversed, but only at the expense of energy elsewhere. A heat "pump" (e.g., a refrigerator) can thus be constructed to remove thermal energy from a cold reservoir to a warmer reservoir as long as work is supplied from outside the system. That work, however, must be derived from a heat engine that in turn discards energy to some other cold reservoir.

These two versions of the second law are equivalent, one simply being the converse of the other. Both recognize that thermal energy naturally (i.e., spontaneously) transfers from bodies of high temperature to those of lower temperature. Energy transfer in the reverse direction does not naturally occur, even when all energy is accounted for. Nature thus appears to operate asymmetrically. Of all the conceivable processes that conserve energy and so "obey" the first law, only some will ever occur spontaneously and be predictable by the second law.

The second law fortunately offers a quantitative way of determining whether a process will occur spontaneously or not. The study of reversible cyclic processes (e.g., the Carnot cycle) teaches us that the ratio of the heating rate of a system to its temperature (Q/T) is preserved when evaluated around the entire cycle. Thus, because

$$\oint \left(\frac{Q}{T}\right)_{rev} = 0, \tag{B.17}$$

the variable combination $(Q/T)_{rev}$ is a state variable, which is related to the entropy Φ of the system through the definition

$$\frac{d\Phi}{dt} \equiv \left(\frac{Q}{T}\right)_{rev}. \tag{B.18}$$

The entropy of a system, Φ_{system}, increases whenever it is warmed ($Q > 0$), and it decreases whenever it is cooled ($Q < 0$). When the process is irreversible, the actual heating rate $Q_{irrev} > Q_{rev}$, so one form of the Clausius inequality can be written as

$$\frac{d\Phi_{system}}{dt} \geq \frac{Q}{T}, \tag{B.19}$$

where Q may be taken as Q_{rev} for a reversible process, in which case Eq. (B.18) is recovered. For a general finite process operating between two arbitrary states a and b, integration of Eq. (B.19) yields

$$\Delta\Phi_{system} \geq \int_a^b \left(\frac{Q}{T}\right) dt. \tag{B.20}$$

When one considers the surroundings of the system and recognizes that the Universe is an isolated system (so $Q = 0$), one finds a robust criterion for spontaneous change to be

$$\Delta\Phi_{system} + \Delta\Phi_{surroundings} \equiv \Delta\Phi_{universe} > 0. \tag{B.21}$$

Thus, if the entropy of the Universe increases for any stated process, then that process can be expected to proceed naturally.

A more convenient criterion for spontaneous change is readily obtainable from the Clausius inequality. Now expressed in terms of the system alone, rather than the entire Universe, Eq. (B.19) becomes

$$\frac{d\Phi}{dt} - \frac{Q}{T} \geq 0, \tag{B.22}$$

where it is understood that the variables all relate to the system. We may now apply the first law in either of two ways, depending on whether we wish to consider changes at constant temperature and constant volume or at constant temperature and constant pressure. For processes at constant volume, it is most convenient to use the energy form of the first law, which from Eq. (B.3) becomes $Q = dU/dt$. After slight algebraic manipulation, Eq. (B.22) becomes

$$\frac{d}{dt}(U - T\Phi) = \frac{dA}{dt} \leq 0. \tag{B.23}$$

The quantity in parentheses is the Helmholtz function $A \equiv U - T\Phi$, a grouping of state variables that must decrease for a spontaneous process at constant volume and temperature. Alternatively, for a constant-pressure process, we use the first law in enthalpy form (Eq. B.7) and find

$$\frac{d}{dt}(H - T\Phi) = \frac{dG}{dt} \leq 0. \tag{B.24}$$

The criterion for spontaneous change at constant pressure for the Gibbs function is therefore $G \equiv H - T\Phi$ to decrease. Entropy of the Universe still maintains its importance as a fundamental driver of change, but we achieve the same end with the Gibbs function expressed in terms of system variables alone.

Third law. The entropy of a substance is zero at zero absolute temperature. This law gives us a "foothold" for obtaining the actual magnitude of entropy. Normally, we require knowledge only of changes in variables, not their integrated values. By making the value of entropy known at one temperature ($T = 0\,\mathrm{K}$), one can obtain the value at any other temperature:

$$\Phi(T) = \Phi_0 + \int_0^T \left(\frac{Q_{rev}}{T}\right), \tag{B.25}$$

where Φ_0 is the "zero-point entropy". The third law claims that $\Phi_0 = 0$ in general, but some molecular solids (including ice) have small, non-zero entropies at 0 K. Empirical values of so-called "third-law entropies" are obtained in practice by calorimetric measurements.

Knowledge of the absolute entropies also provides valuable insights into the physical nature of entropy. Because all matter is made of many molecules each of which can exist in a variety of energy states, entropy can be linked (through statistical mechanics) to the number of ways that the molecular states can be arranged and still yield the same macroscopic (i.e., observable) properties of a given system. Boltzmann showed that if W represents the number of possible microstates for a given macrostate (i.e., the statistical weight), then the entropy of the system is numerically given by

$$\Phi = k_B \ln W, \tag{B.26}$$

where k_B is the Boltzmann constant. Systems naturally strive for states of higher entropy because such states are more probable (have larger values of W). The product $T\Phi$ that appears in both the Helmholtz and Gibbs functions (Eqs. B.23 and B.24) thus becomes identifiable with the store of energy that is unavailable for doing work of any sort, mainly because of the disorganized state of the molecules in the system. Thus, with the recognition that the internal energy (U) and the enthalpy (H) represent the total energies in any system, respectively, at constant volume and constant pressure, the differences expressed by the functions A and G can be seen to represent the amount of energy in the system that is available or "free" to perform work.

Appendix C
Boltzmann distribution

[An abbreviated treatment based on Atkins (1986, Chapter 21)]

The Boltzmann distribution is a function that tells us how the molecules or other particles in a system are distributed among the various possible energy states. Thermodynamics treats only the average behavior of a collection of particles, so it falls to statistical mechanics to specify how the energies of the particles vary about the average. Stated another way, thermodynamics describes the macroscopic, observable behavior of a system, whereas statistical mechanics relates what we observe to the microscopic properties of the system. We must keep in mind that all macroscopic behavior reflects the collective actions of the molecules making up the system. The treatment here is focused on the single question, "How many of the particles in the system have what energy?"

Consider a system containing a large number of particles (such as an Avogadro's number of molecules). Let the system be closed, meaning that the particles are confined to the system; the total number N is therefore constant. These N particles are free to be in any energy state, but the total energy of the system is fixed at the value E. Our objective is to find the most likely arrangement of particles among the many discrete energy states ϵ_i, where $i = 0, 1, 2, \ldots$ Any microscopic arrangement of the system is specified by its configuration, defined as the number n_i of particles in each energy state: $\{n_0, n_1, n_2, \ldots n_{i,} \ldots\}$. In every configuration, the total number of particles is the same and calculated by simple addition:

$$N = \sum_i n_i. \tag{C.1}$$

At any given temperature, the total energy, E, the sum of the particle energies, is also constant:

$$E = \sum_i n_i \varepsilon_i. \tag{C.2}$$

Equations (C.1) and (C.2) are statements of the conservation of mass and energy, respectively; they represent constraints imposed on the system.

The particles in the system continually interact (collide) with each other and so are free to rearrange themselves into many different microscopic configurations for any given set of observable conditions (such as temperature). Each configuration is itself the result of many possible combinations of molecular arrangements, some arrangements being more probable than others. Probability theory gives us the mathematical tools to calculate the number of ways that a given configuration can be achieved. Of our N particles, the first particle can be arranged in N ways, the second in $N-1$ ways, the third in $N-2$ ways, and so forth. If all of the particles were labeled uniquely, we would have $N(N-1)(N-2)(N-3)\ldots = N!$ ways to arrange the N particles. However, our particles (molecules) are not distinguishable, so we would have counted too many ways. We account for such overcounting by dividing by the number of ways the particles in each energy levels can be arranged. The result is given by the so-called statistical weight:

$$W = \frac{N!}{n_0!n_1!n_2!\ldots}. \tag{C.3}$$

Other configurations yield other values for W. The most probable configuration $\{n_0^*, n_1^*, n_2^*, \ldots n_i^* \ldots\}$ is that which yields the largest statistical weight $W^* = \max(W)$. It is this particular configuration that gives us the Boltzmann distribution we seek.

A common way of demonstrating that some configurations are more probable than others is to consider an analogy, the rolling of dice (unloaded, of course). Rolling a single die is just as likely to yield a one or a six, or any number in between. All outcomes are equally probable, so we could say that the statistical weight is uniform across all configurations. No maximum in W exists for a single "particle". Next, roll two dice, letting the sum of dots represent the configuration. Calculating the probabilities, or simply tabulating the frequency with which the various outcomes arise shows that a sum of seven is rolled most frequently. Sums in the mid part of the range of possibilities arise in a greater number of ways than do those at the extremes. For instance, a seven can arise via three indistinguishable combinations of the two dice (1,6; 2,5; 3,4), whereas a two or a twelve can each arise in only one way (two ones or two sixes, respectively). Rolling two dice results in a weak maximum of the statistical weight in the mid range. As the number of dice used increases, the relative strength of the maximum increases rapidly. Rolling 10 dice at a time, for instance, often yields sums near 35 (the mid-point of the range), as shown in Fig. C.1. Even after hundreds of trials, one has very little chance of rolling sums near the ends of the range. The extremes are extremely unlikely, whereas the peak in the statistical weight is pronounced in the middle of the range, where the outcome can be achieved in many ways. The rolling of dice provides insight into the nature of probability distributions and gives us a glimpse of the really large statistical weight that is focused on just a few favored configurations when the number of particles in the system becomes huge, as in most molecular systems.

The mathematical approach used for determining the most probable configuration (that which occurs most frequently) is to maximize the statistical weight W (Eq. (C.3)) subject to the constraints of constant mass and energy (Eqs. (C.1) and (C.2)). The constraints are most conveniently expressed in their differential forms:

$$\sum_i dn_i = 0 \tag{C.4}$$

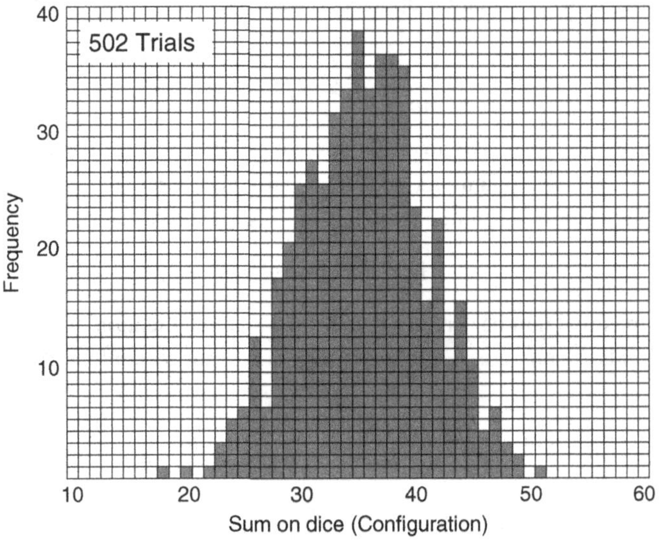

Figure C.1 Outcome of rolling ten dice 502 times. Frequency (ordinate) gives the number of times the indicated sum (abscissa) appeared. Based on an original graph of J. Lamb, used with permission.

and

$$\sum_i \varepsilon_i dn_i = 0. \tag{C.5}$$

Maximizing a function subject to constraints (called "side conditions") is performed using Lagrange's method of undetermined multipliers, which is based on the idea that adding zeros to a function does not change the value of the function. The zeros in question are the constraints (Eqs. (C.4) and (C.5)). Conceptually, we let the configuration numbers (n_i) vary and look for the set of n_i values that give the maximum value for W. Working with factorials of large numbers is difficult, so we use the natural logarithm of W instead; the point at which the maximum of the logarithm of a function lies is the same as that of the function itself. Thus, the criterion for finding the maximum in W is given by

$$d \ln W = \sum_i \left(\frac{\partial \ln W}{\partial n_i} \right) dn_i + \alpha \sum_i dn_i - \beta \sum_i \varepsilon_i dn_i = 0, \tag{C.6}$$

where α and β are the Lagrangian multipliers, constants to be evaluated later. The sum over all energy levels is common to each term, so an equivalent form of Eq. (C.6) is

$$\sum_i \left[\frac{\partial \ln W}{\partial n_i} + \alpha - \beta \varepsilon_i \right] dn_i = 0. \tag{C.7}$$

Every exchange of particles between levels (expressed by dn_i) is independent of every other exchange, so each term of the sum in Eq. (C.7) must be separately equal to zero. Thus, the following equation applies to each value of i and yields the maximum value W^*:

$$\frac{\partial \ln W^*}{\partial n_i} + \alpha - \beta\varepsilon_i = 0. \tag{C.8}$$

The solution to Eq. (C.8) is gained by inserting the function W from Eq. (C.3) and solving for the set n_i^*. The magnitude of N is usually very large, but the task of taking the logarithm of a factorial is simplified by using Stirling's approximation:

$$\ln x! \doteq x \ln x - x, \tag{C.9}$$

where x is any positive number greater than a few tens. Application of this approximation to Eq. (C.3) yields (using j as a dummy index so as not to confuse it with the energy level)

$$\begin{aligned} \ln W &= \ln N! - \sum_j \ln n_j! \\ &\doteq N \ln N - N - \sum_j n_j \ln n_j + \sum_j n_j \\ \ln W &\doteq N \ln N - \sum_j n_j \ln n_j. \end{aligned} \tag{C.10}$$

The derivative of $\ln W$ (needed for Eq. (C.8)) can be simplified by noting that N is constant and that the number of particles in one state does not depend on the number in any other state (meaning that $\partial n_j / \partial n_i \neq 0$ only when $n_j = n_i$):

$$\begin{aligned} \frac{\partial \ln W}{\partial n_i} &= -\sum_j \frac{\partial}{\partial n_i}\left(n_j \ln n_j\right) \\ &= -\sum_j \frac{\partial n_j}{\partial n_i} \ln n_j - \sum_j n_j \cdot \frac{1}{n_j}\frac{\partial n_j}{\partial n_i} \\ &= -\ln n_i - 1 \\ \frac{\partial \ln W}{\partial n_i} &\doteq -\ln n_i. \end{aligned} \tag{C.11}$$

Equation (C.8) now becomes

$$-\ln n_i^* + \alpha - \beta\varepsilon_i = 0, \tag{C.12}$$

which can be rearranged as

$$n_i^* = e^{\alpha} e^{-\beta\varepsilon_i}. \tag{C.13}$$

Evaluation of the coefficients α and β in Eq. (C.13) is achieved by applying the mass and energy constraints (Eqs. C.1 and C.2). The constraint on total number gives

$$N = \sum_i n_i^* = e^{\alpha} \sum_i e^{-\beta\varepsilon_i}, \tag{C.14}$$

from which it follows that

$$e^{\alpha} = \frac{N}{\sum_i e^{-\beta \varepsilon_i}}. \tag{C.15}$$

The denominator here is a weighted sum, known as the "sum over states" or the "partition function",

$$Z \equiv \sum_i e^{-\beta \varepsilon_i}. \tag{C.16}$$

The partition function is a measure of the average number of states occupied by the most likely configuration. Next, the total energy of the system is calculated to be

$$E = \sum_i n_i^* \varepsilon_i = \frac{N}{Z} \sum_i \varepsilon_i e^{-\beta \varepsilon_i}. \tag{C.17}$$

Note that β is an intensive variable that is common to all configurations. Moreover, this variable has inverse energy units, so it should not be surprising that it is closely associated with the thermodynamic temperature T. More detailed treatments justify the relationship

$$\beta = \frac{1}{k_B T}, \tag{C.18}$$

where k_B is the Boltzmann constant. The Boltzmann distribution is therefore

$$n_i^* = \frac{N}{Z} e^{-\frac{\varepsilon_i}{k_B T}}. \tag{C.19}$$

For a given temperature, the number $\left(n_i^*\right)$ of particles in each energy state can thus be calculated.

Often, the number of particles in any one energy state (ε_i) only needs to be compared with that in some other energy state (ε_j). In such cases, it is appropriate to form a ratio and so obtain the relative distribution that depends only on temperature and the energy difference, $\Delta\varepsilon_{ij} \equiv \varepsilon_i - \varepsilon_j$. This population ratio is called the Boltzmann factor,

$$\frac{n_i^*}{n_j^*} = e^{-\frac{\Delta\varepsilon_{ij}}{k_B T}}. \tag{C.20}$$

At low temperatures, when the available thermal energy is small compared with the spacing between neighboring energy levels ($\Delta\varepsilon_{ij} >> k_B T$), relatively few energy levels are occupied. Particles (or molecules) then simply settle into the lower energy states, unable to be kicked into higher energy states. As the temperature increases, the probability of occupying the higher energy states increases, but the total number of particles and the collective energy remain unchanged. The Boltzmann factor, with its exponential character, is useful in many scientific disciplines, including cloud physics.

References

Ahrens, C.D. (1994). *Meteorology Today*. St. Paul, MN: West, 591 pp.

Atkins, P.W. (1986). *Physical Chemistry*. New York: W.H. Freeman/Oxford University Press, 857 pp.

Bailey, M. and J. Hallett (2004). Growth rates and habits of ice crystals between $-20\,°C$ and $-70\,°C$. *J. Atmos. Sci.*, **61**, 514–544.

Baker, M.B. and M. Baker (2004). A new look at homogeneous freezing of water. *Geophys. Res. Lett.*, **31**, L19102.

Baker, M.B., R.G. Corbin, and J. Latham (1980). The influence of entrainment on the evolution of cloud droplet spectra: I. A model of inhomogeneous mixing. *Quart. J. Roy. Meteor. Soc.*, **106**, 581–598.

Beard, K.V. and H.T. Ochs (1984). Collection and coalescence efficiencies for accretion. *J. Geophys. Res.*, **89** (D5), 7165–7169.

Berry, E.X. and R.L. Reinhardt (1974). An analysis of cloud drop growth by collection: Part II. Single initial distribution. *J. Atmos. Sci.*, **31**, 1825–1831.

Böhm, J.P. (1989). A general equation for the terminal fall speed of solid hydrometeors. *J. Atmos. Sci.*, **46**, 2419–2427.

Böhm, J.P. (1992). A general hydrodynamic theory for mixed-phase microphysics. Part II: collision kernels for coalescence. *Atmos. Res.*, **27**, 275–290.

Bowen, E.G. (1950). The formation of rain by coalescence. *Austr. J. Sci. Res., Series A*, **3**, 193–213.

Brasseur, G. and S. Solomon (1986). *Aeronomy of the Middle Atmosphere*. Dordrecht: D. Reidel, 452 pp.

Browning, K.A., J.C. Fankhauser, J.-P. Chalon, *et al.* (1976). Structure of an evolving hailstorm. Part V: synthesis and implications for hail growth and hail suppression. *Mon. Wea. Rev.*, **104**, 603–610.

Burnet, F. and J.-L. Brenguier (2007). Observational study of the entrainment-mixing process in warm convective clouds. *J. Atmos. Sci.*, **64**, 1995–2011.

Burton, W.K., N. Cabrera, and F.C. Frank (1951). The growth of crystals and the equilibrium structure of their surfaces. *Phil. Trans. Roy. Soc. London*, **23A**, 299–358.

Byers, H.R. and R.R. Braham (1949). The thunderstorm. Report of the Thunderstorm Project. Washington DC: US Weather Bureau, Dept. of Commerce.

Chagnon, C.W. and C.E. Junge (1961). The vertical distribution of sub-micron particles in the stratosphere. *J. Meteor.*, **18**, 746–752.

Chen, J.-P. (1994). Theory of deliquescence and modified Köhler curves. *J. Atmos. Sci.*, **51**, 3505–3516.

Chen, J.-P. and D. Lamb (1994a). The theoretical basis for the parameterization of ice crystal habits: growth by vapor deposition. *J. Atmos. Sci.*, **51** (9), 1206–1221.

Chen, J.-P. and D. Lamb (1994b). Simulation of cloud microphysical and chemical processes using a multicomponent framework. Part I: description of the microphysical model. *J. Atmos. Sci.*, **51**, 2613–2630.

Chisholm, A.J. and J.H. Renick (1972). Kinematics of multicell and supercell Alberta hailstorms. Alberta: Research Council of Alberta. Hail Studies Report 72–2, 24–31.

Cooper, W.A. (1974). A possible mechanism for contact nucleation. *J. Atmos. Sci.*, **31**, 1832–1837.

DeMore, W.B., S.P. Sander, C.J. Howard, *et al.* (1997). *Chemical Kinetics and Photochemical Data for Use in Stratospheric Modeling.* Pasadena, CA: NASA, 266 pp.

Dong, X., T.P. Ackerman, and E. Clothiaux (1997). Microphysical and radiative properties of boundary layer stratiform clouds deduced from ground-based measurements. *J. Geophys. Res.*, **102** (D20), 23829–23843.

Doswell, C.A.I. (1985). The operational meteorology of convective weather. Vol. 2 : storm scale analysis. Tech Memo ERL ESG-15, NOAA.

Dye, J.E., J.J. Jones, W.P. Winn, *et al.* (1986). Early electrification and precipitation development in a small, isolated Montana cumulonimbus. *J. Geophys. Res.*, 91 (D1), 1231–1247.

Eisenberg, D. and W. Kauzmann (1969). *The Structure and Properties of Water.* Oxford: Oxford University Press, 296 pp.

Feingold, G. and Z. Levin (1986). The lognormal fit to raindrop spectra from frontal convective clouds in Israel. *J. Climate Appl. Meteor.*, **25**, 1346–1363.

Finlayson-Pitts, B.J. and J.N. Pitts (2000). *Chemistry of the Upper and Lower Atmosphere.* San Diego, CA: Academic Press, 969 pp.

Fleagle, R.G. and J.S. Businger (1963). *An Introduction to Atmospheric Physics.* New York: Academic Press, 346 pp.

Fletcher, N.H. (1962). *The Physics of Rainclouds.* Cambridge: Cambridge University Press, 386 pp.

Fletcher, N.H. (1965). Heterogeneous nucleation of ice crystals. *J. Australian Inst. Metals*, **10**, 101–105.

Fletcher, N.H. (1970). *The Chemical Physics of Ice.* Cambridge: Cambridge University Press, 271 pp.

Fukuta, N. and L.A. Walter (1970). Kinetics of hydrometeor growth from a vapor-spherical model. *J. Atmos. Sci.*, **27**, 1160–1172.

Furukawa, Y. and H. Nada (1997). Anisotropic melting of an ice crystal and its relationship to growth forms. *J. Phys. Chem. B.*, **101** (32), 6167–6170. Amsterdam: Elsevier Science, pp. 559–573.

Goodstein, D.L. (1985). *States of Matter.* New York: Dover, 500 pp.

Grabowski, W.W., 1993: Cumulus entrainment, fine-scale mixing and buoyancy reversal. *Quart. J. Roy. Meteor. Soc.*, **119**, 935–956.

Hallett, J. (1963). Temperature dependence of viscosity of supercooled water. *Proc. Phys. Soc. (London)*, **82** (530), 1046.

Hansen, J.E. and L.D. Travis (1974). Light scattering in planetary atmospheres. *Space Sci. Rev.*, **16**, 527–610.

Hartmann, D.L. (1993). Radiative influences of clouds on Earth's climate. In *Aerosol–Cloud–Climate Interactions*, ed. P.V. Hobbs. San Diego, CA: Academic Press, pp. 151–173.

Heymsfield, A.J., P.N. Johnson, and J.E. Dye (1978) Observations of moist adiabatic ascent in Northeast Colorado cumulus congestus clouds. *J. Atmos. Sci.*, **35**, 1689–1703.

Hicks, B.B. (2005). A climatology of wet deposition scavenging ratios for the United States. *Atmos. Environ.*, **39**, 1585–1596.

Hobbs, P.V., S. Chang, and J.D. Locatelli (1974). The dimensions and aggregation of ice crystals in natural clouds. *J. Geophys. Res.*, *79* (15), 2199–2206.

Hoppel, W.A., J.W. Fitzgerald, G.M. Frick, and R.E. Larson (1990). Aerosol size distributions and optical properties found in the marine boundary layer over the Atlantic Ocean. *J. Geophys. Res.*, **95** (D4), 3659–3686.

Houze, R.A. (1981). Structures of atmospheric precipitation systems: a global survey. *Radio Sci.*, **16** (5), 671–689.

Houze, R.A., M.I. Biggerstaff, S.A. Rutledge, and B.F. Smull (1989). Interpretation of Doppler weather radar displays of midlatitude mesoscale convective systems. *Bull. Amer. Meteor. Soc.*, **70**, 608–619.

Incredio (2010). Insolation top of atmosphere. Available at http://commons.wikimedia.org/wiki/File:InsolationTopOfAtmosphere.png.

Jaecker-Voirol, A. and P. Mirabel (1989). Heterogeneous nucleation in the sulfuric acid-water system. *Atmos. Environ.*, **23** (9), 2053–2057.

Jayaratne, O.W. and B.J. Mason (1964). The coalescence and bouncing of water drops at an air/water interface. *Proc. Roy. Soc. London A*, **280** (1383), 545–565.

Jeffery, C.A. and P.H. Austin (1997). Homogeneous nucleation of supercooled water: results from a new equation of state. *J. Geophys. Res.*, **102** (D21), 25269–25279.

Ji, W. and P.K. Wang (1990). Numerical simulation of three-dimensional unsteady viscous flow past fixed hexagonal ice crystals in the air: preliminary results. *Atmos. Res.*, **25**, 539–557.

Jimenez, J.L. M.R. Canagaratna, N.M. Donahue, *et al.* (2009). Evolution of organic aerosols in the atmosphere. *Science*, **326**, 1525–1529.

Kajikawa, M. (1972). Measurement of falling velocity of individual snow crystals. *J. Meteor. Soc. Japan*, **50**, 577–583.

Knight, C.A. (1979). Ice nucleation in the atmosphere. *Adv. Colloid Interface Sci.*, **10**, 369–395.

Knight, C.A. and N.C. Knight (2005). Very large hailstones from Aurora, Nebraska. *Bull. Amer. Meteor. Soc.*, **86**, 1773–1781.

Knight, N.C. (1981). The climatology of hailstone embryos. *J. Appl. Meteor.*, **20**, 750–755.

Koop, T., G. Luo, A. Tsias, and T. Peter (2000). Water activity as the determinant for homogeneous ice nucleation in aqueous solutions. *Nature*, **406**, 611–614.

Korolev, A.V. (2007). Limitations of the Wegener–Bergeron–Findeisen mechanism in the evolution of mixed-phase clouds. *J. Atmos. Sci.*, **64**, 3372–3375.

Korolev, A.V. and I.P. Mazin (2003). Supersaturation of water vapor in clouds. *J. Atmos. Sci.*, **60**, 2957–2974.

Kuettner, J.P., Z. Levin, and J.D. Sartor (1981). Thunderstorm electrification: inductive or noninductive? *J. Atmos. Sci.*, **38**, 2470–2484.

LaChapelle, E.R. (1969). *Field Guide to Snow Crystals*. Seattle, WA: University of Washington Press, 101 pp.

Lamb, D. (1999). Atmospheric ice. In *Encyclopedia of Applied Physics, Update 1*, ed. G. Trigg. New York: Wiley-VCH, pp. 1–25.

Lamb, D. (2001). Rain production in convective storms. In *Severe Convective Storms*, Vol. 28, ed. C.A. Doswell. Boston: American Meteorological Society, pp. 299–321.

Lamb, D. and R. Blumenstein (1987). Measurement of the entrapment of sulfur dioxide by rime ice. *Atmos. Environ.*, **21**, 1765–1772.

Lamb, D., J. Hallett, and R.I. Sax (1981). Mechanistic limitations to the release of latent heat during the natural and artificial glaciation of deep convective clouds. *Quart. J. Roy. Meteor. Soc.*, **107**, 935–954.

Lamb, D., A.M. Moyle, and W.H. Brune (1996). The environmental control of individual aqueous particles in a cubic electrodynamic levitation system. *Aerosol Sci. Technol.*, **24**, 263–278.

Lamb, D. and W.D. Scott (1972). Linear growth rates of ice crystals grown from the vapor phase. *J. Cryst. Growth*, **12**, 21–31.

Lemon, L.R. and C.A. Doswell (1979). Severe thunderstorm evolution and mesocyclone structure as related to tornadogenesis. *Mon. Wea. Rev.*, **107**, 1184–1197.

Levine, J.S. (1982). The photochemistry of the paleoatmosphere. *J. Mol. Evol.*, **18**, 161–172.

Lew, J.K. and H.R. Pruppacher (1983). A theoretical determination of the capture efficiency of small columnar ice crystals by large cloud drops. *J. Atmos. Sci.*, **40**, 139–145.

Libbrecht, K.G. (2005). The physics of snow crystals. *Rep. Prog. Phys.*, **68**, 855–895.

Low, T.B. and R. List (1982). Collision, coalescence and breakup of raindrops. Part I: experimentally established coalescence efficiencies and fragment size distributions in breakup. *J. Atmos. Sci.*, **39**, 1591–1606.

MacGorman, D.R. and W.D. Rust (1998). *The Electrical Nature of Storms*. New York, Oxford: University Press, 422 pp.

Mader, H.M. (1992). Observations of the water-vein system in polycrystalline ice. *J. Glaciology*, **38** (130), 333–347.

Manahan, S.E. (1993). *Environmental Chemistry*. Boca Raton, FL: CRC Press, 898 pp.

Markowski, P.M. and Y.P. Richardson (2010). *Mesoscale Meteorology in Mid-latitudes*. Chichester, UK: John Wiley & Sons, 406 pp.

Mason, B.J. (1961). The growth of snow crystals. *Sci. Am.*, **204** (1), 120–131.

Mitchell, D.L. (1996). Use of mass- and area-dimensional power laws for determining precipitation particle terminal velocities. *J. Atmos. Sci.*, **53**, 1710–1723.

Mitra, S.K. and A.U. Hannemann (1993). On the scavenging of SO_2 by large and small rain drops: V. A wind tunnel and theoretical study of the desorption of SO_2 from water drops containing S(IV). *J. Atmos. Chem.*, **16**, 201–218.

Mordy, W. (1959). Computations of the growth by condensation of a population of cloud droplets. *Tellus*, **11**, 16–44.

Mossop, S.C. (1985). The origin and concentration of ice crystals in clouds. *Bull. Amer. Meteor. Soc.*, **66**, 264–273.

Ochs, H.T. R.R. Czys, and K.V. Beard (1986). Laboratory measurements of coalescence efficiencies for small precipitation drops. *J. Atmos. Sci.*, **43**, 225–232.

Oraltay, R.G. and J. Hallett (2005). The melting layer: a laboratory investigation of ice particle melt and evaporation near 0 °C. *J. Appl. Meteor.*, **44**, 206–220.

Oseen, C.W. (1910). Über die Stokessche Formel and über die verwandte Aufgabe in der Hydrodynamik. *Arkiv. Mat., Astron. och Fysik*, **6** (No. 29).

Paluch, I.R. and C.A. Knight (1984). Mixing and the evolution of cloud droplet size spectra in a vigorous continental cumulus. *J. Atmos. Sci.*, **41**, 1801–1815.

Pauling, L.C. (1960). *The Nature of the Chemical Bond*. Ithaca, NY: Cornell University Press, 644 pp.

Petty, G.W. (2004). *A First Course in Atmospheric Radiation*. Madison, WI: Sundog Publishing, 443 pp.

Pollack, J.B. (1982). Atmospheres of the terrestrial planets. In *The New Solar System*. ed. J.K. Beatty, and A. Chaikin. Cambridge: Cambridge University Press, pp. 57–70.

Pruppacher, H.R. and J.D. Klett (1997). *Microphysics of Clouds and Precipitation*. Dordrecht: Kluwer Academic Publishers, 954 pp.

Rasmussen, R.M. V. Levizzani, and H.R. Pruppacher (1984). A wind tunnel and theoretical study of the melting behavior of atmospheric ice particles, II: a theoretical study for frozen drops of radius <500 μm. *J. Atmos. Sci.*, **41**, 374–380.

Rasmussen, R.M. and H.R. Pruppacher (1982). A wind tunnel and theoretical study of the melting behavior of atmospheric ice particles, I: a wind tunnel study of frozen drops of radius <500 μm. *J. Atmos. Sci.*, **39**, 152–158.

Raymond, T.M. and S.N. Pandis (2002). Cloud activation of single-component organic aerosol particles. *J. Geophys. Res*, **107** (D24), 4787.

Roberts, P. and J. Hallett (1968). A laboratory study of the ice nucleating properties of some mineral particulates. *Quart. J. Roy. Meteor. Soc.*, **94**, 25–34.

Roehl, C.M. J.J. Orlando, G.S. Tyndall, *et al.* (1994). Temperature dependence of the quantum yields for the photolysis of NO_2 near the disssociation limit. *J. Phys. Chem.*, **98**, 7837–7843.

Rogers, R.R. and M.K. Yau (1989). *A Short Course in Cloud Physics*. New York: Pergamon Press, 293 pp.

Salby, M. L. (1996). *Fundamentals of Atmospheric Physics*. San Diego, CA: Academic Press, 624 pp.

Schaller, R.C. and N. Fukuta (1979). Ice nucleation by aerosol particles: Experimental studies using a wedge-shaped ice thermal diffusion chamber. *J. Atmos. Sci.*, **36**, 1788–1802.

Schmenauer, R.S., J.I. MacPherson, G.A. lsaac, and J.W. Strapp (1980). Canadian participation in HYPLEX 1979. Toronto: Report APRB 110 p 34, 206 pp.

Schwartz, S.E. (1986). Mass-transport considerations pertinent to aqueous phase reactions of gases in liquid-water clouds. In *Chemistry of Multiphase Systems*. W. Jaeschke, ed. Berlin: NATO ASI Series, Springer-Verlag, pp. 415–471.

Schwartz, S.A. and J.E. Freiberg (1981). Mass-transport limitations to the rate of reaction of gases in liquid droplets: application to oxidation of SO_2 in aqueous solutions. *Atmos. Environ.*, **15**, 1129–1144.

Seinfeld, J.H. and S.N. Pandis (1998). *Atmospheric Chemistry and Physics*. New York: John Wiley and Sons, 1326 pp.

Selezneva, E.S. (1966). The main features of condensation nuclei distribution in the free atmosphere over the European territory of the USSR. *Tellus*, **XVIII**, 525–531.

Shaw, R.A. and D. Lamb (1999). Experimental determination of the thermal accommodation and condensation coefficients of water. *J. Chem. Phys.*, **111**, 10659–10663.

Shaw, R.A., W.C. Reade, L.R. Collins, and J. Verlinde (1998). Preferential concentration of cloud droplets by turbulence: effects on the early evolution of cumulus cloud droplet spectra. *J. Atmos. Sci.*, **55**, 1965–1976.

Slinn, W.G.N. (1984). Precipitation scavenging. In *Atmospheric Science and Power Production*, ed. D. Randerson. Springfield: VA: US Department of Energy, pp. 466–532.

Stein, A.F. and D. Lamb (2002). Chemical indicators of sulfate sensitivity to nitrogen oxides and volatile organic compounds. *J. Geophys. Res.*, **107** (D20), 4449.

Stevens, B. and G. Feingold (2009). Untangling aerosol effects on clouds and precipitation in a buffered system. *Nature*, **461**, 607–613.

Szyrmer, W. and I. Zawadzki (1999). Modeling of the melting layer. Part I: Dynamics and microphysics. *J. Atmos. Sci.*, **56**, 3573–3592.

Takashi, T., T. Endoh, G. Wakahama, and N. Fukuta (1991). Vapor diffusimal growth of free-falling snow crystals between −3 and −23 °C. *J. Meteor. Soc. Japan*, **69** (1), 15–30.

Torr, D.G. (1985). The photochemistry of the upper atmosphere. In *The Photochemistry of Atmospheres*, ed. J. S. Levine. Orlando, FL: Academic Press.

Trenberth, K.E., J.T. Fasullo, and J. Kiehl (2009). Earth's global energy budget. *Bull. Amer. Meteor. Soc.*, **90** (3), 311–323.

Twomey, S.A. (1964). Statistical effects in the evolution of a distribution of cloud droplets by coalescence. *J. Atmos. Sci.*, **21**, 553–557.

Twomey, S.A. (1974). Pollution and the planetary albedo. *Atmos. Environ.*, **8**, 1254–1256.

Twomey, S.A. and J. Warner (1967). Comparison of measurements of cloud droplets and cloud nuclei. *J. Atmos. Sci.*, **24**, 702–703.

Turco, R.P. (1997). *Earth Under Siege*. New York: Oxford University Press, 527 pp.

Vonder Haar, T.H. and V.E. Suomi (1971). Measurements of the Earth's radiation budget from satellites during a five-year period: Part I: extended time and space means. *J. Atmos. Sci.*, **28** (3), 305–314.

Walcek, C.J. and H.R. Pruppacher (1984). On the scavenging of SO_2 by cloud and raindrops: I. A theoretical study of SO_2 absorption and desorption for water drops in air. *J. Atmos. Chem.*, **1**, 269–289.

Wallace, J.M. and P.V. Hobbs (2006). *Atmospheric Science An Introductory Survey*. Amsterdam: Academic Press, 483 pp.

Wang, P.K., S.N. Grover, and H.R. Pruppacher (1978). On the effect of electric charges on the scavenging of aerosol particles by clouds and small raindrops. *J. Atmos. Sci.*, **35**, 1735–1743.

Wang, P.K. and W. Ji (2000). Collision efficiencies of ice crystals at low-intermediate Reynolds numbers colliding with supercooled droplets: a numerical study. *J. Atmos. Sci.*, **57**, 1001–1009.

Weidner, R.T. and R.L. Sells (1960). *Elementary Modern Physics*. Boston, MA: Allyn and Bacon, 513 pp.

Whitby K.T. (1972). The aerosol size distribution of Los Angeles smog. *J. Colloid Interface Sci.*, **39** (1), 177–204.

Yokoyama, E. (1993). Formation of patterns during growth of snow crystals. *J. Cryst. Growth*, **128**, 251–257.

Young, K.C. (1993). *Microphysical Processes in Clouds*. New York: Oxford University Press, 427 pp.

Zahn, S.G. (1993). An investigation of warm-cloud microphysics using a multi-component cloud model: interactive effects of the aerosol spectrum. Master's thesis. Pennsylvania State University: 107 pp.

Zhang, Y.-C. and W.B. Rossow (1997). Estimating meridional energy transports by the atmospheric and oceanic general circulations using boundary flures. *J. Climate*, **10**, 2358–2373.

Index

Printed in the United States
By Bookmasters